Environmental Protection in the European Union
Volume 2

Edited by
Michael Schmidt and Lothar Knopp, Cottbus

Environmental Protection in the European Union

Volume 1
M. Schmidt, L. Knopp
**Reform in CEE-Countries with Regard
to European Enlargement**
2004, XII, 205 pages
ISBN 3-540-40259-4

Michael Schmidt · Elsa João
Eike Albrecht (Eds.)

Implementing Strategic Environmental Assessment

With 90 Figures and 61 Tables

 Springer

Professor Dr. Michael Schmidt
Department of Environmental
Planning
Brandenburg University
of Technology (BTU), Cottbus
Postfach/P.O. Box 101344
03013 Cottbus
Germany
schmidtm@tu-cottbus.de

Dr. Elsa João
Graduate School
of Environmental Studies
University of Strathcycle
Level 6, Graham Hills Building
50, Richmond Street
Glasgow G1 1XN
Scotland
elsa.joao@strath.ac.uk

Dr. Eike Albrecht
Department of Environmental Law
Brandenburg University
of Technology (BTU), Cottbus
Postfach/P.O. Box 101344
03013 Cottbus
Germany
albrecht@tu-cottbus.de

Cataloging-in-Publication Data
Library of Congress Control Number: 2004115714

ISSN 1613-8694
ISBN 3-540-20562-4 Springer Berlin Heidelberg New York

This work is subject to copyright. All rights are reserved, whether the whole or part of the material is concerned, specifically the rights of translation, reprinting, reuse of illustrations, recitation, broadcasting, reproduction on microfilm or in any other way, and storage in data banks. Duplication of this publication or parts thereof is permitted only under the provisions of the German Copyright Law of September 9, 1965, in its current version, and permission for use must always be obtained from Springer-Verlag. Violations are liable for prosecution under the German Copyright Law.

Springer is a part of Springer Science+Business Media

springeronline.com

© Springer Berlin · Heidelberg 2005
Printed in Germany

The use of general descriptive names, registered names, trademarks, etc. in this publication does not imply, even in the absence of a specific statement, that such names are exempt from the relevant protective laws and regulations and therefore free for general use.

Hardcover-Design: Erich Kirchner, Heidelberg

SPIN 10972595   64/3130-5 4 3 2 1 0 – Printed on acid-free paper

Foreword

After a long and difficult legislative process, the European SEA Directive finally came into force in July 2001. The European Parliament as co-legislator always supported the idea of SEA and, therefore, of the SEA Directive. The European Parliament was also insistent that the concept of "monitoring" was brought into the SEA Directive. The SEA Directive is a major step forward to sustainable development. The Directive ensures that the environmental consequences of certain plans and programmes are identified and assessed during their preparation and before their adoption. Furthermore, the SEA Directive demands participation and information of the public as well as transboundary consultations. This will lead to better decisions and higher public acceptance of such plans and programmes. It is now crucial that SEA is carried out with quality and this Handbook could be a great help in this process. I welcome the publication of this Handbook as a basis for the necessary discussion on SEA for academics and practitioners, and for the implementation process of further assessment tools which is not finished yet.

Karl-Heinz Florenz, Chair of the Committee on the Environment, Public Health & Food Safety, European Parliament
Strasbourg, October 2004

It is very likely that Strategic Environmental Assessment (SEA) will have an ever-wider implementation in the future. The European SEA Directive, introduced nationally in 2004, represents only one important component in the widespread use of SEA. Developing countries have to be considered in the discussion of the move towards global sustainability, and SEA can be an invaluable tool for this. I particularly welcome this Handbook for including a wide range of points of view on SEA, ranging from developed to developing countries. The United Nations Environment Programme plays an increasingly key role in promoting sustainability and the World Summit on Sustainable Development, in Johannesburg 2002, was one important milestone in this respect. Sustainable Development has to be implemented at both national and international levels in order to change current trade policy and its effects on developing countries. Globalisation can cause environmental impacts, therefore there is a pressing need to do further research on experiences of SEA around the world and for SEA methodologies to be adopted for developing countries. For this to happen, capacity building for SEA is key. The spreading of knowledge, such as the contribution of this Handbook, is crucial in order that SEA can be fully embedded in decision-making processes.

Klaus Töpfer, Executive Director, United Nations Environment Programme
Nairobi, September 2004

Preface

This Handbook is about Strategic Environmental Assessment (or SEA), i.e. the environmental assessment of proposed policies, plans and programmes with the aim of informing decision-making. The main catalyst for this Handbook has been the implementation of the Directive 2001/42/EC, known as the "SEA Directive". By 21st July 2004 the EU Member States have had to implement this new Directive. According to this Directive, before a plan or programme is approved or adopted, the public must be consulted and an Environmental Report must be prepared describing the likely significant effects of the plan or programme and of any reasonable alternatives to the proposal. In addition, from 1st January 2004 United Nations countries – that are members of the United Nations Economic Commission for Europe (UNECE) or have consultative status with the UNECE – may ratify the 2003 Kiev SEA Protocol, with similar aims to the European Directive. It remains to be seen how many countries will ratify the Kiev SEA Protocol, though at the time of going to press there are 37 signatories. Although the SEA Directive is not the start of formal SEA legislation in Europe (e.g. The Netherlands have had SEA since 1987 and the Czech Republic since 1992), it is a fact that there is now a far wider predominance of SEA than ever before. This is why the Handbook was written and the authors feel it is a timely and important addition to the topic of SEA.

The Handbook covers a wide range of SEA-related topics for different sectors, different countries, different natural resources, and from different SEA perspectives: not all chapters are 'pro-SEA'; some also present a critique of SEA. The chapters are written by a variety of different authors: academics, practitioners, civil servants and upcoming SEA researchers. The Handbook is divided into eight Parts. Part I introduces key SEA principles and the legal framework in relation to the SEA Directive. Part II describes the implementation of the SEA Directive in eleven EU Member States. Part III discusses the SEA experience in the USA, in Canada and in New Zealand. Part IV evaluates SEA requirements in developing and fast developing countries by discussing the cases of Kenya, Ghana, Ukraine and China. Part V discusses methodologies for SEA and public participation. Part VI evaluates SEA for soils, water (including links with the European Water Framework Directive) and biodiversity. Part VII discusses SEA links with landscape planning, urban planning, transport planning, agriculture, waste management, and the mining industry. Finally, Part VIII concludes by discussing capacity-building, best practice, and future challenges and possibilities for SEA.

The Handbook has been edited in great detail in terms of content and format. All chapters start with an introduction, and end with a section on conclusions and recommendations for future practice. In tandem with the theory, many of the chapters have a case study in a box that illustrates real-world examples. A key component of the Handbook is the *consolidated list of legislation* that appears at the end of the Handbook. All the legislation that is mentioned in the different chapters is listed in this single list of legislation that is organised by country, by European Union legislation and by International Conventions. In the chapters, legislation is mentioned by a short name (e.g. Treaty of Nice or EIA Directive) rather than by a long legal name and number. The consolidated list of legislation then gives detailed information about each law and regulation. We believe this will greatly increase the usefulness of the Handbook.

The completion of this Handbook leaves us indebted to many people. First of all we wish to thank the 61 contributors, drawn from 18 different countries from all over the world, without whose articles this Handbook would not have been possible. Twenty-one of these contributors participated at a workshop on SEA in Marienthal, Germany, on 23-26 November 2003. We thank the German Academic Exchange Service in Bonn (DAAD – Deutscher Akademischer Auslandsdienst) and the German Federal Foundation for Environment (DBU – Deutsche Bundesstiftung Umwelt) for generously financing the Marienthal SEA workshop that provided such fruitful discussions. These annual scientific workshops are key events organised by the Brandenburg University of Technology (BTU) Cottbus, Germany. They are carried out by an international network for Education and Research in Environmental and Resource Management (ERM) with more than 20 partner universities world-wide working in close co-operation with UNEP.

We are particularly indebted to the excellent work provided by Hendrike Helbron without whom the writing of this Handbook would have been much more difficult. Several other individuals provided special assistance in the final stages of the preparation of the Handbook and we are very thankful for their help at such a crucial time. Among these were Vyacheslav Afanasyev, Heike Bartholomäus, Anna Kulik, Heiko Lübs, Dmytro Palekhov and Susanne Scheil. We are also indebted to the publishers and authors of publications who have granted copyright permission to reproduce extracts from their work for inclusion in the Handbook.

The writing of this Handbook was the result of a collaboration between the Brandenburg University of Technology (BTU) Cottbus in Germany and the Graduate School of Environmental Studies at the University of Strathclyde in Scotland. To all the researchers, academics, practitioners, students, and central and local government officials, we hope you will find this Handbook invaluable in your SEA work and research.

Michael Schmidt, Elsa João and Eike Albrecht
Cottbus and Glasgow, September 2004

Table of Contents

Foreword V

Preface VII

List of Contributors XXI

Part I – Key Principles and Legal Framework for SEA 1

1 Key Principles of SEA 3
Elsa João

    1.1   Introduction 3
    1.2   Environmental Assessment of 'Strategic Actions' 4
    1.3   Tiering and the Link between SEA and Project EIA 4
    1.4   Two Key Principles of SEA 7
    1.5   The Ideal SEA Team 12
    1.6   Conclusions and Recommendations 13

2 Purpose and Background of the European SEA Directive 15
Christian Kläne and Eike Albrecht

    2.1   Introduction 15
    2.2   The SEA Directive and its European Context 15
    2.3   Definition of SEA 22
    2.4   History of the SEA Directive 22
    2.5   Content, Objectives and Purpose of the SEA Directive 25
    2.6   Outlook 27
    2.7   Conclusions and Recommendations 28

3 Legal Context of the SEA Directive – Links with other Legislation and Key Procedures 31
Eike Albrecht

    3.1   Introduction 31
    3.2   Competence of the Community 31
    3.3   Relation to the Aarhus Convention 33
    3.4   Relation to the Espoo Convention 34
    3.5   Harmonisation of Environmental Assessment Procedures 34
    3.6   Relation to other European Acts 35
    3.7   Area of Application 39
    3.8   Procedure of SEA 45
    3.9   Consequences of Mistakes in the SEA Procedure 52
    3.10 Conclusions and Recommendations 52

4 Transposition of the SEA Directive into National Law – Challenges and Possibilities 57
Lothar Knopp and Eike Albrecht

- 4.1 Introduction 57
- 4.2 Connection to Member States' Laws 57
- 4.3 Transposition of the SEA Directive 58
- 4.4 Transposition into the National Legal System 63
- 4.5 Conclusions and Recommendations 66

5 Some Legal Problems of Implementing the SEA Directive into Member States' Legal Systems 69
Jerzy Sommer

- 5.1 Introduction 69
- 5.2 Implementation Problems in Member States 69
- 5.3 Problems in the Implementation in Poland 73
- 5.4 Conclusions and Recommendations 78

Part II – Current Status and National Strategies for the Implementation of SEA in the European Union 81

6 Current SEA Practice in England 83
Thomas B. Fischer

- 6.1 Introduction 83
- 6.2 Current SEA-Type Assessment Practice 84
- 6.3 Practice in Land Use Planning in England 86
- 6.4 Practice in Transport Planning 91
- 6.5 Conclusions and Recommendations 95

7 Implementing SEA in Germany 99
Thomas Bunge

- 7.1 Introduction 99
- 7.2 Intended Amendments to the EIA Act 101
- 7.3 The Amendment to the Building Code of 24 June 2004 105
- 7.4 The Amendment to the Spatial Planning Act of 24 June 2004 108
- 7.5 Links between the new SEA Legislation and Existing Assessment / Planning Tools 108
- 7.6 Methodology 109
- 7.7 Data Requirements 110
- 7.8 Support of the Implementation and Quality Control 111
- 7.9 Conclusions and Recommendations 114

8 Implementing SEA in Italy – The Case of the Emilia Romagna Region 117
Paola Gazzola and Maristella Caramaschi
   8.1   Introduction 117
   8.2   Using the Emilia Romagna Region to Explore the Implementation Status of SEA in Italy 119
   8.3   Context of Planning in Italy 119
   8.4   Relationship between Strategic Planning and SEA 121
   8.5   Planning in the Region of Emilia Romagna 122
   8.6   VALSAT versus SEA Directive 128
   8.7   Implementing the SEA Directive in the Emilia Romagna Region 129
   8.8   A Look at the Bigger Picture: Implementing the SEA Directive in Italy 130
   8.9   Conclusions and Recommendations: Considerations for the Region of Emilia Romagna and Italy as a whole 131

9 First Experiences with implementing SEA Legislation in Flanders (Belgium) 137
Marc Van Dyck
   9.1   Introduction 137
   9.2   Development of a Flemish combined Legislation on EIA and SEA 138
   9.3   Practice of SEA in Flanders 141
   9.4   Common Points of Interest for Future Use of SEA in Flanders 144
   9.5   Conclusions and Recommendations 147

10 Implementing SEA in Austria 149
Ralf Aschemann
   10.1   Introduction 149
   10.2   Legislative Efforts to implement the SEA Directive 150
   10.3   Case Studies 150
   10.4   SEA of the Land use Plan of Weiz 152
   10.5   Accompanying Measures 155
   10.6   Conclusions and Recommendations 156

11 Implementing SEA in Finland – Further Development of Existing Practice 159
Mikael Hildén and Pauliina Jalonen
   11.1   Introduction 159
   11.2   Scope of Application – the EIA Act and the SEA Directive 160
   11.3   SEA in Practice 161
   11.4   Discussion 169
   11.5   Conclusions and Recommendations 172

12 Problems of a Minimalist Implementation of SEA – The Case of Sweden 177
Lars Emmelin and Peggy Lerman

12.1 Introduction 177
12.2 Environmental Assessment in Sweden 178
12.3 The Legislation Implementing the SEA Directive 181
12.4 Understanding the Swedish Mode of Implementation 185
12.5 Conclusions and Recommendations 189

13 National Strategy for the Implementation of SEA in the Czech Republic 193
Lucie Václavíková and Harald Jendrike

13.1 Introduction 193
13.2 Previous Situation Regarding SEA in the Czech Republic 193
13.3 Current Situation Regarding SEA in the Czech Republic 196
13.4 Ensuring the Quality of SEA 199
13.5 Conclusions and Recommendations 200

14 Developments of SEA in Poland 201
Joanna Maćkowiak-Pandera and Beate Jessel

14.1 Introduction 201
14.2 Development of SEA 202
14.3 Compliance of the Polish Environmental Protection Law with the European SEA Directive 204
14.4 Administrative Structure and Procedure of SEA in Poland 206
14.5 Environmental Planning and SEA 209
14.6 Availability of Environmental Data in Poland 210
14.7 Examples of SEA in Poland 211
14.8 Methodological Problems and Quality of SEA 213
14.9 Conclusions and Recommendations 214

15 Experiences with SEA in Latvia 217
Sandra Ruza

15.1 Introduction 217
15.2 Practical Experience Gained with Existing Assessment Instruments and its Legal Provisions 218
15.3 Review of Differences in Applying SEA in Latvia 219
15.4 Quality Assurance 222
15.5 Case Study – SEA for the Latvia Development Plan 223
15.6 Conclusions and Recommendations 224

16 Implementing SEA in Estonia 227
Olavi Hiiemäe

    16.1   Introduction 227
    16.2   Project EIA and SEA in Estonia 228
    16.3   Actors in the SEA Process in Estonia 229
    16.4   SEA in Practice 230
    16.5   The Quality Criteria of the SEA Document 233
    16.6   Conclusions and Recommendations 234

Part III – Experience of SEA in North America and Oceania 237

17 Improved Decision-Making through SEA – Expectations and Results in the United States 239
Jehan El-Jourbagy and Tyson Harty

    17.1   Introduction 239
    17.2   National Environmental Policy Act 239
    17.3   NEPA Analysis: Procedural Process 242
    17.4   Successes and Failures 246
    17.5   Conclusions and Recommendations 249

18 SEA in Canada 251
Norma Powell

    18.1   Introduction 251
    18.2   SEA Legislation and Principles 252
    18.3   Roles and Responsibilities 255
    18.4   Methodologies and Tools 256
    18.5   Documentation and Reporting 261
    18.6   Critical Analysis 261
    18.7   Canadian SEA Case Study 264
    18.8   Conclusions and Recommendations 267

19 SEA of Plan Objectives and Policies to Promote Sustainability in New Zealand 269
Ali Memon

    19.1   Introduction 269
    19.2   Competing Rationales Underpinning Section 32 271
    19.3   The Formative Period (1974-1984) 272
    19.4   A New Era in Environmental Impact Assessment (1991-Present) 275
    19.5   Section 32 Methodological Procedure 279
    19.6   Conclusions and Recommendations 285

Part IV – Requirements of SEA in Developing and Fast Developing Countries 289

20 The Need for SEA in Kenya 291
Vincent Onyango and Saul Namango

    20.1 Introduction 291
    20.2 Kenya and its Formulation of Strategic Decisions 292
    20.3 History of Environmental Assessment of Strategic Decisions in Kenya 293
    20.4 SEA Elements in Kenya and Tools to Implement them 294
    20.5 Institutional Framework for SEA Elements 297
    20.6 Need and Justification for SEA in Kenya 298
    20.7 How SEA will Supplement EIA and Physical Planning in Kenya 300
    20.8 Conclusions and Recommendations 301

21 SEA for Water Resource Management in Ghana 305
Eric Ofori

    21.1 Introduction 305
    21.2 Impacts of Water Resource Developments in Ghana 306
    21.3 Environmental Assessment in Ghana 308
    21.4 Applying SEA and EIA to Water Resource Management in Ghana 310
    21.5 Benefits of SEA for Water Resource Management in Ghana 313
    21.6 Conclusions and Recommendations 315

22 SEA in Ukraine 321
Vyacheslav Afanasyev

    22.1 Introduction 321
    22.2 Historical Background of SEA in Ukraine 321
    22.3 The Existing Conditions of SEA Application in Ukraine 322
    22.4 Obstacles for SEA Application in Ukraine 326
    22.5 Possible Levels of SEA Development in Ukraine 327
    22.6 Conclusions and Recommendations 328

23 Importance of SEA in China – The Case of the Three Gorges Dam Project 331
Cynthia Huang and Jennifer Yang

    23.1 Introduction 331
    23.2 Environmental Situation and Policies in China 331
    23.3 The Case Study of the Three Gorges Project 336
    23.4 The Three Gorges Project and SEA 341
    23.5 Conclusions and Recommendations 344

Part V – Methodologies for SEA and Public Participation 347

24 Tools for SEA 349
Riki Thérivel and Graham Wood

24.1 Introduction 349
24.2 What Makes a Good SEA Tool? 349
24.3 SEA Tools 350
24.4 Impact Matrices 354
24.5 Geographical Information Systems 356
24.6 Causal Effect Diagrams 361
24.7 Conclusions and Recommendations 362

25 Methodological Approaches to SEA within the Decision-Making Process 365
Beate Jessel

25.1 Introduction 365
25.2 Procedural and Methodological Demands on SEA and EIA – a Comparison 366
25.3 Identifying Possible Alternatives for Waste Dumping-Ground Sites 372
25.4 Identification and Assessment of Possible Corridors and Alignments for New Roads 375
25.5 Landscape Planning as an Information Base for SEA .. 380
25.6 Conclusions and Recommendations 382

26 Strategic Level Cumulative Impact Assessment 385
Riki Thérivel

26.1 Introduction 385
26.2 Definitions and Concepts 386
26.3 Principles of Cumulative Impact Assessment 388
26.4 Carrying out Cumulative Impact Assessment 388
26.5 Conclusions and Recommendations 394

27 Handling Transboundary Cumulative Impacts in SEA 397
Tyson Harty, Daniel Potts, Donald Potts and Jehan El-Jourbagy

27.1 Introduction 397
27.2 Definition of Cumulative Effects 398
27.3 Cumulative Effects Analysis in the US 399
27.4 Transboundary Implications of Cumulative Effects 405
27.5 Conclusions and Recommendations 407

28 Cultural Integrity as a Criterion of SEA 409
Engelberth Soto-Estrada, Rina Aguirre-Saldivar and Shafi Noor Islam

28.1 Introduction 409
28.2 The Impact on Cultural Integrity 410
28.3 Consideration of Cultural Elements within SEA 411
28.4 The Concept of Culture 412
28.5 The Cultural Impact Assessment 416
28.6 Cultural Impact Assessment Procedure 417
28.7 Conclusions and Recommendations 419

29 Requirements and Methods for Public Participation in SEA 421
Stefan Heiland

29.1 Introduction 421
29.2 Opportunities and Obstacles of Public Participation in SEA 422
29.3 What Does "Public" Mean? 423
29.4 What Does Participation Mean? 424
29.5 Methods of Public Participation within SEA – Possibilities and Experiences 427
29.6 Conclusions and Recommendations 430

30 Public Participation for SEA in a Transboundary Context 433
Harry Meyer-Steinbrenner

30.1 Introduction 433
31.2 Public Participation in the SEA Directive 434
30.3 Methodological Aspects 435
30.4 Public Participation for the SEA in a Transboundary Context 436
30.5 Pilot Project 436
30.6 Conclusions and Recommendations 440

31 Developing Quantitative SEA Indicators Using a Thermodynamic Approach 443
Jo Dewulf and Herman Van Langenhove

31.1 Introduction 443
31.2 Sustainability and Technology 444
31.3 Thermodynamics – its Basic Laws and the Exergy Concept 447
31.4 Quantitative Information for the SEA of Technology from theThermodynamic Approach 449
31.5 Conclusions and Recommendations 456

32 A Structural and Functional Strategy Analysis for SEA 459
Anastássios Perdicoúlis

32.1 Introduction 459
32.2 Background and Definitions 460

	32.3	The Method of Strategy Analysis 462
	32.4	Contribution of the Strategy Analysis... 466
	32.5	Conclusions and Recommendations 468

Part VI – SEA for Abiotic and Biotic Resources 471

33 Soil Resources and SEA 473
Robert Mayer

    33.1 Introduction 473
    33.2 SEA Links with Environmental and Planning Legislation
         in Germany 475
    33.3 Impacts and Effects – Adequate Methods for Soil Evaluation 478
    33.4 Methods Relating Soil Properties with Functions............................ 482
    33.5 Strength and Weakness of Methodologies, Data Requirements 487
    33.6 Synergies from the Use of a Common Environmental
         Data Base 489
    33.7 Conclusions and Recommendations 491

34 Towards the Implementation of SEA – Learning from EIA for Water Resources 495
Damian Lawler

    34.1 Introduction 495
    34.2 Water in the SEA and EIA Processes 496
    34.3 Water and energy resources of Azerbaijan 498
    34.4 Environmental Impacts Summary... 499
    34.5 Challenges for EIA Procedures in the FSU 499
    34.6 Water Resource Data Appropriateness: General Limitations
         for EA 500
    34.7 Specific Environmental Data Issues in the FSU 504
    34.8 A Revised Water Impact Assessment (WIA) Procedure 504
    34.9 SEA into the Future: Learning from Project EIA 505
    34.10 Conclusions and Recommendations 508

35 Links between the Water Framework Directive and SEA.. 513
Natalia Gullón

    35.1 Introduction 513
    35.2 From a "Hydrological Policy" to a Sustainable "Water Policy"...... 513
    35.3 Strengths of SEA in the Water Sector.. 514
    35.4 Weaknesses of SEA in the Water Sector 516
    35.5 SEA Directive and the Water Framework Directive........................ 517
    35.6 Conclusions and Recommendations 521

36 Assessing Biodiversity in SEA 523
Udo Bröring and Gerhard Wiegleb

36.1 Introduction 523
36.2 Definition of Biodiversity 524
36.3 Measurability of Biodiversity in Theory and Practice 525
36.4 The Value of Biodiversity 528
36.5 Conclusions and Recommendations 533

37 Biodiversity Programmes on Global, European and National Levels Related to SEA 539
Gerhard Wiegleb and Udo Bröring

37.1 Introduction 539
37.2 Determinants or Driving Forces of Biodiversity 540
37.3 Biodiversity in SEA 541
37.4 Available Data for Biodiversity Assessment 546
37.5 Decision-Aid in Biodiversity Protection 547
37.6 A Hypothetical Example 548
37.7 Conclusions and Recommendations 550

Part VII – Implementing SEA in Spatial and Sector Planning 555

38 Co-ordination of SEA and Landscape Planning 557
Frank Scholles and Christina von Haaren

38.1 Introduction 557
38.2 The Intention of the SEA Directive 558
38.3 The Situation in Germany 558
38.4 The Intersection of SEA and Landscape Planning 559
38.5 Conclusions and Recommendations – A Model for Procedural Integration and its Requirements 566

39 Urban Planning and SEA 571
Markus Reinke

39.1 Introduction 571
39.2 Existing Liabilities due to the Federal Building Code 572
39.3 Relation of Landscape Planning and Urban Planning 573
39.4 Analysis of SEA Liabilities Exceeding Previous Requirements 575
39.5 Suggestions for Contents of SEA and the Assessment of Alternatives 578
39.6 The Question of Acceptability 582
39.7 Conclusions and Recommendations 583

40 SEA in Transport Planning in Germany **585**
Wolfgang Stein, Jürgen Gerlach and Paul Tomlinson

    40.1   Introduction 585
    40.2   Screening 586
    40.3   Scoping 587
    40.4   Identifying and Describing the Environmental Effects 590
    40.5   Assessing Alternatives 592
    40.6   Decision-Making Process 592
    40.7   Monitoring 595
    40.8   Conclusions and Recommendations 595

41 SEA for Agricultural Programmes in the EU **599**
Michael Schmidt, Harry Storch and Hendrike Helbron

    41.1   Introduction 599
    41.2   EU Agricultural Policy 1992-2005 600
    41.3   Influence of CAP on Agricultural Land Use 602
    41.4   SEA for Agricultural Plans and Programmes 605
    41.5   Suitability of the German Preliminary Agrarian Structure
           Planning (gAEP) for the implementation of SEA 608
    41.6   Conclusions and Recommendations 614

42 SEA of Waste Management Plans – An Austrian Case Study **621**
Kerstin Arbter

    42.1   Introduction 621
    42.2   Fields of Application 621
    42.3   The SEA for the Viennese Waste Management Plan 622
    42.4   Methodologies and Data 623
    42.5   Process Design and Public Participation 625
    42.6   Conclusions and Recommendations 628

43 Mining Industry and SEA – An Example in Turkey **631**
Evren Yaylaci

    43.1   Introduction 631
    43.2   SEA and the Mining Industry 632
    43.3   Challenges of SEA in Developing Countries 640
    43.4   Conclusions and Recommendations 642

Part VIII – Conclusions **647**

44 Capacity-Building and SEA **649**
Maria do Rosário Partidário

    44.1   Introduction 649
    44.2   Meaning of Capacity-Building for SEA 650

44.3 Current Key Priorities for SEA as a Strategic Impact Assessment Approach 653
44.4 What are the Needs and the Key Drivers in SEA Capacity-Building? 655
44.5 Training for SEA Capacity-Building 659
44.6 Conclusions and Recommendations 661

45 A Critique of SEA from the Point of View of the German Industry 665
Jürgen Ertel

45.1 Introduction 665
45.2 The Role of Industry in the Legislatorial Process 666
45.3 The Priority Level of SEA for the Industry 666
45.4 Involvement of the Industry in the Legislative Process 668
45.5 Conclusions and Recommendations 671

46 Best Practice Use of SEA – Industry, Energy and Sustainable Development 673
Ross Marshall and Thomas B. Fischer

46.1 Introduction 673
46.2 Institutional Links Between SEA and Industry 674
46.3 SEA Application in Industry Planning 675
46.4 SEA – A Strategic Decision-Making Framework for Businesses 677
46.5 Electricity Provision and the UK Transmission Industry ... 678
46.6 The Perceived Benefits of SEA to ScottishPower 680
46.7 The Mid-Wales Case Study – ScottishPower's First Steps to SEA implementation 680
46.8 Applying the Seven Phases of Regional Transmission Network SEA 685
46.9 Conclusions and Recommendations 687

47 SEA Outlook – Future Challenges and Possibilities 691
Elsa João

47.1 Introduction 691
47.2 SEA Barrier 1: Bland Alternatives 691
47.3 SEA Barrier 2: Weak Public Participation 693
47.4 SEA Barrier 3: Lack of the 'Right Data' 694
47.5 SEA Barrier 4: Poor Procedures and Methodology 697
47.6 Conclusions and Recommendations 698

Appendix 1 – The Full Text of the SEA Directive 701

Appendix 2 – Consolidated List of Legislation 715

Subject Index 733

List of Contributors

Vyacheslav Afanasyev is a doctoral student at the department of Environmental Planning at Brandenburg University of Technology (BTU) Cottbus, Germany. His PhD research project deals with monitoring and evaluation techniques, development of indicator systems and environmental assessment.
Brandenburg University of Technology (BTU) Cottbus, P.O. Box 101344, D-03013 Cottbus
Tel: +49 (355) 869 89 79; Email: afanasyev_v@mail.ru

Rina Aguirre-Saldivar is a full-time lecturer at the National University of Mexico (UNAM) since 1980. She has been head of department at the National Institute of Environment and at the Ministry of Natural Resources of Mexico. Main research areas are air pollution and environmental management.
Facultad de Ingeniería, UNAM, Ciudad Universitaria, 04510 México, D.F.,
Tel: +55 (0) 56 22 30 01; Fax: +55 (0) 56 16 10 73; Email: ras@correo.unam.mx

Eike Albrecht is senior lecturer at the Department of Constitutional, Administrative and Environmental Law, Brandenburg University of Technology Cottbus (BTU) since 1999. In 2002 he has received his Doctorate Degree at the University of Leipzig, Germany in soil protection law which is still the main research area. Further research areas are international law and product related civil and administrative law.
Department of Constitutional, Administrative and Environmental Law, Brandenburg University of Technology (BTU) Cottbus, P.O. Box 101344, 03013 Cottbus, Germany
Tel.: +49 (0)355 69 27 49; Fax: +49 (0)355 69 35 02; Email: albrecht@tu-cottbus.de;
Web: www.tu-cottbus.de/Umweltrecht/eike_albrecht.htm

Kerstin Arbter, a landscape planner and an SEA expert, is head of the consultancy "Arbter – SEA Consulting & Research" in Vienna. She has carried out SEA in various sectors and has developed the "SEA Round Table" – an SEA model for pro-active stakeholder participation.
Arbter – SEA Consulting & Research, Vorgartenstraße 145-157/2/16, 1020 Vienna, Austria
Tel/Fax: +43 (0) 1 218 53 55; Email: office@arbter.at; Web: www.arbter.at

Ralf Aschemann, an environmental scientist, is director of the Austrian Institute for the Development of Environmental Assessment (An !dea). One of his main research interests is the application, evaluation, quality insurance and development of environmental assessment instruments, especially SEA.
Austrian Institute for the Development of Environmental Assessment (An !dea), Elisabethstrasse 3/3
8010 Graz, Austria
Tel. +43 (0)316 31 81 98; Fax +43 (0)316 38 46 777; Email: office@anidea.at; Web: www.anidea.at

Udo Bröring is lecturer at the Department of General Ecology, Brandenburg University of Technology (BTU) Cottbus, Germany since 1994. Main research areas are theoretical and applied ecology and zoology.

Department of General Ecology, Brandenburg University of Technology (BTU) Cottbus, P.O. Box 101344, 03013 Cottbus, Germany
Tel: +49 (0)355 69 27 46; Fax: +49 (0)355 69 22 25; Email: broering@tu-cottbus.de;
Web: www.tu-cottbus.de

Thomas Bunge is the head of the section "Environmental Impact Assessment" at the German Federal Environmental Agency (*Umweltbundesamt*), Berlin. He is actively involved, inter alia, in the activities to implement the SEA Directive in Germany at the federal level. He is also an honorary Professor for Environmental and Planning Law at the Technical University of Berlin and co-editor of the EIA Handbook (*Handbuch der Umweltverträglichkeitsprüfung,* loose-leaf collection, Erich Schmidt Verlag, Berlin).

German Federal Environmental Agency, Bismarckplatz 1, 14193 Berlin, Germany
Tel: +49 (0)30 89 03 27 20; Fax: +49 (0)30 89 03 29 06; Email: thomas.bunge@uba.de;
Web: www.umweltbundesamt.de

Jo Dewulf is assistant professor at Ghent University with research and teaching activities in clean and environmental technology. As part of the research group ENVOC he investigates quantitative indicators for the assessment of the environmental sustainability of technology.

Environmental and Clean Technology, Research Group ENVOC, Ghent University, Coupure Links 653, B-9000 Ghent, Belgium
Tel: +32 (0)92 64 59 49; Fax: +32 (0)92 64 62 43; Email: jo.dewulf@ugent.be;
Web: http:www.envoc.ugent.be

Jehan El-Jourbagy is an attorney and environmental advocate. She worked on environmental issues at the Oregon State Public Interest Research Group and at the Western Environmental Law Center. She is a 2003 graduate of the University of Georgia Law School, where she was president of the Environmental Law Association and co-director of the Coalition Against Environmental Racism.

Oregon Bar & Environmental Advocate, 1875 Longview Avenue, Eugene, Oregon 97403 USA
Email: jeljourbagy@yahoo.com

Lars Emmelin is professor of environmental assessment in the Spatial planning programme, Blekinge Institute of Technology at Karlskrona. He has researched and taught on environmental issues in Sweden and Norway, worked with conservation and tourism in the Arctic and Scandinavian mountains and worked for major international organisations such as the Unesco, UNEP and OECD on environmental education. At present he is programme director of a major research programme on tools for SEA funded by the Swedish Environment Protection Agency and directs a MSc on European Spatial Planning.

Spatial planning, BTH, SE 371 79 Karlskrona, Sweden
Email: lars.emmelin@bth.se

Jürgen Ertel is the founder and head of the Department of Industrial Sustainability at the Brandenburg University of Technology, Cottbus since 1994. Previous to his university career he worked for Siemens AG and established relations to various industry associations. His focus is on environmentally benign design of industrial goods, the rating methods and in particular the recycling properties.

Head of the Department of Industrial Sustainability, Brandenburg University of Technology (BTU) Cottbus, P.O. Box 101344, 03013 Cottbus, Germany
Tel: +49 (0)355 69 43 85; Fax: +49 (0)355 69 47 00; Email: ertel@tu-cottbus.de;
Web: www.tu-cottbus.de/neuwertwirtschaft

Thomas Fischer is a senior lecturer in the Department of Civic Design at the University of Liverpool. He previously worked as project manager, administrator and researcher in several private and public institutions on EIA and SEA. He is chair of the SEA section of the International Association for Impact Assessment (IAIA) and member of various SEA and EIA related boards and networks, among which the international advisory board of the EIA Review journal and the BEACON (Building Environmental Assessment CONsensus on the transeuropean transport) network. He has been involved in projects and provided training on SEA for the European Commission, UNEP, UNDP, World Bank and others.

Department of Civic Design, Liverpool University, 74 Bedford Street South, Liverpool L69 7ZQ, England
Tel: +44 (0)151 7 94 31 13; Fax: +44 (0)151 7 94 31 25; Email: fischer@liverpool.ac.uk

Paola Gazzola is a planner and currently undertaking a PhD at the University of Liverpool in England. Her current research project is aiming at identifying the most appropriate form on integrating SEA in different planning systems, focusing particularly on Italy, the UK and The Netherlands.

Department of Civic Design, Liverpool University, 74 Bedford Street South, Liverpool L69 7ZQ, England
Tel : +44 (0)151 7 94 34 53; Fax: +44 (0)151 7 94 31 25; Email: gazzola@liv.ac.uk

Jürgen Gerlach is the head of the Department of Civil Engineering at the University of Wuppertal. His research and lecturing fields are road safety training programmes, road traffic planning and engineering. He is further member of a number of relevant committees related to transport and urban planning.

Department of Civil Engineering, University of Wuppertal, Pauluskirchstraße 7, 42285 Wuppertal, Germany
Tel: +49 (0)202 4 39 40 87 or 40 88; Fax: +49 (0)202 4 39 40 88; Email: jgerlach@uni-wuppertal.de;
Web: www.svpt.de

Natalia Gullón is a Civil Engineer at the Spanish Ministry of Environment. She has been working at the Júcar River Basin Authority, one of the pilot river basins for the application of the Water Framework Directive. One of her main research interests is sustainability and the instruments for its assessment, specially SEA.

Ministerio de Medio Ambiente, Plaza de San Juan de la Cuz s/n, 28071 Madrid, Spain
Tel: +34 (0)9 15 97 60 00; Fax: +34 (0) 9 15 97 59 71; Email: ngullon@mma.es

Christina von Haaren is a full professor and director of the Institute of Landscape Planning and Nature Conservation (ILN) of the University of Hanover. She is also a member of the German Council of Environmental Advisors (SRU).

Christina von Haaren, Institut of Landscape Planning and Nature Conservation, University of Hanover, Herrenhäuser Str. 2, 30419 Hanover, Germany

Tyson Harty is a doctoral student in the Department of Zoology at Oregon State University. He is a National Science Foundation Graduate Research Fellow with primary emphasis on adaptive control and behaviour in biological systems. His research has examined the group dynamics, conservation, and ecology of free-living animal populations in the United States and Kenya.

Department of Zoology, Oregon State University, Corvallis, Oregon 97331, USA
Email:tyson.harty@orst.edu

Stefan Heiland is a landscape planner and responsible for questions on urban ecology at the Leibniz Institute of Ecological and Regional Development (IOER.) One of his main topics is communication and participation in planning processes.

Leibniz Institute of Ecological and Regional Development (IOER), Weberplatz 1, 01217 Dresden, Germany
Tel. +49 (351) 46 79 219; Fax +49 (351) 46 79 212; Email: s.heiland@ioer.de; Web: www.ioer.de

Hendrike Helbron is a scientific research assistant and lecturer at the Brandenburg Technical University of Cottbus, Department of Environmental Planning. The topic of her PhD thesis involves an analysis, evaluation and application of environmental indicators for SEA for regional planning.

Department of Environmental Planning, Brandenburg University of Technology (BTU), Cottbus, P.O. Box 101344, D-03013 Cottbus, Germany
Tel.:+49 (0) 3 55 69 23 52; Fax: +49 (0) 3 55 69 27 65; Email: helbron@tu-cottbus.de

Olavi Hiiemäe is a doctoral student in the Swedish EIA-centre at Swedish Agricultural University (SLU). His research topic is dealing with environmental and socio-economic analysis of landscape and land use changes in Estonia. He is a researcher and lecturer in the Laboratory of Landscape Ecology, Environmental Protection Institute at Estonian Agricultural University.

Environmental Protection Institute, Laboratory of Landscape Ecology, Estonian Agricultural University, 51003 Tartu, Estonia
Tel: +372 (0) 742 74 34; Fax: +372 (0) 742 74 32; Email: olavi@envinst.ee; Web: www.envinst.ee

Mikael Hildén is the programme director of the Finnish Environment Institute. He has been the leader of several national policy assessments and evaluations of environmental policy instruments. He has participated in the development of SEA legislation in Finland and in the European Union.

Finnish Environment Institute (SYKE), Research Department, PO Box 140, FIN-00251 Helsinki, Finland
Tel: +358 (0) 9 40 30 04 01; Fax: +358 (0) 9 40 30 04 90; Email: mikael.hilden@ymparisto.fi; Web: http://www.ymparisto.fi/syke

Cynthia Huang is a graduate student at the Brandenburg Technical University of Cottbus, Germany. She has integrated SEA into the management of World Heritage Sites in her Master Thesis, of which the topic is "Strategic Environmental Assessment and its Implementation Procedure at Programme Level for World Heritage Sites: A Case Study at Wulingyuan Scenic and Historic Interest Area, China".

Juri-Gagarin-Str.3/305, 03046 Cottbus, Germany
Tel : +49 (0) 1794842770; Email: cynthia_huang@hotmail.com

Shafi Noor Islam is a PhD student at the Brandenburg Technical University at Cottbus, Germany. He is interested to implement SEA in his PhD research project: "Cultural Landscapes Changing due to Human Influences and Threats to Ecosystems: A Case Study on the Sundarbans for Sustainable Management".

Chair of Ecosystems and Environmental Informatics. Brandenburg Technical University-Cottbus.
P.O.Box.-131044, D- 03013 Cottbus, Germany
Tel: +49 (0)355 69 28 31; Fax: +49 (0)355 69 27 43; Email: shafinoor@yahoo.com

Pauliina Jalonen is a researcher in the programme for environmental policy of the Finnish Environment Institute. Her research focuses on risk assessments and municipal health programmes.

Finnish Environment Institute (SYKE), Research Department, PO Box 140, FIN-00251 Helsinki Finland
Tel: +358 (0)9 40 30 09 96; Fax: +358 (0)9 40 30 03 91; Email: pauliina.jalonen@ymparisto.fi;
Web: http://www.ymparisto.fi/syke

Harald Jendrike is a lawyer and now the head of the Department "Nature reserves, interventions, landscape planning" at the Saxon State Ministry for Environment and Agriculture. In 2003/2004 he worked as a Pre-Accession Adviser in the field of EIA and SEA at the Czech Ministry of Environment.

Saxon State Ministry of the Environment and Agriculture, Archivstraße 1, 01097 Dresden, Germany
Tel: +49 (0)351 5 64 21 22; Fax: +49 (0)351 5 64 21 30; Email: harald.jendrike@smul.sachsen.de;
Web: www.smul.sachsen.de and www.env.cz/EIA.Web/index.htm

Beate Jessel is full professor for Landscape Planning at the University of Potsdam. She also holds numerous functions in advisory boards, such as the advisory board for regional planning and development at the German Federal Ministry of Transport, Building and Housing and the German Council for Land Stewardship and is the chair of the scientific advisory committee of the Leibniz Institute of Ecological and Regional Development in Dresden. Her special research interests are in the field of ecologically oriented planning, in particular environmental impact assessment, assessment according to the Habitats Directive and watershed management according to the requirements of the Water Framework Directive.

University of Potsdam, Institut for Geoökologie, Postfach 601553, 14415 Potsdam
Tel: +49 (331) 977-2116; Fax: +49 (331) 977-2068; Email: jessel@rz.uni-potsdam.de;
Web: http://www.uni-potsdam.de/u/Geooekologie/index.htm

Elsa João is a lecturer and the director of research of the Graduate School of Environmental Studies (GSES) at the University of Strathclyde, Scotland. At GSES she is responsible for the PhD programme and helps run the successful MSc in Environmental Studies. She has expertise in the areas of SEA, Project EIA, environmental analysis, GIS, scale issues, and spatial data quality. She runs both long and short courses on SEA.

Graduate School of Environmental Studies – GSES, University of Strathclyde, Level 6, Graham Hills Building, 50 Richmond St, Glasgow G1 1XN, Scotland
Tel: +44 (0) 141 54 84 05 6; Fax: +44 (0) 141 5 48 34 89; Email: elsa.joao@strath.ac.uk;
Web: http://www.strath.ac.uk/Departments/GSES/

Christian Kläne is the head of the Department "Company Audit", Tax Administration of Lower Saxony, Germany, since 2003. In 2002 he completed his doctorate thesis, with the support of the German Federal Environmental Foundation, Osnabrück, on SEA in preparatory land use planning at the University of Osnabrück, Germany.

Department "Company Audit", Tax Administration of Lower Saxony, Revenue Office, Eisenbahnweg 4a, D - 49699 Cloppenburg,
Email: dr-christian.klaene@fa-clp.niedersachsen.de

Lothar Knopp is the head of the Department of Constitutional, Administrative and Environmental Law, Brandenburg University of Technology (BTU) Cottbus since 1999 when he was offered a chair at the BTU. His main research areas are administrative, constitutional and environmental law including environmental liability law and environmental management.

Head of the Department of Constitutional, Administrative and Environmental law, Brandenburg University of Technology (BTU) Cottbus, P.O. Box 101344, D-03013 Cottbus, Germany
Tel.: +49 (0)355 69 21 16; Fax: +49 (0)355 69 35 02; Email: umweltrecht@tu-cottbus.de;
Web: www.tu-cottbus.de/Umweltrecht/leitung.htm

Damian Lawler is head of the Hydrogeomorphology Research Group in the School of Geography, Earth and Environmental Sciences at the University of Birmingham. He has published extensively on hydrological and water resource issues, and his principal research interests lie in river process responses, climate change impacts on river flows, urban impacts on river systems, water quality and the development of improved 'third generation' Environmental Assessment protocols, especially as applied to water resources.

School of Geography, Earth and Environmental Sciences, The University of Birmingham, Birmingham B15 2TT, England
Tel: +44 (0) 121 414 5532 / 5543 / 5544 / 6935 International Fax: +44 (0) 121 414 5528; Website: http://www.gees.bham.ac.uk

Peggy Lerman is a legal counsel working in the field of environmental and planning law. She has been assistant judge at The Court of Appeal for southern Sweden. She was formerly head of legal affairs at the National Housing, Building and Planning Agency and national expert in the EU commission expert group for EIA and SEA involved in the preparation of the SEA Directive. She has been a legal expert and consultant to a number of national government or parliamentary commissions on diverse topics such as the development of the Environmental Code, on sustainable development, environmental assessment, disposal of nuclear wastes, infrastructure, biodiversity, and served as coordinator for Nordic cooperation on EIA and planning.

Lagtolken AB, Krakvagen 17, SE 370 24 NATTRABY, Sweden
Tel/Fax: +46 (045) 54 99 98; Email: peggy@lagtolken.se; Web: www.lagtolken.se

Joanna Maćkowiak-Pandera is a scientific employee at the University of Potsdam in the Department of Landscape Planning and a PhD student of the Polish University in Poznań, of Plant Taxonomy. The topic of her thesis is the compliance of EIA methodology for highways in Germany and Poland and working out the good practice for Poland.

Institut for Geo-Ecology, Potsdam University, Postfach 60 15 53, 14415 Potsdam, Germany
Email: mackowia@tlen.pl;
Web: www.uni-potsdam.de/u/Geooekologie/institut/personal/mackowiak.html.

Ross Marshall joined the UK's Environment Agency to set up the National Environmental Assessment Service. NEAS has responsibility for screening all high risk Agency plans and programmes (about 400 per year) for flood risk management, navigation, water abstraction, contaminated land and fisheries, and provides an advisory service to all other internal functions on EIA, SEA, landscape design and archaeology. From July 2004 the Environment Agency will be the UK's largest single obligated preparer of SEA Environmental Reports. He is an EIA/SEA practitioner with over 15 years experience of impact assessment for large scale infrastructure projects in the energy, water treatment, transport, waste and oil/gas sectors.

National Environmental Assessment Service, Environment Agency, Waterside House, Waterside North, Lincoln LN2 5HA, England
Tel: +44 (0) 1522 78 58 19; Email: ross.marshall@environment-agency.gov.uk

Robert Mayer is professor for Landscape Ecology and Soil Science, University of Kassel. His research focuses in environmental cycling and impacts upon soil, water and vegetation; soil science in planning and soil protection.

University of Kassel, Departement of Architecture, Urban and Landscape Planning, Gottschalkstrasse 28, 34109 Kassel
Tel: +49 (0) 5 61 80 42 350; Fax: +49 (0) 5 61 80 43 558; Email: rmayer@uni-kassel.de;
Web: www.uni-kassel.de/fb6/fachgebiete/bodenkunde.htm

Ali Memon is professor of Environmental Management and Planning at Lincoln University in Canterbury, New Zealand. He has published widely on several areas of environmental policy and planning.

Environment, Society & Design Division, Lincoln University, Canterbury, New Zealand
Email: memona@lincoln.ac.nz; Web: www.lincoln.ac.nz/esdd

Harry Meyer-Steinbrenner is EU officer in charge of project and research coordination at the Saxon Ministry of the Environment and Agriculture in Dresden, Saxony. He is responsible for a pilot project for implementing the SEA Directive in regional planning.

Sächsisches Staatsministerium für Umwelt und Landwirtschaft (SMUL), Referat Grundsatzfragen Umwelt und Landwirtschaft/EU-Angelegenheiten, Archivstr. 1, 01097 Dresden, Germany
Email: Harry.Meyer-Steinbrenner@smul.sachsen.de

Saul Namango is a lecturer in the Department of Chemical and Process Engineering Moi University, Eldoret Kenya, presently on study leave as a doctoral research candidate at the Brandenburg Technical University Cottbus. He is currently researching on the chemistry of building materials.

Department of Chemical and Process Engineering, Moi University, P.O Box 3900 Eldoret, Kenya
Email: snamango@yahoo.com; Web: http://www.tu-cottbus.de/altlasten/

Eric Ofori has an MSc in Environmental and Resource Management from the Brandenburg Technical University. His area of interest is SEA and Project EIA in the West African Sub-region with emphasis on water resources management.

Juri-Gagarin Str. 1/213-2, 03046 Cottbus, Germany
Email: oforieric@hotmail.com

Vincent Onyango has completed his MSc in Environment and Resources Management with a thesis on the development of an SEA approach for assessing General Agreement on Tariffs and Trade (GATT) / World Trade Organisation (WTO) rules in developing countries. He will test the efficacy of the SEA approach for international trade during his PhD studies.

Wilhelm-Kulz Str. 50, 03046 Cottbus, Germany
Email: vin_onyngo@yahoo.com

Maria do Rosário Partidário is professor of environmental planning, sustainability and impact assessment at the New University of Lisbon, Portugal. Her expertise is on strategic assessment approaches to sustainable development in the public and private sector. She is a member of SEA international panels and networks, a consultant developing guidance and capacity-building programmes and a professional trainer in national countries and international organizations such as the World Bank, USAID, Dutch Aid, UNDP, UNEP, IDB, European Commission. She is the co-editor of two books on SEA. Maria was president of the International Association for Impact Assessment in 1997-98 and was awarded the IAIA Individual Award in 2002 for contributions to the development of SEA.

DCEA, FCT-UNL, Quinta da Torre, 2829-516 CAPARICA, Portugal
Tel/Fax: +35 (0) 12 12 94 96 16; Email:mp@fct.unl.pt; Web:campus.fct.unl.pt/mp/

Anastássios Perdicoúlis is an assistant professor at the University of Trás-os-Montes e Alto Douro, Portugal. He holds a BSc from the University of Washington (1990) and a PhD from the University of Salford, UK (1997). His research interests are on Sustainable Development, with particular emphasis on SEA, EIA, and Environmental Management Systems. He is currently leading the Environmental Dynamics Special Interest Group of the System Dynamics Society (SDS) and the Impact Dynamics discussion list of the International Association for Impact Assessment (IAIA).

Department of Biological and Environmental Engineering, University of Trás-os-Montes e Alto Douro, Apartado 1013, 5000-911 Vila Real, Portugal
Tel. +351 259 350 728; Fax: +351 259 350 480; Email: tasso@utad.pt; Web: home.utad.pt/~tasso

Daniel Potts is a doctoral student in the Department of Ecology and Evolutionary Biology at the University of Arizona. As a member of the National Science Foundation science and technology center for Sustainability of semi-Arid Hydrology Riparian Areas (SAHRA), his primary emphasis is on water resources research and public outreach. His MSc in Range Management is also from the University of Arizona (2002).

Department of Ecology & Evolutionary Biology, University of Arizona, Tucson Arizona 85721 USA
Email: dlpotts@email.arizona.edu

Donald Potts is professor of Watershed Management in the Department of Forest Management at the University of Montana. He is Associate Director of the Montana University System Water Center and a past-president of the American Water Resource Association. His research interests focus on wildland hydrology and assessment techniques for determining cumulative watershed effects resulting from fire and resource management activities.

Department of Forest Management, University of Montana, Missoula, Montana 59812, USA
Email: dpotts@forestry.umt.edu

Norma Powell is a wildlife biologist for an environmental consulting firm with offices in Victoria and Vancouver, British Columbia. She specialises in environmental impact assessment for land development related to infrastructure, residential developments, and recreational developments, among others.

ENKON Environmental Ltd., Unit 201 - 2430 King George Highway, Surrey, British Columbia, Canada, V4P 1H8
Email: powell_norma@hotmail.com

Markus Reinke is a landscape planner and responsible for evaluation and development of environmental protection instruments at the Leibniz Institute of Ecological and Regional Development (IOER). His research interests focuses on the possibilities to reach environmental safe land use planning.

Institut für ökologische Raumentwicklung e.V. (IÖR), Weberplatz 1, 01217 Dresden, Germany
Tel : +49 (0)351 46 79 272; Email: m.reinke@ioer.de; Web:www.ioer.de

Sandra Ruza is a senior environmental expert at the Ministry of Environment in Latvia. She was involved in the process of drafting Latvia EIA and SEA legislation and was working as national expert in the Finnish – Latvian project for conducting SEA of Latvia Single Programming Document and its Complement. She was actively involved in the process of developing the Kiev SEA Protocol. She is Vice-chair to the Working Group on EIA under the Espoo Convention.

Ministry of Environment, Peldu 25, LV-1494, Riga, Latvia
Tel +371 (0) 7026526; Fax: +371 (0) 7820442; Email: Sandra.Ruza@vidm.gov.lv

Michael Schmidt is vice president and the head of the Department of Environmental Planning at the Brandenburg University of Technology (BTU) Cottbus. His scientific research and lecturing fields include environmental planning, environmental assessment, strategies for sustainable development, techniques for combating desertification as well as monitoring and evaluation.

Head of department of Environmental planning, vice president for international affairs, Brandenburg University of Technology (BTU) Cottbus, P.O. Box 101344, D-03013 Cottbus, Germany
Tel: +49 (0) 3 55 69 24 54; Fax + 49 (0) 3 55 69 27 65; Email: schmidtm@tu-cottbus.de;
Web: www.tu-cottbus.de/environment

Frank Scholles is a senior lecturer at the Institute of Regional Planning and Regional Science (ILR) of the University of Hanover. He is also the chairman of the German EIA Association.

University of Hanover, Institut of Regional Planning and Regional Science, Herrenhäuser Str. 2, 30419 Hanover
Tel.: +49 (0) 5 11 762 26 17; Fax: +49 (0) 5 11 762 52 19; Email: scholles@laum.uni-hannover.de;
Web: www.laum.uni-hannover.de/mitarbeiter/scholles.html

Jerzy Sommer is professor at the Institute of Law Studies of the Polish Academy of Sciences, Warsaw. He is the head of the Research Group on the Environmental Law of the Institute, located in Wrocław. He is specialised in environmental legislation and policy development of Poland and of the EU. He is the President of the Polish Environmental Law Association. He is also the member of: the Commission on Environmental Law of the IUNC, the International Council of Environmental Law, the National Council of Environmental Protection and the Commission on Environmental Protection of the Polish Academy of Sciences.

Editor-in-chief of the quarterly: Ochrona Œrodowiska. Prawo i Polityka (Environmental Protection. Law and Policy), Zespół Prawa Ochrony Środowiska, ul. Kuźnicza 46/47, 50-138 Wrocław, Poland
Tel: +48 (0) 713 28 46 01; Fax: +48 (0) 713 28 00 28; Email: sommer@prawo.uni.wroc.pl

Engelberth Soto-Estrada is a PhD student at the BTU Cottbus, Germany. He is carrying out his PhD research project on "Cultural Integrity as a Criterion of Strategic Environmental Assessment (SEA)."

Erich-Weinert-Str. 6 Zi. 614/3, 03046 Cottbus, Germany
Email: engel_soto@yahoo.com.mx

Wolfgang Stein is an officer for EIA at the State Enterprise for Roads in the federal state North-Rhine-Westphalia, Germany. He is a member of the working group on SEA within the German EIA Association.

Landesbetrieb Straßenbau NRW, Mindener Str. 2, D-50679 Cologne
Tel: +49 (0)221 801 91 26 91; Fax: +49 (0)221 801 91 - 2267, Email: wolfgang.stein@strassen.nrw.de

Harry Storch is assistant lecturer at the Department of Environmental Planning, Brandenburg Technical University of Cottbus, Germany. Main research areas are geographical and environmental information systems for environmental planning.

Department of Environmental Planning, Brandenburg University of Technology (BTU), Cottbus, P.O. Box 101344, D-03013 Cottbus, Germany

Tel.:+49 (355) 69 21 22; Fax: +49 (355) 69 27 65; Email: storch@tu-cottbus.de; Web: http://www.tu-cottbus.de/BTU/Fak4/Umwplang

Riki Thérivel is a partner of Levett-Thérivel sustainability consultants and a visiting professor at Oxford Brookes University's School of Planning. She specialises in SEA and environmental impact assessment. She has advised a wide range of organisations on SEA, has written some key guidance documents on SEA, and has (co-)authored three books on SEA. She is the 2002-3 recipient of the International Association for Impact Assessment's Individual Award for Contribution to Impact Assessment.

28A North Hinksey Lane, Oxford OX2 0LX, England
Tel/Fax: +44 (0) 18 65 24 34 88; Email: riki@levett-therivel.fsworld.co.uk; Web: www.levett-therivel.co.uk

Paul Tomlinson is head of Environmental Assessment & Policy in the Centre for Sustainability at the Transport Research Laboratory (TRL Limited). He has over 20 years experience in environmental assessment and is an internationally recognised expert in SEA. He leads the Environmental Planning and Protection Network of the Royal Town Planning Institute (RTPI) and is an advisor to the European Conference of Ministers of Transport (ECMT) on transport appraisal. He is the editor of the joint TRL and ECMT Newsletter on SEA & Transport Planning.

TRL Limited, Crowthorne House, Nine Mile Ride, Wokingham, Berkshire, RG40 3GA, England
Tel: +44 (0) 13 44 77 08 00; Fax: +44 (0) 13 44 77 03 56; Email: ptomlinson@quista.net; Web: www.trl.co.uk and www.sea-info.net

Lucie Václavíková worked at the Ministry of Environment of the Czech Republic for five years in the field of EIA and SEA. She is now working at a private company called Nordicpharma.

Email: lucie.vaclavikova@nordicpharma.cz

Herman Van Langenhove is professor in environmental chemistry and technology at Ghent University. Within the research group ENVOC he investigates quantitative indicators for the assessment of the environmental sustainability of technology.

Environmental and Clean Technology, Research Group ENVOC, Ghent University Coupure Links 653, B-9000 Ghent, Belgium
Tel: +32 (0) 92 64 59 53; Fax: +32 (0) 92 64 62 43; Email: herman.vanlangenhove@ugent.be; Web: www.envoc.ugent.be

Marc Van Dyck has over 12 years of experience in advising the Flemish and Belgian authorities on the development of methods for SEA and on other policy development and evaluation methods. He was an independent expert in the drafting of the Flemish SEA legislation and is involved as SEA coordinator in the SEA projects for the Masterplan Mobility Antwerp, the SIGMA plan and the Long Term Vision for the Flemish ports. He gives strategic policy advice to Belgian and international government institutions on mobility, environment and nature, land use planning and regional economics.

Resource Analysis NV, Wilrijkstraat 37, B-2140 Antwerp, Belgium
Email: marc.van.dyck@resource.be; Web: www.resource.be

Gerhard Wiegleb is professor of General Ecology at BTU Cottbus. His current research focuses on ecological and socio-economic driving forces of biodiversity change in disturbed landscapes. He is a head of the working group of Restoration Ecology of the German Ecological Society.

Department of General Ecology, Brandenburg University of Technology (BTU) Cottbus, Postfach 101344, D-03013 Cottbus
Tel: +49 (0)355 69 22 91; Fax: +49 (0)355 69 22 25; Email: wiegleb@tu-cottbus.de;
Web: www.tu-cottbus.de

Graham Wood is a senior lecturer in the Department of Planning at Oxford Brookes University. His main research interests cover environmental assessment techniques and methods including GIS, EIA and SEA, with particular emphasis upon the evaluation and communication of environmental information, impact significance and uncertainty in environmental decision making. He has undertaken research, consultancy and training activity for a wide variety of bodies including the Economic and Social Research Council, the European Commission and the Asian Development Bank. He is course leader on the MSc in Environmental Management and Technology.

Department of Planning, School of the Built Environment, Oxford Brookes University, Oxford. OX3 0BP, England
Tel: +44 (0) 1865 483942; Fax: +44 (0) 1865 483559; Email: gjwood@brookes.ac.uk

Jennifer Yang is a PhD student at the department of General Ecology at Brandenburg University of Technology (BTU) Cottbus, Germany. She is carrying out her doctoral research project on 'Ecotourism and Natural Resource Management – A Case Study of Jiuzhaigou Biosphere Reserve in Sichuan'. The topic of her PhD research project is the implementation of SEA into the ecotourism plan in China.

Juri-Gagarin-Str.1/104C, 03046 Cottbus, Germany
Tel:+49 (0)355 8696838; Email: jennifercottbus@hotmail.com

Evren Yaylaci is a mining engineer. He is currently taking the international master programme "Environmental and Resource Management" at Brandenburg University of Technology (BTU) in Cottbus. His main interest areas are mineral resource management in developing countries, tools for sustainable energy projects development and specific requirements of SEA for the mining industry.

Erich-Weinert Str. 6/615-3, 03046 Cottbus, Germany
Tel: +49 (0)179- 2599715; Email: evrenyaylaci@yahoo.de

Part I – Key Principles and Legal Framework for SEA

The year 2004 was when Strategic Environmental Assessment (SEA), the environmental assessment of strategic actions, gained more predominance than ever before. The 21st of July 2004 was the deadline by when all 25 EU member states had to implement the Directive 2001/42/EC, referred to in this Handbook as the "SEA Directive" and beyond the European Union other changes have occurred. From the 1st of January 2004, United Nations countries – that are members of the United Nations Economic Commission for Europe (UNECE) or have consultative status with the UNECE – may ratify the Kiev SEA Protocol (created in May 2003). So far there are 37 signatories of this SEA protocol (36 countries plus the European Community itself). Part I of the Handbook introduces the key principles of SEA and the legal framework for SEA, with particular regard to the new legislation. Part I creates the basis for much of the following seven Parts of the Handbook.

Chapter 1 describes the main ideas underlying SEA: that SEA should *improve* rather than just *analyse* the strategic action, and that SEA should consider different alternatives (or options) and compare them in an assessment context. Chapter 2 goes "behind the scenes" to analyse the purpose and background of the European SEA Directive: an advanced piece of legislation, which has had more than two decades of discussions (remarkably, early plans for the creation of SEA at the European level started back in 1975). In Chapter 3 the *legal* context of the SEA Directive is evaluated, giving particular emphasis to links between SEA and other legislation (such as the Aarhus Convention, the Habitats Directive, the EIA Directive and the Water Framework Directive). Chapter 4 investigates the potential benefits, and as important, the problems of transposing the SEA Directive into national law, revealing the increased complexity of the situation of the federal as compared to non-federal systems. Finally, Chapter 5 uses the example of Poland to present a critical evaluation of legal deficiencies concerning the implementation of the SEA Directive into the legal systems of the member states.

1 Key Principles of SEA

Elsa João

Graduate School of Environmental Studies, University of Strathclyde, Scotland

1.1 Introduction

This chapter sets out the general philosophy and explains the fundamental principles related to Strategic Environmental Assessment (SEA) as used in this Handbook. The Handbook is being published at a time when the European SEA Directive is being implemented across all 25 European Union member countries (see Chap. 2). According to Thérivel (2004), by late 2003 about 20 countries worldwide had established legal requirements for SEA, so the uptake of SEA due to the SEA Directive is significant at a global scale. This is despite the fact that the SEA Directive does not mark the start of SEA legislation in Europe. For example, the Netherlands have had SEA since 1987 (see Sadler and Verheem 1996) and the Czech Republic since 1992 (see Chap. 13). However, it is the case that there is now a wider predominance of SEA than ever before. A powerful contributor to the introduction of SEA is also the 2003 UNECE (United Nations Economic Commission for Europe) Kiev Protocol on SEA (see Chap. 3), which has similar aims to the SEA Directive. From 1 January 2004 United Nations countries – that are members of the UNECE or have consultative status with the UNECE – may ratify this SEA Protocol. It remains to be seen how many countries will ratify the Kiev SEA Protocol (at the end of September 2004 there were 37 signatories), but the impact of this Protocol is likely to be significant and might even "require the European Union to amend the SEA Directive" (Robinson and Elvin 2004 p.1028).

This chapter starts by explaining exactly what SEA is. The full set of possible strategic actions is considered rather than the more restricted definition used by the SEA Directive (which only considers plans and programmes). It then discusses the issue of tiering, and addresses the link between SEA and Project EIA, including the role of SEA in cumulative impact assessment. The chapter then proposes two key principles of SEA: that SEA must evaluate alternatives in an assessment context and that SEA must improve (and not just analyse) the strategic action. This is followed by an evaluation of the ideal SEA team, which is closely related to the need to improve the strategic action. The chapter concludes by evaluating SEA's key advantages in minimising the damage to our environment.

Implementing Strategic Environmental Assessment. Edited by Michael Schmidt, Elsa João and Eike Albrecht. © 2005 Springer-Verlag

1.2 Environmental Assessment of 'Strategic Actions'

SEA is the process of evaluating the environmental impacts of proposed policies, plans or programmes, in order to *inform decision-making*. An example of a proposed policy could be a national transportation policy; a proposed plan could be a local development plan; and a proposed programme could be a co-ordinated series of dams. More precisely, a policy can be defined as an inspiration and guidance for action, for example, whether or not to promote the development of nuclear power in a particular country. A plan can be defined as a set of linked proposed actions, with a specific timeframe, that implement the policy, such as how much nuclear power to produce by 2020. Finally, a programme can be defined as a set of proposed projects in a particular area that will implement the plan, for instance the proposal of four new nuclear power stations with X capacity in the area Y by the year 2020. (Definition based on Wood (1991) and example taken from Oxford Brookes University (2004).)

However, in practice these distinctions and stages are not necessarily very clear-cut and the terminology is not consistently used. For example, many UK local councils talk about policies that are a sub-set of local plans rather than the overarching policy meant in the previous definition. Also different countries might use a different terminology altogether. For example, the top 'policy' level is named 'strategies' by the Scottish Government (Scottish Executive 2003) while in the Czech Republic strategies, policies, plans and programmes are jointly called 'conceptions' (see Chap. 13). A misunderstanding due to different terminology can cause problems such as in the implementation of the SEA Directive (see Chap. 5 and Robinson and Elvin 2004), and reminds us on how important it is to clarify how different strategic actions are defined.

In this Handbook, in order to solve this lack of consistency in SEA-related terminology across different countries, the term 'strategic actions' will be used when jointly referring to policies, plans, programmes, strategies or conceptions. Individual terms (e.g. policy, plan, programme, strategy, conception) will still be used when relevant, for example when a SEA methodology or case study refers to plans alone. Ultimately, whatever an individual strategic action is called, what is important to recognize is that SEA deals with a hierarchy of strategic decision-making where 'more strategic' actions affect other strategic actions and, ultimately, affect what projects are built on the ground (Oxford Brookes University 2004). This hierarchy might differ in detail from country to country but is nevertheless there, and is a key aspect of what SEA is all about. This concept is closely linked to the idea of 'tiering' described in Sect. 1.3.

1.3 Tiering and the Link between SEA and Project EIA

As mentioned in the example used in Sect. 1.2, the proposal of four new nuclear power stations with X capacity in the area Y by 2020 is what is called a 'programme' and the impacts of such programme would be covered by SEA. On the other hand, each individual proposed nuclear power station in this example is called a 'project' and its impacts would be dealt with by the Environmental Impact Assessment (EIA) process for individual projects. (In the European Community, EIA is done according with the implementation by individual Member States of the EIA Directive.) EIA, the environmental assessment of individual proposed projects (such as a new road bypass), is now also commonly called 'Project EIA'. This new term came about exactly because of the advent of SEA as it helps clarify its specific area of application to *projects* alone.

It is important to bear in mind that the three levels of SEA are linked to Project EIA. The SEA of policy will affect and inform the SEA of plans, which in turn will affect and inform the SEA of programmes, which in turn will affect and inform the EIA of projects. This linkage is called 'tiering' (see also Fig. 26.1 in Chap. 26). Tiering means that aspects of decision-making and SEA carried out at one level do not necessarily need to be subsequently revisited at 'lower' levels, so that "tiering can potentially save time and resources" (Thérivel 2004 p.13). In California, for example, the SEA experience has been that certain aspects of *subsequent* projects have not then needed to be assessed in detail (Bass 1991, cited in Wood 1995). Each level is considered a 'tier' exactly because there should not be unwanted duplication between the different stages. Bottom tiers can also link up with and influence top tiers in terms of future decisions and planning. Even what is found at Project EIA could, and should, affect later decisions at more strategic levels. This is illustrated in Fig. 1.1.

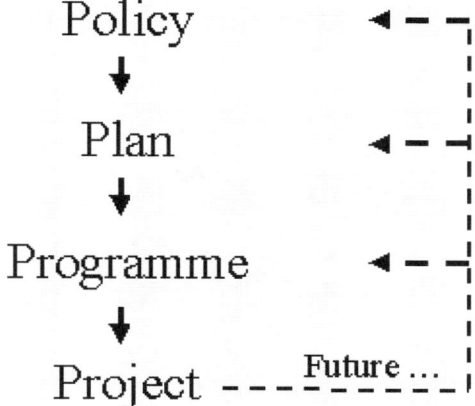

Fig. 1.1. The links between the different tiers of policy, plan, programme and project

Although it can be argued that SEA was born together with Project EIA back in 1969 with the National Environmental Policy Act or NEPA (see Chap. 17), until recently Project EIA has been the dominant environmental assessment process around the world.

Often SEA has had a latent existence in the shadow of Project EIA. For example, discussions about a European SEA Directive started back in 1975 when it was intended that one *single* Directive would cover both projects and strategic actions, but, when the EIA Directive was approved in 1985, its application was restricted to projects only (Thérivel 2004). However, as more and more Project EIA were implemented around the world, there was an upsurge of disillusionment with the capacity of Project EIA to assist, *as a single tool,* for "sound environmental decision-making" (Partidário 1999 p.60). Part of this disillusion had to do with the timing of Project EIA. Essentially, Project EIA enters decision-making too late - when decisions at policy or planning level, that could influence the type and amount of projects that are actually built on the ground, have already been taken (Partidário 1999).

The other difficulty of Project EIA is its inability to deal with cumulative impacts (see Glasson et al. 2004). By contrast, it is often argued that one of the main justifications for SEA is that it can deal with cumulative impacts in a more effective way (see Chaps. 27 and 28). A very good example of this was presented by Thompson et al. (1995) in relation to the farming of Atlantic salmon in Scotland. Normally, if fish farms are below a certain size (as specified in EIA legislation) and are not located in an ecologically sensitive area, then they might not even require a Project EIA. But even if they do, each one will be evaluated on a one-to-one basis and the cumulative impacts of fish farming in general is not necessarily considered. This is particularly pertinent in the case of the West Coast of Scotland where there are a large number of fish farms (see Fig 1.2) whose cumulative impacts are not easily evaluated at project level but, according to Thompson et al. (1995), could ideally be done with SEA.

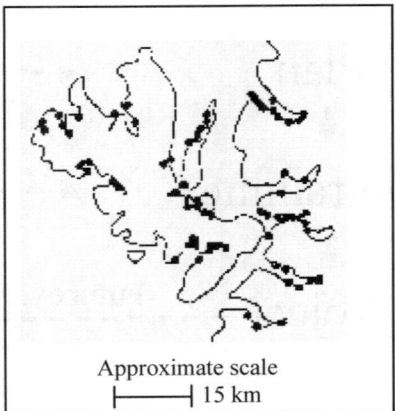

Fig. 1.2. Distribution of Salmon fish farms in the Island of Skye and adjacent mainland on the West Coast of Scotland in 1989 (map extract taken from Thompson et al. 1995 – each point represents one Salmon farm lease)

1.4 Two Key Principles of SEA

There are two fundamental principles to implementing SEA. First, that SEA must clearly identify feasible policy, plan or programme options (or alternatives) and compare them in an assessment context. Second, that SEA must *improve*, rather than just *analyse*, the policy, plan or programme. The two principles are related, as it can be argued that only by considering alternatives can the best strategic action be created. Sects. 1.4.1 and 1.4.2 discuss both of these key principles in detail.

1.4.1 SEA must evaluate Alternatives in an Assessment Context

Just like Project EIA, the concept of SEA should be associated with options or alternatives. SEA (as well as Project EIA) should *not* be used to mitigate environmental impacts of actions that have *already been decided upon*. Instead the environmental assessment process should be used to inform the decision of what action should take place, i.e. what alternative should be chosen. There are different types of alternatives at strategic level, such as (based on Thérivel and Partidário 1996):

- 'Do nothing' or 'continue with present trends' option.
- Demand reduction, e.g. reduce the demand for water through water metering.
- Different locational approaches, e.g. build new houses in existing towns versus in new towns.
- Different types of development that achieve the same objective, e.g. produce energy by coal or wind.
- Fiscal measures, e.g. toll roads or congestion charges.
- Different forms of management, e.g. waste management by recycling or incineration.

The SEA Directive, for example, is very clear on the need for alternatives and requires the consideration of "reasonable alternatives (to the plan) taking into account the objectives and the geographical scope of the plan". In addition the SEA Directive requires that the environmental report should provide "an outline of the reasons for selecting the alternatives".

What is important to note is that alternatives at strategic level are not just about different types of development which achieve the same objective (e.g. produce energy by coal or wind), they are also about demand reduction (e.g. reduce the demand for energy production by insulating buildings). In other words SEA alternatives are also about *obviating development*, e.g. making new power stations unnecessary. The draft guidance on implementing the SEA Directive in England (ODPM 2002) addresses obviation of development in particularly robust terms:

> "It is no longer enough just to consider different possible locations for development. The shift from predict and provide to plan, monitor, manage means that alternative ways to meet needs or respond to development demands should also be considered, including dif-

ferent types of development, and ways of obviating development, e.g. better local amenities or services might make some journeys unnecessary.

Obviation is not the same thing as restricting or thwarting demands. It should be seen as looking for different, more sustainable means to, achieve human quality of life ends. For example, obviating journeys should be seen as providing people with access to the things they want with less need for mobility."

The English guidance on implementing the SEA Directive (ODPM 2003) – the first guidance to be published in Europe – includes a useful diagram that stages the design of different alternatives (see Fig. 1.3). The hierarchy of alternatives starts by deciding if development can be obviated, if not, then it considers how should it be done in a more environmentally or sustainable way, only after this stage does it consider alternatives of where should the development go and decides about timing and detailed implementation.

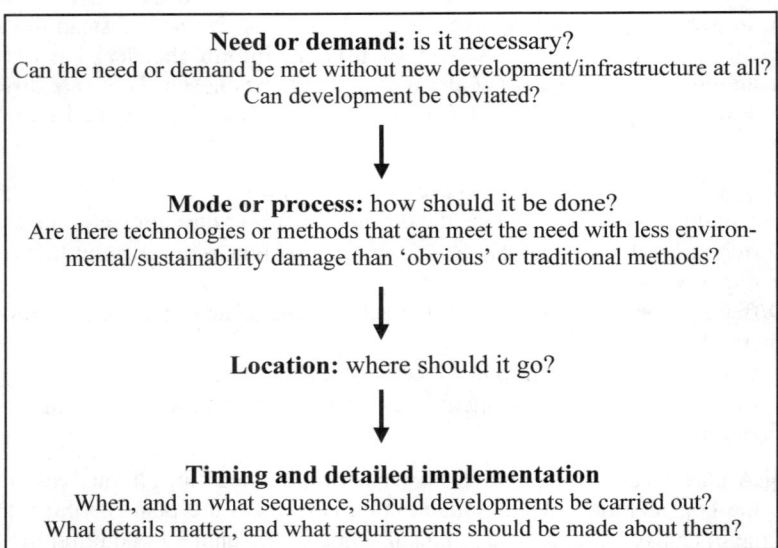

Fig. 1.3. 'Hierarchy' of alternatives: from obviation to detailed implementation (ODPM 2003)

Developing and comparing alternatives (or options) allows the decision-maker to determine which is the best option, although this is not necessarily straightforward. The reason being that, in strategic decision-making, options (or alternatives) can be not only "either-or" alternatives (e.g. nuclear vs. renewable energy) but also "mix-and-match" options (e.g. recycle and reduce waste) (see Thérivel 2004 p.12). Alternatives can also be "broad alternatives, leading to a choice of preferred alternatives, in turn leading to more detailed statements of how the preferred alternative will be implemented" (Thérivel 2004 p.12).

1.4.2 SEA must improve (and not just analyse) the Strategic Action

The second key principle of SEA is that SEA must *improve*, rather than just *analyse*, the policy, plan or programme. The emphasis should be on incorporating SEA in the formulation of the strategic action: looking at the evolving strategic action afresh and being willing to change and improve it in the light of the SEA findings. SEA must contribute to the decision-making process!

In order to evaluate how SEA could improve the strategic action it is helpful to see the different ways in which SEA could interact with strategic-decision making. Fig. 1.4 shows (in a symbolic way) four different ways in which SEA stages can interact (or not) with different strategic-decision stages.

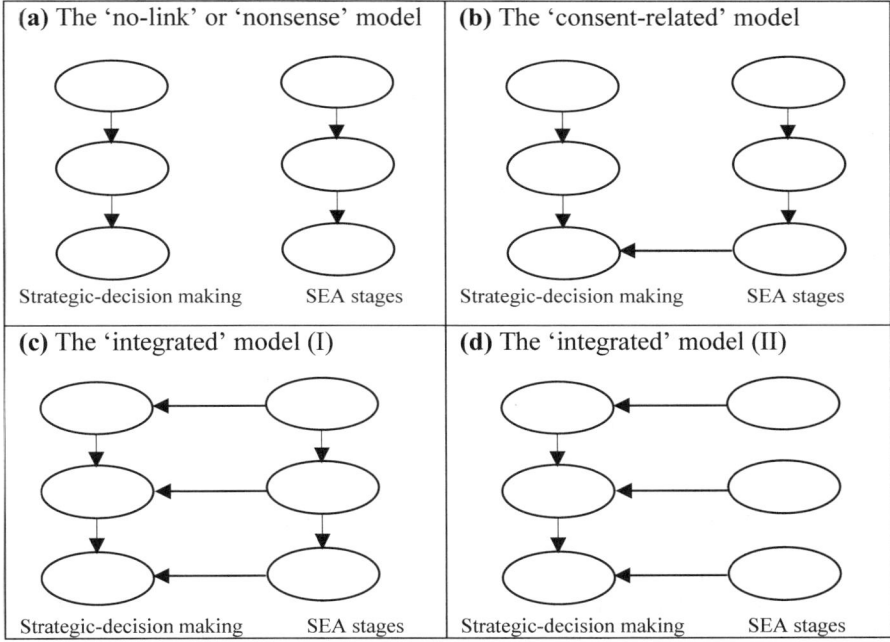

Fig. 1.4. Links (or not) between strategic decision-making stages and SEA stages - circles symbolize the stages and arrows indicate the links: (a) and (b) are *not* good-practice models; while (c) and (d) are considered 'best practice' models

The 'no-link' or 'nonsense' model (Fig. 1.4 a) is a very poor practice model. It assumes that there are no links between SEA and the strategic-decision making, and therefore SEA is done just as a 'tick-box' exercise. The 'consent-related' model (Fig. 1.4 b) only adjusts strategic-making to include an SEA stage that informs the strategic action's decision-making stage (Thérivel and Partidário 1996). Again this is not considered a best practice model.

The best practice models in Fig. 1.4 are the two integrated models (Fig. 1.4 c and d). The integrated model assumes that strategic actions are subject to multiple stages of decision-making and attempts to integrate SEA into each of these decisions (Thérivel and Partidário 1996). The difference between the two integrated models is a subtle one. In the integrated model (I) (Fig. 1.4 c) there are links between the SEA stages, while in the integrated model (II) (Fig. 1.4 d) the emphasis is on the flow between the strategic-decision making stages and the SEA stages exist only to inform decision-making. The integrated model (II) shows more than any other how the ultimate aim is *not* to carry out the SEA *per se* but to use the SEA to achieve the best strategic decision possible.

Using the integrated model (II) as a basis, Fig. 1.5 shows the main strategic-decision making stages and how they are affected by SEA. Interestingly, SEA's influence occurs right from the start in the improvement of the objective of the strategic action. The role of SEA in this respect is to try and reshape the strategic action objective so that it includes environmental and sustainability issues. Oxford Brookes University (2004) provides a good example for this: an original objective for a transport strategy was "to ensure the free flow of all forms of transport and improve the county's economic base whilst minimizing the environmental harm associated with transport". This objective was found to have problems. It was likely to increase traffic, reduce environmental quality, and would not necessarily improve people's ability to access necessary facilities or services. The new, improved objective proposed thanks to the SEA process was: "to improve accessibility and reduce the need to travel".

The above re-wording of the strategic action can be considered a mitigation measure in SEA (see Thérivel 2004). "SEA mitigation measures do not look like those of EIA" (Thérivel 2004 p.167). Mitigation measures in Project EIA might concentrate on location or design for example. In contrast, SEA mitigation can include (Thérivel 2004): development of new options (e.g. by combining the best aspects of existing options), or requirements and terms of reference for EIA of certain types of projects (this is associated with tiering – see Sect. 1.3). Importantly, SEA contributes to the improvement of the strategic action not only by mitigating *negative* impacts but also by enhancing *positive* impacts (stage D, Fig. 1.5).

SEA is therefore a tool for improving the strategic action, irrespective of the fact the strategic action has mainly negative or mainly positive impacts. "SEA is not a post-hoc 'snapshot'. This means that the SEA should be started early, be integrated in the decision-making process, and focus on identifying possible alternatives and modifications to the strategic action" (Levett and Thérivel 2003). If SEA is integrated in the decision-making process in this way, then by the time the Environmental Report (the report that documents the SEA process) is being written most of the job is already done. Therefore, the following quote makes sense: "the preparation of an SEA report is probably the least important part of the SEA" (Oxford Brookes University 2004). This is because the real value in SEA is as "a creative tool in the design cycle of the strategic action" (Oxford Brookes University 2004).

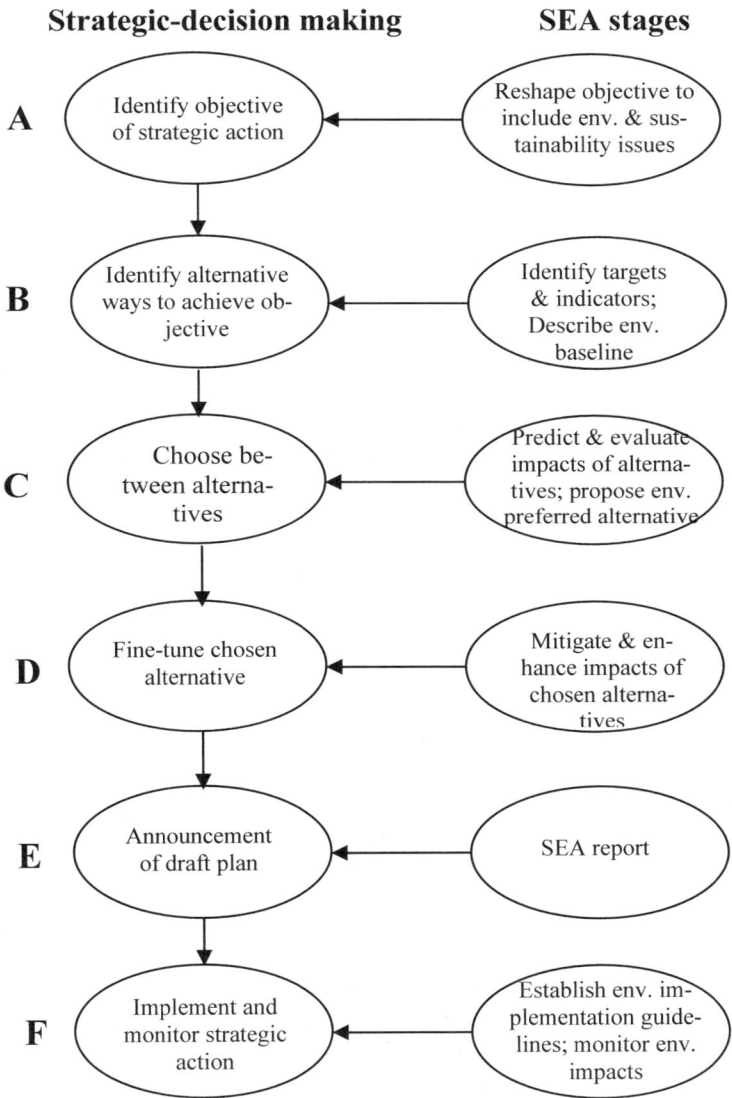

Fig. 1.5. Key strategic-decision making stages and how they are influenced by SEA

It is also important to describe the importance of public participation in the SEA process (see Chaps. 24, 29 and 30). According to the SEA Directive, decisions on the scope and level of detail of the SEA (stage B of Fig. 1.5) require consultation of authorities with environmental responsibilities. While the announcement of the draft plan and accompanying environmental report (stage E of Fig. 1.5) requires consultation of authorities with environmental responsibilities *and* the public.

Finally, in order for the interaction between strategic-decision making and SEA to work, the decision-makers need to be involved in the SEA process in some active capacity, as only they can ensure that the SEA findings are fully taken into account in the decision-making process. This in turn affects the composition of the ideal SEA team (discussed in Sect. 1.5). It is also crucial that SEA interacts at the 'appropriate points' in the decision-making process. The ANSEA Team (a team of international SEA experts that devised the "analytical SEA methodology", or ANSEA) calls these appropriate points, when critical choices are made which have environmental consequences, "decision windows" (Dalkmann et al. 2004).

1.5 The Ideal SEA Team

The make-up of the ideal SEA team is closely associated with the key principle discussed in Sect. 1.4.2 that SEA should improve rather than just analyse the strategic action. This means that the decision-maker needs to be involved in the SEA process in some form or another. It can be argued therefore that it is impossible for an external organization to carry out the whole SEA in the name of a responsible authority (i.e. the authority responsible for the strategic action and therefore responsible for the SEA) as in this case it will be difficult, if not impossible, to use the SEA to improve the strategic decision. That said it can be helpful to involve in the SEA process *both* people who are directly involved in producing the strategic action and others, either within the responsible authority or from outside, who can contribute a more detached and independent view to the exercise. The 'outsider' could be posing the awkward questions of 'why can't it be done differently' (i.e. in a more environmentally friendly and sustainable manner), rather than falling into the trap of doing things as they always have been done in the past.

ODPM (2003) suggest that either an individual or a team can undertake an SEA. However a team approach is more likely to give results that are based on shared values and a common understanding. ODPM (2003) further recommends that if a team is not used in the assessment, someone who was not directly involved in the plan making process should ideally critically review the draft results.

Closely linked to setting up a SEA team is the issue of expertise required on carrying out an effective SEA (see also Chap. 44). Thérivel (2004) suggests that planners, engineers, consultants, environmentalist and politicians all need SEA training. For people actually carrying out the SEA, training is needed on SEA requirements and ideas on SEA best practice. "Also given the amount of group work, consensus-building and community involvement required in many SEA, training on negotiation skills is also likely to be useful" (Thérivel 2004 p.207). Even decision-makers will need to understand legal SEA requirements and find out how best to use SEA results in decision-making.

1.6 Conclusions and Recommendations

The SEA process aims to minimize any negative impacts of the preferred strategic action until they are no longer significant, as well as maximize positive impacts and generally enhance the environment where possible. The key aim of SEA is therefore to improve the strategic action. The major advantage of SEA over Project EIA is that it allows the consideration of a much wider range of mitigation and enhancement measures, particularly measures to avoid impacts at an earlier stage of decision-making. The choice and assessment of alternatives (or options) is crucial to this. Part of being *innovative* in strategic-decision making is reflected in the kind of alternatives that are considered (including 'obviation of development').

According to the draft English SEA guidance (ODPM 2002) strategic decision making has always been a process of making choices between different potential ways to handle different (sometimes conflicting) issues (e.g. housing, transport, employment demands, environmental constraints). However, to date, "relatively few authorities have formally appraised options related to their plan and documented them, and plans have often been criticized for failing to identify or give fair consideration to options which might have been better than those chosen" (ODPM 2002). ODPM (2002) suggests that SEA should remedy this situation by ensuring a wide-ranging, systematic and transparent consideration of options.

In summary, Fischer (2002) considers that there are five main potential benefits of SEA. First, SEA allows for a wider consideration of impacts and alternatives. Second, SEA is a pro-active tool that can be used to support strategic action formulation for sustainable development. Thirdly, SEA can increase the efficiency of tiered decision-making (including strengthening of Project EIA). Fourthly, SEA allows for a systematic and effective consideration of the environment at higher tiers of decision-making, and finally, more consultation and participation of the public should take place. Thérivel (2004 p.18) expresses the potential benefits of SEA in a more succinct way: "SEA can help to achieve, clearer, more environment-friendly and more publicly acceptable strategic actions that are approved more quickly." The key recommendation for this to become a reality is that SEA should start as *early* as possible within the decision-making process.

References[1]

Bass RE (1991) Policy, plan and programme EIA in California. EIA Newsletter 5:4, 5 (EIA Centre, Department of Planning and Landscape, University of Manchester) (cited in Wood 1995)

Dalkmann H, Jiliberto-Herrera R and Bongardt D (2004) Analytical strategic environmental assessment (ANSEA) developing a new approach to SEA. Environmental Impact Assessment Review 24(4):385-402

---

[1] Note that legislation mentioned in the chapter is listed at the end of the handbook in the consolidated list of legislation (Appendix 2).

Fischer TB (2002) Strategic Environmental Assessment in Transport and Land use Planning. Earthscan, London
Glasson J, Chadwick A, Thérivel R (2004) Introduction to Environmental Impact Assessment, 3rd edn. Spon Press, London
Levett-Thérivel (2003) Web site of the consultancy 'Levett-Thérivel sustainability consultants'. Internet address: http://www.levett-therivel.fsworld.co.uk, last accessed 28.09.2004
ODPM – Office of the Deputy Prime Minister (2002), Draft guidance on the Strategic Environmental Assessment Directive. Proposals for practice guidance on applying Directive 2001/42/EC "on the assessment of the effects of certain places and programmes on the environment" to land use and spatial plans in England. 23 October 2002. Draft guidance prepared for the Office of the Deputy Prime Minister by Levett-Thérivel Sustainability Consultants
ODPM – Office of the Deputy Prime Minister (2003) The SEA Directive: Guidance for Planning Authorities. Practical guidance on applying European Directive 2001/42/EC 'on the assessment of the effects of certain plans and programmes on the environment' to land use and spatial plans in England. ODPM, London
Oxford Brookes University (2004) Web-based distance-learning training packages on SEA. Internet address: http://www.brookes.ac.uk/schools/planning/SEAmicro/PPP%20defs.html!, last accessed 28.09.2004
Partidário MR (1999) Strategic Environmental Assessment – principles and potential. In: Petts J (ed) Handbook of Environmental Impact Assessment, vol. 1. Blackwell, Oxford, pp 60-73
Robinson J, Elvin D (2004) The Environmental Assessment of Plans and Programmes. Journal of Planning and Environmental Law, August 2004
Sadler B, Verheem R (1996) Strategic Environmental Assessment – Status, Challenges and Future Directions. Ministry of Housing, Spatial Planning and the Environment of the Netherlands
Scottish Executive (2003) A Partnership for a Better Scotland. A partnership coalition agreement between Labour and the Liberal Democrats, Internet address: http://http://www.scotland.gov.uk/library5/government/pfbs-05.asp, last accessed 28.09.2004
Thérivel R (2004) Strategic Environmental Assessment in Action. Earthscan, London.
Thérivel R, Partidário MR (1996) The Practice of Strategic Environmental Assessment. Earthscan Publications Ltd, London
Thompson S, Treweek J, Thurling D (1995) The potential application of SEA to the farming of Atlantic salmon (Salmo salar L.) in mainland Scotland. Journal of Environmental Management, 45:219-229
Wood C (1991) EIA of policies, plans and programmes. EIA Newsletter 5:2-3
Wood C (1995) Environmental Impact Assessment – A comparative review. Longman Scientific & Technical, Harlow

2 Purpose and Background of the European SEA Directive

Christian Kläne[1] and Eike Albrecht[2]

1 Department "Company Audit", Tax Administration of Lower Saxony, Revenue Office Cloppenburg, Germany
2 Department of Constitutional, Administrative and Environmental Law, Brandenburg University of Technology (BTU), Cottbus, Germany

2.1 Introduction

After more than 20 years of discussion, the SEA Directive entered into force on 21 July 2001 (see Appendix 1 for the full text of the SEA Directive). This Directive is an advanced piece of legislation that will change how environmental matters are dealt with in the 25 Member States of the European Union. All these Member States have had to transpose the Directive into their own national legislation since 21 July 2004. This chapter links the SEA Directive to European Environmental policies and their major concepts. It gives introductory information about the term "Strategic Environmental Assessment", the historical background of the SEA Directive and describes the legislative process of the SEA Directive. Furthermore this chapter describes the purpose and objectives of the Directive. Finally this chapter closes with an outlook on the future of SEA and related instruments, such as Sustainability Impact Assessment (SIA), in the European Union.

2.2 The SEA Directive and its European Context

There is a long history of the use of SEA or similar instruments all over the world. First steps have been taken in USA in the Seventies with the National Environmental Policy Act (NEPA), which stated in Sec. 102 para 2 c), that proposals for legislation and other major federal action significantly affecting the environment had to include a detailed statement on the environmental impacts (for details see Chap. 17).

Implementing Strategic Environmental Assessment. Edited by Michael Schmidt, Elsa João and Eike Albrecht. © 2005 Springer-Verlag

The first European countries which enacted SEA were The Netherlands (included in the EIA Act) in 1987 and Denmark in 1993. But there was – besides the discussion in the wake of the legislative process for the EIA Directive – no original Community SEA legislation. The reasons for this and the development of the European environmental policy and legislation will be described in this section with a special focus on the historical, legal, political and systematic background of the SEA Directive.

2.2.1 The Development of European Environmental Law

Today, the relevance of European environmental law for the Member States' law is significant, as partly the whole national legal system of environmental law has changed (Rengeling 2003 p.VI). Though European Environmental Law has – from today's view – quite long tradition, environmental issues were not subject of European legislation in the beginning of the process of European unification in 1952. The starting point of European environmental law are the years 1971 (Bulletin EC 9/10-1971 p.63) and 1972 (OJ 1972 C 52, p.1, 8) when the Commission communicated that environmental protection was an important task of the Communities (see also Schröder 2003 p.201). For a full understanding of the legal and political background of the SEA Directive the process of the European Unification should be briefly explained.

Start of the European Unification

The formal start of the European Unification after World War II was the foundation of the European Coal and Steel Community (ECSC) in 1951, which came into force in 1952. This community had six Member States (Belgium, France, Germany, Italy, Luxembourg and The Netherlands) and was the core of further integration measures. In 1957 the Treaties of Rome, first the Treaty on Foundation of the European Economic Community (EEC) and second the Treaty on Foundation of the European Atomic Energy Community (EAEC) were signed. Both treaties came into force at the beginning of 1958. The main subject of all three treaties was economical matters. Whereas the ECSC and the EAEC had mainly sectoral character – only a certain sector, on one hand coal and steel industries and on the other hand atomic industries – the EEC had a broader approach and regulated economical and trade questions in general and especially the creation of the "Common Market".

In the beginning, the EEC was mainly a Tariff-Union but non-economical questions got, step-by-step, more relevance. Nevertheless, still all mandatory environmental measures from the European institutions had to be justified with economical reasons, because this was the only competence basis for any binding acts (see Sheate 2003 p.333; Kloepfer 2004 p.675).

Environmental Competences in the European Single Act

In 1973 Denmark, Ireland and the United Kingdom had joined the Community, followed by Greece 1981. But it was in 1986 (the same year that Portugal and Spain joined) that the European Single Act, which was a treaty to amend the already existing Treaties (entering into force at 1 July 1987), widened the competences of the Community on environmental subjects (Art. 130r ff. of the EC Treaty). This meant, that now environmental action could be taken from the European institutions for environmental reasons only, and, no economical impact has to be proved anymore (Kloepfer 2004 p.676). The Commission, which has the role of a kind of governmental or head-executive organ, used this new competence provisions for a huge number of environmental activities, especially the enactment of environmental directives.

Environmental Law and the European Union

The foundation of the European Union with the Treaty of Maastricht from 1992 was an important step. For the now existing political Union which went far over the original purpose of an economical community, the legal and institutional conditions and guarantees had to be organised. To the already existing communities (ECCS, EAC and EEC, with the Treaty of Maastricht renamed into European Community – EC) two political sectors were added, the Common Foreign and Security Policy (Art. 11-28 EU Treaty) and the Police and Judicial Cooperation in Criminal Matters (Art. 29-42 EU Treaty). So, the main competence for environmental measures still was based on Art. 130r ff. EC Treaty. In 1995 Austria, Sweden and Finland and in 2004 the Eastern European Accession States joined the European Union. The European Union has now 25 Member States.

The European Union is composed of the three already existing Communities and two political fields where close cooperation is agreed in the EU Treaty (see Fig. 2.1).

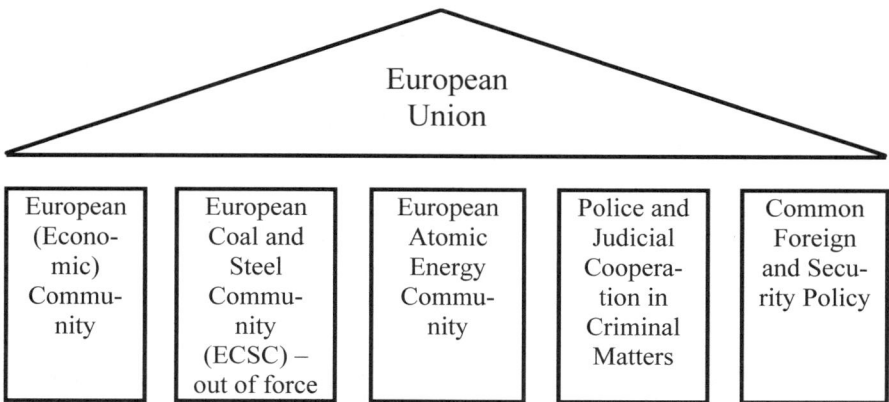

Fig. 2.1. The Organisation of the European Union

Whereas the newly added political sectors, the Common Foreign and Security Policy and the Police and Judicial Cooperation in Criminal Matters, are still developing – well noticed to the varied positions between the different European Member States in respect to the participation at the, from a legal point of view probably illegal, attack of the Iraq – the fields which are regulated in the EC Treaty have a long and constituent law and administrative tradition.

The Treaty of Amsterdam from 1997 established a new counting, so that Art. 130r ff. were renamed into Art. 174 ff. The Treaty of Nice from 2001 on the amendment of the European Treaties included smaller changes in respect to the competence provisions for environmental measures. With the new Art. 175 para 2 of the EC Treaty a fitting competence basis for the SEA Directive is laid down in European law.[1]

2.2.2 Focal Points of European Environmental Law

Though European environmental policy had throughout the years different focal points, some central principles did not change. Here, the principle of precaution and prevention, laid down in Art. 174 para 2 of the EC Treaty, the principle of sustainable development, laid down in Art. 2 of the EC Treaty *and* Art. 2 of the EU Treaty, furthermore laid down in Art. 6 of the EC Treaty, and the idea of integrative environmental protection, has to be mentioned.

Additionally, Art. 6 of the EC Treaty is of importance. Environmental protection requirements have to be integrated into the definition and implementation of the Communities' policies and activities. This does not mean that environmental matters automatically have priority in relation to other matters, but the effects of any European measure to the environment has to be considered (Callies and Ruffert 2002, Art. 6 EC Treaty, rec. 6).

Principle of Precaution and Prevention

Art. 174 para 2 EC Treaty mentions the precautionary principle and the principle that preventive action should be taken. Though there is a discussion in literature, if there is a difference between both principles (for details see Callies and Ruffert 2002, Art. 174 EC Treaty, rec. 26; Kloepfer 2004 p.685), there is a common feature of both principles. Environmental measures should be taken as far as possible to prevent the environment from negative effects (Callies and Ruffert 2002, Art. 174 EC Treaty, rec. 28). Furthermore, the current generation has an obligation to future generations to leave sufficient stocks of social, environmental and economic resources for them to enjoy levels of well being at least as high as our own. Therefore these principles are connected with the principle of sustainability

---

[1] What has to be noticed is that the SEA Directive is based on Art. 175 para 1 of the EC Treaty (see entering phrase of the SEA Directive; for details see Kläne 2003, p.142) because the new Art. 175 para 2 of the EC Treaty came into force after the SEA Directive was enacted. Therefore this provision could not been used as a competence norm.

(Callies and Ruffert 2002, Art. 174 EC Treaty, rec. 29; see also OPOCE 2002, p.53). Additionally, though the principles of precaution and prevention are solely laid down in Art. 174 para 2 EC Treaty, the European Court of Justice regards these principles not only as environmental principles but as general political principles of Community's activities, for example in respect to the protection of consumers against bovine spongiform encephalopathy, better known as "Mad Cow Disease" (European Court of Justice, Case 157/96, European Court reports 1998, p.I-2211; (64) Commission/United Kingdom; Case 180/96 European Court reports 1998, p.I-2265; (100) Commission/United Kingdom).[2] Therefore it can be said, that the precautionary principle and the principle, that preventive action should be taken, are fundamental European principles.

Principle of Sustainable Development

The concept of Sustainable Development was first given real political momentum in the United Nations Brundtland Commission report of 1987 ("Our Common Future") and widely propagated at the "Earth Summit" in 1992 in Rio de Janeiro, adopting the Rio Declaration on Environment and Development, as well as the Agenda 21. In European policy, Sustainable Development is defined as a method to meet the needs of the present generation without compromising those of future generations (OPOCE 2002 p.9). This vision of progress links economic development, protection of the environment and social justice. A Sustainable Development perspective helps to clear the fact that many current policies often do not pay enough attention to long term issues, or to the connections between different policy areas and their complexity. Achieving Sustainable Development therefore means improving the quality of policy making. Hence, Sustainable Development is closely linked to Governance, Better Regulation and Impact Assessment.

As the first action field for European Union's strategy for Sustainable Development the improvement of policy coherence is mentioned (OPOCE 2002 p.26). One of the activities to meet this target is the assessment of the potential economic, environmental and social benefits and costs of action or lack of action, both inside and outside of the EU will be proposed (OPOCE 2002 p.27). However, the principle of sustainability is laid down in Art. 2 and 6 of the EC Treaty *and* in Art. 2 of the EU Treaty (for details see Kloepfer 2004 p.675).

Integrative Environmental Protection

Integrative environmental protection means that measures for the protection of environmental goods should not be seen only in the specific sector, but in an overall view all positive and negative effects to the environment as a total should be taken into consideration. This principle is negatively defined in Recital 7 of the IPPC Directive: different approaches to control emissions into the air, water or soil separately could lead to the shifting of pollution between the different environ-

---

[2] More information under: http://viswiz.imk.fraunhofer.de/~steffi/madcow/madcow.htm, last accessed 24.8.2004.

mental media rather than protecting the environment as a whole. Positively formulated is this principle in recital 8: the objective of an integrated approach to pollution control is to prevent emissions into air, water or soil wherever this is practicable, taking into account waste management, and, where it is not, to minimize them in order to achieve a high level of protection for the environment as a whole (see also Wolf 2002 p.390; Kloepfer 2004 p.675). The continuation of the integrative environmental policy is finally stated in sections 3.2 and 5.3 of the Sixth Environmental Action Programme.

Public Participation

Principle 10 of the Rio Declaration recognizes that public participation in environmental issues is a basic condition for sustainable development. Also, public participation in environmental matters is a major task of European environmental policy also and is a very important part of the Fifth Environmental Action Programme (see EEA 1999 p.435 for further information). The first major step towards public participation was the enactment of the Directive on Access to Environmental Information. Without the right to get access to official information on the environment, the public cannot be involved in environmental relevant decision making processes. This supporting of the engagement of the private sector in the decision making process was continued with a number of directives and is an important subject of the SEA Directive. Furthermore, the European Union signed the Aarhus Convention and enacted some directives to transpose the objectives of this Convention into European Union law. Central provisions of the SEA Directive are the regulations on public participation and consultation.

Environmental Action Programmes

The Community set up meanwhile six Environmental Action Programmes. Until the Fourth Environmental Action Programme (OJ C 328, p.1), one plan embraces a period of five years. These five-years-plan had different focal subjects. Starting with the First Environmental Action Programme in 1973 (OJ C 112, p.1) the Community laid main emphasis on the taking environmental impacts into account at an early stage (see especially sections 9, 10 and 11). In respect to the SEA Directive, the Third Environmental Action Programme from 1982 to 1986 (OJ C 46, p.1) should be mentioned, because this was the political impetus for the EIA Directive in 1985 (see section II, No. 11). The enactment of EIA provisions was difficult enough because the Communities had no original competences for environmental activities at that time. The EIA Directive could be justified with competences for competition rules in the EEC Treaty. For a Directive on the assessment of plans, programmes and policies as well, then obviously, this competence norm would have been overstressed (Sheate 2003 p.333).

The Fifth Environmental Action Programmes (OJ C 138, p.1), valid from 1993 to 2000, is – regarding the SEA Directive – the most important Environmental Action Programme. This becomes clear alone with the title "towards sustainability".

But in addition, Part 1, Section 7.3 of this Programme provided a rationale for the SEA Directive in explicit terms.

The Sixth Environmental Action Programme from 22 July 2002 (OJ L 242, p 1), valid now for ten years from 2002 until 2012 is based on the fundamental environmental principles of European environmental law. This Programme was enacted politically in the co-decision procedure (Art. 251 EC Treaty) and is based on the (new) competence provision of Art. 175 para 3 EC Treaty which was amended with the Treaty of Amsterdam in 1997. The Sixth Environmental Action Programme is agreed with all Member States' governments (via the Council), with the European Parliament and with the Commission. It is the first Environmental Action Programme which is legally binding.[3] This is a difference to most of the Member States' environmental policy programmes and gives the programme a quite strong weight. The Sixth Environmental Action Programme is in relation to the previous programmes not as precise and ambitious, but now, the European institutions are bound to this programme.

One of the mentioned political focal points of the Sixth Environmental Action Programme is the integration of environmental policy into other political fields (section 7.1). Therefore the SEA Directive, though enacted before the enforcement of the Sixth Environmental Action Programme, fits perfect to this subject, and it can be assumed that further political fields will be provided with different assessment methods.

Inclusion into the SEA Directive

At first, the importance of these above mentioned principles is identified in the recitals at the beginning of the SEA Directive. Here the Community describes the reasons for the Directive and the – legal, political and economical – background. The precautionary principle is mentioned in Recital 1 as a central reason for the enactment of the SEA Directive. The principle of sustainability (referred to in the cross section clause in Art. 6 of the EC Treaty) is also mentioned in the first Recital. The principle of sustainability is additionally cited in relation to the Fifth Environmental Action Programme "Towards Sustainability" in Recital 2. Therefore these two principles are the political and legal fundament for the SEA Directive. The topic public participation and consultation is mentioned in Recital 17 and 18. Especially Recital 18 states, that the Member States have to make sure that the public is informed and relevant information is made available to them. Furthermore, all these principles are directly transposed in the Directive's provisions. So most of the articles can be directed to one or more environmental principles, for example Art. 6 (Consultations) and 7 (Transboundary Consultations) to the public participation principle.

---

[3] Therefore the Sixth Environmental Action Programme is now published in section "L" (for *législation*) and not anymore in section "C" ("*Communications et informations*").

2.3 Definition of SEA

In the scientific literature at least about ten, more or less different, definitions have been used, to describe what SEA is (for details see Arbter 2003, p.175 with further references). The importance of these definitions became rather low with the progress of European Community's SEA law and the development of a characteristic definition.

The Commission, which first presented and then accompanied the SEA Directive from the very beginning though all stages, defined SEA in 1999 as "a similar technique to environmental impact assessment (EIA) but normally applied to policies, plans, programmes and groups of projects. SEA provides the potential opportunity to avoid the preparation and implementation of inappropriate plans, programmes and projects and assists in the identification and evaluation of project alternatives and identification of cumulative effects. SEA comprises two main types: sectoral SEA, applied when many new projects fall within one sector, and regional SEA, applied when broad economic development is planned within one region" (EC 1999, Integrating environment concerns into development and economic cooperation).

No wonder the procedure regulated in the SEA Directive is not new to people familiar with the EIA. The concept of SEA, as defined in the SEA Directive, is based to a big extent on the existing EIA experience and related legal requirements. At the moment, major projects in the Member States likely to have an impact on the environment must be assessed under the EIA Directive. SEA is an important tool both for harmonisation of policies and for integration of environmental considerations into economic and other sectoral decisions. It combines the precautionary principle with principles of public participation and of preventing environmental damage. However, this assessment takes place at a stage when options for significant change are often limited. Decisions on the site of a project, or on the choice of alternatives, may already have been taken in the context of plans for a whole sector or geographical area. The SEA Directive plugs this gap by requiring the environmental effects of a broad range of plans and programmes to be assessed, so that they can be taken into account while plans are actually being developed, and in due course adopted.

2.4 History of the SEA Directive

The SEA Directive is a rather advanced piece of legislation, which has gone through more than 20 years of discussions. Very early plans for the creation of SEA on the European level started in 1975 when a more integral concept was in the discussion for plans, programmes, policies and projects. In this year – at this time the European Communities had no competence for action on the environmental sector – the Commission carried through studies on the feasibility of SEA in Europe (for details see Platzer-Schneider 2004 p.15). The Commission intended a long time to supplement the Council Directive on the EIA for projects of 1985 with a Council Directive on SEA, but because of general critique from the Council the Commission abstained from it (Kläne 2003 p.131). The discussion of the implementation of SEA in the European legislation was continued during the preparation of the Habitats Directive where Art. 6 para 3 requires an environmental assessment. Also the Structural Funds Regulation (expired) followed this concept in Art. 8 para 4 subpara 3 of the SEA Directive where an assessment of the environmental impacts has to be carried through. The successor Structural Fund Regulation requires an "ex-ante" assessment, similar to the SEA (Art. 41). Furthermore, SEA was discussed during the establishment of the Trans-European Transport Network.[4] Here, Art. 8 para 2 obliged the Commission to develop suitable methods of analysis for the carrying through of strategically evaluating the environmental impact of the whole network.

Besides these more sectoral approaches, the Commission tried several times to prepare proposals for the implementation of SEA, based on the 4th and 5th Environmental Action Programme (Platzer-Schneider 2004 p.16-17). Between 1989 and 1996 about six different alternatives for a proposal were discussed in the Commission.

2.4.1 Early Proposals

The first proposal for a Council Directive of August 1990 was barely discussed and then given up by the Commission. After five years the Commission intensified the efforts to introduce the SEA in European legislation and published in April 1995 a second and in September 1995 a third proposal for a Council Directive on SEA. The wide scope of these proposals included plans, programmes and policies. This caused partial opposition of some Member States, especially Germany. Therefore the Commission changed the strategy. The scope was limited, especially SEA of policies was given up (for details see Jacoby 2000 p.154).

2.4.2 Development of the SEA Directive

The basic proposal was amended on 4 December 1996 and limited to spatial planning (OJ C 129 from 25.04.1997, p.14). This proposal included the examination of alternatives, certain measures for prevention, minimization and compensation of environmental impacts. Not mentioned was the problem of duplication of the assessment in a plan or programme hierarchy (now regulated in Art. 4 para 3 of the SEA Directive).

On 25 March 1997 this proposal was communicated to the Council, and from there to the Economic and Social Committee, to the Committee of Regions and to the European Parliament. The Economic and Social Committee supported gener-

---

[4] Decision No 1692/96/EC of the European Parliament and of the Council of 23 July 1996 on Community guidelines for the development of the trans-European transport network, OJ L 228, p.1.

ally the proposal with opinion from 29 May 1997 (OJ C 287 from 22.09.1997, p.101). The Committee of Regions welcomed the direction of the proposal but considered an examination of the need for an additional directive to the EIA Directive and doubted the competence provision of Art. 175 para 1 EC Treaty/ex Art. 130s para 1 EC Treaty (OJ C 64 from 27.02.1998, p.63).

On 20 October 1998 the European Parliament discussed the proposal in the First Reading positively and demanded broader area of application, because the advantages of such an instrument should be used as far as possible (OJ C 341 from 09.11.1998, p.18). The European Parliament proposed 29 amendments (Schmidt et al. 2002 p.357; critical Ziekow 1999 pp.290-291). Some of them were accepted by the Commission. The amended text of the proposal (OJ C 83 from 25.03.1999, p.13) served as the basis for negotiations at the Council level with 15 Member States participating during 1999.

The Committee of Regions submitted its opinion on 15 September 1999 (OJ C 374 from 23.12.1999, p.9), the European Parliament approved the opinion from the First Reading on 16 September 1999 (OJ C 54 from 25.02.2000 p.76)

Still, the discussions on Council level were difficult and led to an orientation debate on the level of the Member States' Environment Ministers (Kläne 2003 p.134). In December of 1999, the Environment Ministers agreed on a text for the future Directive as a "Common Position". This result was not the "big point" which was expected from and corresponding with the position of the Commission, the European Parliament and some Member States. Rather, the lowest common denominator was found. At the last minute, those plans and programmes, the ones mentioned now in Art. 3 para 8 of the SEA Directive were left out of the area of application (Kläne 2003 p.135). Anyway, this trimmed version reached unanimous consent of all Member States and was formally adopted on 30 March 2000 as Common Position (OJ C 137 from 16.05.2000 p.11).

On 6 September 2000, the European Parliament approved the Common Position subject to the amendments voted on at its plenary session during the Second Reading (OJ C 135/155 from 07.05.2001, p.155). The Commission formulated its opinion on the amendments to the Common Position on 16 October 2000 (COM (2000) 636 from 16.10.2000). In the conciliation process, opened on 27 February 2001, most of the Council's positions were carried through, and a version of the directive was drafted which was acceptable to all Member States (Kläne 2003 p.137). But the delegation of the European Parliament succeeded in some point also, in particular, the idea of monitoring (Art. 10 SEA Directive), which was proposed by the European Parliament with the argument, that conservation of the environment is often agreed upon, but the realisation is rarely controlled. This success is owed the entry into force of the Amsterdam Treaty which strengthened the position of the European Parliament in co-decision procedures in comparison with the co-operation procedure being in force before (Feldmann and Vanderhaegen 2001, p.120).

On 31 May 2001 the European Parliament (Bulletin EU, 5-2001, no. 1.4.51) and on 5 June 2001 the Council formally adopted the SEA Directive 2001/42/EC (Bulletin EU, 6-2001, No. 1.4.44). The text of the Directive was published in the Official Journal L 197 of 21 July 2001, page 30. Therefore, the Directive entered into force at the same day, Art. 14 SEA Directive. Art. 13 of the SEA Directive obliges the Member States to implement the content of the Directive by 21 July 2004 (see Chap. 4). The implementation period falls within a small window of time; however, provisions for the transposition of the SEA Directive are already under way.

2.4.3 Coordination of the Member States' Transposition

To coordinate the transposition process between the Member States, a European Group of experts on EIA and SEA was set up. These experts met twice a year. Following a Commission's proposal, in one of the meetings – Vaxholm 2001 – it was decided to install a Drafting Group under the leadership of Finland and UK for the creation of an SEA Guidance. The draft was discussed in these expert meetings and is now available in its actual version at the Commission's homepage (http://europa.eu.int/comm/environment/eia/030923_sea_guidance.pdf).
This Guidance discusses mainly questions of interpretation of terms, definitions and provisions and not the practical carrying through of an SEA (Platzer-Schneider 2004 pp.25-26). So, this guidance is a kind of official commentary on the SEA provisions. If the Member States follow this guidance in their transposition process it is likely that the concept of SEA is realized in the same way throughout the European Union.

2.5 Content, Objectives and Purpose of the SEA Directive

Before describing the content, the objectives and the purpose of the SEA Directive, one important matter should be mentioned. The provisions of the SEA Directive are mainly of a procedural nature. Only in a few provisions the Directive directs material rights to the respective addressees. Material rights mean that the Directive gives substantial rights to a natural or juridical person or to any other body. Especially participation and consultation rights have to be mentioned in this matter. On the other hand, procedural law means that the Directive rules the formalities, especially procedural steps, requirements and consequences. Most of the provisions are about the SEA procedure, starting from the screening and ending with the monitoring.

2.5.1 Content of the SEA Directive

The SEA Directive consists of 15 articles and two annexes. 20 Recitals with consideration points are put in front of the text. Art. 1 describes the objectives, Art. 2 consists of definitions of relevant terms, Art. 3 rules the area of application. In Art. 4 to 9 the SEA procedure is regulated. Art. 10 contains the monitoring, Art. 11 details provisions for the relationship to other European legislative acts, especially to the EIA Directive. Art. 12 regulates information, reporting and review obligations and measures of both Commission and the Member States. The first Commission's report on the application and the effectiveness of the SEA Directive has to be sent to the European Parliament and the Council before 21 July 2006. Art. 13 consists of provisions on the transposition of the Directive through the Member States, to which the Directive is addressed by Art. 15. Art. 14 of the SEA Directive rules the entry into force with publication in the Official Journal. Annex I consists notices on the information referred to in Art. 5 para 1, the environmental report. Annex II contains criteria for determining the likely significance of effects referred to in Art. 3 para 5.

2.5.2 Objectives of the Directive

The objectives of the SEA Directive, described in Art. 1 of the SEA Directive, are to provide for a high level of environmental protection and to contribute to the integration of environmental considerations into the preparation and adoption of plans and programmes with a view to promoting sustainable development. For this, the Directive should ensure that an environmental assessment is carried out of certain plans and programmes which are likely to have significant effects on the environment.

These objectives link the Directive to the general objectives of the Community environmental policy as laid down in the EC Treaty (Art. 174). Art. 6 of the EC Treaty lays down that environmental protection requirements must be integrated into the definition and implementation of Community policies and activities, in particular with a view to promoting sustainable development. Art. 1 is to be read in conjunction with Recitals 4 to 6 of the SEA Directive. They describe the aims of the directive to ensure that such effects of implementing plans and programs are taken into account during their preparation and before their adoption (Recital 4). Further, to benefit undertakings by providing a more consistent framework in which to operate by the inclusion of relevant environmental information into decision making. The inclusion of wider set of factors in decision making should contribute to more sustainable and effective solutions (Recital 5). Finally to provide for a set of common procedural requirements necessary to contribute to a high level of protection of environment (Recital 6).

2.5.3 Purpose of the SEA Directive

The purpose of the SEA Directive is to ensure that environmental consequences of certain plans and programmes are identified and assessed during the preparation stage and before they get adopted. Environmental consequences should be considered at the planning stage and not at the time at which projects are already implemented, which often turns out to be too late to take environmental matters into account. The SEA Directive supplements the EIA Directive.

The public and environmental authorities must be involved and may give their opinion. The results are taken into account and integrated during the planning procedure. After the adoption of the plan or programme, the public will be informed of the decision and how it was made. If significant transboundary effects are likely, other Member States and its public must be informed and should be given the opportunity to provide comments that must also be integrated into the national decision-making process. The SEA Directive, therefore, is a procedural law.

To achieve sustainable development, which is mentioned as a major principle in Art. 2 and 6 of the EC Treaty *and* furthermore in Art. 2 of the EU-Treaty, SEA will contribute to more transparent planning by involving the public and by integrating environmental considerations. The high level of protection referred to in Art. 1 SEA Directive is a policy aim that is written down in Art. 174 Para 2 EC Treaty. In addition, the Directive provides for an enforcement mechanism to achieve environmental aims for the European Union, Art. 175 Para 1 EC Treaty. According to this mechanism, the Council may enforce legal actions in accordance with a procedure referred to in Art. 251 EC Treaty that is called the co-decision procedure.

2.6 Outlook

The commission has launched two communication papers on additional provisions on impact assessments in June 2002 (COM (2002) 276 final; COM (2002), 278 final) to start a new initiative towards more integration with further impact assessments (IA). The aim of these papers is the improvement of the quality and coherence of the policy and legislative development processes. An IA should be carried out for all major initiatives, whether strategies and policies, programmes or legislation. Though this strategy is directed to Community's acts, it should cause pressure for the application of IA to various policies, building on previous analysis of trade policy. Additionally this strategy could influence activities of the Member States towards the implementation of IA measures to their national activities.

This kind of IA is called Sustainability Impact Assessment (SIA). The intention of such Impact Assessments is to help analyse the impacts of such initiatives in terms of the three pillars of sustainable development, the economic, the social and the environmental field (OPOCE 2003 p.293, Box 13.15; see also OPOCE 2002, p.26). With the IA, the process of assessing major initiatives should be simplified, by incorporating the key elements of several existing evaluation methodologies and superseding them. Besides SEA, these include Business Impact Assessment (BIA), Regulatory Impact Assessment (RIA) and Health Impact Assessment (HIA). Again, the key question is how these aims can be transformed into practice (OPOCE 2003 p.293, Box 13.15.).

At the national level, there are almost no requirements to use IA yet. Only Finland, The Netherlands and Great Britain are using sectoral or similar procedures. Finland currently carries out SIA through the use of adapted SEA. In The Netherlands a range of coordinated tests is applied, e.g. the environment test (E-test) and business test (B-test), furthermore feasibility and enforceability tests. The Integrated Policy Appraisal (IPA) is currently promoted in the United Kingdom as a new tool. Furthermore RIA is adopted as a standard and integrated approach to policy-making (OPOCE 2003 p.293, Box 13.15.). But national approaches using (sectoral) RIA, SIA, BIA or – what is also mentioned – sustainable development assessment (SD) or sustainability appraisal (SA), will continue, although a broader use of IA is estimated as more likely (OPOCE 2003 p.293, Box 13.15).

So the assessment of environmental effects will be on the table for a while to support the Sixth Environmental Action Programme and the European Union Strategy for Sustainable Development (OPOCE 2002; see also Feldmann and Vanderhaegen 2001 p.121). Explicit the assessment of effects of policies on sustainable aims is one of the major aims of the European Union Strategy for Sustainable Development (OPOCE 2002 p.26; see also Art. 3 no. 3 of the Sixth Environmental Action Programme). Therefore assessment of the effects on the environment is one of the important tools for the Community's sustainable and precautionary environmental policy.

2.7 Conclusions and Recommendations

The transposition process is still not finished and there are no, or almost no, experiences from real life, yet. Depending on the results of SEA, the Commission will probably launch amendments to the SEA Directive or a new directive on Environmental Assessments to further fields. This would be logical because the Commission and the European Parliament strongly supported the inclusion of policies and strategies into the application area of the SEA Directive. Furthermore, endeavours for implementing provisions for SEA in policies and strategies were agreed at the 5th Ministerial Conference "Environment for Europe" at an extraordinary meeting of the Parties to the Espoo Convention in Kiev (Ukraine) on 21 May 2003 (see Chap. 3). So it is very likely that this endeavour obligation of the Kiev SEA Protocol will result in further Commission's activities.

References[5]

Arbter K (2003) SEA and SIA – Two Participative Assessment Tools for Sustainability. In: Kopp U et al. (eds) Conference proceedings of the EASY ECO 2 Conference, May 15-17, 2003, Vienna, pp 175-181. Internet address: http://www.arbter.at/pdf/paper_Arbter.PDF, last accessed 03.08.2004

EEA – European Environmental Agency (1999) Environment in the European Union at the turn of the century – Environmental assessment report No 2. Internet address: http://reports.eea.eu.int/92-9157-202-0/en/tab_abstract_RLR, last accessed 12.08.2004

---

[5] Note that legislation mentioned in the chapter is listed at the end of the handbook in the consolidated list of legislation (Appendix 2).

EC – European Commission (1999) Integrating environment concerns into development and economic cooperation, cited after European Environmental Agency (EEA) multilingual environmental glossary http://glossary.eea.eu.int/EEAGlossary/S/strategic_environmental_assessment, last accessed 03.08.2004.

Feldmann L, Vanderhaegen M (2001) The SEA Directive and its context. UVP-report 15(3):119-122

Jacoby C (2000) Die Strategische Umweltprüfung (SUP) in der Raumplanung (SEA in spatial planning). Erich Schmidt, Berlin

Kläne C (2003) Strategische Umweltprüfung (SUP) in der Bauleitplanung (SEA in land use planning). Dr. Kovac, Hamburg

Kloepfer M (2004) Umweltrecht (Environmental law), 3rd edn. CH Beck, Munich

OPOCE (2002) A European Union Strategy for Sustainable Development, Luxembourg 2002

OPOCE (2003) Europe's Environment: the Third Assessment (Environmental report No 10), Chapter 13. Internet address: http://reports.eea.eu.int/environmental_assessment_report_2003_10/en/kiev_chapt_13.pdf, last accessed 03.08.2004

Platzer-Schneider U (2004) Entstehungsgeschichte, Funktion und wesentliche Inhalte der Richtlinie zur strategischen Umweltprüfung sowie der Koordination der mitgliedstaatlichen Umsetzung (Development, function and main contents of the SEA Directive and the coordination of Member States transposition measures). In: Hendler R, Marburger P, Reinhardt M, Schröder M (eds) Die strategische Umweltprüfung (sog. Plan-UVP) als neues Instrument des Umweltrechts. Erich Schmidt, Berlin, pp 15-26

Schmidt M, Rütz N, Bier S (2002) Umsetzungsfragen bei der strategischen Umweltprüfung (SUP) in nationales Recht (Problems of transposing the SEA Directive into national (German) law). DVBl. 6:357-363

Schröder M (2003) Umweltschutz als Gemeinschaftsziel und Grundsätze des Umweltschutzes (Environmental protection as an aim of the Community and the environmental principles). In: Rengeling H-W (ed) Handbuch zum europäischen und deutschen Umweltrecht, 2nd edn. Carl Heymanns, Köln, pp 199-238

Sheate WR (2003) The EC Directive on Strategic Environmental Assessment: A Much Needed Boost for Environmental Integration. European Environmental Law Review 12(12):331-347

Wolf J (2002) Umweltrecht (Environmental law). CH Beck, Munich

Ziekow J (1999), Strategische Umweltverträglichkeitsprüfung – ein neuer Anlauf (SEA – a new try). UPR 8:287-294

3 Legal Context of the SEA Directive – Links with other Legislation and Key Procedures

Eike Albrecht

Department of Constitutional, Administrative and Environmental Law, Brandenburg University of Technology (BTU), Cottbus, Germany

3.1 Introduction

The purpose of the SEA Directive is to ensure that environmental consequences of certain plans and programmes are identified and assessed during the preparation stage and before they get adopted (Recital 4 of the SEA Directive). Environmental consequences should be considered at the planning stage and not at the time at which projects are already implemented, which often turns out to be too late to take environmental matters into account. Art. 1 of the SEA Directive states the objective of the Directive as follows: "[…] to provide for a high level of protection of the environment and to contribute to the integration of environmental considerations into the preparation and adoption of plans and programmes with a view to promoting sustainable development, by ensuring that, in accordance with this Directive, an environmental assessment is carried out of certain plans and programmes which are likely to have significant effects on the environment."

This chapter describes the connection between the SEA Directive and, first, certain international agreements and, second, other European Acts. Furthermore this chapter discusses the basic terms of the SEA Directive and describes the objectives, the area of application and the procedure of SEA from a legal point of view. Finally, this chapter closes with an outlook.

3.2 Competence of the Community

The Community can only act if the Treaties direct competences for the respective act to the Community (for details see Kloepfer 2004 p.676; Kläne 2003 p.141).

Implementing Strategic Environmental Assessment. Edited by Michael Schmidt, Elsa João and Eike Albrecht. © 2005 Springer-Verlag

The competence for the SEA Directive is based on Art. 175 para 1 (former Art. 130s) EC Treaty. Art. 175 para 1 EC Treaty gives the Council the competence to act in order to achieve the objectives referred to in Art. 174. This provision states that the environmental policy of the Community is to contribute to the preservation, protection and improvement of the environmental quality, the protection of human health and the prudent and rational utilisation of natural resources. The Community's environmental policy is to be based on the precautionary principle, Art. 174 para 2 EC Treaty. Additionally Art. 6 of the EC Treaty provides that the environmental protection requirements are to be integrated into the definition of Community policies and activities, in particular with a view to promote sustainable development.

3.2.1 Critique in Regard to the Competence Norm

Concerning the competence provision in the EC Treaty, it was discussed if Art. 175 para 1 or para 2 EC Treaty provides the Community with the necessary competence to act in regard of the SEA Directive (Kläne 2003 pp.142-145; Spannowsky 2004 pp.201-203; Stüer 2004 p.89 with further references in footnote 15). The main difference is, that Art. 175 para 2 EC Treaty requires a unanimous decision of all Member States. Anyway, there was a unanimous vote for the proposal of the SEA Directive (for details see Chap. 2), so this discussion is more or less academic. Furthermore, Art. 308 EC Treaty, which directs supplementary competence to the Community, was discussed as a competence norm (for details see Spannowsky 2004 pp.201-202). But Art. 308 is subsidiary if there is a special provision in the Treaties (Calliess and Ruffert 2002, Art. 308 EC Treaty, rec. 40 ff.). In respect to the SEA Directive this is Art. 175 para 1 EC Treaty (Spannowsky 2004 p.202).

Furthermore, the competences of the Community for the Directive have been doubted, because of its effects on the areas of urban planning statutes, framework development plans or preparatory land-use plans for which there was no competence in the EC Treaty for the Community to act. For example, the German *Laender* voted against the Proposal during consultations in the *Bundesrat* and demanded that the Federal Government object to the Proposal (*Bundesrat*-Parliamentary Reports, BR-Drs. 277/97 (*Beschluss*/Decision, p. 3; for further information see Kläne 2003 p.144). Because all other Member States seemed to accept the Proposal, the *Laender* could not assert themselves against the Federal Government, which did not want to be isolated with a negative vote.

With the Treaty of Nice, which came into force later than the SEA Directive, the provision of Art. 175 para 2 EC Treaty was changed. Now there would be a fitting Community competence for measures affecting town and country planning, but – as a compensation for the loss of national competence – with the requirements of a unanimous vote.

3.2.2 Community Competence for Procedural Provisions

The SEA Directive is of a procedural nature. This is confirmed in Recital 9 of the SEA Directive. This could cause problems in respect to the compatibility of the material requirements of the SEA Directive within Art. 249 para 3 EC Treaty. Art. 249 para 3 of the EC Treaty states that a directive is binding, as to the result to be achieved, upon each Member State to which it is addressed, but leaves to the national authorities the choice of form and methods. But the Directive contains some very specific provisions and leaves no room for a transposition which regards national legal traditions and structure. This problem occurs under the term "detailed directive", especially in respect of technical provisions. In the case of technical provisions the accordance with European Treaties and therefore their legality is mostly undisputed (Calliess and Ruffert 2002, Art. 249 EC Treaty, rec. 45). But, if a directive contains procedural requirements, there are some voices of reservation in juridical literature, for example regarding Art. 3 para 2 of the Water Framework Directive, (Knopp 2003 pp.28-29; Reinhard 2001a pp.145-148; Reinhard 2001b p. 126; Breuer 1998 pp.1001-1003). Similar objections are raised in respect to the EC Directives for the transposition of the Aarhus Convention (Danwitz 2004 pp.276-278). However, the European Court of Justice ruled that there is no violation of the EC Treaty, if procedural requirements are regulated in a directive as far as the "effet utile" requires such provisions (European Court of Justice, Case 184/97, European Court reports 1999, p. I-7837, (55) and (58), Commission/Germany, also printed in ZfW 2000, p. 171, 176-177; see also Ekardt and Pöhlmann 2004 pp.129-131).

3.3 Relation to the Aarhus Convention

The UN ECE Convention on Access to Information, Public Participation in Decision-making and Access to Justice in Environmental Matters was adopted on 25 June 1998 in the Danish city of Aarhus at the Fourth Ministerial Conference of the "Environment for Europe" process. The Aarhus Convention entered into force on 30 October 2001 according to Art. 20 para 1 (for details see Danwitz 2004 pp.272-274). The European Community and all Member States have signed the Convention though not all Member States have ratified the convention.[1] To meet the two fundamental subjects of the Aarhus Convention, the European Union created first as the main instrument to align Community legislation with the provisions of the Aarhus Convention on public access to environmental information, the Environmental Information Access Directive. The instrument to support European law within the provisions of the Aarhus Convention on public participation is the Public Participation Directive. This Directive also covers the third pillar of the Aarhus Convention, the access to the courts.

---

[1] For details see http://www.unece.org/env/pp/ctreaty.htm, last accessed 17 August 2004.

Additionally, provisions to force public participation in environmental decision-making are to be found in the SEA Directive and also in the Water Framework Directive. So, the SEA Directive is one of the first relevant acts of European legislation to support the aims of the Aarhus Convention (Feldmann 2004 p.30). Therefore the SEA is consistently left out of the area of application of the above-mentioned directives for the transposition of the Aarhus Convention into European law.

3.4 Relation to the Espoo Convention

The SEA Directive mentions in Recital 7 the UN ECE Convention on Environmental Impact Assessment in a Transboundary Context of 25 February 1991, signed in Espoo/Finland, the so called Espoo Convention. At the second conference of the parties to this convention in Sofia at the end of February 2001, it was decided to prepare a legally binding protocol on a strategic environmental assessment to supply the existing provisions on the environmental impact assessment in a transboundary context (see Jendroska and Stec 2003 pp.106-107).

Before the parties of the Espoo Convention met at the Fifth Ministerial Conference "Environment for Europe" in Kiev, Ukraine, on 21 May 2003, to sign a binding supplement on SEA, the (EC) SEA Directive came into force. Therefore, the frame for supplements to the Espoo Convention was set through the already existing European Directive (Feldmann 2004 p.31; Jendroska and Stec 2003 p.108). Thus, the content of this supplement is very close to the content of the SEA Directive. Only requirements concerning health protection are going further than the SEA Directive (Art. 2 para 7, Art. 7 para 2 or Art. 11 para 2 of the Kiev SEA Protocol (Feldmann 2004 p.32; Jendroska and Stec 2003 pp.109-110) and the protocol contains an article on policies and legislation (Art. 13 of the SEA Protocol). Art. 13 obliges the parties to endeavour the consideration and integration of environmental concerns in political and legislative processes (for details see Jendroska and Stec 2003 p.109).

3.5 Harmonisation of Environmental Assessment Procedures

There are already a lot of provisions in the Member States which oblige the authorities to carry through a kind of environmental assessment in a more or less detailed form. These different environmental assessments are often not coherently regulated. To use an example from German law, for Land-Use-Plans, based on the Building Code, or for Regional Plans, based on the Spatial Planning Act, a parallel system of special nature protection plans exists (see Chap. 38), such as Landscape Programmes, Landscape Master Plans (Art. 15 German Federal Nature Conservation Act 2002) and Landscape Plans (Art. 16 German Federal Nature Conservation Act 2002).

One aim of the SEA Directive (Recital 19 of the SEA Directive) is to point out possibilities for the Member States to harmonise and to connect different environmental assessment procedures. Harmonisation should not be understood as unification of procedures but as an elimination of contradicting evaluations. Therefore, the Member States should firstly take the chance to clear parallel assessments and procedures. Secondly, they could use the transposition process to harmonise their assessment provisions. For Germany it is considered that the SEA procedure, to harmonise it with the EIA, could be opened even for certain plans and programmes, e.g. plans based on the Building Code, which do not necessarily fall under the SEA provisions (Stollmann 2004 p.82; Krautzberger 2004 p.242).

Secondly, it could be reasonable to harmonise EIA provisions with requirements of the Habitats Directive. Therefore, an obligation for an EIA is proposed in all cases when an assessment in respect to the Habitats Directive is required (Bunge 2004 pp.210-211). The EIA provisions could contain alternatives, as is required by the SEA Directive for an SEA (Bunge 2004 p.202). Additionally, a monitoring procedure could be implemented into current EIA law and lastly, the public should be involved in procedures concerning the Habitats Directive (Bunge 2004 pp.203-204). As a minimum the Member States should harmonise the SEA provisions with procedures which are required by earlier European legislation. For example, monitoring systems based on the Water Framework Directive or the IPPC Directive could be used for monitoring measures required by the SEA Directive (Roder 2004 pp.236-237).

3.6 Relation to Other European Acts

There are overlaps between the SEA Directive and other European legislation. The SEA Directive specifies that certain plans and programmes require an assessment in accordance with its provisions. Some of these plans and programmes are required by other European acts which themselves demand different kinds of environmental assessments from that laid down in the SEA Directive. Art 11 of the SEA Directive is the basic provision for the understanding of the relationship between the SEA Directive and other European legislation. Art. 11 para 1 of the SEA Directive seems to be very clear and states that an SEA has no prejudice to any requirements under the EIA Directive and to any other Community law requirements.

Further requirements are laid down in Art. 3 para 2 b), Art. 3 para 9, Art. 5 para 3 and Art. 12 para 4. The relationship between the SEA Directive and other European legislative acts should nevertheless be pointed out because the national legislative power has to decide whether the SEA Directive requires additional assessment elements. If so, the Member States could either implement a cumulative procedure law or they could join the existing requirements with the new SEA-elements to one procedure. For example, in German building law it seems that the different procedural requirements will bind together into an amended Building Code. In respect to other fields, German legislative power, or better, the competent ministry officials of the Federal Ministry for the Environment, Nature Conservation and Nuclear Safety, are tending to create additional SEA provisions to the EIA Act (for details see Chap. 4 and 7).

Regardless which form of transposition the Member States choose, they have to notify their adopted measures under Art. 13 para 1 of the SEA Directive, and therefore it will be helpful to point out the SEA elements in current material law even though no transposition measures have been estimated as necessary.

3.6.1 EIA Directive (85/337/EEC)

Normally the EIA Directive and the SEA Directive should not overlap. The EIA Directive rules concrete projects and the SEA Directive rules plans and programmes. Although the area of application seems to be clearly divided, there are possibilities of overlaps. For example, Art. 2a of the German Federal Building Code in conjunction with Art. 17 of the German EIA Act, requires an EIA for the resolution of certain Legally Binding Land-Use Plans (for more examples see Bunge 2004 pp.200-201). Plans like this are typical subjects for an SEA, therefore with the transposition of the SEA Directive into German law there could be an overlap of SEA and EIA. But besides these – maybe only German – specialities, both Directives tend in the same direction, so that the areas of application for EIA and SEA should harmonize.

3.6.2 Water Framework Directive (2000/60/EC)

There could be overlaps between the Water Framework Directive and the SEA Directive in two points. Firstly, Art. 11 of the Water Framework Directive requires a Programme of Measures which has to be prepared by authorities. An SEA is only required if a Programme of Measures sets the framework for the future development consent of projects.

Secondly, Art. 13 of the Water Framework Directive introduces a River Basin Management Plan to coordinate water quality-related measures within each river basin. Like the Programme of Measures laid down in Art. 11 of the Water Framework Directive, the decision as to whether an SEA is necessary or not has to be decided case by case. If the River Basin Management Plan sets the framework for future projects, an SEA has to be done. On the other hand, there is no SEA obligatory if the River Basin Management Plan only summarizes what has already been laid down in the basic Programme of Measures.

3.6.3 Nitrates Directive (91/676/EEC)

Art. 5 of the Nitrates Directive requires action programmes for areas with high nitrate pollution. However, these action programmes are normally directed towards certain agricultural practices and are not intended to set the framework for certain projects. Where such programmes refer to agricultural practices only, an SEA is not mandatory. But in certain cases it could be that such action programmes set the framework for the future development consent of projects. In this case, such an action plan is considered as a programme with respect to the SEA Directive.

3.6.4 Waste Framework Directive (75/442/EEC)

Art. 6 of the Waste Framework Directive requires waste management plans to be developed from the Member States. The main elements of such waste management plans are regulated in Art. 6 of the Waste Framework Directive. Additional elements are applied by the Hazardous Waste Directive and the Packaging and Packaging Waste Directive. One objective of waste management plans could be to identify suitable disposal sites or installations. Insofar they set the framework for future licensing of waste disposal facilities which are covered by Annex I (No. 9 and 10) and Annex II (No. 11 b) of the EIA Directive, and therefore an SEA is mandatory (Lindemann 2004 p.66; Beckmann 2004 pp.137-139).

Additionally waste management plans could set certain criteria for disposal facilities or could delegate this task to plans on a lower level. In both cases these waste management plans set frame for subsequent development consents. Therefore it would be necessary to require an SEA for them as well. Only if waste management plans do not set the framework for future licensing, which could be the case, if the current disposal capacities are not exceeded, they are not covered by the SEA Directive (Lindemann 2004 p.66).

3.6.5 Air Quality Framework Directive (96/62/EC)

Art. 8 para 3 and 4 of the Air Quality Framework Directive requires the preparation or implementation of plans and programmes, if the level of one or more pollutants exceeds certain limits. In these plans and programmes the Member States should lay down measures concerning how the level of the relevant pollutants could reach the limit values. The main objective of these plans is to improve air quality. Though these plans have effects on many sectors, they do not necessarily fall under the alternative of Art. 3 para 2a) of the SEA Directive (Sander 2003 p.338).

But they do if such plans or programmes meet the criteria laid down in Art. 3 para 4 of the SEA Directive and set the framework for the future development consent of projects and the Member States determine them likely to have significant environmental effects. Such plans and programmes for the improvement of air quality normally have significant environmental effects (Sander 2003 pp.337-

338). Because the SEA Directive has no restriction in respect to only negative environmental effects, an SEA has to be done also if positive environmental effects are expected (Hendler 2004 p.107 with reference to Annex I, lit. f. of the SEA Directive).

3.6.6 Wild Birds Directive (79/409/EEC)

When assessments are required by the Wild Birds Directive, Recital 19 of the SEA Directive proposes joined or harmonised procedures. On the other hand, the Wild Bird Directive does not require the preparation of plans and programmes which could fall under the SEA Directive.

3.6.7 Habitats Directive (92/43/EEC)

The aims of the Habitats Directive and the SEA Directive are different. An SEA should be carried through for better planning decisions. The assessment according to Art. 6 and 7 of the Habitats Directive should protect the European network "Natura 2000" against negative effects through projects and plans (Bunge 2004 pp.204-205; Hösch 2004 p.212). The Habitats Directive aims at setting up a coherent European ecological network of special areas of conservation. Therefore the Member States have to transmit a list of ecologically worthy areas to the Commission as a proposal.

Normally this proposal does not set a framework for future development, therefore the procedures under the Habitats Directive do not fall under the SEA Directive. In opposite manner, Art. 3 para 2b) of the SEA Directive also requires an SEA for plans and programmes which, in view of the likely effect on sites, have been determined to require an assessment pursuant to Art. 6 or 7 of the Habitats Directive, transformed into German national law by Art. 34 and 35 of the Federal Nature Conservation Act from 2002.

However, these provisions are not sufficient with respect to the SEA Directive (Hendler 2004 p.117). Therefore the provisions for the assessments on the base of the Habitats Directive and the SEA Directive should be harmonised, e.g. by joining procedures to avoid parallel assessments, as Recital 19 of the SEA Directive advices. Difficulties with respect to harmonization are not expected, because the Habitats Directive does not contain detailed procedural provisions and the aims are different (Bunge 2004 p.206).

3.6.8 Plans and Programmes under the Structural Fund and under the European Agricultural Guidance and Guarantee Fund

Art. 3 para 9 of the SEA Directive states, that the SEA Directive does not apply to Plans and Programmes under the Structural Fund and under the European Agricultural Guidance and Guarantee Fund under the current respective programming periods. The reason is, that the Directive is addressed only to the Member States and does not apply to the institutions of the Community. The Commission has introduced a procedure for assessing the impact of its own proposals, the "communication on impact assessment" of 5 June 2002 (COM (2002) 276 final). Because of Art. 6 of the EC Treaty a kind of an environmental assessment is mandatory anyway for any measures of the Community (Calliess 2004 pp.157-158).

3.7 Area of Application

Plans and programmes are defined in the SEA Directive to mean "plans and programmes, including those co-financed by the European Community, as well as any modifications to them:

- which are subject to preparation and/or adoption by an authority at national, regional or local level or which are prepared by an authority for adoption, through a legislative procedure by Parliament or Government, and
- which are required by legislative, regulatory or administrative provisions."

This definition is remarkable because its grade of precision is rather low. It is also remarkable, that the term "plan" and "programme" is used without distinction. So the provisions of the SEA Directive concern both plans and programmes without any differences. In respect to German law, the terms "plans" and "programmes" are not used in a standardized way. It depends more on regional matters, if a certain "act" is called a plan or a programme by the respective law,[2] though there is a tendency that plans are more detailed in relationship with programmes (Hendler 2003 p.229).

Only plans and programmes, which are required by legislative, regulatory or administrative provisions, are mandatory subject to an SEA. If these conditions are not met, the SEA Directive does not apply. Such voluntary plans and programmes usually arise because legislation is expressed in permissive terms, e.g. "the authority may prepare a plan", instead of "the authority shall prepare a plan", or when there is no regulation about the preparation of a plan.

The obligation for a preparation or adoption of a plan or programme could be based inside the plan or programme itself. This concerns plans and programmes which have to be continued (Hendler 2004 pp.104-105). The decision, if a certain plan or programme is required by law, is not always easy. There are plans, e.g. plans for the need of federal highways on the base of the German Federal Highways Act, which are not directly required by law, but the existence of the plan is the condition for the further development plans and measures. So such plans should be assessed by an SEA (Hendler 2004 p.105) because otherwise the basic

---

[2] E.g. Art. 7, 9 Hessian Planning Law mentions a "Development Plan for the Land" and "Regional Plan", Art. 4, 6 Regional Planning Act for Lower Saxony names the same act with the same function as "Comprehensive Regional Planning-Programme for the *Land*" and "Regional Comprehensive Regional Planning-Programme".

primary plan is not subject to an SEA but the later (and depending on the primary) plan has to be assessed (Hendler 2003 p.231, with a hint to Recital 9 of the SEA Directive; see also Chap. 40). Therefore, the Member States should draw a wide circle of plans and programmes which have to be assessed, to avoid a violation against the SEA Directive by not including all relevant plans and programmes to the transposition acts.

3.7.1 Policies

Only plans and programmes are mentioned in the SEA Directive. This means that policies are not subject to the SEA Directive though in early proposals to the SEA Directive policies have been included (see Chap. 2 and for extra information Hendler 2004 p.101). The reason was that general political concepts normally stand on the top of the decision hierarchy. On the other hand, policies or political concepts are often not concrete enough to have direct and future effects to the environment. But, the Member States are free to open the SEA requirements to policies as well.

3.7.2 Strategies

Whenever strategies are to be counted as something different from policies, there is the same result to be found as above. On the other hand, strategies could be something very similar to a programme. Then, of course the Member State have to be very careful in their decision whether a strategy is a programme in respect to the SEA Directive or if it is more like a general guideline without direct effects and therefore more a policy. In the first case an SEA is mandatory for strategies (if the other conditions are fulfilled), in the second, there is no obligation for the Member State. But, of course, the respective organ could open the transposition of the SEA Directive to strategies. For example, in Scotland strategies will also be the subject to an SEA according to the forthcoming SEA Bill that goes beyond the requirements of the SEA Directive (see Chap. 47).

3.7.3 Legislation

It can be controversially disputed if legislative acts fall under the SEA Directive (for details see Hendler 2003 p.230). Especially laws have often direct or indirect effects on the environment, e.g. income tax law, which seems to have no contact to environmental matters: If, for example, the estimated amount of travel costs for the use of private cars are raised, more people would move to the suburbs and take the car, with all negative effects on nature, city centres and infrastructure.

The term "through a legislative procedure by Parliament or Government" leads to the argument that even laws could be subject to an SEA. One the other hand, normally, laws are not identical to plans and programmes. So, if laws were meant to be included into the SEA Directive, this should have been mentioned in the text of the Directive. And as the history of the SEA Directive shows, the exclusion of legislative acts was intended. An additional clue is the reference in Recital 7 to the Kiev SEA Protocol, where legislation and policies are mentioned separately.

Furthermore, it can be doubted that the EC Treaty provides EC organs with the necessary competence to regulate national legislative procedures. Then, an obligation for an SEA for legislative acts would cause more problems than it would solve. The typical legislative organ is the Parliament, but Parliaments normally cannot revert to executive resources. If for legislative proposal of the Parliament, a certain number of Members of the Parliament, parliamentary parties or parliamentary groups, depending on the relevant constitution or parliamentary standing orders, an SEA is required, the ability to introduce laws to parliamentary procedures is very reduced (in this direction Manegold, contribution to the discussion on the paper of Hendler, printed in: Hendler/Marburger/Reinhardt/Schröder (Eds.), Die strategische Umweltprüfung (sog. Plan-UVP) als neues Instrument des Umweltrechts 2004 p.123). On the other hand this argument does not exclude an obligation for other legislative proposals, e.g. proposals introduced into the parliamentary system from the Government. In these cases, an SEA could be done, because the resources of the executive power, especially the Ministries and subordinated authorities could be used (Ginzky 2002 p.48; Hendler 2004 pp.103-104). But to avoid different types of legislative proposals – some with an SEA (from the Government), some without (from inside of the Parliament) – the area of application should not be exceeded on legislative acts.

Finally, usually there are very limited obligations for the adoption of laws. So the requirement of Art. 2a), 2nd alternative of the SEA Directive ("which are required by legislative, regulatory or administrative provisions") is usually not fulfilled.

3.7.4 Licenses or Permits

Concrete licenses or permits are not subject to an SEA. Though this is not directly written in the SEA Directive it is presupposed. Additionally, the combination of Art. 3 para 2a) and para 4 of the SEA Directive in conjunction with the EIA provisions leads to the result that plans and programmes have to be different from licenses (Hendler 2004 p.102). Therefore Legally Binding Land-Use Plans fall under the SEA Directive, Decisions on Planning Approvals (Art. 74 German Administrative Procedures Act) do not (Hendler 2004 p.102).

3.7.5 Modifications to Plans and Programmes

Modifications to plans and programmes fall under the definition of Art. 2a) of the SEA Directive. Therefore the provisions of the directive have to be followed not only for the preparation or adoption of plans and programmes but also for modifications (Hendler 2003 p.229). However, for modifications, the exception rule in

Art. 3 para 3 of the SEA Directive has to be observed. If there are only minor modifications, the Member States could exclude them from an obligation for an SEA, if they are not likely to have significant environmental effects.

3.7.6 Private Plans and Programmes

Plans and programmes in respect to the SEA Directive require the preparation or the adoption of an authority. Therefore private plans and programmes are typically not subject to the SEA Directive. This excludes all possible private plans and programmes, e.g. business plans or economical programmes inside of a company, societies or private households and finally the "environmental programme" referred to in Art. 2 h) of the EMAS II-Regulation, as a part of the implementation of an EMAS environmental management system.

On the other hand, plans or programmes developed from private parties could be subject to an SEA, if they are adopted from authorities. This is not unusual, especially in the area of German Federal Building Law, Land-Use-Plans are often prepared from private architecture-offices (Hendler 2004 p.103). Another example is the Remediation Plan pursuant to Art. 14 para 1 of the German Federal Soil Protection Act, normally prepared from private experts (Albrecht 2003a, p.231). It is difficult to decide whether a private Remediation Plan referred to in Art. 13 para 1 of the same act has to contain an SEA or not. The private Remediation Plan is required and has to be – under certain conditions – accepted from the competent authority which normally declares the plan as binding. This has the same effect as a binding plan from the authority, therefore an SEA should be required.

So it has to be clarified, what "authority" means in the context of the Directive. The European Court of Justice has given a large scope of what it summarizes under an authority. It can be defined as a body, whatever its legal form and regardless the extend (local, regional or national) of its powers, which has been made responsible, pursuant to a measure adopted by the State for providing a public service under the control of the State, and it has for that purpose special powers beyond those which result from the normal rules applicable in relations between individuals (European Court of Justice, Case C 188/89, European Court reports 1990, p. I-3313 (20), Foster et al./British Gas plc). Therefore privatized utility companies could be treated as authorities, if they carry out tasks or duties which in not privatized regimes would be carried out by public authorities, e.g. obligation for private water-suppliers for long term plans to ensure water resources.

3.7.7 International Plans and Programmes

Art. 11 of the Water Framework Directive requires Programmes of Measures. Therefore each Member State shall ensure the establishment for the part of an international river basin district within its territory. Art. 3 para 3 of this Directive obliges Member States to ensure that a river basin covering the territory of more than one Member State is assigned to an international river basin district. The

Member States shall ensure the appropriate administrative arrangements, including the identification of the appropriate competent authority, for the application of the rules of the Water Framework Directive.

The same is the case where a river basin district extends beyond the territory of the Community. Then the Member State(s) concerned shall endeavour to establish appropriate coordination with the relevant non-Member States, with the aim of achieving the objectives of the Water Framework Directive throughout the river basin district. In both cases, the programme of measures could be prepared from an supranational body or authority with the relevant competences, as Art. 3 para 6 of the Water Framework Directive shows (Hendler 2003 p.230). As far as states are concerned which are not obliged by the SEA Directive, the assessment of environmental effects could only be supported by the Member States. But, in case of programmes of measures for international river basins situated only in Member States, e.g. the rivers Maas, Odra or Elbe there is an obligation for an SEA.

On the other hand, if the international coordination is not organised from a supranational body or authority, there is no SEA required by the SEA Directive. This could be the case, if the programme of measures is agreed in an international treaty between the relevant Member States. Theoretically, an SEA could be done before the relevant Parliament adopts the treaty. But environmental effects would be assessed too late to influence the treaty. Thus, it is considered that there is no obligation for an SEA for the adoption of international treaties (Hendler 2003 p.230). The same result has to be stated with respect to River Basin Management Plans (Art. 13 of the Water Framework Directive) (see also Chap. 35).

3.7.8 Plans and Programmes for Defence or Civil Emergency Purposes

Art. 3 para 2 of the SEA Directive identifies the plans and programmes that are subject to SEA; moreover, in Art. 3 para 8 of the SEA Directive, those plans and programmes are identified which are not subject to SEA. This is relevant for "plans and programmes the sole purpose of which is to serve national defence or civil emergency." In comparison to the EIA provisions, the SEA Directive contains stricter conditions. So the EIA provisions do not apply to projects serving national defence purposes. On the other hand, an SEA is not necessary for plans and programmes the *sole* purpose of which is defence or civil emergency. So if a plan or programme serves national defence and civil emergency purposes as well as other purposes, e.g. regional land use plans, an SEA is mandatory, if the other conditions are fulfilled. Generally spoken, the derogation should be construed narrowly. Therefore, plans and programmes that address floods or other natural catastrophes fall under the SEA Directive, if further purposes are included in the plan or programme, which normally is the case. If a certain plan or programme just sets out what action should be taken if a flood were to occur, an SEA is not obligatory.

3.7.9 Financial or Budget Plans or Programmes

Additionally, no SEA is required for financial or budget plans and programmes. Budgetary plans or programmes, referred to in Art. 3 para 8 of the SEA Directive, are the annual budgets of authorities at the national, regional or local level. Financial plans and programmes could include ones which describe how some projects or activity should be financed, or how grants or subsidies should be distributed, regardless if they have significant effects on the environment or not. For example, certain subsidies to avoid the use of fertilizer, based on a plan for improving groundwater quality could have positive effects on the environment. Anyway, an SEA is not required in these cases.

3.7.10 Plans for the Use of Small Areas at Local Level

Plans and programmes which only determine the use of small areas at a local level are principally excluded from any obligation for an SEA, but the Member States can provide an SEA for these plans and programmes. The main criterion for the application of the Directive is not the size of area covered, but whether the plan or programme would be likely to have significant environmental effects. So if a plan or programme, even if it refers to a small area on local level, has significant environmental effects, an SEA has to be done. It cannot exactly be defined what "small area" in this context means. Differences between the Member States have to been taken into account. So the assignment has to be decided case by case. A similar problem contains the phrase "local level". The Directive does not establish a clear link with local authorities, but implies a contrast with national or regional levels.

3.7.11 Plans and Programmes as Base for Certain Decisions

Art. 3 para 2a) of the SEA Directive requires an SEA for plans and programmes which are prepared for "agriculture, forestry, fisheries, energy, industry, transport, waste management, water management, telecommunications, tourism, town and country planning or land use and which set the framework for future development consent of projects listed in Annexes I and II to Directive 85/337/EEC", the EIA Directive. Having a closer look to these categories, two more general sectors can be found (town and country planning/land use) and ten specialized categories (Hendler 2004 p.110). So, normally more than one category is concerned.

According to the above provision, an SEA must be prepared for plans and programmes in two situations: First, the nature of a plan or programme can be directed to one of the mentioned sectors; second, the plan or programme has to set the framework for licensing of certain projects, which are listed in Annexes I and II of the EIA Directive, e.g. thermal power stations and other combustion installations with a heat output of 300 megawatts, construction of motorways, express roads and lines for long-distance railway traffic and of airports with a basic runway length of 2100 m or more, or the extraction of coal and lignite by open-cast mining.

3.7.12 Plans and Programmes with Effects on Habitats

Art. 3 para 2b) of the SEA Directive requires an SEA also for plans and programmes "which, in view of the likely effect on sites, have been determined to require an assessment pursuant to Art. 6 or 7 of Directive 92/43/EEC," the Habitats Directive. If an SEA is required for such plans and programmes which fall also under the Habitats Directive, a single habitat's assessment could be avoided. Recital 19 of the SEA Directive advices coordinated or joint procedures. Of course, the special and additional requirements of the Habitats Directive have to be fulfilled (Bunge 2004 p.223).

3.7.13 Other Plans and Programmes

Art. 3 para 4 of the SEA Directive obliges the Member States to determine whether plans or programmes, which are not listed in the catalogue of Art. 3 para 2, but set the frame for future development consent of projects, are likely to have significant environmental effects. This provision opens the scope of the Directive. Contrary to Art. 3 para 2 it does not automatically deem certain plan and programmes to have significant effects on the environment, but forces the Member States to make a specific determination. This includes both plans or programmes in sectors which are not mentioned in Art. 3 para 2 and projects which are not listed in the annexes to the EIA Directive.

3.8 Procedure of the SEA

An environmental assessment has to be carried out during the preparation of a plan or programme and before its adoption or submission to the legislative procedure. This should support the early integration of environmental aims in planning, decision, and licensing procedures. Art. 2b) of the SEA Directive defines the term "environmental assessment". This definition includes the "preparation of an environmental report, the carrying out of consultations, the taking into account of the environmental report and the results of the consultations in decision-making and the provision of information on the decision in accordance with Articles 4 to 9" of the SEA Directive. If an environmental assessment has to be carried out because of other directives or respective transpositions into national law, the Member States could join procedures or state coordination. This is relevant especially for the Wild Birds Directive, the Habitats Directive or the Water Framework Directive (see Recital 19 of the SEA Directive) (see also Chaps. 36 and 37).

Where plans and programmes are part of a hierarchy, the Directive advises against duplication of environmental assessment (Recital 19 and Art. 4 para 3 of the SEA Directive). The SEA Directive is of procedural nature (see Recital 9 of the SEA Directive) and provides an administrative procedure with several different steps (Calliess 2004 p.160; Schmidt et al. 2002 p.359).

3.8.1 Screening

The first step is screening. This is not required by the Directive and is only mentioned in Art. 3 para 5 of the SEA Directive. This pre-selection, for which plans and programmes an SEA should be carried through, can be decided either case by case or determined by general provisions, e.g. a law (Calliess 2004 p.167). For plans and programmes where no determination exists, a screening is useful to identify which plans have to be assessed and those which have not to be. For example, plans and programmes which determine the use of small areas at local level (Art. 3 para 3 of the SEA Directive) could be asserted out or – the other way round – all plans and programmes which require an SEA could be listed in a "positive list" (Ginzky 2002 p.49).

3.8.2 Scoping

The next procedural step to follow is scoping. Here, the extent to which details will be included in the assessment is the next step (Art. 5 para 4 SEA Directive). Therefore, scoping is seen as the central factor for the quality of the environmental assessment (Ginzky 2002 p.51; Calliess 2004 p.167). Art. 6 para 3 of the SEA Directive states that authorities that have been designated by the Member States for their specific environmental specialities (e.g. for Water and Water Management, for Nature Conservation or for Emission Control) should be consulted. But, Member States' laws could open this participation for NGOs and the public in general as well, though this is not required by the SEA Directive (Calliess 2004 p.168).

3.8.3 Environmental Report

The environmental report is seen as the main element of the SEA procedure (Platzer-Schneider 2004 p.22; Calliess 2004 p.168). The production of an environmental report is mandatory. Therefore this is one of the main differences to the EIA Directive. In the EIA provisions only information had to be provided somehow, but not necessarily in a separate report. The reason was, that the Commission and the Member States tried to avoid an obligatory requirement for a single environmental document (Sheate 2003 p.344). This was subject to critique which is today more justified than it has been when the EIA provisions were enacted in 1985. Now there is a competence for European environmental acts in the EC Treaty which did not exist in 1985 (Sheate 2003 p.331; see also Chap. 2).

Definition

The environmental report is "the part of the plan or programme documentation containing the information required in Art. 5 and Annex I." This includes an outline of the contents, main objectives of the plan or programme and relationship with other relevant plans and programmes, the environmental characteristics of areas likely to be affected, the likely significant effects on the environment, including issues such as biodiversity, population, human health, fauna, flora, soil, water, air, climatic factors and some more. Details are stated in Annex I to the SEA Directive.

The environmental report is not a final description of the environmental situation (Calliess 2004 p.168). It is, on the contrary, the opening report for the whole planning procedure. The content of the environmental report has to be considered, but not amended or improved, during the planning procedure (Schmidt-Eichstaedt 2004 p.87).

Position inside the SEA Procedure

The Environmental report is the central element of the SEA and forms the basis for monitoring the significant effects of the implementation of the plan or programme (Calliess 2004 pp.168-169). The report requires an identification and description of the plan or programme as well as an evaluation of the likely significant effects on the environment of an implementation of the plan or programme. It also should describe reasonable alternatives which have to be described at the same level as the original plan or programme. So, differently to the EIA, the main task of the Environmental Report is the finding and description of alternatives (Calliess 2004 p.169; Schink at a conference on SEA, 28 April 2003, Rostock, Germany, cited after Hönig 2003 p.1193). One of the alternatives is the non-pursuit of a plan or programme, the so called zero-variant (Calliess 2004 pp.169-170).

Responsible Authority

The Environmental Report must be prepared by the authority that is competent to perform the respective planning. Because of the specified content (Art. 2c) of the SEA Directive) the Environmental Report should be a coherent text. It also may be helpful to structure the report, as far as possible, on the titles used in the Annex I. In any case, there should always be a non-technical summary of the information provided under these titles, so that non-technical professionals, like lawyers, could understand the main targets of the report (ODPM 2003 p.26, Fig.7).

Integrated or Separate Information

The SEA Directive does not regulate, whether the Environmental Report should be integrated in the plan or programme itself or in a separate document. But, if it is integrated, it should be clearly distinguishable as a separate part of the plan or programme, so that it could be easily found and assimilated by the public and authorities.

Content of the Environmental Report

Annex I of the SEA Directive specifies the minimum information that is to be provided in the Environmental Report. The Member States could introduce provisions on the content that go further than the requirements of the SEA Directive. Anyway, the report should concentrate on issues related to the significant effects on the environment (so Art. 5 of the SEA Directive), because an overload with information on insignificant effects or irrelevant issues makes the report difficult to digest and might lead to important information being overlooked.

The following information has to be provided as a minimum and should include secondary, cumulative, synergistic, short, medium and long-term permanent and temporary, positive and negative effects (for details see Lindemann 2004 pp.74-76):

1. an outline of the contents, main objectives of the plan or programme and relationship with other relevant plans and programmes;
2. the relevant aspects of the current state of the environment and the likely evolution thereof without implementation of the plan or programme;
3. the environmental characteristics of areas likely to be significantly affected;
4. any existing environmental problems which are relevant to the plan or programme including, in particular, those relating to any areas of a particular environmental importance, such as areas designated pursuant to Directives 79/409/EEC and 92/43/EEC;
5. the environmental protection objectives, established at international, Community or Member State level, which are relevant to the plan or programme and the way those objectives and any environmental considerations have been taken into account during its preparation;
6. the likely significant effects on the environment, including on issues such as biodiversity, population, human health, fauna, flora, soil, water, air, climatic factors, material assets, cultural heritage including architectural and archaeological heritage, landscape and the interrelationship between the above factors;
7. the measures envisaged to prevent, reduce and as fully as possible offset any significant adverse effects on the environment of implementing the plan or programme;
8. an outline of the reasons for selecting the alternatives dealt with, and a description of how the assessment was undertaken including any difficulties (such as technical deficiencies or lack of know-how) encountered in compiling the required information;
9. a description of the measures envisaged concerning monitoring in accordance with Art. 10;
10. a non-technical summary of the information provided under the above headings.

3.8.4 Consultations

The information of the public and transboundary consultations are subject to international environmental conventions, like the Aarhus Convention for public information and participation or the Espoo Convention on transboundary environmental impact assessment. Especially the Espoo Convention, and especially the Kiev SEA Protocol, which was agreed at the second COP 2001 in Sofia (and signed 2003 in Kiev), had a big influence on the international consultation process and is seen as a basis for transboundary consultations in SEA-procedures (Platzer-Schneider 2004 pp.29-30). The SEA Directive includes these international concepts in Art. 6 and 7. Art 6 requires a second consultation with the public and those authorities that have been involved in the definition of the scope.

Definition of "the Public"

The public, under the SEA Directive, is "one or more natural or legal persons and, in accordance with national legislation or practice, their associations, organisations or groups." This definition follows that of the Aarhus Convention. The term "natural or legal person" is clear. But the following "associations, organisations or groups" leave a wide range for national determination (Calliess 2004 p.171). Art. 6 para 4 of the SEA Directive obliges the Member States to identify the public for the participation steps, mentioned in Art. 6 para 2 of the SEA Directive. This contains an early and effective opportunity within appropriate time frames to express their opinion on the draft plan or programme and the environmental report before the adoption (Ginzky 2002 p.51). So only those parts of the public have privileged participation rights, which are provided with by the Member States. This includes as a minimum those, who are affected or likely to be affected by, or having an interest in the decision-making subject to the SEA Directive. This also includes, according to Art. 6 para 4 of the SEA Directive, relevant NGOs, especially those promoting environmental protection (Calliess 2004 p.171).

An example related to German law is particularly valid: In Germany NGOs in the environmental sector have the right to participate in different environmental procedures, e.g. Art. 60 para 2 Federal Nature Conservation Act 2002 (former Art. 29), or Art. 12a Waste Law of *Baden-Wuerttemberg* (see Albrecht 2003b) § 12a, Nr. 3.1.), Art. 51 Nature Protection and Conservation Law of Saxony (see in detail Wilrich 2002 pp.392-395). For this reason these groups have to be involved in the SEA, because national legislation or practice provides them with participation rights.

The strengthening of public participation is a steadily ongoing process, which is supported from a lot of actions from the Community. So the new Directive on the Access to Environmental Information opens most of the official information and data to the public. In some Member States, like the Scandinavian countries, there is a culture of open official information. In Germany, on the contrary, there is a deeply rooted "arcane culture". On the federal level there are only certain areas of official information opened for public access, e.g. the documents of the state security service of the (former) GDR or, as mentioned, information on the environment (for details see Partsch and Schurig 2003 pp.483-484).

Additionally to the Aarhus Convention and their transposition Directives (see Sect. 3.3), there are further provisions to support public participation. Besides European directives, like the Water Framework Directive, the before mentioned Espoo Convention (see Sect. 3.4) and the Kiev SEA Protocol (see Sect. 3.4) has to be pointed out, to which the SEA Directive refers to in Recital 7 of the SEA Directive (see for details Jendroska and Stec 2003 pp.108-109).

Consultation Procedure

The procedure of consultations should meet certain requirements. A consultation is at first a dialogue between the authorities and the interest circles (authorities, domestic and foreign public). Therefore the use of the internet is an important part of delivering information to the public, but is not sufficient as a single measure, because certain parts of the public could be excluded from the participation process. So additional measures have to be done. The procedure of public participation in licensing procedures according to Art. 10 of the German Federal Immissions Control Act[3] gives a hint what measures could be done (for details see Wolf 2002 pp.424-428; Sparwasser et al. 2003 pp.728-735). So the competent authority shall give public notice of the project in its official gazette and, additionally, in any local daily newspapers as are widely read in the area where the plan and programme are to be established. The notification should indicate the date and place of when and where the documents will be laid out for public inspection. The documents should be made available for public inspection for a reasonable time period following such notice, e.g. the period of one month. Any objections raised against the plan or programme may be lodged with the competent authority in writing, e.g. until the end of two weeks after expiry of the inspection period, which should be sufficient. There also could be introduced a public discussion as a part of the consultations though this is not mandatory (Lindemann 2004 p.72). The Member States are free to determine the duration of the time frame, as long as it meets the requirement to give an early and effective opportunity for responses. Different time frames could be set for different types of plans and programmes.

The procedure of public participation could be organised differently. Of course there is the possibility general to open all SEA to public participation. On the other hand, this might overstress a lot of planning procedures, so that a minimum of public participation in form of the participation of NGOs or environmental groups could be arranged (Lindemann 2004 pp.71-72). To meet the requirements of a wide public participation and on the other hand the needs for a planning procedure which is not overloaded, a graded procedure is provided. So a general public participation on the lower and therefore concrete level, especially on the local level is suggested and a participation of NGOs on the higher, and therefore more abstract level (Lindemann 2004 pp.71-72). Problem here could be that the text of

---

[3] The English translation of the text of the German Federal Immission Control Act is printed – besides other relevant federal environmental laws – in Mulloy et al. 2001.

the SEA Directive does not direct privileged public participation rights to NGOs. On the contrary, Art. 6 para 4 of the SEA Directive tends to treat NGOs as parts of the public like others (Lindemann 2004 p.71).

International Consultation

Additionally, Art. 7 of the SEA Directive obliges the planning Member State to consult with other Member States if the Member State foresees likely significant effects on the environment in the other Member States. Art. 7 of the SEA Directive also provides those Member States a right to apply for consultation, if the Member State will likely be affected by the respective plans and programmes in the other Member State.

3.8.5 Decision Making

Art. 8 of the SEA Directive obliges the planning authority to take into account the environmental report, the opinions expressed pursuant to Art. 6 of the SEA Directive (authorities and the public), and the results of transboundary consultations when preparing the plan or programme before adoption or submission to the legislative procedure.

3.8.6 Information on the Decision

Art. 9 of the SEA Directive states that the respective authority should inform the public and Member States, if they have been consulted on

- the adoption of the plan or programme,
- a statement summarizing how environmental considerations, alternatives and the opinions have been considered and integrated, and
- the monitoring measures; Art. 9 para 1a) to c) of the SEA Directive (for details see ODPM 2003 p.26).

The information could be transported via internet, but additionally other media should be used, like information in local or regional newspapers, to avoid the exclusion of people without access to the internet. If appropriate, even individual information may be required. The notification procedure to the public is similar to the EIA, so the Member States could use the experiences made with the EIA provisions.

3.8.7 Monitoring

Finally, monitoring the implementation of a plan or programme is required in respect to significant environmental effects (Art. 10 para 1 of the SEA Directive). Para 2 of the same article allows the use of existing monitoring arrangements to avoid duplication of measures. The Directive contains only little requirements about the monitoring measures and leaves the decision to the Member States. But the monitoring has to cover significant environmental effects. The SEA Directive requires not necessarily a direct monitoring. An indirect monitoring could be sufficient, e.g. through pressure factors or mitigation measures. Further, if monitoring can be integrated in the regular planning cycle, this could be sufficient and no separate monitoring step has to be carried out.

3.9 Consequences of Mistakes in the SEA Procedure

The following main mistakes could occur in respect to the SEA. At first, no SEA will be carried through for a plan or programme which falls under the SEA Directive. And second, formal or procedural mistakes will happen in the carrying through of an SEA. Both questions cannot be answered according to the SEA Directive because there are – consequently in respect to the subsidiarity principle in Art. 5 EC Treaty – no provisions for this subject. One the other hand, this principle obliges the Member States to provide a sufficient and effective executive system to take care that Community law is carried out effectively (European Court of Justice, Case C-8/88, European Court reports 1990, p. I-2321, (36), Commission/Germany; see also Kahl 1996 p.349).

Therefore, the answer has to be found in Member States' law. If no SEA has been carried through or if there occurred mistakes in the SEA procedure this is a formal or procedural mistake, which always happen in administrative procedures. So it depends on the national provisions which consequences therefore arise. It is conceivable that minor procedural mistakes could be healed in a later step of the SEA procedure. If no SEA is carried through at all, though necessary, this could either lead to voidness or to invalidity of the plan, or as a weaker consequence to contestability before the competent authority or court. But this is subject to the respective national legal system. And, especially in respect to the EIA provisions, a consistent jurisdiction exists in (probably) all Member States (in regard to German and French law see Kloepfer 2004 pp.376-377).

3.10 Conclusions and Recommendations

To give a personal evaluation of possible future development of SEA, it can be stated, that the obligation for an SEA may help to avoid negative environmental effects in cases when plans and programmes determine the future use of the environment. A lot of laws require such assessments for plans and programmes already, but others do not. Therefore the SEA Directive could force the legislation to enact a consistent and standardized procedure for any plan or programme with likely significant environmental effects. The further consultation of the public could be the major benefit of the SEA Directive, because external expertise could be gained for better decisions and future problems could be avoided or minimized.

On the other hand, another procedure and maybe another procedural law has to be followed which makes planning procedures more bureaucratic. Faster and more efficient procedure provisions could be counteracted, double consideration could result, and new procedural hurdles could be established. Furthermore, there is often a long time between the SEA and the concrete project. So the results of the assessments are in a lot of cases not up to date, when – maybe years later – the project starts. If the monitoring measures in the Member States do not work, the expected positive effects for the environment might not be realised.

Besides this, it has always to be observed very critically, when EU organs are directing typical procedure law via a directive to the Member States and wearing out EU competences. Though action of European organs is in a lot of cases useful, the Member States should avert the hollowing out of central European principles, e.g. the principle of subsidiarity (Art. 2 para 2 EU Treaty and Art. 5 EC Treaty). Ironically, Recital 8 of the SEA Directive states that the principle of subsidiarity is regarded, but doubts, if this is more than just a political statement, have to be entertained.

References[4]

Albrecht E (2003a) Die Kostenkonzeption des Bundes-Bodenschutzgesetzes (The cost's conception of the (German) Federal Soil Protection Act). Peter Lang, Frankfurt /M

Albrecht E (2003b) Commentary on Art. 12a Waste Act Baden-Wuerttemberg. In: Kretz K, Knopp L, Michler H-P (2003) Das Abfallrecht in Baden-Württemberg (Waste Law in Baden-Wuerttemberg), 2nd supplementary sheets. Kommunal- und Schulverlag, Walluf/Wiesbaden

Beckmann M (2004) Umweltverträglichkeitsprüfung im Abfallrecht (EIA in waste law). In: Erbguth (ed) Die Umweltverträglichkeitsprüfung: Neuregelung, Entwicklungstendenzen. Nomos, Baden-Baden, pp 121-140

Breuer R (1998) Der Entwurf einer EG-Wasserrahmenrichtlinie (The proposal of the Water Framework Directive). NVwZ 17(10):1001-1010

Bunge T (2004) Zur Harmonisierung von UVP, SUP, FFH-Verträglichkeitsprüfung und Raumverträglichkeitsprüfung (To the harmonisation of the EIA, SEA, Habitats Assessment and impact assessment based on the Federal Spatial Planning Act). In: Hendler R, Marburger P, Reinhardt M, Schröder M (eds) Die strategische Umweltprüfung (sog. Plan-UVP) als neues Instrument des Umweltrechts. Erich Schmidt, Berlin, pp 191-224

Calliess C, Ruffert M (2002) Kommentar zu EU-Vertrag und EG-Vertrag (Commentary on EU Treaty and on EC Treaty), 2nd edn. Neuwied, Kriftel

Calliess C (2004) Verfahrensrechtliche Anforderungen der Richtlinie zur strategischen Umweltprüfung (Procedural requirements of the SEA Directive). In: Hendler R, Marbur-

---

[4] Note that legislation mentioned in the chapter is listed at the end of the handbook in the consolidated list of legislation (Appendix 2).

ger P, Reinhardt M, Schröder M (eds) Die strategische Umweltprüfung (sog. Plan-UVP) als neues Instrument des Umweltrechts. Erich Schmidt, Berlin, pp 153-178

Danwitz T v (2004) Aarhus-Konvention: Umweltinformation, Öffentlichkeitsbeteiligung, Zugang zu den Gerichten (Aarhus Convention: environmental information, participation of the public, access to the courts). NVwZ 3:272-282

Ekardt F, Pöhlmann K (2004) Die Kompetenz der Europäischen Gemeinschaft für den Rechtsschutz – am Beispiel der Aarhus-Konvention (The competence of the EC for the right for access to courts – example of the Aarhus Convention). EurUP 3:128-133

Erbguth W (2004) Die Umweltverträglichkeitsprüfung: Neuregelung, Entwicklungstendenzen (EIA: new laws, developments). Nomos, Baden-Baden

Feldmann L (2004) Die strategische Umweltprüfung im Völkerrecht – SEA-Protokoll zur Espoo-Konvention (SEA in international law – SEA Kiev Protocol to the Espoo Convention). In: Hendler R, Marburger P, Reinhardt M, Schröder M (eds) Die strategische Umweltprüfung (sog. Plan-UVP) als neues Instrument des Umweltrechts. Erich Schmidt, Berlin, pp 27-36

Ginzky H (2002) Die Richtlinie über die Prüfung der Umweltauswirkungen bestimmter Pläne und Programme (the SEA Directive). UPR 2:47-53

Hendler R (2003) Zum Begriff der Pläne und Programme in der EG-Richtlinie zur strategischen Umweltprüfung (The Definiton of Plans and Programmes in the EC Directive on Strategic Environmental Assessment). DVBl. 4:227-234

Hendler R, Marburger P, Reinhardt M, Schröder M (2004) Die strategische Umweltprüfung (sog. Plan-UVP) als neues Instrument des Umweltrechts (SEA as a new environmental instrument). Erich Schmidt, Berlin

Hönig D (2003) Die Umweltverträglichkeitsprüfung: Neuregelungen, Entwicklungstendenzen (EIA: new laws, developments), conference-report, conference on SEA, 28 April 2003, Rostock, Germany. DVBl. 18:1193-1194

Hösch U (2004) Die FFH-Verträglichkeitsprüfung im System der Planfeststellung (The Habitats assessment in the system of plan decision procedure). NuR 4:210-219

Jendroska J, Stec H (2003) The Kiev Protocol on Strategic Environmental Assessment. Environmental Policy and Law 33(3-4):105-110

Kahl W (1996) Europäisches und nationales Verwaltungsorganisationsrecht (European and national law on the organisation of authorities). Die Verwaltung 31:341-384

Kläne C (2003) Strategische Umweltprüfung (SUP) in der Bauleitplanung (SEA in land use planning), Dr. Kovac, Hamburg

Kloepfer M (2004) Umweltrecht (Environmental law), 3rd edn., CH Beck, Munich

Knopp L (2003) Flussgebietsmanagement und Verwaltungskooperation (Management of River Basins and administrative co-operation). In: Erbguth (ed) Änderungsbedarf im Wasserrecht – zur Umsetzung europarechtlicher Vorgaben (Needs for changes in water law – to the transposition of European requirements). Nomos, Baden Baden, pp 27-41

Krautzberger M (2004) Europarechtsanpassungsgesetz Bau – EAG Bau 2004: Die Neuregelungen im Überblick. UPR 7:241-246

Kretz K, Knopp L, Michler H-P (2003) Das Abfallrecht in Baden-Württemberg (Waste Law in Baden-Wuerttemberg), 2nd supplementary sheets. Kommunal- und Schulverlag Walluf/Wiesbaden

Lindemann J (2004) Die Richtlinie zur strategischen Umweltprüfung aus gliedstaatlicher Sicht (The SEA Directive from federal states'/provinces' perspective). In: Hendler R, Marburger P, Reinhardt M, Schröder M (eds) Die strategische Umweltprüfung (sog. Plan-UVP) als neues Instrument des Umweltrechts. Erich Schmidt, Berlin, pp 61-79

Mulloy M, Albrecht E, Häntsch T (2001) German Environmental Law, Erich Schmidt, Berlin

Office of the Deputy Prime Minister – ODPM (2003) The Strategic Environmental Assessment Directive: Guidance for Planning Authorities. Internet Address: http://www.odpm.gov.uk/stellent/groups/odpm_planning/documents/page/odpm_plan_025198.pdf, last accessed 17.08.2004

Partsch C, Schurig W (2003) Das Informationsfreiheitsgesetz von Nordrhein-Westfalen – ein weiterer Schritt aus dem Entwicklungsrückstand Deutschlands (The freedom of information act in North Rhine-Westphalia). DÖV 12:482-488

Platzer-Schneider U (2004) Entstehungsgeschichte, Funktion und wesentliche Inhalte der Richtlinie zur strategischen Umweltprüfung sowie der Koordination der mitgliedstaatlichen Umsetzung (Development, function and main contents of the SEA Directive and the coordination of Member States transposition measures). In: Hendler R, Marburger P, Reinhardt M, Schröder M (eds) Die strategische Umweltprüfung (sog. Plan-UVP) als neues Instrument des Umweltrechts. Erich Schmidt, Berlin, pp 15-26

Reinhard M (2001a) Wasserrechtliche Richtlinientransformation zwischen Gewässerschutzrichtlinie und Wasserrahmenrichtlinie (The transposition of water law related directives). DVBl. 3:145-154

Reinhard M (2001b) Deutsches Verfassungsrecht und Europäische Flussgebietsverwaltung (German constitutional law and European River Basin Management). ZUR Extra edition:124-128

Roder M (2004) Monitoring nach Art. 10 SUP-Richtlinie (monitoring according to Art. 10 of the SEA Directive). In: Hendler R, Marburger P, Reinhardt M, Schröder M (eds) Die strategische Umweltprüfung (sog. Plan-UVP) als neues Instrument des Umweltrechts. Erich Schmidt, Berlin, pp 225-251

Sander A (2003) Strategische Umweltprüfung für das Immissionsschutzrecht? (SEA for immissions control law?). UPR 9:336-342

Schmidt M, Rütz N, Bier S (2002) Umsetzungsfragen bei der strategischen Umweltprüfung (SUP) in nationales Recht (Problems of transposing the SEA Directive into national (German) law). DVBl. 6:357-363

Schmidt-Eichstaedt G (2004) Die Richtlinie zur strategischen Umweltprüfung aus kommunaler Sicht (The SEA Directive from a local communities' perspective). In: Hendler R, Marburger P, Reinhardt M, Schröder M (eds) Die strategische Umweltprüfung (sog. Plan-UVP) als neues Instrument des Umweltrechts. Erich Schmidt, Berlin, pp 80-98

Sheate WR (2003) The EC Directive on Strategic Environmental Assessment: A Much Needed Boost for Environmental Integration. European Environmental Law Review 12(12):331-347

Spannowsky W (2004) Rechts- und Verfahrensfragen einer „Plan-UVP" im deutschen Raumplanungssystem (Legal and procedural questions of SEA in the German spatial planning law). UPR 6:201-210

Sparwasser R et al. (2003) Umweltrecht (Environmental law), 5th edn. Hüthig, Heidelberg

Stollmann F (2004) Umweltverträglichkeitsprüfung im Baurecht (EIA in building law). In: Erbguth (ed) Die Umweltverträglichkeitsprüfung: Neuregelung, Entwicklungstendenzen. Nomos, Baden-Baden, pp 63-83

Stüer B (2004) Umweltverträglichkeitsprüfung in der Verkehrswegeplanung (EIA in the transport planning). In: Erbguth (ed) Die Umweltverträglichkeitsprüfung: Neuregelung, Entwicklungstendenzen. Nomos, Baden-Baden, pp 88-100

Wilrich T (2002) Verbandsbeteiligung im Umweltrecht (Participation of NGOs in environmental law). Nomos, Baden-Baden

Wolf J (2002) Umweltrecht (environmental law). CH Beck, Munich

4 Transposition of the SEA Directive into National Law – Challenges and Possibilities

Lothar Knopp and Eike Albrecht

Department of Constitutional, Administrative and Environmental Law, Brandenburg University of Technology (BTU), Cottbus, Germany

4.1 Introduction

This chapter discusses how the SEA Directive could be transposed into national law and describes its challenges and possibilities. After some introductory words about the connection of the SEA Directive to Member States' law, the main problems about the transposition process are discussed. The section starts with a description of the addressees and the content of the Directive. The next two sections discuss the transposition instruments and the transposition measures. Here the – from a legal point absolutely important – question is discussed, whether the Community has the competence to rule procedural matters and – following the European Court of Justice's ruling – if so, what kind of legal quality the result of the national transposition act must have to fulfil the Court's requirement on legal certainty and preciseness. Finally, the difficulty of implementing European law into federal systems, like Austria, Germany or Belgium, and the different possibilities to link the contents of the SEA Directive to the respective national legal system are discussed.

4.2 Connection to Member States' Laws

According to Art. 14 the SEA Directive went into force on 21 July 2001. The time period for the transposition into national law is three years as stated in Art. 13 of the SEA Directive. Therefore the Member States were obliged to change and amend their national provisions to the content of the SEA Directive by 21 July 2004 at the latest.

Implementing Strategic Environmental Assessment. Edited by Michael Schmidt, Elsa João and Eike Albrecht. © 2005 Springer-Verlag

The SEA Directive is a fundamental change for environmental law. Currently, only projects of a certain size and with likely effects on the environment have to be assessed in an EIA procedure based on the EIA Directive and its national transpositions. This assessment takes place at a time when the possibilities for changes or the outlining of alternatives are quite restricted. Normally the decision for a certain site for such an EIA requiring projects and possible alternatives is fixed through previous planning decisions of a binding nature. The SEA Directive closes this gap for a broad range of plans and programmes with the introduction of preliminary assessments of the effects on the environment. Environmental aspects can be considered when plans and programmes are developed and adopted (Calliess 2004 p.154).

Additionally, the public has to be involved through the SEA Directive during the stage where plans and programmes are developed and in the creation of the environmental report. The opinions of the public have to be considered. A stronger involvement of the public, e.g. through participation of NGOs, environmental groups or organizations and the citizens, is a major aim of European policy, especially of European environmental policy. This is clarified through a Directive for the transposition of the Aarhus Convention, the Public Participation Directive. This Directive is a parallel directive to the SEA Directive (see Art. 2 para 5 of the Public Participation Directive), which firstly guarantees certain information and participation rights for the public and secondly guarantees admittance to the courts (see Chap. 3).

4.3 Transposition of the SEA Directive

European law contains an obligation to transpose directives into national law. It is discussed if this obligation is laid down in Art. 249 para 3 of the EC Treaty (Case C-129/96, European Court reports 1997 p. I-7411 (40), Commission/Belgium; Calliess and Ruffert 2002, Art. 249, Rec. 43) or in Art. 10 in conjunction with the respective provision in the directive (see for details Westbomke 2004 p.122 with further reference in footnote 5). Art. 10 of the EC Treaty describes in general that the Member States have to take the necessary measures to fulfil their obligations in respect of the Treaty. So, Art. 249 of the EC Treaty is the more specific norm with respect to a special obligation, the obligation to transpose directives into national law. The solution is Art. 30 of the Vienna Treaty Convention which describes some basic legal principles, for example the principle "lex specialis derogat lex generalis".[1] In comparison with Art. 249 of the EC Treaty, Art. 10 is the more general provision. So Art. 249 of the EC Treaty is the relevant norm to oblige the Member States for transposition.

---

[1] This means, the special provision substitutes the general provision.

4.3.1 Addressees of the Directive

According to Art. 249 para 3 EC Treaty, a directive is addressed to the Member States. A Directive is binding as to the result to be achieved, but leaves the choice of form and methods to the relevant national authorities (Kloepfer 2004 p.680). The institution or organ responsible for the transposition within a Member State is determined according to the relevant competence provisions of that Member State and not to any specific European law. (European Court of Justice, Case 96/81, European Court reports 1982, 1791 (12), Commission/The Netherlands); for the whole topic see also Calliess and Ruffert 2002, Art. 249 EC Treaty, rec. 59). Problems regarding the responsible organ or institution inside a Member State could primarily arise in Member States that are organised under a federal system, such as Germany, Austria or Belgium. As regards the Community, the Member State has the obligation for proper transposition at the federal level. Insofar Art. 249 para 3 of the EC Treaty is clear. Therefore the federal level is also responsible in cases when a directive was not transposed in time, even if the Constitution addresses the competence for the transposition of certain Community Laws to the constituent state. Therefore, if some or all of the *Laender*[2] in Germany do not transpose a directive on topics that are not addressed to the federal level, e.g. procedural provisions which are regularly in the competence of the *Laender*, the federal level has to face an infringement procedure, with the consequences described in Art. 226 ff. of the EC Treaty, especially the consequence of paying a lump sum or a penalty payment according to Art. 228 para 2 subpara 2 of the EC Treaty.

4.3.2 Content of the Transposing Provisions

Because a directive is binding as to the result to be achieved, the Member States have a broad range with respect to the choice of form and methods. This concerns either the choice of the instrument for the transposition, and the question of which measures should be taken to implement the content of a directive into national law. The Member States should have the possibility to implement European law in such a manner, that it fits as far as possible to their national legal system.

European legislation normally leaves enough room for autonomous provisions to the Member States, merely to fulfil the principle of proportionality, laid down in Art. 5 para 3 of the EC Treaty (Calliess and Ruffert 2002, Art. 5 EC Treaty, rec.

---

[2] According to Art. 20 para 1 of the German Basic Law, Germany is a Federal State. The principle of federalism is a key factor within the German legal system, because along this line the relationship between the *Bund* (federal level) and the 16 *Laender* is organized. The German Basic Law directs principally the competences in legislative, executive and juridical matters to the 16 *Laender*, and only as an exception to the federal level. Though this principle-exception relation is in fact turned around, there still exist many competences for the *Laender*, especially in respect of executive powers and laws to that matter (for details see Freckmann and Wegerich 1999, pp.61-62). Because the SEA Directive consists of many procedural provisions, *Laender* competences are highly affected.

49 ff.).[3] But there are exceptions. If a directive is strict and binding in certain areas, no room is left for the Member States' decision on how to transpose the content of the directive. Then the Member States have to follow the content, the structure or the procedures required from the directive. Therefore they have no choice as regards the form and methods. To this type of directive the SEA Directive has to be counted, because in parts it consists of very detailed and specific provisions and therefore significantly intrudes into the transposition competences of the Member States.

4.3.3 Transposition Instruments

Deciding which normative measures should be given for the transposition of directives is a controversial problem. "Measure" in this respect does not mean a certain measure of legal regulation, but the legal form of the measure for the transposition into national law. In this respect the question occurs of whether the Community is allowed at all to prescribe certain compulsory transposition measures, like a licence, plan, charge, or forms of civil action to mention a few (Reinhard 2001 p.149; Knopp 2003 pp.28-29; for a different opinion see Calliess and Ruffert 2002, Art. 249 EC Treaty, rec. 51 with further references). Because the SEA Directive prescribes and defines certain procedural steps for the carrying through of an SEA, e.g. Environmental Assessment according to Art. 2b) of the SEA Directive or Environmental Report (Art. 5; defined in Art. 2c of the SEA Directive), this question could be picked out as a central topic. But the European Court of Justice generally does not share these objections (European Court of Justice, Case 184/97, European Court reports 1999, p. I-7837, (55) and (58), Commission/Germany, also printed in ZfW 2000, pp.171, 176-177). Besides, the SEA Directive leaves enough room for the Member States to transpose the form of the prescribed procedural elements according to their national legal tradition, system and structure.

4.3.4 Transposition Measures

Furthermore, it has to be asked, which form of transposition into national law is sufficient in respect to Community law. This is a well known dispute, if it may be sufficient to enact appropriate provisions, of whatever legal quality they are, or if it may even be sufficient to refer to an established jurisdiction. In the meantime the European Court of Justice has established its jurisdiction on this matter.

---

[3] The principle of subsidiarity is not relevant for this problem (Art. 5 para 2 of the EC Treaty). This principle is relevant for the question whether the Community is allowed to act in general, see Calliess and Ruffert 2002, Art. 5 EC Treaty, rec. 17 ff.

General Requirements

Following the European Court of Justice's ruling the transposition of directives into domestic law does not necessarily require that its provisions have to be incorporated formally and verbatim in (expressly) specific legislation. A general legal context may, depending on the content of the directive, be adequate for the purpose, provided that it does indeed guarantee the full application of the directive in a sufficiently clear and precise manner so that, where the directive is intended to create rights for individuals, the persons concerned can ascertain the full extent of their rights and, where appropriate, rely on them in national courts (European Court of Justice, Case 29/84, European Court reports 1985, p. 1661, (23) and (28), Commission/Germany; Case C-131/88, European Court reports 1991, p. I-825, Commission/Germany; Case 363/85, European Court reports 1987, p. 1733, (7), Commission/Italy; Case 247/85, European Court reports 1987, p. 3029, (9), Commission/Belgium; Case 361/88, European Court reports 1991, p. I-2567 (16 ff.), Commission/Germany, also printed in DVBl. 1991, p. 869 and NVwZ 1991, p. 866; see also Calliess and Ruffert 2002, Art. 249 EC Treaty, rec. 51 with further references). Whereas the second point of this ruling occupies a secondary position with respect to the transposition of the SEA Directive – only participation rights could be counted as provisions for favouring individuals – the first point is of large importance. Due to the European Court of Justice's ruling, the relevant national legislative power has to enact appropriate provisions to ensure that the authorities in charge of carrying out an SEA completely implement the requirements of the Directive. Therefore not only the complete registration of those plans and programmes for which an SEA shall be carried out is necessary, but also the establishment of sufficiently precise provisions in national law about the different procedural steps for the carrying out of an SEA.

Parliamentary Acts and Ordinances

Following the Court of Justice's requirements, parliamentary formal acts and ordinances are sufficient in respect to clearness and preciseness of the provisions to be transposed. Parliamentary acts and ordinances are enacted in a formal procedure. Because they are published in the Member States law gazettes everybody has the chance to get knowledge about the content.

Contracts and Agreements

Contracts or agreements between state authorities and privates can be deemed adequate if they are sufficiently binding and the respective individual rights in the national field are sufficiently guaranteed (Communication from the Commission to the Council and the European Parliament on Environmental Agreements, COM (96) 561 final; No. 31 ff., printed in *Bundesrat*-Parliamentary Reports BR-Drs. 20/97, p.96). Art. 249 para 3 of the EC Treaty directs the obligation to transpose directives to respective national authorities in the Member States. Therefore, the transposition through a communal statute is sufficient because a communal statute is a binding law for externals (Calliess and Ruffert 2002, Art. 249 EC Treaty, rec. 54 with further references). However, a transposition through either contracts and agreements or communal statutes is not likely, because administrative bodies or authorities are subject to the obligations of the Directive. The content of such public obligations, like the obligation to carry out an environmental assessment or to consult the public or neighbour states, is normally laid down to the respective authority or body by law, and not through contracts or statutes.

Constant Administrative Practice

On the other hand, a constant administrative practice which concurs with a directive is not sufficient in the opinion of the Court because administrative practice could change (European Court of Justice, Case 102/79, European Court reports 1980, p. 1473, (10 f.), Commission/Belgium; Case 96/81, European Court reports 1982, p. 1791, (12), Commission/The Netherlands; Case 97/81, European Court reports 1982, p. 1819, (12), Commission/The Netherlands; Case C-131/88, European Court reports 1991, p. I-825, (8), Commission/Germany; Case C-358/98, European Court reports 2000, p. I-1255, (17), Commission/Italy). Also not seen as sufficient – by the Court – is the transposition of a directive through administrative circulaires and rules, insofar as individual rights are concerned. However, this judgement was subject to vehement critique (see Calliess and Ruffert 2002, Art. 249 EC Treaty, rec. 56 ff.; Westbomke 2004, p. 123 with further references).

In the SEA Directive, procedural provisions are mainly laid down by law, but not only. A few provisions, though of subordinate importance, provide individual rights in respect to participation and consultation in the procedure of carrying out an SEA. Therefore the transposition of the SEA Directive by administrative circulaires or rules does not meet the Court's requirements regarding the reliability and preciseness of provisions.

Referral to a Directive in National Law

Finally a transposition technique should be closer examined, which could be the easiest way to transpose European acts into national law: a simple referral to the Directive in national law. The European Court of Justice accepted at least referrals to European directives in national law under certain conditions (Case C-96/95, European Court reports 1997, p. I-1653, (40), Commission/Germany). A simple referral is not sufficient. As a minimum standard at least a precise referral in the national law to the European Act is required, so that the affected person in question could easily find the provision. If a directive directs certain rights to individuals, these individuals must have a chance to get knowledge of their rights.

In respect to the SEA Directive this means that the Member States can transpose the directive into national law with a referral in their national law gazette, but this referral must be precise and easily perceptible, because certain participation rights are directed to individuals.

4.4 Transposition into the National Legal System

There are different possibilities to transform the SEA Directive into national law. So the SEA Directive could be transposed with a special SEA Act, maybe just with a word-by-word implementation or just with a national legal reference to the Directive, as far as the Court of Justice's requirements are followed. Another alternative is a supplement of SEA provisions to either EIA law or to special laws with planning duties in the Member States. Which method of transposition is used is connected with existing experiences with SEA. In a number of European States SEA is already practiced.

4.4.1 Use of SEA in Europe

Before enacting the SEA Directive, SEA was already used in the Community, but large variations existed across the EU. Some countries have an established history of SEA of plans or policies, others are moving towards systematic SEA. In western European countries, there is a lot of experience with application of SEA, and some countries have working systems. On the other hand, SEA is often carried out in an *ad hoc* way. In most of the cases the SEA is confined to specific sectors, particularly land-use plans and transport plans. But some of the countries have their only experience of SEA being through the assessment of regional plans as required by the European Council regulation for structural funds (2081/93, now superseded by 1260/1999). This indicates a lack of guidance as the main problem in the implementation-process of legal or administrative requirements. Therefore the situation has been described as patchy (OPOCE 2003 p.292; *Bundesrat-Parliamentary Reports BR-Drs. 277/97 p.6*).

Much more difficult to estimate is the situation in Central and Eastern European (CEE) countries. For some countries it can be stated that SEA provisions are working well. There has been a series of capacity-building programmes which have helped to achieve compliance with the requirements of European directives. In other countries new legislation is still in the process of development, so no judgement could be done. But some problems can be mentioned: First, there is a lack of systematic coverage of content requirements, second, a lack of enforcement provisions and third, a lack of application of SEA to any sector other than land use. For those CEE countries which have – with the exception of Bulgaria and Romania following in 2007 – already joined the European Union in May 2004, the SEA Directive, as a part of the *acquis communautaire* will have a beneficial effect because those countries are subject to the requirements of the European SEA Directive (OPOCE 2003 p.293; Feldmann and Vanderhaegen 2001 p.121).

Besides the new Member States of the European Union, how is the situation about SEA in former Soviet Union states, being candidates for the accession to the European Union? In those states, EIA systems are primarily based on the State Ecological Review (SER) or the State Ecological Expertise (SEE) and assessment of ecological impacts (OVOS)[4] systems which are inherited from the former Soviet Union. The SER/SEE and OVOS systems theoretically cover the requirements of an SEA. There are mentioned some major problems in the operation of these systems, mainly examined in respect to the EIA. Because of certain structural parallels between EIA and SEA, the results are transformable.

First problem mentioned, is, that the citizens are not aware of their rights and duties and therefore they do not participate. Furthermore, transboundary impacts are not taken into consideration. Finally, there are no or only a few EIA specialists and administrative fines or penalties for non-compliance is inadequate. Another common problem is that the transition of legislation from the former SER/OVOS system to a more 'western' style of assessment-procedures in some countries is leading to a situation where two systems are operating in parallel and therefore causing confusion (OPOCE 2003 p.292).

In practice, some more problems have been identified, again most related to EIA, but the problems for SEA are estimated as similar to those for EIA (OPOCE 2003 p.293). Though countries have been through a capacity-building programme (inclusive new legislation in place) there remain operational problems through a lack of training of responsible officers, or a lack of organisations with suitable experience to be able to carry out EIA. These problems have to be expected for SEA procedures as well. In conjunction with the EIA implementation, other issues are cited, which probably will be relevant for the SEA process as well. First, the quality of environmental statements is not very good and that specific guidance needs to be developed. Furthermore the process of EIA takes too long under the administrative procedures adopted and exceeded the duration of one year in some cases. Both could be imagined for SEA procedures as well. As a main problem, the lack of baseline environmental data was identified (OPOCE 2003 p.293). Because of the financial situation in those countries, it is not very likely that there are fast progresses in the former Soviet Union states. So capacity-building assistance, similar to the new Member States of the European Union, should be considered, because this has been successful in many cases (OPOCE 2003 p.293) (see Chap. 44).

4.4.2 Transposition with a Special SEA Act

To implement the SEA into national law, there is the possibility to enact a special SEA act. This could resemble an EIA Act, as a transposition of respective European EIA Directives. In addition, details could be regulated within an ordinance to the SEA law, like it has been done, e.g. in Germany, in respect to an ordinance on the EIA Act. As mentioned above, an ordinance fulfils the Court of Justice's requirements on legal certainty. The benefit of a special law on SEA would be the emphasis of environmental aims in the process of the creation of certain plans and programmes. Another positive impact would be that provisions on the environmental effects in other laws, like the environmental assessment of agricultural

---

[4] Ozenka Vozdejstviya na Okruzhayushuyu Sredu (for details see Chap. 22).

planning or in the planning of federal traffic routes (for details see Hendler 2004 pp.85-87) could be centralised in one act.

On the other hand, bodies with limited but regulated planning abilities which fall under the SEA Directive, e.g. communal bodies, responsible for Preparatory Land-Use plans, could have difficulties to handle a mixture of different laws (Kläne 2003 p.191). This difficulty does exist in respect to the EIA law for two reasons: First an EIA must be carried out only for projects with a certain size and weight, so most of the projects do not fall under the EIA Act; second, the EIA has to be applied to the authorities from the person who wants to carry out the project, so the applicant is responsible for the EIA and not the authority. Responsible for the SEA is the planning body that is typically an authority.

4.4.3 Supplement to the EIA Act

Another possibility of transposition into national law could be a supplement to the respective EIA acts where existent. There are a lot of parallels between the EIA and the SEA, but there are differences as well. The EIA Act was amended in 2001 with the provisions on "screening"[5] that is a relevant part of the SEA Directive too.

One the other hand, some instruments have to be amended which are not known in the EIA law yet. For example, "monitoring" is a new instrument and was not mentioned in the EIA Directive and therefore not obligatory to transpose into national law (Roder 2004 p.226). Additionally, the EIA law would have to be amended to include rules for assuring that the SEA seeks out competent authorities, and not, as the current EIA-Acts states, the applicant for a project.

Actually, there is a broad discussion about the best way for transposition, and for example in respect to German law, most of the transposition work is being done with amendments to special planning regulations. Nevertheless, the EIA Act has to be amended in respect to general provisions, especially provisions about the area of application and provisions on the procedure (for details see Chap.7).

4.4.4 Supplements to Special Laws with Planning Duties

The SEA Directive is relevant for a lot of different types of plans and programmes. So, a lot of different laws are affected, as far as there is no Environmental Code which includes all SEA relevant plans and programmes. Therefore it might be helpful to implement the relevant SEA provisions into the respective law (in respect to land use planning see Kläne 2003 pp.172-185). This procedure meets both the requirements of each Member States' legal tradition and the needs of the planning practice. Those steps of the SEA which are not part of the national legislation yet could be included to the already existing parts. The respective planning or programme-creating authority could find all relevant provisions in one ap-

---

[5] Act of 27. July 2001, Fed. Law Gazette I, p. 1950.

plicable act. So mistakes and deficient execution of European SEA law could be better avoided. On the other hand, this proceeding is quite laborious because every exiting regulation has to be assessed for the correspondence with and divergence from the SEA Directive.

Additionally in Member States with a federal system with different legislative competences between the federal and the *Laender* level, e.g. Germany, this procedure demands for a high collaboration and coordination effort. For this, the three year transposition time limit (Art. 13 of the SEA Directive) is not long enough.

4.5 Conclusions and Recommendations

The transposition of the SEA Directive into national law has to be done by the Member States. According to the ruling of the Court of Justice on cases of similar procedural provisions in directives, there are no effective arguments against the correctness of the underlying competence of this European act, although there have been numerous criticisms. Therefore the Member States have to do the transposition in time, namely before 21 July 2004 (Art. 13 para 1 SEA Directive).

To transpose the content of the SEA Directive into national law, the Member States could enact a special SEA Act. This would be the easiest form to transpose the Directive, though accordance to the national legal system might be questionable. Other alternatives are amendments to either EIA law or to special planning law. A transposition like this might have the benefit of considering national legal systems and traditions, but is more complicated and therefore more time consuming. Then the risk has to be faced, that the transposition could not be carried through in time. In some Member states, the federal system could double the difficulties in respect of transposition done in time.

According to the ruling of the Court of Justice the transposition of the SEA Directive will be done by laws or ordinances. As long as the requirements of the Court of Justice on preciseness and legal reliability, as described above, are met, other forms of transposition are possible, but not very likely. So, for example, Germany has to face an infringement procedure. Theoretical consequences of no or incorrect transposition are described in Art. 226 ff. of the EC Treaty (payment of a lump sum) and in Art. 228 para 2 subpara 2 of the EC Treaty (penalty payment).

References[6]

Calliess C (2004) Verfahrensrechtliche Anforderungen der Richtlinie zur strategischen Umweltprüfung (procedural requirements of the SEA Directive). In: Hendler R, Marburger P, Reinhardt M, Schröder M (eds) Die strategische Umweltprüfung (sog.

---

[6] Note that legislation mentioned in the chapter is listed at the end of the handbook in the consolidated list of legislation (Appendix 2).

Plan-UVP) als neues Instrument des Umweltrechts. Erich Schmidt, Berlin, pp 153-178

Calliess C, Ruffert M (2002) Kommentar zu EU-Vertrag und EG-Vertrag (Commentary on EU Treaty and on EC Treaty), 2nd edn. Neuwied, Kriftel

Freckmann A, Wegerich T (1999) The German Legal System. Sweet and Maxwell, London

Hendler R (2004) Die Bedeutung der Richtlinie zur strategischen Umweltprüfung für die Planung der Bundesverkehrswege (the meaning of the SEA Directive on Federal Highway planning). EurUP 2:85-93

Kläne C (2003) Strategische Umweltprüfung (SUP) in der Bauleitplanung (SEA in land use planning). Dr. Kovac, Hamburg

Kloepfer M (2004) Umweltrecht (environmental law), 3rd edn. CH Beck, Munich

Knopp L (2003) Flussgebietsmanagement und Verwaltungskooperation (Management of River Basins and administrative co-operation). In: Erbguth (ed) Änderungsbedarf im Wasserrecht – zur Umsetzung europarechtlicher Vorgaben. Nomos, Baden-Baden, pp 27-41

Reinhard M (2001) Wasserrechtliche Richtlinientransformation zwischen Gewässerschutzrichtlinie und Wasserrahmenrichtlinie (the transposition of water law related directives). DVBl. 3:145-154

Roder M (2004) Monitoring nach Art. 10 SUP-Richtlinie (monitoring according to Art. 10 of the SEA Directive). In: Hendler R, Marburger P, Reinhardt M, Schröder M (eds) Die strategische Umweltprüfung (sog. Plan-UVP) als neues Instrument des Umweltrechts. Erich Schmidt, Berlin, pp 225-251

Westbomke K (2004) Die Umsetzung von EU-Richtlinien in nationales Recht (the transposition of EU Directives into national law). EurUP 3:122-127

5 Some Legal Problems of Implementing the SEA Directive into Member States' Legal Systems

Jerzy Sommer

Institute of Legal Studies of Polish Academy of Sciences, Research Group on Environmental Law, Poland

5.1 Introduction

The implementation of European directives such as the EIA Directive or the SEA Directive may be analysed from several points of view. Only one of those being the legal point of view. This chapter is devoted to legal problems connected with the implementation of the SEA Directive. The main point of interest is the scope of the Directive and procedures of its preparation and adoption. This is a very important problem because it affects the successful implementation. The SEA Directive defines programmes and plans which must be submitted to environmental assessments but, on the other side, national legal systems have their own definitions of programmes and plans as well as procedures to their preparation and adoption, and their own environmental assessment.

This chapter presents the problems connected with the SEA Directive, with particular reference to the problem of its implementation in Poland. This example is interesting because the legislation on SEA was introduced in Poland before enactment of the Directive. The chapter starts by discussing implementation problems in Member States in general and then it discusses particular problems in the implementation of the SEA Directive in Poland.

5.2 Implementation Problems in Member States

The directives on EIA (Directive 85/337/EEC; Directive 97/11/EC) create difficult implementation problems. They are, from the legal point of view, procedural provisions and try to introduce a uniformed regulation, whereas in Member States the laws in this respect are highly differentiated.

Implementing Strategic Environmental Assessment. Edited by Michael Schmidt, Elsa João and Eike Albrecht. © 2005 Springer-Verlag

It is worth to mention that four years after the deadline for the implementation of the last modification of the EIA Directive, only one country, Denmark, has *not* been subject to infringement proceedings or action before the Court of Justice for non-transposition, non-communication or poor application of the Directive. At the other extreme, Spain has the worst record with 81 complaints and/or proceedings pending (Europe Environment 2003, No. 637). This situation does not necessarily reflect a lack of willing of the Member States, but shows instead the real trouble with this Directive and others of the same legal character. It is suggested in this chapter that the same will happen in respect to the SEA Directive. This might be the result of the reasons described in this section.

5.2.1 Undefined Legal Terms

First, general difficulties connected with the implementation of EC law have to be mentioned. The EC law has a partial character and does not create a coherent system as national laws. This is a result of its history. EC law has developed as the answer to specific problems of creation of the internal market and therefore it has some shortcomings. One of these shortcomings is the use of terms in European law which have not a direct equivalent in national legal languages. Often these terms are not sufficiently clear and precise. There are many undefined legal terms which often are interpreted in different ways in different legal systems. The European legislation very often uses terms which have no legal value because they have a technical meaning without national equivalent (e.g. reference methods of measurement). The structure of EC law is in certain degree strange: the so-called daughter's directives and mother's or father's directives.

But above all there is one feature of the EC law that creates most problems in the process of transposition. It is the concept of "directives". According to Art. 249 para 3 of the EC Treaty, directives are binding Member States only to the results to be achieved, but the Member States ought to decide upon the forms and methods of realisation of the prescribed aims. This is connected with the competence of the Member States to establish their political and administrative structure. The concept of the directive only appears clear, but in reality, as may be deemed from the activity of the European Court of Justice, there are many legal discussions around it. It seems that the concept of directives present in the EC Treaty is based on very fragile construction namely that the notion of aim in law is sharp and clear. Yet that may be seriously doubted. In reality there are two interests: the Community and the Member States (or only even one), and very often they are in opposition. So the legal rules established by the European Court of Justice to resolve the conflicts around the implementation of the directives (Pernice 1994 pp.328-340) may be only supported if the supremacy of EC law is taken as the supreme necessity, but not according to their legal values. Generally, we have two legal systems: national legal systems and community legal systems. In the sphere of European Union competence the community legal system has a priority on national ones, but this priority does not denote that the Community law suspends or derogates national laws, except for regulations. Only the national legislator is obliged to create the conformity between the Community and national legislation. This is a very difficult task and it is dubious whether it may be fulfilled to the full extent.

5.2.2 Different Legal Tradition

The EIA Directives and the SEA Directive are procedural provisions (see Chap. 2). Member States' procedural regulations are sophisticated but also differentiated legal constructions. It is not easy to find the proper way of implementation, which would not destroy the national systems and law traditions of each individual country. Certain compromises must be found (Ladeur and Prelle 2001 pp.197-198). For example, there is some vagueness of the SEA Directive, e.g. the definition of the subject matter of SEA: plans and programmes are not clear in every aspect (Hendler 2003 p.227). This subject matter has no uniform equivalent in national laws. There are many plans and programmes in the legal turnover of the national laws, and they have differentiated legal character and normative meanings. This is in addition to the differentiated procedures of preparation and adoption of plans and programmes. Therefore it was difficult to incorporate requirements of the EIA Directives to national laws and, probably, it will be more difficult to introduce the SEA Directive to national legal systems, taking into account its unclear terminology and other legal deficiencies.

5.2.3 Procedural Requirements

Procedural provisions are of minor importance in comparison to substantial requirement especially in the view of courts. The Directives on EIA and SEA contain no material and substantial requirement though they demand that an influence of a planned undertaking on the environment will be identified, described and assessed. But this question cannot be answered without the identification, description and assessment of influences on the environment, which is not the subject of these Directives. So the EIA and SEA Directives should support substantial material values though they are only procedural provisions (Radecki 2001 pp.28-40, see also Chap. 3 for further details).

5.2.4 Scope of the SEA Directive

There is a problem with the scope of the SEA Directive: what plans and programmes are included in this procedure. The SEA directive names three criteria: the first is mentioned in Art. 2 a) SEA Directive and last two are laid down in the Art. 3 SEA Directive. The problem is to explain the meaning of "set the framework for future development consent" or vague criteria of environmental significance comprised in Annex II (see also Chap. 2).

5.2.5 Normative Value of Plans and Programmes

In national laws different procedural requirements on the plans and programmes exist. These procedural requirements take into account the normative value of the plans and programmes. Some plans have *real* normative value, like local land-use plans or water management plans in Poland. Some plans and programmes have *partial* normative value, when they are binding only to the subordinated authorities, like regional and national land-use plans or waste-management plans in Poland, but some plans and programmes are only political documents. Accordingly, the regulation of preparation and the requirements of the context of plan are differentiated. The more normative value a plan has, the more its preparation and contents are regulated by law.

5.2.6 Legal Value of the Environmental Report

There is a problem with the legal value of the environmental report. The environmental report is a kind of proof or evidence with the same value as other proofs, which are subjected to estimation of authority and court and it is not binding for both. The authority and court must consider the SEA but the same must be done with all other evidence. The only real requirement is that this evidence must be presented and taken into account in the preparation of plans and programmes. There is uncertainty in this respect. Art. 8 of the SEA Directive demands only that the environmental report and prescribed consultation (including any transboundary consultation as relevant) will be taken into account in a stage of the preparation of the plan or programme. But Art. 9 of the SEA Directive demands that public and the authorities be informed on the plan or programme adopted and how the environmental consideration, environmental report and the results of the consultation were considered in the decision of the plan or programme. It is no problem when the responsibility for the preparation and the adoption of the plan or programme is placed on the same authority. However, but there are potential problems when one authority is preparing the plan or programme, and another authority, normally a superior to the first, adopts it. The decision on the plan or programme and its reasons may be different in both instances. The draft plan or programme may be changed in the final stage, and the decision may have very different contents in comparison to its draft.

Further, it must be remembered that, according to the Directive, the environmental report is prepared by the authority responsible for the preparation of the plan or programme. What may denote that the authority takes into account the environmental report that it has prepared by itself? The quality of the environmental report is of minor importance. Whether it is sufficient quality or not it is at the discretion of the authority because of the vague substantial requirements comprised in the Directive. For example, it is impossible to evaluate the national energy plan in full accordance to this criterion.

Also it must be taken into account that the authority which prepares the plan or programme makes the environmental report. In this situation it is difficult to imagine that the environmental report will reject the main ideas of the plan or programme or the main methods and tools of its realisation. If not, we shall have to do with a kind of "dual personality".

5.2.7 Transposition in the Accession States

It must be taken into account that the problems with implementation of EC law are much more serious for the Accession States than for the Member States. Accession States must implement many normative acts in a very limited time. These acts comprise the legal institution alien to the hitherto existing law (Damohorsky 2002 pp.20-23). Poland at the end of 2001 enacted 271 Parliamentary acts aiming at harmonization with EU legislation. In the same time the Polish Parliament enacted 619 acts in total. So the harmonization acts amounted to 43 % of all enacted acts. 78 % of acts were enacted on the initiative of Minister's Council, 14% on initiative of Members of Parliament, 7% on initiative of Parliamentary Commission and the rest were enacted on initiative of the Senate and the President of Poland. From these acts – 106 regulated matters at the time not regulated in Polish law (39 %), 165 acts regulated matters already regulated in Polish law: 67 acts were new acts (25 %) and 98 acts amended the existing legislation (36 %). The acts on environmental protection amounted to about 27, i.e. 10 % of the total number of acts aiming at harmonization with EU legislation. Seven acts were only amendments of existing legislation, the rest were regulating the matters at the time not regulated in Polish law or were regulating the matter anew by replacing the old regulation (Sommer 2004 p.37). It must be remembered that this regulation must be integrated with the rest of law.

5.3 Problems in the Implementation in Poland

The requirements of the SEA Directive were introduced to the Polish law by the Act of 09.11.2000 on Public Access to Information and Environmental Impact Assessments. This Act was binding only ten months, then the provisions of this Act were amended to the Act on Environmental Protection Act of 27.04.2001. Both Acts were based on the draft Directive on SEA.

5.3.1 Basic Provisions in the SEA Directive

At first, it seems valuable to recall basic provisions of SEA in this respect. The SEA Directive defines plans and programmes submitted to an environmental assessment procedure in Art. 2 a) and in Art. 3 para 1-4 and 8 and 9 of the SEA Directive. According to Art. 2 a) of the SEA Directive plans and programmes are the documents which are subject to preparation and/or adoption by an authority at national, regional or local level, or which are prepared by an authority for adoption, through a legislative procedure by Parliament or Government. The authority must have a legal duty to prepare the plans and programmes, and the duty is established by legislative, regulatory or administrative provisions (Hendler 2003 p.231). Art. 3 of the SEA Directive narrows the scope of the plans and programmes, subordinated to the Directive, to plans and programmes which "are likely to have significant environmental effects". The plans and programmes, fulfilling the requirements of Art. 2 a), will be subordinated to the Directive only where the Member States determine that they are likely to have significant environmental effects. Art. 3 para 2 introduces two presumptions. First, all plans and programmes which are prepared for specific sector of social life and which set the framework for future development consent of projects listed in Annexes I and II of the EIA Directive are likely to have significant environmental effects and therefore demand environmental assessment. Second, the plans and programmes that have been determined to require an assessment pursuant to Art 6 or 7 of the Habitats Directive, in view of the likely effect on sites due to the site's conservation objectives, they also are obligatory subordinated to environmental assessment according to the SEA Directive.

Art. 3 para 3 allows to dismiss from the requirement of environmental assessment whose plans and programmes referred to in Para 2 which determine the use of small areas at local level or represent a minor modification to plans and programmes referred to in Para 2. The Member States may decide to subordinate such plans and programmes only where they determined that they are likely to have significant environmental effects. According to Art. 3 para 4 of the SEA Directive the Member States shall determine whether plans and programmes, other than those mentioned in Para 2 and which set the framework for future development consent of projects, are likely to have significant environmental effects.

5.3.2 Legal Situation in Poland

This section will now evaluate the issues discussed in section 5.2 with regards to the Polish situation. The implementation of the SEA Directive in Poland in 2004 will need to take into consideration that Poland has had SEA legislation since 1995 (see also Chap. 14).

(1) Environmental Assessment in Polish Law: The Polish Environmental Protection Act determines the scope of plans and programmes submitted to environmental assessment procedure but this is different to the SEA Directive. The EPA determines two groups of plans and programmes which ought to be subordinated to environmental assessment procedure. One group is comprised of a draft national land use development outline, draft land use plans, and draft regional development strategies. The second group consists of the draft plans and of the draft programmes on industry, energy, transport, telecommunications, water management, waste management, forestry, agriculture, fisheries, tourism and land use (this specification is the same as in the SEA Directive) that are prepared by the central and regional authorities and their preparation is provided for by the laws.

The Act uses also the terms policies and strategies, but in Polish law these terms do not have a distinct meaning as compared with plans and programmes. It is worthy to mention that the regulation must be comprised only in a Parliamentary Act and second that the connection with future development consent is not necessary. So the regulation of the Polish EPA, on the one side, is broader in comparison with the SEA Directive because it does not demand that the plans and programmes must create the framework for the development consent, but on the other side it is narrower because only a Parliament Act may be the basis for the programmes and plans which should be submitted to EA procedure.

(2) Preparation by Authorities: The Polish EPA determines plans and programmes subordinated to environmental assessment procedure, and does not use terms like "preparation by authority of national, regional and local level" and "required by legislative, regulatory or administrative provisions". Instead of these terms it uses the names of the drafts of the SEA Directive (first group of plans and programmes). As for the second group of plans and programmes, the local authorities are excluded and the meaning of central and regional authorities is not in accordance with the SEA Directive. According to the Polish text the central and regional authorities are used in a specific sense. The term "central administrative authorities" has its specific legal meaning. In Poland it is discerned three kinds of authorities on the central level: main (political) administrative authorities, central administrative authorities and state organisational units. The main administrative authorities are those which create, according to Constitution (Art. 147), the Ministers' Council. The main administrative authorities are mainly Prime Minister and ministers. The central administrative authorities are the administrative authorities subordinated to main administrative authorities. The main characteristic features of these two groups are: they possess their own competence and task distinct of other authorities, and they have decision rights (legislative or administrative). The state organisational units are auxiliary administrative structures situated in the Ministers' Council or in ministries and they do not have decision rights. So according to EPA, only documents prepared by the central administrative authorities (in the above meaning) are submitted to environmental assessment (EA) procedure envisaged in the Act. The documents prepared by the main administrative authorities (by ministers and Prime Minister) are not subjected to EA procedure as well as the documents prepared by the state organisational units. A draft energy policy is prepared by the Minister of Economy and the draft of national land use plan is prepared by the Government Centre of Strategic Studies which is the state organisational unit.

Also there are the problems with identifying regional authorities whose plans and the programmes are subordinated to EA procedure (in Polish text the word is used that takes into account only specific authorities acting on regional level, namely these acting on the territory of region or province). Further, there are also the supra-regional authorities, e.g. in water management, and they have competence to prepare and adopt the plans that fulfilled the characteristic of the SEA Directive. There are also the sub-regional authorities. So the EPA excludes the plans and programmes prepared by the main administrative authorities, state organisational units (kind of central authorities), local authorities, supra-regional and sub-regional authorities. Finally, the SEA Directive does not use the term "administrative authorities" but only "authorities" which has a broader sense. In reality, other acts could be subject to environmental assessment procedure, also documents which are prepared by excluded authorities. This is especially true for the land use management and water management, and the national energy plan. In that respect the requirement of Art. 4 para 2 of the SEA Directive demand that the requirements of the Directive shall either be incorporated into existing procedures in Member States for the adoption of plans and programmes, or incorporated in procedures established to comply with this Directive. But, Poland uses both methods simultaneously.

(3) Significant Environmental Effects: Next deciding feature whether to submit a plan or programme to the environmental assessment procedure or not, is that it is likely to have significant environmental effects. That requirement is comprised in the EPA only as the possibility for exclusion of a document from the EA procedure. The conditions of exclusion are the same as provided in Art. 3 paragraph 3, but the paragraph 4 is not transposed. The EPA does not contain an equivalent for the Annex II to the Directive.

(4) Framework for Further Development: Also the EPA does not contain any reference to the next characteristic feature of plans and programmes subordinated to the Directive mentioned in Art. 3 para 3 a). According to the Directive plans and programmes must create the framework for future development consent. The term "development consent" is not clear, because in many countries there are various decisions that may be treated as the development consent. Practically every project to be realised must obtain a series of decisions. More helpful is the definition of development consent comprised in EIA Directive: "Development consent means the decision of the competent authority or authorities that entitles the developer to proceed with the project". In turn the EIA Directive defines the competent authority as follows: "The competent authority or authorities shall be those which the Member States designated as responsible for performing the duties arising form this Directive" (Art. 1 para 2, the third indents and para 3 of the EIA Directive with amendments). Translating this definition to the Polish legal system it seems that, in most cases, the equivalent of development consent it will be a location decision. In Polish law only the local land-use plans create such a framework for location decisions. They also create the framework for other decisions that may be treated as the equivalent of development consent.

The preparation and adoption of local land use plans is regulated almost exclusively in Land Use Planning and Management Act of 27.03.2003. Almost exclusively because the EPA contain in this respect the delegation for the Environmental Minister to regulate in Executive Order the details of the environmental report of local land use plans. The competence to enact the local land use plans is carried out almost exclusively by the local government.

The local land use plans of marine territory are the only exceptions to the rule that the local government, with main role of commune council, is responsible for the enactment of the local land use plans. The preparation and adoption of the local land use plans which are pertaining to inner marine waters, territorial waters or exclusive economic zone are regulated separately in the Act on the Sea Areas and Maritime Administration of the Polish Republic of 21.03.1991. A Director of Sea Office is responsible for the preparation of draft land use plan with the environmental report. The Minister of Building and Housing acting together with the Minister of Defence accepts the plan.

(5) Harmonization with the Habitats Directive: The new Nature Protection Act of 16.04.2004 introduces the category of Nature 2000 areas and establishes the protective regime for these areas, the same as in the Habitats Directive. But the EPA does not comprise the equivalent of Art. 3 para 2 b) of the SEA Directive. Now, there is a draft in preparation which ought to remedy the situation.

(6) The Procedure of the Environmental Assessment: In other matters the transposition of the SEA Directive is much better. According to EPA, public administrative authorities responsible for the preparation of certain plans and programmes are obliged to carry out the environmental assessment during the preparation of a plan or programme and before its adoption. The environmental assessment procedure consists of the preparation of the environmental report, of the communication to the public of a draft document together with the environmental report, of the consultation of public and of the competent authorities on the draft document. (Art. 19 para 2, No. 1 and No. 2, Art. 34 and Art. 40 of EPA). The right of the public to be consulted comprised in EPA is taken very broad. Everybody has a right to take part in consultation. The SEA Directive demands that the Member States shall identify the public that will have right to consultation (Art. 6 para 4). The provision formulates three criteria: public affected or likely to be effected or having an interest in the decision-making. There is a question whether these criteria refer also to non-governmental organisation. The text of the SEA Directive is not clear in this point. The provisions on organisations are very vague. Taking the decision on the plans or programmes, the authority must take into account the opinion expressed during the consultation. The EPA does not demand from the competent authority to publish the decision on the adoption of the plans or programmes. It is contrary to Art. 9 of the SEA Directive. According to this article the Member States are obliged to ensure that the consulted public, authorities and other Member States and their public and authorities (Art. 7 of the SEA Directive) will be informed on the adopted plan or the programme. A statement ought to be enclosed how the environmental consideration and the environmental report and opinion expressed in consultation, according to Art. 6 and 7 of the SEA Directive, were taken into account in the decision on the adoption of the plan or programme. It must be mentioned that the requirements for the environmental report, comprised in Art. 41 para 2 of the EPA, are almost the same as comprised in Annex 1 to the SEA Directive.

The provisions on transboundary consultation comprised in Art. 7 of the Directive are only partially transposed by the EPA. The EPA regulates transboundary consultations only to plans and programmes comprised in Art. 40 para 2 No. 2 and not in Art. 40 para 2 No. 1. It denotes that the plans and programmes on the land use management are excluding, contrary to the SEA Directive, from transboundary consultation. These matters are also not regulated in other Acts.

5.4 Conclusions and Recommendations

The discrepancies between the SEA Directive and the Polish EPA, and in certain respect other Acts (especially on transboundary consultations), are a result not only of the fact that Polish laws were enacted before the Directive. The reasons are more serious. They are connected with legal nature of plans and programmes on one side and the procedural regulation of and adoption of the plans and programmes on the other side.

In Polish law only limited number of plans have normative value. They are treated as normative acts, as a source of generally binding norms. To this group belong local land use plans and water management plans, and the procedure of their preparation and adoption are regulated at length. Some programmes and plans are not treated as normative acts, comprising generally binding norms, but they are binding only indirectly via other plans and programmes or they have only a political character. But it does not denote that they are without importance. To such plans belong, for example the plans on industrial policies. From these reasons the regulation ought to be very precise, going in line with the SEA Directive. It is impossible to regulate all plans and programmes with the same intensity.

According to the legal character of the plans and the programmes the procedural regulation pertaining their adoption is developed. Most developed provisions are comprised in the Land Use Planning and Management Act of 27.03.2003. They pertained to the local and regional land use plans. The provisions pertaining to the national land use development outline are comprised in Land Use Planning and Management Act but also in Ministers' Council Act of 08.08.1996, in the Resolution No 49 of 19.03.2002 comprising Rules of the Ministers' Council and in Act on the Sections of Governmental Administration of 04.09.1997. These last three normative acts regulated the procedure of preparation and adoption of all other plans and programmes prepared by all authorities acting on central level. The procedures of preparation and adoption of plans and programmes on regional level are comprised in the Regional Local Government Act of 05.06.1998 and in the Statutes of Local Government as well as in the acts regulating specific sectors of social life, e.g. Water Law Act. The procedure of the preparation and the adoption of the plans and programmes are comprised in the many acts of differentiated legal character therefore it is very difficult to regulate the procedure in specific act. The procedure must take into account the specific nature of the plan or the programme. At present time there are doubts to the range of the plans and programmes which ought to be subjected to the requirements of the SEA Directive. It is the result of the method of regulation of this subject. It would be a lot better if the regulation would be for each kind of plan or programme. But such a method is more difficult and time consuming.

Nevertheless the SEA Directive, as well as acts which have or will implement it, have some value. It forces the authority to include the environmental requirements on the preparation of plans and programmes, and obliges them to include in the preparation the public and authority responsible for the environmental protection. This may have influence on the quality of the plans and programmes but surely it shall have the influence on the people that are active in the field of environmental protection also, as well as, on the environmental conscience overall.

References[1]

Damohorsky M (2002) Główne problemy harmonizacji czeskiego prawa środowiska z legislacją Wspólnoty Europejskiej (The Main Problems of Harmonization of Environmental Law of Czech Republic with the EC Law). Ochrona Środowiska. Prawo i Polityka (Environmental Protection. Law and Policy) 4:19-28

Hendler R (2003) Zum Begriff der Pläne und Programme in der EG-Richtlinie zur strategischen Umweltprüfung (The Definiton of Plans and Programmes in the EC Directive on Strategic Environmental Assessment). DVBl. 4:227-334

Ladeur K-H and Prelle R (2001) Environmental Assessment and Judicial Approaches to procedural Errors: A European and Comparative Law Analysis. Journal of Environmental Law 13(2):185-198

Pernice I (1994) Kriterien der normativen Umsetzung von Umweltrichtlinien der EG im Lichte der Rechtsprechung des EuGH (Criteria of the Normative Transposition of EC Environmental Directives with regard to the Ruling of the European Court of Justice). Europa Recht 3:325-341

Radecki W (2001) Charakter prawny raportu oddziaływania na środowisko (The Legal Character of Environmental Report). Ochrona Środowiska. Prawo i Polityka (Environmental Protection. Law and Policy) 2:28-40

Sommer J (2004) The organizational and legal instruments available for harmonizing Polish environmental law with EC environmental law. In: Schmidt M, Knopp L (eds) Reform in CEE-Countries with Regard to European Enlargement. Springer Berlin, Heidelberg, pp 29-52

---

[1] Note that legislation mentioned in the chapter is listed at the end of the handbook in the consolidated list of legislation (Appendix 2).

Part II – Current Status and National Strategies for the Implementation of SEA in the European Union

Part II of the Handbook focuses on discussing the status of SEA in the European Union (EU) in 2004 and evaluates different national strategies employed by eleven individual EU Member States in order to implement the binding SEA Directive (the deadline for introduction was 21 July 2004). Interestingly, many of the EU Member States already had SEA legislation in place, or used tools similar to SEA, before the SEA Directive came into force. So part of the challenge in the implementation of the SEA Directive was how it would relate to existing SEA legislation that, in some cases, had even broader requirements than the SEA Directive (e.g. the original national SEA legislation might cover policies and strategies, while the SEA Directive does not).

Chapter 6 considers the situation within England (for SEA legislation, England is separate from the other countries of the UK). England has had a simplified version of SEA, called Environmental Appraisal, since 1992 and was the first country in the EU to produce guidance for the implementation of the SEA Directive. Chapter 7 outlines the German progress in implementing SEA. Germany makes a particular interesting case study because of its federal system and how that affects the introduction of the SEA Directive. This chapter forms an important basis for some of the later methodology chapters (in Parts V, VI and VII of the Handbook) that use Germany as their case study.

Chapter 8 discusses the difficulties in the implementation of the SEA Directive in Italy at the *national* level and concentrates on one of the Italian *regions* where SEA requirements has been introduced: the Emilia Romagna Region. Italy is one of the EU member countries where the national implementation of the SEA Directive has *not* occurred on time. Because of the lack of a national implementation of SEA this has lead the regions (such as the one described in this chapter) to develop their own SEA legislation.

The first experiences and case studies in the Flanders region of Belgium are described in Chapter 9, where the implementation of the SEA legislation has been progressing since 2002. Chapter 10 evaluates the implementation of SEA in Austria, by describing some voluntary SEA and SEA-like pilot studies that have been undertaken since 1998. Chapter 11 discusses the further development of the implementation of SEA in Finland taking into account existing practice. Finland introduced a general requirement to assess the environmental effects of policies, plans and programmes in the Environmental Impact Assessment Act of 1994. In contrast, Chapter 12 discusses the problems of a minimalist approach with the implementation of SEA in Sweden.

The last four chapters of Part II describe the implementation of the SEA Directive in some of the new Member States that joined the European Union on the 1st of May 2004. These East European countries have problems that are very specific to their background, but interestingly some of these countries already had SEA legislation before the advent of the SEA Directive. Chapter 13 describes the national strategy for the implementation of the SEA Directive in the Czech Republic, which has had SEA legislation since 1992. One of the interesting aspects of the established SEA process in the Czech Republic is the use of a system of licensed environmental assessment experts in order to help ensure that the SEA process is done with quality.

Chapter 14 describes the developments of SEA in Poland, which started in 1995. Currently, Polish law considers SEA as an environmental assessment procedure to be carried out for plans, programmes, policies and strategies. Interestingly, although the SEA Directive only refers to plans and programmes, SEA in Poland applies to all these four fields of action. Chapter 15 describes the experience with SEA in Latvia. Despite the fact that SEA experience is still at a relatively early stage, the chapter discusses how the SEA Directive is seen as an important tool for assessing environmental impacts in Latvia. Finally, Part II concludes by describing the implementation of SEA in Estonia, where there has been SEA legislation since 2001 (Chapter 16).

6 Current SEA Practice in England

Thomas B. Fischer

Department of Civic Design, University of Liverpool, England

6.1 Introduction

SEA type assessments have been applied widely in the UK for a number of years. Furthermore, UK SEA type assessment guidance has been among the earliest in Europe. Regarding current practice, environmental and sustainability appraisal of land use development plans and regional planning guidance are widely known, both nationally and internationally (Fischer 2004). Other SEA related experience exists in transport, energy and economic development planning, water management and EU structural and rural planning. Besides practical SEA type application, there is also some considerable research experience and UK based authors have had a high input to the international SEA literature. Gazzola and Fischer (forthcoming), for example, found that out of a sample of 45 international SEA related books and conference proceedings, one third were either written or edited by UK-based authors.

In 2002, the Royal Society for the Protection of Birds and the Transport Research Laboratory (RSPB et al. 2002) identified those plans and programmes (PP) that might be subject to formal SEA application in the UK, following the requirements of the European SEA Directive. This was based on a list of 81 statutory and non-statutory PP from different sectors, including town and country planning, water management, marine environment, bio-physical environment, tourism, forestry, economic development, EU, rural development and agriculture planning, transport, energy, telecommunications, defence and sustainable development. Twenty-seven PP were identified to clearly fall under the requirements of the SEA Directive. Among these, 12 were spatial or land use PP, five were transport PP, three were energy PP, two were water management PP, one was an economic PP and one was an environment PP. There were also three rural development and agriculture PP that are not UK specific, but prepared EU wide in the context of the structural funds. Thirty-five PP were said to potentially require formal SEA in the future. Only two of the 81 plans and programmes were said to clearly not require formal SEA after 21 July 2004, the deadline for Member States to formalise SEA.

Implementing Strategic Environmental Assessment. Edited by Michael Schmidt, Elsa João and Eike Albrecht. © 2005 Springer-Verlag

Furthermore, the status of eight PP was said to be unclear. Considering these numbers, the UK faces a huge task in developing specific SEA approaches, particularly as there are currently no indications that any substantial new resources will be made available in public decision making for SEA. Generally speaking, in the UK, most people involved in SEA practice or research recognize that SEA needs to be applied according to the specific circumstances of a sector or administration, i.e. there is a 'no one fits all' attitude to SEA (see, for example, Tomlinson 2002b). Whilst SEA type assessments have been conducted in various sectors for more than a decade, current practice will need to change substantially in order to meet the new SEA Directive requirements. What this will mean for land use and transport planning is discussed in Sects. 6.3 and 6.4.

This chapter portrays and discusses current SEA type assessment practice, focusing on land use and transport planning. Furthermore, future challenges for SEA, following the implementation of the SEA Directive are identified. A land use sustainability appraisal for the Oldham Unitary Development Plan (UDP) and the transport appraisals according to the Guidance on Multi-Modal Methodologies (GOMMMS) and the New Approach to Trunk Roads Appraisal (NATA) are evaluated based on five SEA Directive core elements. These include the scope of SEA application itself, proper tiering, consultation, the consideration of a range of substantive aspects in report preparation and follow-up. It is concluded that while current practice can provide a good basis for formal SEA, some amendments are needed. These particularly include improved baseline descriptions of the environment, a wider consideration of alternatives, more effective public participation, better quantification of impacts and monitoring.

6.2 Current SEA-Type Assessment Practice

This section is divided into three parts. Firstly, current strategic assessment instruments are introduced and their relevance for future formalised SEA is discussed. Subsequently, SEA related guidance, currently in use or being developed, is presented. Finally, recent publications on SEA related experiences in the UK are identified.

6.2.1 Strategic Assessment Instruments

In the UK, a wide range of assessment instruments are applied and formally promoted at strategic decision making levels (UK government strategy unit 2002; Levett-Thérivel 2002). These include:

- *Environmental and Sustainability Appraisal* for land use development plans
- *Sustainability Appraisal* for Regional Planning Guidance (RPG)
- *Appropriate Assessment* for development activities that might have a significant impact on areas falling under the EU Habitats Directive

- *Health Impact Assessment* for policies, plans, programmes and projects with likely impacts on health and safety
- *Regulatory Impact Assessment* for policies with likely impacts on businesses
- *Transport Scheme Appraisal* for local transport plans (LTPs) and trunk roads proposals
- *Multi-Modal Studies* for defined corridors/ infrastructure improvement areas
- *Rural Proofing* for policies with possible impacts on rural areas
- *Gender Impact Assessment* for policies with likely impacts on equal treatment
- *Consumer Impact Assessment* for policies that may affect costs for consumers
- *Risk assessment* to identify risks of policies, plans, programmes and projects, often within other impact assessment instruments
- *E-commerce Impact Appraisal* for policies with likely impacts on E-commerce

Five of these instruments explicitly aim at supporting the consideration of the biophysical environment in planning, namely environmental and sustainability appraisal of land use development plans, sustainability appraisal of RPG, appropriate assessment, transport scheme appraisal and multi-modal studies. Environmental appraisal and more recently sustainability appraisal has been applied for over a decade in land use development plan making. Furthermore, sustainability appraisal has been applied to Regional Planning Guidance since 2000. Environmental aspects have also been considered in transport scheme appraisal for local transport plans and the trunk roads programme for over two decades. Multi-modal studies represent a new approach to appraisal. There is currently comparatively little reported evidence of appropriate assessment application, which may consist of merely a few pages on expected impacts within project, programme, plan or policy proposals.

6.2.2 Guidance

A range of guidance documents for environmental assessment at strategic levels of decision making have been prepared over the last decade, including:

- Policy appraisal and the environment (DoE 1991)
- Environmental Appraisal of Development Plans – A Good Practice Guide (DoE 1993)
- Good Practice Guide on Sustainability Appraisal of Regional Planning Guidance (DETR 2000a)
- Flood and coastal defence project, appraisal guidance, strategic planning and appraisal (DEFRA 2001)
- Guidance on Multi-Modal Methodologies (GOMMMS – DfT 2000, 2003)
- New Approach to Trunk Roads Appraisal (NATA – DETR 1998a)
- Guidance on the SEA Directive (ODPM 2003, 2004)

Furthermore, SEA guidance is currently prepared by the Environment Agency for England and Wales on general SEA application. English Nature, the Environment

Agency, the Countryside Council for Wales and the Royal Society for the Protection of Birds (RSPB) work on guidance on biodiversity considerations in SEA. Furthermore, the Department for Transport (DfT) is preparing guidance for local transport plans and the Office of the Deputy Prime Minister (ODPM) for local development frameworks and regional spatial strategies, two new types of development plans to be introduced in the second half of 2004 (see Sect. 6.3).

6.2.3 Recent SEA-Related Publications in the UK

More recently, several authors have reported on SEA type experiences in the UK. For spatial/land use planning, these include Thérivel (1998), Curran et al. (1998), Russel (1999), Fischer (2002b, 2003) and Short et al. (2004). For transport planning, authors include Tomlinson (1999), Fischer (2002b) and Sheate et al. (2004). In other sectors, Marshall (2003) reported on SEA related energy planning and Thames Water (2000) on voluntary water management SEA. Furthermore, the Department for Trade and Industry (dti) has done work on SEA for offshore oil and gas facilities (2003a) and offshore wind energy (2003b).

Subsequently, practice in the two areas where SEA application is most extensive, i.e. land use and transport planning, are discussed. In order to avoid confusion, the focus is on England, as there are a number of – growing – differences in UK SEA practice between Scotland, Wales, Northern Ireland and England.

6.3 Practice in Land Use Planning in England

This section is sub-divided into three parts. Firstly, an overview of the current situation is provided. This is followed by a summary of the development of SEA type assessment practice at the beginning of the 1990s. Finally, the sustainability appraisal of the UDP Oldham is evaluated in terms of the requirements of the SEA Directive.

6.3.1 Overview of Current Situation

SEA type experiences in England are most extensive in land use planning. Practice follows requirements introduced in 1992 by national Planning Policy Guidance (PPG) 12 (last amended 1999, see DETR 1999). More recently, there has also been some considerable experience at the regional level with sustainability appraisal of Regional Planning Guidance (RPG), following PPG 11 (DETR 2000c). In England, different types of land use plans are prepared, including:

- Unitary development plans (UDPs) in unitary authority areas (single tier structure),
- Structure plans at county level and local plans at local levels in non-unitary authority areas (two tier structure).

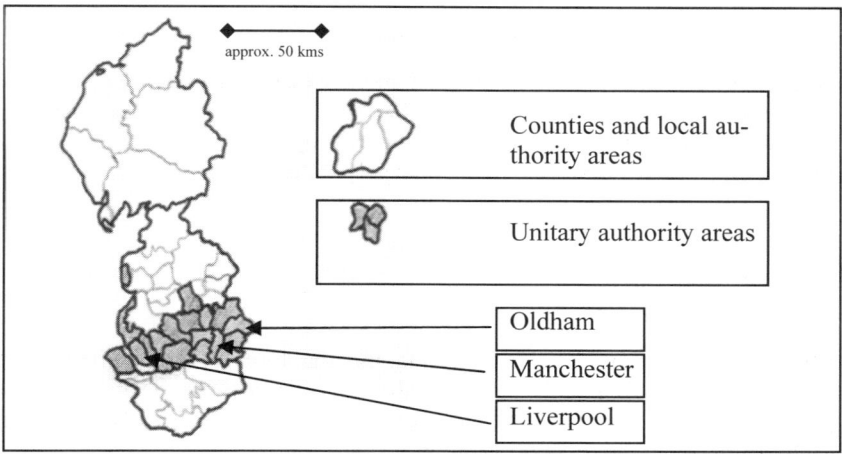

Fig. 6.1. North West England – administrative structure (up until mid-2004)

Currently, unitary authorities are responsible for preparing UDPs – consisting of two main parts and dealing with strategic choices and concrete development policies. Local authorities prepare local plans that need to take objectives of county wide structure plans into account. Fig. 6.1 shows the administrative structure in the region of North West England, consisting of counties and their local authority districts and unitary authority areas. The locations of Manchester and Liverpool, the two biggest cities in the region, and of Oldham, subject to the evaluation in Sect. 6.3.3 are also shown.

The land use planning system in England will change in mid-2004, following the Planning and Compulsary Purchase Bill from 2003. A uniform hierarchy of plans will be introduced throughout the country, consisting of regional spatial strategies and local development frameworks, comprising portfolios of local development plan documents, local development scheme, supplementary planning documents and a statement of community involvement. This poses additional challenges to the formalised application of SEA, as assessors are faced with a new and untested planning system. Whilst SEA guidance for the new system is currently in preparation, it remains to be seen how effectively new SEA requirements can be implemented within a new planning system.

6.3.2 Development of Land Use SEA Type Assessment Practice

Environmental appraisals for structure plans, UDPs and local plans were first conducted at the beginning of the 1990s. The earliest related government guidance 'environmental appraisal of development plans – a good practice guide' was released in 1993 (DoE 1993). Since then development plan appraisal practice has developed in several stages. These may be summarised as follows (Fischer 2003):

- Beginning of the 1990s: Environmental appraisal taking the form of qualitative 'matrix evaluation' (see, for example, Thérivel et al. 1992); this was usually focusing on ecological aspects only and was done by one person quasi ex-post for the four main stages of development plan preparation:

    - evaluation of the old plan
    - consideration of development options
    - UDP deposit draft
    - UDP final version

- Mid-1990s: Environmental appraisal taking the form of qualitative 'matrix evaluation'; now frequently also considering socio-economic aspects; several persons were often involved in the quasi ex-post assessment of the four main stages of development plan preparation.
- End of the 1990s: Environmental appraisal taking the form of qualitative 'matrix evaluation'; usually also considering socio-economic aspects in a quasi ex-ante appraisal, conducted by several persons for the four main stages of development plan preparation, often including external consultation; environmental appraisal is increasingly transformed into sustainability appraisal.
- Beginning of the 2000s: Environmental and sustainability appraisals are conducted in an objectives-led process, referring to sustainable development strategies (see, for example NWRA 2003). In this context, increasingly the term 'integrated appraisal' is also used. This has been encouraged by the government not just as part of its efforts to promote more sustainable patterns of governance (DEFRA 2002) but also in response to its wider modernising government agenda (Prime Minister and Cabinet Office, 1999; see also Kidd and Fischer, forthcoming).
- 21 July 2004: SEA to be formally conducted, based on existing environmental appraisal and sustainability appraisal practice. Necessary amendments and changes are discussed in Sect. 6.3.3.

Fig. 6.2 shows a typical impact matrix used in environmental and sustainability appraisal, assessing development plan policies in terms of 15 criteria, falling into three overall categories; 'global sustainability', 'national resources' and 'local environmental quality'. This kind of matrix approach has proved to be a very efficient way of assessing impacts in a qualitative manner in objectives-led assessment. However, following the SEA Directive requirements for baseline-led assessment, this approach on its own will be insufficient and other assessment techniques will be needed, including, for example, GIS and overlay mapping.

Criteria	Global sustainability								Natural resources				Local environmental quality		
	1	2	3	4	5	6	7	8	9	10	11	12	13	14	15
Proposed Policies/Action	Transport energy efficiency	Transport trips	Housing energy efficiency	Renewable Energy potential	CO2 fixing	Wildlife Habitats	Air quality	Water conservation	Soil quality	Minerals conservation	Landscape	Rural environment	Cultural heritage	Public access to parks	Building quality
Urban regeneration	✓	✓	✓	✓	✓	✓	✓	✓	×?	•	✓	•	✓	✓?	✓
Improved trams	✓	✓	?	✓?	✓	•	✓	•	•	•	•	•	✓	?	✓
Use of brownfield sites	•	•	•	✓?	✓	×?	•	•	×?	✓	✓	?	✓	✓	✓

•	No relationship or insignificant impact
✓	significant beneficial impact
✓?	likely but unpredictable beneficial impact
?	uncertainty of prediction or knowledge
×?	likely but unpredictable adverse impact
×	significant adverse impact

Fig. 6.2. Environmental assessment impact matrix

6.3.3 Sustainability Appraisal of the UDP Oldham

Development appraisal performance in England varies widely. This has been shown, for example by Short et al. (2004). Subsequently, a recent case – the sustainability appraisal of the UDP Oldham – is evaluated in terms of the future EU SEA Directive requirements. In the region of North West England, this is widely considered a good practice case (see Fischer 2003). Table 6.1 identifies the extent to which the SEA Directive requirements are currently met.

There are currently a number of shortcomings that will need to be addressed in order to comply fully with future requirements. This was confimed by a number of authors, including Levett-Thérivel (2002); Sheate et al. (2004); Short et al. (2004) and Russel (1999). Positive aspects in the examined case include the consideration of environmental protection objectives and the qualitative assessment of the likely impacts. Furthermore, the relationships with other plans and programmes were acknowledged and the public was able to comment on the underlying plan and an EA report was published. Current shortcomings include:

- no description of the baseline environment; whilst the plan is appraised in terms of environmental protection objectives, the baseline environment is not described (in this context see also Fry et al. 2002).
- no consideration of different development alternatives; development plan policies are only 'optimised' in terms of their likely significant effects on the environment, without considering development alternatives
- no rigorous quantitative predictions; the appraisal is entirely based on qualitative evaluations and no quantifiable information is provided.
- no mitigation and compensation measures; currently, the focus is purely on optimising policies and not on mitigating or compensating for remaining impacts.
- no description of impacts in the appraisal report. Currently, the report describes the process, but is quiet on the overall impacts

Table 6.1. Sustainability appraisal Oldham and SEA Directive requirements (Fischer 2003)

	requirement met
TIERING – VERTICAL AND HORIZONTAL	
Is the assessment focusing on those issues appropriate to the hierarchical level of the plan?	Yes
Is the assessment being carried out as part of a wider procedure to avoid duplication within a tiered system of decision making?	Indirectly, according to land use planning
Have existing monitoring arrangements been consulted and utilised to avoid unnecessary duplication?	Indirectly, in underlying plan
COMMUNICATION, CONSULTATION & PARTICIPATION	
Is there public consultation/participation?	Indirectly, in underlying plan
Is there expert consultation?	Yes
Have relevant environmental authorities been involved in determining the scope of the appraisal?	No
Did the consultees receive a draft version of the plan or program and an accompanying Environmental Report (ER)?	Partly, draft version of plan
Is it clear how the opinions collected in the consultation processes, influenced the preparation of the plan or program?	No
Has adequate consideration been given to publication and advertisement of the plan or program once formally adopted?	Yes
SUBSTANTIVE – REPORT AND ALTERNATIVES	
Is there an environmental report (ER)?	Yes, only describing process
Does ER/appraisal section outline the relationship with other PP?	Yes
Does it include baseline data for the state of the environment?	No
Does it include the environmental characteristics of areas likely to be significantly affected?	No
Is there information on existing environmental problems and anything relating to areas of particular environmental importance?	No

Table 6.1. (cont.)

Have reasonable alternatives been identified, described & evaluated?	No
Does it outline the reasons for selecting alternatives?	No
Does it consider environmental protection objectives?	Yes
Does it consider the significant effects on the environment?	No
Does it include information on mitigation?	No
Does it include a description of how the assessment was undertaken?	Yes
Is there a description of measures concerning monitoring?	No
Is there a non-technical summary of the ER/appraisal section results?	No
FOLLOW-UP: MONITORING AND REVIEW	
Are the significant environmental effects resulting from the implementation of the plan monitored?	So far unclear
Are the measures envisaged within the monitoring process included within the ER/appraisal section?	So far unclear
Does monitoring include the identification of progress towards the plan and SEA objectives?	So far unclear

6.4 Practice in Transport Planning

In England, besides land use planning, there is some extensive SEA type experience in the transport sector. This particularly includes practice in programme type scheme appraisals, conducted in local transport 'plan' making and in the national Trunk Roads Programme. Furthermore, more recently, multi-modal study appraisals have also been conducted. On the one hand, these were prepared in a national context and supported by the Department for Transport (DfT) at local, regional and cross-regional scales. On the other hand, there have also been pilot corridor assessments conducted in the context of the Trans-European Transport Networks (TENs). Generally speaking, the need for impact assessment in the transport sector is widely acknowledged. In this context, Tomlinson (2002b p.1), for example, suggested that:

> Single infrastructure transportation projects at best act to delay the problem rather than provide the solution. It is increasingly recognized that a multi-dimensional approach is needed to adequately address transportation problems.

Subsequently, practice of local transport plan appraisal, trunk roads appraisal and multi-modal appraisals is discussed. This is followed by an evaluation of the current approaches based on the requirements of the SEA Directive.

6.4.1 Local Transport Plan Appraisal

Local transport plans (LTPs) were introduced in 1998 by the UK government white paper 'a new deal for transport' (DETR 1998b). In 2000, the DETR released guidance on local transport plan appraisal, according to which local authorities need to assess the quality of LTPs and produce annual progress reports (DETR

2000b). Where major projects of over £5M (5 Mio Pounds) are included in an LTP, the plan must be appraised with and without the project. LTPs consider a range of SEA type elements (Levett-Thérivel 2002). Most importantly, these include:

- agreement on objectives
- identification of constraints
- consideration of alternatives
- monitoring
- appraisal

Scheme appraisal practice in LTPs is very similar to that in the national trunk roads programme. This is further explained below.

6.4.2 Trunk Roads Appraisal

The national Trunk Roads Programme has been prepared for several decades and includes scheme appraisal based on cost-benefit analysis. This also considers environmental aspects. In the 'new approach to trunk roads appraisal' (NATA), released in 1998 (DETR 1998a), guidance was provided on how to identify and assess policy options and compare them – using appraisal summary tables. This includes five objectives; the environment, safety, economy, accessibility and integration. Furthermore, NATA identified three dimensions for addressing current transport problems. The first dimension revolves around the consideration of all feasible options. The second dimension asks for options to be realistic and the third dimension requires options to deliver improvements.

6.4.3 Multi-modal Studies

The need for multi-modal studies (MMS) was firstly formulated in the UK government white paper 'a new deal for trunk roads in England' (DETR 1998b). Similarly to the NATA approach, MMS focus on integration, safety, economy, environment and accessibility. The Guidance on Multi-Modal Methodologies (GOMMMS) was released by the government in 2000 (DfT 2000; see also Tomlinson 2002a). To date, more than 20 MMSs have been prepared. These cover very different scales, including multi-regions wide MMSs (London to the South West and South Wales), regional MMSs (London to Ipswich) and local MMSs (South East Manchester). The GOMMMS methodology includes 13 main aspects, as follows (Source: DfT 2003):

1) Establish local, regional or study-specific objectives
2) Develop understanding of current situation in terms of:
   - Current transport and other policies impacting transport
   - Opportunities and constraints
   - Current travel demands and levels of service

- Current transport related problems
3) Understanding the future situation in terms of:
- Future committed land use and policies
- Future committed transport system changes
- Future travel demands and levels of service
- Future transport-related problems
- Undertake consultations, participation and information assembly
4) Identification of potential solutions
5) Defining the information needed to apply the appraisal framework
6) Establishing the methods by which information is to be provided through
- Transport model or land use/transport interaction model
- EIA procedure
- Cost-benefit analysis (CBA) procedure
- Geographic information systems (GIS)
7) Implementation costs, operation maintenance & enforcement to be defined
8) Option testing and appraisal
9) Distillation and comparison of options
10) Consultations on options
11) Study outputs
12) Review funding sources to confirm option feasibility
13) Formulation of implementation programme

This methodology reflects an overall tiered approach to decision making, starting with the establishment of objectives and projections of the likely future situation. Potential solutions need to be identified, information needs are established and appropriate methods are to be chosen. Costs and options testing, consultation and review precede the final formulation of an implementation programme.

6.4.4 Evaluation of NATA and GOMMMS Practice

Both, NATA and GOMMMS have the potential to integrate SEA Directive requirements. However, whilst both apply to plans, their approach is scheme – ie project – based and it is currently unclear how strategic level alternatives are to be dealt with. This means that a more strategic approach may in effect be needed in order to conduct SEA effectively. Table 6.2 identifies the extent to which current strategic transport planning in England meets the requirements of the SEA Directive, combining GOMMMS and NATA approaches.

Similarly to town and country planning, transport appraisal is dealing well with addressing those aspects appropriate to the tier of application. Furthermore, there are fairly widespread communication and consultation requirements. However, it appears that practice will have to open up to more extensive public consultation. Whilst transport appraisal requires different alternatives to be considered, baseline conditions are currently insufficiently considered, similarly to land use planning. There are usually sections on mitigation and monitoring, but no non-technical summaries. An additional problem is an apparent overlap of GOMMMS and NATA appraisals, ie they deal with similar substantive issues. Generally speaking, transport planning practice in England has the potential of dealing well with aspects of assessment tiering. In this context, eight different levels have been suggested by Tomlinson (2002a) that can be allocated to a systematically tiered SEA system consisting of four tiers, as proposed by Fischer (2002a). This consists of the preparation of regional transportation strategies, multi-modal studies, evaluation of alternatives, implementation programmes, design and EIA within consent processes (including public inquiry and implementation). Table 6.3 illustrates the four systematic levels in the left column, whilst possible transport planning stages are shown in the right column. As only a properly tiered system is likely to fully result in all the SEA benefits advertised in the academic literature (see Fischer 2000 and Jansson 2000), the introduction of the eight planning stages would be a big step towards achieving more beneficial SEA.

Table 6.2. GOMMMS and NATA and SEA Directive requirements

	requirement met
TIERING – VERTICAL AND HORIZONTAL	
Is the assessment focusing on those issues appropriate to the hierarchical level of the plan?	Yes
Is the assessment being carried out as part of a wider procedure to avoid duplication within a tiered system of decision making?	Unclear
Have existing monitoring arrangements been consulted and utilised to avoid unnecessary duplication?	No
COMMUNICATION, CONSULTATION & PARTICIPATION	
Is there public consultation/participation?	Indirectly, by underlying plan
Is there expert consultation?	Yes
Have relevant environmental authorities been involved in etermineing the scope of the appraisal?	No
Did the consultees receive a draft version of the plan or program and an accompanying Environmental Report (ER)?	Yes
Is it clear how the opinions collected in the consultation processes, influenced the preparation of the plan or program?	Unclear
Has adequate consideration been given to publication and advertisement of the plan or program once formally adopted?	Partly, varies
SUBSTANTIVE – REPORT AND ALTERNATIVES	
Is there an environmental report (ER)?	No, only section in plan
Does ER/appraisal section outline the relationship with other PP?	Yes
Does it include baseline data for the state of the environment?	No
Does it include the environmental characteristics of areas likely to be significantly affected?	No
Is there information on existing environmental problems and anything relating to areas of particular environmental importance?	No

Table 6.2. (cont.)

Have reasonable alternatives been identified, described & evaluated?	Partly, scheme based
Does it outline the reasons for selecting alternatives?	Yes
Does it consider environmental protection objectives?	Yes
Does it consider the significant effects on the environment?	No
Does it include information on mitigation?	Yes
Does it include a description of how the assessment was undertaken?	Yes
Is there a description of measures concerning monitoring?	Yes
Is there a non-technical summary of the ER/appraisal section results?	No
FOLLOW-UP: MONITORING AND REVIEW	
Are the significant environmental effects resulting from the implementation of the plan monitored?	No
Are the measures envisaged within the monitoring process included within the ER/appraisal section?	Yes
Does monitoring include the identification of progress towards the plan and SEA objectives?	Partly, plan objectives

Table 6.3. Tiering in transport planning in England

Systematic tiers	Proposed transport planning stages	
Policy	1 + 3:	Regional Transportation Strategy (to be completed only after multi-modal studies at the plan level have been conducted)
Plan	2	Multi-modal studies
	4	Evaluate design and implementation alternatives of transportation measures
Programme	5	Select preferred transportation measures/implementation programme
Project	6	undertake transportation design, environmental assessment and consent processes
	7	public inquiry and announcement
	8	transportation measure design and tender/implementation process

6.5 Conclusions and Recommendations

There is some considerable experience with the application of SEA-type assessment instruments at strategic decision making levels in the UK. Whilst practice is most widespread in land use and transport planning, other sectors have also undertaken SEA related assessments. These include waste and water management, energy and economic development planning and the EU wide practiced structural and rural planning. Table 6.4 summarises the most important SEA-type assessment instruments in land use and transport planning. A detailed description of how practice compares with the future SEA requirements is provided in Sects. 6.3 and 6.4.

Table 6.4. SEA-type assessment techniques in land use and transport planning

SEA-type assessment	year it started	reference link
Environmental appraisal in land use development planning	1992	DoE 1993
Sustainability appraisal in RPG	2000	DETR 2000
NATA	1998	DETR 1998a
GOMMMS	2000	DfT 2000

The evaluation of the sustainability appraisal of the land use Oldham UDP and the transport GOMMMS and NATA show that whilst current practice can be a solid basis for the formal application of SEA, certain shortcomings will need to be addressed in order to fully comply with the SEA Directive requirements. Future challenges particularly include a better description of the baseline environment, a more comprehensive consideration of alternatives, more effective participation, a better quantification of impacts and monitoring. In addition to the task of formalising SEA in 2004, another challenge is connected with the introduction of a new planning system, consisting of regional spatial strategies and local development documents. Whilst the SEA guidance recently released by central government deals mainly with the 'old' planning system, guidance for use in the new system is currently prepared. However, how planning officers will manage to deal with two new parallel challenges remains to be seen. In transport planning, current debates at the national level suggest that there is a good chance for assessment tiering to become more comprehensive and effective than has been practiced to date.

References[1]

Curran JM, Wood C, Hilton M (1998) Environmental appraisal of UK development plans: current practice and future directions. Environment and Planning B: Planning and Design 25:411-433

DEFRA – Department for Environment, Food and Rural Affairs (2001) Flood and Coastal Defence Project Appraisal Guidance. Strategic Planning and Appraisal. London. Internet Address: www.defra.gov.uk/environ/fcd/pubs/pagn/default.htm, last accessed 18.12.2003

DEFRA – Department for Environment, Food and Rural Affairs (2002) Sustainable development in Government: First Annual Report 2002. Internet Address: www.sustainable-development.gov.uk/sdig/reports/ar2002/index.htm, last accessed 13.01.2003

DETR – Department of Environment, Transport and the Regions (1998a) New Approach to Appraisal. DETR, London

DETR – Department of Environment, Transport and the Regions (1998b) A New Deal for Transport. DETR, London

---

[1] Note that legislation mentioned in the chapter is listed at the end of the handbook in the consolidated list of legislation (Appendix 2).

DETR – Department of Environment, Transport and the Regions (1999) Planning Policy Guidance Note 12: Development Plans. DETR, London
DETR – Department of Environment, Transport and the Regions (2000a) Good Practice Guide on Sustainability Appraisal of Regional Planning Guidance. DETR, London
DETR – Department of Environment, Transport and the Regions (2000b) Guidance on full local transport plans. DETR, London
DETR – Department of Environment, Transport and the Regions (2000c) Regional Planning Guidance. DETR, London
DfT – Department for Transport (2000) Guidance on Multi-modal Methodologies. DfT, London
DfT – Department for Transport (2003) Applying the multi-modal new approach to appraisal (GOMMMS) to highway schemes. Internet Address: www.dft.gov.uk, last accessed 17.12.2003
DoE – Department of the Environment (1991) Policy Appraisal and the Environment. HMSO, London
DoE – Department of the Environment (1993) Environmental Appraisal of development plans – a good practice guide. HMSO, London
dti – Department of Trade and Industry (2003a) Strategic Environment Assessment Area North and West of Orkney and Shetland. Consultation Document September 2003, London. Internet Address: www.offshore-sea.org.uk, last accessed 17.12.2003
dti – Department of Trade and Industry (2003b) Environmental Report: SEA (Phase 1) for Offshore Wind Energy Generation. London. Internet Address: www.og.dti.gov.uk/offshore-wind-sea/process/envreport.htm, last accessed 17.12.2003
Fischer TB (2000) Lifting the Fog on SEA. Towards a ctegorisation and ientification of some major SEA tasks. In: Bjarnadottir H (ed) Environmental Assessment in the Nordic Countries. Nordregio, Stockholm, pp 39-46
Fischer TB (2002a) Towards a more systematic approach to policy, plan and programme environmental assessment – some evidence from Europe. In: Marsden S, Dovers S (eds) SEA in Australasia, Sydney Place Federation Press, pp 99-113
Fischer TB (2002b) Strategic Environmental Assessment in Transport and Land use Planning. Earthscan, London
Fischer TB (2003) Impact Assessment of UDP Oldham: Evaluating a positively perceived process (Die Folgenprüfung zum Entwicklungsplan Oldham: Ein positiv wahrgenommenes Verfahren auf dem Prüfstand). UVP-report 1:29-33
Fischer TB (ed) (2004) Progress towards meeting the requirements of the European SEA Directive. European Environment Journal, special issues 14(2):55-134 and 14(3):135-200
Fry C, McColl V, Tomlinson P, Eales R, Smith S (2002) Analysis of Baseline Data Requirements for the SEA Directive – Final Report. TRL Ltd, Crawthorne
Gazzola P, Fischer TB (forthcoming) Procedural flexibility – a valid SEA effectiveness criterion for Italy? EIA Review
Jansson AH (2000) Strategic Environmental Assessment for transport in four Nordic countries. In: Bjarnadottir H (ed) Environmental Assessment in the Nordic Countries. Nordregio, Stockholm, pp 81-88
Kidd S, Fischer TB (forthcoming) Integrated Appraisal in North West England. Environment and Planning C
Levett-Thérivel (2002) Implementing the SEA Directive: Analysis of existing practice. Report to the South West Regional Assembly

Marshall R (2003) SEA and Energy. Conference paper presented at the annual IAIA meeting in Marrakech, June
NWRA – North West Regional Assembly (2003) Implementing Action for Sustainability: An Integrated Toolkit for the North West. NWRA, Wigan
ODPM – Office of the Deputy Prime Minister (2003) The SEA Directive: Guidance for Planning Authorities. Practical guidance on applying European Directive 2001/42/EC 'on the assessment of the effects of certain plans and programmes on the environment' to land use and spatial plans in England. ODPM, London, www.planning.odpm.gov.uk
ODPM – Office of the Deputy Prime Minister (2004) A Draft Practical Guide to the SEA Directive. Welsh Assembly Government, Scottish Executive, Northern Ireland Department of the Environment. ODPM, London
Prime Minister and Cabinet Office (1999) Modernising Government White Paper. Command Paper 43 (HMSO, London). Internet address: www.archive.official-documents.co.uk/document/cm43/4310/4310-02.htm, last accessed 12.06.2003
RSPB – Royal Society for the Protection of Birds, CPRE and TRL – Transport Research Laboratory (2002) Implementing the SEA Directive: experts workshop on 23 October 2002 – background material. RSPB, London
Russel S (1999) Environmental appraisal of development plans. Town Planning Review 70(4):529-546
Sheate WR, Byron HJ, Smyth SP (2004) Implementing the SEA Directive: Sectoral Challenges and Opprotunities for the UK and EU. In: Fischer TB (ed) Progress towards meeting the requirements of the European SEA Directive. European Environment Journal 14(2):73-93
Short M, Carter J, Baker M, Jones C, Wood C (2004) Current practice in the SEA of development plans in England. Regional Studies 38(2):177-190
Thames Water (2000) Planning for Future Water Resources: Bets Practicable Environmental Programme. Final Report prepared by Consultants in Environmental Science, Land Use Consultants, OXERA and Wessex Archaeology
Thérivel R (1998) Strategic environmental assessment of development plans in Great Britain. EIA review 18:39-57
Thérivel R, Wilson E, Thompson S, Heaney D, Pritchard D (1992) Strategic Environmental Assessment. Earthscan, London
Tomlinson P (1999) Strategic Environmental Assessment for Transport in the UK. Paper presented at the OECD/ECMT conference on SEA for Transport, Warsaw 14-15 October 1999. TRL Ltd, Crawthorne
Tomlinson P (2002a) New approaches to environmental assessment of multi-modal transport schemes. paper presented at the IHT Alan Brant National Workshop, 16 April 2002. Internet address: www.trl.co.uk/static/environment/env_IHT02.pdf, last accessed 17.12.2003
Tomlinson P (2002b) Multi-modal environmental assessment. Internet address: www.trl.co.uk/static/ environment/env_Waterfront01.pdf, last accessed 17.12.2003
UK Government Strategy Unit (2002) Impact assessment and appraisal: guidance checklist for policy makers. Internet address: http://policyhub.gov.uk, last accessed 17.12.200

7 Implementing SEA in Germany

Thomas Bunge

German Federal Environmental Agency, Germany

7.1 Introduction

This chapter gives an overview of the activities to adapt German law and practice to the requirements of the SEA Directive and reflects the author's own personal view. Much of this work is still in progress; however, as far as local development planning and spatial planning are concerned, the necessary legislation at the federal level has recently been passed. Another important set of legal provisions, the amendment to the (Federal) EIA Act, only exists, as yet, in the form of a government bill. It is expected, though, that this will also be adopted during the next months.

As in many other fields of environmental policy, the implementation of the SEA Directive in Germany is faced with a special situation: the country's federal structure. Under the Federal Constitution (Basic Law), the federation is not entitled to transpose all the requirements of the Directive into national law. Rather, only part of the subject-matter falls within its legislative competence. As far as the rest is concerned, the 16 federal states ("*Laender*") will have to develop their own legal provisions. This is due to the broad range of plans and programmes covered by the SEA Directive which can be grouped in three categories: some of them, e.g. local development plans, may be dealt with comprehensively in federal legislation (e.g. in the federal Building Code); for others, like spatial plans at *Land*-wide and regional level or water management plans, the federation may only enact framework legislation, while the details must be left to *Laender* provisions. A third category of plans and programmes, such as general planning and route determination of *Laender* highways and roads, falls into the exclusive legislative competence of the federal states. Consequently the implementation of the SEA Directive requires a lot of co-ordination and co-operation between the federal and the *Laender* levels.

Implementing Strategic Environmental Assessment. Edited by Michael Schmidt, Elsa João and Eike Albrecht. © 2005 Springer-Verlag

In addition, the areas of planning mentioned in Art. 3 of the Directive are dealt with, in Germany, in a number of different acts,[1] so that it was discussed, as one option, to adapt each of these "special" provisions to bring them in line with the European requirements. However, finally, the federal government decided to take a slightly different approach: it is now intended, basically, to amend the federal EIA Act, adding to it, in particular, provisions on the range of application of SEA, the procedural steps, the content of the environmental report, the role of the SEA results in the decision-making context, and monitoring.

The EIA Act will, however, not be the only one that deals with SEA in federal law. In its present state, it already contains a "subsidiarity" clause stating that it will only be applicable as far as there are no *other* legal rules laying down the same or more stringent requirements for EIA. That has made it possible, during the 1990s, to develop specific EIA rules for particular subject areas, for instance for industrial projects or for mining activities. In future, the act will include a similar subsidiarity clause referring to strategic assessment. Consequently, other acts may also contain legal provisions on SEA. These will, however, only be valid if they do not fall below the standard laid down in the EIA Act. Such special rules will then leave no room for applying the EIA Act simultaneously. The most important provisions of this kind have been laid down, in June 2004, in the Building Code and the Spatial Planning Act (see Sects. 7.3 and 7.4).

The *Laender* will probably proceed in a similar way. Most of them have adopted their own acts on EIA (which were necessary for particular kinds of projects mentioned in the EIA Directive, such as *Land* highways, cable railways, specific agricultural projects, etc., for which the federation has no legislative competence), but in substance these are fairly similar to the federal EIA Act. It is therefore likely that the future amended text of that last-mentioned act will again serve as a model for the *Laender* Acts on EIA. In addition, however, the *Laender* will also have to modify their Acts on Spatial Planning. The federal Spatial Planning Act lays down only framework rules which need specification at the level of the individual federal states.

As yet, however, no draft of such *Laender* provisions is available. This chapter will therefore (in Sects. 7.2 to 7.4) be confined to the federal level and give a brief outline of the intended amendments to the EIA Act, of the recent changes in the Building Code and those in the Spatial Planning Act. This legislation will not be described in full, but the overview will only cover rules that either specify the European provisions or lay down more stringent requirements. Subsequently, the

---

[1] For example in the federal Highways Act, the Federal Immissions Control Act, the Waste Management Act, the federal Spatial Planning Act, the *Laender* Spatial Planning Acts, the federal Building Code, the federal Water Management Act and the *Laender* Water Acts. Although in Germany plans or programmes exist for most of the sectors mentioned in Art. 3 para 2 of the Directive, there are some of these sectors (such as telecommunications) for which no planning requirements have been laid down in legal or administrative provisions. These subject areas may, however, be dealt with in comprehensive planning (*Land*-wide and regional plans and local development plans).

implementation of the SEA Directive at the level of planning practice will be addressed in Sects. 7.5 to 7.8.

7.2 Intended Amendments to the EIA Act

7.2.1 Plans and programmes requiring SEA

The draft amendment of the EIA Act[2] contains rules, first, on the plans and programmes for which SEA will be mandatory. In this context it follows the approach of the SEA Directive. A plan or programme will have to be subject to SEA if it is required by legislative or administrative provisions, and

- either belongs to one of the categories set out in Art. 3 para 2 of the SEA Directive (agriculture, water management, waste management etc.) and sets the framework for future development consent of projects for which an EIA is required,
- or will be subject to an assessment under the Habitats Directive.

The draft amendment specifies, however, what it understands by "setting the framework" for future development consent. This clause does not only refer to plans and programmes that are strictly binding for the authorities, but also those that (only) have to be *taken into consideration* when deciding on the development consent. This is important in the context of German law since many projects requiring an EIA are being authorised in a so-called plan approval procedure (*"Planfeststellungsverfahren"*) where the competent authority is granted far-reaching discretionary powers. In these cases the final decision on the project is only outlined vaguely in the relevant legal provisions, and usually not all plans and programmes relating to the project area will be strictly mandatory. Even if they are, they do not, as a rule, stipulate very detailed requirements. Non-binding plans, on the other hand, set out preferable conditions for development that will have to be weighed and balanced against other interests and concerns when deciding on the authorization of a project. SEA will be mandatory for both these plan categories.

Under German law, there are some plans and programmes that *may* contain clauses setting the framework for future development consent of projects, but do not necessarily do so in each case. These instruments will only be subject to SEA if the competent authority intends to include "framework-setting" provisions in a specific plan or programme.

In a number of fields, different levels of plans exist. As an example the transport planning system may be mentioned, as it poses some special problems. For the whole of Germany, a *Federal transport plan* lays down which transport projects (in particular, highways, waterways, railways and airports) should be realised

---

[2] Federal Government Bill to Introduce Strategic Environmental Assessment – Act to Implement Directive 2001/42/EG, Parliamentary reports (*Bundestags-Drucksache*), 15/3441, Art. 1.

during the next years, provided they can be financed. The question whether this kind of plan should be subject to SEA was discussed rather controversially during the legislative process. Since there is, under German law, no *legal* obligation to prepare Federal transport plans, it was argued that the SEA Directive did not cover them. Eventually, though, it was decided to include this category in the list of plans and programmes for which the draft amendment to the EIA Act will make SEA mandatory. The Federal Ministry for Transport will, however, be entitled to lay down, by ordinance, *special rules* for SEA in these cases. Those may deal with the assessment procedure, the environmental report, the way to inform others of the decision on the plan, and monitoring.

Before any projects mentioned in a Federal transport plan can be constructed, there will be usually another two-tiered procedure: First, the route or site of the specific project will be determined in rather broad terms. In the case of federal highways and waterways this determination is binding in principle, while for other projects it will be taken in a special regional planning procedure ("*Raumordnungsverfahren*") and its result will only be recommendatory. Secondly, a plan approval procedure will follow, dealing with the project in detail and finally deciding whether or not to authorise the project. This second procedure will thus lead to a "development consent" in the terms of the EIA Directive. On the other hand, the question whether the first should now be seen as a planning procedure falling within the range of SEA has led to some argument. In the past, German law has required, as a rule, an *environmental impact assessment* for these route determinations or site recommendations because they precede development consent decisions. This was due to the fact that at the time there was no legal basis for SEA as a plan-related procedure. Now, under the SEA Directive, it might be more appropriate to subject route determinations as well as site recommendations to SEA. The draft amendment, however, has not taken up this recommendation (which, anyway, would not change much in terms of the actual assessment requirements). Instead it modifies the legal basis for EIA in special regional planning procedures (*Raumordnungsverfahren*) to some extent: At present the decision whether or not to make EIA mandatory in such procedures cases rests with the *Laender*. In future, the obligation to carry out an EIA at this level will be laid down in the federal EIA Act.

As to the plans and programmes referred to in Art. 3 paras 3 and 4 of the SEA Directive, the draft amendment to the EIA Act will make an SEA mandatory for them only if they are likely to have significant environmental effects. This will have to be determined, on a case-by-case basis, in a screening process.

7.2.2 Procedural Requirements

One of the political principles underlying the implementation of the SEA Directive in Germany – and stressed by both the federal and the *Laender* governments – is that, as in the case of EIA, no new or "separate" administrative procedures for SEA should be developed. Rather, the existing procedures laid down in German law for planning and programme-making should be amended in order to include the steps required by the SEA Directive. The draft amendment to the EIA Act therefore only specifies which additional steps will be necessary in future when applying such procedures. In that it follows the SEA Directive closely. However, it specifies a number of provisions which have been left rather general at the European level. The more important of these are outlined below.

Public Participation: Under Art. 6 para 4 of the SEA Directive, Member States "shall identify the public for the purposes of paragraph 2", i.e. for public participation. Under the envisaged amendment to the EIA Act, the draft plan or programme, the environmental report, and possibly other information will have to be laid open for inspection by the *general public*,[3] while *the public whose interests will be concerned* by the plan or programme may submit their opinions. This will allow for rather a broad participation of the public; still, anyone wishing to send comments to the authority will have to state his or her special (legal, economic, or other) interests which may be affected. The draft amendment here extends the rules on public participation in EIA which have existed since 1990 to the plan and programme level. Similarly to the EIA provisions, it also requires that an appropriate time – at least one month – must be granted for such comments in the SEA procedure.

SEA and Decision-Making: Furthermore, the draft amendment to the EIA Act adds some detail to the provision that the information gathered during the procedure is to be "taken into account" in the final decision on the plan or programme (Art. 8 of the SEA Directive). This has to do with the fact that under the existing EIA Act the competent authority will, in EIA cases, have to take the environmental impacts of the project into account in two steps: It will have (1) to prepare a *summary of the environmental impacts* based on the information submitted by the proponent and the comments received from the public and from other authorities, and (2) to *evaluate these impacts* using the yardstick of the relevant environmental legislation (e.g. the Federal Immissions Control Act or the Water Act of the respective *Land*). Art. 12 of the EIA Act expressly demands to carry out this evaluation "with a view to an effective environmental precaution".

The draft amendment follows this pattern also in the case of SEA. However, in the context of EIA, practitioners have always had their doubts whether the clear distinction made between the description of the facts ("summary") and its subsequent evaluation can be realised in practice. As far as SEA is concerned it is seen as impossible to divide the planning process (which necessarily includes evaluative elements from the preparation of the planning concept to the final decision on the plan) into a "descriptive" and a "normative" phase. Consequently, the environmental report will in each case already contain value statements. Once the involvement of the public and other authorities has come to an end, therefore, it will be the task of the competent authority not to begin the evaluation, but to *finalise* it

---

[3] The term "the public" includes also associations, organisations and groups (sec. 2 para 6 of the draft amendment).

before the decision to adopt or reject the plan or programme can be taken. The draft amendment to the EIA Act is worded accordingly.

In SEA, the basis for the evaluation is to be found again in the "relevant legal provisions". These have also to be applied with a view to an effective environmental precaution. This is important since the German legislation on aims, principles and content of plans and programmes is formulated rather broadly, only stating general requirements and leaving much leeway for interpretation by the competent authorities.

Once the environmental effects of the plan or programme have been identified, described and evaluated they will be weighed and balanced with other concerns in the decision-making process. Regarding this step, the draft amendment uses the same terms as Art. 8 of the SEA Directive in stating that the SEA results will have to be taken into account. It does not require expressly applying the precautionary principle here as well. However, in a number of cases the relevant special legislation (e.g. the Water Management Act) lays down as a general objective that all measures envisaged (including plans or programmes) should aim at furthering a sustainable development. In addition, more specific provisions may apply describing "precautionary" aims of the plans and programmes in question.

7.2.3 Tiering

Where plans and programmes form part of a hierarchy, Art. 4 para 3 of the SEA Directive allows for special provisions on tiering "with a view to avoiding duplication of the assessment". This topic is also addressed in the draft amendment to the EIA Act: The competent authority will, as a rule, have to determine during the scoping phase at which particular level of the overall planning process specific environmental effects should mainly be assessed. At subsequent planning levels, the respective SEA procedures may then deal only with "additional and other relevant environmental impacts", besides updating the information from previous SEAs and discussing it in a more detailed way. In such cases, also, the EIA required at a later stage in any development consent procedure for a project may be simplified in the same way.

Since the competent authority will have to address the topic of tiering already at an early stage – when determining the scope of SEA – it is in a position to discuss the matter with the other concerned planning authorities. This may even lead to an agreed "division of work" between them. However, each particular SEA will, also in such cases, have to conform to the European requirements.

Under Art. 5 para 3 of the SEA Directive, "relevant information on environmental effects of the plans and programmes which has been obtained at other levels of decision-making" may be used for preparing the environmental report. This option, too, has been included in the draft amendment to the EIA Act, which however adds that such information can only be utilized if it is as specific as required for the actual SEA, and furthermore is up to date so that it still gives a valid picture.

7.2.4 The Environmental Report

As to the environmental report, the draft amendment to the EIA Act expressly mentions all the items laid down in Annex I of the SEA Directive. In this context, a procedural simplification is worth noting: The amendment allows the authority to include the environmental report also, at a later stage, in the reasons and considerations on which the plan or programme is based. This leads to some reduction of work since German planning law usually states that, in the preparation procedure, a draft of these reasons and considerations will already have to be laid open for public inspection. However, if the authority acts in this way it will have to reconsider and probably amend the environmental report after the public and the other authorities have given their comments. If the environmental report is kept, on the other hand, as a separate document no such updating will be necessary. In any case, the draft reasons and considerations for the plan or programme will have to be brought in line with the final decision taken by the planning authority.

7.3 The Amendment to the Building Code of 24 June 2004

As mentioned, the Building Code and the Federal Spatial Planning Act have been modified in June 2004[4] so that both now include provisions on SEA. The new requirements have become effective on 20 July 2004.

The amendment to the Federal Building Code is a bit more specific than the draft to bring the EIA Act in line with the European requirements. This is due to the fact that the Building Code deals with only two kinds of plans at the local level: the zoning plan (*Flächennutzungsplan*) which covers the whole area of a town or city, gives an outline of the future land use, and binds only public authorities, and the building scheme (*Bebauungsplan*) which is prepared in most cases only for a particular area of the municipality, sets out the land use in detail, and is legally binding for everyone. Both these plans fall under the categories "town and country planning" and "land use" mentioned in Art. 3 para 2 of the SEA Directive. The Building Code does not contain any "subsidiarity clause" relating to SEA so that its provisions are in each case binding for the municipalities.

7.3.1 Range of Application of SEA

The Building Code makes SEA mandatory basically for all zoning plans and building schemes. The only exception are specific plans not affecting the outline of planning as it already exists, i. e. plans only laying down the status quo. For these, an SEA is not necessary

---

[4] Act to Adapt the Building Code to EU Directives of 24 June 2004, Federal Law Gazette, part I, p.1359, Arts. 1 and 2.

- if they do neither set the framework for projects requiring an EIA nor authorize such projects, and
- if there is no clue that a special protection area listed under the Wild Birds Directive or a special area of conservation listed under the Habitats Directive may be impaired.

Due to these provisions there will be no need for a full screening process in local development planning. In this respect, the amendment took into account the criticism of municipalities which saw a screening test as rather costly and time-consuming in relation to its outcome, and preferred to carry out SEA in practically all cases.

7.3.2 SEA Procedure

The procedural requirements of SEA will also be "added on" to the existing planning procedure as laid down in the Building Code. This procedure is basically the same for zoning plans and building schemes. The new provisions follow the SEA Directive closely, specifying only some of its rather general requirements.

In local development planning, the draft plan and the environmental report will be laid open for public inspection for one month, and comments will have to be submitted to the planning authority within this time. Everyone is entitled not only to inspect these documents, but also to comment on them. These provisions giving each individual the right to participate in the process have not been developed in the context of the SEA Directive, but have been in place, basically, for nearly thirty years. They are somewhat broader than those in the draft amendment to the EIA Act which allows comments only from those whose interests are concerned by the plan.[5]

It is interesting to note that the Building Code expressly mentions the possibility for the authority to use electronic information technologies, e.g. the internet. Due to the fact that many people have no access to these technologies, though, this can be used just as a *supplementary* way to provide information, and it will still be indispensable to prepare, as well, paper copies of the draft plan and the environmental report which the public can inspect.

7.3.3 The Environmental Report

Under the Building Code, the environmental report will be prepared, first, in a draft form for public inspection as well as for the participation of other authorities and, if relevant, other countries whose environment may be affected by the plan. After these procedural steps have been completed, the planning authority will have to revise, and possibly modify or supplement, the report. This will then be in-

---

[5] C.f. above in 7.2.2 on *Public Participation*. The draft amendment to the EIA Act allows, however, provisions in special legislation that award the right of comments to the general public.

cluded in the reasons and considerations on which the proposed plan is based. Thus it will be subject to the final decision on the plan by the municipality. Consequently it will then have to be reconsidered and, if necessary, amended once again before it is made available to the public, the involved authorities and the other affected countries.

7.3.4 Tiering

The new legislation also deals with tiering. Local development planning forms, in Germany, a part of the "comprehensive" land use planning system (which also includes spatial planning at the *Land*-wide and regional levels). Thus, in carrying out an SEA for a zoning plan or a building scheme, the municipality will, as a rule, have to use the SEA results of any other relevant comprehensive plan, and will have to restrict its own assessment work to those environmental effects that have not been addressed sufficiently in these other results. Consequently, when preparing a zoning plan, the municipality is usually obliged to utilize the information gathered in the SEA procedures of the *Land*-wide spatial plan, the regional plan, and any building schemes. Moreover, it will have to ascertain to what extent relevant information is available from project-related EIA.

Tiering requirements apply, as well, to cases where two comprehensive plans are being elaborated simultaneously. This may happen especially at the local level. For instance, if a building scheme is envisaged which deviates from the zoning plan it is necessary to modify the latter as well, in order to avoid conflicting planning provisions. In all such "parallel" cases (concerning different comprehensive plans at any level) the Building Code now makes it necessary for the competent authority or authorities to determine at an early stage which of the procedures will include the full SEA, and which will use the relevant information gathered in that process.

7.3.5 Monitoring

The provision on monitoring in the Building Code closely resembles that of the SEA Directive, and is similarly vague. Thus the ways and means as well as the frequency of monitoring are left to the *Laender* and the municipalities. The only specification is that the authorities will be *obliged* to use existing mechanisms for monitoring purposes. In order to support the competent agency in this respect, other authorities are required, once the planning procedure has been completed, to inform it of any significant environmental effects caused by the plan.[6] This clause may possibly be useful, in particular, as far as monitoring of unforeseen adverse effects is concerned.

---

[6] This will not make it necessary for the other authorities to begin their own monitoring activities as well. Rather, they will only have to provide any information they already may have to the municipality.

7.4 The Amendment to the Spatial Planning Act of 24 June 2004

As the federal Spatial Planning Act belongs to the category of framework legislation, the amendment to this act restricts itself to rather general requirements which the *Laender* will then have to implement in their respective Planning Acts. It does not specify the provisions of the SEA Directive. Still, SEA will be included, as at the local level, in the procedures laid down for *Land*-wide and regional planning.

One special aspect worth noting concerns spatial planning in the German exclusive economic zones of both the North Sea and the Baltic. In these areas, there is considerable interest at present to construct huge wind farms. They are increasingly being used for other purposes as well, e. g. for gas pipelines. As a response to these developments, the Spatial Planning Act makes it mandatory to draw up particular spatial plans at the federal level for the exclusive economic zones. In this context an SEA will have to be carried out.

7.5 Links between the new SEA Legislation and Existing Assessment / Planning Tools

The new legislation on SEA will make planning processes more transparent, and will furthermore provide a procedural framework for existing tools such as the rules on "impairments to nature and landscape"[7] as well as the assessment under the Habitats Directive. Both these instruments deal with impacts of projects or plans on nature and landscape. Contrary to SEA they do not set out a procedural basis for the assessment of such effects, but substantial rules for the decision on the activities: Adverse impacts will have to be prevented as far as possible. If that cannot be done in a sufficient way, the authorities will not be able, in some cases, to authorize the project or adopt the plan. In other cases they will have to require the proponent or planning agency to adequately compensate the likely impacts (by improving the environmental quality of other areas); if that is not possible, the project authorization or approval of the plan cannot be granted.[8]

The amendment to the federal EIA Act allows the authorities to apply these tools *within the framework of SEA*; the Building Code even makes such a "comprehensive" application obligatory. It is important, however, to bear in mind that SEA cannot replace the other tools, but should allow for their proper application within the assessment procedure: In particular, it will be indispensable for the environmental report under Art. 5 of the SEA Directive to give a *separate account* of

---

[7] The so-called *Eingriffsregelung* under the Federal Nature Protection Act and the Building Code. As far as plans and programmes are concerned these rules are applicable, as yet, only in local development planning.

[8] The features of the "impairment" rules and of the assessment under the Habitats Directive are similar to some extent, but differ in many details. They cannot be described here (on this topic, see, for instance, Bunge 2004).

the effects the plan may have on nature and landscape, and on Natura 2000 habitats. This is necessary in order to ensure that the decision on the plan or programme is adequately based (also) on the specific provisions mentioned.

7.6 Methodology

Implementation of the SEA Directive means, of course, not only adapting the national legislation to the European requirements. In addition, each member state will have to develop the structures, capacity and methods needed for an effective SEA.

7.6.1 General

At a first glance it seems that it should not prove too difficult for the authorities to assess the environmental effects of plans and programmes, since under the existing planning system they have had to do just that for many years. The relevant legislation requires them to take all relevant concerns into account that are likely to be affected by the plan, and to weigh and balance them carefully against each other. This principle, if taken seriously, makes it indispensable also to predict and evaluate any significant environmental impacts likely to be caused by the draft plan or programme.

Looking more closely at the actual practice, however, one will find that SEA methodology has as yet not been developed very far in Germany. The ways and means in which environmental concerns are being taken into account at present differ widely. The methods applied in EIA, i. e. at the project level, can only be used for SEA to a limited extent, as the data available in the planning process are usually less detailed, the plan covers a comparatively large area, and it will often take considerable time before it will have been fully implemented. Also, the consideration of alternatives plays a different role in SEA: At the planning level, other (more general) kinds of options will have to be addressed than at the project level. For instance, in general transport planning it is important to discuss different modes of transport (e.g. by road, by rail, or by water), while in the procedure to authorize a particular section of a highway, alternative locations within a narrow geographical area and details of road design and construction will have to be addressed.

Thus, in order to implement the SEA Directive effectively it will be important to adapt the existing EIA methodology to the requirements of SEA, or to develop new methods. Work of that kind has been in progress for some years now in various areas of planning and programme-making (see Sect. 7.8). In particular, a German research institution concerned, inter alia, with highway planning and construction, "Society for Research in the Field of Roads and Transport (*Forschungsgesellschaft für Straßen- und Verkehrswesen*)" is preparing general recommendations for SEA in transport planning.[9] These also contain methodological hints (see as-well Chap. 40). However, in many other areas SEA methodology is still less advanced.

7.6.2 Tiering

The relationship of the assessments carried out at various levels of a planning system is another matter where not much experience has been gained as yet in Germany. Thus the SEA legislation on tiering (cf. Sects. 7.2.2 and 7.3.4) poses some practical problems. Although it seems possible, in principle, to avoid duplication of work by taking over the SEA results from a previous level, it is important that each SEA should deal adequately with all environmental impacts relevant for the specific plan or programme it is intended for. Considering that such plans and programmes are not always drawn up in the order envisaged in the relevant legislation, but "lower" level plans frequently precede those at "higher" level, and that they will possibly be amended more often than the latter, it is doubtful to what extent the new tiering rules will save time and resources.

7.6.3 Monitoring

In actual practice, the question is basically still open how the environmental effects of the realisation of a plan or programme can best be monitored. In this respect it is unclear, too, how far it will really be efficient to use existing mechanisms, as the available general monitoring systems (e.g. for air quality or water quality purposes) often do not provide the specific data necessary for the purposes of the SEA Directive. In particular, it will frequently be difficult to establish whether an observed effect has been caused by implementing the plan or programme in question, or whether it can be traced back to other causes. Some planning authorities have already developed specific processes of their own for the monitoring activities required under the SEA Directive (cf. Frenk 2004).

7.7 Data Requirements

At present it cannot be ascertained to what extent the authorities will require "new" or additional information for SEA. As mentioned, they are already obliged under German planning law in its present form to predict and evaluate the relevant environmental impacts of the plan or programme. In each planning process they

---

[9] Instructions on strategic environmental assessment of plans and programmes in the transport sector (*Merkblatt zur Strategischen Umweltprüfung von Plänen und Programmen im Verkehrssektor*), draft, 2004.

will therefore have to collect and adapt a large amount of data, even if they do not conduct a "formal" SEA.

Nevertheless, it is doubtful whether the planning authorities, in actual practice, deal adequately with all aspects of the environment – especially with items such as impacts on biodiversity and on human health,[10] and with changes to the interrelationship of the environmental factors.[11] In addition it will not be always easy for them to specify the relevant environmental objectives as required in Annex I, lit. e of the SEA Directive, since this clause comprises many different provisions, principles, resolutions, etc. It is by no means certain that each planning authority is aware of all pertinent objectives laid down at international and European Community level.

It remains to be seen, therefore, how well the authorities will be able to cope with the data requirements set out in Annex I of the SEA Directive. Those that have already carried out SEA on a voluntary basis in the past[12] will probably be in a better position.

7.8 Support of the Implementation and Quality Control

Both the development of legal rules for SEA and the practical application of this tool have been and still are being supported, inter alia, by various research projects. Of these, the following may be mentioned:

In two comprehensive studies (Spannowsky et al. 2000; Ginzky and Kraetzschmer 2001), proposals for the legal implementation of the SEA Directive in Germany were elaborated.[13] Also, an expert commission set up by the Federal Ministry of Transport, Building and Housing provided detailed advice on adapting the Building Code to the SEA Directive requirements (*Bundesministerium für Verkehr, Bau- und Wohnungswesen* 2003). Several other research projects concerned themselves with specific legal matters, e.g. the question for which plans and programmes under German law the Directive requires an SEA (Hendler 2002), or the definition of the term "environmental protection objectives" in Annex I, lit. e of the SEA Directive (Sommer et al. 2002). Eberle et al. (2004) developed detailed proposals for legal rules and administrative provisions implementing the SEA Directive, as well as for (recommendatory) guidance. These are based on experience gathered by applying the SEA Directive's requirements to three actual regional planning cases (for a brief overview see Box 7.1).

---

[10] However, in local development planning, health effects caused by air pollution and noise are probably dealt with in all relevant cases.

[11] In German EIA practice between 1990 and 1998, Wende (2000 pp.192-197) noted that in many cases not all environmental impacts have been addressed properly.

[12] In Germany, some 200 towns and cities have decided since the 1980s to apply the principles of EIA in their own communal activities, in particular in local development planning.

[13] Both these projects were completed before the adoption of the Directive in 2001; their results are thus based on draft texts of the Directive.

The results have been used in developing the amendments to the EIA Act set out in Sect. 7.2. In the context of local development planning, the German Institute for Urban Studies (*Deutsches Institut für Urbanistik*) conducted an "experimental planning game" for one and a half years in order to find out how the municipalities would be able to cope with the new SEA requirements set out in the (then intended) amendments to the Building Code. In this game, eight towns, cities, and districts participated. The final report of this game (Bunzel et al. 2004), besides stating the participants' favourable view of the proposed legislation in general, also includes a number of recommendations to modify particular provisions. Many of these suggestions have subsequently been accepted by Parliament in the Act of 24 June 2004 (cf. Sect. 7.3).

Another important field of research is the development of suitable methods for SEA. Various projects have dealt with this subject in relation to regional planning (Schmidt et al. 2004), spatial planning for offshore wind power utilization (Koeppel et al. 2003) and transportation planning (Koeppel et al. 2004). As to a particular aspect, Giegrich and Knappe (2004) have specified monitoring indicators and criteria in the context of waste management and water management planning. Also, Barth and Fuder (2002) have elaborated recommendations for monitoring.

The quality of implementing the SEA Directive is being controlled at present mainly in the framework of co-operation between the Federal and the *Laender* environmental ministries. In particular, working drafts of federal legislation are being discussed at this level. Also, at the request of any *Land,* its own legislation may be dealt with in this context. In preparing legal rules for SEA, the EU Commission's *Guidance on the Implementation of Directive 2001/42/EC on the assessment of the effects of certain plans on the environment* (2003) is being taken into account as well. Furthermore, the co-operation of the EU Member States at Commission level allows comparing the various countries' approaches in adapting their legislation and planning practice.

Quality control of the application of SEA in actual practice will probably be left mainly to the competent planning authorities. At present the legislation mentioned contains no provisions on this subject.

Box 7.1. German Case studies

In 2002, the Federal Environmental Agency commissioned a research project in order to ascertain whether, and to what extent, the requirements of the SEA Directive were met already in actual regional planning. For requirements not laid down in the existing spatial planning acts, the project should develop proposals for new legislation, for administrative regulations or for "informal" guidance. This was to be done on the basis of empirical data from planning activities in different *Laender*.

The project was carried out by *D. Kraetzschmer* (Hanover), *Ch. Jacoby* (Munich), *D. Eberle* (Tübingen), and *A. Naeckel* (Rostock). The procedures for preparing three regional plans,

- *Regionales Raumordnungsprogramm* Braunschweig (Lower Saxony),
- *Regionalplan* Mittelfranken (Bavaria),
- *Regionaler Raumordnungsplan* Westpfalz (Rhineland Palatinate)

served as test cases. As each of these procedures was in a different stage, SEA requirements could be considered in various planning phases. The Braunschweig planning process had only just begun, whereas the Westpfalz plan was already rather far advanced. In the Mittelfranken case the planning activities were delayed to some extent (due to unforeseen problems) so that it could only partially be used for the purposes of the research project; however, it was possible to include the updating of the plan chapters "Water" and "Preferential sites for wind power installations" in the research project.

It was found that the authorities in all three cases had no major difficulties with applying the Directive's requirements. Still, in relation to a number of procedural and substantial details, they saw future additional guidance as helpful. Based on the experience gathered from the cases, the research project resulted in a large number of recommendations for conducting SEAs. These refer to

- the scoping process,
- tiering,
- the assessment of alternatives,
- the environmental report (methodological suggestions, content items, utilization of the report when preparing the reasons and considerations on which the plan is based),
- details of public participation,
- the decision on the plan, and
- the requirement to make the plan as well as other information publicly available, and
- monitoring.

The researchers recommend that only some of these proposals should be implemented in the form of legal or administrative provisions. For the others, non-binding "rules of good planning practice" should be developed.

Details can be found in the research report by Eberle et al. (2004)

7.9 Conclusions and Recommendations

Although there has been, and still is, some controversy regarding the legal implementation of the SEA Directive in Germany, two of the relevant federal acts have now been amended in order to meet these European requirements. The additional federal and *Laender* legislation necessary to make SEA mandatory for all relevant plans and programmes will probably be adopted in 2004 and 2005.

It may be somewhat more difficult to ensure an ambitious and effective implementation also at the practical level. In spite of the fact that German planning law has made it mandatory for many years now to predict and evaluate, inter alia, the relevant environmental consequences of plans and programmes, problems and shortcomings in this context are likely, since the SEA Directive requires the planning authorities to deal with the effects on the environment in a special way: These have to be assessed separately from other relevant considerations, and it is important, as well, to look at the impacts not just individually, but also comprehensively. Consequently, methods for predicting and evaluating environmental impacts at the plan and programme level will have to be further developed. As far as evaluating the effects on the environment is concerned, *landscape planning* (in which rather much experience has been gathered in Germany) may possibly play an important role as each landscape plan is required under the existing legislation also to evaluate both the existing state of nature and landscape and its expected development. Landscape plans and programmes have to be elaborated, in principle, at the *Land*-wide, regional, and local levels for the whole area of Germany.

In addition, it will be useful to prepare SEA recommendations and guidelines for the planning authorities. These should address, in particular, details of scoping, the content of the environmental report, and monitoring requirements. Some work of this kind has already begun.

Another important item for research, as well as for practical guidance, is the role the assessment results play in the final decision-making process. The value of SEA as an instrument of environmental policy depends on how far its results will actually influence the decision to adopt a plan or programme. Thus the planning system should warrant in each case that the competent authorities consider the environmental concerns identified through SEA in an adequate way, from a cautionary approach and with a view to promoting sustainable development. Consequently, the implementation of SEA Directive at the practical level should not only guarantee an efficient and transparent SEA procedure leading to a reliable prediction and an appropriate evaluation of environmental impacts, but should also make sure that the likely environmental impacts of plans and programmes will be seen as important factors in each decision.

Finally, as far as the relationship between SEA and decision-making is concerned, there is an obvious difference between EIA and SEA: While at the project level the proponent and the competent authority are, as a rule, different subjects, at the plan and programme level the competent authority itself will also be responsible for the assessment and its results. As yet, this problem has not received much attention during the implementation process. However, it will have to be tackled if

SEA is expected to lead to results that are as impartial as possible, and that play a decisive part in the preparation and adoption of plans and programmes.

References[14]

Barth R, Fuder A (2002) Implementing Art. 10 of the SEA Directive 2001/42/EC. Research report commissioned by the EC Commission's IMPEL network. Darmstadt

Bundesministerium für Verkehr, Bau- Wohnungswesen (ed) (2003), Novellierung des Baugesetzbuchs. Bericht der Unabhängigen Expertenkommission (Amending the Building Code. Report of the Independent Expert Commission)

Bunge T (2004) Zur Harmonisierung von Umweltverträglichkeitsprüfung, strategischer Umweltprüfung, FFH-Verträglichkeitsprüfung und Raumverträglichleitsprüfung (On harmonising the instruments of environmental impact assessment, strategic environmental assessment, the assessment under the Habitat Directive, and spatial impact assessment). In: Hendler R (ed) Die strategische Umweltprüfung (sog. Plan-UVP) als neues Instrument des Umweltrechts (SEA as a new instrument of environmental law). Erich Schmidt, Berlin, pp 191-224

Bunzel A et al. (2004) Planspiel BauGB 2004. Bericht über die Stellungnahmen der Planspielstädte und Planspiellandkreise (Experimental planning game 'Building Code 2004'. Report on the comments by the participating towns, cities and districts). Deutsches Institut für Urbanistik, Berlin

EC – Commission of the European Union (2003) Guidance on the Implementation of Directive 2001/42/EC on the assessment of the effects of certain plans on the environment. Brussels

Eberle D (ed) (2004) Umweltprüfung ausgewählter Regionalpläne – Praxistest (SEA of selected regional plans –Test of Practical Application). Research report commissioned by the Umweltbundesamt

Frenk J (2004) Monitoring of the Environmental Effects of Local Development Planning – Approach of the City of Leipzig (in German). In: Bunzel A et al (ed), Monitoring und Bauleitplanung – neue Herausforderungen für Kommunen bei der Überwachung von Umweltauswirkungen (Monitoring and local development planning – new challenges for the municipalities in monitoring of environmental effects). Deutsches Institut für Urbanistik, Berlin, pp 87-98

Giegrich J, Knappe F (2004) Entwicklung geeigneter Indikatoren/Kriterien für die Überwachung der Umweltwirkungen bestimmter Pläne und Programme und Prüfung der Nutzbarkeit bestehender Überwachungsmechanismen (Development of suitable indicators/criteria for the monitoring of certain plans and programmes, and an examination of the possibility to use existing monitoring mechanisms). Research report commissioned by the Umweltbundesamt

Ginzky H, Kraetzschmer D (2001) Vorschläge für rechtliche Grundlagen und methodische Hilfen für die Umweltverträglichkeitsprüfung von Plänen und Entwicklung von Vorschlägen zur rechtlichen Verzahnung von UVP in vorgelagerten Verfahren mit UVP in nachgelagerten Verfahren (Proposals for legal bases and methodological aids for

---

[14] Note that legislation mentioned in the chapter is listed at the end of the handbook in the consolidated list of legislation (Appendix 2).

environmental assessment of plans and programmes, and development of proposals to link EIA in prior procedures with EIA in subsequent procedures). Research report commissioned by the Umweltbundesamt

Hendler R (2002) Der Geltungsbereich der EG-Richtlinie zur strategischen Umweltprüfung (The range of application of the EC Directive on SEA). Research report commissioned by the Federal Ministry for Environment, Nature Conservation and Nuclear Safety

Koeppel J (ed) (2003) Anforderungen an die Strategische Umweltprüfung im Rahmen der Ausweisung von Eignungsgebieten für Offshore-Windenergienutzung (Requirements for SEA in the framework of designating suitable areas for utilization of offshore wind energy). Research report commissioned by the Federal Ministry for Environment, Nature Conservation and Nuclear Safety (Ökologische Begleitforschung zur Windenergie-Nutzung im Offshore-Bereich der Nord- und Ostsee – Teilprojekt: Instrumente des Umwelt- und Naturschutzes: Strategische Umweltprüfung, Umweltverträglichkeitsprüfung und Flora-Fauna-Habitat-Verträglichkeitsprüfung, vol. IV)

Koeppel J (ed) (2004) Anforderungen der SUP-Richtlinie an Bundesverkehrswegeplanung und Verkehrsentwicklungsplanung der Länder (Requirements of the SEA Directive for federal transportation planning and transportation development planning at *Laender* level). Berlin, Umweltbundesamt (Texte No. 13/04)

Schmidt C (ed) (2004) Die Strategische Umweltprüfung in der Regionalplanung am Beispiel Nordthüringens (SEA in regional planning, taking Northern Thuringia as an example). Research report commissioned by the Federal Ministry of Education and Research

Sommer K (2002) Machbarkeitsstudie für ein Behördenhandbuch „Umweltschutzziele" in Deutschland (Feasibility study concerning a handbook for authorities on environmental objectives). Research report commissioned by the Umweltbundesamt

Spannowsky W (2000) Praxisuntersuchung und Expertise zu einer Umsetzung der europarechtlichen Umweltverträglichkeitsrichtlinien in das Raumordnungsrecht (Investigation of planning practice and expertise concerning the implementation of the SEA Directive in spatial planning law). Kaiserslautern

Wende W (2000) Praxis der Umweltverträglichkeitsprüfung und ihr Einfluss auf Zulassungsverfahren (The practice of EIA and its influence on licencing procedures). Nomos, Baden-Baden

8 Implementing SEA in Italy – The Case of the Emilia Romagna Region

Paola Gazzola[1] and Maristella Caramaschi[2]

1 Department of Civic Design, University of Liverpool, England
2 Consultant at Pordenone, Italy

8.1 Introduction

In Italy, attempts to start implementing the SEA Directive started well before the 2004 deadline. In March 2002 Law No. 39/2002 established that the Directive was to be introduced in the Italian legislative system by 26 of March 2003. Another Italian law (Law No. 284/2002) postponed this deadline to the 31 of December 2003. This deadline has again not been respected. The continuous postponing of the implementation of the SEA Directive at a national level is perhaps due to two reasons. Firstly, to the chronic delay with which Italy has started to deal with the environment; and secondly, to the willingness of the politicians of the centre-right government, to *empty* the entire existing legislation (concerning the environment) of its meaning and of its effectiveness. This may also be confirmed by the recent environmental legislations approved (in a hurried manner), which cumulatively are likely to make the application of the SEA Directive, ineffective at a national level (see Sect. 8.8). This situation has therefore decreased the level of general trust towards a national implementation of the SEA Directive and has motivated the regions to hurry the development of their own SEA legislations.

Since 1996, through an Act of Instruction and Coordination (Dpr 1996) EIA competences have been delegated to the regional levels. Regions received full EIA competences, after directly implementing the 1997 EIA Directive, amending the 1985 one (regions have also been delegated full planning competences in 1970s). Therefore, consistency between the European Union and Italy only exists through the regional governments. A national EIA framework is in fact still missing (Gazzola 2002). Furthermore, in 2001 the new Art. 117 of the Italian Constitution has been approved. According to this article, all levels of government (state, regional, provincial and local) are directly subjected to European acts.

Implementing Strategic Environmental Assessment. Edited by Michael Schmidt, Elsa João and Eike Albrecht. © 2005 Springer-Verlag

Regions (and the two Autonomous Provinces of Trento and Bolzano) in particular, are now allowed to participate in decisions concerning the development of EU legislation and to implement international agreements as well as EU acts. Subsequently, regions will directly implement the SEA Directive and legislation will therefore differ from region to region. Today, most of the Italian regions have developed their own EIA law. Some of these have also introduced SEA provisions anticipating the forthcoming SEA Directive. SEA requirements have in fact been introduced so far in Italy or through urban regional laws or through regional EIA laws. This has occurred in the regions of Emilia Romagna, Puglia, Liguria, Tuscany, Calabria, Umbria, Friuli-Venezia Giulia, Valle d'Aosta and in the two Autonomous Provinces of Trento and Bolzano (Chitotti 2002).

Analysing the implementation status of SEA in Italy, in theory means analysing the implementation status of SEA in the country's 20 different regions and in its two Autonomous Provinces (Trento and Bolzano). However, by looking at one of Italy's regions, i.e. Emilia Romagna, an attempt to describe national trends will be made. Fig. 8.1 shows a map of Italy and the location of the Emilia Romagna region.

A brief introduction of why the region of Emilia Romagna has been chosen as a case study to discuss the implementation status of SEA in Italy follows. Subsequently, the general planning context of Italy and of the region of Emilia Romagna, with references to the relationship between strategic planning and SEA and the existing planning instruments, will be presented. The case of the region of Emilia Romagna will be explored through an analysis of the region's planning law (i.e. Law No. 20/2000) and its environmental assessment procedure for plans and programmes (i.e. VALSAT). Finally, the implementation of SEA in the region of Emilia Romagna and in Italy will be considered, highlighting the opportunities and barriers for effective SEA application. Subsequently, the conclusions will be drawn and recommendations will be provided.

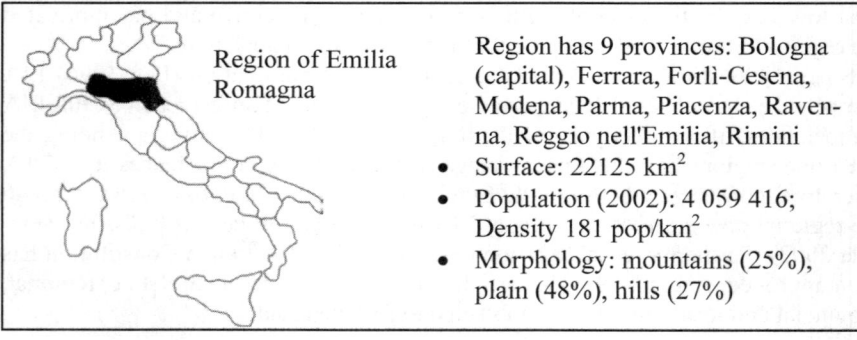

Fig. 8.1. Map of Italy and location of the region of Emilia Romagna

8.2 Using the Emilia Romagna Region to Explore the Implementation Status of SEA in Italy

The region of Emilia Romagna is one of the first regions that introduced the concept of environmental sustainability and strategic planning in its legislative planning system. It also defined environmental sustainability as the main principle upon which, territorial and urban planning processes must be developed and implemented. For these reasons, it is considered one of the most advanced regions at a national scale, and will therefore be used as a case-study in this chapter. Furthermore, the implementation of the SEA Directive represents a stopping point for the region of Emilia Romagna, where multiple directions for the introduction of the new EU regulation, can be taken. Currently, the region, the provinces and the municipalities must submit their plans to a "preventive environmental and territorial sustainability assessment (VALSAT), in consistency with national and community legislations" (Art. 5 of Law No. 20/2000). Governments must undertake the assessment throughout the plan's formulation and approval procedures, identifying the effects likely to arise from its implementation. This chapter also aims to analyse the relationship between the VALSAT and the SEA Directive, testing the performances and consistency of the first against the latter.

8.3 Context of Planning in Italy

Generally speaking, the Italian planning system is defined by the National Urban Act No. 1150, 1942 (promulgated by the King Vittorio Emanuele III), which is still in force today. Other laws and decision-making levels (regions in the '70s and provinces in the '90s) have been introduced, but without affecting the validity of the 1942 urban act and the planning principles upon which, the Italian system has been developed (Salzano 2003). Fig. 8.2 presents a simplified summary of the Italian planning system, with the different decision-making levels and their relative planning competences.

As shown in Fig. 8.2, Italy has an established planning system with different tiers of decision-making: national, regional, provincial and local. Each tier has planning competences and defined responsibilities. With the introduction of the regions, planning competences have been delegated to the regional level, whilst the national one – through the Ministry for the Environment and Territorial Protection, the Ministry for Public Works and the Ministry for Cultural Goods and Activities – still carries the responsibility for policies, sectoral plans-programmes and projects of a national interest, and for defining "acts of direction and coordination". Regions, however, are the government level with the main planning competences. As mentioned previously, each region has developed their own planning law and has distributed planning competences through the lower decision-making tiers. As a consequence, planning is practiced in different ways according to the region and relative regional law, taken in consideration.

Furthermore, despite the "ordinary" (state, regional, provincial and local) decision-making levels, in Italy there are also a number of sectoral agencies (e.g. ANPA, ARPA, APPA, park agencies or river basin agencies, etc.), which also have planning competences. These may be at a national level, as well as regional and provincial. For both systems – ordinary and sectoral – the Italian planning approach follows the hierarchical bond and the subsidiarity principles (Salvia and Teresi 1992, Salzano 2004). Whilst the first principle, links the planning instruments of different decision-making levels in such a way that, the plan of a lower tier develops the strategies and objectives defined in the plan of the upper-tier, but it cannot derogate, correct or amend it; according to the second principle, decisions should be taken at the lowest level consistent with effective action, decentralising the vertical power sharing (Jordan and Jeppesen 2000)

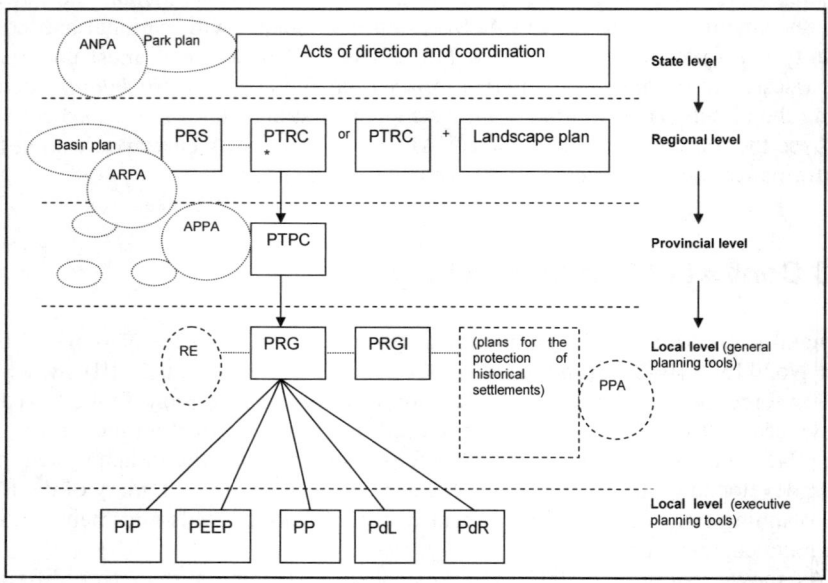

Fig. 8.2. Planning competences and responsibilities in the Italian system (simplified)

Regional level: PTRC*: territorial coordinated regional plan with a landscape component; PTRC + Landscape plan: like above, but in two different plans; PRS: regional development plan that comprehends an economic programming component
Provincial level: PTPC: territorial coordinated provincial plan;
Local level: RE: building regulation; PRG: general regulator plan; PRGI: intermunicipality general regulator plan; PPA: multi-annual implementation programme.
PIP: plan for productive settlements; PEEP: economic community building plan; PP: detailed plan; PdL: division plan; PdR: recovery plan
Sectoral agencies: e.g. ANPA: National agency for environmental protection; ARPA: Regional agency for environmental protection; APPA: Provincial agency for environmental protection

8.4 Relationship between Strategic Planning and SEA

The introduction of SEA requires the practice of strategic planning, which has been first introduced in Italy in 1995, with the Tuscany's Regional Urban/Planning Law No. 5/1995 (Caramaschi et al. 2003). The first generation of urban regional laws (after the institution of the regions in 1972) simply followed the structure of the National Urban Act, No. 1150, 1942. As mentioned previously, it was only in the mid 1990s that things have started to change (Salzano 2004). By taking into account the cultural debate on sustainable development, regional laws have started to include new approaches for managing urban and territorial transformations. This is the case for example, of the region of Tuscany in 1995, Liguria in 1997, Basilicata in 1999, Lazio in 1999 and Emilia Romagna in 2000.

One of the basic requirements for undertaking an SEA, rather than an EIA, is that the policy, plan and programme (PPP), must be strategic or include a strategic component. Otherwise, an SEA cannot be applied. Strategic planning and SEA are therefore two processes that are strongly linked: distortions or lacking elements of the first, consequently impact the latter (Caramaschi et al. 2003). For example, if a plan is lacking articulation according to a hierarchical scale of strategic objectives, but has rather indicated single project level actions, the SEA will only be able to consider the impacts of those actions on the environment. The potential and scope of SEA will be limited to an EIA. Submitting therefore "traditional" plans and programmes to a SEA, means simply adding a bureaucratic approval to the plan, e.g. a technical attachment. Finally, strategic planning and SEA must work in a synergetic way. SEA can only be applied, if strategic planning is intended as (Caramaschi et al. 2003):

- An instrument for structuring policies and for constructing consensus: through consensus, strategic planning may guide the research and development of alternatives, delimiting the range of possible consequences;
- An instrument for tracing future auspicial scenarios, emphasising trade-offs: strategic planning does not identify decisions, but represents a framework for decisions.
- A structured process open for interactions and capacity building, in which actions are developed through dynamic and interactive processes.

However, strategic planning implies a hierarchical conception of the role of public institutions; and practices and experiences have proved its inefficiency. Strategic planning was originally introduced in Italy between the 1950s and 1970s, during a phase of economic and urban quantitative growth, with the purpose of regulating land use (Ciciotti et al. 1997). From the 1990s new approaches to planning were being formulated, on the basis of models already experimented and applied outside of Italy, but which had originally been introduced in the country in the 1950s-70s. These new approaches therefore, acquired new definitions and considerations for research and experimentation and, by the end of the decade several Italian urban regional laws were developed, taking inspiration from strategic planning. The main features of strategic plans are not so much the consideration of "changing prevailing objectives [...] but rather the identification of specific objectives somehow verifiable" (Avarello 2000); and the guidance for action that the administration should provide, in order to achieve the objectives. Strategic planning, as considered in the (few) regional laws approved, is inspired from the following principles:

- The preparation of comprehensive frameworks with long-term perspectives (of 40-50 years): these must take into account complex variables and phenomenon, in order to provide a reference for the subsequent elaboration and implementation of structured plans with short-term perspectives.
- The central importance of interaction, transparency and participation, from the earliest phases of plan making, i.e. definition of the plan's objectives.
- The necessity to subject plans to constant analyses, predictions, assessments and monitoring.
- A new orientation and approach towards regulating policies and the development and implementation of policies within a context and vision of sustainability.

However, where the traditional planning (non-strategic) approach is still practiced and rooted in the system, the decision-making context is profoundly changing: the hierarchical and sectoral approaches are now being substituted or developed in parallel, with new forms of negotiated programmes – societies for urban transformation, territorial deals, programmes agreements, conferences of services and open administrations (*societa' di trasformazione urbana, patti territoriali, accordi di programma, conferenze di servizi e sportelli unici*) – with the purpose of coordinating public decision-making in a systematic way (Stanghellini 1999).

8.5 Planning in the Region of Emilia Romagna

Through its planning Law No. 20/2000 (general discipline for the protection and use of the territory) and the operating of its institutional asset, the region of Emilia Romagna has explicitly committed itself to establish with its citizens, a relationship based on the following principles: subsidiarity, institutional and administrative simplification agreements (horizontal integration and simplification of administration procedures). Law No. 20/2000 also defines objectives, contents and innovative instruments that the governments within the region, must pursue through planning. Furthermore, in order to regulate territorial transformations, the planning law acknowledges and distinguishes baseline information from forecast information. Law No. 20/2000 aims in fact at developing a planning system able to guarantee coherence between the characteristics and status of the territory and the plan's predictions, to achieve a balanced and agreed relationship between development and protection. Baseline elements as well as assessment ones will therefore be provided, supporting the plan's decision-making process. Every choice or decision within a plan causes or is subjected to impacts or to a chain of impacts. The law therefore aims to stress the need to preliminarily diagnose the state of the environment, in order to effectively assess the pressures caused or subjected to the environmental and anthropic components by the plan's choices. Suitable or compatible responses may then be elaborated.

With this new law, the region provides provincial and local governments with a land-use instrument, through which, the sustainability of environmental and territorial planning choices can be defined. The preventive environmental and territorial sustainability assessment (VALSAT) is to be considered as fundamental for all territorial and urban planning processes, i.e. general and sectoral planning:

- General planning: Art. 10 of Law No. 20/2000 defines general plans as instruments through which public governments discipline the protection and land-use of the entire area of their competence (e.g. region's are responsible for the entire regional area, provinces for the entire provincial area, municipalities for the entire municipal area).
- Sectoral planning: Art. 10 of Law No. 20/2000 defines sectoral plans as instruments through which, public governments and agencies with specific planning competences (only when expressed by law), discipline the protection and land-use of the territory, according to the profiles related to their specific functions.

The planning law of the region introduces two main innovations: strategic planning and the VALSAT (*Environmental and territorial sustainability assessment*).

- *Strategic planning*: by defining objectives, contents and innovative instruments in order to regulate territorial transformations, it recognizes and distinguishes baseline information, forecast information, assessment and monitoring information. Strategic planning assumes that "[...] baseline and assessments must set the foundations for all of the urban and territorial planning processes, therefore for all of the general and sectoral instruments of the region (Regional Territorial Plan and Regional Territorial Landscape Plan – PTR and PTPR), Provinces (*Provincial Territorial Coordinated Plan – PTCP*) and Municipalities (*Municipality Structure Plan and Executive Plan– PSC and POC*) [...]"
- *VALSAT*: integrated assessment procedure for plan-making. It is defined as an iterative process, which must be applied throughout the whole plan-making, in order to be effective. It must aim at sustainability, therefore it must include objectives of "environmental, territorial and social, health and safety, landscape qualification and environmental protection established by the upper-tiers of legislation and planning". It must be based on broad baseline knowledge of the territory, as it must "acquire, through baseline information, the status and the evolving tendencies of natural and anthropic systems, and their interactions as well".

8.5.1 Levels of Planning Instruments for the Region of Emilia Romagna

There are three levels of planning instruments, which correspond to the three main tiers of decision-making: regional, provincial and local. The following three sections describe these three planning levels and related planning instruments for the region of Emilia Romagna (see also Fig. 8.2).

Regional Planning Instruments: Through the regional territorial plan (PTR or PTPR when landscape plan is integrated), the region defines the objectives for development and social cohesion and for the increase of regional competitiveness, in consistency with European and national development strategies. The aim is to guarantee the reproducibility, qualification and valorisation of the social and environmental resources, by identifying the limits and conditions for sectoral and planning instruments of lower tiers.

Provincial Planning Instruments: The provincial territory is managed through the Provincial Territorial Coordinated Plan (PTCP). The PTCP must implement those interventions defined at a national and regional level – such as those relating to the infrastructural system and major projects concerning areas of provincial development – and define the asset and land-use directions for the localisation and dimensioning of structures and services of a provincial or super-municipal interest. The PTCP must identify criteria for defining the vulnerability, thresholds and potentials of natural and anthropic systems. Furthermore, it must define the balance of territorial and environmental resources for the territorial and environmental sustainability of the lower-level (i.e. municipalities) planning predictions, establishing thresholds, conditions and limits for land-use.

Local Planning Instruments: According to Law No. 20/2000, structural contents are separated from programmatic ones, resulting in the two planning instruments at the local level:

1. The municipality structure plan (PSC), which defines strategies for the territorial asset and development, in order to ensure protection of the physical integrity, of the environment and of the cultural identity.
2. The municipality executive plan (POC), which identifies and disciplines the interventions of territorial protection, valorisation, organisation and transformation. It has a length of five years.

The municipality structure plan is composed by:

- Baseline information framework (*Quadro Conoscitivo*): it must re-construct the state of art of the territory; the analysis of the evolving dynamics in act, the technical and discretionary assessment of resources, as well as opportunities and limiting factors.
- Preliminary document (*Documento Preliminare*): it defines the plan's strategic contents; sets the general objectives and policies towards which, the plan must aim its actions; it is subjected with both, the VALSAT and the baseline information framework, for examination to the "planning conference": the only institutionalised stage of public participation (see Sect. 8.6).
- Structural plan: concluding document resulting from the implementation of the preliminary document's observations, amendments and integrations expressed by the planning conference.

Therefore, on one hand the PSC (of an indeterminate length) defines the major strategies and projects of a significant scope, the rules for the physical transformations, the definition of compatible uses according to the territorial characteristics identified, providing directions and prescriptions in order to better satisfy needs for the protection of the environment. On the other hand, the POC (five year length) establishes prescribed uses and the physical urban and building transformations to implement, within the corresponding administrative mandate (each political mandate also has a five year length).

8.5.2 Principles and Objectives of Law No. 20/2000 of Emilia Romagna

The regional law of Emilia Romagna has had a major impact on the planning system. Despite the novelties described previously concerning the contents and specific objectives of each planning instruments, Law No. 20/2000 defines the general (sustainability) objectives that planning must aim to achieve. According to Art. 2, planning instruments must:

- Promote an organised territorial development
- Achieve compatibility between land-use transformations and the protection of physical integrity and the cultural identity of the territory
- Improve the quality of life and of urban settlements
- Reduce the pressure of settlements on natural and environmental systems, also through appropriate impact mitigations
- Improve the environmental, architectural and social quality of the urban territory through re-qualifications of the existing built estate
- Allow land consumption only when no other alternative options (reorganisation and re-qualification of existing settlements) are possible.

To guarantee a major coherence between the strategies of the different planning instruments and to define a common vision of sustainable development, Law No. 20/2000 introduces "planning conferences" (Art. 14): procedural stage of the plan's approval process and only moment of "participation" required by the VALSAT, in which forms of institutional agreements (*concertazione istituzionale*) are realised.

Law No. 20/2000 establishes that sustainability inputs and elements that make reference to a sustainable development may occur in two moments of the planning process: while assessing the plan's objectives with those defined by Art. 2 (described earlier in this section) and during the approval procedure of the plan, therefore when different public agencies and socio-economic representatives are involved in the preventive environmental and territorial sustainability assessment (VALSAT). Within this context, a Regional Territorial Plan (PTR) for example, should provide a unique vision of sustainable development for the entire regional area, expressed through objectives. These objectives, in consistency with those listed in Art. 2 of Law No. 20/2000 should layout the future strategies for the regional as well as the provincial and local development. The objectives should also represent a valid support for the sustainability assessment.

8.5.3 The Environmental and Territorial Sustainability Assessment: VALSAT

Law No. 20/2000 defines the environmental and territorial sustainability of plans, as a balanced relationship between development and territorial protection. Planning processes can only start after a thorough knowledge of the territory, therefore after an analysis of its main features, of its state of art and of its evolving processes (Art. 4). All planning instruments, general and sectoral, territorial and urban, of a regional, provincial and local level, are therefore subjected to "knowledge" (baseline information) and assessment activities, subsequently resulting in two technical documents, both constituting elements of the approved plan: the baseline information framework (*Quadro Conoscitivo*) and the VALSAT (*Valutazione preventiva della sostenibilita' ambientale e territoriale*). The approach is to be considered as a preventive assessment of the effects that the plan is likely to have on the territorial systems (Art. 5).

The VALSAT starts during the elaboration of the plan. It must identify the effects deriving from the implementation of the plan's choices and select the alternatives, also identifying the planning measures to mitigate, avoid, reduce or compensate the impacts likely to occur. In order to guarantee effective and coherent procedures, the region provides the provinces and municipalities with minimum elements to be included in the VALSAT. The sustainability assessment consists of a six step linear process articulated through the different phases of the plan-making process. It is to be considered as an integrated step of the plan's approval procedure. Fig. 8.3 represents a description of the VALSAT's tasks in the plan-making process.

The VALSAT aims to verify the consistency of the plan's choices with the general planning objectives (defined in Art. 2 of Law No. 20/2000) and with the sustainability objectives for territorial development defined in the general and sectoral plans, of a community, national, regional and provincial level. The VALSAT therefore, wishes to represent a guaranteeing instrument for the decisions made in the planning process. As an instrument integrated in the planning process, the VALSAT is applied by the same administration developing the plan.

During the subsequent elaboration phases of the plan, the administration must also integrate the findings of the VALSAT in the preliminary document. Whilst institutional and representative stakeholders participate to the process through forms of institutional co-operation and agreements (i.e. planning conferences), citizens are informed and consulted through more traditional approaches, such as the presentation of observations (see Sect. 8.6). The planning conference aims to develop an agreed baseline information framework of the territory concerned and of the limits and necessary conditions for sustainable development. Furthermore, it aims to express preliminary assessment of the planning objectives and decisions formulated in the preliminary document (Art. 14).

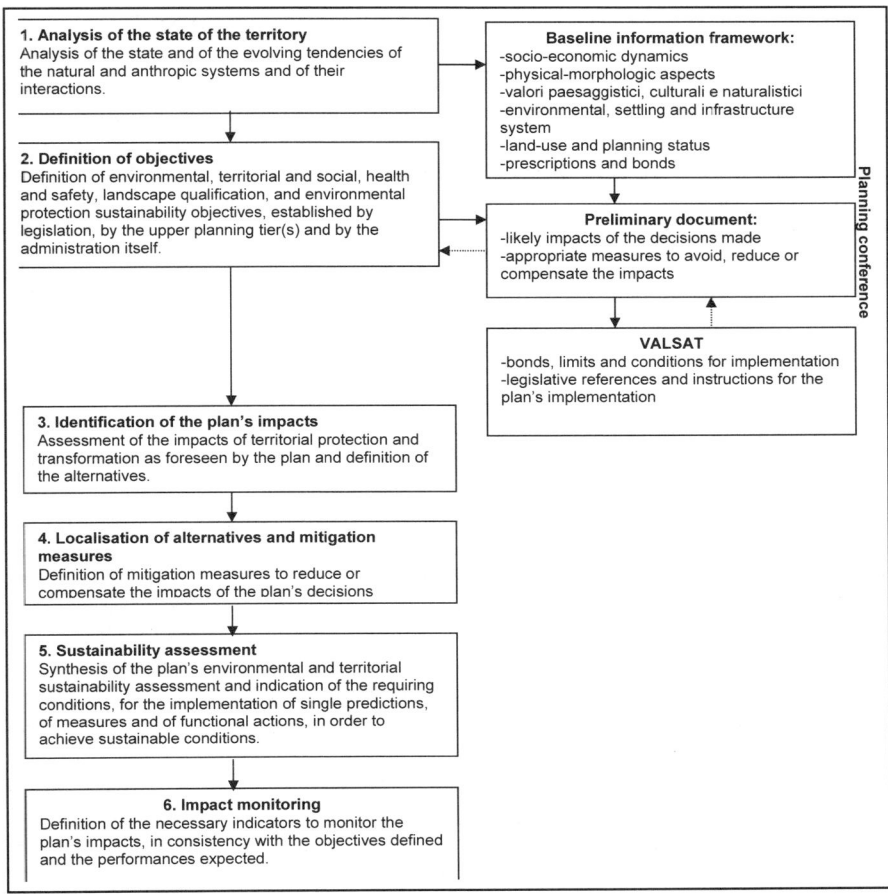

Fig. 8.3. The VALSAT in the planning process

The preliminary document is, therefore, instrumental to the planning conference. Territorial and administrative agencies identified for each plan, as well as those administrations entitled to provide opinions in the respect of agreements, participate to the planning conference. Social and economic associations do not participate to the conference, but are invited to set agreements with the conference itself. Provinces and municipalities may define common objectives and strategic decisions through territorial agreements, aiming to coordinate the implementation of planning instruments (Art. 15).

Although Law No. 20/2000 does not provide an explicit definition of sustainability, it is implicitly assumed that those described as general objectives in Art. 2 could be intended as sustainability objectives. Law No. 20/2000 does not provide a common set of objectives and targets to fulfil the VALSAT's purpose of achieving a balanced relationship between development and territorial protection. Furthermore, the scoping stage for the entire planning process is not included in the VALSAT procedure. Despite assessing the coherence between planning choices and the definition of mitigating measures, the assessment of the degree of compatibility between the sustainability objectives and targets and the strategic policies is not foreseen, as well as the assessment of the impacts of the proposed alternatives. Besides identifying limits, conditions and bonds for the plan's implementation in terms of achieving sustainability, the VALSAT procedure should also emphasise objectives and targets, according to which, policies may be stressed, amended or introduced *ex-novo* to achieve sustainable development, i.e. an environmental and territorial sustainability. A comparison of the VALSAT and of the SEA Directive follows in Sect. 8.6.

8.6 VALSAT versus SEA Directive

The environmental and territorial sustainability assessment process of the region of Emilia Romagna and the European SEA process for plans and programmes, differ in terms of their aims and procedures. Whilst the SEA Directive aims to ensure that the environmental consequences of plans and programmes are identified and assessed during the preparation and before their adoption, the VALSAT aims to assess the consistency of the plan's choices with the planning's general objectives and with the sustainability objectives defined by the general and sectoral plans, and by provisions of community, national, regional and provincial levels. Both procedures emphasise the availability of territorial and environmental data as a starting point, as a benchmark against which, the assessment of the different alternative's performances could be made.

For certain aspects the SEA Directive goes beyond the provisions of the VALSAT. The Directive in fact, requires the designation of environmental authorities and their consultation on the scope of the assessment and on the contents of the environmental report; the public must also be consulted on the environmental report and the information on how the environmental considerations have been integrated in the final plan or programme, must be provided. Furthermore, the SEA Directive requires the final reports to meet quality requirements and assumes a distribution of competences for quality control. The VALSAT instead, neither requires the designation of environmental authorities nor makes explicit references to it. The planning and the assessment processes are both a competence of the governmental administration. The VALSAT requires the planning conference, as a moment of institutional consultation and participation in which the interested and relevant administrations (with planning functions), integrate their competences and search for common objectives in the preliminary document (not on the scope of the assessment and not on the plan). If compared to the European SEA, public participation in the VALSAT is relatively reduced: more attention is given to institutional participation (i.e. horizontal integration) and to the involvement of the declared socio-economic forces, than to the public. Law No. 20/2000 defines the contents of plans and of the preliminary document, but does not provide the respect of quality criterion during their preparation. Quality control is not required for the VALSAT's concluding statement (the preparation of an environmental report is not required).

Both SEA stress the importance of an *iterative assessment* process with the plan and/or programme making process. Furthermore, whilst the SEA Directive provides the adopted plan or programme to be adequately published and advertised, requiring also a synthesis in the environmental report describing the way in which environmental considerations have been integrated in the plan or programme (Art. 9 of the SEA Directive), Law No. 20/2000 demands different measures. According to Law No. 20, after being adopted, each plan (regional, provincial and local) is deposited in the competent territorial offices for sixty days and is published in the Official Bulletin of the region. During the sixty days, citizens are entitled to take vision of the plan and present "observations". These can be accepted or rejected by the interested public administration (the region for regional plans, the province for provincial ones and the municipality for local ones).

As for the concluding statement of the VALSAT, the president of the planning conference must only transfer a copy of it to all actors convoked to the conference.

8.7 Implementing the SEA Directive in the Emilia Romagna Region

The comparative analysis of the VALSAT with the SEA Directive in the previous section showed how the first does not fulfil all the requirements of the second. However, the innovative elements introduced by Law No. 20/2000 of the region of Emilia Romagna, must be acknowledged, in particular, the introduction of *strategic planning*. Fig. 8.4 presents a summary of the elements of the SEA Directive that must be introduced in the VALSAT in order to achieve compliance between the region of Emilia Romagna and the European Union.

The VALSAT was introduced in the planning practices in March 2000. Despite four years of implementation – in planning terms this is not very long and this partly explains why its application has not yet demonstrated valid results – it has not yet become a consolidated and well-structured practice in the region.

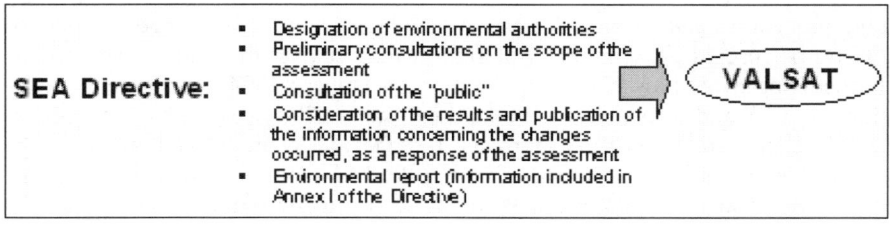

Fig. 8.4. Transposing the SEA Directive in the region of Emilia Romagna's VALSAT

Sani and Di Stefano (2001) recognized four barriers:

- Difficulties related to the integration of the VALSAT's steps in the plan's decision-making process.
- Difficulties related to the introduction of new procedures in an existing planning system, with approaches well-structured and consolidated.
- Difficulties related to the availability of trained personnel and professionals and
- To the distribution of competences and responsibilities.

The integration of the SEA Directive with the VALSAT in one single procedure for the assessment of plans and programmes, could imply the distribution of the assessment responsibilities and an extension of the participation input of environmental authorities and more in general, of the public. The object of the participation process would be not only the content of the assessment itself, but the final environmental report as well.

8.8 A Look at the Bigger Picture: Implementing the SEA Directive in Italy

The legislative and cultural framework for the implementation of SEA in Italy is still uncertain. Despite an intense debate surrounding the argument, shortcomings due to the way in which the environment is tackled and to the practice of project level assessments in Italy, are still present (Bettini 1995; Schmidt di Friedberg 1996, 1997; Del Furia and Wallace-Jones 2000; Bettini and Gazzola 2001; Gazzola et al. 2004). However, particular tendencies concerning how the introduction of SEA will occur are being formalised. The main one is to follow the SEA approach applied for the Structural Funds, also adopting the same methodology (Caramaschi et al. 2003). A brief description of elements considered key for an effective introduction of a SEA system in Italy, follows.

8.8.1 Evolving Environmental Legislative Framework

The legislative framework for the Italian planning system is changing. This concerns in particular, a reorganisation process aiming at the simplification of planning procedures and the introduction of new legislations. The changes introduced by the Berlusconi government, will significantly impact the effectiveness of the forthcoming SEA. Law No. 5 approved on 16 January 2004 defines new regulations on EIA and formally institutionalises "super-EIA": simplified and accelerated EIA, also known as "EIA shortcuts" (see Gazzola et al. 2004). Furthermore, it diminishes the number of experts forming the EIA Commission (commission of twenty nominated experts that have the task to provide opinions on EIA and to verify and assess the EIA submitted, as requested by the Ministry of the Environment) and formulates new requirements for the authorisation procedure for projects of a strategic importance. The impacts of the new Art. 117 of the Italian Constitution (as described in Sect. 8.1) are also relevant here.

8.8.2 Problems with Implementing Vertical Integration in the Italian Planning System

As shown in the simplified representation in Fig. 8.2, the Italian planning system is highly complex. There are numerous decision-making tiers and sectoral agencies with different planning competences and responsibilities. Therefore, it often is difficult to balance conditions defined by top-down policies with the willingness to establish local level decision-making processes. The subsidiarity principle has the merits for making citizens closer to the decision-making process. However, by not diminishing the number of institutions present on the territory (with planning competences), it has often multiplied the conflicts through those same institutions. Increasing difficulties for local governments are also due to the incapacity or impossibility to socially manage the public participation processes introduced by the new planning instruments (i.e. the forms of negotiated programmes mentioned in Sect. 8.4) and by Local Agenda 21. The implementation of vertical integration appears to be problematic even at the higher tiers, i.e. provincial and regional. If on one hand, more importance is currently being given to local issues, on the other their implementation seems to significantly extend the timings. In some cases, the conditions necessary for achieving local needs are not even implemented. Important decisions (i.e. those that are really strategic, such as those for the infrastructure system) are taken at the upper tiers (provincial and/or regional and/or national) without the local governments being thoroughly consulted and in the absence of EIA. An important ring of the chain of strategic objectives is therefore missing, since the necessary conditions for implementing vertical integration through a tiered process cannot be possible at the local level (Caramaschi et al. 2003). At this tier, due to the planning principles practiced in Italy, the plan is obliged to implement the provisions of the upper level.

8.9 Conclusions and Recommendations: Considerations for the Region of Emilia Romagna and Italy as a whole

In implementing the SEA Directive in Emilia Romagna, the region is likely to follow the trends currently being formulated in the national debate, which is to equalize the SEA Directive with the SEA process for the structural funds. During an SEA seminar organized by the Italian Ministry for the Environment and Territorial Protection (2001), representatives of the region of Emilia Romagna said the following:

> "[…] in implementing the Community Directive, and in order to develop and approve sustainable and environmentally compatible plans and programmes, the region of Emilia Romagna considers necessary to favour an approach that guarantees the best results.

Therefore, the best option is to pursue an SEA approach oriented towards the SEA for structural funds, as defined in community legislation [...]" (Di Stefano and Sani 2001)

According to the national debate, it is assumed that the region of Tuscany will follow the same direction too.

Since the implementation process of the SEA Directive started, the Italian Ministry of the Environment and Territorial Protection has been very active in organising and attending conferences concerning the introduction of SEA in Italy and in arranging roundtables with representatives of the Italian regions (see Ministry of the Environment and Territorial Protection 2004). SEA methodologies, according to the indication of the Directive, are also taking place, on different types of plans and programmes in order to test the existence of procedural difficulties in decision-making processes and to test the possibility of applying common procedures. Outcomes of these experiments include: checklist for technicians (with related guidelines), checklist for administrators, the environmental report and a glossary on SEA (Polizzy 2003). The goal of the Ministry is to provide the regions, guidance and directions for implementing the SEA Directive in their own legislative framework. Several regions, research institutes (such as *Fondazione Eni Enrico Mattei* (2002)) and universities have been organising conferences on SEA as well (IUAV 2001, 2003).

Whilst, in theory the cultural debate seems to be heading towards more innovative directions (e.g. Law No. 20/2000 of the Region of Emilia Romagna), in reality planning is responding in a different way. In practice, the updating or amendment of known schemes are preferred to more radical choices. It could be said that Italy is therefore, more keen on mitigating the externalities of the implementation process of the SEA Directive. In the latest generations of plans – the so-called strategic ones – the consideration of environmental parameters is in fact more articulated and systematic, occupying a relevant part of the analytical analysis of the whole plan. But the consideration of environmental aspects is not reflected in the articulation of appropriate and adequate sustainability objectives. These are still defined in a generic manner and are not articulated in an organic way, neither at a horizontal level (per function), nor at a vertical one (per sub-objectives and targets). Furthermore, with the enlargement of the debate on sustainable development to the consideration of the city (e.g. urban sustainability) and on the definition of indicators to measure the city's sustainability, the same has not occurred for the non-urban territory. In practice, concepts, definitions and methods for elaborating urban sustainability, have been mechanically transposed to the non-urbanised territory. Finally, in relation to the adoption of a SEA approach and methodology it is highly likely that a new methodology will not be developed. Instead, and based on past experience in Italy, it is likely that the existing SEA methodology for the structural funds will be applied to the environmental assessment of plans and programmes, once the SEA Directive is implemented in Italy.

The implementation of the SEA Directive could have been an important occasion for Italy to develop a less fragmented and a more complete framework for the environmental assessment of plans and programmes (and of projects as well). The SEA Directive could have created a unique opportunity for facilitating the transition from the *traditional planning* approach to the *latest generation planning* based on strategic and sustainability principles. Italy overall has instead decided to mitigate the impact of the implementation of the SEA Directive and has mostly chosen to simply update the existing planning system. And even the more innovative regions, such as the Emilia Romagna region described in this chapter, are likely to follow the same approach. The most advanced regional governments, conscious of the fact that the future implementation decree (at a national level) of the SEA Directive was not going to have significant strategic contents or the sustainability perspective wished by the EC's Directive, did not have sufficient legitimating strength to pursue and adopt the novelties introduced by the SEA Directive. Therefore, when the regions introduced SEA on a procedural level in their own regional laws, the shift towards a more strategic approach to planning could not be completed. This is a typical "chicken and egg" situation, further complicated by the country's general attitude towards the environment and its cultural and professional delay in environmental and sustainability assessment and planning.

References[1]

Avarello P (2000) Il Piano Comunale (The municipal plan). Il sole 24ore S.p.a. Milan

Bettini V (1995) L'Impatto Ambientale. Tecniche e Metodi (The environmental impact – techniques and methods). Cuen, Naples

Bettini V, Gazzola P (2001) La valutazione in campo ambientale: da strumento ancillare a componente strategica (Evaluation in the environmental field: from maid instrument to strategic component). In Stame N (ed) Valutazione 2001. Lo sviluppo della valutazione in Italia (Evaluation 2001 – the development of evaluations in Italy). Franco Angeli, Milan

Caramaschi M, Gazzola P, Sirocco L, Zanin D (2003) Il contributo della Valutazione Ambientale Strategica (VAS) alla sostenibilità: il caso del Piano Strutturale di Sorbolo (Parma) (The contribution of strategic environmental assessment (SEA) to sustainability: the case of the Structural Plan of Sorbolo (Parma). Master thesis, University Institute of Architecture of Venice (IUAV)

Chitotti O (2002) La valutazione ambientale strategica in Italia (Strategic environmental assessment). Notizie dal Centro V.I.A. Italia 2002:20

Ciciotti E, Florio R, Perulli P (1997) Approcci strategici alla pianificazione territoriale (Strategic approaches to territorial planning). In Perulli P (ed) Pianificazione strategica (Strategic planning). Conference proceedings of the seminar held in Venice May 31 1996. Daest, Venice

Del Furia L, Wallace-Jones J (2000) The effectiveness of provisions and quality of practices concerning public participation in EIA in Italy. Environmental Impact Assessment Review 20:457-479

Di Stefano AM, Sani MM (2001) La Valutazione di Sostenibilità Ambientale e Territoriale (VALSAT) dei piani territoriali ed urbanistici prevista dalla legge regionale

---

[1] Note that legislation mentioned in the chapter is listed at the end of the handbook in the consolidated list of legislation (Appendix 2).

dell'Emilia Romagna n.20/2000 Disciplina generale sulla tutela e uso del territorio (The environmental and territorial sustainability assessment (VALSAT) of urban and territorial plans according to law 20/2000 general disciplines on the protection and use of the territory of the region of Emilia Romagna). Paper presented at the seminar on Strategic Environmental Assessment, Ministry for the Environment and Territorial Protection. November 28-29 2001, Rome

Fondazione Eni Enrico Mattei (2002) Prospettive di sviluppo della Valutazione Ambientale Strategica in Italia (Development prospective for strategic environmental assessment) Conference proceedings, February 5 2002, Milan

Gazzola P (2002) Prospettive di implementazione della direttiva VAS in Italia (Implementation prospective for the SEA Directive in Italy). Archivio di Studi Urbani e Regionali 74:77-92

Gazzola P, Caramaschi M, Fischer TB (2004) Implementing the SEA Directive in Italy. Opportunities and barriers. European Environment 14 (3):188-199

IUAV – Planning Department, University Institute of Architecture of Venice (2001) La valutazione ambientale strategica (VAS) e la nuova Direttiva Europea (Strategic environmental assessment (SEA) and the new European Directive). Conference proceedings, October 26 2001, Venice

IUAV – Planning Department, University Institute of Architecture of Venice (2003) L'integrazione della VAS nei processi di piano. Riflessioni sulle esperienze in corso in vista del recepimento della Direttiva 42/2001/CE (The integration of SEA in planning processes. Reflections on ongoing experiences while awaiting for the implementation of Directive 42/2001/EC). Conference proceedings

Jordan A, Jeppesen T (2000) EU environmental policy: adapting to the principle of subsidiarity? European Environment 10:64-74

Ministry of the Environment and Territorial Protection (2004) Web page of the *Ministero dell'Ambiente e della Tutela del Territorio*. Interent address: http://www.minambiente.it, last accessed 25.03.2004

Ministry of the Environment and Territorial Protection (2001) Seminario sulla Valutazione Ambientale Strategica (Seminar on strategic environmental assessment). Conference proceedings, November 28-29 2001, Rome

Polizzy L (2003) L'attivita' di sperimentazione del Ministero dell'Ambiente e della Tutela del Territorio (Experimenting activities of the Ministry of the Environment and Territorial Protection). Paper presented at the conference held in Venice, 20.11.2003

Province of Bologna (2003) Valutazione Ambientale Strategica di piani e programmi – Esperienze italiane a confronto (Strategic environmental assessment of plans and programmes – comparing Italian experiences). Workshop organised by the Province of Bologna, November 11 2003, Bologna

Region of Emilia Romagna (2000) Disciplina generale sulla tutela ed uso del territorio (General discipline on the use and protection of the territory). Law of March 3 2000 n.20

Salvia F, Teresi F (1992) Diritto Urbanistico (Urban law), Cedam

Salzano E (2003) Fondamenti di Urbanistica (Urban fundaments). Edizioni La Terza

Salzano E (2004) Web page of *Prof. Edoardo Salzano*. Interenet address: http://www.eddyburg.it, last accessed 25.03.2004

Schmidt di Friedberg P (1996) Recent EIA Developments within the European Union. Environmental Assessment in Italy. In: Wood C, Barker AJ, Jones CE (eds) EIA Newsletter 13, EIA Centre, Department of Planning and Landscape, University of Manchester, UK

Schmidt di Friedberg P (1997) Recent EIA Developments within the European Union. Environmental Assessment in Italy. In: Wood C, Barker AJ, Jones CE (eds) EIA Newsletter 15, EIA Centre, Department of Planning and Landscape, University of Manchester, UK

Stanghellini S (1999) Riforma urbanistica e domanda di valutazione (Urban reform and demand for evaluation). In Lombardi P, Micelli E (eds) Le misure del piano. Temi e strumenti della valutazione nei nuovi piani (The measures for plans. Evaluation themes and instruments in the new plans). Franco Angeli, Milan

9 First Experiences with Implementing SEA Legislation in Flanders (Belgium)

Marc Van Dyck

Resource Analysis NV, Belgium

9.1 Introduction

This chapter evaluates the first experiences in the implementation of the new SEA legislation in the region of Flanders in Belgium. Belgium consists of three regions. The environmental legislation has been developed differently in these three Belgian regions. In the Flemish region or Flanders, environmental legislation including environmental reporting has been implemented since 1989, in Walloon since 1985 and in the Brussels Capital region since 1992. The main elements of the EIA legislation are the same in the three regions. However, in practice, EIA procedures differ slightly.

This chapter will not address the differences in the legislation between the Belgian regions, but will evaluate the legislation and practice of SEA for strategic actions (plans and programmes) in Flanders, where the implementation of the SEA legislation has been developed since 2002. SEA legislation had neither been developed in the Walloon region nor in the Brussels Capital region. Due to the fact that SEA has been implemented the furthest in Flanders, and good practice cases are known, focus lies in this chapter on the experiences in this region.

In the next section of this chapter the development process of the new legislation is described. At this moment the implementation phase of SEA in Flanders has started up in 2002 with a few pilot projects. A small number of cases are presented in Sect. 9.3. Sect. 9.4 will address problems with the practical implementation of the SEA Directive and describe a few common points of interest for the implementation of SEA. The pilot projects illustrate some of the implementation problems. A short final section presents a summary of conclusions and recommendations.

9.2 Development of a Flemish Combined Legislation on EIA and SEA

This paragraph describes how the legislative framework in Flanders developed from the introduction of EIA to that of SEA for strategic actions. First the former EIA legislation is explained. In the second part, the development process of the SEA legislation is highlighted and, in the third part, the current procedure for SEA is presented.

9.2.1 The Old EIA Legislation

The Flemish legislation on EIA for projects dates from the Decree[1] of 23 March 1989, the EIA Decree. The required elements of the contents of the EIS were taken from the EIA Directive. In addition to the EC requirements, an employment report and a description of the planned investment and the type of the produced goods and services are part of a Flemish EIA. Commercial and confidential industrial data need not be added to the EIA.

In Flanders the quality control of the EIS is guaranteed by a unique system: accredited EIA experts are the only persons allowed to write the chapters of an EIS. These experts have to be independent from the project initiator and are personally accredited for a period of five years by a ministerial Decree. This system guarantees that experienced people write the chapters and perform the investigation into the environmental impacts. In practice, this system also has disadvantages because no other systems of quality control during the process were put into place and the report authors still are not fully independent of the project initiators, because these experts get their assignments from them. However, if they do not deliver an objective and scientifically correct impact evaluation, the renewal of their accreditation could be refused.

The EIA procedure in itself is very simple and only includes two formal steps: the confirmation of the team of accredited experts in an initial dossier and the validation of the final EIS. An informal procedure with discussions on scoping and on the draft version of the EIS is developed over the years by the Flemish administration – EIA Unit. The EIA Unit from the Flemish Community government is the competent authority and controls the quality of the EIA and SEA processes.

9.2.2 Development of an SEA Legislation and Update of the EIA Legislation

As soon as the EC decided to implement a new guideline on SEA, an internal working group was established in 2001, including members of the Flemish administration on Environment and a few independent EIA experts and academics.

---

[1] A law of the Flemish community is called a Decree.

This working group started the drafting of a text for the new adapted Decree on EIA, including at the same time new SEA approach. Almost two years pass before a final consensus on the basic text is reached. Difficult points were the inclusion of a procedural step to have a public enquiry before the EIS is drafted, and the text for an annex that defines the types of plans to which the SEA legislation will apply. Finally the public enquiry was accepted in the scoping phase and additional regulations on the communication of EIA results with the broad public were to be determined later on.

The EIA and SEA Decree introduces all elements of the SEA Directive (Risse et al. 2003). The Decree combines the regulation for project EIA and SEA. In Flanders, SEA is called "EIA for plans and programmes" and has a short and simple procedure. For project EIA a new and more elaborate procedure is installed in this new legislation. The system of the accredited EIA experts as report authors remains operative and also the contents of the EIA remain the same. For the contents of the SEA ("EIS of plans and programmes") the SEA Directive was the main source of inspiration.

The Decree is tuned with other Flemish environmental legislation and legislation on land use planning, integrated water management and nature conservation. A new item in the procedure is that also the SEA or EIA co-ordinator[2] has to be accredited. He or she will lead a team of experts that can be accredited EIA experts or experts that are put at his of her disposal by the initiator of the strategic action. New is also the exemption clause that states that projects, that have a detailed SEA for the plan they belong to, can be exempt from the obligation to draft a project EIA, when no new data or information can be expected from this EIA procedure, or when the project belongs to the list of Annex II (list B in the Flemish Decree) and the check with the criteria shows that no significant environmental impacts can be expected.

Table 9.1. Overview of Flemish plan types to subject to or exempt from SEA obligation (Annex I to the Regulation 12/03/2004)

Obligatory subjection to SEA	Exempt from SEA
The Flemish Manure Action Plan, 1991	The Primary Minerals Exploitation Plan, 2003
The Policy plan for Gravel exploitation, 1993	The Land Use Execution Plans, Decree of 1999
The General Water Sanitation plans, Decree of 1971	The Nature Design Plans, Decree of 1997
	The Re-allotment plans, Decree of 1970

---

[2] The co-ordinator of the EIA or the SEA is the expert who takes on the co-ordination of the team of accredited experts. He is responsible for the coherent drafting of the EIS and all communication with the competent authorities.

The EIA and SEA Decree, completed with a few articles and a few annexes by a government Draft Resolution to the EIA and SEA Decree, will become operative from the 1 July 2004. The Resolution states the new project type lists A and B for the Flemish community. These lists are similar to the Annexes I and II of the SEA Directive. The criteria for the accreditation of the EIA experts are listed and the annexes 1a through 1c of the Resolution clarify the area of application of the SEA legislation.

The plans in Table 9.1, mentioned in the Draft Resolution to the EIA and SEA Decree, will be subjected to an SEA or will be exempt, based on the legal framework. These plans will not be evaluated by the EIA Unit prior to the decision to subject the plan to an SEA. All other policies, plans or programmes can be subjected to SEA, based on an evaluation of the characteristics of the plan or of the area involved. Characteristics or criteria on which the evaluation of the SEA obligation is performed are (Annex 1 to the Regulation 12/03/2004):

Characteristics of the plan or programme itself:

- The plan or programme is the frame for projects that involve the area, the volume, the type and/or the exploitation conditions as well as the distribution of natural resources.
- The plan or programme influences other plans or programmmes
- The plan's relevance to the integration of environmental aspects, focusing on-sustainable development
- The plan has environmental consequences
- The plan or programme influences the application of the EU environmental legislation

Characteristics of the involved areas:

- Frequency, reversibility, cumulative character and cross-border character of the impacts
- Risks for human health and safety and for the environment
- Magnitude of the impacts and area of influence
- Value or vulnerability of the areas involved
- Impacts on specially protected areas

At this moment a few strategic actions have already been subjected to an SEA: these can be considered as pilot projects. Sect. 9.3 will describe a few of these pilot studies. Methodologically, the SEA research still has be developed, within the Flemish geographical and policy context. In the near future a guideline book for SEA will be drafted.

9.2.3 Procedural Steps in SEA

The procedures for EIA and SEA have almost the same structure. The first step is the project definition and scoping, that is described in the so-called "notification dossier". This document includes also information on the team of accredited experts. It is submitted to the municipalities, involved in the plan or programme, and is available for consultation by the public for the period of one month. The public enquiry also can include other communication activities that are not specified in the legislation. The notification dossier includes a description of the strategic action, data on the cross-border exchange of information, previously executed research, the scope of the study and the description of plan alternatives and the team of experts.

The response of the public enquiry and the additional guidelines of several administrations will be the basis for the EIA Unit to develop guidelines for the drafting of the Strategic EIS. These guidelines will address the required contents of the Strategic EIS, including the alternatives that need to be investigated as well as the investigation methods and the level of detail needed in the Strategic EIS for the plan or programme.

In the following phase the Strategic EIS is drafted and in a final phase it will be controlled on completeness and quality (Berube and Cusson 2002). For this purpose, the EIA Unit will seek advice from other administrations, competent in the different environmental aspects (nature, water, soil, etc). After an approval is given by the EIA Unit, the document becomes public and is available for consultation at the Flemish administration and on the official website of the EIA Unit (www.mervlaanderen.be).

The Flemish government can decide on more active forms of communication about the plan and/or the SEA for the plan.

9.3 Practice of SEA in Flanders

Before the development of the SEA legislation, several location studies and feasibility studies had already been drafted for the larger government funded infrastructure developments. For most of the plans, however, an environmental impact assessment only was made at the time when the projects were defined up to such a level of detail, that technical project alternatives could not be brought into the evaluation. The legislation of 1989 did not demand the evaluation of more alternatives than the project proposal and the reference situation (i.e. the case when no project is developed). This is why many projects have been developed without looking at possibly more environmentally friendly alternatives. The new SEA legislation allows evaluating plans and programmes on environmental aspects before the projects, that arise from these plans or programmes, are (fully) defined (Thérivel 2004; Partidário 2000).

Because of the fact that it took a long time to define the types of strategic action to which the legislation applies, some large programmes and a few plans have been subjected to the procedure of SEA, before the legislation became operative. For these first pilot processes the EIA Unit have had frequent discussions with the individuals responsible for carrying out the SEA, about the methodology used to describe the impacts at a plan level, about the methods used to develop plan alternatives, and about the way the public enquiry can be organised.

A few cases will be briefly described in this section. Many of these cases are still going on at the moment of writing, so definitive conclusions are not yet available. However, the pilot projects are important as they allow some first comments to be made on the implementation problems of the SEA procedure in Flanders.

Four cases have been chosen for this chapter: the mobility plan of Flanders that had an SEA in 2001, the port policy plan of Flanders that has an SEA in 2004, the river flood management programme for the Scheldt river (SIGMA-plan) that has an SEA in 2003-2004 and finally the master plan mobility of Antwerp that had an SEA in 200-2004. The reason why these case studies have been chosen is because they are the first SEA processes being developed during the period in which the new Flemish SEA legislation has come into effect and because they represent a variety of strategic actions. Furthermore these SEA processes have been almost finalized at the moment of writing, so that there is enough case information available for the analysis.

9.3.1 Mobility Plan Flanders – SEA 2001

In the Flemish policy plan for the mobility a large number of concrete policy actions and infrastructure projects has been defined. This list of measures has been subjected to an environmental assessment. The assessment report is composed of two reports: a methodology report and the actual assessment report. In the methodology report a set of criteria for the evaluation of the environmental impacts is described (Ministry of the Flemish Community 2001).

First the plan was developed and afterwards the SEA assessment was made, comparing two scenarios: the trend scenario (i.e. continue with present trend situation) and the sustainable scenario, in which a set of achieved targets on mobility policy would have been attained. Scenario development process was not during the policy plan development but only later on during the drafting of the Strategic EIS.

The description of environmental impacts was split into a qualitative evaluation of indicators for each of the two scenarios and a description of the specific impacts of the concrete policy measures. Most of the impacts of these measures (other than infrastructure projects) were described in a very general manner or it was indicated that no impacts could be determined in this stage of the plan evaluation. Neither a participative process nor a public enquiry was organized.

This was the first SEA performed in Flanders and there were still a number of drawbacks in the used methodology, as well as a lot of communication problems between involved parties.

9.3.2 Port Policy Plan Flanders – SEA 2004

The policy plan for the Flemish ports is being drafted at the moment of writing. At the same time inclusion of environmental aspects is performed during the plan development. To be able to perform this parallel reporting, the policy plan will be developed in two main steps: definition of the policy strategies on a level with little detail, which will be subjected to an SEA, and a second version of the plan with more details on the policy actions which will include the results of the environmental assessment. The plan is developed in an interactive way with all involved stakeholders in a participative process: environmental impacts are integrated in the plan development process. The SEA document will be in fact the evaluation of the draft plan, but will probably already have a minimum of environmental impacts. A public enquiry will be organized with a sound board of representative organizations (Ministry of the Flemish Community 2004a).

A point of interest in this SEA is that many of the policy measures are not yet concrete enough to determine the exact impact on the environment. Organizational measures e.g. change in legislation, can have indirect environmental impacts but it is difficult to describe exactly the whole chain of causes and effects. The final decision making process on this policy plan will be fully supported by an environmental assessment.

9.3.3 River Flood Management Programme for the Scheldt River (SIGMA-plan) – SEA 2003-2004

The programme aims at flood safety and is being developed in close co-operation with all involved parties (Ministry of the Flemish Community 2004b).

The SEA of this programme evaluates different packages of infrastructure measures on the river to control floods now and in the future, taking into consideration that the climate change will lead to a rise in sea level. The SEA compares different strategies: e.g. the heightening and strengthening of dikes, "space for the river" (development of controlled inundation areas), a stormsurge barrier, and the possibility to create a new channel between the Eastern and Western Scheldt. Also combinations of these strategies are evaluated.

The programme is developed on a sufficiently detailed level, so that all environmental impacts can be evaluated. At the same time a definite choice for one of the strategies is not yet made. The development of the SIGMA-plan and the SEA are performed together so that the final version of the plan will include the results of the environmental assessment. Also the decision making process seems to follow all necessary steps to have the environmental aspects support this process.

9.3.4 Master Plan Mobility for Antwerp: an Infrastructure Programme – SEA 2002-2004

The master plan mobility for the city and port of Antwerp is an infrastructure programme. The SEA addresses both the different options for the infrastructure projects (road, rail, bus, tram, waterways) and a plan alternative where infrastructure development is replaced by economic measures (toll and road pricing, to limit the growth of the mobility on the highways around the city) (Management Company for the Mobility of Antwerp 2004).

The problem with this SEA is that is came into the picture at a very late stage in the decision making process. Previously taken decisions have an impact on the objective evaluation of environmental impacts at different levels. However, it seems that the decisions taken seem to lead to the probably most environmentally acceptable solution for the mobility problem around Antwerp. One could consider the whole process of discussions between government officials, technical experts and environmental experts as a fully integrated SEA process, even before it officially was called an SEA process. Involvement of the public during the whole SEA process is actively organised, through information encounters with the broad public.

The infrastructure projects in the programme were developed in a different level of detail (Van Dyck 2002a, 2002b). For the tramway lines concrete project descriptions were already available and at the same time the alternatives of the combined use of the international ring road and a city ring road were only developed as general concepts. This creates difficulties in evaluating plan alternatives on a set of evaluation criteria. SEA research therefore was performed both on a plan level covering the whole Antwerp region and on the level of separate areas within the Antwerp region.

9.4 Common Points of Interest for Future Use of SEA in Flanders

In this section, comments on these first experiences of SEA in Flanders are presented. The remarks are the personal opinion of the author.

9.4.1 Plans are not the same as Programmes

It seems that there is a need for a different approach for plans and for programmes. A policy plan includes the development of a general policy vision for the long term and results in a set of policy strategies (Fischer 2001). To evaluate the environmental impacts of policy strategies will demand expert judgment rather than detailed modelling of environmental impacts. Most of the time, these policies or plans are inspirational for political intentions. Political discussion will lead to the definition of concrete programmes with a defined set of policy measures and project combinations. It will be easier to determine the magnitude of the environmental impact of these programmes on a more quantitative basis via modeling and calculations. In a programme evaluation choices will be made and priorities will be set. Table 9.2 shows the differences in characteristics. In the Flemish context, it seems there is need of a different SEA approach to general policies and plans and the concrete programmes, which are derived from these plans.

Table 9.2. Differences in characteristics in SEA for plans and for programmes (Marc Van Dyck 2002a)

Characteristics	Policies/plans	Programmes
Political decision process	Giving direction; policies are inspirational for political intentions	Making choices
Alternatives	Visions (policies) / strategies (plans)	Project combinations / location and timing
SEA process	Integrated; open to all alternatives solutions; scenario development	Only cumulative impacts; limited choices
Evaluation process	Iterative; open to new options; vague	Not fully iterative; specific evaluation qualitative and quantitative
Stakeholder involvement	In a political process with as much scientific basis as possible	In a scientific evaluation process

9.4.2 The Status of the Strategic Action Decision is not always what it should be

It seems that there is a different situation for all strategic actions concerning the status of the decision that is already taken, before an SEA procedure is started up. In an ideal situation no decision is taken at all and all options are open. Most of the time, however, a few principal decisions already have been taken on a political level before the real evaluation of impacts can begin. In the case of an infrastructure programme, there is sometimes already a principal decision, before the first evaluation is made. The reason is that this political decision can have a positive short term impact on the public, who often is waiting for a concrete solution for a societal problem.

In Flanders, where the structured planning and decision making philosophy, including the use of EIA and SEA as an evaluation tool, has only entered into policy preparation about five to ten years ago, there is still a transitional phase going on towards a more structured form of planning and decision making processes.

9.4.3 Tiering of Plans and Programmes in a Decision Making Process still is a new Item

Normally a policy plan should be developed before a concrete programme is determined. In practice most programmes are solutions to practical problems in society, e.g. mobility problems or economic problems, and there is no time to make a general policy plan to evaluate all policy options, before a concrete programme is developed. Planning in advance seems to be the solution, but the faster societal changes do not always allow long planning processes.

Political decisions and policy preparation still need to be made more compatible in Flanders and Belgium, also for fast changing societal situations. Use of scenarios in the development of plans and in SEA could help to realize more robust policy preparation.

9.4.4 There are still Problems with the Level of Detail of the SEA for a Strategic Action: Where do we Draw the Line?

This seems to be a common problem in all SEA implementation processes: where is the thin line between a project, a programme, a plan or a policy? And what does it mean for the level of detail that is needed in the SEA research (Fischer 2004)?

In Flanders discussions are going on between the EIA experts and the Flemish administration about the level of detail that is needed in each phase of the decision making process, going from general policies and plans to concrete projects, which need a building permit or an environmental permit. At this moment there still seems to be a strong demand to have all the detailed information in the first evaluation step – the SEA. For the higher level of decision making on general policy options, this level of detail is not needed. When the SEA is filled with detailed information, the main policy choices to be made and the questions to be answered might get lost in the details: in that way SEA risks becoming an inappropriate instrument for the intended decision making level.

9.4.5 Should we Involve the Stakeholders from the Beginning or later on?

In the first experiences with SEA, the early public inquiries led to great protest because the broad public wanted to have more detailed information about the plan. Also the study of plan alternatives is sometimes not accepted. Protest arises, even from mentioning the different options in a strategic policy action (Mathiesen 2003). Because of the limited experience in Flanders with sound boards and public enquiries during the process of plan development (Devuyst et al. 2000), some unsuccessful measures have been taken to steer the plan or programme or the SEA evaluation process. For the broad public the difference between "different optional strategies" and "a final decision" is not always clear.

In Flanders there is a need for more experience in the handling of controversial public enquiries and to respond appropriately to the protest that arises. Most of the time, the protest of a few "bigmouths" is dominant in the discussions and the risk arises that this will influence the quality of the SEA evaluation processes. Involvement of the public should be organized during the full length of the decision making process. Political interference should be avoided as much as possible during the preparatory study phases and in the public communication of the SEA results.

9.4.6 There is a Need for Generally Accepted Criteria for the Evaluation of Environmental Impacts

The development of a set of generally accepted criteria (Kornov and Thissen 2000), derived from the general policy orientations in the Flemish society, is still an ongoing process. At this moment, it is difficult for the SEA author to evaluate environmental impacts because the policy goals are still unclearly defined. The main cause of this problem is the lack of existing policies and plans in different policy fields. The only solution to this problem, in this pilot phase of the development of the instrument SEA in Flanders, is to determine the strategic action goals in participative way with all stakeholders. If the goals have a sufficiently general character, they could be used in future SEA processes.

9.5 Conclusions and Recommendations

The implementation of SEA in Flanders is being realized at this time in different strategic actions. The pilot cases allow presenting a few common points of interest to enhance smooth implementation of SEA.

At the same time the government administration in Flanders has understood that a good planning process is needed to develop plans and projects, and that involvement of all parties can be an advantage, rather than a disadvantage.

The general conclusion is that the transitional phase towards a structured planning process with public enquiries has been stimulated and supported by the development of EC legislation and the EC Directives. The SEA Directive still needs to be linked better to all other existing planning processes, but a big leap forward has been taken in the last five years. However, the implementation on a national level still encounters a few difficulties.

References[3]

Berube GG, Cusson C (2002) The environmental legal and regulatory frameworks. Assessing fairness and efficiency. Energy Policy 30(14):1291-1298

Devuyst D, Van Wijngaarden T, Hens L (2000) Implementation of SEA in Flanders: Attitudes of key stakeholders and a user-friendly methodology. Environmental Impact Assessment Review 20(1):65-83

Fischer TB (2001) Practice of environmental assessment for transport and land-use policies, plans and programmes. Impact Assessment and Project Appraisal 19(1):41-51

Fischer TB (2004) Transport policy making and SEA in Liverpool, Amsterdam and Berlin – 1997 and 2002. Environmental Impact Assessment Review 24(3):319-336

---

[3] Note that legislation mentioned in the chapter is listed at the end of the handbook in the consolidated list of legislation (Appendix 2).

Kornov L, Thissen WAH (2000) Rationality in decision- and policy-making: Implications for strategic environmental assessment. Impact Assessment and Project Appraisal 18(3):191-200

Management company for the Mobility of the City of Antwerp (2004) SEA for the masterplan Mobility Antwerp. Resource Analysis, work in progress

Mathiesen AS (2003) Public participation in decision-making and access to justice in EC environmental law: The case of certain plans and programmes. European Environmental Law Review 12(2):36-51

Ministry of the Flemish Community – Department of Environment and Infrastructure – Administration Environment, Nature, Water and Land management – Unit Air Policy (2001) Environmental Impact Assessment for the draft Mobility plan Flanders using SEA. University of Antwerp – Institute for Environmental management, 22nd of September 2001

Ministry of the Flemish Community – Department of Environment and Infrastructure – Administration of Waterways and the Marine – Unit Policy Ports and Waterways (2004a) Long Term Vision for the Flemish ports including SEA. Resource Analysis, work in progress

Ministry of the Flemish Community – Department of Environment and Infrastructure – Administration of Waterways and the Marine (2004b) SEA of the SIGMA-plan – flood management programme on the river Scheldt. Resource Analysis, work in progress

Partidário MR (2000) Elements of an SEA framework – Improving the added-value of SEA. Environmental Impact Assessment Review 20(6):647-663

Risse N, Crowley M, Vincke P, Waaub J-P (2003) Implementing the European SEA Directive: The Member States' margin of discretion. Environmental Impact Assessment Review 23(4):453-470

Thérivel R (2004) Strategic Environmental Assessment in Action. Earthscan, London

Van Dyck M (2002a) What to do when SEA comes later into the decision making process? – case: the Antwerp Mobility Masterplan. Proceedings of the conference of IAIA'02, The Hague

Van Dyck M (2002b) SEA process of the Mobility Masterplan Antwerp, Proceedings of the SEA congress of the Province of Northern Italy, Milan

10 Implementing SEA in Austria

Ralf Aschemann

Austrian Institute for the Development of Environmental Assessment (An !dea), Austria

10.1 Introduction

This chapter deals with the actual state and development regarding the transposition of the requirements of the SEA Directive in Austria. In the revision of the EIA Act 2000 there are no provisions for SEA. However, beginning with the mid 1990s numerous SEA related activities can be named, such as legislative efforts (see Sect. 10.2) and case studies (see Sect. 10.3).

In order to understand these measures it is necessary to briefly illustrate the Austrian political system. The legislative and executive competences are split between the federal and the provincial level (in all there are nine provinces in Austria). For example mining matters, water management and forestry are federal competences, whereas spatial planning and nature conservation are responsibilities of the provinces. Consequently, Austria's activities to implement the SEA Directive involve actors both at federal and provincial level. The main ones are:

- Federal Ministry of Agriculture and Forestry, Environment and Water Management
- Federal Ministry of Transport, Innovation and Technology
- relevant departments of the nine administrations of the provincial governments

Currently, an "Austrian Convention", intending to finish its task at the end of 2004, is working on renewing Austria's constitution and its administrative system. Its work might lead to some changes in this division of competences. In the Convention, all relevant groups of the Austrian society are represented.

The structure of the chapter is as follows: At first, legislative efforts to implement the SEA Directive are presented, see Sect. 10.2. Afterwards, eight Austrian SEA case studies will be briefly tackled (see Sect. 10.3), whereas the SEA of the land use plan of Weiz will be discussed in more detail (see Sect. 10.4).

Implementing Strategic Environmental Assessment. Edited by Michael Schmidt, Elsa João and Eike Albrecht. © 2005 Springer-Verlag

Furthermore, the text focuses on some accompanying measures in the context of implementing the SEA Directive (see Sect. 10.5). Finally, some conclusions and recommendations will finish this chapter, see Sect. 10.6.

10.2 Legislative Efforts to Implement the SEA Directive

The Federal Government amended the Water Management Act 2003, in order to integrate the requirements of the SEA Directive (and other pieces of European legislation like the Water Framework Directive) and it came into force on 22 December 2003. With the same intention, the province of Salzburg amended its Spatial Planning Act, which became effective on 1 May 2004. Hence, only two pieces of Austrian legislation reached the deadline of the SEA Directive and came into force before 21 July 2004. For an environmental progressive country like Austria this is quite disappointing.

Further legislation transposing the SEA Directive is on its way, for example there is a proposal regarding amendments of the Federal Waste Management Act. Moreover, there are ongoing preparations for adapting federal laws concerning clean air, transport and noise protection in order to comply with the SEA Directive. At the provincial level, proposals exist to amend the Spatial Planning Act of Lower Austria and to create a new Carinthian Act on Environmental Planning. In Styria, the competent committee of the provincial parliament is dealing with a first proposal amending the Styrian Spatial Planning Act. This means that the introduction of SEA in Austria will mainly be realized through amending a lot of existing pieces of legislation instead of creating new ones (the only exception is the mentioned Carinthian proposed act).

In order to support provincial governments implementing the SEA application for land use plans in a more homogeneous way, two guidance documents have been published. One is dealing with process and implementation issues (Vorschläge 2003), the other focuses on methodological issues (Methodenpapier 2004). Both of them are not binding, but a good source of advice.

10.3 Case Studies

Amongst the described legislative tasks there exist various activities to prepare implementing the requirements of the SEA Directive. The elaboration of SEA case studies is to be highlighted in this context. A number of voluntary SEA and SEA-like pilot studies have been undertaken in Austria since 1994 in order to gain as much practical experience as possible. These cover different geographical areas and planning sectors. Table 10.1 gives an overall view of all eight case studies.

Five of the eight SEA listed can be judged as being full or comprehensive SEA, two (Danube corridor, Lower Austria) as partial SEA, and one (Graz local energy plan, that started back in 1994) was not designed as an SEA process, but main parts of it could be interpreted as one. The most recent SEA of the waste management plan for the Salzburg province started at the beginning of 2003. Its environmental report was finished on 18 February 2004 and is available via the project's website (Salzburg Province 2004), unfortunately only in the German language.

With the exception of the SEA of the Local Energy Plan of Graz City, all other Austrian SEA case studies have been elaborated within the last six years (1998-2004). This is a result of the publication of the proposal of the SEA Directive in 1996 (COM (96) 511 final). This document acted as an initiating starting point for many Austrian SEA activities beside the case studies, see Sect. 10.5.

It is important to state that the Austrian SEA pilot projects were undertaken under some specific circumstances. They are likely to be different from future routinely applied SEA to a certain extent because of the following reasons. All pilot SEA received additional resources in terms of staff and money.

Table 10.1. Austrian SEA case studies (1994-2004)

Case study	Type of SEA	Who was responsible for the SEA	Period for conducting the SEA	Further information (all in German language, websites last accessed 26 August 2004)
Local Energy Plan of Graz City	p	Graz City Council	1994-1996	graz.at/umwelt/kek.htm
Land use plan of Weiz	f	Weiz City Council	1998-1999	See section 10.4
Regional development plan for the Danube area in Lower Austria	p	Provincial government of Lower Austria	1999	Final report is unpublished and not available
Regional programme of Tennengau	f	Provincial government of Salzburg	1999-2001	www.salzburg.gv.at/stra_tennengau
Demonstration study Danube corridor (transport)	p	Federal Ministry of Environment	2000	www.bmvit.gv.at/sixcms_upload/media/231/band004.pdf
Waste management plan of Vienna City	f	Vienna City Council	2000-2001	See Chap. 42 of this book
Urban and transport development in the North-East of Vienna City	f	Vienna City Council	2001-2003	www.wien.gv.at/stadtentwicklung/supernow/
Waste management plan of Salzburg Province	f	Provincial government of Salzburg	2003-2004	www.salzburgerabfall.at

p: partial SEA, contains only some SEA elements
f: full SEA, contains all or nearly all SEA elements

With these additional means, the SEA were able to involve many experts, including spatial and landscape planners, transport professionals and communication experts. This would have not been possible under normal circumstances, because there is no possibility to provide a considerable extra amount of money for the future "daily SEA practice", considering the recent context of rethinking and redefining public tasks and shrinking public budgets. Nevertheless, all future Austrian SEA will have to comply with the requirements of the SEA Directive.

For more details on these case studies, two recent publications are particularly useful: A special issue of the magazine of the Austrian Society for Spatial Planning was dedicated to SEA (Forum Raumplanung 2004). Another publication highlights lessons learned from Austrian SEA (Aschemann 2004). This last publication focuses on three main issues: the involvement of stakeholders and the public, the methods in the assessment stage, and the impact of SEA on decision-making.

The practical experiences demonstrated that all Austrian pilot SEA improved the planning process itself (for example relating the consideration of alternatives, the systematic analysis of its environmental consequences or the documentation of its likely significant environmental effects). To a certain degree the pilot projects also contributed to a better environmental quality and to the quality of the adopted strategies (Bass et al. 2003).

10.4 SEA of the Land use Plan of Weiz

The case study dealt with the third revision of the land use plan of the municipality of Weiz, a regional capital with approximately 9 300 inhabitants and a total area of 507 hectares, located in the Northern part of the Austrian province of Styria. The legal basis for this plan revision is the Styrian Spatial Planning Act. The planning preparations started in 1997, the revision was adopted in 1999. This SEA might be seen as the Austrian pioneering one, because it was the first comprehensive SEA in the country (see Table 10.1).

This voluntary SEA was commissioned by the Austrian Ministry of Environment, Youth and Family Affairs and the Department of Spatial Planning of the Styrian provincial government. Its approach was to integrate the SEA into the spatial planning process, using for example one common public participation phase, both for the SEA and the planning. The key actors of this SEA were the City Council of the Municipality of Weiz as the responsible authority for the elaboration and adoption of the plan; the Styrian provincial government, both as final approving authority for the land use plan and as environmental authority for the SEA; and the public concerned. A consultation with possibly affected Member States was not necessary, because the SEA did not identify significant transboundary environmental effects. After three scoping meetings a scoping document has been developed. A crucial issue of any scoping document is to decide on the indicators, to be used for assessing the likely significant effects of the plan and its alternatives on the environment. A selection of these indicators is summarized in Table 10.2.

Table 10.2. Selected indicators for the assessment of the land use plan Weiz (Translated from SEA report (Aschemann 1999))

Issues	Indicators
Human beings	Emissions and their impact on human health, noise
Fauna and flora	Biodiversity, sensitive areas, special forests
Soil	Agricultural areas, contaminated sites, pollutants in soils
Water	Quality of groundwater, wastewater standards
Air	Sources and trends of emission, ambient air quality, extent of smell
Climatic factors	Greenhouse gas emissions, frequency of smog situations
Landscape	Valuable areas, aesthetical factors, protected areas
Material assets and cultural heritage	Archaeological sites, protected buildings

Beside the assessment of (a) the no action alternative and (b) the (old) land use plan of 1994, two realistic and feasible planning alternatives have been developed by a planner, who was commissioned by the municipality and was involved in the SEA process, too. One alternative (c) is dealing with the intentions of the municipality of Weiz and (d) the other is proposed as environmentally-friendly planning option.

Table 10.3. Scheme of an assessment matrix (blank) (Translated from SEA report (Aschemann 1999))

Area No. x	(a) no action alternative	(b) old land use plan	(c) land use plan municipality driven	(d) environmentally friendly land use plan
General description				
Environmental indicators				
Socio-economic indicators				
Weighting process				
Recommendations, mitigation measures, comments				

Note: The second and third line of the matrix has to be filled to represent the magnitude of the impact on the environment: 1 (very positive impact), 2 (positive impact), 3 (neutral), 4 (adverse impact) and 5 (very adverse impact).

The assessment was done in a comprehensive way, evaluating the environmental (see Table 10.2) and the socio-economic impacts (using six indicators for components such as local economy, settlement, social infrastructure or technical infrastructure) of the four options mentioned, focusing on 25 key areas with intended important land use changes. These areas have been chosen by the planner mentioned in co-operation with the municipality. The assessment used a matrix for each of those areas (Table 10.3).

Beside the judgement of the environmental effects of all 25 areas mentioned of the plan and its alternatives, such a matrix approach offers another opportunity. It allows identifying potential conflicts between environmental and socio-economic planning interests (laid down in the line 'weighting process' of Table 10.3). Those observations serve as valuable support and crucial input for the decision-makers and lead for example to propose appropriate mitigation measures.

The assessment described above led to a preliminary environmental report (including maps and assessment matrices) for the draft plan. The draft plan and the draft environmental report have been made available to the stakeholders (relevant authorities and interest lobbies as laid down in the Styrian Spatial Planning Act such as the Chamber of Commerce, the Chamber of Labour or surrounding municipalities) and the general public for a period of eight weeks, offering the possibility to comment on both documents. In all, 15 comments have been received, but disappointingly none from the public. Additionally, two meetings on the drafted plan and on the SEA have been organized and a non-technical summary of the SEA issues has been published in the local newspaper of the municipality, which is free of charge for every Weiz household.

Unfortunately, the influence of the SEA of the Weiz land use plan on the decision-making was limited. The City Council adopted a final land use plan, which led to some changes compared to the recommendations of the environmental report. These changes can be judged as a step towards a degradation of environmental quality in terms of soil protection, climate, landscape protection, recreation and noise (Aschemann 1999). This demonstrates that SEA serves as supporting tool for the decision-making bodies, but these bodies can disregard the recommendations of the SEA, when other overruling interests are stronger.

Nevertheless, the SEA of the land use plan of Weiz reached its expected objectives and therefore gained experiences regarding methodological, procedural and communication issues of an SEA. These experiences served as a valuable source for the following Austrian case studies (see Sect. 10.3). The main lessons learned are summarized in the following (Aschemann 1999):

- Start the SEA as soon as possible in order to prevent "informal" planner's work creating unchangeable framework conditions
- The integration of SEA into the planning process improved its quality
- The communication and co-ordination processes within the planning process are as much important as procedural and methodological issues
- Public participation could be more successful by taking care for professional public relation measures

- Some research needs have been identified such as analysis and assessment of indirect and cumulative effects, dealing with uncertainties at the strategic level and the necessity of practical guidance material

10.5 Accompanying Measures

The preparation process for implementing the SEA Directive in Austria has not only included the pilot SEA mentioned, but involved a range of other accompanying activities. These include the elaboration of studies, the organization of training courses and other measures. Some of them are briefly described in the following:

- A review of international and national SEA experiences and approaches was done in 1995 and 1996, commissioned by the Federal Ministry of Environment, Youth and Family Affairs (Jorde et al. 1997). It collected and analysed SEA experiences in three European countries (Germany, the Netherlands and United Kingdom) and North America (Canada, USA), based on comprehensive interviews with 58 SEA experts. Moreover, it gave first hints for an SEA introduction in Austria by analysing its strategic actions both at federal and provincial level.
- A study on the environmental assessment of policies and legislation was elaborated, focusing on Styria (Aschemann 2001). It collected experiences in this field, analysing approaches especially in Denmark, Finland, Norway and the Netherlands. They were used as a basis for designing an assessment process of policies and legislation in Styria.
- Recently, studies on the range of potential plans and programmes that will be subject to the SEA Directive (for Austria and for the City of Vienna) have been completed. Moreover, a proposal for SEA screening procedure and criteria has been elaborated. Finally, there are ongoing studies on the SEA assessment methodology and the SEA monitoring. Both are expected to be completed at the end of 2004.
- Two series of training workshops for administration staff have been held in 1997 and 2001/2002. Furthermore, three bigger SEA conferences have been organized in 1996, in 1998 during the Austrian EU Council Presidency, and the most recent one in April 2003. Different working groups are meeting regularly, e.g. the one dealing with "SEA and transport". Last but not least, a number of national and international presentations and papers and a handbook on SEA (Bass et al. 2003) illustrate various aspects of the Austrian SEA activities.

For some of the case studies, evaluation studies have been commissioned in order to identify SEA benefits and obstacles and to conclude some lessons learned. One is dealing with the SEA of the land use plan of Weiz (see Sect. 10.4), another one analyses the experiences of the SEA of the Viennese waste management plan, see Chap. 42. The third one focuses on lessons learned with the SEA of the urban and transport development in the North-East of Vienna City, it is expected to be finished at the end of 2004.

10.6 Conclusions and Recommendations

Austria is on its way to fully transpose the requirements of the SEA Directive. Before the deadline of transposing the SEA Directive only two pieces of legislation have been amended, but numerous others are in the pipeline. Beside these legislative tasks, Austria gathered a lot of experience by undertaking seven SEA case studies since 1998. Additionally, a lot of studies have been elaborated in the SEA context, and workshops and conferences have been organized.

Due to the relatively advanced state of the Austrian planning system with regard to data availability, know-how of the authorities and a well developed environmental consciousness, there have been neither serious problems concerning the methodologies used for the assessment nor regarding the existence of baseline data, because there is an established monitoring system for the main environmental indicators. Every three years, the Federal Environment Agency is reporting on the "State of the Environment in Austria", collecting comprehensive data related to Austria on the following issues (State of the Environment in Austria 2002):

- Population and growth of built-up area
- Emissions and air quality
- Global climate change
- Stratospheric ozone depletion
- Water
- Soil
- Forest
- Nature protection
- Agriculture
- Transport
- Industries
- Eco-audit according to the EMAS regulation
- Waste
- Contaminated sites
- Energy
- Noise
- The safe handling of chemicals
- Pesticides and other biocidal products
- Gene technology
- Radioecology

Proceeding to the future aspects, one has to think about the emerging Sustainability Impact Assessment (SIA), see for example Thérivel (2004). SIA analyses, identifies and assesses the environmental, economic and social effects of a proposed action. Similar to other countries and following the Communication from the European Commission on Impact Assessment (CEC 2002) Austria is exploring the possibilities of SIA of policies and legislation. The Federal Ministry of Agriculture and Forestry, Environment and Water Management has commissioned a study and its results are expected at the end of 2004.

Here we are in the centre of one of the current SEA discussions. Is sustainability appraisal really "an alternative to the environmental focus of SEA" (Sheate et al. 2003 p.15)? And how much "care is needed to ensure the environment is not diminished in decision-making as a consequence of taking a more 'integrated' approach through sustainability appraisal" (Sheate et al. 2003 p.15)? Gibson (2002 p.157) is asking a similar question, whether "a shift to sustainability-based assessment could reduce the attention given to ecological considerations". The future SEA application will show, whether SEA is developing into some kind of strategic sustainability assessment or not.

The future development and application of SEA in Austria has to be waited for, but the potential of SEA to integrate environmental concerns into all sectors of strategic planning seems to be appropriate to react to an OECD report regarding Austria's environmental performance (OECD 2003). This report recommends inter alia to "further integrate environmental concerns into spatial plans at provincial level and into planning and zoning decisions at municipal level" and to "ensure that nature conservation objectives are more systematically incorporated into spatial planning at provincial level, and planning and zoning at municipal level (OECD p.17, 22)." The regular SEA application as consequence of the SEA Directive seems to be a very promising approach in order to overcome the deficiencies mentioned in the OECD report.

References[1]

Aschemann R (1999) Endbericht zum Pilotprojekt "Strategische Umweltprüfung (SUP) des Flächenwidmungsplanes 3.0 (FWP) der Stadtgemeinde Weiz". Erstellt im Auftrag des Bundesministeriums für Umwelt, Jugend und Familie (Abt. I/1) und des Amtes der Steiermärkischen Landesregierung (Fachabt. Ib) (Final report for the pilot project "SEA of the third revision of the land use plan Weiz". Commissioned by the Federal Ministry of Environment, Youth and Family Affairs and the Provincial Government of Styria). Arbeitsgruppe Wissenschaftsladen Graz/Büro Arch. DI Hoffmann, Graz

Aschemann R (2001) Umweltbeurteilung von Gesetzen und Verordnungen – Schwerpunkt Steiermark. Gefördert vom Bundesministerium für Land- und Forstwirtschaft, Umwelt und Wasserwirtschaft, Umweltanwalt des Landes Steiermark (Environmental assessment of policies and legislation, focusing on Styria. Commissioned by the Federal Ministry of Agriculture and Forestry, Environment and Water Management and the Environment Ombudsman of Styria). Österreichisches Institut für die Entwicklung der Umweltfolgenabschätzung, Graz

Aschemann R (2004) Lessons learned from Austrian SEA. European Environment 14(3):165-174

Bass R, Thérivel R, Arbter K (2003) Handbuch Strategische Umweltprüfung. (Handbook on Strategic Environmental Assessment). 2. überarbeitete, stark erweiterte Auflage, 2. Aktualisierungs-Lieferung Juni 2003. Institut für Technikfolgen-Abschätzung der Österreichischen Akademie der Wissenschaften, Wien

EC – Commission of the European Union (2002) Communication from the Commission on Impact Assessment. COM 2002 (276) final, Brussels, 5 June 2002

---

[1] Note that legislation mentioned in the chapter is listed at the end of the handbook in the consolidated list of legislation (Appendix 2).

Forum Raumplanung (2004) Nützt die Strategische Umweltprüfung (SUP) der Stadt- und Regionalplanung? (What is the use of strategic environmental assessment for city and regional planning?). Österreichische Gesellschaft für Raumplanung, Wien

Gibson RB (2002) From Wreck Cove to Voisey's Bay: the evolution of federal environmental assessment in Canada. Impact Assessment and Project Appraisal 20(3):151-159

Jorde T, Aschemann R, Hittinger H (1997) Umweltprüfung für Politiken, Pläne und Programme, Untersuchung der Umsetzungsmöglichkeiten in Österreich. Teil 1: Erhebung in- und ausländischer Ansätze, im Auftrag des Bundesministeriums für Umwelt, Jugend und Familie (Strategic environmental assessment – An analysis of possibilities for its implementation in Austria. Part 1: An exploration of experiences and approaches in selected foreign countries, commissioned by the Federal ministry of Environment, Youth and Family Affairs). Österreichisches Ökologie-Institut, Wien

Methodenpapier – Methodenpapier zur Umsetzung der Richtlinie 2001/42/EG in die Raumplanungspraxis Österreichs (2004) Endfassung vom 6.2.2004, Österreichische Raumordnungskonferenz (Methodological paper on implementation of the SEA Directive into Austria's spatial planning practice, final version from 6 February 2004, Austrian Spatial Planning Conference). Österreichische Raumordnungskonferenz, Wien

OECD (2003) Environmental Performance Review Austria, Paris

Salzburg Province (2004) Website of the SEA of the waste management plan of Salzburg province. Internet address: http://www.salzburgerabfall.at, last accessed 26.08.2004

Sheate WR, Dagg S, Richardson J, Aschemann R, Palerm J, Steen U (2003) Integrating the Environment into Strategic Decision-Making: Conceptualizing Policy SEA. European Environment 13(1):1-18

State of the Environment in Austria (2002) Edited by the Federal Environment Agency Ltd., Vienna

Thérivel R (2004) Strategic Environmental Assessment in Action. Earthscan, London

Vorschläge der Länderexperten für die Umsetzung der SUP-Richtlinie (2003) Der zugehörige Bericht wurde von der Landesamtsdirektorenkonferenz am 29.10.2003 angenommen (Proposals of the provincial experts for implementing the SEA Directive. The corresponded report has been adopted by the Conference of the Heads of the administrations of the Austrian provincial governments on 29 October 2003). Landesamtsdirektorenkonferenz, Wien

11 Implementing SEA in Finland – Further Development of Existing Practice

Mikael Hildén and Pauliina Jalonen

Finnish Environment Institute, Finland

11.1 Introduction

Finland introduced a general requirement to assess the environmental effects of policies, plans and programmes, henceforth strategic actions, in the Environmental Impact Assessment Act of 1994 (EIA Act, 468/1994), but the requirement is limited to one section in the EIA Act (Section 24). The Act does not specify how assessments are to be carried out, but mandated the Ministry of the Environment and the Council of State to issue guidelines. Such guidelines have subsequently been issued on the assessment of policies, plans and programmes (Ministry of the Environment 1998) and separately for government bills (Council of State 1998). The Ministry of the Environment has also supported environmental assessments of plans and programmes through training and demonstration activities and expert advice. The Land Use and Building Act of 1999 (132/1999) includes explicit requirements to carry out environmental assessments of land use plans and these requirements are linked to the procedural stages of the preparation of land use plans. Single sections demanding the assessment of environmental effects of plans and programmes can also be found in other Acts, for example, in the Forest Act (1093/1996). In the Regional Development Act (602/ 2002) there is a general demand to assess development plans and programmes.

As a consequence of the legislation, a number of environmental assessments of strategic actions have been carried out in different branches of the administration in Finland. This chapter will, in the light of these experiences, address key challenges in implementing the SEA Directive. The legislation for the full implementation of the SEA Directive is in preparation.

According to normal practice in Finland the legislative work related to the SEA Directive was initiated by appointing an ad hoc ministerial working group in December 2001 to do the preparatory work (Working group for environmental assessment of plans and programmes, henceforth SEA WG).

Implementing Strategic Environmental Assessment. Edited by Michael Schmidt, Elsa João and Eike Albrecht. © 2005 Springer-Verlag

The group was appointed and chaired by the Ministry of the Environment, but included as members representatives of all key Ministries, Interior Affairs, Finance, Agriculture and Forestry, Trade and Industry and Transport and Communications. The Association of Finnish Local and Regional Authorities and one Regional Environment Centre were also represented. In addition, the senior author of this Chapter participated as an expert representative of the Finnish Environment Institute. The SEA WG completed its work by submitting a proposal for SEA legislation in June 2003 (SEA WG 2003). After that the Ministry of the Environment has carried the work further. The proposal has been reviewed by stakeholders and the Ministry of the Environment will submit a bill to the Parliament in early autumn of 2004 with the purpose of establishing a SEA Act that meets the requirements of the SEA Directive and also the SEA Protocol of the UN ECE Convention on Environmental Impact Assessment in a Transboundary Context (the Espoo Convention).

The chapter is organised as follows. After a general discussion on the scope of the SEA obligations, Finnish SEA practice is reviewed in the light of the demands of the SEA Directive. This is followed by a general discussion highlighting some of the key issues in the implementation. The last section presents conclusions and recommendations.

11.2 Scope of Application – the EIA Act and the SEA Directive

The Finnish EIA Act (Section 24) requires that an environmental assessment be made for all policies, plans and programmes that may have significant effects on the environment. This means that the Finnish requirement to carry out environmental assessments of strategic actions is clearly broader than that of the SEA Directive. The authority responsible for drafting the policy, plan or programme is responsible for determining the need for an assessment.

The two key restricting conditions of the SEA Directive are that the plan or programme has to be based on law, decree or administrative order and that the plan or programme should "set the framework for the consent of projects". Material compiled by the SEA WG has shown that these two conditions will restrict the application of the SEA Directive to a fraction of all policies, plans and programmes that are being prepared at different levels in the administration from municipalities to ministries (SEA WG 2003; Hildén et al. 2004b). The most difficult concept has been that of an administrative order, because the concept does not appear as such in any existing legislation and it can affect the scope of the application of the SEA Directive considerably.

Land use plans clearly fall under the SEA Directive. They also differ from most other strategic actions in that the Land Use and Building Act of 1999 specifies procedures for the preparation and the assessment of the plans. The procedural requirements cover regional land use plans, local master plans and the local detailed plans. The Act also requires the preparation of general national land use guidelines, which guide more detailed planning. One could argue that the national guidelines are a general expression for the land use policy and that they do not as such set the framework for individual decisions on consent, although the Land Use and Building Act requires Government authorities to take national land use guidelines into account, promote their implementation and assess the impact of their actions on local structure and land use (Section 24, The Land Use and Building Act).

Land use planning is also linked to road planning. In 1999 the Public Roads Act was revised by strengthening this link between road planning and land use planning. The Public Roads Act now includes an explicit demand that road planning must be harmonised with land use planning and also with the national land use guidelines (Public Roads Act, Section 10; the amendment 133/1999). In the SEA WG's proposal (2003) the national land use guidelines are included among plans and programmes requiring assessments according to the SEA Directive, because the document may also include fairly detailed requirements that can influence decisions on consent directly.

The Finnish experiences of the assessment of strategic actions have been sufficiently positive to maintain the general requirement to carry out assessment, but have not provided strong arguments for establishing additional fixed procedures. Therefore the SEA WG's (2003) proposal suggests that the general requirement of Section 24 in the EIA Act should be maintained in the new SEA Act. In this respect the scope of application of the Finnish SEA Act will be broader than that of the SEA Directive by including policies as well as plans and programmes that do not set the framework for the consent of projects. The procedural requirements of the SEA Directive will, however, be applied only to the subset of plans and programmes that are covered by the Directive.

11.3 SEA in Practice

The following discussion of Finnish experiences focuses on eight key features of SEA practice, which also roughly accord with the headings of the Art. in the SEA Directive: timing, tiering, consideration of alternatives, assessment methods and approaches, reporting, public participation, use of SEA in decision making, and monitoring. They illustrate likely challenges in implementation of the Directive, but highlight also more generally issues that are relevant in the assessment of policies, plans and programmes.

11.3.1 Timing

The introduction of the requirement for the environmental assessments of strategic actions in the EIA Act of 1994 revealed that the timing of the assessment is problematic. In some cases arguments about the need to carry out an assessment delayed the initiation of the assessment. Other delays in this regard were caused by arguments over whether or not an assessment can be made before a policy,; plan or programme is available, at least in draft form. Delays in the initiation of the assessment clearly affected negatively the possibilities to use the assessment results. This is also in line with international findings in transport planning (Hildén et al. 2004a).

For land use plans the question of timing has, in principle, been solved already at the level of legislation. The Land Use and Building Act requires a specific participation and assessment scheme to be prepared at the beginning of the planning. The monitoring of the implementation of the Building and Land Use Act which has shown that the meaning and role of such a scheme has remained somewhat unclear (Kangas et al. 2002), which demonstrates that legislation alone cannot solve timing problems.

11.3.2 Tiering

The requirement of Art. 4(3) of the Directive, namely that "Member States shall, with a view to avoiding duplication of the assessment, take into account the fact that the assessment will be carried out, in accordance with this Directive, at different levels of the hierarchy" includes the simple idea that one should not duplicate work. It is, however, more problematic to specify what kind of assessments should be carried out at a particular hierarchical level and to what extent an assessment at one level can exclude the need for assessments at another.

In Finnish regional development programmes there has been tiering within regions and links to other plans and programmes that have direct implications for an assessment. For example, Agenda 21 programmes and municipal policies have been used as a reference, since the driving force in the development of regional programmes is the regional council, which in Finland is formed by municipalities, i.e. a 'bottom – up' structure (Ministry of the Interior and Ministry of the Environment 1999). However, funding for regional development is channelled through regional government authorities, which represent a 'top – down' structure. These conflicting pressures have to be taken into account, although they do not represent an orthodox view of what tiering is about. Avoiding duplication in data collection is relatively easy, but the main issue and the time consuming phase of assessments relate to the discussions that take place in the preparation of the policy, plan or programme when several discussions will be "duplicated". This appears to be unavoidable, given the nature of the planning task.

Land use planning has incorporated a vertically tiered approach to SEA. In horizontal planning there should be links between regional scheme, regional development plan (without procedural requirements) and regional land use plan (with procedural requirements). However, the plans do not follow one another in a temporal sequence and therefore an ideal tiering that takes the assessment from the general broad brush picture to the detailed assessments of specific development activities is rarely possible. Links with other branches of the administration complicate the picture further.

A particular example of the difficulties related to tiering is the linking of land use planning and transport planning. Establishing working links is complicated due to the separate traditions and planning paces in these sectors. The interagency collaboration between regional councils, regional environmental authorities, municipalities and the Finnish Road Administration is difficult. It is also uncertain how detailed transport plans should proceed in land use planning. This shows that there is also tiering across different sectors and not only hierarchical tiering within sectors. Similar "cross-tiering" may occur within general regional development efforts and sector specific programmes.

11.3.3 Consideration of Alternatives

The Finnish Guidance on SEA (Ministry of the Environment 1998) has stressed the importance of alternatives and alternatives have generally been considered in Finnish SEA. Implementing the requirement to consider alternatives according to the SEA Directive is therefore not likely to be problematic. The Finnish experiences show, however, that alternatives can play very different roles: exploratory and visionary alternatives map possible worlds; variations on a single theme prepare the ground for a compromise; and demonstrative alternatives serve to prove that the chosen solution is the only possible or clearly the best alternative (Hildén and Jalonen 2003).

In land use planning explicit comprehensive alternatives are seldom drafted. Neither are they required by the Land Use and Building Act. This reflects in part the nature of land use planning, which differs from the preparation of many other policies and programmes. Land use planning is a process that produces new alternatives and excludes old ones continuously. Different plans have different temporal and spatial scales. Alternatives are produced and presented rather in individual projects or in local detailed plans. This also explains why there is really no definitive state of completed impact assessment. The assessment is, as the planning itself, a never-ending story, although parts of the land used get gradually fixed through incremental decisions. Sufficient agreement on the impact assessment needs to be found during the process for each incremental decision. The whole planning process is thus characterised by constant negotiation enabling certain trading between separate projects. This negotiated process is accentuated in large municipalities that undergo dynamic change due to expansion of population or economic development. It also explains why there are frequently difficulties in linking transport planning with land use planning: transport planning has traditionally focused on large projects with very concrete alternatives whereas land use planning is concerned with struggles over future options. Similar problems in matching land use planning with infrastructure planning have been encountered in the planning of other "linear" projects such as high voltage electricity lines.

11.3.4 Assessment Methods and Approaches

The SEA Directive does not specify the type of methods or approaches that should be used in an assessment of the issues listed in Appendix I of the Directive. Finnish assessments have used a wide range of methods, from quantitative modelling to collection of qualitative information and expert opinions. In fact, common to all is a combination of methods and approaches. Moreover, the assessment experiences so far indicate that it is not possible to undertake detailed quantitative analyses of all relevant aspects. Several assessments, such as the assessment of the programme for renewable energy resources, have hinted at the need to carry out a life cycle analysis (LCA) of some aspects of the policy, plan or programme, but so far no assessment have applied this approach. Given the available resources, it is unlikely that a LCA could be done as part of the assessments.

Another key methodological issue concerns the scope of assessment, in particular the combination of economic analyses with the environmental assessment. Most plans and programmes have links to economic activities and economic consequences, and for some there are established procedures for carrying out the economic appraisals. In the assessments of the Natura 2000 programme (Hildén et al. 1998) based on Habitats Directive and the National Forest Programme 2010 (Hildén et al. 1999), economic calculations were part of the assessment. In these cases, the combined assessment clearly increased the interest in and weight of the whole assessment. In the assessment of the National Climate Strategy (Box 1), the environmental assessment was carried out separately but co-ordinated with the economic analyses (Hildén et al. 2001; Forsström and Honkatukia 2001; Kemppi et al. 2001). This arrangement caused some confusion. The main interest was clearly in the economic assessment, but some issues, such as the views of different interest groups, could probably have been handled better, through a truly combined assessment.

All assessments have to deal with uncertainties and data deficiencies. These uncertainties have been noted, and in some of the assessments the uncertainties have been systematized. The uncertainties can also be exploited by various groups in conflicts related to the subject of the policy, plan or programme. The validity of various pieces of information becomes an issue of power and authority: who can say what and on what grounds, whose information is accepted and who will have to produce extraordinary evidence in order to convince others?

In order to achieve a systematic treatment of some uncertainties the assessment of the major transport infrastructure plan for southern Finland called the Nordic Triangle, separated the issues influencing the development into background variables and decision variables. In this way it was possible to highlight more clearly what the assessment was about (Valve 1999). In several assessments analyses of strengths, weaknesses, options and threats (SWOT) have been made and synthesised into best and worst case scenarios. These SWOT analyses have helped to diversify the view of what the strategic action actually is about.

The Land Use and Building Act sets certain standards on planning. These include general objectives in land use planning as well as specific content requirements for the various plans concerning these objectives. The general objectives include sustainable development. Regional planning emphasises regional and community structure (ecological sustainability of land use and environmental and economically sustainable arrangements of transport and infrastructure, for instance).

Box 11.1. Environmental Assessment of the Finnish National Climate Strategy (Hildén et al. 2001; Kumari et al. 2003)

Process: A group of key Ministers agreed on three basic scenarios (a baseline and two alternatives) for the national climate strategy. The assessment framework was developed in a steering group with representatives of all key Ministries (Environment, Agriculture and Forestry, Transport and Telecommunications, Trade and Industry). The assessment linked environmental, technical, economic and social aspects. Stakeholder participation was an integral part of the assessment and provided information on perceived characteristics of the scenarios and also on risks and opportunities associated with the scenarios. All assessment plans and results were made public.

The alternatives: The baseline scenario was developed assuming a steady economic growth, including growth in production industries. Population growth was assumed to be low. Assumptions were also made concerning the price of oil and the price of natural gas. The net import of electricity was assumed to decline. The alternative scenarios included a program supporting the development of renewable energy resources and a program aiming at saving energy in buildings and households. In one scenario an additional 1300 MW nuclear power plant was assumed whereas the other included an explicit prohibition to use coal in the production of electricity. The difference between the two alternative scenarios amounted to alternative ways of producing additional energy and to relatively small differences in the energy taxation.

Results: The technical, economic and environmental assessments provided an analysis of the energy use, greenhouse gas emissions, costs and environmental effects of the different scenarios until 2020. A synthesis of the available information was produced using the SWOT (strengths, weaknesses, opportunities, and threats) approach. A general observation was that alternative scenarios would be generally beneficial compared to the baseline. However, the differences between the alternative scenarios were small when done over a 10 year period, and would increase slightly over 20 years, but would still be limited.

The SWOT analysis further confirmed that the two alternatives did not make a great difference; but rather the factors assumed constant in the model (such as level and structure of energy taxation, electricity imports) would change the course of developments more than the measures assumed. Technical and economic assessments were linked and thus different aspects of the climate strategy could be subject to a simultaneous and balanced public review instead of dealing with one issue (environment, technology, economics, social aspects) at a time.

Conclusions: The assessment revealed that the scenarios were variations on a theme rather than explorative of distinctly different situations. The assessment concluded that the scenarios were myopic and too narrow in scope, and unable to capture all concerns and arguments on possible energy futures – thus limiting the scope for a broad public discussion. The SWOT analysis overcame some of these problems and the Parliament made extensive use of it in deliberations on the strategy. The Parliament requested some widening of the scope of the analyses. The present government is in the process of revising the climate strategy, thus showing the repetitive nature of planning and assessment.

The local master plan, in turn, must be drafted considering the community structure and living environment, i.e. functionality, economy and ecological sustainability of the community structure and opportunities for a safe and healthy living environment, which takes different population groups into equal consideration. How these contents requirements are assessed in the planning process has been the key question in land use. While it is too early to assess the fulfilment of general sustainability and living environment quality objectives, there have been some consequences stemming from the specific contents requirements (Kangas et al. 2002). Due to improved consultations, different content requirements are emphasised relative to those emphasised in the previous legislation. New impact assessment needs have also arisen. Specific guidance on how to carry out assessments have been produced, for example Söderman (2003) provides advice on biodiversity assessments in the context of spatial planning.

Spatial planning has also changed with the introduction of Geographical Information Systems (GIS) and other numerical tools (Irjala and Hallin 2002). GIS-based tools (see Chap. 24) have for example been developed for spatial planners in order to aid the combination of information on the natural environment and socio-economic factors. Box 11.1 gives an example of a SEA of the Finnish National Climate Strategy.

11.3.5 Reporting

In Finland public reporting of decision making material is mandatory and has a background in the constitution. Assessment findings have thus been reported in the actual documents for the policy, plans or programme. In some cases, the assessments have also been published as separate reports. There are also examples of "tiered reporting". The National Climate Strategy document itself included a brief mention of the environmental effects (Council of State 2001); the background document of the strategy devoted a chapter to the various assessments that were made (Ministry of Trade and Industry 2001); and, finally, the detailed assessments were published separately, as well as the sector specific material that was used in the preparation of the strategy (Hildén et al. 2001, Forsström and Honkatukia 2001, Kemppi et al. 2001). In land use planning the draft plan itself also works as assessment report although separate assessments are frequently published.

The most common form of publication has been that of printed reports. The more recent assessments have used the Internet as one important channel for the distribution of information, including progress with implementation. A further advantage of the Internet relates to its cost effectiveness as a way of presenting and distributing drafts of the policy, plan or programme and the assessment of the proposal. For example, the National Climate Strategy and its assessment were posted in draft form on the Internet and these www-pages have been retained and developed further (Ministry of Trade and Industry 2004).

The contents of assessment reports have varied from brief descriptions of likely effects that have been identified to more extensive analyses of consequences and alternatives. The original National Forest Programme 2010 contained a one page expert opinion based overview of environmental effects (Ministry of Agriculture and Forestry 1999), but the Chancellor of Justice did not consider this to be a sufficient assessment of the likely effects, which led to a detailed assessment that dealt with environmental, economic and social effects (Hildén et al. 1999).

The Land Use and Building Act has a stronger emphasis on the reporting than its predecessor. Providing information is nevertheless challenging. Stakeholder surveys have indicated that about 50 percent of the stakeholders consider the available information insufficient for active participation in the planning process (Roininen et al. 2003).

11.3.6 Public Participation

According to the Finnish constitution, everyone has the right of access to documents and recordings produced by, or in the possession of, the public authorities. The Act on the Openness of Government Activities (621/1999) makes it clear that the public authorities should inform citizens not only of their decisions but also of the preparatory work leading to decisions. This applies also to policies, plans and programmes and thus gives strong backing for public participation in the environmental assessments.

In the structural funds programmes, participation has been based on the concept of partnership. In practice, this has meant assembling a broad group of stakeholders, but not providing direct access for individual citizens. Similar broad groups with members representing various interests have been used in other assessments as well, but many have also provided broader access for the public (Hildén and Jalonen 2003). In a few assessments, attempts have been made to reach the general public by newspaper advertisements. These attempts, in the programme for renewable energy resources, for instance, have not been particularly successful. The response was limited and the general interest in the programme was low.

One of the innovations of the Land Use and Building Act is the specific participation and assessment scheme that has to be prepared for plans in the preparatory phase of planning (Section 63). In the scheme the target and objectives of planning, the key impacts to be assessed and the participants to be consulted are defined. Evaluations of the land use and building act suggest stakeholders have had some difficulties in grasping the role of the assessment scheme (Kangas et al. 2002).

In 2000-2001 negotiations with the regional environment centre on the adequacy and implementation of the scheme had been organised eight times out of 16 suggested implying that negotiations are not used very often. Deficiencies in the schemes were identified with respect to participation, studies and impact assessments. The municipalities and regional environment centres generally agree on sufficient participation possibilities (uncertainty on the sufficiency caused oversized participation arrangements in some cases). The sufficiency of studies and impact assessment has been more often than not been a significant cause of disagreement between the municipalities and the regional environment centre.

The continuous nature of land use planning as well as its long timeframe constrains public participation because the public may have difficulties in understanding the indefinite target of planning. Correspondingly the scope of the impact assessment can be the subject of never-ending discussions as new participants and new aspects enter during the planning process. Due to this organic character of land use planning, participation is also a continuous process with frequent openings as there are always new planning processes about to begin.

None of the strategic environmental assessments have systematically analysed the input provided by the participatory processes. Neither is there any requirement in the Land Use and Building Act to do so. This suggests that public participation has been seen as an expression of democracy and an expansion of democratic rights, but little consideration has been spent on the conditions for improving its effect on planning and on how one can develop analytical and systematic dialogues based on participation.

11.3.7 Use of SEA in Decision-Making

In principle assessments affect the final documentation and decision-making concerning policies, plans and programmes. It is, however, often difficult to verify in detail the actual influence on the choices made. Due to the nature of the decision-making process, individual findings seldom have a clear-cut effect on the outcome but instead may exert a gradual influence. In order to detect this influence, a separate evaluation that can clarify both the direct and indirect effects of the assessment would be necessary.

An evaluation of the assessment of the Helsinki Metropolitan Area Transport System Plan 1998 indicated a lack of strong immediate influence on decision-making, primarily because the basic agenda was set using criteria other than those dealt with in the assessment. This does not, however, mean that the assessment would not influence the planning, only that the influence is slower and more indirect than a rationalistic view of such assessments would otherwise indicate. (Kaljonen 2000). In transport planning the general emphasis has shifted from viewing assessments as a way of producing specific information to a more integrated view that makes the assessment a part of the preparatory process (Jansson 2000).

In land use planning the national land use guidelines and regional land use plans are general policies that are concretised in local master plans. Regional land use plans have more direct use in decision making in rural communities whereas in densely populated communities the solutions from regional plans are always specified in more detail.

Local land use planning is an interesting case since the local planning authority has a certain freedom to choose how strongly the plan should affect future development. According to the Land Use and Building Act it is possible to draft a local master plan without legal consequences (Section 45) making the plan a policy document instead of a plan. The choice is also likely to affect the nature of the impact assessment and the use of the impact assessment, but there is as yet insufficient empirical material to analyse these effects.

Land use planning is also a consistent example of representative democracy in action, as the final decisions on local plans are made by local councils. The Land Use and Building Act requires that reasoned responses are given to parties that have expressed opinions on the plan. The new planning procedure in the new Act aims at a consensus-like planning posture and emphasises the questions arising from the coexistence of participative and representative democracies and monitoring of the participation has been developed (Roininen et al. 2003). In most other plans and programmes the control exerted by the institutions of representative democracy is often indirect and weak, except for those that are approved at a high political level by the Government, some of which may be subject to considerations by the Parliament such as the National Climate Strategy (Council of State 2001).

11.3.8 Monitoring

The SEA Directive emphasises monitoring and some form of monitoring generally exists for most policies, plans and programmes in Finland (Paldanius and Tallskog 2003). It has therefore also been natural to include monitoring as a part of the proposed SEA legislation (SEA WG 2003). The Land Use and Building Act already specifically sets requirements for keeping plans up-to-date (for regional plans, master plans and local detailed plans Art. 26, 36 and 51 respectively). For local detailed plans a special monitoring obligation is also specified (Art. 60 and 61). These requirements were introduced in response to problems caused by the previous legislation under which an obsolete plan could become activated or used as an argument for projects that otherwise could no longer be justified.

The monitoring is part of the general policy cycle and includes the compilation of information, analysis, conclusions and use in the drafting of new strategic actions. The weakest part has often been the analysis (Paldanius and Tallskog 2003). The underlying causes of the observations made in the monitoring are not fully explored and thus the monitoring results provide less support for the revision of the plans than they potentially could. This is related to a broader need to develop evaluation practices in the environmental field (Mickwitz 2003) and is not limited to SEA. The problems are also reflected in the general problems that have been observed in the preparation of bills (Ervasti et al. 2000).

11.4 Discussion

Are all Policies, Plans and Programmes similar from an Assessment Point of View?: Fischer (2001) argues that it is possible to distinguish between "policy SEA", "plan SEA" and "programme SEA" and that assessments within each group are similar. In his classification, policy assessments are broad, scenario-driven and examine broadly different types of effects, which are fully integrated into the formulation of the policy itself. The SEA Directive excludes the policy level, but what about the plans and programmes? One particularly interesting comparison is that between land use planning and other plans and programmes, and also policies.

One key aspect is the continuous nature of land use planning and its long time frame. The incremental nature of planning is accentuated in land use planning and this is also reflected in the development and treatment of alternatives and also in certain aspects of the impact assessment. For example a partial master plan raises certain issues that are important within the scope of the partial plan itself, but then there are other issues that are relevant and that arise from the relationship between the partial master plan and the whole municipality.

The role of public participation is more widely emphasized and obvious in land use planning. The Finnish constitution gives everyone the right to influence decisions that concern their own living environment. The spirit of the Land Use and Building Act requires reflexivity in planning, and it was also expected to make planning culture more open and interactive (Roininen et al. 2003). The new planning procedure has clearly emphasized the role of participation and collaboration although impact assessments have already had a specific role in land use planning before the new Act came into force. In the formulation of other policies and programmes there is less tradition for public participation and the legal backing is also absent except for the general constitutional demands. Developing public participation for other plans and programmes can therefore be expected to be more challenging.

There are also methodological differences between land use planning and other plans and programmes. Land use planning deals with the physical location of activities and their interaction. The rapid development of numerical tools has benefited land use planning (Irjala and Hallin 2002), which is also supported by quantitative monitoring of community change and infrastructure. These techniques and approaches are more difficult to use in plans and programmes, for example, in plans for regional development or support for renewable energy resources, where the spatial aspect is less prominent. It is nevertheless obvious that most plans and programmes will benefit from better information management, but the role of numerical and quantitative information will depend on the context.

One can thus conclude that strategic actions are obviously sensitive to context. Attempts to standardize procedures, approaches and methods may fail miserably, if they are based on the idea of a fixed and specific role for the environmental assessment. Empirical research strongly suggest that the preparation of strategic actions are often best understood as expressions of social struggles over problem definitions, objectives and acceptable means (Hildén et al. 2004a). Under such circumstances flexibility and sensitivity to context is a key to successful SEA.

Key Challenges in Implementing the SEA Directive: Several authors have argued that the environmental assessment of policies, plans and programmes should lead to more sustainable societal development. Sheate (2003) argues that SEA "may be seen as an important counter balance to the onward march of weaker sustainability approaches to assessment" (p.347). Noble (2002) concludes that "the effectiveness of SEA in achieving sustainability objectives will only be realized when a structured and systematic methodological framework is adopted" (p.14). Does the SEA Directive represent such a framework for Finland and should one put such strong normative demands on the environmental assessment?

The SEA Directive provides a general framework, but it does not, and in the light of the Finnish experiences, should not, specify a methodological framework in any detail. The methodological approaches are context specific and although it may be helpful to develop such frameworks for certain sectors, they should not be specified at the level of legislation. Legally specified methodological frameworks easily become straightjackets that may lead to a waste of resources and a loss of legitimacy for SEA in real planning situations (see also Partidário 2000; Nilsson and Dalkmann 2001, for a more critical account Fischer 2003).

The thought that SEA could dictate sustainable solutions is at odds with Finnish experiences and also international experiences in, for example, transport planning (Hildén et al. 2004a). The assessment may contribute to open and transparent preparation of strategic actions, but it cannot dictate the outcome, although interest groups are likely to make such demands.

One of the main challenges in implementing the SEA Directive will be in fitting it into the existing system of preparing plans and programmes. This is affected by its scope. The Finnish experiences suggest that many types of strategic actions will be affected by assessment obligations. Although policies are explicitly excluded from the procedures according to the SEA Directive, they are often indirectly linked to plans and programmes. The dividing line between policies on one hand and plans and programmes may also turn out to be difficult to specify.

A second challenge will be the integration of the assessment with the preparation of the plan or programme in such a way that it improves the preparatory work. It will take political will to initiate and use assessments and such a will does not develop overnight (Hildén et al. 2004b). In Finland, the translation of the basic idea of SEA into more or less accepted practice has taken nearly a decade, and it is still not completed. To what extent the procedural requirements of the SEA Directive will speed up this integration is still open to debate. There is a clear analogy with the experiences of EIA in a transboundary context. Rules and procedures set the scene, but real changes require a development of the culture and practice (Hildén and Furman 2001).

The Land Use and Building Act guides SEA in terms of land use planning. The planning itself is largely formalized through procedures specified by the Act (Chapter 8 of the Act) and the assessment is an integral part of the process. Thus the Land Use and Building Act illustrates one way of integrating assessment with the planning process, including broad participation possibilities. The implementation of the act has only been preliminarily evaluated (Kangas et al. 2002). It may also be questionable to use land use planning as a model for the preparation of other policies, plans and programmes. The formalism of land use planning and the fact that there is a large body of experts devoted to the task of planning may make it easier to adopt formalised approaches to environmental assessments. In other strategic actions there may be a greater need for flexibility.

The third main challenge lies in developing public participation. In part, this challenge is practical, for example, how does one reach potentially interested stakeholders and organize their input in such a way that it can be used in the assessment. On a deeper level issues of representative democracy arise. One must ask, for instance, which groups can act as spokespersons for the public and to what extent the participatory processes are introduced to truly influence public decisions instead of simply giving access to information.

11.5 Conclusions and Recommendations

In the light of Finnish experiences one can conclude that the SEA Directive can provide a general framework for the search of sustainable solutions in plans and programmes, but SEA cannot dictate sustainable solutions. If sustainable strategic actions are to emerge (whatever sustainability happens to mean), they will be the result of a development of the whole preparatory process, and not the result of a separate specified "SEA procedure".

One of the main conclusions and recommendations based on the Finnish experiences is that the requirements of the SEA Directive should be fitted into the existing system of preparing plans and programmes instead of becoming a new add-on. Although policies are explicitly excluded from the SEA Directive, they are often indirectly linked to plans and programmes in such a way that issues related to the tiering of assessments will arise. This stresses the need for flexibility.

The assessment should further be integrated with the preparation of the plan or programme in such a way that it improves the preparatory work from a technical information standpoint. This is, however, not sufficient. The assessment should also improve the possibilities for democratic input into the preparatory processes.

Environmental assessments can enhance public participation, but the task is demanding with both practical and general democracy issues arising. Only when sufficient attention has been paid to both aspects, it is possible to claim that the SEA Directive and its implementation have contributed positively to the development of democracy in Europe.

Acknowledgements

This chapter is a further development of research on SEA supported by the Ministry of the Environment in Finland, and a joint Nordregio project. Thanks are due in particular to Ulla-Riitta Soveri of the Ministry of the Environment for many discussions on the issues dealt with here.

References[1]

Council of State (1998) Säädösten ympäristövaikutusten arviointi (Guidelines for the environmental assessment of Government bills). Edita, Helsinki

Council of State (2001) Kansallinen ilmastostrategia: valtioneuvoston selonteko eduskunnalle (The National Climate Strategy: the Council of State's report to Parliament). Ministry of Trade and Industry, Publications 2, Helsinki

Ervasti K et al. (2000) Lainvalmistelun laatu ja eduskunnan valiokuntatyö, in Finnish (The quality of the preparation of legislation and work of the Parliamentary Committees). The National Research Institute of Legal Policy, Publications 172, Helsinki

Fischer T (2001) Practice of environmental assessment for transport and land-use policies, plans and programmes. Impact Assessment and Project Appraisal 19:41-51

Fischer T (2003) Strategic environmental assessment in post-modern times. EIA Review 23:155-170

Forsström J, Honkatukia J (2001) Suomen ilmastostrategian kokonaistaloudelliset kustannukset, in Finnish (The economic costs of the National Climate Strategy). The Research Institute of the Finnish Economy Discussion Papers 759

Hildén M, Furman E (2001) Assessment across borders – stubling blocks and options in the practical implementation of the Espoo Convention. EIA Review 21:537-551

Hildén M, Jalonen P (2003) Key issues in strategic environmental impact assessment. In: Hilding-Rydevik T (ed) Environmental assessment of plans and programs. Nordregio R 2003(4), pp 41-73

Hildén M, Tahvonen O, Valsta L, Ostamo E, Niininen I, Leppänen J, Herkiä L (1998) Natura 2000-verkoston vaikutusten arviointi (Impacts of the Natura 2000 network in Finland). Finnish Environment 201

Hildén M, Kuuluvainen J, Ollikainen M, Primmer E (1999) Kansallisen metsäohjelman ympäristövaikutusten arviointi (Assessment of the National Forest Programme 2010). Ministry of Agriculture and Forestry

Hildén M, Attila M, Hiltunen M, Karvosenoja N, Syri S (2001) Kansallisen ilmastostrategian ympäristövaikutusten arviointi (Environmental Impact Assessment of the National Climate Strategy). Finnish Environment 482

Hildén M, Furman E, Kaljonen M (2004a) Views on planning and expectations of SEA: the case of transport planning. EIA Review 24:519-536

Hildén M, Kosola M-L, Jalonen P (2004b) Suunnitelmien ja ohjelmien ympäristövaikutusten arviointi (Assessment of policies, plans and programmes: a summary of Finnish experiences). Finnish Environment, in press

Irjala A, Hallin L (2002) Numeerinen maakuntakaava (Numerical regional plan). Finnish Environment 572

Jansson A (2000) Strategic environmental assessment for transport in four Nordic countries. In: Bjarnadóttir H (ed) Environmental Assessment in the Nordic Countries – Experience and Prospects. Nordregio R 2000(3) pp. 81-96

Kaljonen M (2000) The role of SEA in planning and decision-making: the case of the Helsinki Metropolitan Area Transport System Plan 1998. In: Bjarnadóttir H (ed) Envi-

---

[1] Note that legislation mentioned in the chapter is listed at the end of the handbook in the consolidated list of legislation (Appendix 2).

ronmental Assessment in the Nordic Countries – Experience and Prospects. Nordregio R 2000(3) pp. 107-116

Kangas P, Haapanala A, Laitio M, Tulkki K, Vastamäki J (2002) Maankäyttö- ja rakennuslain toimivuus (How does the Land Use and Building Act work? Assessment of experience). Finnish Environment 565

Kemppi H, Perrels A, Lehtilä A (2001) Suomen kansallisen ilmasto-ohjelman taloudelliset vaikutukset (The economic effects of the Finnish national climate strategy). Government Institute for Economic Research. VATT-Research Reports 75

Kumari K, Watson R, Gitay, H, Bosquet B, Harley M, Hildén M, Lantheaume F (2003) Selected case studies: harmonisation of climate change mitigation and adaptation activities, with biodiversity considerations. In: Secretariat of the Convention on Biological Diversity (ed) Interlinkages between biological diversity and climate change. Convention on Biological Diversity Technical Series 10, pp 111-134

Mickwitz P (2003) A framework for evaluating environmental policy instruments – context and key concepts. Evaluation 9:415-436

Ministry of Agriculture and Forestry (1999) The National Forest Programme 2010. Ministry of Agriculture and Forestry Publications 2/1999

Ministry of the Environment (1998) Guidelines for the environmental assessment of plans, programmes and policies in Finland. Ministry of the Environment, Helsinki. (also at: http://www.vyh.fi/eng/orginfo/publica/electro/eia/planprog.pdf)

Ministry of the Interior and Ministry of the Environment (1999) Alueellisten kehittämisohjelmien ympäristövaikutusten arviointi (Guidelines for the environmental assessment of regional development programmes). Ministry of the Interior/Ministry of the Environment, Helsinki

Ministry of Trade and Industry (2001) Kasvihuonekaasujen vähentämistarpeet ja mahdollisuudet Suomessa: Kansallisen ilmastostrategian taustaselvitys (Background report for the National Climate Strategy). Ministry of Trade and Industry Publications 4, Helsinki

Ministry of Trade and Industry (2004) National Climate Strategy and international negotiations. Internet address: http://www.ktm.fi/index.phtml?menu_id=164&lang=3&fs=10, last accessed 06.08.2004

Nilsson MA, Dalkmann H (2001) Decision making and strategic environmental assessment. Journal of Environmental Assessment Policy and Management 3:305-327

Noble BF (2002) The Canadian experience with SEA and sustainability. EIA Review 22:3-16

Paldanius J and Tallskog, L (2003) Julkishallinnon suunnitelmien ja ohjelmien seuranta – periatteita ja toimintamalleja seurannan kehittämiseen (Monitoring and evaluation of plans and programmes in the public sector: principles and models). Finnish Environment 663

Partidário MR (2000) Elements of an SEA framework – improving the added-value of SEA. EIA Review 20:647-663

Roininen J, Horelli L, Wallin S (2003) Osallistuminen ja vuorovaikutus kaavoituksessa (Participation and interaction in land use planning. Frame of reference and methods for monitoring and evaluation). Finnish Environment 664:1-84

SEA WG (2003) Suunnitelmien ja ohjelmien ympäristövaikutusten arviointityöryhmän mietintö (Report of the Working group for environmental impact assessment of plans and programmes). Finnish Environment 643

Sheate WR (2003) The EC directive on strategic environmental assessment: A much-needed boost for environmental integration. European Environmental Law Review 12:331-347

Söderman T (2003) Luontoselvitykset ja luontovaikutusten arviointi (Biodiversity impact assessment in regional planning, environmental impact assessment and in Natura 2000 assessment) (abstract in English). Environmental Guide 109

Valve H (1999) Frame conflicts and the formulation of alternatives: environmental assessment of an infrastructure plan. EIA Review 19:125-142

12 Problems of a Minimalist Implementation of SEA – The Case of Sweden

Lars Emmelin[1] and Peggy Lerman[2]

1 Department of Spatial Planning, Blekinge Institute of Technology, Sweden
2 Legal counsel in environmental planning law, Lagtolken AB, Sweden

12.1 Introduction

Sweden has taken a clearly minimalist approach to the implementation of the SEA Directive. The mode of implementation and some of the problems of the Swedish approach will be discussed. In an international perspective it is interesting to examine why Sweden, a self-proclaimed international leader in environmental policy, has chosen this approach. The discussion is based on a detailed analysis of the legislation and an outline of history of EA in Sweden. Three general types of approaches to implementation of European Directives on environment and planning are outlined: minimalist, intentionalist and environmentalist. Paradoxically one of the more ambitious approaches might have led to fewer problems.

Discussion in this chapter is restricted to the legal system of EA and what can be said from a comparison of the legislation and the Directive. This is the only part of the implementation open for analysis at this stage. There are limitations of such an approach; indeed one of the authors has elsewhere discussed this in some detail (Emmelin 1998a). Arguably, however, the theoretical systems aspects in the Swedish case give a good indication of coming problems in the practice of SEA. The legislation lacks elementary clarity on what the concept of SEA, as opposed to EIA, is to be. The decision to use the same term for EIA and SEA in Swedish, the integration of SEA regulations into the existing EIA legislation and the lack of analysis in drawing up the legislation add to the confusion. This confusion is likely to severely hamper development of sound SEA practice in Sweden.

The concern here is thus with the issues of the systems structure's contribution to handling that aspect of practice that has been termed "the dark side of planning theory" i.e. the power struggles and conflicts of interest that shape the actual plans and programmes and the outcome of planning (Harris 2002).

Implementing Strategic Environmental Assessment. Edited by Michael Schmidt, Elsa João and Eike Albrecht. © 2005 Springer-Verlag

There are Swedish examples of ambitious attempts at developing and applying SEA methods, for instance in local comprehensive land use planning, as well as individual examples from policy making and sector planning. This "bright side" of SEA development in Sweden has been reviewed well by Bjarnadóttir and Åkerskog (2003).

12.2 Environmental Assessment in Sweden

The need for information on impacts on the environment has long been recognized in a general way in Swedish legislation. However environmental assessment as a reasonably formalized process was introduced relatively late in Sweden if seen in an international perspective. A broad and compulsory demand for an EIS in several sectors including spatial planning was established in 1991. Since no screening procedures were introduced it lead to a wide variety of documents, depending on the nature of the project or plan. The only criterion for the document was that the EIS should make possible an overall assessment of the impacts on health, environment and management of resources. Procedures for scoping and specifications for the contents of the EIS were regulated in 1998.

Compliance with the EIA Directive in Sweden has been contested by the EU commission – in a note in April 2003 in accordance with Art. 226 of the EC Treaty – for a specified number of types of development including major issues such as infrastructure and heavy tourism development. It is highly debatable whether Sweden does in fact have a system which conforms to a reasonable degree with what is internationally considered to be good EA practice (Emmelin 1998b). On the one hand there are general requirements for EA in a large number of laws. On the other hand the legislation specifies few of the requirements that distinguish EIA from any other planning process. The mandatory EIS in environmental permit processes for projects in several sectors makes some form of knowledge of impacts available to much of decision making, which may be positive. The rarity of high quality EIS seems to be the price of this mass information.

Provisions for EA of spatial plans at the local level, both the comprehensive and the detailed and legally binding plans, have thus existed in Sweden prior to the implementation of the SEA Directive but not at higher, strategic levels in decision-making or spatial planning. The spatial planning system is in Sweden essentially untiered, with a local planning monopoly and limited planning at higher levels. The lack of tiering and strategic planning in some cases lead to conflict between levels. The traditional planning and policy making system in Sweden is also characterised by a lack of transparency on one crucial EA issue. International good practise which requires that an EIS should be reviewed, updated and available in complete and often officially approved form for decision making does not apply in Sweden. It is thus difficult or impossible to judge what material is in fact used by the decision maker. Westerlund's (1997) criterion that a complete EIS should be available at the time of the decision is not necessarily met.

12.2.1 Resistance to EA

Implementation of the SEA Directive in Sweden can only be understood in the light of a longer development of the spatial planning system and of the struggles to introduce environmental assessment over several decades. This process and the implementation of the EIA Directive as part of the background to the implementation of the SEA Directive are discussed by Emmelin and Lerman (2004). It can be seen as a professional struggle between a spatial planning tradition and profession and the emerging environmental professions. The planning tradition in Sweden is based on architecture and on politics as the ultimate arbitrator in the adoption of plans. The professional culture of the environmental administration on the other hand is based in natural science with a clearly rationalist understanding of planning. A discussion of the effects on EIA development in the Nordic countries can be found in (Emmelin 1998a, 1998b).

One of the major arguments against EIA in Scandinavia was a general fear, enhanced by initial US experiences, of a complex procedure overloaded with information. Critics in Sweden focused on the existing, advanced planning system and on-going development of spatial planning at the national level. Proponents on the other hand focused on the inability of this system to make large scale planning decisions in a rational way and the need for less *ad hoc* approaches to large projects with major policy implications and environmental impacts. Thus the early debate on EIA in Sweden dealt with problems now identified within the realm of SEA. EIA as introduced in NEPA created a new working order in which US federal agencies found a mechanism for negotiating answers to conflicts, even if these could not be resolved. To do so it was necessary to have a structured process with explicit requirements and mechanisms for quality control which both stimulated and regulated entrepreneurial competition on "the analysis market" (Wandesford-Smith and Kerbavaz 1988). These are precisely the characteristics of EIA that have been resisted in Sweden.

Much of the early resistance to EA was in the spatial planning sector and concerned a classic problem of strategic planning and decision making. Strategic decisions, including decisions on large projects that may be what Etzioni (1967) calls "contextuating" i.e. project that shape subsequent development and in retrospect therefore appear to be strategic, operates within the dilemma of "weighting versus daring". Weighting is the rational deliberation on alternatives and impacts whereas daring characterises the intuitive political acting. At strategic levels the Swedish spatial planning system favoured daring over weighting.

12.2.2 Experiences of SEA

The Swedish planning system comprises two main types: sector plans and municipal spatial plans. There is experience of EA in both types but mostly in road planning and in spatial planning at the local, municipal level. The infrastructure sector contains much planning activity and there is planning and programming at national, regional and local levels. At national and regional level issues on funding and priorities between competing projects are the primary focus. An attempt at policy SEA in national transport policy making (SOU 1997) is so far the most ambitious attempt at a strategic, national level. At the local planning level project alternatives are developed. National and regional sector agencies have tried to contribute with their specific issues, but found the lack of information at the appropriate level of detail and precision a major obstacle. Some sectors focus information on local level and at a detailed scale but have little experience in describing strategic issues. Another experience is that early integration is hard to achieve and "add on assessments" are therefore common. The local plans leading on to a project such as a road can benefit from project EIA experiences and are better developed than national and regional planning and programming. The difficulty to find the right level of detail and address the questions at the best level is more of a general than an EA specific problem. These problems notwithstanding, infrastructure planning in Sweden can be said to be tiered, which spatial planning is not.

Cross sectoral spatial or land use planning is done by the municipalities. They have a planning monopoly for the development of built areas and the state can intervene only in certain cases, for example if national interests or public health are threatened. A comprehensive land use plan, covering the whole of the municipality, showing the present land use and sector interests such as nature, cultural heritage, tourism and roads is mandatory. The plan gives recommendations on suitable land use but is not binding. A detailed development plan regulates building and covers rather small areas. Both types of plans are prepared in an elaborated consultation procedure which involve both authorities and the public. The County Administrative Board, which is the regional state authority representing all the national sectoral interests, has an important role to assist the municipality in the planning process, especially with data and information on these interests. This role is mainly fulfilled in dialogue and the state gives only recommendations to the municipality, in respect of the planning monopoly. Sweden is thus one of the few EU countries essentially without a regional spatial planning level.

Development and application of SEA in spatial planning so far has occurred in comprehensive planning in a small number of municipalities (Bjarnadóttir and Åkerskog 2003). Otherwise experiences of EA in municipal spatial planning are mostly at the detailed level. The uncertainties of what the concrete implementation of a plan will be are addressed with a "maximum approach". The EA is based on a situation where all possible rights to build contained in the plan are assumed to be used. Uncertainties concerning what kind of development there will be in an industrial area are addressed by a "typical approach", describing characteristic types of impact of the permitted land use but leaving the specific impacts of a certain industry to the coming development consent. A study of a sample of detailed municipal plans with EA indicates a highly variable quality of both work process and substance; the general picture being one of relatively low quality and substantial uncertainty over relevant issues to include (Oscarsson et al. 2003). This in spite of the application of EIA methodology and the lower level of uncertainty which is characteristic of detailed and shorter term planning. Part of the explanation may be attributed to lack of competence and resources in smaller municipalities. However systems problems such as a lack of division of responsibility between planning and assessment or failure to make use of available environmental expertise are important factors. As Asplund and Hilding-Rydevik (1996) note EA in municipal comprehensive planning is "a daily struggle to keep environment on the agenda" in competition with acute and short term interests with higher political visibility. Formal support of binding requirements for quality in procedure and content would be important in such a struggle but are not provided by the minimalist mode of implementation of the SEA Directive.

Studies of economic development programming have been made in four regions and a manual with guidance for SEA in such cases was produced (Balfors and Schmidtbauer 2002) but does not seem to have been applied. The experience of how environmental matters have in fact been incorporated in project screening and decisions on funding have not been systematically gathered but an ongoing study indicates that practice may not be living up to either national expectations or the programme rhetoric (Nilsson and Emmelin 2004).

A threat to effective SEA is the trend to regard SEA as a declaration of goal attainment. This interpretation of SEA may take the focus away from an important function of EA: the examination of alternatives and the uncovering of negative side effects and impacts. As Sager (2001) notes in a Nordic review, many large development projects show the problems for EA to "prevent the development of a risk-blind and self-reassuring management culture on all large projects". In the typical SEA case in Sweden there will not be the adversarial situation which is the institutional context of project EIA used for environmental permits. The need in strategic planning for "contingency planning" (Sager 2001) and "worst case analysis" (Emmelin 1996) is hardly smaller than for major projects.

12.3 The Legislation Implementing the SEA Directive

The preparation of the Swedish legislation was made in great haste and the law was issued in mid June 2004, giving no time to produce guidelines for rules that come into force one month later. Guidelines will be of crucial importance, since the law is only a vague framework leaving the details to coming regulations. Such regulations will not be issued before the law comes into force. Since screening is not regulated in the law, which leaves practise in almost total uncertainty concerning what plans and programmes that need to initiate the formal SEA now required by law.

The main regulations on procedures and documents are given by the Environmental Code, but will be supplemented with specific regulations in the separate sector legislations and for spatial planning. This split regulation makes it necessary to look at the whole decision making system when analyzing the function of SEA in Sweden. Every legislation has its own special demands, that will influence both procedures and decision making despite common rules. This may be seen as an adaptation to the decision making at hand, in line with the Directive. An adaptation that makes it possible for SEA to work in a certain sector in accordance with the Directive may however require several changes to sector legislation and guidelines. The minimalist approach contains no such changes but merely adds a reference to the Environmental Code. Analysis that supports that there is no need for change in any sector law was not presented in the preparation of the implementation.

12.3.1 Objectives of SEA

The common objective for SEA is given by the Environmental Code, but it contains only a selection of the issues stressed by the SEA Directive. Sustainable development is a shared overall goal. But while the SEA Directive focuses on the integration of environmental considerations into planning, the Environmental Code only mentions the plan or programme as such as a tool for sustainability. The given procedure makes integration into planning possible, but the wording with focus on the document does not promote integration. Lack of integration may be a serious threat to another objective in the Directive: a high level of protection. This objective is not mentioned in the Environmental Code. It may be deduced from other regulations but is not spelled out. The proposal made by the committee (SOU 2003) used the exact words of the Directive, but for reasons unknown they have been taken away in the legislation passed by parliament. Is this simply a mistake or is it a result of a conscious minimalist choice?

Another goal implied in the Code is the contribution to transparent decision making. There are many Swedish rules with the purpose to promote transparency, within and outside the Environmental Code. Since transparency helps ensuring that information is both comprehensive and reliable, according to the SEA Directive, it is questionable whether this is sufficient or if this purpose of SEA should have been clearly stated

Reliable and full information are important parts of quality assurance and review. The SEA Directive underlines this and explicitly request Member States to ensure that environmental reports are of sufficient quality and to communicate to the Commission measures taken. There are no regulations on this goal in the amendments to the Environmental Code, nor any indications on measures to be taken. The explicit reason given for this is that authorities are presumed to loyally apply the legislation and that general supervision is sufficient to ensure this. While this may be true, the important goal of transparency and the concomitant problem of quality are only implied. The minimalist approach gives the impression of according low importance to these issues even if this is not intended.

Another objective given in the SEA Directive is to promote more efficient solutions, by assessments providing a wider set of factors and fuller background material in decision making. Finally the Directive points out that a more consistent framework is provided to the benefit of companies, by planning based on relevant environmental information. Neither of these objectives can be considered to be implied in the Swedish Environmental Code.

Thus the SEA Directive shows more clearly than the Swedish legislation what to have in mind in preparing a SEA. That gives a good foundation for the application of SEA, whereas what may possibly be implied in the Environmental Code needs rather enthusiastic practitioners to be found at all, let alone applied. This may be a partly unwanted consequence of the chosen minimalist approach, possibly not intended, but still negative.

12.3.2 The Scope of the SEA Directive

The SEA Directive gives clear recommendations on addressing questions at the right level and avoiding duplication. The several versions of documents in which Swedish implementation has been proposed contain a selection of plans and programmes, but explicitly only as examples and no full list of plans and programmes that might be covered by the SEA Directive in the Swedish system. Combined with the fact that there are no screening provisions, it is impossible to understand the precise scope of application. Two examples are possible to discuss: road planning and municipal spatial planning, representing two different solutions. Spatial planning for roads will not be covered by the Swedish SEA regulations, but will still be under the project EIA regulations. Municipal spatial planning is, on the other hand, defined as planning under the Directive. Until screening provisions are issued it is not possible to know the extent: Are all detailed plans and comprehensive plans covered or only a selection, and if so using what criteria? A question of duplication arises as these plans will also fall under older regulations about impact assessment of comprehensive plans and project EIS for detailed plans. This can not be further elaborated here, but performance of the Swedish systems to date afford little reassurance (Emmelin and Lerman 2004).

Looking closer at road planning further problematic points can be seen. Road planning starts with initial studies to decide whether a road is needed or not. In practice there are four planning steps in such a study, where other measures to meet the goal are investigated before the solution "new road" can be accepted. If the answer is a new road and there is more than one possible corridor, a localization study follows with the purpose to select the best corridor. The next step is a development plan that is a necessary basis to acquire the land where the road line is placed; this final stage is described as a development consent. The development plan is not the only form of consent, as separate parts of the construction such as bridges, tunnels, drainage and so on, need special development permits. Road planning is thus a mix of several steps, from program and planning to several development consents. The background documents for the new Swedish SEA legislation, contrary to this, concludes that road planning is a part of the development consent process and should therefore not fall under the Directive. This is a simple way of not implementing the Directive, but it may prove problematic in the end. The separate steps in road planning already have an EIA procedure and an EIS, but these requirements do not comply with the SEA Directive. Present practice in project EIA shows great difficulties to find a suitable level of detail in the planning documents. The strategic planning stages tend increasingly to look like the final, detailed planning level, on par with a development consent. A proper SEA might have been a remedy to this, giving the procedure a chance to develop the benefits of planning and leaving the details to the coming consents process.

12.3.3 Notable Lack of Clarity on Several Points

Since the public is not defined in the new Swedish regulations on SEA, it is not possible to evaluate how the participative ambitions of the Directive will be fulfilled. The Directive states that consultation with the public shall be early as well as effective and within appropriate time frames. The Swedish translation of the Directive has left out "effective", but an early proposal for Swedish legislation caught the meaning by adding that it should be possible to take the opinions into account. However, the legislation passed by Parliament only states there shall be reasonable time to express an opinion. The time from the given opinion until the decision is not regulated, nor are there any criteria. It is, once again, only implied that time is needed to make it possible to take the opinions into account. Procedures for direct public participation of the type used in local, detailed spatial planning is not easily translated into methods at strategic levels.

The criteria for selecting the relevant material for the document are very much like those of the SEA Directive. The Swedish legislation has however a very good addition, stating that the interests shown by the public also should be a part of the scoping. This requires an early consultation however, that is so far not regulated. The listing of the substance of the document is not identical to the specifications in the Directive and since no reasons are given only speculations can be made here. "Significant" is omitted in relation to the current state of the environment, possibly indicating a broader demand than the SEA Directive. "Characteristic" is omitted when describing the affected area, with the same possible purpose. The description of existing problems in the environment is limited to, or at least focused on, their relevance for Natura 2000, possibly omitting other areas of particular importance. This may make exempt areas of cultural value from the scope.

One of the confusing aspects of present Swedish EA legislation is the inconsistent use of terminology: the same concept can be covered by different terms while identical or similar concepts covered by different terms. The definition of "environment" in different legislations is a case. Whereas the Directive distinguishes the document resulting from an EIA and a SEA these documents are called by the same name in the Swedish legislation. Equating a SEA document with an EIS in project EIA, with the motive that there are several similarities between the two and that the contents of an SEA may be used for a project EIS, seems to conflate SEA and EIA and add to confusion and potentially to resistance to SEA.

The Swedish regulations on decision making makes it unclear whether there is supposed to be any emphasis on integration into planning or focus only on a document. The Directive says that SEA and consultations should be taken into account during the preparation and before adoption, in Sweden shortened to "before adoption". The statement is regulated in the same manner, with demands only for a description of how SEA and consultations have been taken into account, omitting "during"; the word leading to actual integration. Minimal changes in words but possibly major impacts.

12.4 Understanding the Swedish Mode of Implementation

The implementation taking shape in Sweden can partly be understood as a conscious choice of approach. Three basically different approaches to implementation of European directives on environment are possible.

Minimalist: whenever a Directive is a minimum Directive the relevant question is how a national system can implement it in the specific national context with a minimum of disruption. Implementation needs to be based on formal analysis of the minimum requirements of compliance. Such analysis takes as its point of departure any formal criteria contained directly in a Directive and neglects the intentional information of the preamble. Based on formal criteria one would decide what has to be covered, what elements constitute a risk of being seen as non-compliance and what can safely be left out. Criteria for effectiveness or efficiency of the system would play a minor role in such an analysis.

Intentionalist: the object here is to align national legislation with the intentions of the Directive, making the appropriate adaptations to achieve the goals and objectives. The analysis needed is an examination of the background and stated purposes of the Directive, especially with regard to such concrete purposes as what substantive areas are to be covered, indicated by the listing of sectors in the Directive. Any formal criteria narrowing the scope of the Directive would be seen as a means of making it efficient. For example restricting the types and numbers of plans covered by the SEA Directive would be regarded as a means of making SEA efficient by restricting requirements of exhaustive and formal SEA procedures to those areas of greatest potential environmental impact. From the EU point of view the intentionalist approach would be the natural and preferred alternative: attempting an appropriate transformation of a Directive into national legislation so as to achieve the goals that the Directive aims for.

Environmentalist: here a Directive is used as lever to change national policy. The questions become clearly normative and the analysis should uncover needs related to environmental policy making and substantive national issues rather than the signals implied in a Directive. Typically one might ask: What do we want to achieve and how can it be justified using the Directive?

One problem in evaluating implementation may be that the approach taken is not as clear-cut as the three types above indicates. A minimalist approach may be obfuscated by environmentalist statements and rhetoric. Ambitious but vague intentionalist implementation at the legal systems level may not be followed up at the level of implementation structures. The problem of how the latter may happen due to professional and organisational culture in EIA has been discussed based on empirical material from the Nordic countries by Emmelin (1998a, 1998b).

This said, it is however abundantly clear that the approach taken in Sweden has been consciously minimalist. The committee originally proposing the amendments to the Environmental Code as well as the government notes that the approach has been to implement the Directive with as little disruption to present Swedish legislation as possible (SOU 2003). The ambition is defined as a compliance which causes no formal complaints from the EU.

A basic problem with a minimalist approach is that defining the minimum requirements may be difficult. In the first proposal for Swedish legislation it was noted that the review of what Swedish plans are in fact covered by the Directive may not be exhaustive. In a minimalist implementation this would be necessary in order to guarantee compliance. Thus the implementation in Sweden suffers from a fundamental problem of lacking in assurance that implementation meets minimum requirements.

12.4.1 A Lack of Clarity on what is Implemented

In a developing field such as EA methodology and doctrine it is natural and appropriate that different views and definitions are used. SEA covers a wide range of EA of plans and programmes that have different geographic scope, substance, legal, institutional and professional contexts etc. However in attempting to regulate a phenomenon in legal form a certain amount of conceptual clarity is desirable. The technical solution chosen in Sweden of implementing the Directive directly alongside EIA regulation and with considerable terminological overlap, confusion and inconsistency seems to equate SEA with EIA. This begs the question of why a special Directive has been produced rather than an amendment to the EIA Directive.

The implementation of SEA in Sweden lacks elementary clarity on what the concept of SEA as opposed to EIA is to be. The decision to use the same term for EIA and SEA in Swedish, the integration of implementation of the Directive into the existing EIA legislation and the lack of analysis in drawing up the legislation add to the confusion. This is likely to contribute significantly to both resistances in the individual cases, where *ad hoc* decisions are to be made on whether an SEA is necessary, and to the problems of understanding what it is to be. The committee originally proposing the legislation argues that the distinction between EIA and SEA should not be drawn up in a law. Against this one could argue that different concepts should not be covered by a term hitherto specific to one of them; a generic concept would then be appropriate as well as clear indications that the distinction needs to be made in guidelines. Discussing SEA in a context related way would also help clarifying how major types of programmes and plans could be treated differently within a framework of SEA.

Several approaches to SEA can be discerned. In Sweden there are basically two competing modes of approaching SEA. One can be called the "EIA mode" and the other the "planning mode" (Emmelin 1996). These approaches can be said to take two differing stances on the role and functioning of SEA. Working in the "EIA mode", the role and functioning of SEA can essentially be seen as characterised by proposal assessment. The consideration of alternatives is made at a discrete, set point in time and often the aim of an assessment will be to increase the acceptability of a proposal. The SEA will, in these cases, be undertaken outside the planning or programme and policy preparation process and will be a separate assessment procedure that takes place following the submission of a proposal. In contrast, SEA operating in the "planning mode" can be seen as integrated into a planning or policy process, and therefore alternatives are considered in succession. This mode is evolutionary and incremental. One of the main aims of the "planning mode" SEA would be to identify and highlight areas of problems and conflicts and to serve as guidance to future planning. Widely used international definitions seem to emphasise the two modes somewhat differently: Thérivel et al. (1992) leaning towards the EIA mode, Sadler and Verheem (1996) towards the "planning mode" and Partidario (1999) attempting, at the cost of considerable vagueness, to unite them. Partidario (1999 p.72) questions the need for SEA of the EIA mode. In the Swedish context policy and programme SEA could however well be of this type. The traditional method of examining policy options and policy instruments in parliamentary committees used in Sweden would be suited to such assessment methodology. The case for scenario based impact assessment of policy and policy instruments at national level has been made and methodology developed (Emmelin 1996).

In most cases however SEA must be something other than EIA. A useful point of departure could have been the simple but clarifying distinctions that can be made between assessment of sectoral programmes, regional development plans, spatial plans outlining land use policy, municipal development plans, legally binding local spatial plans etc. One such clarification is proposed by Partidário (1999). Each of the possible categories would have distinct requirements for different approaches to SEA. SEA may rightly serve the purpose of both attempting to outline the consequences of a given development or merely identify future environmental conflicts. In the spirit of Art. 3 of the SEA Directive such identification could be followed up by indications of how future environmental conflicts would be assessed and policy statements for how arbitration could be made.

12.4.2 Why a Minimalist Approach

There are a number of reasons that contribute to the minimalist approach taken in Sweden. In the absence of studies the discussion must remain tentative. The authors posit that there are several groups of reasons:

The timing of implementation. Both the environmental and the spatial planning legislation are at present under review. Part of the overt explanation for the minimalist approach is the lack of time to lay a foundation for anything more ambitious. Time could however have been gained by an earlier start; as it was work was begun a year after the passing of the Directive. SEA is in an unclear position midway between the environmental and the spatial planning legislation in the Swedish system. Arguably an approach entailing as little change as possible to the two laws is rational at a time of change, at least from the strict technical point of view. However this could have been better handled by an analysis of how the Directive could be utilised to promote the changes needed in both legislations. An intentionalist or even environmentalist implementation would have served as part of the ongoing process of revision of the legislations.

Lack of clarity of what SEA is. As noted above the two contending views of SEA complicate a reasonably clear interpretation of how SEA should be applied in practice to different plans and programmes. The seeming conflation of SEA and EIA which may seem a good solution for the law maker but shifts the problems to practitioners; probably at great cost in argument over whether SEA is needed.

Resistance to EA in spatial planning . Much of the resistance seems to be a relatively unreflected reaction to perceptions of EA as a cumbersome, separate process needing expertise that planners do not have and threatening the professional position of planners. The previous lack of clarity on methods and requirements, which is rather increased than otherwise by the legislation implementing the SEA Directive, aggravates this problem. Sweden is reaping the fruits of not sorting out the "integration confusion" i.e. the relationship between planning and EA and between a plan and an EIS (Emmelin 1996). Given the complex, inconsistent and vague nature of EA in combination with requirements for EA on a vast number of issues in Sweden the resistance to environmental assessment can however not simply be explained away as anti-environmental.

The issue of whether the Directive applies to municipal comprehensive planning is important in this respect. A large number of Swedish municipalities have not produced reasonably updated comprehensive plans and this planning level has come into question in recent years. The Union of Swedish Municipalities in commenting on the initial proposal for legislation claimed that the Directive can not be interpreted as applying to municipal comprehensive planning. Initially the proposal was that the Directive should not apply to this plan type. The legislation however finally came to include municipal comprehensive planning. The lack of discussion of what SEA appropriately should be at this planning level is likely to increase resistance to both this plan type and to SEA. Some of the technical features of the legislation are likely to add argument to this; a comprehensive plan involving a Natura 2000 reserve would for example seem to require at least three different and separate environmental assessments as a result of the totality of the Swedish legislation.

Participative aspects of EA as a burden on strategic decision making. There are indications that the formal participative elements in both spatial planning and EA are increasingly seen as a hindrance to strategic decision making. The conflation of SEA and EIA in the legislation is a rational reason for decision makers concerned with this to be sceptical of SEA. Formalised participation may furthermore be relatively ineffective also in achieving the democracy goals of participation.

EA seen as formalistic. In spatial planning the incremental and *ad hoc* nature of processes has been seen to contrast with precise and formal requirements of project EIA aimed at supporting formalised consent procedures. To meet this objection clarification that SEA in for example comprehensive planning can legitimately be many things would have been helpful. The identification of environmental problems and some simple guidance to lower levels charged with the actual planning or permit giving is an important function of SEA at programme and higher planning levels.

Impact assessment and environment seen as too restricted. With recent stress on the application of the so called "environmental quality goals" and the emphasis on sustainable development and sustainability assessment EA has been described as reactive. While SEA has sometimes been termed pro-active (Sadler 1996) the notion that wider concerns than environment need to be the focus of assessment has been frequent. This is not the place to argue the issue as such, although it seems unclear who has the mandate and competence to do this within the framework of EA. In the spatial planning sector the argument would seem to serve well in the professional struggle between planners and environmentalists (Emmelin and Kleven 1999); planners claiming holism as their professional role. A minimalist implementation may superficially seem to leave this issue open.

12.5 Conclusions and Recommendations

To approach SEA with a minimalist intention may seem rational, given that the implementation coincides with revisions of both environmental and planning legislation in Sweden. While the minimalist approach may be partly caused by this situation, the lack of argument as well as of analysis of the consequences of choosing this approach is problematic. The adding on of another component to a system already complex, inconsistent and patently not well functioning in several respects, is not likely to be successful. The lack of attempt at clarifying what SEA is, as opposed to project EIA, is likely to cause problems in the spatial planning sector, especially at the level of comprehensive municipal planning. As it is, municipalities are not doing a very good job of EA even at the much more concrete level of building plans. A majority of Swedish municipalities at present do not have a reasonably updated comprehensive plan and resistance to this type of planning is not likely to diminish if clear guidelines are not issued making SEA at this level meaningful and manageable.

The work on guidelines has begun in the responsible agencies. Much can be done to remedy the situation, although all of the problems can not be remedied by good intentions in guidelines and education. The choice of the Environmental Code as the base for the system means that binding detailed regulations for spatial planning will be issued by the environmental agency rather than by the central planning agency. Competence in planning as well as other parts of the environment, such as culture, health and risk, will be mainly outside the regulatory authority. The lack of time may prevent co-operation with other competent national authorities, although such co-operation has started. A partial way out of this dilemma would be rapid development of sector guides, which has been the practise so far in Sweden with new legislation on planning and environment. The lack of clarity in the legislation and the lack of supporting analysis and guidance from the proposals will be a problem in this case, inviting different sectors to take their own stand. Sweden may expect a fairly long period of application without guidance or guidance with a limited perspective.

Quality assurance is likely to remain a major problem in Sweden, especially with regards to all forms of municipal land use planning. Again, guidance would go some way to ameliorate the problem. The lack of mechanisms for assuring quality is however a serious problem in the Swedish system with neither the possibility of appealing a plan on the grounds of poor quality EA or any formal body charged with assuring quality. The lack of conceptual and terminological clarity or of specificity with regards to content of EA naturally makes quality assurance problematic.

The lack of clear political ambitions for the implementation can however not be remedied by guidelines. This means that the decisive factor will be the ambitions of each and every sector that will have to apply SEA and the different types of professions that represent the decision making powers there. The lack of expressed ambition for SEA in regional development programming is likely to be the most obvious major arena where sustainability policy will lack even the support of a SEA requirement.

There have been various examples of development of SEA in Sweden reported in the literature. Along with other examples of environmental policy and practice such as Local Agenda 21, these would seem to paint a brighter picture than the one sketched in this chapter. To this the present authors would say that a minimalist implementation lacking in both clarity and direction involves the risk of discouraging and hindering the ambitious facing opposition, as well as providing no stimulus to the undecided or any coercion on the recalcitrant.

References[1]

Asplund E, Hilding-Rydevik T (1996) SEA: Integration with Municipal Comprehensive Land-use Planning in Sweden. In: Thérivel R, Patidário MR (eds) The Practice of Strategic Environmental Assessment. Earthscan, London

Bjarnadóttir H, Åkerskog A (2003) Sustainable development and the role of SEA in municipal comprehensive planning in Sweden. In: Hilding-Rydevik A (ed) Environmental assessment of Plans and Programs. Nordregio R 2003(4), pp 73-98

Balfors B, Schmidtbauer J (2002) Swedish Guidelines for Strategic Environmental Assessment (SEA) for EU Structural Funds. European Environment 12(1):35-48

Emmelin L (1996) Landscape Impact Analysis: a systematic approach to landscape impacts of policy. Landscape Research 21(1):13-35

Emmelin L (1998a) Evaluating Environmental Impact Assessment Systems – Part 1: Theoretical and Methodological Considerations. Scandinavian Housing and Planning Research 15:129-148

Emmelin, L (1998b): Evaluating Environmental Impact Assessment – Part 2: Professional Culture as an Aid in Understanding Implementation. Scandinavian Housing and Planning Research 15:187-209

---

[1] Please find legislation mentioned in the chapter at the end of the handbook in the consolidated list of legislation (Appendix 2).

Emmelin L, Kleven T (1999) A paradigm of Environmental Bureaucracy? Attitudes, thought styles, and world views in the Norwegian environmental administration. NIBR Pluss Series 5-99, Oslo

Emmelin L, Lerman P (2004) Implementing the SEA Directive in Sweden – adding to the confusion. Forthcoming as BTH Research report. Blekinge Institute of Technology, Karlskrona

Etzioni A (1967) Mixed-scanning: A 'Third' Approach to Decision-making. Public Administration Review

Harris N (2002) Collaborative Planning: From Theoretical Foundations to Practice Forms. In: Allmendinger P, Tewdwr-Jones M (eds) Planning Futures, New Directions for Planning Theory. Routledge, London New York, pp 21-43

Nilsson J-E, Emmelin L (2004) Practice and rhetoric in integration of environmental considerations in structural fund decisions: a case study. Forthcoming as BTH Research report. Blekinge Institute of Technology, Karlskrona

Oscarsson A et al. (2003) MKB för detaljplan – användning och kvalitet (EIA for detailed plans – application and quality). Internet address: www.boverket.se/novo/filelib/arkiv03/mkbdetaljplan.pdf, last accessed 15.03.2004

Partidário MR (1999) Strategic Environmental Assessment – principles and potential. In: Petts J (ed) Handbook of Environmental Impact Assessment. Vol I, Blackwell, London, pp 60-73

Sadler B (1996) Environmental Assessment in a changing world: evaluating practice to improve performance. International Study of the Effectiveness of Environmental Assessment: Final Report. Canadian Environmental Assessment Agency, International Association for Impact Assessment.

Sadler B, Verheem R (1996) Strategic environmental assessment – status, challenges, and future directions. Ministry of Housing, Spatial Planning and the Environment, Amsterdam, pp 105-168

Sager T (2001) A planning theory perspective on the EIA. In: Hilding-Rydevik T (ed) EIA, large development projects and decision making in the Nordic countries. Nordregio R 2001(6)

SOU (2003) Miljöbedömningar avseende vissa planer och program (EA on certain plans and programmes). Report of the national commission to revise the spatial planning legislation 2003(70)

SOU (1997) Ny kurs i trafikpolitiken (New Directions for Traffic Policy). Swedish national commission report 1997(35)

Therivel R et al. (1992) Strategic environmental assessment. RSPB/Earthscan, London

Wandesford-Smith G, Kerbavaz J (1988) The co-evolution of politics and policy: elections, entrepreneurship and EIA in the United States. In: Wathern, P (ed), Environmental Impact Assessment: Theory and Practice. Unwin Hyman, London, pp 161-191

Westerlund S (1997) Genuine Environmental Impact Assessments (EIA) and a Genuine EIA Concept. In: Basse M (ed) Miljökonsekvensbeskrivning i ett rättsligt perspektiv. Nerenius and Santérus, Stockholm

13 National Strategy for the Implementation of SEA in the Czech Republic

Lucie Václavíková[1] and Harald Jendrike[2]

1 NORDIC Pharma Ltd., Czech Republic
2 Saxon State Ministry for Environment and Agriculture and in 2003/2004 Pre-Accession Adviser at the Ministry of Environment of the Czech Republic

13.1 Introduction

This chapter discusses the national strategy for the implementation of SEA in the Czech Republic, one of the new Member States that joined the European Union on 1 May 2004. What is particularly noteworthy is that the Czech Republic has more than ten years experience with the practice of SEA. The SEA legislation in the Czech Republic exists since 1992. It therefore is one of the oldest SEA legislations in Europe. The Czech Republic also has an interesting system of licensed EIA experts, also used for the SEA process, that ensures the SEA process is done with quality. This chapter starts by explaining the previous situation regarding SEA in the Czech Republic and it then analyses what changes to the national legislation had to take place, namely regarding the requirements of the SEA Directive, due to joining the European Union. The chapter concludes with a description of the Czech system to ensure the quality of the SEA process.

13.2 Previous Situation Regarding SEA in the Czech Republic

SEA had been regulated in the Czech legislation for more than ten years by Art. 14 of the Act No. 244/1992 Coll., as amended by Act No. 132/2000 Coll. and Act No. 100/2001. This "Act on Environmental Impact Assessment" originally covered both EIA and SEA.

Implementing Strategic Environmental Assessment. Edited by Michael Schmidt, Elsa João and Eike Albrecht. © 2005 Springer-Verlag

However, from 1 January 2002 EIA has been regulated in the new Act No. 100/2001 Coll. (before 1 May 2004, this Act covered only the assessment of projects – constructions, activities, and technologies defined in Annex No. 1 to the Act). Only Art. 14 and Annex No. 3 about the assessment of plans and programmes remained in Act No. 244/1992 Coll., (the name of the Act therefore changed into "Act on Environmental Impact Assessment of Development Conceptions and Programmes"). On 29 January 2004 a new Act No. 93/2004 Coll. was adopted by the Czech Parliament, which came into force on 1 May 2004 and which ensures the implementation of the SEA Directive (see Sect. 13.3).

Art. 14 of Act No. 244/1992 Coll. required that conceptions (this was the name for plans and programmes in the Czech Republic) submitted and approved on the level of the central authorities of state administration in the fields of energy industry, transportation, agriculture, waste management, mining and processing of minerals, recreation, tourism and land use planning as well as Water Management Plans had to be assessed on the basis of the following procedure:

- the proponent of the conception had to elaborate a documentation on environmental assessment of the conception. The main features of the documentation on environmental assessment of the conception pursuant to Parts C. III and C. IV. of Appendix No. 3 of Act No. 244/1992 Coll., as amended, were a complex description of expected environmental impacts and an estimation of their significance as well as a description of provisions proposed for the prevention, elimination, minimization and potentially compensation of the effects on the environment
- the draft of the conception, together with the documentation on environmental impacts of the conception, was made available to the public (on notice boards, in local newspapers, leaflets etc.) by the proponent for a period of time of 60 days for comments
- the proponent had to send the draft of the conception, revised with consideration of possible comments of the public, to the competent authority (Ministry of Environment), which had to issue its statement on the conception within 30 days from receiving the draft of the conception
- the statement of the Ministry of Environment was used as a basis for the approval of the conception – the conception could not be approved without this statement.

The procedure described above concerned national development conceptions approved on the level of central authorities of the state administration. Individual development conceptions submitted and approved on lower level, e.g. the level of regional authorities did not necessarily have to be assessed pursuant to Act No. 244/1992 Coll., unless it was a land use planning conception. When the Ministry of Environment issued its statement, it evaluated whether the proponent of the conception ensured:

- description and evaluation of trends in the development of the environment quality, which is related to the existing development of the said territory or sector

- consideration of all key conception goals of environment protection, which are defined in the State environmental policy and related programme documents
- accord of the conception with limits for the site utilisation set in the relevant territorial planning documentation
- analysis of specific environmental impacts of the conception
- publication of the draft of the conception and SEA documentation, according to the system agreed with the Ministry of Environment
- analysis of the key comments of the public on the conception and SEA documentation
- proposal of a system for the monitoring of the real environmental impacts of the conception
- continuous analysis of the background documentation for the preparation of the conception and supplementing points of view of the environment protection, which should be considered in the drafts of the working versions of the conception.

Also a "Methodology of environmental impact assessment of regional development conceptions", was produced by the Ministry of Environment in 2001. The material interprets and recommends steps of SEA and main features of an efficient SEA like: timeliness, integration into the elaboration of the regional development conception, and ensuring comments by the public.

From the beginning of the validity of Act No. 244/1992 Coll. until its abolition on 1 May 2004, the Ministry of Environment assessed and issued statements on altogether 57 conceptions, out of those 29 land use plans of large territorial units. The overwhelming majority ended with a positive statement without reservation or with conditions, which did not fundamentally change the conception. For only four conceptions negative statements were issued:

- In 1997, the plan for organizing the Ski World Championship 2003 in the region of Liberec was rejected, because all proposed variants would have further affected the forests and their ecological functions in the mountain range of Jizerské hory already seriously damaged by the air pollution of the 1970s and 80s.
- In 1999, the proposal for a national energy policy was rejected, mainly because it did not include any variant approaches like the use of renewable energies.
- In 2001, the national strategy for depositing radioactive waste and spent nuclear fuel was rejected, because, among others, alternatives to depositing nuclear waste in deep geological layers, to the proposed ways of processing nuclear fuel and to the use of nuclear energy in general were not taken into consideration in the conception. Instead of depositing nuclear waste, the Czech Republic is now planning interim storages on the sites of the two existing nuclear power plants.
- In 2003, another proposal for a national energy policy was rejected, because, among others, with regard to the use of brown coal, it was not in compliance with the national policies on raw materials and on the environment and the necessity for the proposed building of two new blocks of nuclear power plants was not proved taking into consideration the use of renewable energies.

Compared to the number of SEA actually carried out, the number of discussed conceptions including concepts of land use plans was nevertheless more than twice as much. But many of them were not completed for various reasons: they failed due to a lack of money, changes of relevant authorities, time etc.

In 2003, on the other hand, applications for SEA of conceptions were submitted, which at that time were not subject to the assessment by law, such as regional energetic conceptions (South Bohemia Region), conceptions for air protection (Plzeň Region), prognosis of regional development of the South Moravian Region (Regional Authority Brno), Waste Management Plan of the Czech Republic and updated State Policy of the Environment (both in the responsibility of the Ministry of Environment). In those cases the Ministry of Environment does not issue a statement pursuant to Art. 14 of the Act, but gives only an opinion on it.

13.3 Current Situation Regarding SEA in the Czech Republic

As explained in Sect. 13.2, the Czech Republic has prepared and adopted a new SEA law by Act No. 93/2004 Coll., which became a special part of the current EIA Act No. 100/2001 Coll. This new SEA law is compatible with the SEA Directive and the Kiev SEA Protocol (the Czech Republic signed this Protocol but has not yet ratified it). The amendment of Act No. 100/2001 Coll. was approved by the Czech Parliament on 29 January 2004 (see Act No. 93/2004 Coll.) and came into force on 1 May 2004.

The field of application are strategies, policies, plans and programmes (defined as "conceptions", see Art. 3 para b) on national, regional and local level in the areas of agriculture, forestry, hunting, fishery, surface or ground water management, energy, industry, transport, waste management, telecommunications, tourism, territorial planning, regional development and environment including nature protection, furthermore conceptions, for which, in view of their possible effect on the environment, the necessity of their assessment follows from a special legal regulation, and finally conceptions co-financed by European Community funds (see Art. 10a para 1).

The competent authorities for executing SEA are either the Ministry of Environment (if the affected territory comprises the whole territory of the Czech Republic or the whole territory of a Region or if it concerns the territory of several Regions or the territory of a National Park or a so-called Protected Landscape Area, see Art. 21d, or the Regional Authorities *(Krajské úřady)* in all other cases (see Art. 22b).

The procedure includes the stages of notification of an intended conception to the competent authority (see Art. 10c), the fact-finding by the competent authority (i.e. screening and scoping combined, see Art. 10d), elaboration of an evaluation (i.e. the environmental report pursuant to Art. 5 of the SEA Directive) by an authorized person (see Art. 10e, for the authorized persons see Chap. 13.4 below), submission of the draft conception together with the evaluation to the competent authority (see Art. 10f) and issuing of a statement by the competent authority (Art. 10g).

The final statement of the competent authority has to be taken into account, when the conception is adopted, but it is not binding. However, the body competent for approving the conception has to justify why it did not or only partly take the statement into account. This justification has to be published (see Art. 10g para 4).

Public participation is obligatory at all stages of the procedure. The public is informed of the notification of a conception (see Art. 10c para 2), of the conclusion of the fact-finding procedure (see Art. 10d para 6), of the selection of the authorized person elaborating the evaluation (see Art. 10e para 1), of the draft conception (see Art. 10f para 2), of the time and place of the public hearing (see Art. 10f para 3) and of the final statement of the competent authority (see Art. 10g para 3). People can submit written viewpoints on the notification (see Art. 10c para 3) as well as on the draft conception (see Art. 10f para 5) and they can participate in the public hearing (at least one public hearing about the conception is obligatory, see Art. 10f para 3 and 4).

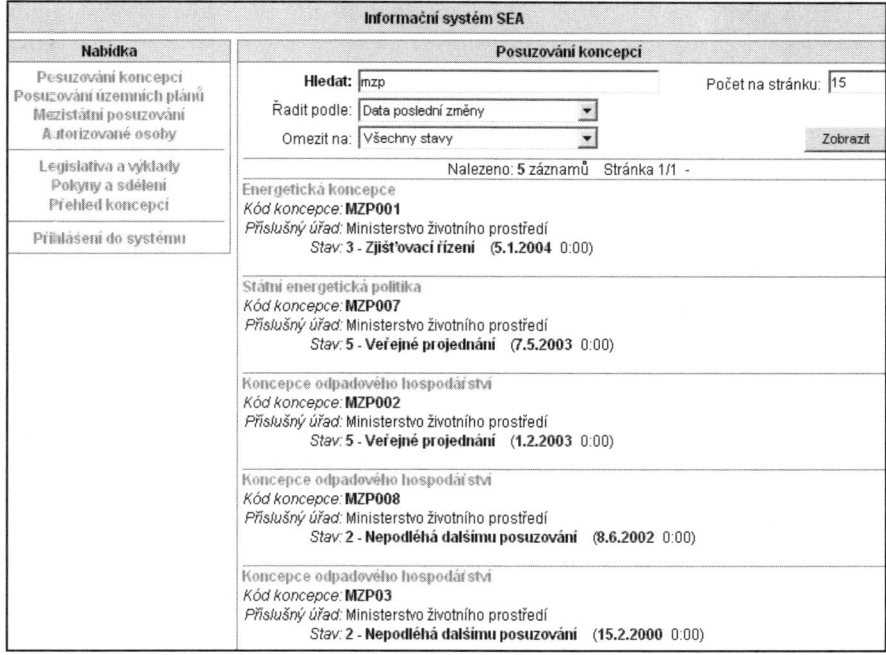

Fig. 13.1. SEA Information System (*Informační Systém SEA*) (Internet: www.ceu.cz/ EIA/SEA/UPD/Default.aspx)

The Internet is widely used. Basically all information provided by the proponent of the conception (see Art. 10c para 2 and Art. 10f para 2) as well as the conclusions of the competent authority (see Art. 10d para 6 and Art. 10g para 3 in connection with Art. 16 para 3) have to be published on the Internet in its full wording. An electronic database of all EIA projects with original documents, links to relevant legislation, contact dates of relevant people and bodies, a list and contact dates of authorized persons etc. already exists since 1992. Based on this system a comprehensive information system for SEA is available on the website of the Czech Ecological Institute *(Český ekologický ústav)*. The new system can be found on the website www.ceu.cz/EIA/SEA/UPD/Default.aspx. A screenshot of the SEA information system is shown in Fig. 13.1.

Transboundary assessment is ensured for the Czech Republic both as State of origin and as affected State (see Art. 14a for conceptions implemented within the territory of the Czech Republic and Art. 14b for conceptions implemented outside the territory of the Czech Republic).

A special article deals with SEA for land use plans to ensure compatibility with the Building Act (see Art. 10i). Another special provision deals with monitoring (see Art. 10h), but it is not very detailed.

The new Czech SEA legislation differs from the SEA Directive mainly in the following points:

- environmental assessment is to a great extent carried out by an authorized person (see Art. 10e). This comes from the long experience with this institution in the Czech Republic in the field of EIA. The authorization is granted by the Ministry of Environment for environmental and by the Ministry of Health for health issues (see Art. 19)
- the competent authority issues the statement something "above" the environmental assessment (see Art. 10g), because it summarizes all the viewpoints submitted by the affected bodies and the public, in addition to the findings of the environmental assessment
- due to the very complicated land use planning procedure regulated in the Building Act a special SEA procedure had to be created in order to connect SEA and land use planning procedure (see Art. 10i)
- another special procedure was created for cases where the proponent of the conception is another ministry (see Art. 10j). This article was added in the course of the legislative procedure for political reasons, when some ministries wanted to carry out SEA by themselves, but in Parliament this intention was changed to the current wording. Now the assessment of these conceptions must also be organized by the Ministry of Environment, but some special rules ensure more influence of the proponent
- a public hearing is obligatory

The new Czech SEA legislation differs from the previous Czech SEA legislation mainly in the following points:

- not only plans and programmes, but also strategies and policies are mentioned as subjects of SEA, but this does not really mean a difference to the previous law, because the terms "policy" and "strategy" usually just appear in the name of the conception, but actually do not differ from the terms "plan" or "programme" (see EC 2003, where it is said in section 3.3 that the name alone will not be a sufficiently reliable guide and that the characteristic of the material is the crucial point)
- not only conceptions on the national level, but also conceptions on the regional and local levels have to be assessed. Therefore the competent authority is not any longer only the Ministry of Environment, but also Regional Authorities
- the assessment must be carried out by an authorized person
- a screening step was introduced
- certain deadlines were added
- precise definition of the term "conception"
- changes of the conception are also subject to the assessment
- regulations for cross-border SEA were added

The methodology for the environmental assessment of regional development conceptions (Czech Ministry of Environment 2004) has been adapted to the new SEA legislation. This will help to ensure the efficiency of the new SEA system in terms of timeliness, integration of SEA into the elaboration of conceptions, and ensuring comments by the public.

In short, everything was done to fulfil the requirements of the SEA Directive. It is hard to say, if the new Czech SEA legislation differs from the *ideal* SEA legislation. Every EU Member State must try to fulfil the requirements of the SEA Directive as well as possible according to its legal and administrative traditions and framework. Only time and practice will show what the ideal legislation or ideal SEA may be.

13.4 Ensuring the Quality of SEA

The Czech Republic uses a system of authorized EIA experts for the SEA process as well to ensure that the SEA process is done with quality. This is organized in the following way:

- The experts (only natural persons) must fulfil certain requirements like appropriate professional qualification, practice in the field for a period of at least three years, the lack of a criminal record and full legal capacity (see Art. 19 para 3 of Act No. 100/2001 Coll.).
- The professional qualification has to be proved by documents and is verified by a written and oral examination (Art. 19 para 4).
- This examination is carried out by a special commission established by the Ministry of Environment.

- Authorization is granted by the Ministry of Environment and in the case of health by the Ministry for Health (see Art. 19 para 1). It is valid for a period of five years and can – repeatedly – be prolonged by another five years on request of the holder (Art. 19 para 7). It can in certain cases be withdrawn by the Ministry of Environment, e.g. if the holder seriously or repeatedly infringes his/her obligations (Art. 19 para 9).
- A list of all authorized persons is published on the Internet (see website www.ceu.cz/EIA/is/osoby.asp)

Only an authorized person is allowed to elaborate the evaluation of the possible environmental impacts of the intended conception (Art. 19 para 1). Thereby it should be ensured that no weak SEA will be elaborated. The authorized person uses the same vocabulary and the same sources of information. If specialists are necessary for certain questions, the expert knows whom to involve. In the case of a bad performance the authorization can be withdrawn (Art. 19 para 9). The system works since 1992 in the field of EIA and until now the Ministry of Environment is quite satisfied. That is the reason why the Ministry chose the system for SEA as well. The environmental assessment elaborated by the authorized person is controlled by the affected authorities and the public in the course of the procedure (Art. 10f para 3). All comments must be taken into account (Art. 10f para 5).

13.5 Conclusions and Recommendations

Since 1 May 2004 Czech legislation on SEA is in compliance with the SEA Directive. However, the links between SEA legislation and the legal regulations the various conceptions are based on, have still to be adjusted, especially with regard to Art. 9 of the SEA Directive concerning information of the public of the decision. The experience of more than ten years of SEA and EIA in the Czech Republic has been used for the legal transposition of the SEA Directive and can now be used for its practical implementation. The system of authorized experts will play an important role in the quality of this process. These experts as well as the responsible staff at the Regional Authorities should therefore intensively be trained on the requirements of the new law.

References[1]

EC – Commission of the European Union (2003) Guidance on the Implementation of Directive 2001/42/EC on the assessment of the effects of certain plans on the environment. Brussels

---

[1] Note that legislation mentioned in the chapter is listed at the end of the handbook in the consolidated list of legislation (Appendix 2).

14 Developments of SEA in Poland

Joanna Maćkowiak-Pandera[1,2], Beate Jessel[1]

1 University of Potsdam, Germany
2 Adam Mickiewicz University Poznań, Poland

14.1 Introduction

The chapter deals with the development of SEA in Poland and summarizes some essential experiences gained by this instrument so far. In Poland SEA started in 1995 with the Land Use Act from 1994 and was then applied to land use plans (see Table 14.1). The implementation of the SEA Directive and thus the adoption of European legal requirements took place in 2000. However, the existing SEA regulations were not adapted to the new EU recommendations and thus strategic action is divided now into two separate systems: one for land use plans and another one for programmes, strategies, policies and other plans which are summarized here as "sectoral plans".

The Polish law considers SEA as an EIA procedure that shall be carried out for plans, programmes, policies and strategies. Although the SEA Directive only refers to plans and programmes, SEA in Poland applies to all four fields of action but without giving any further legal definition either of SEA itself, or of these components (Tyszecki and Behnke 2002, for definitions of plans, programmes and policies see Chap. 5).

In the first parts of this chapter the development of the SEA in Poland and legal aspects of its implementation are pointed out (see Sects. 14.2 and 14.3). The administrative structure and procedure of SEA in Poland is described, comparing the differences in application to land use plans and to sectoral planning (Sect. 14.4). A survey of available environmental data in Poland is given and the relations of SEA to environmental planning are briefly outlined (Sects. 14.5 and 14.6). Finally the chapter illustrates the practice of SEA application based on the Polish National Development Plan and discusses methodological problems connected with the implementation of strategic actions (Sects. 14.7 and 14.8).

Implementing Strategic Environmental Assessment. Edited by Michael Schmidt, Elsa João and Eike Albrecht. © 2005 Springer-Verlag

14.2 Development of SEA

The first legal basis, which stated the necessity of environmental assessment for proposed projects, having significant impacts on the environment in Poland, dates back over 20 years. Legal background was the Environment Protection and Management Act of 1980. The environmental impact assessment (EIA) system was not adequate in quality and effectiveness for environmental impact assessment reports. The reason was the lack of specific standards for the EIA procedures, mostly due to the fact that Poland was then one of the first European countries which applied environmental assessment.

The European integration process played an important role for the Polish government, forcing it to pass a difficult test. Poland was obliged to implement hundreds of legal acts from the beginning of the nineties until December 2002 (the end of official negotiations). With such an amount of new legal instruments, the integration and coordination of the new regulations with the existing ones became a big challenge.

Further in 1999, there were substantial changes in the Polish administration regarding the organizational structures and, in addition of this, the legal basis. This also interfered with the responsibilities for EIA and SEA which to a great extent were delegated to the regional and local levels. Along these provisions, the EIA and SEA also had to be adopted to the Polish law in a form, that could be reconciled with the European guidelines. The European models of these two environmental planning tools were taken over by the EIA Act 2000. This act implemented the following regulations:

- EIA Directive 85/337/EC amended by Directive 97/11/EC
- Espoo Convention from 1991 on Environmental Impact Assessment in a Transboundary Context
- Aarhus Convention from 1998 on access to information, public participation and to justice in environmental matters
- Common positions of the SEA Directive (later Directive 2001/42/EC).

The EIA Act 2000 was in force for only ten months; subsequently the Environmental Protection Act from 27 April 2001 (EPA 2001) came into effect, which is the current regulation for all environmental issues in Poland (see Chap. 5). The Land Use Planning and Management Act of 27 March 2003 (Land use Planning and Management Act 2003) brought about further modifications for EIA and SEA concerning land use plans particularly pertaining to public participation and the procedure. The history of developments of environmental assessment in Poland is introduced in Table 14.1.

The current legal basis for EIA and SEA in Poland are thus the EPA 2001, the Land Use Planning and Management Act 2003 and the Ordinance of 14 November 2002 on detailed criteria of SEA prognosis for land use plans. The SEA prognosis is the Polish term for environmental report.

Table 14.1. Overview about important legal EIA and SEA acts in Poland and their effects

Legal act	Effective	Commentary
Environmental Protection and Management Act 1980	September 1980	First mention of EIA
Toll Motorways Act, 1994	December 1994	EIA for highways was introduced
Land Use Act, 1994	January 1995	First legal demand for SEA
Ordinance of 09.03.1995 on the requirements to be met by the environmental prognosis of land use plans.	March 1995	First demands for environmental report in SEA for land use plans
Ordinance of 13.05.1995 on developments particularly harmful to the environment and human health and terms to be met by EIS.	June 1995	Naming of projects with mandatory EIA
Ordinance of 05.06.1995 on the terms to be met by the assessment of the impact of highways on the environment, agricultural and forest land and cultural heritage.	June 1995	Demands on environmental impact report of highways
EIA Act 2000	January 2001	Implementation of European-EIA regulations
Environmental Protection Act (EPA), 2001	October 2001	Amendment of EIA, SEA and public participation procedure
Ordinance of 24.09.2002 on the types of projects which may have significant impacts on the environment and detailed criteria for project screening for EIA.	October 2002	List of type of projects EIA mandatory based on EU law
Ordinance of 14.11.2002 on detailed criteria of SEA prognosis for land use plans.	December 2002	Demands on SEA for land use plans
Land Use Planning and Management Act 2003	2003	Changes in functioning of EIA and SEA

14.3 Compliance of the Polish Environmental Protection Law with the European SEA Directive

Despite these legal amendments, there are still some significant differences left between European SEA regulations and current Polish implementation. They are introduced in Table 14.2.

Table 14.2. Differences between the SEA Directive and the Polish EPA 2001

SEA Directive	EPA 2001
Art. 1	It shall be mandatory to carry out the assessment not only for draft plans and programmes, but also for draft policies and strategies (see Art. 40 EPA 2001). The word "adoption" is not to be found in Art. 40-45 EPA 2001 concerning SEA; it is not so clearly emphasized that SEA has to be part of the several steps of a decision-making process.
Art. 2 a)	Mandatory SEA for plans, programmes, policies and strategies only on national and regional, not on local level.
Art. 3 para 3	Except land use plans where SEA is mandatory for communal, voivodeship (i.e the regional level) and national level (i.e. for all levels mentioned in the EC Directive).
Art. 3 para 2b)	No reference to the Habitats Directive and Wild Birds Directive.
Art. 3 para 3	The administration authority which prepares the draft plan, programme, policy or strategy, may decide not to carry out the EIA procedure if it determines, that the implementation of these documents would not have a significant impact on the environment. This also seems to be a departure from the SEA Directive, which (in Art. 3, para 3) only allows an exception from SEA in case of land use plans and programmes for small areas on local level which would have no significant impact on the environment.
Art. 3 para 5	EPA 2001 defines types of sectors with mandatory SEA, but cites no types of plans, programmes, policies and strategies. As a result there is a lot of ambiguity about which document shall be the subject of SEA procedure.
Art. 3 para 7	For local land use plans the Land Use Planning and Management Act 2003 does not guarantee the participation of the public in preparation of the document.
Art. 3 para 8, 9	EPA 2001 does not explicitly exclude financial or budget plans, programmes and projects for national defence and civil emergency. These plans and programmes are yet exempt from public participation and public access to environmental information.
Art. 4 para 3 Art. 5 para 2, 3	No provisions about tiering.
Art. 10	No explicit regulation for monitoring the significant environmental effects of the implementation of the plan or programme
Annex I	No normative statements which parts of the environment have to be assessed, including cumulative aspects and interrelationship between the factors (cf. Art. 41, para 2 EPA 2001).

While Polish regulations for EIA are highly elaborated and meet nearly all the requirements of the European Directive, some discrepancies between the Polish law and the European SEA Directive remain. Table 14.2 enumerates the details. The main discrepancies are (see also Céwe et al. 2003a, b):

- *The role of SEA within the decision-making process:* Art. 1 of the SEA Directive states, that its goal is integrating environmental consideration into the preparation and adoption of plans and programmes, whereas Art. 44 EPA 2001 specifies that those who are preparing or modifying a document "shall take into account the findings of the environmental impact prognosis". Thus the character of SEA of being integrated into the decision-making process is not emphasized as clearly as it is in the SEA Directive. The Polish law does not specify any provisions about "tiering", i.e. how to deal with SEA, when plans, programmes, policies and strategies have a complex structure and have to be carried out in several steps (see e.g. Art. 4 para 3 of the SEA Directive).
- *Area of application:* The scope of the Polish regulations does not fully cover the European requirements. Art. 2a para 3 of the SEA Directive refers to plans and programmes affected by legislative, regulatory or administrative provisions. For example it is necessary to conduct the SEA procedure for "II State Ecological Policy" (i.e. the national ecological policy for the years 2000-2003), but in fact it has not been accomplished for any policy yet. There is no specific methodology or approach for application of SEA to policies or strategies in Poland. Art. 40 para 1 No. 2 EPA 2001 mentions the term "law" in the Polish sense "ustawa", which only includes laws adopted by the parliament. That means many plans and programmes will not fall within the SEA Directive. Further, the EPA does not contain any determinations for SEA of sectoral plans and programmes on the local level. It should be mentioned that the Polish law on land use imposes an obligation to apply SEA or at least an assessment similar to SEA since 1994. However, there is no obligation to carry out an SEA on local level, such as e.g. for plans or programmes to local waste management. Further on it is unclear which document is SEA mandatory. The formulation that every plan, programme, strategy and policy provided by law falls into the SEA is too general (Kowalczyk and Starzewska-Sikorska 2003).
- *Monitoring:* Art. 10 of the SEA Directive states, that the Member State shall monitor the significant environmental effects of the implementation of plans and programmes. Art. 41 para 2 No. 10 EPA 2001 only says that the environmental report should include information on the methods used for monitoring the implementation. This formula gives merely the underlying message that monitoring should be carried out, however there is no explicit claim for it as in the SEA Directive.
- *Consultation and public participation:* For the plans and programmes, the authority preparing the draft document shall obtain approval of the environmental authority about the level of detail which must be included in the environmental report and it should also secure the possibility of public participation (see Art. 42 para 1 and Art. 43 para 2 EPA 2001). The Land Use Planning and Management Act 2003 regulates it in Art. 17 for local land use plans and in Art. 41 for

regional land use plans. According to these provisions the use of land by the state is excluded from public participation.
- *Environmental factors as subject maters in SEA*: Art. 41 para 2 EPA 2001 itemizes the necessary aspects the environmental report has to deal with. This article just uses the word "environment" and does not specify both the different environmental factors and natural assets and furthermore the secondary, cumulative, synergistic, and other effects like Annex I of the SEA Directive does. The same applies to the interrelationships between the factors which are also mentioned in Annex I of the Directive. Thus it does not become clearly evident that all these factors have to be deliberately considered within an SEA.

14.4 Administrative Structure and Procedure of SEA in Poland

As mentioned before the first framework for SEA in Poland was created by the Land Use Act in 1994. According to this Act, it was mandatory to deliver the environmental report *after* the preparation of the local land use plan. This meant that the environmental report did not have sufficient impact on the content of a land use plan and could not really be taken into account during the decision-making process. Now strategic assessment issues are regulated by EPA 2001. The new model assumed, with moderate changes, the old strategic assessment regulations of 1995, and additionally created a new SEA system. Because of the lack of integration between the old and the new systems and the existing tools of SEA for land use plans from 1995 Poland currently has two kinds of procedures – one for land use plans and one for sectoral plans which differ in the course of their proceedings.

The section provides a survey of important administrative and procedural aspects of SEA in Poland such as area of application, competent authorities, different procedures for SEA for land use plans and SEA for sectoral plans and public participation.

14.4.1 Area of Application of SEA

According to Art. 40 of EPA 2001 the SEA procedure shall be carried out for the following plans, programmes, policies and strategies:

- Sectoral plans: Draft policies, strategies, plans, and programmes in the fields of industry, transport, energy, telecommunications, water and waste management, forestry, agriculture, fisheries, tourism and land use, where their preparation by the national or voivodeship administration authorities is provided by law.
- Land use plans: The draft concept of the national land use policy, draft land use plans and draft regional development strategies,
- Modifications of these documents after their adoption.

By mentioning also policies and *strategies* as being submitted to an SEA, the Polish law goes beyond the requirements of the SEA Directive. However, as a concise definition is missing, this field of application stays rather dim. As for plans and programmes no single types but only the general fields of application are mentioned in the EPA 2001. Thus the provisions leave open to which documents exactly an SEA shall be applied to.

14.4.2 Competent Authority

The decision on the results of a strategic assessment, the so-called "opinion", is prepared on the following administrative levels:

1. For plans, programmes, policies, and strategies: by national and regional public administrative authorities, i.e
   - the Ministry of the Environment (Department of Environmental Protection Tools) with Chief Sanitary Inspector for the national stage.
   - the voivodeship (Department of environmental protection) for the regional stage.
2. For land use plans on the local, regional and national level: by the corresponding level of administration, i.e.
   - the commune (a department or person in charge for environmental protection),
   - the voivodeship (Department of environmental protection),
   - or the Ministry of the Environment (Department of environmental protection tools) with Chief Sanitary Inspector.

14.4.3 Procedure: SEA for Land Use Plans vs. SEA for Sectoral Plans

SEA in Poland means an own proceeding with an independent decision as a result, i.e. unlike SEA models in other European countries it does not merely form part of a comprehensive admission procedure. SEA and EIA employ different procedures. As mentioned above, there are two kinds of SEA processes in Poland – for land use plans and for sectoral plans – which differ substantially. Figs. 14.1 and 14.2 illustrate the different SEA procedures for sectoral plans and for land use plans.

A major difference between SEA for land use plans and for sectoral plans is the level at which the decisions are taken (see also Sect. 14.4.2): Only for land use plans SEA is mandatory for each administrative level. Strategic assessment for plans, programmes, policies or strategies is carried out on either regional or national level. Another substantial difference is that for plans and programmes it is possible to apply the screening and skip the SEA if there are no significant impacts to be likely on the environment according to the prepared sectoral plan.

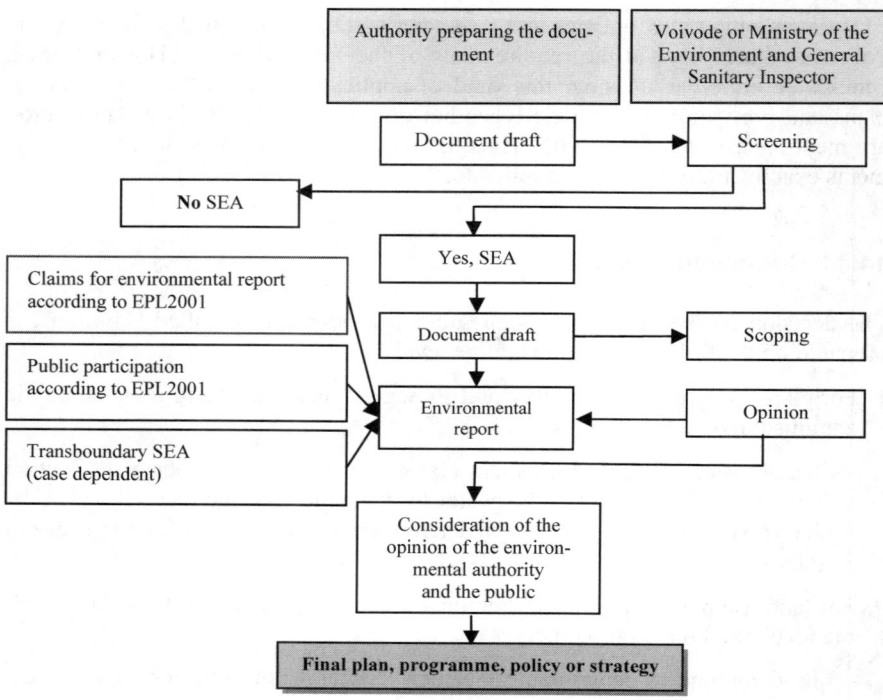

Fig. 14.1. Simplified diagram of the SEA procedure for plans, programmes, policies and strategies in Poland according to the EPA 2001

For land use plans according to the Land Use Planning and Management Act no scoping is provided for obligatory. The explanation given is that there is a separate order on the scope of the environmental report from 2003 (Tyszecki and Behnke 2002) compared to the plans and programmes for which it is mandatory to state the scope of the report with authorities.

14.4.4 Public Participation

According to Art. 10 of the EPA 2001 everyone should have the right to participate in the procedures relating to:

- granting of decisions in the scope of environmental protection,
- adoption of draft policies, strategies, plans or programmes for development or restructuring as well as draft land use studies and plans

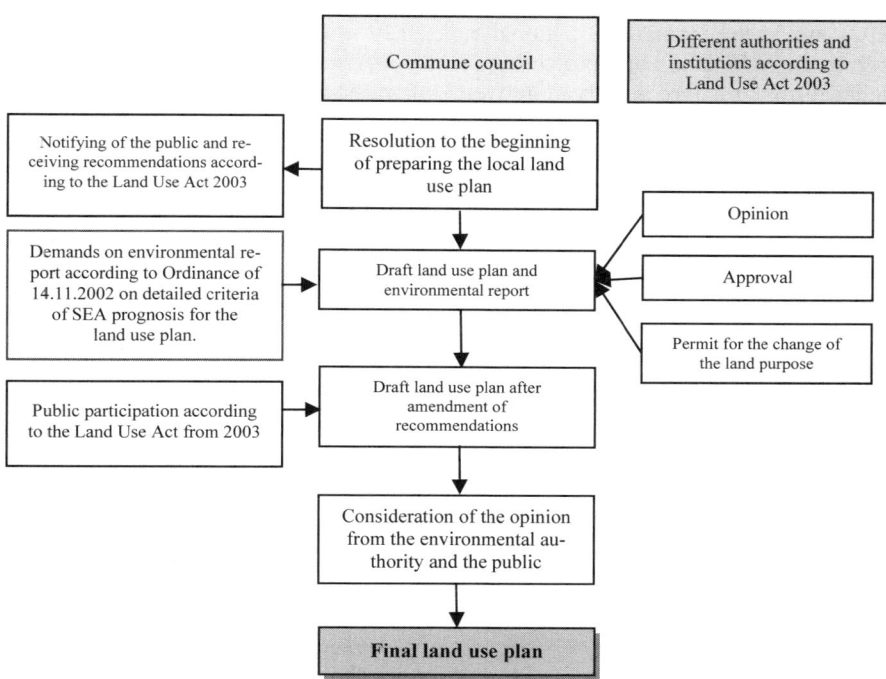

Fig. 14.2. Simplified diagram of the SEA procedure for local land use plans according to the Land Use Planning and Management Act 2003

The public participation for SEA is regulated separately on two legislative bases: For land use plans according to Land Use Planning and Management Act 2003 and for plans and programmes according to EPA 2001. In both cases the public has to be notified about the beginning of the procedure. In the second stage, when the environmental report is already prepared, the public has 21 days (SEA for secoral plans or local land use plans) or three months (for voivodeship land use plans) to submit the recommendations, which should be considered by the responsible authority.

14.5 Environmental Planning and SEA

The term environmental planning is not used in Poland. Instead Poland applies environmental management (which includes EIA, SEA, ISO 14000, 14001), nature preservation and its protective measures (e.g. protection plans for national and landscape parks) and land use planning. It is necessary to add, that the nature preservation and its tools in Poland have a long history supported with good experiences – the first National Park (Białowieski National Park) was established in 1921 and at the moment 23 national parks, 120 landscape parks, 1345 nature reserves and 412 landscape protection areas do exist in Poland, which cover with other forms of nature preservation over 31% of the area of Poland (Brodowska et al. 2003; Froehlich-Schmitt et al. 2000).

Since the mid nineties, EIA has become more and more important, which can be credited to the fact that it is the only instrument capable to influence the environmental aspects in the planning process. The current SEA model was prepared under EU adjustment pressure. As a result no detailed regulations, which consider the specific requirements of SEA have been compiled so far and there is no integration between nature conservation, land use planning or EIA and SEA.

14.6 Availability of Environmental Data in Poland

For an effective use of SEA it is necessary to have an easy access to relevant baseline data. This is valid not only for traditional environmental data as on biotopes, air and water quality, but also for data on environmentally related health issues. Tables 14.3 and 14.4 present the data commonly available in Poland.

Table 14.3. Availability of basic data about the environment useful for SEA

Type of data	Available at	Scale
"Ecophysiographic studies" – information about natural resources. Basic information about environment in a municipality.	Commune, voivodeship	1:5 000 1:10 000 1:25 000
Topographic maps	Central Agency for Geodesy and Cartography	1:10 000 1:25 000 1:50 000
Air photographs	Central Agency for Geodesy and Cartography	1: 26 000
"Nature inventory" – biotic and abiotic nature elements in a municipality	Commune, voivodeship	1:10 000 1:25 000
Soil – agricultural maps	Commune or poviat	1:5 000 1:25 000 1:100 000
Hydrographic maps	Central Agency for Geodesy and Cartography	1:50 000 (for particular parts of Poland)
Forest maps	Regional directory for State Forest	1:5 000 1:20 000

Table 14.4. Other data about the environment in Poland

Kind of information	Available at
Natura 2000	Ministry of the Environment
Climate data	Institute of Meteorology and Water Management
Nature preservation	State Board for National Parks
Statistic data – figures and indicators about population, economy industry, environment, economy development and the like.	Central Agency for Statistic Data
Control and protection of the state of the environment	Central Agency for Environmental Protection

The main problem with the application of environmental maps is the lack of information where to obtain the data. There are also no requirements inducing consultants responsible for preparing environmental reports to use them.

14.7 Examples of SEA in Poland

Even if the implementation of SEA is not yet complete, there are some examples in which the authors are searching for good solutions in SEA procedure and a sound environmental report. The SEA for the National Development Plan (NDP) (see Box 14.1), which was one main pilot project for SEA can be considered as one of these examples. For Poland the NDP is the strategic blueprint for planning domestic and European activities. This document will explain how Poland is going to use the structural funds from the EU and will organize its own budget in the years 2004-2006. The document is divided into the following sectors:

1. Increasing of activity in the manufacturing sector
2. Strengthening of human potential
3. Restructuring of the food sector and development of rural areas
4. Fishery
5. Development of road infrastructure
6. Environmental protection and water management.

One main goal of the pilot project "SEA for the National Development Plan" was to consider how the concerns of environmental protection were represented in particular sectors of NDP. The project started at the end of March 2002 on an initiative of Regional Environmental Center for Central and Eastern Europe (REC) and on behalf of the Ministry of Economy. An independent expert team was appointed. The work on the environmental report ran at the same time as the composition of the assessed document.

Box 14.1. SEA for the National Development Plan (Koziarek and Rzeszot 2003)

The proceeding ran in the following stages:

1. Analysis of about 90 Polish, European and international legal instruments (laws, conventions, directives, strategies, resolutions etc.).
2. As a result of the analysis, selection of 14 basic legal acts and formulation of 250 detailed criteria (representing ecological goals from legal acts).
3. Synthesis in 50 criteria according to the sectors, as: transport, power industry, agriculture, nature and landscape protection, forestry, fishery, and general aspects.
4. Comparing the criteria with the tasks of the NDP, in order to identify the important fields for evaluation. Consulting the result with individual, sectoral teams from the Ministry.
5. Formulation of evaluation criteria:

    a) 11 specialized criteria for the evaluation of the NDP-contents and proposed activities and their impact on the environment, such as:

    I. Will the planned activities contribute to the effective use of the natural resources, to the change of consumer and production behaviour and or will they manage the demand for these resources?
    II. Will the planned activities support the development of renewable energy?
    III. Will the planned activities positively affect the conditions of the following parts of the environment: air, water and groundwater, earth's surface?
    IV. Will the planned activities reduce the health threats, which are an effect of the environmental pollution?

    b) 12 formal criteria for the evaluation of the construction and cohesion of the document and its consistence with principles of sustainable development, such as:

    V. Does NDP cover the environmental goals?
    VI. Do the intended activities agree with the "Strategy for sustainable development of the European Union", "VI EU Program for environmental protection" and "II Polish ecological policy"?
    VII. How and to what extend were the significant impacts on the environment considered?
    VIII. Was the document submitted to broad consultation and were the results taken into account?

6. Evaluation of every sector in a matrix. The criteria were evaluated in a 4-level-classification (from 0 to 3).
7. In the final result 60 recommendations were suggested. Most of them were considered in the last version of the NDP.
8. Transmission of the work version to the Ministry of Economics; public consultation by placing of the report on the Polish web-site of Regional Environmental Center for Central and Eastern Europe (REC) and creating a forum on the server of one NGO.
9. Consideration of the opinions and recommendations of SEA in the last version of the National Development Plan.

The main conclusions coming from the example in Box 14.1 are:

- The environmental impact report was integrated in the preparing process of NDP and it was feasible.
- The recommendations of the SEA team were implemented in the NDP.
- The Ministry of the Economy was cooperative (Koziarek and Rzeszot 2003).

One of the problems was that the public did not care for participation in the procedure. Furthermore the assessed document – the National Development Plan – as a working version was still changed during the SEA process by the main Author – the Ministry of Economics. The NDP has an interdisciplinary character, so it was difficult to set general evaluation criteria conceiving the whole document. Furthermore no formal criteria for the evaluation of the environmental report were developed. At least there was a lack of good practice for provision of implementation of SEA recommendations in the document (Koziarek 2003).

14.8 Methodological Problems and Quality of SEA

The discussion about implementing SEA has been mainly focused on legal and procedural aspects so far (Kowalczyk and Szulczewska 2002). This is true not only for Poland, but for most of the EU Member States (see Chap. 26). Hitherto the experiences about SEA in Poland are rather restricted when talking about which methods could be used for prognosticating and assessing the environmental impacts on a strategic level (Céwe et al. 2003a). The lack of methods suitable for the strategic level may be one reason that so far SEA is not used in Poland in the majority of cases to make substantial inputs to decision-making processes, but is dealt with just as a formal requisite that has to be added to the documents for the approval of a project.

Generally speaking, SEA in Poland was formed based on the methodology of EIA. There are some good procedural guidances for EIA and SEA of land use plans from the publisher Eko-Konsult, from the Institute for Cities Development or Ministry of the Environment (Kowalczyk and Starzewska–Sikorska 2002; Wiszniewska et al. 2002; Kowalczyk and Szulczewska 2002; Czerwieniec et al. 2002; Tyszecki and Behnke 2002; Tvevad et al. 2002).

There is also a lack of not even procedural, but specific methodological approaches to the SEA for policies, strategies, programmes and plans. It would be essential to work out specific methods for SEA procedures, to secure the public participation and to set standards on the environmental impact report. The regulations from EPA 2001 about strategic assessment are frequently not even applied to policies, strategies, plans and programmes with mandatory SEA provisions. Concrete examples are some waste management plans and restructuring programs of the industrial sectors cited by Kowalczyk and Starzewska-Sikorska (2002). One main reason is the lack of knowledge and good practice. Céwe et al. (2003a) also claim the elaboration and publishing of pilot studies to provide examples for an exemplary SEA practice that are easy accessible.

The most important methodological problems with SEA in Poland are related with the integration of the environmental report within the procedure of preparing the plan, programme, policy or strategy and the stage on which the SEA shall be considered (Kistowski 2003). The experiences with the National Development Plan (see Sect. 14.7) show that the solution to integrate the environmental report into the preparation of the plan or programme and to influence it would be most reasonable because it guarantees the consideration of environmental issues within the permit procedure of the document. Another problem is the influence of the environmental report on the result of the SEA procedure or how its statements have to be taken into account when the decision about a project is worked out.

14.9 Conclusions and Recommendations

Poland has made a tremendous progress in environmental protection and in approaching the structures of the EU in recent years. However as one of the biggest countries in Europe which became EU participant in May 2004 it has not finished its integration process yet. SEA already existed in Poland before the SEA Directive came about. The implementation of the European requirements in 2000 was not effective enough; therefore it is still necessary to amend the EPA 2001 to meet all the requirements of the SEA Directive. As the "old" SEA system (for land use plans) and the new one (for the other fields of action) differ in their areas of application, their courses of proceedings and in demands on public participation, the lack of integration between these two systems proves as a stumbling block to come to a common comprehension of what is "SEA". Nevertheless, beyond all the formal requirements, it will be *SEA practice*, that will prove, what SEA can really bring about. Thus, it becomes important to develop coherent or even standardized procedures for land use plans and for the different kinds of sectoral plans. This may give good examples that can be taken as standards of good SEA practice and to set up the guidelines that induce administration and planners to implement SEA. Another way to increase the quality and the effectiveness of SEA in the decision-making process will be to encourage the participation of the general public, in which Poland is short on experience so far.

References[1]

Brodowska M, Czajka J, Dygas Ciołkowski L, Fornal B, Gruszecki P, Jaworski R, Kasprowicz H, Krajewski Z, Kureczko B, Miłoszewski A, Szatkowska - Konon H, Rudlicka A, Wolnicki Z, Wróblewska D, Zrałek E (2003) Raport o stanie środowiska w Polsce (Report about the state of the environment in Poland). Biblioteka Monitoringu Środowiska, Warszawa

---

[1] Note that legislation mentioned in the chapter is listed at the end of the handbook in the consolidated list of legislation (Appendix 2).

Céwe T, Fischer T, Heuberger C, Jessel B, Thierfelder H (2003a) Final Report on Implementation of EIA and SEA Directives in Poland. Twinning Project "Strengthen Environmental Impact Assessment" PL/2000/IB/EN/01, Report on Activity 2.3.1

Céwe T, Hartlik J, Heuberger C, Jessel B, Thierfelder H (2003b) Recommendations for Implementing EIA and SEA Directive in Poland, Final Status Paper. Twinning Project "Strengthen Environmental Impact Assessment" PL/2000/IB/EN/01, Activity 1.7.2

Czerwieniec M, Pawłowska K, Schmager M, Słysz K, Zgud K (2002) Podstawy metodyczne sporządzania strategicznych ocen oddziaływania na środowisko dla potrzeb planowania (Methodological principles for preparing the strategic environmental assessment in planning processes), Kraków. Internet Address: www.mos.gov.pl, last access 05.05.2004

Froehlich-Schmitt B, Mitlacher G, Wróbel J (2000) Deutsch-polnisches Handbuch zum Naturschutz, Polsko-niemiecki podręcznik ochrony przyrody (German-Polish manual for the nature conservation). Bundesamt für Natuschutz, Bonn, Ministerstwo Środowiska, Warszawa

Kistowski M (2003) Metody sporządzania strategicznych ocen oddziaływania na środowisko przyrodnicze (Methods of the strategic environmental assessment), Problemy Ocen Środowiskowych. Eko-Konsult (2)21

Kowalczyk R, Starzewska Sikorska A (2002) Strategiczne oceny oddziaływania na środowisko w układach sektorowych (Strategic environmental assessment in sectoral planinng). Eko-Konsult

Kowalczyk R, Szulczewska B (2002) Strategiczne oceny oddziaływania na środowisko do planów zagospodarowania przestrzennego (Strategic environmental assessment for the land use plans). Eko-Konsult

Koziarek M, Rzeszot U (2003) Ramowa Strategiczna Ocena Oddziaływania na Środowisko Narodowego Planu Rozwoju na lata 2004-2006 (Strategic environmental assessment framework for the National Development Plan (years 2004-2006), Ogólnopolska Konferencja z cyklu Instrumenty Zarządzania Ochroną Środowiska, Problematyka Ocen Środowiskowych w przededniu wstąpienia Polski do Unii Europejskiej, Akademia Górniczo Hutnicza, Kraków

Koziarek M (2003) Doświadczenia i metodologia pracy nad Ramową Strategiczną Oceną Oddziaływania na Środowisko Narodowego Planu Rozwoju na lata 2004-2006 (Experiences and methodology of the work on strategic environmental assessment of the National Development Plan). Internet address: www.rec.org.pl, last access 05.05.2004

Tyszecki A, Behnke M (2002) Analiza systemu ocen oddziaływania na środowisko w obszarze planów i programów-dokumentów strategicznych (Analysis of the SEA system for plans and programmes – strategic documents). Internet address: www.mos.gov.pl, last access 05.05.2004

Tvevad A, Farr J, Jendrośka J, Szwed D (2002) Udział społeczeństwa w postępowaniu w sprawie oceny oddziaływania na środowisko (Public participation in the environmental impact assessment procedure), Ministerstwo Środowiska, Warszawa

Wiszniewska B, Farr J, Jendrośka J (2002) Postępowanie w sprawie oceny oddziaływania na środowisko planowanych przedsięwzięć (Environmental assessment procedure for planned projects). Ministerstwo Środowiska, Warszawa

15 Experiences with SEA in Latvia

Sandra Ruza

Ministry of Environment, Latvia

15.1 Introduction

The integration of environmental protection requirements into planning documents of other sectors have over the time been recognized in Latvia as a significant principle for ensuring adequate level of environmental protection. Several new legal requirements for environmental policy integration have been established within the past few years, in recognition of a number of issues where insufficient assessment of environmental concerns have been identified. Those include: stricter requirements for land use planning in the Baltic Sea coastal area and other sensitive areas, and preservation of biological diversity. However, with an increasing number of cross-cutting issues such as climate change, nature conservation and the protection of biological diversity, a specific new tool like SEA has become necessary to identify likely negative effects of proposed planning documents. While experience with SEA and continuous discussions regarding its better application are still at the relatively early stage in Latvia, the implementation of the SEA directive in 2004 is to be seen as an important tool to create a continuous and stable process of assessing environmental impacts. This chapter reflects the author's own personal view on this implementation in Latvia.

It has been also recognized that an effective SEA system will require to overcome existing difficulties regarding the understanding of the nature of SEA and technicalities. Discussions on whether the actual instrument will increase the effectiveness of integrating the environmental policy into other sectors are to be seen as continuously ongoing. However it may change with time when more useful practical examples of SEA will be known. It is also important to show to planners and authorities responsible for other sectors that SEA cannot be effective if not built into the planning process. Therefore already developed processes for taking decisions in areas like, transport, agriculture, industry and forestry should overcome some modifications and be more open in introducing SEA.

Implementing Strategic Environmental Assessment. Edited by Michael Schmidt, Elsa João and Eike Albrecht. © 2005 Springer-Verlag

It seams that the only area where SEA can be integrated quite easily into existing planning and decision-making process is territorial planning. This is mostly because many elements of SEA are part of the procedure for developing territorial plans. When in 2001 work on developing a Latvia SEA system took place, the practice gained from specific planning and decision-making processes (e.g. development of territorial plans, regional development strategies) was analysed beforehand. It showed that some of the fundamental elements of the SEA Directive are incorporated already in our legal system. However analyses of current practice also showed that we do not have concrete provisions set by the law on how to proceed with SEA in other sectors besides territorial planning.

Before defining and explaining specific elements to the Latvia SEA process, a brief summary of the existing assessment tools will be given in Sect. 15.2 to see whether today's practice can be a part of the SEA process. Sect. 15.3 provides an introduction to the legislative context of SEA and a review of differences in applying SEA in Latvia. Sect. 15.4 focuses on activities regarding quality assurance of the SEA process. Sect. 15.5 discusses case study that provides information about pilot SEA project.

15.2 Practical Experience Gained with Existing Assessment Instruments and its Legal Provisions

As a starting point for introducing an SEA system, information was gathered and analyzed to identify what administrative and legal frameworks were already in place for applying SEA. The findings of the study on introducing the status and challenges of SEA processes and practice (Ruza 2001) indicated that territorial planning requirements in Latvia have many common procedural steps with regard to the SEA Directive. Those include:

- identification of environmental issues and related impacts;
- evaluation of impacts likely to be significant and important for particular areas and people living there;
- preparation of the planning documentation, including environmental impacts;
- consultations with relevant state and municipal authorities during the scoping stage and planning documentation review stage;
- public participation with public involvement within the scoping stage, public review of the draft documentation and final draft before submitting for approval;
- submission of the documentation to decision-makers.

However there are also some procedural steps that are missing in the territorial planning system and those relate to identification and evaluation of alternatives, requirements for assessing cumulative and transboundary effects, requirements for mitigation measures, requirements for monitoring, requirements for managing the total quality of the process, including decision-making process and communication of the decision passed to the public. As regards to the issue of integration, it has been found that new legal provisions shall be introduced on how to integrate environmental assessment results into the decision-making process. Procedure for taking the final decision of territorial plan also requires improvements to be in line with provisions of SEA Directive.

The above-mentioned study on introducing the status and challenges of SEA processes and practice (Ruza 2001) also showed that considerable progress has been reached in preparing regional development strategies. With few procedural exceptions, mostly those ones already mentioned for the territorial planning, the final regional development strategies intend to be called as test examples where an integrated approach to environmental impact analysis was used (Dusik 2001).

As a result, a conclusion was reached that legal provisions and practice in place for assessing environmental impacts of territorial plans shall be improved to comply with the SEA directive. However existing experience in general was considered to be positive, especially in meeting requirements for involving the public when drafting and assessing territorial plans and regional development strategies. This conclusion was taken into account when developing the new SEA system. A procedural link is therefore made between SEA legislation and territorial planning legislation to be able to keep current public participation practice during the plan preparation process. This also shows how important it is to try to integrate SEA elements into existing procedures.

15.3 Review of Differences in Applying SEA in Latvia

In Latvia, the SEA Directive is implemented through the Environmental Impact Assessment (EIA) Act 1998 and new secondary legislation (e.g. Cabinet of Ministers Regulations). Procedural framework for conducting SEA for plans and programmes has been transposed by amending the EIA Act on 26. February 2004. Cabinet of Ministers Regulations on procedure for conducting SEA was approved by the Cabinet of Ministers on 23 March 2004 and includes more detailed provisions on application and screening requirements, scoping stage, public participation procedure, environmental reports content, its drafting and evaluation procedure, consultation and monitoring requirements.

15.3.1 Definitions

The definition of SEA in the Latvia legislation refers to the type of environmental assessment for plans and programmes as required by the SEA Directive. The difference is to be found when issues arise about *applicability*. The SEA Directive applies to both plans and programmes, but neither of these terms is defined in the Directive. In Latvia the term "planning document" is used instead, which covers not only plans and programmes but also other strategic documents. According to the Regulations on Rules of Procedure for Cabinet of Ministers, when a set of planning documents is defined, the SEA applicability shall be in line with the above-mentioned Regulations. It has been decided that SEA will apply to the following types of planning documents: plans, programmes, conceptions and strategies. Since the meaning of these terms is a matter for country specific determination the following definitions are provided below:

Plan: Timely organised schedule of commitments or activities in a particular area, that implements a policy or programme.

Programme: A set of co-ordinated priorities, timed objectives, tasks and measures for the implementation of the policy in a particular area.

Conception: A set of necessary activities to be undertaken for solving a particular issue or problem. Conception is to be elaborated before initiating a new legal act.

Strategies: (no definition is provided)

15.3.2 Requirements for SEA Application

In considering further the differences of SEA application in Latvia it is important to mention that the area of application in relation to specific sectors has been widened. In addition to the sectors already mentioned in the Art. 3 para 2 of the SEA directive, SEA will be carried out also for planning documents, which are prepared for regional development, for extraction of mineral resources and for harbour development plans. Extractive industry has been especially emphasized due to special legal provisions by which this sector is regulated. SEA will also be mandatory for all territorial plans drafted for so called major towns of Latvia and for all districts. In Latvia we do have seven major towns and 26 districts.

However certain types of planning documents have been exempt from the rule to have obligatory SEA. These are territorial plans at local level (e.g. parish and municipality) and also so-called detailed plans. Territorial plans covering other planning levels (e.g. national and regional) are required to have SEA.

15.3.3 Competent Authority for SEA

It has been decided that the State EIA bureau, which is responsible for co-ordinating the EIA procedure in Latvia will also be responsible for the SEA process, especially for deciding on SEA application through case-by-case examination, for deciding on the scope and level of detail of the information to be included in the environmental report and for evaluating the environmental report. The approach for introducing one central institution directly responsible for certain SEA process elements associated with project EIA, where a similar approach has been introduced mostly to ensure the appropriate quality of the EIA report and the EIA procedure itself. Whether the approach with central competent authority will show positive signs mostly depends on the skills and competencies of those few experts becoming responsible for the SEA subject in the State EIA bureau.

15.3.4 Preliminary Assessment and SEA Application Form

The preliminary assessment is introduced for those planning documents, which are assessed through case-by-case examination. It is used as a filter to help the State EIA bureau to identify those planning documents that will be subject to SEA. Bases for the preliminary assessment are SEA application form submitted by the authority responsible for developing the planning document in question and consultation results with state and local authorities responsible for environmental protection and health issues. Professional associations having an interest in a particular sector to be covered by the planning document and environmental non-governmental organizations shall also be consulted prior to submitting the application form to the State EIA bureau.

The opportunity given to environmental organizations to be partners in the consultation process can lead to better acceptance of SEA results. Through the involvement of these organizations it is more likely that public concerns about a proposed planning document are identified and communicated to the competent authority at an early stage of decision making.

When the consultation process for identifying potential significant environmental effects has been concluded, the State EIA bureau, taking into account the results of the preliminary assessment, is responsible for deciding on whether the particular planning document is subject to SEA or not.

However in a system where case-by-case approach is used in determining the need for SEA some difficulties may arise for authorities responsible for developing planning documents. The existing system makes it difficult to know when the planning document is subject to preliminary assessment. Therefore further guidelines are needed to provide more detailed explanation of preliminary assessment procedure and to indicate those types of planning documents that might be subject to this assessment.

15.3.5 Information about a Decision to Prepare a Planning Document

The authority responsible for developing the planning document, which by the rules has been made subject to obligatory SEA, shall communicate to the State EIA bureau information on the decision to start to prepare a planning document. This information should be submitted to the State EIA bureau at an early stage of developing the planning document and before consultations are taking place on the scope and level of information to be included in the environmental report.

15.3.6 Review of the Draft Environmental Report

The main purpose of the review is to assess whether the draft environmental report prepared by the responsible authority is in compliance with conditions set by the State EIA bureau in the scoping stage and also with legal requirements for SEA. This review is undertaken by the State EIA bureau. The actual review procedure includes also consultations with authorities and public to ensure that the draft environmental report has been adequately prepared.

This pre-decision review process can help in dealing with information gaps, accuracy and credibility in presenting certain data and can help to set conditions needed for the monitoring. The formal review process is to be seen as an instrument for ensuring better quality of the environmental report and planning document. However further guidelines are still needed for the personnel working in the State EIA bureau to know more precisely what are the review criteria and techniques to be used for reviewing. Otherwise the quality standards required will be hard to reach.

15.4 Quality Assurance

For reaching better quality of the whole SEA process it is still necessary to disseminate the principles and methods of SEA into traditional sectors and to continue the communication process within and between institutions becoming involved in the SEA application. However some activities regarding the quality assurance of the SEA process within Latvia took place during 2001 and 2003 and a number of initiatives have helped to contribute to better understanding of the SEA process.

Box 15.1. Contents of the Guidance document: Introducing the status and challenges of SEA processes and practice

1. SEA definition and coverage
2. Advantages of SEA application
3. New SEA Directive – requirements and analyses
4. Environmental assessment at the European level (Habitats Directive, Wild Birds Directive, Water Framework Directive, pre-Structural funds requirements)
5. SEA limitations
6. Legal bases for the implementation of the SEA Directive in Latvia (analyses of existing laws and regulations)
7. Evaluation of SEA related plans and programmes in Latvia
8. Proposals for SEA legislation establishment in Latvia

    Annex 1   Examples of plans and programmes for which SEA could be undertaken
    Annex 2   Criteria to determine significance of proposed plan or programme
    Annex 3   Issues to be included in the SEA environmental report
    Annex 4   Questionnaire that could be used for assessing the environmental impacts of proposed plan or programme
    Annex 5   Case Study: Experience on the integration of environmental assessment in the development process – Regional Development Plan for the Latgale Region
    Annex 6   Key references

The most important initiatives were:

- Publication of the Guidance document: *Introducing the status and challenges of SEA processes and practice* published in November 2001. It has been disseminated (also via workshops) to the main stakeholders in order to give them needed information on SEA in the national language. Box 15.1 gives some information on the contents.
- The pilot SEA project, which aimed at assessing likely significant environmental impacts of the Latvia Development Plan, was undertaken during June 2002 and February 2003. Information about this SEA project is presented in Sect. 15.5.

15.5 Case Study – SEA for the Latvia Development Plan

Strategic Environmental Assessment for the Latvia Development Plan (Single Programming Document and Programme Supplement) was carried out during June 2002 to February 2003 within the framework of Latvian – Finnish bilateral assistance project. The purpose of this project was to increase knowledge of the state and municipal authorities and various organizations (including non-governmental organizations) on issues related to the SEA by involving them into a practical SEA project aimed at assessing the likely significant environmental impacts of the Latvia Development Plan (LDP).

The main objective of this project was to prepare an SEA environmental report for the LDP and it was reached by using the following approach. Simultaneously with the SEA training workshops, SEA was carried out using "learning by doing" approach. The joint Latvian and Finnish expert group, who organized the SEA training course, also participated but mostly as moderators of the SEA process.

SEA took place simultaneously with the drafting process of the LDP. During the SEA process, the working group, which consisted of members of the SEA training course, evaluated five drafts of the LDP and participated in five workshops. In between the workshops several meetings with representatives from responsible authority, the Ministry of Finance, were held. The public discussion on the Draft SEA environmental report took place simultaneously with the public discussion on the draft LDP in September 2002. Each member of the public was able to access the draft environmental report by visiting the Internet site of the Ministry of Finance. It also included a specially designed form for submitting comments. Every single assessment result prepared by the SEA working group, including the final version of the SEA environmental report, was submitted to the Steering Committee of the LDP. Thus the current version of the LDP was constantly being updated and improved taking into consideration the results of the SEA working group.

The SEA process was definitely a complicated one because the working group had to assess every newly drafted version of the LDP and had to submit proposals, within a limited period of time. The following issues were addressed in the final SEA environmental report:

- Information of the SEA procedure
- The description of the SEA principles and methodology of work is provided
- The analysis of the current situation
- The assessment of the quality and sufficiency of information provided in the draft LDP that is required to identify the possibilities and main solutions for Latvia to achieve the strategic objectives
- Compliance assessment of the strategy objectives
- The analyses of the compliance of the Strategic objectives, which are set in the LDP with the objectives of every priority and measure, as well as conformity to the objectives of sustainable development of Latvia and the EU
- Environmental assessment of priorities and measures included in the LDP
- Environmental assessment of the effects from implementing the measures of the LDP, assessment of both positive and negative impacts; and finally
- Recommendation for the monitoring of the implementation of the LDP

15.6 Conclusions and Recommendations

The implementation of the SEA Directive in Latvia is to be seen as an important opportunity to create a continuous and stable process of assessing environmental impacts and increase environmental policy integration. Procedural framework for conducting SEA for plans and programmes has been transposed by amending the EIA Act on 26 February 2004 and new secondary legislation.

Through the application of legal instruments Latvia has chosen a traditional approach to handle the environmental integration issues. However, there is a need for using different instruments (flexible), such as non-legislative to overcome the weakness of the SEA system. Further guidelines are needed to provide more detailed explanation of preliminary assessment procedure and to indicate those types of planning documents that might be subject to this assessment. Other guidelines are needed for the personnel working in the State EIA bureau to know more precisely what are the review criteria and techniques to be used for reviewing the SEA environmental report. Otherwise it will be hard to reach the quality standards required.

From the existing SEA case studies evaluated (e.g. SEA for the Latvia Development Plan – see Sect. 15.5) it is still an issue to be argued – how to apply the SEA framework within diverse sectors with specific decision making systems and how to measure the effectiveness and success of SEA? These two questions should be an issue for further development.

References[1]

Dusik J (2001) Approaches to strategic environmental Assessment in Central and Eastern Europe. Background documents for an informal workshop of the Sofia EIA Initiative. Szentendre, April 9-10, 2001, (unpublished report)

Finnish Neighbouring Area Co-operation Bilateral Project (2003), Developing monitoring system to Latvia in the area of Regional Development: SEA Component, Final Report, Riga

Ruza S (ed) (2001) Latvijas – Somijas projekta skaidrojošs materiāls: Stratēģiskais vides novērtējums (Latvian-Finnish project Guidance document: Introducing the status and challenges of SEA processes and practice), Riga

---

[1] Note that legislation mentioned in the chapter is listed at the end of the handbook in the consolidated list of legislation (Appendix 2).

16 Implementing SEA in Estonia

Olavi Hiiemäe

Estonian Agricultural University, Estonia

16.1 Introduction

The interest in SEA has grown significantly over the last several years, as its relevance has been recognized more broadly. It is now generally understood that the prediction of possible damages made to the environment is easier and cheaper to avoid rather than to deal with the consequences in the future. In Estonia there is SEA legislation since 2001. This chapter discusses the current state of the SEA process in Estonia and tries to analyze the positive aspects and shortcomings of the present law and existing SEA reports. This will be done also taking into account the requirements of the new SEA Directive, which Estonia will need to comply with following integration within the European Union in May 2004.

This chapter explains the current status of the SEA directive and SEA process in Estonia. First, in Sect. 16.2, an introduction to the main ideas of project EIA and SEA according to the existing national law of "EIA and Environmental Auditing" is given. Accession to the European Union (EU) and the need to adopt the SEA Directive that requires all new EU Member States to introduce the laws, regulations and administrative provisions necessary for its implementation has accelerated the development of Estonian environmental legislation. In May 2004 the new draft of EIA and Environmental Management legislation, that include detailed introduction to the SEA as well, was passed to the Parliament. In Sect. 16.3 the chapter gives an overview of the main actors in the SEA process, and the results of a survey of SEA actors carried out by Peterson (2004) are introduced in Sect. 16.4. Finally, the quality criteria of the existing SEA documents are explained and general recommendations to improve the decision-making process via strategic planning documents are presented.

Implementing Strategic Environmental Assessment. Edited by Michael Schmidt, Elsa João and Eike Albrecht. © 2005 Springer-Verlag

16.2 Project EIA and SEA in Estonia

In Estonia both Project EIA and SEA are regulated by the national law on "Environmental Impact Assessment and Environmental Auditing" (EIA and EA). The law was enforced from 1 January 2001, and became the main driving force for the introduction of SEA in Estonia. It is stated in Art. 22 of this law that "SEA should go in parallel with the drafting process of the policy documents and that likely environmental impacts arising from the implementation of policies, plans and programmes should be taken into account" (Peterson 2003). According to the law, the idea behind the SEA is to simplify the decision making process in early stage of the policies, plans and programmes, to ensure that all proposed alternatives are adequately assessed, all impacts are considered, the public is fully consulted, and the decisions has made keeping in mind the concept of sustainable development and "Good Practice".

Project EIA has a longer history and practice in Estonia (national regulation was adopted in 1992), whereas SEA is more recent (first mentioned in 2001) and thus is little practiced yet. Accession to the EU has accelerated the development of national legislation on environmental assessment and transposition of corresponding EU legislation (the draft of the new law was passed to the parliament on 31 May 2004). The requirement for mandatory SEA on national plans, programmes and spatial plans (the requirement of mandatory SEA on spatial plans was removed from the law in 2003) has brought about a forceful introduction of SEA into different sectors and public administration in a short period of time. The public administrations, either responsible for carrying out SEA alongside with the drafting process of policy documents or supervising the process, were not prepared for the implementation of the SEA Directive and much of the today's experience lies on learning-by-doing practice (Peterson 2003).

Followed by the enforcement of the law, there have been seven SEA of programmes and plans at national level, and probably hundreds of spatial plans developed at county or municipal level in the period between 1 January 2001 and 1 August 2003. Usually these SEA documents have been accompanied by only a short description of one or two pages on the possible environmental impacts (Peterson 2003). SEA was conducted (by the listed proponents and adopting bodies) for the following national strategic documents in Estonia in 2000-2003 (Peterson 2004):

- Single Programming Document (Ministry of Finance/ Government)
- Forestry Development Plan (Ministry of Environment/ Parliament)
- Fuel and Energy Development Plan (Ministry of Economy and Communication/ Parliament)
- National Programme on Minimisation of Emissions of Greenhouse Gases (Ministry of Environment/ Government)
- National Development Plan Sustainable Estonia 21 (Ministry of Environment/ Parliament)
- Rural Development Plan (Ministry of Agriculture/ Government)

- Planning Permission for the Central Test Ground for National Defence Forces (Ministry of Defence/ Government)

Since July 2004 the SEA Directive requires all new EU Member States, including Estonia, with the duty to introduce the laws, regulations and administrative provisions necessary for its implementation. It means that Estonia is approaching the third phase in its EIA development. In May 2004 a new draft of EIA legislation passed to the Parliament, that includes a detailed introduction to SEA. In the draft the principle of SEA is defined as the systematic assessment of impacts of strategic planning documents, policies and programmes. The assessment is made before the decision, made available to the decision maker and the public, and should demonstrate the impacts of both implementation and operation. The goal is to ensure that the environmental issues are taken into account when a decision is reached about whether and under which conditions a planning or political document should be allowed to proceed.

16.3 Actors in the SEA Process in Estonia

There are five main groups of actors in the SEA process in Estonia:

1. Parliament/Government, which adopts and approves of the policy documents,
2. Public authority, which is responsible for drafting of the policy document (a programme or a plan) and assessing the potential environmental effects arising from the implementation of the programme or plan,
3. Environmental authority (environmental competent authority), which reviews the SEA documentation and approves of the SEA programme and report,
4. SEA expert(s), who is (are) contracted by the public authority to conduct the SEA and
5. Stakeholders, who have interest in the issues, either those from the policy document and/or from the SEA programme and report (see also Fig. 16.1).

The public authorities developing programmes and plans (PP) are usually ministries appointed by the Government to develop policy documents in their sector of authority. The environmental authority in the Estonian case is the Ministry of Environment, which performs the quality control of SEA process (see Sect. 16.5) and approves the SEA programmes and reports.

Due to the legal requirement in the Estonian law, only licensed environmental experts can conduct environmental assessments, including SEA. SEA experts are specialists on certain research area, who hold a valid license, issued by the special commission next to the Ministry of Environment. Legal bodies can be subcontracted for SEA by public authorities only if they employ licensed experts. It should be noted that there is no distinction being made in registration of experts performing project-level EIA and policy level SEA.

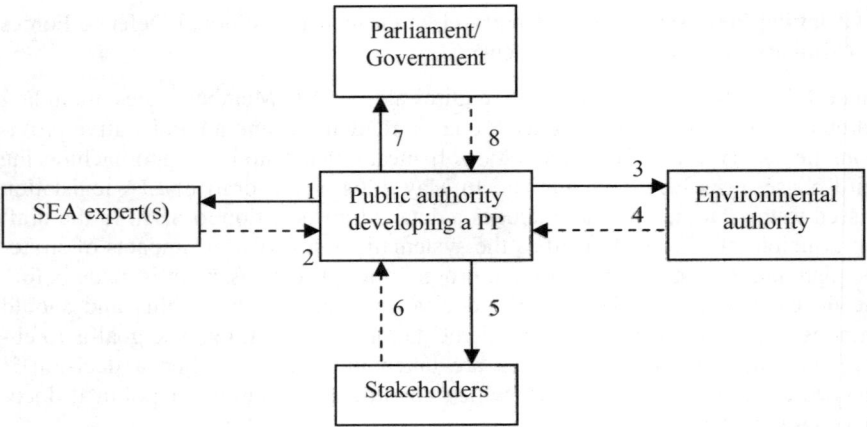

Fig. 16.1. Actors and their roles, and circulation of documents in the SEA process in Estonia (full line – flow of documents and information, dashed line – feedback) (Peterson 2003)

Flow of documents and information, and feedback of the SEA process in Estonia:

1. Terms of reference (ToR) for SEA;
2. Draft SEA programme and report submitted to be approved;
3. Draft SEA programme and report transferred to environmental authority for review and approval;
4. Approval of SEA programme and report;
5. Draft SEA programme and report put on public display, stakeholders are consulted, public meetings are held;
6. Stakeholders give comments and propose amendments;
7. Programme or plan submitted for adoption;
8. Feedback on decision – adopted or rejected (including amendments).

"Stakeholders" are either general public or interest groups in the respective policy area or, more commonly, environmental NGOs or other SEA experts that are not directly involved in this particular SEA but have either content or professional interests in it. According to the EIA and SEA law, the SEA document must be available for all stakeholders for 30 days to give comments and make replacements in the document. After that the expert together with public authority will organize the public meeting. All received information must be included in the report, and arguments for considering or not considering must be presented.

16.4 SEA in Practice

The national law in Estonia does not specify how, when and to what extent the two processes (drafting of plan or programme and SEA) should interact. The main idea of SEA – to identify, describe and evaluate the likely significant effects on the environment of implementing a plan or programme and reasonable alternatives – gets usually lost in formal bureaucratic process and eventually has limited impact on the policy document. Moreover, SEA is much regarded as a waste of time, creating unnecessary problems, which result in delays of the drafting and adoption of policy documents and thereby increases the total cost (Peterson 2003).

Even, if it stated in the law that SEA is mandatory and should be carried out in parallel to the drafting process of the policy document, still SEA is usually launched later than the drafting process starts. In the majority of cases this results in situations where SEA experts find themselves with limited options to consider, since important policy decisions have been made in the earlier stages of the process.

Once the scope and, sometimes, the objectives are fixed it leaves limited maneuver, if any at all, for environmental experts and stakeholders to propose possible amendments and alternative settings to the outline of the policy document. This will increase the likelihood of facing delays in the completion of the draft policy document and of excessive costs borne by the public authorities, if later charges are forced, either by environmental experts or stakeholders (Peterson 2003).

On the other hand, since the law does not provide any guidelines and only a few models are available on how to carry out SEA, there is general a lack of information about SEA. All SEA documents are prepared according to the SEA Directive, but with self-developed methods interpreted by each expert differently. So far, in practice, experts pay very little attention to the analysis of alternatives and the consideration of cumulative environmental impacts of policies, plans and programmes. This makes environmental impact statements (EIS) more formal documents with detailed overview of the background data and current situation, and less analytical documents. For example, there is a case where one 77 pages long SEA document has only one and half pages dedicated to the description of alternative options and the comparison of those alternatives, and three pages only for the overview of mitigation measures.

Sometimes the EIS are written in a way that the suggestions and analysis do not provide clear understanding, and can be interpreted in many ways. Totally missing descriptions on how results were defined. Difficulties arise in the interpretation of results and suggestions when they are implemented. Reports are too liberal and sometimes even confusing, and give decision makers the opportunity to weigh up the implications of findings of environmental impacts with their own interests. As a results of these factors, SEA documents might turn into fragmented, less well understood reports, and can often have unintended and unpredictable outcomes.

In practice, SEA has currently much bigger effect on the identification of stakeholders and public involvement into the policy drafting process than on setting environmental criteria and forecasting of environmental impacts. The educational aspect of SEA is much appreciated by public authorities, once different aspects of the environmental protection are brought to their attention over the course of the SEA. At same time many authorities and experts still underestimate the benefit of the public participation, and see it just as a formal annoying procedure that causes delay of planning process and increases total costs of the project. In the last years, the attitude seems to have change. Recently more attention has been paid to communicating with the public, and is rarer to find projects where "public manipulation" has been taken place. The public and stakeholders are starting to be actively involved into the SEA process from the early stages of the project, and there are more cases where the decision makers have accepted public opinion. According to the existing legislation, the public and stakeholders must be informed and consulted at least after the scoping stage and when finalizing the draft SEA report. In practice, many experts have realised that it is much easier to manage the process when public is involved already in early stage of the project. Usually, most actively the public is involved while making the scoping.

All seven SEA reports were analysed by Peterson (2004), where the study has explored the role of SEA in Estonia on the bases of 26 evaluations by three groups of respondents – public authorities, SEA experts and environmental NGOs. The respondents were asked questions, which explored the following main questions:

- When was the SEA started in relation to the start of drafting policy documents?
- What was the expected role of SEA and its impact on the policy document?
- What was the actual impact of SEA on the policy document?
- What were the major difficulties experienced in the SEA project?

The results of the survey showed, that in many cases the meaning of the SEA is still misunderstood. Often it was much due to the unclear definitions in the law what gives a possibility for manipulation and dual interpretation for those definitions. According to the Peterson (2004) "the law requires that SEA is applied to national policies, plans and programmes, but not for other types of policy document in Estonia. The need for such definition is obvious, given the diversity of terms currently used in Estonia, such as 'a (national) development plan', 'a (national) programme', 'a strategy' and 'an action plan'. Diverse interpretations of these terms have made it difficult for public authorities and courts to decide whether a SEA of a particular document is needed".

In six out of seven cases the SEA was started later than the process of developing the policy document. The delay varied from one month to one year and was explained by saying that at the beginning of the drafting process "there was nothing to assess yet" (Peterson 2004). Peterson (2004) concluded "this is a sign of the attitude to SEA as a 'reactive' assessment tool, rather than a 'proactive' planning aid. The late start of the SEA process result that there are only a limited number of options to consider, since important policy decisions have already been made".

All SEA actors who have participated in the survey agreed that expectations and objectives of SEA were to identify the potential activities proposed by the policy documents that have significant effects on the environment, to identify the possible conflicts between the objectives of the policy document and national environmental objectives, and reaching a public consensus on them. However, there was no clear agreement between different actors in SEA process on question that what should be the overall aim of SEA. Peterson concluded that "if the objectives of SEA are not agreed among the actors of the process at the beginning of the process, it may lead to different, even contradictory, expectations about the outputs and outcomes" (Peterson 2004 p.162).

The participants in the survey were asked about the overall impact of SEA and its effects on specific aspects of the strategic planning process and SEA documents. All respondents believed that SEA had expanded the number of stakeholders (public meetings and increased communication between all actors) involved in the policy drafting process, better coverage of environmental issues in the policy document and growth of personal environmental education through communication between different actors involved in SEA process were just some main positive impacts pointed by the respondents. The difficulties in managing public consultation, public meetings were delaying the planning process and increased the cost were comments from the negative side of the SEA process. However, "only three out of twenty six respondents reported that SEA had no impact at all, while 21 regarded it as limited, and two considered it significant" (Peterson 2004 p.162).

Only SEA experts and public authorities were asked the question regarding difficulties in carrying out SEA. The major difficulty experienced by both responding groups was to fit SEA activities as consultation with public, public hearings, processing of comments and providing feedback into the given timeframe. The lack of established methods and approaches for identifying and assessing significant impacts, the lack of sectoral environmental objectives and criteria, and lack of qualified SEA experts were main difficulties named out by SEA experts and public authorities.

16.5 The Quality Criteria of the SEA Document

Thissen (2000 p.119) argues that, policies will get better as the available information on problems, causes, alternatives and impacts is better, more scientific, and more balanced. The core objective of analysis is to generate and present such information. Further in his article, he argues about the "quality" and "effectiveness" criteria of SEA reports. He defines a "quality criteria" as a criterion "used to assess the goodness of institutional arrangements, methods and other inputs", while "effectiveness" is related to "direct and indirect outcomes".

In Estonia, the quality of the SEA documents is estimated using similar criteria as the one used for the EIA documents:

- Are all relevant topics considered or not considered?
- Are all analyses realistic or unrealistic?
- Are suggestions implementable in practice or not?
- Are the recommendations, analysis and suggested alternatives transparent or not?
- Was there any improvement compared to the previous policy/programme?
- Does it help decision makers to reach better decisions?
- Does it help plans, policies and programmes to be more proactive and direct the process?

The quality of the SEA report is still based much on personal skills of experts and requisition of decision maker, made in the beginning of the agreement. It is up to the decision maker (in practice often used involvement of expert(s) in early stages of SEA preparation stage.) to set up detailed list of questions that SEA report must come up with. In the end, it is up to the decision maker to accept the report or to request the expert for an update or for more information.

16.6 Conclusions and Recommendations

The interest in SEA has grown significantly over the last years, as its relevance has been recognized more broadly and practical experience has accumulated. The authorities and experts have recognized that the prediction of possible damages made to the environment is easier and cheaper to avoid in advance than to deal with consequences in the future. Therefore, the new SEA documents are prepared in parallel with plans and policies, with tight cooperation between SEA experts and policy makers. During the last years the authorities and decision makers started to pay more attention to the required procedure and to the quality of SEA reports.

EIA and SEA both originated from a need to foster the attention for and weight given to environmental considerations in decision-making. It has, however been felt that EIA has its shortcomings, especially while talking about environmental concerns in planning and policymaking. "Applied at the project level, an EIA can usually only affect decision making in a reactive way, attempting to block or modify a project proposal if it is environmentally harmful. SEA came into being as a reaction to these perceived shortcomings of EIA. Its purpose is to more structurally introduce and safeguard the attention to environmental concerns in the policy making and planning phases of decision making" (Thissen 2000 p.114).

The new draft 2004 law that adopts the requirements of SEA Directive into the Estonian context, recommends the ideological structure of SEA, and sets up the quality assessment criteria for SEA reports. For the first time in Estonia, the law provides a comprehensive basis for the appraisal of policies, plans and programmes, including the provision of an environmental report, consideration of alternatives, consultation and public participation, inclusion of recommendations, and monitoring and review mechanisms.

However, the SEA Directive has some shortcomings too. The main problem faced in the SEA process is to identify whether SEA is required or not. It has been argued that at the moment there is no obligation to conduct SEA since the current law in Estonia does not specify the types of programmes and plans subject to SEA. The need for legal definition of programme and plan is obvious, giving the diversity of terms currently used in Estonia, e.g.: 'arengukava' – development plan, 'riiklik arengukava' – national development plan, 'programm' – programme, 'riiklik programm' – national programme, 'strateegia' – strategy; 'tegevuskava' – action plan. The lack of legal definitions of various policy documents has confused the screening process of SEA, i.e. whether SEA is needed or not and whether and why some programmes and plans are subject to SEA and others are not (Peterson 2003).

There is still no clarity as to what level of detail the SEA document is expected and what is a hierarchy of different political documents in Estonia. This results in different interpretations of what sort of alternatives should be considered and what should be the level of detail of mitigation measures, which eventually leads to the different expectations regarding to SEA.

In addition, the methods for assessment used in SEA and the quality criteria need better clarification. Only after methods of assessment and quality criteria are clarified, SEA can become a systematic process during the preparation of a plans and strategies, and SEA can achieve the environmental, economic and social objectives by which sustainable development is defined. However, it can be concluded that the importance of SEA is rapidly growing as a tool for forecasting environmental impacts and integrating environmental issues into the strategic decision-making process.

References[1]

Peterson K (2003) Role of SEA in complex Decision Making in Estonia. Tallinn
Peterson K (2004) The role and value of strategic environmental assessment in Estonia: stakeholder's perspectives. Impact Assessment and Project Appraisal 22(2):159-165
Thissen W (2000) Criteria for Evaluation of Strategic Environmental Assessments. In: Partidário M, Clark R (eds) Perspectives on Strategic Environmental Assessment. CRC/Lewis Boca Raton, pp 113–127

---

[1] Note that legislation mentioned in the chapter is listed at the end of the handbook in the consolidated list of legislation (Appendix 2).

Part III – Experience of SEA in North America and Oceania

Throughout the world there is a long history on the use of SEA or similar instruments that precedes the new European SEA Directive. The first steps were taken in the United States in 1969 with the National Environmental Policy Act (NEPA), which required the environment be considered at *every level* of federal decision-making. Therefore it can be argued that *both* Project EIA and SEA started with NEPA in 1969. Chapter 17 discusses the improved decision-making that is possible with SEA, in relation to the expectations and results of NEPA in the United States. There is also a lot to learn from the Canadian system (Chapter 18) and from the New Zealand approach (Chapter 19). Both Canada and New Zealand have a legal requirement for SEA since early 1990s and both put a strong emphasis on the links to promoting sustainability. Analysing the successes and failures of long-established SEA systems such as these is very helpful so that legislators, academics and practitioners know how to best incorporate SEA into strategic-decision making.

In Canada, although there has been a commitment to assess the potential environmental implications of federal policies since 1984, SEA as we know it began in 1990. More recently, in 1999, a revision was done in order to strengthen the role of SEA by linking environmental assessment to the implementation of sustainable development strategies. New Zealand was one of the first countries to codify sustainability into law by enacting the RMA (Resource Management Act of 1991). New Zealand has had SEA procedures in place since 1991. One of the main objectives of the SEA procedures is to ensure that resource management plans (prepared within the framework of the RMA) achieve the sustainability objective of the Act.

Part III – Experience of SEA in North America and Oceania

17 Improved Decision-Making through SEA – Expectations and Results in the United States

Jehan El-Jourbagy[1] and Tyson Harty[2]

1 Attorney, Member of the Oregon Bar & Environmental Advocate, USA
2 Department of Zoology, Oregon State University, USA

17.1 Introduction

The National Environmental Policy Act (NEPA), one of the first legislative efforts in the world to address SEA and the first to require Environmental Impact Statements (EIS), was a landmark initiative by the United States government that required environmental considerations at every level of federal decision-making. In approaching a procedural framework in which to integrate SEA, the European Union may find useful an examination of a similar system in the United States. As the United States balances federalism with states' rights, NEPA only address federal actions; therefore, major actions by state agencies are governed at the discretion of state legislatures. Because of this dichotomy, NEPA's reach is less extensive then that of the EC Directive because every Member State agreed to abide by the rules whereas the states in the United States are under no such obligation. By first offering a brief background on NEPA – the United States' procedural law addressing environmental assessment – followed by a discussion of what triggers NEPA analysis, this chapter focuses on NEPA's successes and failures in an effort to assist future lawmakers, academics, and advocates on how to best incorporate SEA into law and policy.

17.2 National Environmental Policy Act

If principle is good for anything, it is worth living up to. Benjamin Franklin

NEPA, passed in 1969 and signed into law by President Richard Nixon on January 1, 1970, was one of the first federal laws created during a period in American legislation known as the "environmental decade" (Pub. L. 91-190, codified at 42 USC §§ 4321-4370a).

Implementing Strategic Environmental Assessment. Edited by Michael Schmidt, Elsa João and Eike Albrecht. © 2005 Springer-Verlag

The Clean Air Act of 1963 (amended in 1970), the Clean Water Act of 1972 (originally called the Federal Water Pollution Control Act), the Marine Protection, Research, and Sanctuaries Act of 1972 (also known as the Ocean Dumping Act), the Endangered Species Act of 1973, the Safe Drinking Water Act of 1974, and the Resource Conservation and Recovery Act of 1976 were all either created or significantly overhauled within this ten-year period from 1969 to 1979.

NEPA includes two main provisions: first, the law outlines procedural requirements for federal agencies when taking major actions that may significantly affect the human environment, and second, the law created the Council on Environmental Quality (CEQ). NEPA has not changed significantly since implementation though its application has evolved through litigation and subsequent case law. The procedural requirements are described in detail in the next section and are followed by a discussion on the successes and drawbacks of the rules. As a three-member panel under the Executive Office of the President, the CEQ is charged with compiling and publishing annual Environmental Quality Reports. The CEQ is also responsible for coordinating agency compliance with NEPA, and it maintains a website at *http://www.whitehouse.gov/CEQ* that provides a wide range of information about NEPA. The CEQ recently appointed a NEPA task force that provides recommendations to the CEQ. The task force meets this charge by reviewing NEPA documents and determining the level of detail required for environmental assessments, categorical exclusions, and other products of the NEPA process. More information regarding the taskforce may be found at the webpage *http://ceq.eh.doe.gov/nepa/nepanet.htm*.

The United States was groundbreaking in requiring an environmental impact statement. NEPA is also useful in garnering environmental assessment for all legislation and federal action and is therefore somewhat similar to the SEA Directive. NEPA requires a discussion of environmental impacts, adverse effects, and alternatives any time an action may significantly affect the environment – just as the SEA Directive requires a discussion of environmental effects and reasonable alternatives. The balance of this section introduces the purpose of NEPA, its implementing governmental organizations, and what type of action triggers NEPA.

17.2.1 Purpose

NEPA is codified in the United States Code (USC) and is actually quite broad; more specific guidelines promulgated by the CEQ may be found in the Code of Federal Regulations (CFR) (40 CFR pts. 1500-1517). The basic purpose of NEPA is to provide federal agencies, such as the United States Forest Service and the Army Corps of Engineers, with information on which to base environmentally sound decisions. The purpose found in Section 2 of NEPA is as follows:

> "To declare a national policy which will encourage productive and enjoyable harmony between man and his environment; to promote efforts which will prevent or eliminate damage to the environment and biosphere and stimulate the health and welfare of man; to enrich the understanding of the ecological systems and natural resources important to the Nation; and to establish a Council on Environmental Quality" (42 USC § 4321).

The purpose asserted by the CEQ in the federal regulations is to make decisions "based on understanding of environmental consequences and take actions that protect, restore, and enhance the environment" (40 CFR § 1500.1(c)). Information presented pursuant to NEPA must be of high quality, and the law requires an agency, such as the National Park Service, to perform an environmental analysis *before* making a decision (40 CFR § 1500.1(b) (emphasis by the authors)).

17.2.2 Implementing Government Bodies

All United States government agencies are required to comply with NEPA. Examples of agencies that often engage in NEPA analysis include the United States Forest Service, the Bureau of Land Management, the United States Park Service, United States Fish and Wildlife, and the Department of Energy. All documents produced under NEPA are considered public documents, which means that they must be accessible and available to the general public. Federal agencies often post NEPA-related documents on websites and also allow the members of the public to acquire such documents and make copies at their offices.

As aforementioned, NEPA is a federal law enacted by the United States legislature, so only federal (or United States) agencies are required to comply; however, some states have enacted a similar law that applies to state agencies and municipalities. For instance, the state of Washington enacted the State Environmental Policy Act (SEPA) that is similar to NEPA (RCW 43.21C.030). A key difference between Washington's SEPA and NEPA is that the Washington law only applies to adverse environmental impacts – not both beneficial and adverse impacts as is the case under NEPA (RCW § 43.21C.031 (2002); WAS 197-11-330(1)(b)). In contrast to NEPA, the SEA Directive applies to every Member State instead of just applying to EU-initiated plans and programmes.

17.2.3 Triggering NEPA: "Major Federal Action"

The NEPA process is triggered at the proposal stage for any "major federal action significantly affecting the quality of the human environment" (42 USC § 4332). NEPA does not address non-federal actions even if such action affects several states. Unlike in the Member States, where all plans and programmes trigger environmental assessment and consultation with other Member States is required upon a finding of potential transboundary impacts, non-federal initiatives in the United States that impact several states may only be addressed through other environmental legislation such as the Clean Water Act or Clean Air Act. The law calls for baseline data and consistent methods and procedures (*Id*). Every environmental report must include a discussion on environmental impacts, unavoidable environmental impacts, alternatives, relationship of short-term to long-term productivity, and any irreversible/irretrievable resources (42 USC § 2332(C)(i)-(v)). These requirements are very similar to what is required in an environmental report under the SEA Directive, which requires the inclusion of likely significant environmental effects and reasonable alternatives (Art. 5).

According to the EC Directive on SEA, all "plans and programmes" fall under the ambit of the procedural guidelines (Art. 1 of the EC Directive). The language used in NEPA is slightly different, with "major actions" triggering the procedural requirements (42 USC § 4332(C)). Federally, "major" is read to enforce the term "significant" – in other words, the size of the proposal alone does not impact NEPA analysis and other issues that affect significance (such as frequency, reversibility or probability of the impacts) are also considered (40 CFR § 1508.18 Major Federal action).

"Action" is the more determinative word. "Action" is defined as "new and continuing activities, including projects and programs entirely or partly financed, assisted, conducted, regulated, or approved by federal agencies; new or revised agency rules, regulations, plans, policies, or procedures; and legislative proposals" (40 CFR § 1508.18(a)). The SEA Directive states that environmental assessments are required for plans and programmes prepared for agriculture, forestry, fisheries, energy, industry, transport, waste management, water management, telecommunications, tourism, town and country planning, or land use (...)" (Art. 2 para 2a)). NEPA's guidelines do not list specific areas of applicability. However, unlike the EU Directive, NEPA's procedural regulations apply to more than just plans and programmes. Federal proposals, permits, funding, and policies may also trigger the procedures. In contrast, in the EU this will only happen in the cases where EU countries view the SEA Directive as a "floor" instead of a "ceiling." At minimum, plans and programmes will trigger SEA. However, a country may choose to apply SEA to other types of actions. For instance, Scotland is planning to include policies in applying SEA (see Deasley 2003). Also, the Directive states that Member States may make a list of types of plans and programmes that are likely to cause significant environmental effects and/or they may address plans and programmes on a case-by-case basis (Art. 3). In the United States, all actions are dealt with on a case-by-case basis, but Member States will likely create lists of types of plans and programmes to assist in application.

17.3 NEPA Analysis: Procedural Process

The execution of the laws is more important than the making of them. Thomas Jefferson

To understand how NEPA procedurally operates, reviewing the typical decision-making process step by step is illustrative (see Box 17.1). First, an agency will see if a major action is already addressed under existing NEPA coverage. For instance, the action may already be encompassed under an Environmental Assessment (EA) or an Environmental Impact Statement (EIS) (see later in this section for a more detailed explanation of EA and EIS).

17 Improving Decision-Making through SEA in the United States   243

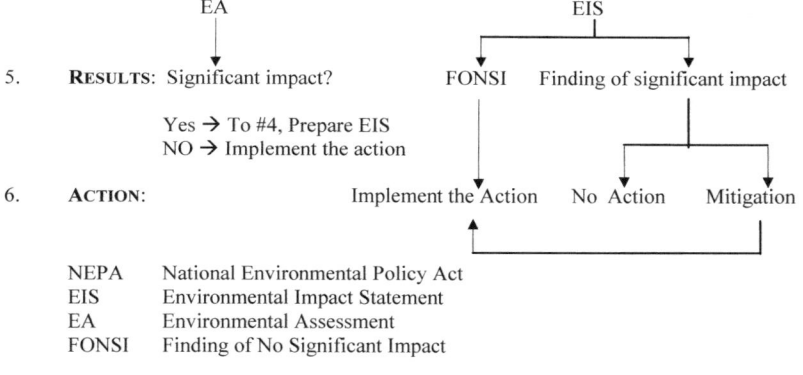

Box 17.1. A Step-by-step Flowchart of NEPA Decision Making Process

1. Is action **ALREADY COVERED** in a previous NEPA document?

    Yes → If new information/changes, then prepare supplemental EIS (to # 4)
       If no new information, IMPLEMENT according to previous findings
    NO → To # 2

2. Does a **CATEGORICAL EXCLUSION** apply?

    Yes → IMPLEMENT
    NO → To # 3

3. Will the action result in **SIGNIFICANT ENVIRONMENTAL IMPACTS**?

    Yes → To # 4, Prepare EIS
    Not sure → To #4, Prepare EA

4. **REPORT PREPARATION**

                EA   EIS

5. **RESULTS**: Significant impact?   FONSI   Finding of significant impact

    Yes → To #4, Prepare EIS
    NO → Implement the action

6. **ACTION**:   Implement the Action   No Action   Mitigation

    NEPA   National Environmental Policy Act
    EIS   Environmental Impact Statement
    EA   Environmental Assessment
    FONSI   Finding of No Significant Impact

If the action is covered under a NEPA document, then the preparer should review the action to see if there are any substantial changes or new information. If so, then a supplemental EIS should be prepared.

The second step is to determine if a categorical exclusion applies. A categorical exclusion is a previously determined finding that the action will result in no significant impact (40 CFR § 1508.4). For instance, trail reconstruction in a popular area, such as the Grand Canyon, may fall within a categorical exclusion because the trail is used so often and is often maintained. In 1985, the city of Dupont, Washington proposed that a new interchange be built to address current and projected growth (West vs. Secretary of the Department of Transportation, 206 F.3d 920, 922 (9th Cir. 2000)). After a review of potential environmental impacts, the Washington State Department of Transportation concluded that the project would not result in any significant environmental impact and further stated that the project fell within a categorical exclusion (*Id.* at 924). However, the Court of Appeals for the Ninth Circuit held that "an entirely new, $18.6 million, four-lane, 'fully-directional' interchange constructed over a former Superfund site and requiring 500,000 cubic yards of fill material, 30,000 tons of crushed surfacing, and 32,000 tons of asphalt concrete" did not fit within the documented categorical exclusion constituting "approvals for changes in access control" (*Id.* at 928).

The third step in the NEPA procedure is to determine if a significant environmental impact will occur. Significant impact is evaluated by considering both context and intensity (40 CFR § 1508.18(a)). An evaluation of context considers "society as a whole, the affected region, the affected interests, and the locality" (*Id.* § 1508.27(a)). Intensity may be either beneficial or adverse (*Id.* § 1508.27(b)(1)). This is similar to the EU Directive in that both "positive and negative effects" are triggers (Annex I, n. 1). Other intensity factors include impact on "public health or safety"; unique characteristics (proximity to historic or cultural resources, park lands, scenic rivers, etc.); controversial factors; unique/unknown risks factors; future impacts; cumulative effects (see Chap. 27); impact on infrastructures and other scientific, cultural or historic resources; impacts to endangered species; and whether the action may violate laws (40 CFR § 1508.27(b)).

If the agency is certain that there will likely be a significant environmental impact, then they must prepare an Environmental Impact Statement (EIS). An EIS is a "detailed written statement concerning the environmental impacts of the proposed action and any adverse environmental effects which cannot be avoided" (42 USC § 4332(2)(C)). If the agency is uncertain as to the impact, then they must prepare an Environmental Assessment. An Environmental Assessment (EA) is a "concise public document . . . that serves to briefly provide sufficient evidence and analysis for determining whether to prepare an environmental impact statement or a finding of no significant impact" (40 CFR § 1508.9(a)(1)). An EA is basically a briefer version of an EIS, a preliminary review to see if there are indeed significant environmental impacts. An environmental assessment serves as a systematic, interdisciplinary approach to reviewing environmental consequences. Another key difference is that public involvement and scoping is not required to be incorporated into the EA (although it is highly recommended), while public comments must be addressed in an EIS (40 CFR § 1503.4).

Typical contents of an EA and EIS include title page, glossary, executive summary, purpose and need for the proposed action, description of the affected environment, description and analysis of environmental impacts of the proposed action and reasonable alternatives, applicable environmental permits and regulator requirements that would need to be obtained; list of agencies and persons consulted, list of preparers, and references. All that is required under the regulations is a brief discussion of the need for the proposal, alternatives, environmental impacts of the proposed action and alternatives, and a listing of agencies and persons consulted (40 CFR § 1508.09(b)). Alternatives are at the heart of an EA and EIS (40 CFR § 1502.14 Alternatives including the proposed action). A range of alternatives should be offered, but courts have permitted a minimum of two alternatives: the proposed action and no action (see Box 17.2).

> **Box 17.2.** Case Study: Timber Sale in Ochoco National Forest, Oregon, USA
>
> In 1998, the United States Forest Service (USFS) submitted a Finding of No Significant Impact (FONSI) and a corresponding Environmental Assessment (EA) on a proposed action to remove timber in the Ochoco National Forest. The proposed action involved commercially harvesting 215 acres, precommercially thinning 131 acres, and burning 600 acres, as well as removing some existing roadway. Located on the northeastern slopes of Lookout Mountain in central Oregon, the site is within the Lookout Creek watershed, which flows into the North Fork of the Crooked River.
>
> The EA submitted by USFS was insufficient in two major ways. First, the EA failed to adequately develop reasonable alternatives. Agencies are required to "use the NEPA process to identify and assess ... reasonable alternatives" (40 CFR § 1500.2(e)). Second, the EA failed to adequately address environmental impacts. NEPA requires that agencies shall "identify environmental effects and values in adequate detail" *(Id.* § 1501.2(b)).
>
> USFS only presented two options: (1) no action, or Alternative I, and (2) the proposed action, identified as "Alternative II," or the preferred alternative. Though a comparison was made between Alternatives I and II, USFS did not provide adequate choices. All USFS offered was its choice or no action. Applying the "rule of reason," a good comparison would contrast the proposed action to several alternatives that attempt to carry out the purpose and need of the project. The no action alternative did not fulfill the purpose of the action because it did not provide wood products (such a wood for paper or building) nor did it return stands back to historic levels.
>
> USFS failed to "rigorously explore" all reasonable alternatives *(Id.* § 1502.14(a)). The Service could have identified at least one viable alternative to their proposed action, but instead, USFS simply went through the motions of compliance without adhering to the letter of the law. USFS did include reasons for eliminating an alternative, as required in 40 CFR § 1502.14(a), but that alternative was completely contrary to one of the purposes of the proposed action – to return stands to historic viability. The eliminated alternative sought to remove large and old stands that are below historic levels, an action contrary to the interim directive found in Amendment II of the Ochoco Forest Land and Management Plan.
>
> An example of USFS's inadequate study of environmental effects pertains to its cursory treatment of air quality. The USFS only mentioned minor smoke in the spring and fall. In addition, though USFS identified potential impacts to soil via compaction and displacement, the EA did not discuss how a timber harvest may be particularly detrimental. USFS found timber harvesting had the greatest impact on cultural resources, but in the EA, USFS did not seem overly concerned. USFS did not really discuss indirect effects, though the proposed action may lead to further harvests because of accessibility from the self-maintaining roads; however, this may be unlikely and was not the greatest error. The most glaring omission in the EA was USFS's failure to discuss cumulative effects. Evidence of cumulative effects was apparent throughout the EA but was not discussed in a comprehensive manner. Discussions on cumulative effects must include past, present, and reasonably foreseeable actions as well as actions on adjacent lands. USFS mentioned some past actions on the land, but did not discuss the past effects in tandem with present and future impacts.

Such failure to put reasonable effort in considering all alternatives could likely be challenged on the grounds that no action is not a viable alternative. Environmental impacts or effects encompass ecological, historic, cultural, economic, social or health impacts, as well as direct, indirect and cumulative impacts (40 CFR § 1508.8 Effects). For instance, with regard to a timber sale, the direct impact would be cutting the trees. An indirect impact would be global warming. Cumulative impacts, often underdeveloped in most plans, would include the effect on a watershed, surface runoff in a large area, etc.

Upon an EA being completed, the agency may decide that there are significant environmental impacts and thus complete an EIS. An EIS is an in-depth study on environmental impacts and includes all of the provisions required in an EA (40 CFR § 1508.11 Environmental impact statement). A more formal review process, an EIS identifies environmental consequences and alternatives for pursuing a proposal. An EIS also includes adverse environmental effects that cannot be avoided, alternatives, the "relationship between local short-term uses of man's environment and the maintenance and enhancement of long-term productivity," and "any irreversible and irretrievable commitments of resources" (42 USC § 102(2)(C)).

Two main results ensue: (1) a finding of significant impact, which then requires mitigation or a decision to not go forth with the action, or (2) a finding of no significant impact (FONSI) (40 CFR § 1508.13). A FONSI may be challenged under the Administrative Procedure Act (APA) (5 USC § 706(2)(A)). The agency must prepare a Record of Decision (ROD), stating the decision, alternatives, and mitigation methods, and the ROD is published in the Federal Register (40 CFR § 1505.2). In the case of the Ochoco Forest timber sale, the USFS decided to proceed with an EA instead of an EIS, and after making a cursory effort at evaluating environmental impacts, they arrived at the conclusion that a FONSI was appropriate. Even if an EA is inadequate, the resulting document allows the public to evaluate impacted resources and challenge the findings. If a FONSI is found, an agency may proceed with the action unless complaints from the public lead the agency to voluntarily hold off, or the agency may be forced to discontinue any action upon the issuance of an injunction pending a legal challenge to the FONSI. A court may rule that the agency's FONSI is "arbitrary and capricious, an abuse of discretion, or not in accordance with law" (APA 5 USC § 706(2)(A)). The procedure requires attention to detail, consideration of the environment at every step, and a resulting document that is available to the public. No matter what the outcome, NEPA, similar to the SEA Directive, requires evaluation of significant environmental impacts and public participation in decisions that effect the environment.

17.4 Successes and Failures

Example is more efficacious than precept. Samuel Johnson

Perhaps the best way to learn from United States environmental assessment law is to identify its successes and failures, thus enabling emulation of its strengths and avoidance of its weaknesses.

The CEQ submitted a report on the 25th anniversary of NEPA, documenting the areas that have proved most useful and the areas that need to be improved upon (CEQ 1997). The CEQ identified five key areas of strategic planning, public information and input, interagency coordination, interdisciplinary place-based approach to decision-making, and science-based and flexible management approaches (CEQ 1997). All of these key areas will be addressed in this section.

17.4.1 Strategic Planning

First, in regard to strategic planning, many people who implement actions covered by NEPA have said that the law has helped them to do their jobs better (CEQ 1997). By requiring a rigorous evaluation system, poor proposals are quickly discarded if not discouraged altogether. Strategic planning also reduces documentation at later stages (CEQ 1997) – though a main complaint is that documents are sometimes too long and too technical. The requirement of alternatives supports innovation and creative problem solving. Another benefit of strategic planning is that it results in comprehensive reporting. In other words, by looking at environmental factors, plan preparers are more likely to consider cumulative impacts.

Sometimes the NEPA process is triggered too late to be fully effective (CEQ 1997). For instance, an agency may decide to sell burned timber to a logging outfit and then later cater the EA or EIS to suit their purpose (see Case Study in Box 17.2). This type of use has been termed "post-hoc rationalization." The documents that result from a NEPA study may also be very long or incredibly technical. The rules do state that the reports should be understandable (40 CFR § 1502.1); however, scientists are often charged with preparing the documents and are sometimes guilty of overusing scientific jargon.

Another problem is that training for agency officials is often inadequate. Many agencies are now offering more guidance, hosting training sessions and providing information on the agency websites; in addition, state bar associations and law institutes are also offering courses on how to prepare an EA or EIS. Unfortunately, some preparers view the NEPA process as a compliance requirement instead of a tool for better decision-making. Another abuse of the process occurs when agencies break down an action into several pieces, thus avoiding the result of a larger plan causing significant environmental impact.

17.4.2 Public Information and Input

Second, public information and input is viewed as a critical tool in the NEPA process. NEPA's outreach to the public is viewed as a framework for collaboration. The CEQ requires that agencies invite comments immediately after preparing a draft EIS (40 CFR 1503.1a)) (an EA does not require public input although it is recommended and is the best practice). Agencies must seek comments from agencies with jurisdiction over the potential environmental impact; from appropriate state and local agencies; potentially impacted Indian tribes; and the general public (40 CFR § 1503.1(2)-(4)).

Methods of reaching out to the public include group list emailing, newspaper advertisements, signs at the site, and websites. Public input is invaluable to the decision-making process and assists in recognizing potential environmental consequences. Comments should be specific and they must be incorporated in an EIS. Responding to the comments, the agency may modify or add alternatives, make corrections, or explain why the comment does not warrant a response (40 CFR § 1503.4 Response to comments). Not only does public input improve agency effectiveness, but the sharing of information also increases the public's knowledge and awareness (CEQ 1997).

A difficulty with public input is that some agencies find it challenging to reach out systematically to interested parties. When evaluating public comment, agencies look for comments that are most persuasive and written by those most affected. In the US, agencies have been criticized for their failure to adequately respond to comments. For instance, the whole purpose of inviting public comment is to assist in the planning stage, but many agencies simply obtain the opinions and then disregard them. Instead, planners should modify the plans and add alternatives in response to suggestions as advised by the CEQ in 40 CFR § 1503.4(a). Citizens are sometimes viewed as adversaries instead of potential partners; moreover, agencies are unable to reach all of the public (CEQ 1997). Usually, the agencies send out notices to groups that have expressed their interest in the issue. Agencies also rely on citizen initiative, and the public is not always able to handle the analysis.

17.4.3 Interagency Coordination

Third, interagency cooperation has led to more efficient environmental planning. By working across agencies, conflicts can be resolved and duplication is reduced or avoided (CEQ 1997). Interagency cooperation also has drawbacks in that it may require more time to coordinate, which may lead to delays. A framework should be in place before attempting to mesh to agencies together with specifically designated contacts. Different agencies also have different timetables (CEQ 1997), so again, the agencies should attempt to coordinate their schedules.

17.4.4 Interdisciplinary, Place-Based Approach to Decision-Making

Fourth, by combining different disciplines, more wide-ranging environmental data may be obtained. Another advantage of place-based approaches is a greater attention to cumulative effects (CEQ 1997). Environmental indicators serve as an excellent tool to evaluating environmental consequences, and by bringing different scientists, such as geologists, hydrologists, and ecologists, more environmental effects are addressed. A major problem with interdisciplinary studies is that there is a lack of baseline data, thus hampering the requisite comparison of alternatives (CEQ 1997).

17.4.5 Monitoring and Adaptive Management

Finally, science-based and flexible management approaches have led to effective monitoring. As mitigation of environmental consequences is integral to achieving long-term success, monitoring is vitally important. Instead of being overly detailed at the planning stage, monitoring can serve as an excellent mitigating effect. Generally, no legal requirement exists in the United States to require monitoring, though some state acts require monitoring. Moreover, many agencies include monitoring in their mandates. The problem with monitoring is that there has been a failure to collect long-term data, so checking for increased levels of a specific chemical may be difficult (CEQ 1997).

17.5 Conclusions and Recommendations

When addressing SEA, planners should start as early as possible. Often people are intimidated by what they perceive to be a lengthy study (CEQ 1997), but they should not be scared off. Instead, planners should first evaluate the project using broad strokes. Entities charged with implementing SEA should view the process as a positive step to arriving at ecologically sound programming. Within a government body, leadership should encourage integrating SEA early in the planning process (CEQ 1997) and they should foster a culture of enthusiasm. SEA should be viewed as a useful tool, not a hindrance.

Also, by introducing SEA to large plans and programmes – for instance, for railroad plans or flood control – many agencies can work together and seek more public input. Public comment is not required when preparing an EA in the United States; however, some agencies do provide a comment period when preparing an EA (CEQ 1997). The EU has an excellent opportunity to insist upon public participation with SEA.

When seeking public input, the government should provide as much background information as possible. The public should be informed from the very start so that their input is more valuable. By providing the public with information on environmental issues, the government is empowering the public to become more involved and to feel as though they have a stake in what happens to the environment. To interact with the public, hearings should be held and documents should always be readily accessible (CEQ 1997). Alternatively, other public outreach mechanisms may be used, such as round table discussions or workshops (CEQ 1997).

EU countries are in excellent position to perform SEA with the full cooperation of the public. They are also in the unique position of working across borders and should take advantage of planning on an ecosystem scale (rather than being restricted by administrative boundaries). By encouraging public input and through techniques such as monitoring, the EU could become the world leader in environmental planning and conservation.

References[1]

CEQ – Council on Environmental Quality (2004) Website. Internet address: http://www.whitehouse.gov/CEQ, last accessed 07.02.2004

Council on Environmental Quality NEPA Website, accessed on February 7, 2004. Internet address: http://www.ceq.eh.doe.gov/nepa/nepanet.htm, last access 07.02.2004

Deasley N (2003) Strategic Environmental Assessment. SEPA View – The magazine of the Scottish Environment Protection Agency 16:16-17

CEQ – Council on Environmental Quality, Executive Office of the President (1997) The National Environmental Policy Act: A Study of Its Effectiveness After Twenty-Five Years, January 1997

---

[1] Note that legislation mentioned in the chapter is listed at the end of the handbook in the consolidated list of legislation (Appendix 2).

18 SEA in Canada

Norma Powell

ENKON Environmental Ltd, Canada

18.1 Introduction

Canada is a Federal State with a Federal Government, ten Provincial and three Territorial Governments. There is a constitutional division of legislative authorities and a resulting division of policy authorities. Although not a complete list, the Federal government administers trade and commerce, taxation, criminal law, public debt, fisheries, currency and coinage, banks and banking, and First Nations (i.e. indigenous peoples) and First Nations' lands. The Provinces administer, among other issues, the sale of non-Federal lands, hospitals, municipal institutions, local works and undertakings and matters of a local or private nature in the Province. Jurisdiction in some areas, such as environment, is shared.

The Federal Government may utilize a number of policy development tools such as: legislation and regulations (e.g. Canadian Environmental Assessment Act), guidelines and codes of practice (e.g. Greening Government Operations), funding programs (e.g. National Infrastructure Program) and management plans (e.g. Parks Management Plans). Developing policy options includes, where necessary, Strategic Environmental Assessment (SEA).

In theory, Canada has been committed to assessing the potential environmental implications of Federal policies since 1984, the Environmental Assessment Review Process Guideline Order defined a "proposal" to include "any initiative, undertaking or activity for which the Government of Canada has a decision making responsibility" (Noble 2002). However, SEA as we know it began in Canada in 1990 when the Federal Government's Cabinet of Departmental Ministers directed their respective Departments to consider environmental concerns at the strategic level of decision-making. This Cabinet Directive was revised in 1999 to strengthen the role of SEA by clarifying obligations of Departments and Agencies and linking environmental assessment to the implementation of sustainable development strategies.

Implementing Strategic Environmental Assessment. Edited by Michael Schmidt, Elsa João and Eike Albrecht. © 2005 Springer-Verlag

This chapter outlines SEA in Canada. Sect. 18.2 summarises Canadian SEA regulations and the principles that guide SEA in Canada. The roles and responsibilities of the Federal Government for SEA are highlighted in Sect. 18.3. Sect. 18.4 sets out methodologies and tools in Canadian SEA, and SEA in general, and Sect. 18.5 discusses the role of documentation and reporting for Canadian SEA. A critical analysis of the Canadian SEA system and a Canadian case study of SEA implementation are outlined in Sects. 18.6 and 18.7, respectively. The author's conclusion and recommendations regarding SEA in Canada concludes the chapter.

18.2 SEA Legislation and Principles

18.2.1 Federal Direction

In Canada, like in the European Union, Strategic Environmental Assessment (SEA) is a systematic, comprehensive process of evaluating the environmental effects of a policy, plan or program and its alternatives. The process is intended to improve decision-making by providing managers, Ministers and Cabinet members at the Federal level with information relating to the positive and negative environmental effects of a policy, plan or program and the means to accentuate the positive effects, and reduce or avoid the negative ones.

According to the Canadian Environmental Assessment Act (2004), Ministers expect a strategic environmental assessment of a policy, plan or program proposal to be conducted when the proposal is submitted to an individual Minister or Cabinet for approval and when implementation of the proposal may result in important positive or negative environmental effects.

Departments and Agencies are also encouraged to conduct strategic environmental assessments for other policy, plan or program proposals when circumstances warrant. An initiative may be selected for assessment to help implement Departmental or Agency goals in sustainable development, or if there are strong public concerns about possible environmental consequences.

The information summarised in this chapter was primarily obtained from CEAA (2004) and Environment Canada (2000).

Enforcement without Legislation

In lieu of exacting SEA legislation in Canada how does Canada ensure the implementation of SEA? It is in the hands of each Federal Department and Agency; to meet key obligations in order to comply with Federal policy. There are five areas of policy to consider with regard to SEAs in Canada.

1. The Federal Government's commitment to sustainable development by way of the *Auditor General Act* requires the completion of SEA. Under the Act, the Commissioner for Environment and Sustainable Development will hold Departments accountable for "greening" their policy, plans and programs and will review progress in implementing the strategies.

2. The Cabinet Directive on SEA, which was discussed in the previous section, requires Departments to complete SEAs.
3. Regulatory impact assessment obligations by way of the *Canadian Environmental Assessment Act*,

   SEA also fits into the regulatory impact assessment requirements for proposed Federal regulations. Under the government's regulatory policy, Federal Departments and Agencies must include environmental implications and risks in the benefit-cost analysis conducted on a proposed regulation. When regulations address health, social, economic or environmental risks, the analysis must demonstrate that the regulatory effort is being expended where it will do the most good.

4. Each Department's commitment to develop environmentally sound policy, plan and program proposals.
5. Canada's commitment to certain international agreements.

   SEA is a key component of Canada's obligations under the UN Convention on Biological Diversity. Article 14(b) of the convention calls for signatories to "introduce appropriate arrangements to ensure that the environmental consequences of policies, plans and programs likely to have adverse effects on biodiversity are duly taken into account." Additionally, the convention on Environmental Impact Assessment in a Transboundary Context sets out protocols whereby neighbouring jurisdictions can contribute to assessments of initiatives of their neighbours where there may be transboundary effects. Canada contributed to the United Nations Economic Commissions for Europe (UN ECE) discussions and is a signatory to the convention.

Potential future amendment to the Canadian Environmental Assessment Act may develop from the Members of the House of Commons Standing Committee on Environment and Sustainable Development's report to the Federal Government, Beyond Bill C-9: Toward a New Vision for Environmental Assessment (SCESD 2003). If action is taken to embrace the recommendations given by the aforementioned Standing Committee, the Prime Minister would have to direct the Privy Council Office to develop legislation, in consultation with the Minister of Environment that establishes a legal framework for mandatory Strategic Environmental Assessment.

18.2.2 Provincial Level

Each Canadian provincial government has enacted its own regulatory legislation to guide environmental assessments for projects that do not fall under the auspices of the CEAA (e.g., project is not on Federal lands and is not Federally funded). Findings to date indicate that there exists considerable variability in the practice of SEA amongst Canadian provinces, with only Saskatchewan, Ontario and Quebec identifying recent SEA practice experience (Noble 2004a). Like, the Federal level, the main barriers to effective implementation are the lack of legislative requirements for SEA, and the limited understanding of the nature and benefits of higher order impact assessments (Noble 2004a).

18.2.3 Guiding Principles

In 1999, the Canadian Environmental Assessment Agency published guidelines on implementing the SEA Directive. The guidelines are based on the principles outlined in Box 18.1.

Ministers expect the SEA to consider the scope and nature of the likely environmental effects, the need for mitigation to reduce or eliminate adverse effects, and the likely importance of any adverse environmental effects, taking mitigation into account.

Box 18.1. Principles of the Canadian SEA	
Early integration	Consideration of environmental effects should begin early in the conceptual or planning stages of the proposal, before irreversible decisions are made.
Examine alternatives	Evaluate and compare different options for the policy, plan or program. The comparison will help identify how modifications or changes to the policy, plan or programme can reduce environmental risk.
Flexibility	The guidelines set out by the Canadian Environmental Assessment Agency are advisory, not prescriptive. Departments and Agencies are encouraged to adapt and refine analytical methodologies and decision-making tools appropriate to their circumstances
Self-assessment	Departments are responsible for applying SEAs and determining how their SEAs should be conducted, how they carry them out and how they report their own results.
Appropriate level of analysis	The scope of the analysis of potential environmental effects should be commensurate with the level of anticipated effects.
Accountability	Strategic environmental assessment should be part of an open and accountable decision-making process within the Federal Government. Accountability should be promoted through the involvement of affected individuals and organisations, when appropriate, and through documentation and reporting mechanisms.
Use of existing mechanisms	In conducting an SEA, Departments and Agencies should use existing mechanisms to conduct any analysis of environmental effects, involve the public if required, evaluate performance and report the results. Such mechanisms shall also be used to report statements of environmental effects.

SEA should contribute to the development of policies, plans and programs on an equal basis with economic or social analysis; the level of effort in conducting the analysis of potential environmental effects should be commensurate with the level of anticipated environmental effects. The environmental considerations should be fully integrated into the analysis of each of the options developed for consideration, and the decision should incorporate the results of the strategic environmental assessment. Departments and Agencies should use, to the fullest extent possible, existing mechanisms to involve the public, as appropriate. Departments and Agencies shall prepare a public statement of environmental effects when a detailed assessment of environmental effects has been conducted through a strategic environmental assessment. This will assure stakeholders and the public that environmental factors have been appropriately considered when decisions are made.

18.3 Roles and Responsibilities

All Ministers are responsible for assessing environmentally relevant policy, plan and program initiatives for their effect on the environment. In the context of their sustainable development goals, objectives and policies, individual Ministers are accountable for ensuring that assessments of relevant policy, plan and program proposals are conducted, and that they take into account how initiatives might contribute to environmental and sustainable development goals.

The Minister of the Environment has a lead role in establishing the guidance in establishing the environmental framework for Canada and promoting the application of environmental assessment to policy, plan and program proposals. The Minister is also responsible for advising other Ministers on the potential environmental effects of policy initiatives before Cabinet decisions are taken and for advising on environmentally appropriate courses of action. This does not constitute either a veto or an approval role.

Departmental and Agency officials initiating a policy, plan or program proposal to be submitted for consideration by Ministers must ensure that, when appropriate, an assessment of the environmental effects is completed. The objective is to ensure that senior managers or Ministers who approve policy initiatives are properly briefed.

Environment Canada will support the Minister by consulting other Departments and Agencies and provide expert policy, technical and scientific analysis and advice on sustainable development and the potential environmental effects of policy, plan and program initiatives.

The Canadian Environmental Assessment Agency will promote the application of environmental assessment to policy, plan and program proposals of the Federal government.

The Commissioner for the Environment and Sustainable Development is tasked with overseeing the government's efforts to protect the environment and promote sustainable development.

The public may also be involved. Stakeholder concerns are a key consideration for any environmental assessment, and the 1999 Cabinet Directive encourages public notification and consultation for an SEA. Making preliminary information available often facilitates public understanding of the ramifications of the proposed initiative and leads to more constructive input. Although confidentiality of some aspects of policy development may preclude full public consultation, any effort to understand stakeholder concerns will improve the quality and credibility of the SEA and the policy itself. It is important to do what is possible within the existing limitations. All consultations should be documented, to show that interested parties' concerns have been considered. Methods for involving the public are varied and can be tailored to the circumstances.

18.4 Methodologies and Tools

There is no single "best" methodology for conducting an SEA of a policy, plan or program proposal (Environment Canada 2000). In fact, the Canadian Environmental Assessment Agency's review of the Canadian Cabinet Directive on SEA suggests that SEA application since the introduction of the Cabinet Directive has been *ad hoc* and inconsistent at best (CEAA 1999). Federal Departments and Agencies are encouraged to apply appropriate frameworks or techniques, and to develop approaches tailored to their particular needs and circumstances.

Based on current, proven, good practices within Federal Departments and Agencies the Cabinet Directive requires that SEA be *flexible* in that SEA can be applied in a variety of policy settings, *practical* in that SEA does not necessarily require specialist information and skills, or a substantial commitment of resources and time and *systematic* in that SEA is based on logical, transparent analysis.

Preliminary Scan

In Canada, as early as possible in the development of a proposal, the analyst, through the use of a decision flowchart as illustrated in Fig. 18.1, should determine whether the proposal requires ministerial or cabinet approval and then whether important environmental considerations are likely to arise from implementing the proposed policy, plan or program. The focus should be on identifying strategic considerations at a relatively general or conceptual level, rather than evaluating quantitative, detailed environmental impacts as in a project-level assessment.

To conduct a scan of the proposal, the analyst may use a variety of tools, including available matrices, checklists and experts both within the Department and from other Departments. The following considerations may also be of assistance in conducting the preliminary scan:

1. The proposal has outcomes that affect natural resources, either positively or negatively.
2. The proposal has a known direct or a likely indirect outcome that is expected to cause considerable positive or negative impacts on the environment.
3. The outcomes of the proposal are likely to affect the achievement of an environmental quality goal (e.g., reduction of greenhouse gas emission or the protection of an endangered species).

4. The proposal is likely to affect the number, location, type and characteristics of sponsored initiatives which would be subject to project-level environmental assessments, as required by the Canadian Environmental Assessment Act or an equivalent process.
5. The proposal involves a new process, technology or delivery arrangement with important environmental implications.
6. The scale or timing of the proposal could result in significant interactions with the environment.

An SEA generally addresses questions such as: What are the potential direct and indirect outcomes of the proposal and how do these outcomes interact with the environment? What is the scope and nature of these environmental interactions? Can the adverse environmental effects be mitigated and can the positive environmental effects be enhanced? What is the overall potential environmental effect of the proposal after opportunities for mitigation have been incorporated?

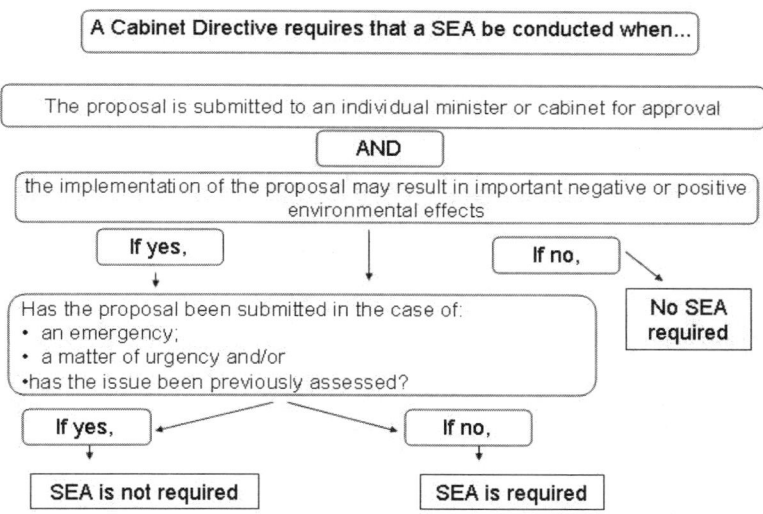

Fig. 18.1. Preliminary scan to determine if an SEA must be conducted (Environment Canada 2000)

The real challenge to policy and program officers preparing to conduct an SEA of a policy, plan or program proposal is to think more broadly about the proposal about the kinds of activities it may trigger (intended or not) and the interaction of those activities on the environment. Above all, policy and program officers should keep in mind that the strategic environmental assessment is not an add-on process, but one linked with the ongoing economic and social analyses under way on the proposal. An effective SEA cannot be done in isolation or after the fact it must identify the direct and indirect outcomes associated with implementing the proposal and consider whether these outcomes could affect any component of the environment.

If the preliminary scan identifies the potential for important environmental considerations, or if there is a high level of uncertainty or risk associated with the outcome, then a more detailed analysis of the environmental effects should be conducted through a strategic environmental assessment. If the scan does not identify the potential for important environmental considerations, no further analysis of environmental effects is required.

Analysing Environmental Effects

SEA is best done using an iterative approach based on six steps (Fig. 18.2) when considering policy options. In concert with the development of the proposal, each step narrows the policy options to identify the most feasible, environmentally sound course of action. This ensures thorough consideration of the pros and cons of each option. The final recommendation should be informed by the results of the SEA.

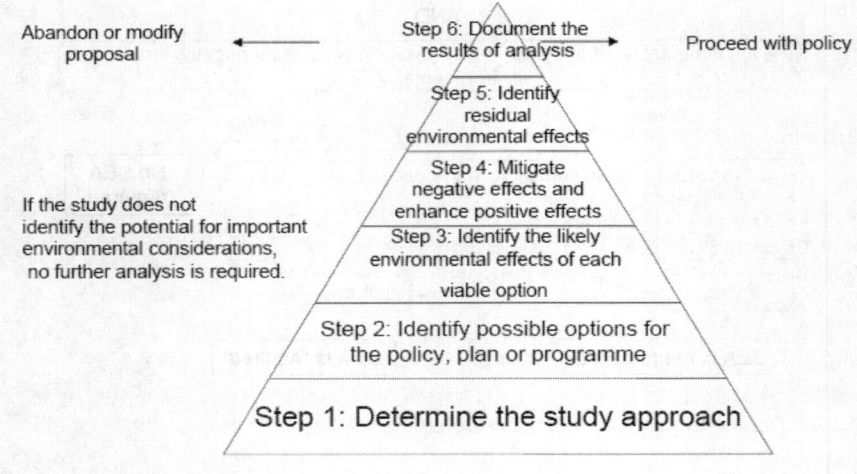

Fig. 18.2. The Canadian SEA approach (Environment Canada 2000)

The SEA should address the s*cope and nature of potential effects*. The analysis should build on the preliminary scan to describe, in appropriate detail, the scope and nature of environmental effects that could arise from implementing the proposal. Environmental effects, including cumulative effects, could result from the use of or changes in, atmospheric, terrestrial or aquatic resources, physical features or conditions. The analysis should identify positive as well as adverse environmental effects.

The SEA should also address *the need for mitigation or opportunities for enhancement*. Analysts should consider the need for mitigation measures that could reduce or eliminate potential adverse environmental consequences of the proposal. Similarly, they should also consider opportunities, where possible, to enhance potential environmental benefits. Mitigation or enhancement could include, for example, changes in the proposal, conditions that may need to be placed on projects or activities arising from the proposal, or compensation measures.

The analysis should describe, in appropriate detail, the *scope and nature of residual effects*, meaning the potential environmental effects that may remain after taking into account mitigation measures and enhancement measures. The SEA should also consider the need for *follow-up measures* to monitor environmental effects of the policy, plan or programme or to ensure that implementation of the proposal supports the Department's or Agency's sustainable development goals.

Lastly the SEA should consider *public and stakeholder concerns*. The analysis should identify for decision makers, where appropriate, concerns about the environmental effects among those likely to be most affected, and among other stakeholders and members of the public.

Appropriate Level of Effort

The level of effort committed to the SEA should be commensurate with the level of environmental effects anticipated from implementation of the proposed policy, plan or program. The factors described in Box 18.2 should assist analysts in assessing potential environmental effects and gauging the appropriate level of effort for the analysis.

Federal Departments and Agencies are encouraged to develop their own sources of information and analytical tools such as relevant literature; previous strategic environmental assessments of policy, plan or program proposals; expert advice from other branches within Departments and other expert Federal Departments; checklists; matrices and modelling; scenario building; and simulation analysis.

Box 18.2. Factors to consider when determining the level of effort in SEA	
Frequency and duration	What is the anticipated scale of the effect? Will it be local regional, national or international in scope?
Location and magnitude	Will the effect be a one-time only occurrence? Will it be a short-term or long-term effect?
Timing	Is the effect likely to occur at a time that is sensitive to a particular environmental feature?
Risk	Is there a high level of risk associated with the effect, such as exposure of humans to contaminants or pollution, or a high risk of accident?
Irreversibility	Is the effect likely to be irreversible?
Cumulative nature	Is the effect likely to combine with other effects in the region in a way that could threaten a particular environmental component?

Public Concerns

The analysis of potential environmental effects should indicate, where appropriate, concerns about these effects among those likely to be most affected, and among other interested stakeholders and members of the general public.

Understanding public concerns can strengthen the quality and credibility of the policy, plan, or program decision. By involving interested parties, decision makers can, at an early stage, identify and address public concerns about a proposal that could otherwise lead to delays or the need for further analysis later in the process. Stakeholders and the public can also be an important source of local and traditional knowledge about likely environmental effects. Due to divergent public interest, decision makers may need to build a consensus among different or opposing interests. Through public participation, decision makers can help develop the credibility and trust in the decision-making process in order to help achieve group consensus. In late stages of the open participation process, the public can help communicate the results of the process and decisions back to the stakeholders. In addition, a visible commitment to understanding and responding to public concerns can help build a sense of public trust and credibility in the decisions of the Department or Agency.

The involvement of the public should be commensurate with public involvement on the overall development of the policy, plan or program proposal itself, and should make use of any public involvement activities that may be under way as part of the proposal. If public documents are prepared for use in a consultation exercise, it is advisable to incorporate them into the results of the strategic environmental assessment to address potential environmental concerns.

18.5 Documentation and Reporting

Reporting is important for ensuring an open and accountable process. Departments and Agencies shall prepare a public statement of environmental effects when a detailed assessment of environmental effects has been conducted through an SEA. Departments will determine the content and extent of the public statement according to the circumstances of each case. The purpose of the statement is to demonstrate that environmental factors have been integrated into the decision-making process. Such statements should be integrated into existing reporting mechanisms to the fullest possible extent. Separate reporting of SEA is not required.

When reporting on sustainable development goals in Sustainable Development Strategies, Reports on Plans and Priorities, or Departmental Performance Reports, Departments and Agencies should report on the extent and results of their environmental assessment practices. Similarly, other corporate documents that summarise organisational effectiveness, or plans to implement Sustainable Development Strategies, would benefit from describing how frequently policies, plans and programs had been assessed, and any impact these assessments had on reaching organisational goals in sustainable development.

For some proposals, such as those involving significant adverse effects or serious public concerns, Departments and Agencies may choose to release a public document that discusses the environmental effects in detail, in addition to any public statement of environmental effects. This document will help demonstrate that environmental factors have been integrated into the decision-making process.

When a policy, plan or program proposal has been assessed for potential environmental effects, the documentation outline in Box 18.3 is recommended.

18.6 Critical Analysis

Noble (2000) argues that a serious problem with SEA, in Canada and elsewhere, lies in the fact that we lack a common understanding of its definition and characteristics. The majority of the definitions proposed in the literature (e.g. Kessler and Toornstra 1998; Tonk and Verheem 1998; Borrow 1997) define SEA with little explanation as to what "strategic" means and therefore what makes it strategic. To establish a common understanding of SEA it is important that these uncertainties are clarified. Noble (2000) suggests the following definition:

> SEA is the proactive assessment of alternatives to proposed or existing PPPs, in the context of a broader vision, set of goals, or objectives to assess the likely outcomes of various means to select the best alternative(s) to reach desired ends.

Box 18.3. Recommended SEA documentation

Submissions to Cabinet and Treasury Board	Documentation should discuss any strategic environmental assessments, public concerns, and the outcomes of this analysis, as an integral part of examining the options presented. The analysis section of the Memorandum to Cabinet should report on potential significant environmental effects of each of the options proposed for consideration, and mechanisms to mitigate potential adverse effects. The statement should specify how the policy, plan or program affects or relates to the Department's or Agency's Sustainable Development Strategy. If a separate public document detailing the assessment has been prepared, it should be appended to the Memorandum to Cabinet, and Cabinet should be requested to approve its release to the public.
SEA submission exempt from the Cabinet Directive	If a policy, plan or program does not require Cabinet approval, but is still assessed, the findings of the assessment should be reported in any relevant decision documents. Departments and Agencies shall prepare a public statement of environmental effects when a detailed assessment of environmental effects has been conducted through a strategic environmental assessment. Such statements should be integrated into existing reporting mechanisms to the fullest possible extent.
Regulatory Impact Analysis Statement	If a Regulatory Impact Analysis Statement is prepared on an initiative, Departments and Agencies should reflect the findings of the strategic environmental assessment. Departments and Agencies shall prepare a public statement of environmental effects when a detailed assessment of environmental effects has been conducted through a strategic environmental assessment. Such statements should be integrated into existing reporting mechanisms to the fullest possible extent.
Reporting in Sustainable Development Strategies, Reports on Plans and Priorities, or Departmental Performance Reports	Departments and Agencies may also report on strategic environmental assessments as part of their reporting in Sustainable Development Strategies, Reports on Plans and Priorities, Departmental Performance Reports, and other documents that concern organisational practices and effectiveness.
SEA summary reports circulation	The strategic environmental assessment should be forwarded to Departmental evaluation and review officers so that future evaluations of the policy, plan or program initiative can incorporate the outcome of the analysis into the evaluation framework.

Another problem with the Canadian SEA experience, like other countries, has been the inability to establish consensus on the methodological principles of SEA (Noble 2002; Noble and Storey 2001; Stinchcombe and Gibson 2001) and how to make SEA principles and characteristics operable while maintaining a practical and effective approach to SEA. The Cabinet Directive offers broad guidance for conducting SEA, but it does not outline a specific methodology for its application.

In fact, the Cabinet Directive states that there is no single best methodology for conducting SEAs. It suggests that "Federal departments and agencies are encouraged to apply appropriate frameworks and techniques, and to develop approaches tailored to their particular needs and circumstances."

Canada has had successes and failures with regard to implementing SEA into the development of policy, plans and programmes. One failure that illustrates an inability to meet the SEA requirement of early integration is the evaluation of the North American Free Trade Agreement (NAFTA) (Noble 2002). The objectives of the NAFTA terms of reference were "to ensure that environmental considerations were taken into account during all stages of the negotiating process; and to conduct and document a review of the potential environmental effects of NAFTA on Canada." The review committee that was charged with the task of reviewing the NAFTA agreement, identify the potential environmental effects on Canada, and submit the review to the Federal Cabinet by no later than the signing date of NAFTA itself (Hazell and Benevides 1998). By the time the assessment was requested, the policy document was already in place and many decision options were already forgone. The NAFTA SEA did not contain information suggesting policy alternatives were given consideration, and because the policy document had already been prepared prior to the SEA, the assessment process made few contributions to sustainability and had only minimal influence on the trade agreement outcome.

Often within Canadian SEA, there is an insufficient use of integration across political decision-makers and disciplinary specialists. There is a need for collaboration within government and beyond in achieving sustainable development because no single institution has the competency, authority or resources to tackle multi-disciplinary issues. To become an effective tool, SEA requires increased integration; order and congruity through cooperative decision-making and improved communications between agencies and specialists; and a process to audit the effectiveness of the SEA and its influence on the decision output and outcomes (Noble 2003). In Canada, the self-assessment process, where the agency that screens and conducts an SEA is also the agency that develops and advocates the policy under study, can result in conflicts of interest. Under this self-assessment process, decision-makers may, for simplicity, avoid or curtail SEA requirements (Stinchcombe and Gibson 2001).

From the Cabinet Directive, an SEA is required when: 1) a proposal is submitted to an individual Minister or Cabinet for approval and 2) the implementation of the proposal may result in important environmental effects. An SEA is not required for other PPPs, which do not meet these requirements. Comparing compliance of the Cabinet Directive with Federal legislation such as the *Farm Income Protection Act* (*FIPA*) highlights the lack of enforcement with respect to the Cabinet Directive. Under the *FIPA*, environmental assessments are required for all programmes established under the Act, whereas SEAs under the Directive are carried out only by matter of policy. Given, that SEAs are triggered by a Memoranda to Cabinet, results are rarely available to the public (Noble 2003).

With regard to quality assurance and decision consistency, little guidance for practitioners exists concerning the use and treatment of expert judgment in SEA processes. As more complex evaluation and techniques are used, the assessment panel is playing a more important role. Several guidelines and considerations are outlined by Noble (2004b) for the use of assessment panels in SEA. First, there is no best method for selecting an assessment panel; it should depend on the SEA objectives, the socio-political assessment context, data and information requirements, available time and resources, and the setting and situation. Second, the value of the expert judgment is often overrated; the input of local parties and stakeholders might be more valuable. Third, making accurate predictions is more difficult at the strategic-levels of decision-making and is not a sufficient measure of SEA performance. Fourth, consensus is neither a necessary nor a sufficient condition for SEA decision-making, and should not be an indicator of quality. Lastly, consistency is not an accurate reflection of expertise, rather of informed decision-making. Analysts should not assume that expert judgment is more credible than that of the non-expert.

Canadian provincial legislation, like Federal legislation, lacks definitive requirements for environmental assessments of policy, plans and programmes. In addition, at the provincial level there is limited understanding of the nature and benefits of higher-order impact assessments, there is a provincial lack of practitioner training and guidance on assessment tools and techniques, and the provinces lack institutional planning to develop organizational arrangements necessary to facilitate coordination across government departments and agencies (Noble 2004b).

18.7 Canadian SEA Case Study

Given the SEAs are triggered by Memoranda to Cabinet, and given that the Cabinet Directive does not require separate reporting of an SEA, results of assessments are rarely made available to the public (Noble 2003). However, the following section outlines an SEA that identifies and evaluates the potential environmental effects of the proposed initiatives in the Core Area Concept for Ottawa, Canada's national capital city. The Core Area Concept is the second phase of a three step planning process to develop the Core Area Sector Plan for the heart of Ottawa, which will provide long-range direction for the development of the downtown areas of Ottawa, Ontario and nearby Hull, Quebec (Du Toit Allsopp Hillier – Delcan Corporation 2000). The SEA analysis includes a brief description of the future conditions in the downtown areas of Ottawa and Hull that may affect each environmental component; a description of the potential effects resulting from the Concept; and recommended mitigation measures that should be considered as part of the future work for the downtown area plan.

Table 18.1. Potential environmental effects from the proposed Core Area Concept for the downtown core in Ottawa and Hull, Canada, and proposed mitigation measures

Potential Environmental Effect	Proposed Mitigation Measure
Water	
Potentially positive effects on shorelines may result through shoreline improvements as part of the development of shoreline sites. Negative effects on the desired objective of naturalised shorelines may occur in areas where support facilities for boating are constructed at the water's edge and through the construction phase of the locks.	Where construction is proposed on waterfront properties, boat launch site rehabilitation and naturalisation of shorelines should be included in plans and boat facilities should be sited to avoid negative effects on shoreline and fish habitat.
Potentially negative effects on water quality are largely possible from the following initiatives in the Core Area Concept: erosion and sedimentation, soil compaction, increase boat use on local rivers, and construction at the water's edge that may result in toxic spills.	Construction mitigation will require stormwater management. Potential effects of increased boating in affected areas can be mitigated through the use of boating Best Management Practices and adherence to the Fisheries Act.
Land	
Addition of greenspace and required realignment of roadways may affect the ecological health of existing greenspace.	Realignment of roadways should avoid negative effects on greenspace and planning should identify greenspace connections to maintain linkages between greenspaces and parks. This will require a more detailed study of park area supply and utilisation to determine if the proposed additions will meet future demand.
Protection of escarpment lands through naturalization and physical work to avoid slumpage.	The Core Area Concept supports the rehabilitation of the escarpment with the study area.
Air	
Road redesign have the potential to lead to decreases in auto traffic as road redesign favours pedestrians, cyclists and public transit, which should improve air quality and help conserve energy.	A transportation study should consider the use of High-Occupancy-Vehicle lanes and road redesign that fosters non-vehicle travel.

Table 18.1. (cont.)

Flora

Landscape improvements with more urban native plantings in parks and greenspace and where streetscape improvements have been made. Re-naturalisation of escarpment land will utilise native plants. Loss of vegetation is likely to occur in areas where there are road extensions and where there is proposed development.	Road extensions should be designed to minimize the negative effects on existing vegetation and include compensation for removed vegetation. Streetscaping should be carefully planned with appropriate native species and should be included in road design.
Naturalisation of shorelines will potentially improved natural area linkages to areas outside of the Ottawa and Hull downtown core area.	No mitigation identified

Fauna

Protection of shorelines and escarpment lands may provide benefits for urban mammals and birds.	No mitigation identified
Sedimentation and erosion into watercourses, and proposed shoreline development may negatively affect fish habitat.	Ensure docking facilities and shoreline developments do not interfere with fish spawning grounds. Stormwater Management Plans should be prepared for re-development sites.

Through the SEA process it was determined that the implementation of the Core Area Concept for Ottawa and Hull would likely have the environmental effects outlined in Table 18.1 and should take into account the accompanying considerations.

In conclusion, the proposed Core Area Concept for Ottawa and Hull will likely have the following effects: urban intensification; reduced automobile dependency; addition of park space within the Core Area; and naturalisation and rehabilitation of shorelines and escarpments. Socio-economic effects relate to heritage buildings and heritage character of some development areas, street improvements to some locales and an expanded public transportation network.

The SEA for the Ottawa Core Area Concept does not propose or evaluate alternative options. The emphasis is on identifying the potential impacts of a predetermined set of planning actions and the implications for implementing a national capital plan. Public comments were received on the Core Area Vision and Concept through public consultation and open houses. The SEA does not put forth follow-up monitoring recommendations and frameworks, such as an environmental monitoring system based in integrated resource management.

18.8 Conclusions and Recommendations

Canada has had successes and failures with respect to SEA implementation. It has made significant advances in the development of SEA principles, especially with respect to the nation's commitment to sustainability, but suffers from mixed success in SEA application. More guidance is required on how to make these principles operational within the context of practical and effective SEA approaches. Integral to Canada's problem is the lack of a specific methodology for SEA application. Canada requires a structured, systematic framework for assessment, where a variety of methods and techniques can be applied at all levels of decision-making across all sectors. With Canada's need for a structured methodology also comes a need for national legislation.

References[1]

CEAA – The Canadian Environmental Assessment Agency (1999) Cabinet Directive on the Environmental Assessment of Policy, Plan and Program Proposals. Government of Canada, Ottawa

Du Toit Allsopp Hillier – Delcan Corporation (2000) Core Area Concept of Canada's Capital Strategic Environmental Assessment. National Capital Commission, Ottawa

Environment Canada (2000) Strategic Environmental Assessments at Environment Canada: How to Conduct Environmental Assessments of Policy, Plan, and Program Proposals. Prepared by the Environmental Assessment Branch

Hazell S, Benevides H (1998) Federal strategic environmental assessment: towards a legal framework. Journal of Environmental Law Practice 5(1998):13-24

Noble BF (2000) Strategic environmental assessment: what is it and what makes it strategic? Journal of Environmental Assessment Policy Management 2(2):203-224

Noble BF (2002) The Candian experience with SEA and sustainability. Environmental Impact Assessment Review 22(2002):3-16

Noble BF (2003) Auditing strategic environmental assessment practice in Canada. Journal of Environmental Assessment Policy and Management 5(2):127-148

Noble BF (2004a) A state-of-practice survey of policy, plan and program assessment in Canadian provinces. Environmental Impact Assessment Review 24(3):351-361

Noble BF (2004b) Strategic environmental assessment quality assurance: evaluating and improving the consistency of judgements in assessment panels. Environmental Impact Assessment Review 24(2004):3-25

Noble BF, Storey K (2001) Towards a structured approach to strategic environmental assessment. Journal of Environmental Assessment Policy and Management 3(4):483-508

SCESD – Standing Committee on Environment and Sustainable Development (2003) Beyond Bill C-9: Toward a New Vision for Environmental Assessment

---

[1] Note that legislation mentioned in the chapter is listed at the end of the handbook in the consolidated list of legislation (Appendix 2).

Stinchcombe K, Gibson RB (2001) Strategic Environmental Assessment as a means of pursuing sustainability: Ten advantages and ten challenges. Journal of Environmental Assessment Policy and Management 3(3):343-372

19 SEA of Plan Objectives and Policies to Promote Sustainability in New Zealand

Ali Memon

Lincoln University, New Zealand

19.1 Introduction

Although human settlement in New Zealand has been relatively recent, the environment has demonstrated a dramatic response to human impact. Environmental change has been particularly manifest during the last 150 years. However, it is only during the last three decades that attempts have been made to develop appropriate institutional arrangements for effective environmental management. One of the more recent initiatives has been the institutionalisation of SEA procedures as an integral part of the statutory planning process.

New Zealand has had SEA procedures in place since 1991. Section 32 in the Resource Management Act 1991 (henceforth, the RMA) requires evaluation of a proposed statutory planning instrument prepared under the Act (a proposed district or regional plan, proposed regional policy statement, a national policy statement or New Zealand coastal policy statement), or a change or variation to it, before it is publicly notified for adoption or when a regulation is made (see Box 19.1). One of the main objectives of these procedures is to ensure that resource management plans prepared within the framework of the RMA achieve the sustainability objective of the Act. New Zealand was one of the first countries to codify sustainability into law by enacting the RMA. A Section 32 evaluation must examine:

- the extent to which each objective in the plan is the most appropriate way to achieve the sustainability purpose of the RMA; and
- whether, having regard to their efficiency and effectiveness, the proposed policies and methods are the most appropriate way for achieving the objectives;
- the benefits and costs of proposed policies and methods; and
- the risk of acting or not acting if there is uncertain or insufficient information about the issues being addressed (the precautionary principle).

Box 19.1. SEA provisions in the RMA 1991 (as amended in 2003)

32. Consideration of alternatives, benefits, and costs –

1. In achieving the purpose of this Act, before a proposed plan, proposed policy statement, change, or variation is publicly notified, a national policy statement or New Zealand coastal policy statement is notified under Section 48, or a regulation is made, an evaluation must be carried out by –

(a) the Minister, for a national policy statement or regulations made under Section 43; or
(b) the Minister of Conservation, for the New Zealand coastal policy statement; or
(c) the local authority, for a policy statement or a plan (except for plan changes that have been requested and the request accepted under clause 25(2)(b) of Part 2 of Schedule 1); or
(d) the person who made the request, for plan changes that have been requested and the request accepted under clause 25(2)(b) of Part 2 of the Schedule 1.

2. A further evaluation must also be made by –

(e) a local authority before making a decision under clause 10 or clause 29(4) of the Schedule 1; and
(f) the relevant Minister before issuing a national policy statement or New Zealand coastal policy statement.

3. An evaluation must examine—

(g) the extent to which each objective is the most appropriate way to achieve the purpose of this Act; and
(h) whether, having regard to their efficiency and effectiveness, the policies, rules, or other methods are the most appropriate for achieving the objectives.

4. For the purposes of this examination, an evaluation must take into account—

(i) the benefits and costs of policies, rules, or other methods; and
(j) the risk of acting or not acting if there is uncertain or insufficient information about the subject matter of the policies, rules, or other methods.

5. The person required to carry out an evaluation under subSection (1) must prepare a report summarising the evaluation and giving reasons for that evaluation.

6. The report must be available for public inspection at the same time as the document to which the report relates is publicly notified or the regulation is made.

32A. Failure to carry out evaluation –

1. A challenge to an objective, policy, rule, or other method on the ground that Section 32 has not been complied with may be made only in a submission under Schedule 1 or a submission under Section 49.

2. SubSection (1) does not preclude a person who is hearing a submission or an appeal on a proposed plan, proposed policy statement, change, or variation, or a submission on a national policy statement or New Zealand coastal policy statement, from taking into account the matters stated in Section 32.

The evaluation report must be available for public information before a plan or plan change is formally adopted by the statutory national,[1] regional[2] or territorial local government[3] planning authority. Failure by a planning authority to comply with the Section 32 directive can provide the basis for formally contesting an objective, policy or rule in a proposed plan.

The present Section 32 provision in the Act to subject proposed policy statements and plans to public scrutiny is a 2003 amendment of the comparable, but less clear, directive in the RMA when it was first enacted in 1991. A local authority is also required under section 35 in the Act to monitor the efficiency and effectiveness of policies in its policy statement or plan and methods of implementation and to prepare a five yearly review report on the results of this monitoring.

This chapter is divided into three parts. The first part provides a historically informed critique of the competing sustainability, economistic and rational decision making rationales that have led to the inclusion of this innovative provision in New Zealand's environmental planning legislation. It is important to have this wider perspective in judging the potential effectiveness of this provision. This critique is followed by an analysis of the Section 32 procedural requirements. The chapter concludes with observations on implementation experience and effectiveness of Section 32 as a means to operationalise the sustainability purpose of the RMA.

19.2 Competing Rationales Underpinning Section 32

It is argued here that the justification for inclusion of this innovative provision in the Act reflects the combined influence of three factors:

- as a logical extension of the decision to incorporate project based EIA procedures within the RMA as a new integrated planning statute, based on the rational planning model.[4]
- as a reflection of the influence of the emerging global sustainability discourse in shaping environmental policies in New Zealand during the 1980s.
- as a reflection of the influence of the New Right discourse in New Zealand to subject intervention by the state to rigorous economic assessment as a means to deregulate the land and property market and in order to minimise transaction (or compliance) costs for business (Memon 1993).

---

[1] Ministry for Environment; Department of Conservation.
[2] Regional councils and unitary authorities.
[3] City and district councils.
[4] Even though, from a statutory perspective the project based EIA requirements for planning consent applications in the Act are separate and distinct from the SEA requirements under Section 32 to subject proposed plans and policies to scrutiny, it could be argued that conceptually both of these procedures require environmental assessment to underpin decision-making.

The development of SEA in New Zealand reflects institutionalisation of rational planning and evaluation methodologies within the framework of the RMA. This statute was enacted in 1991 for integrated management of land, water, air and geothermal resources and based on the sustainability purpose. At the same time, the design of the Act reflects dominance of neo-liberal influences in shaping policy reforms in New Zealand during the last two decades.

The most notable of the rational evaluation methodologies in the RMA is the practice of EIA. Briefly, there have been two phases in the development of EIA in New Zealand (Memon and Morgan 2001). The first phase, between 1970 to 1984, constituted in a sense, a search for an effective EIA model that was compatible with the long-standing role of central government as a major developer in a society economically reliant on international trade. The second phase, from 1991 to the present, has been dominated by a change in the course of the country's development from a highly protected economy to an open economy of deregulated market forces integrated into the global economy, decline of the welfare state, devolution of responsibility for environmental management from central government to regional and local government and adoption of "weak sustainability"[5] as the basis of the new development paradigm to reconcile tensions between biophysical environmental carrying capacity and economic growth.

19.3 The Formative Period (1974-1984)

19.3.1 Impact Assessment as a Centralised Process

As in a number of other Western countries, the growing concern with environmental issues during the 1970s led to reviews of the institutional structures for making public policy. However, these structural reforms, including the development of EPEP (Environmental Protection and Enhancement Procedures), were not the result of a wide ranging review of institutional arrangements for environmental management; rather they appear to have evolved in an *ad hoc* manner. While it was recognized that environmental problems were interrelated, and could not be tackled by individual departments, the notion of a single strong environmental agency – a ministry for the environment – proved politically too radical (Memon 1993).

The institutional framework, which evolved in a piecemeal fashion during the decade, included setting up the Commission for Environment as a secretariat for the Minister for the Environment. The Commission had responsibility to coordinate environmental policy, to advise the government of environmental issues, and to administer environmental impact procedures for government-funded projects and for private-sector projects which required government consent. At the same

---

[5] The term "weak sustainability" signifies marginal changes to mainstream production policies and consumer behavior patterns in a capitalist, consumer oriented society to promote sustainability objectives in contrast to radical changes ("strong sustainability").

time, the Crown was the largest land owner and developer in the country and the Town and Country Planning Act (TCPA) and the Water and Soil Conservation Act (WSCA) were seen as ineffective environmental management measures. There was a measure of uncertainty regarding the extent to which the Crown was bound under the planning legislation. These were important factors in the creation of the EPEP. Despite its relatively wide mandate, the activities of the Commission came to focus on the administration of these.

The decision to create an "arms length" agency, rather than charge an existing central government department with environmental assessment responsibility, was a reflection of the lack of confidence in the ability of existing departments, such as the Ministry of Works and Development, to perform this function satisfactorily. This separation was also expected to affirm the Commission's independence in the assessment process. Such expectations subsequently proved to be unrealistic, as discussed below.

19.3.2 The EPEP

In contrast to a number of other countries which adopted EIA during the 1970s (including the United States), environmental impact assessment was first introduced in New Zealand without a legislative base (Morgan 1983). The EPEP became operative in 1974 and had some of the attributes of SEA. EPEP introduced a system of environmental impact assessment for the works and management policies of all government departments, and for proposals that required licenses or permits from government agencies, or needed government funding. Proposals were to be screened for significant environmental effects, and if such effects looked likely, an environmental impact report (EIR) would be prepared. A set of screening questions was provided in the Procedures to help determine significance; a number of these were concerned with wider than site-specific impacts.

In hindsight, while the Commission had the mandate, it did not have the legislative power or executive status to do the job effectively. The large state development agencies (the Lands and Survey Department, the Forest Service, the Ministry of Works and Development and the Ministry of Energy) did not see that environmental assessment of their activities was really necessary, except where they saw themselves doing it. At the same time, the lack of apparent and real coordination between the EPEP and the parallel decision-making processes under the Town and Country Planning Act and the Water and Soil Conservation Act came to be perceived by developers as unnecessarily expensive and bureaucratic in the face of increasing difficulties to remain competitive in the global economy, and involving much duplication of effort.

19.3.3 A System in Crisis

The constraints on the effectiveness of the Commission became manifest during the late 1970s and early 1980s, a period during which New Zealand experienced a major fiscal crisis and the beginnings of significant structural economic changes. These changes were precipitated by rising international oil prices and the loss of traditional international markets when Britain joined the European common market.

The government response to these problems was a series of large scale resource development projects (known as the 'Think Big' strategy), based on imported technology, overseas borrowing, and greater multinational involvement in the New Zealand economy[6]. The Commission incurred the wrath of the government when it pointed to shortcomings of the Think Big strategy. The Government moved to constrain the role of the Commission and public opposition to large-scale resource development projects by enacting the National Development Act (NDA) in 1979. The Act established a 'fast-track' procedure for those projects deemed to be of national importance by the government and specific time constraints were placed on the decision making process[7].

The fast-track procedure dramatically speeded up the consent process for these major projects, In the event, only three proposals went through the fast-track procedure of the NDA before it fell into abeyance in 1982-1983 (it was only repealed formally in 1986). However, the Think Big policies stimulated a major social debate in New Zealand about the wise use of resources, the need for a national energy policy and a number of other related strategic issues. The NDA was seen by interest groups as a cynical government move almost smacking of totalitarianism to curtail opportunities to scrutinise proposals and thereby suppress debate.

The government amended the NDA and the EPEP in 1981 to narrow the scope for public debate, and remove grounds for judicial review of NDA impact reports, by setting a minimum requirement for an EIR that effectively excluded indirect and cumulative effects (normally central to strategic impact assessments), and relegating social considerations to being of secondary importance after the biophysical impacts (Morgan 1988). The changes to the contents of the impact reports imposed further constraints on the independence of the Commission. The Commission was not allowed to discuss policy issues or include in the audit public submissions that discussed policy issues.

---

[6] Projects constructed or anticipated included a hydro dam, an aluminium smelter, a synthetic petrol plant and an ammonia-urea fertiliser plant.

[7] The Act combined in one hearing applications for all consents in front of the Planning Tribunal (an appeal body under the Town and Country Planning Act 1977 and subsequently the RMA 1991; its name was changed in 1996 to the Environment Court). It overrode 22 other statutes, and restricted rights of appeal. Government was to have the right of final decision.

19.4 A New Era in Environmental Impact Assessment (1991-Present)

19.4.1 Wide Ranging Environmental Policy Reforms

The enactment of the Resource Management Act (RMA) in 1991 marked a fundamental change in the New Zealand system for impact assessment. In particular, EIA was decentralised to district and regional councils, with minimal direct participation by any central government agency. This remarkable change in the institutional arrangements for EIA has been precipitated by a number of inter-related factors that led to wide ranging environmental administration reforms (see Memon 1993 for a fuller discussion).

Underlying the pressure for a major restructuring of institutional arrangements for environmental management during 1984 to 1990, and the role of EIA in this context, was a desire to ensure compliance of the environmental management statutes with the rapidly changing direction of the role of the state in the New Zealand society. Under the aegis of the fourth Labour government, these changes were conceived by a small elite group of politicians, businessmen and government officials, whose primary motivation was to increase the competitiveness of the economy in the global economic order to avert a major economic crisis. Wide ranging restructuring was achieved within a remarkably short period between 1985 and 1990 by deregulating the production and financial sectors of the economy and dismantling the welfare state.

In the context of this New Right political economy setting, the appropriateness of EIA procedures as a form of central government intervention in a market-led economy was questioned, as was the role of the central government in administering these procedures. Treasury – a very influential agency in shaping national policy during that era, in particular, saw the procedures being more appropriately administered by local government, with the role of the new Ministry for Environment explicitly limited to monitoring the effectiveness of these procedures (Memon 1993; Memon and Morgan 2001).

The justification for environmental assessment procedures, as an additional requirement, separate from the statutory planning process, was another reason which ultimately led to a decentralisation of EIA from central to local government. It appeared more logical to incorporate the EIA procedures into the planning process within a new integrated environmental planning statute so that proposals had only one consent process to comply with.

19.4.2 The Resource Management Act 1991

The RMA provides a framework for apportioning environmental planning responsibilities between central, regional and local government agencies, and establishes appropriate decision making procedures for resolving conflicts related to resource use. SEA and project based EIA have been accorded a central role in this process.

In significant respects, the Act is a product of a conventional, theoretical economic analysis of resource management issues. Structured within the school of institutional economics, a key aim of the government was to design an act to address the problem of environmental externality in the context of property right arrangements. Treasury's overriding objective throughout this policy development exercise was to limit the scope of the proposed Act as a means of controlling externalities. It was, for instance, critical of the broader purposes of the proposed legislation, such as sustainability, and the needs of future generations, which it regarded as being based on vague values and could therefore lead to arbitrary decisions by local authorities and the Planning Tribunal (now the Environment Court). The view of the RMA as an 'effects based' statute has been a strong driving force on the part of central government in directing its implementation by local government.

An equally distinctive feature of the RMA is that it provides a statutory framework for a relatively integrated approach to environmental planning, based on the rational planning model It has replaced a large number of separate and in some respects inconsistent and overlapping statutes concerned with the use of natural resources (land, air, water).

The sole purpose of the Act is defined in Section 5 (Part II) in terms of the principle of sustainable management of natural and physical resources:

5. Purpose

(1) The purpose of this Act is to promote the sustainable management of natural and physical resources.

(2) In this Act, "sustainable management" means managing the use, development, and protection of natural and physical resources in a way, or at a rate, which enables people and communities to provide for their social, economic, and cultural wellbeing and for their health and safety while –

  (a) Sustaining the potential of natural and physical resources (excluding minerals) to meet the reasonably foreseeable needs of future generations; and

  (b) Safeguarding the life – supporting capacity of air, water, soil and ecosystems and

  (c) Avoiding, remedying, or mitigating any adverse effects of activities on the environment.

Although sub-Section 5(2)(c) is often interpreted as the provision requiring impact assessment, in fact, the whole Section 5 requires assessment as a basis for decision-making to promote the sustainable management purpose of the Act. Thus, environmental assessment is an integral part of the Act.

Manifestly, sustainable management, as defined in the RMA, is akin to 'weak' sustainability, underpinned by a neo-liberalist political economy stance, to reconcile competing environmental protection and economic growth objectives in an open market economy. Nevertheless, the Section 5 well as the provisions of Sections 6, 7 and 8 in Part II (Purpose and Principles) have significantly broadened the values that have to be taken into account in impact assessment, including those of the Mäori people and protection of the bio-physical environment.

A number of Māori environmental concepts have been incorporated in the provisions of Part II of the RMA. While the Act does not give Māori a right of veto over alternative uses of resources on their ancestral lands and disputed territories, it does give their values priority. It requires those who exercise functions and powers under the Act to recognise and provide for the relationship of Māori and their culture and traditions with their ancestral lands, water, sites, *waahi tapu* (sacred sites) and other *taonga* (Section 6) as matters of national importance; they are also required to have particular regard to *kaitiakitanga* (Section 7) and take into account the Principles of the Treaty of Waitangi (Section 8).

Several other features of the Act are important from a SEA perspective:

- At the sub-national level of government, the environmental management functions are now exercised within a hierarchical, two-tier system of directly elected multi-purpose regional councils and territorial local authorities (city and district councils, and unitary authorities) (Box 19.2 and Fig. 19.1).

Box 19.2. The Resource Management Act: Functions by levels of government

CENTRAL GOVERNMENT
- Overview role
- Develop national policy statements and national environmental standards
- National aspects of coastal management

REGIONAL COUNCILS
- Integrated management of regional resources
- Water and soil management
- Regional aspects of coastal management
- Manage geothermal resources
- Natural hazards mitigation
- Regional aspects of hazardous substances use*
- Air pollution control

TERRITORIAL LOCAL AUTHORITIES
- Control of effects of land use and subdivision
- Noise control
- Controls for natural hazards avoidance and mitigation*
- Local control of hazardous substance use

*Allocation of responsibilities between regional and territorial authorities for these functions is decided on a regional basis.

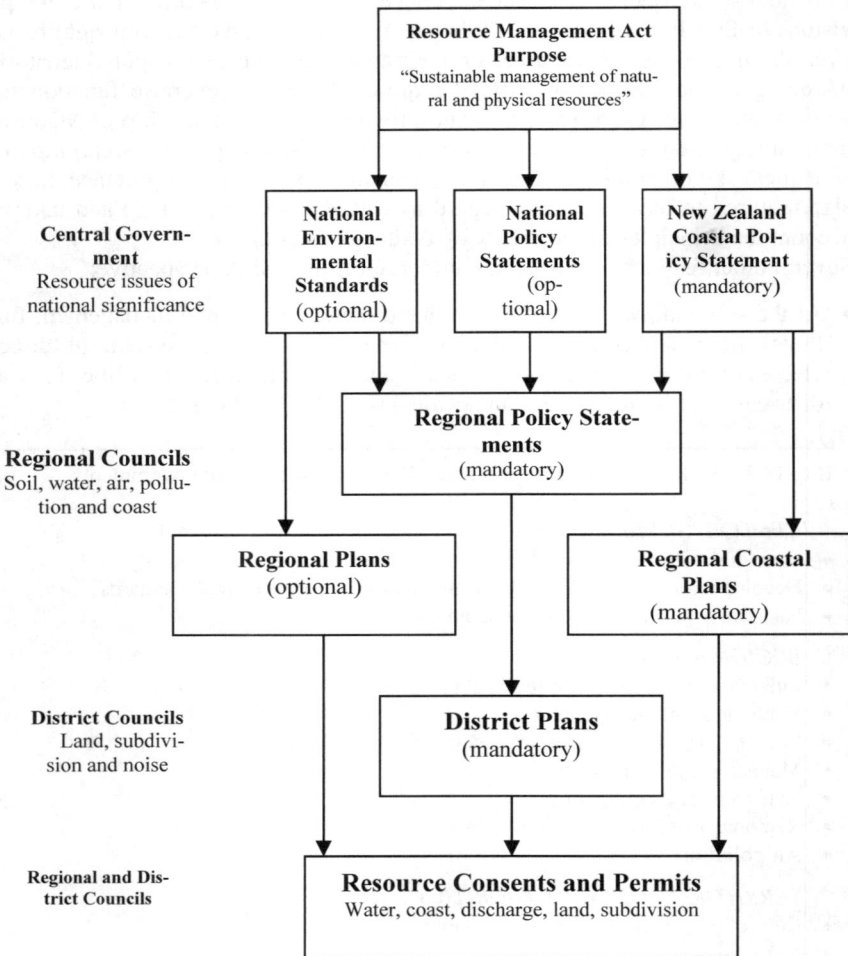

Fig. 19.1. The RMA Hierarchical Planning Framework

This allocation is based on the assumption that decisions should be made as close as possible to the appropriate level of community of interest where the environmental effects and benefits arise.

- National policy statements and standards articulate national priorities. Regional policy statements provide a strategic and holistic overview of resource management issues in each region and policies and methods for their integrated management within a sustainability framework. The more specific objectives and policies and methods for implementing these are formulated in regional and

district plans. These documents provide the basis for evaluating and making decisions on resource consent applications and to monitor policy effectiveness. This highlights the respective roles of SEA and project based EIA in the new planning framework. Objectives, policies and methods in plans are subject to SEA while development applications are subject to EIA.
- The Act provides more flexible administrative processes for resource consents and stronger monitoring and enforcement provisions to manage environmental impacts. The councils are required to be cognizant of cumulative and interactive effects and cross boundary issues, and have a duty to consider alternatives and assess their benefits and costs in carrying out their functions. Regional policies and plans are binding on district councils and stakeholder user groups.
- There are relatively generous provisions for public participation under the Act.

19.5 Section 32 Methodological Procedure

In this Section of the chapter, I will review issues related to the procedural requirements for undertaking a Section 32 analysis in the wider context of the Act. The iterative sequence of steps in a Section 32 analysis is illustrated diagrammatically in Fig. 19.2. Very briefly, decision makers should scope and analyse each environmental issue or problem under the jurisdiction of the agency; determine the extent to which new objectives, policies and methods are needed to deal with the issue and evaluate the appropriateness of these. This requirement applies to all RMA policy statements and plans prepared by central government and district and regional councils (including unitary authorities), as noted earlier.

19.5.1 Wider Significance of Section 32 Provisions

As reviewed earlier, the RMA is underpinned by multiple rationales on the part of Government. Section 32 manifests these contesting rationales. This tension may compromise the effectiveness of the SEA process.

Given the sustainable management purpose of the RMA, the Section 32 requirement echoes the growing international recognition of the value of assessing the environmental impacts of policies, plans and programs. SEA can better cope with cumulative impacts than project assessment. It is pro-active, occurring before development proposals are made, provides guidelines to ensure development is within sustainable levels, enables integration across an area or region and enables cumulative effects to be dealt with (Thérivel et al. 1999 p.24). The inclusion of Section 32 into the RMA could be seen to represent these aspirations.

By the same token, Section 32 introduced a check in the plan preparation stage for the first time (Fookes 2000). It imposes discipline on policy makers so as to ensure that any objectives, policies or methods adopted under the RMA are supported by good policy analysis (MfE 2000). It imposes a rigour on decision-makers by requiring them to evaluate their objectives, policies and methods in meeting the purpose of the Act. The aim is to promote greater accountability for decision-making and greater transparency of decision making.

The inclusion of Section 32 in the RMA was also, in part, motivated by a New Right ideological attack on regulation starting in the mid-1980s. Compared to the preceding Town and Country Planning Act, the RMA is 'permissive' by nature rather than 'regulating' legislation, particularly with reference land use planning. Section 32 is subject to Part II of the RMA (Purpose and Principles). Part II of the Act anticipates a degree of regulation but Section 32 requires regulation be explicitly limited to promoting the purpose of the Act. The implicit objective here is to reduce the burden of excessive local government regulation on land use in order to keep transactions costs low for business.

This 'check', requiring explicit justification for the adoption of objectives policies and rules, represents a significant change from the preceding Town and Country Planning Act regime in New Zealand. In the past councils were not required to undertake an analysis of the plans they created from an environmental sustainability or cost benefit analysis stance. More to the point, planning in the past had been used by government (national and local), albeit quite legitimately, to address social or economic issues eg zoning of land for industry in conjunction with economic incentives from central government. Apparently, according to some observers, including the former Minister for Environment, councils should no longer use the RMA to achieve economic and social ends indirectly (Memon 1993). From the Minister's perspective, restriction of private property rights under the ambit of the RMA was only justifiable to address adverse impacts on the biophysical environment. The logic of such a restrictive New Right economic rationale for planning under the RMA is questionable in the wider context of the broader sustainability purpose which underpins the Act and has been rejected by the Environment Court.[8]

19.5.2 Undertaking a Section 32 Analysis

Section 32 of the RMA sets out a process for councils and government departments to test the *appropriateness, efficiency and effectiveness* of any proposed provisions for national, regional and district plans. The methods for applying these criteria are likely to be contested consequent on the amendment to Section 32 in 2003 and may have to await the development of Environment Court case law for greater clarity. The appropriateness criterion applies to objectives, policies and methods. The effectiveness and efficiency criteria apply only to policies and methods, not to objectives.

---

[8] This restrictive bio-physical "environmental bottom-line" view of the RMA has been challenged by a number of recent decisions of the Environment Court (see Skelton and Memon 2002). A number of Judges have argued that the method of applying Section 5 involves an overall broad judgment of whether a proposal would promote the sustainable management of natural and physical resources.

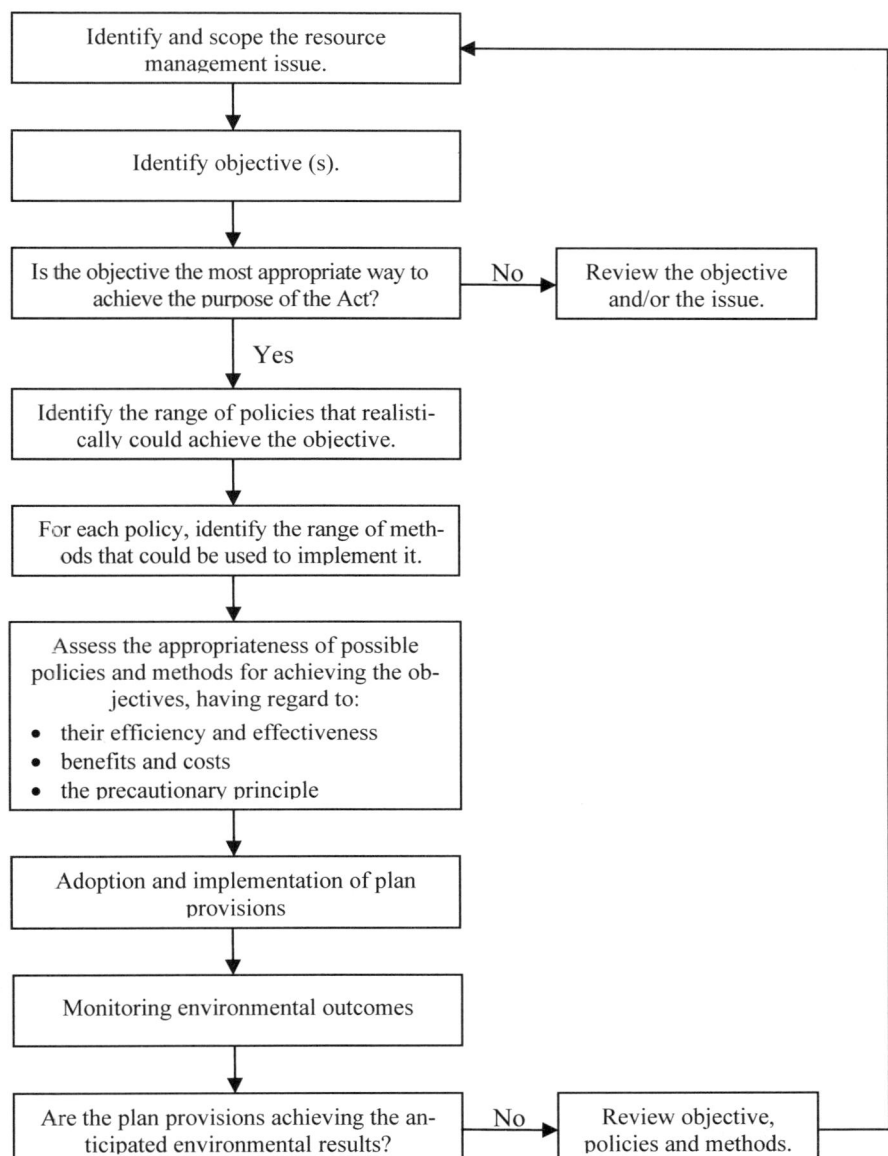

Fig. 19.2. Sequence of steps in a Section 32 analysis

In the context of Section 32, the above terms may be interpreted as follows[9]. *Appropriateness* means the suitability or acceptability of any objective, policy or method in achieving its purpose. To assist in determining whether a proposed policy or method is appropriate, then the *effectiveness* and *efficiency* of the option should be also considered in relation to other options, including 'take no action' option. Effectiveness means how successful a particular option is in addressing the issues in terms of achieving the desired environmental outcome in comparison to other alternatives. The efficiency test is based on a comparison of the benefits to costs (bio-physical environmental benefits minus costs compared to social and economic benefits minus costs). For example, regulation by means of rules in plans is just one method that councils can use to implement policies and objectives. Sometimes methods other than regulation (e.g. economic instruments, information and education) may be more effective and/or efficient.[10] Section 32 explicitly requires decision makers to consider those methods as well.

The evaluation of objectives, policies and methods further must also take into account the monetary and non-monetary benefits and costs of policies and methods and the risk posed by inadequate information about an issue or uncertainty about how to address the issue. For example, a relatively 'blunt' tool may be acceptable to address an environmental issue if the risks posed by not taking any action are deemed high.

19.5.3 The Benefits of Section 32 Analysis

The expected benefits of undertaking a Section 32 analysis from a central government and council perspective have been summarised as follows (MfE 2000 p.6):

1. Better environmental outcomes

- Measures are targeted at achieving the purpose of the RMA
- Plan provisions achieve desired environmental outcomes in effective and efficient ways
- The Council's decision –making is better informed
- The Council can have confidence that its proposed plan provisions are soundly based

2. Costs to the community are kept to a minimum

- The costs borne by affected parties are the least practicable costs consistent with achieving the purpose of the RMA
- The Council's own transaction costs can be factored into decision-making

---

[9] This is based on the information accessed from the Ministry for Environment's Quality Planning website (www.qualityplanning.org.nz).

[10] Regulatory and non-regulatory methods to achieve each policy include: research, monitoring, education, information and training, council services, incentives, voluntary agreements, land purchase, levying charges, and rules.

3. More robust provisions

- The needs of stakeholders are met better
- The acceptability of the plan provision is improved, and the number of submissions opposing plan provisions is reduced
- The number of appeals to the Environment Court is likely to be reduced
- The Council has a sound basis for its case if/when plan provisions are appealed to the Environment Court

4. Provides a basis for future monitoring

- The assessment of the effectiveness of plan provisions provides the information required to set the anticipated environmental results

5. Reporting assists plan implementation and review

- Council and other parties have easy and assured access in the future to authoritative information on why a particular provision is in the plan and why other possible provisions are not.
- A report assists Council decision–makers to exercise their discretion in an appropriate way, and assists councillors and staff when they are required to justify plan provisions to plan users.
- A report assists the Council when it is considering plan variation or plan reviews.

19.5.4 Reflections on Implementation Experience

An in-depth study of the implementation of the RMA has concluded that central government adopted a radical and somewhat sophisticated environmental mandate but has subsequently failed to adequately support its implementation. Consequently, most regional and district council have produced only fair to poor environmental plans, due mostly to limited governance capabilities (Ericksen et al. 2002). Hutchings has observed that: "it was certainly the attitude of the day that local government should be left to fend for itself" (cited in Young 2001 p.36).

Initially, it was unclear as to how Section 32 was to be implemented as it was a new provision and not clearly worded in the Act. Even after its recent amendment, it is likely that the application of Section 32 will continue to pose challenges to practitioners and councils. The quality of Section 32 analysis is largely determined by the ability of those undertaking Section 32 analyses to carry out the task. Recent comments would suggest that practice is variable and that some councils may not "be bothered fully determining what other reasonable alternatives to relatively draconian land restriction are available" (McShane 1998 p.38).

Comments by two Environment Court Judges tend to confirm this. As commented by one Judge:

> "Although Section 32 of the Act is couched in terms which indicate that interference by means of plan provisions should be kept to a minimum, Councils have, to a degree, brushed this to one side and produced plans running to many volumes of unbelievable

complexity [...]. Some plans are so complex that the average man in the street cannot understand them. To produce such a plan is inexcusable" (cited in McShane 1998 p.10).

According to another Environment Court Judge, the Section 32 analyses carried out by councils often fall short of the responsibility imposed. There is often little actual identification, quantification and analysis of the costs and benefits of alternative means to a proposed provision. Even at a basic level, benefits and costs are not adequately identified (Kenderdine 1999 p.5).

It is pertinent to note that the above observations on RMA practice reflect a limited perception of the purpose of Section 32 on the part of these critics and a lack of understanding of its wider intent as a tool for strategic planning. It would appear the equally important significance of Section 32 as a means to operationalise the sustainability purpose of the Act has been marginalised or is not understood. Such differences in perceptions of the underlying purpose of Section 32 reflect the contesting rationales that led to its inclusion in the Act, as reviewed earlier.

Notwithstanding these constraints, Section 32 has encouraged a few councils to identify other methods of achieving the purpose of the Act. The use of alternatives methods of implementation, such as strategies implemented through the Annual Plan of a local authority, is seen in some instances as more flexible and responsive to change than a regional or district plan. There are councils that have applied Section 32 analysis to their policy/plan process and, as a result, have chosen to use alternative methods e.g. Manawatu-Wanganui Regional Council (Browne et al. 1998). In the cited instance, the council chose to prepare non-statutory strategies as an alternative to regional plans after concluding from the Section 32 analysis that the necessary management strategies would be more efficient and effective than a regional plan.

19.5.5 Perceived Challenges in Practice

The following are concerns reported to be experienced by New Zealand planning practitioners undertaking Section 32 analysis.[11] These concerns are evidently also limited to narrow procedural matters, at the expense of the sustainability vision which underpins the RMA.

Recording and reporting Section 32 analysis: Keeping clear and concise records of Section 32 analysis is a critical component of the implementation process. It helps to make the development of any plan transparent. A Section 32 report which summarises the process they went through in deciding on the most appropriate method to deal with a particular environmental issue, as is often the practice, is not adequate. Deciding on how to present and store formation on Section 32 analysis can be a difficult task, especially when dealing with large quantities of information and significant and/or particularly contentious issues. The financial costs as-

---

[11] This Section is based on information obtained from the Ministry for Environment Quality Planning website (www.qualityplanning.org.nz).

sociated with this can also be significant and councils can find it challenging to finance good Section 32 analysis and record keeping.

It is important that the Section 32 records and/or reports are constantly readdressed throughout the planning process. This is to ensure that any changes or decisions are appropriately recorded and reasoned. This can be difficult to achieve when there are tight time constraints and limited finances available.

Recording and reporting Section 32 analysis is also critical where a considerable amount of information held within a council may be institutional knowledge which is not already documented. Where changes in staff occur this important information may be lost if there has been no record of the knowledge held.

Choosing suitable methods: It can be difficult to decide what methods, particularly non-regulatory ones, are suitable to address an issue when they have not been previously tested. Section 32 does not prescribe what methods should be used. However, councils may find it easier to use regulatory methods, as their performance can be better measured through previous implementation. Non-regulatory methods tend to be monitored to a less extent even though they may be a more plausible option for dealing with some issues.

Amount and type of consultation: Consultation assists the production of a sound and transparent Section 32 assessment. It can be difficult though to determine what type and how extensive the consultation should be. The extent and type of consultation should be relative to the nature and scale of the issue and the level of public interest. Wide ranging consultation does not necessarily equate to any less potential for any new provisions to be challenged.

Using monitoring in Section 32 analysis: Monitoring existing activities that are subject to resource consents and state of the environment monitoring can be useful in determining environmental issues but also in assessing the effectiveness of existing plan provisions. Sound monitoring can be a key mechanism in determining the need for future plan revisions. Inadequate monitoring and recording results can be detrimental to the entire process and could lead to the development of inappropriate methods for dealing with an issue.

19.6 Conclusions and Recommendations

New Zealand's earlier experience to institutionalise SEA practices within central government demonstrates what happens in capitalist societies facing deep fiscal crisis when environmental impact assessment, as public processes, arrive at conclusions at odds with government policy. In pre-1984 New Zealand, within the dominant paradigm, the solution to the fiscal crisis was to promote large-scale resource development, deemed by government and industrial concerns to be 'in the national interest'. Critics argued that from a more strategic perspective, these policies involved huge risks, created relatively few jobs, entailed massive subsidies to private interest and posed serious threats to the physical environment. The manner in which the role of the Commissioner for the Environment was successfully circumscribed is illustrative of the limits of strategic policy assessment if it challenges conventional wisdom and the status quo.

In 1991 Section 32 institutionalised SEA in New Zealand's devolved environmental planning legislative mandate, but SEA continues to struggle to gain proper recognition in planning practice. The New Zealand discourse about Section 32 during the last decade has been dominated by narrow instrumental concerns limited to contesting necessity for environmental regulation under the RMA in order to minimise transaction costs. The normative SEA intent of Section 32, as a means to design and fine tune strategic planning objectives and policies to promote the sustainability purpose of the Act has, unfortunately, gone unrecognised. The extent to which the recent amendment to better clarify Section 32, coupled with more guidance by central government, will address this criticism remains to be seen.

Acknowledgements

I would like to thank my colleagues Associate Professor Peter Skelton and Dr Geoff Kerr for their comments on an earlier draft.

References[12]

Browne V, Quayle R, McNeill J (1998) Strategies for Sustainable Management: The Use of Non-statutory Documents. Planning Quarterly, NZPI 129:14-17

Ericksen N, Berke P, Crawford J, Dixon J (2002) Planning for sustainability: New Zealand under the RMA. Waikato University, Hamilton, New Zealand

Fookes T (2002) Environmental assessment under the Resource Management Act. In: Memon, P.A. and Perkins, J. (eds) Environmental Planning and Management in New Zealand. Dunmore Press, Palmerston North, New Zealand, pp 80-83

Kenderdine S (1999) RMA The Best Practicable Option? Accentuate the Positive, Eliminate the Negative. Resource Management Journal 1(VII March 1999):5

Ministry for Environment (2000) What are the options? A guide to using Section 32 of the Resource Management Act. Wellington, New Zealand

McShane O (2003) Think piece two. Internet address: http://www.qualityplanning.org.nz/uploads/best_practice/pubs/other/3809.pdf, last access 29.09.2004

McShane O (1998) Land Use Control under the Resource Management Act. Ministry for Environment, Wellington, New Zealand

Memon PA (1993) Keeping New Zealand Green. Otago University Press, Dunedin, New Zealand

---

[12] Note that legislation mentioned in the chapter is listed at the end of the handbook in the consolidated list of legislation (Appendix 2).

Memon PA and Morgan R (2001) Social assessment in New Zealand resource management. In: Dale A, Taylor N, Lane M (eds) Social Assessment in Natural Resources Management Institutions. CSIRO Publishing, Collingwood, Victoria, pp 61-73

Morgan R (1988) Reshaping EIA in New Zealand. EIA Review 8:293-306

Skelton P and Memon PA (2002) Adopting sustainability as an overarching environmental policy: a review of Section 5 of the RMA. Resource Management Journal 1:1-10

Thérivel R, Wilson E, Thompson S, Heaney S, Prichard D (1992) Strategic environmental assessment. Earthscan, London

Thérivel R, Partidário M (1996) The practice of strategic environmental assessment. Earthscan, London

Young D (2001) Values as Law: The History and Efficacy of the Resource Management Act. Institute of Policy Studies, Wellington

Part IV – Requirements of SEA in Developing and Fast Developing Countries

The impact of SEA in improving strategic-decision making, and in helping to achieve sustainability, will only be felt at a global scale when SEA is also embraced wholeheartedly by developing and fast developing countries. Part IV of the Handbook discusses the requirements and importance of SEA in four different countries: Kenya, Ghana, Ukraine and China. Chapter 20 discusses the need for SEA in Kenya. The chapter discusses how Kenya's Project EIA can be supplemented by SEA and gives recommendations for implementing SEA considering the socio-economic needs and environmental context of the country. This chapter attempts to justify the need for SEA in a 'biodiversity-rich' developing country such as Kenya, whose priority is to reduce poverty by embracing free trade as an engine of economic development. Chapter 21 evaluates the importance of SEA in the case of Ghana, with particular regard to Water Resources Management. The chapter looks at the impacts of large dam projects and the suitability for SEA in water resources management in Ghana. The chapter suggests how SEA could be implemented taking into account Ghana's system of governance.

Chapter 22 describes SEA in Ukraine, one of the Newly Independent States of Countries in Transition. The chapter presents an overview of some of the main obstacles that hinder the effective application of SEA in Ukraine in relation to transparent and environmentally sound strategic actions, and possible future European Union integration. Finally, chapter 23 discusses the importance of SEA in China with particular regard to the controversial case of the Three Gorges Dam Project. The chapter uses the Three Gorges Project as an example to support the idea that SEA implementation has great significance for wise decision making in the field of cultural and archaeological conservation. The chapter is largely speculative in nature arguing how things might have turned out very differently if the Three Gorges Dam Project had used SEA.

Part IV – Requirements of SEA in Developing and Fast Developing Countries

20 The Need for SEA in Kenya

Vincent Onyango[1] and Saul Namango[2]

1 Brandenburg University of Technology (BTU), Cottbus, Germany
2 Department of Chemical and Process Engineering, Moi University, Kenya

20.1 Introduction

Developing Countries (DCs) in Sub-Saharan Africa have never been more challenged in their national strategic decisions and actions to eradicate poverty, cope with recurrent calamities and steer into greater levels of economic development (Reid 2001 pp.26-66). The task is harder given the requirement that this development needs to be sustainable over the long term, balancing economic, social and environmental concerns, and encompassing a greater definition of Human Rights. Although DCs differ in environmental, economic and socio-political aspects from countries where SEA was developed (Modak and Biswas 1999 p.viii), this chapter suggests that SEA can nevertheless aid strategic decision making in DCs and significantly facilitate sustainable development (Wood 2003; Sadler and Verheem 1996, Thérivel et al. 1992; Buckley 1998).

This chapter attempts to justify the need for SEA in a "biodiversity-rich" developing country such as Kenya (Reid 2003 p.196), whose priority is to fast reduce poverty by embracing "free trade" as an engine of economic development (GOK 2003). The chapter gives an insight into the general strategic decision-making process, reviews how strategic decisions are regulated by law and discusses how the environment has historically been integrated into the decision-making process. The chapter analyses key SEA elements already being implemented, highlights how Kenya's Project EIA can be supplemented by SEA and finally gives recommendations for implementing SEA within the socio-economic needs and eco-environmental context of the country. Kenya has been chosen to reflect the need for SEA in a typical Developing Country in Sub-Saharan Africa.

Implementing Strategic Environmental Assessment. Edited by Michael Schmidt, Elsa João and Eike Albrecht. © 2005 Springer-Verlag

20.2 Kenya and its Formulation of Strategic Decisions

Kenya is a developing East African country with an agriculturally dominated economy, a population of 30 million and a recent annual GDP growth rate of 1.1% in 2001 (GOK 2003). Being both biodiversely rich and a Least Developed Country, there is urgency to reduce poverty through greater economic development by following a path of sustainable development that does not harm the environment (NARC 2002). To achieve this goal while protecting the "nature-base" that most Kenyans rely on for subsistence, this chapter attempts to justify how the SEA process can aid policy makers to embed development strategic decisions within a paradigm of sustainable development. SEA will not only predict environmental impacts, but provide an effective process to aid decision-making through evaluation and selection of development alternatives, and offer best combination of long-term sustainability alternatives integrated with environmental protection (Wood 2003; Sadler and Verheem 1996; Thérivel et al. 1992; Fischer 2002).

Kenya is a unitary republic, with policies, plans and programmes administered at national, regional (provincial) and local (district) levels (GOK 2004). Historically, strategic development decisions were formulated at national levels in a top-down manner with *ad hoc* input from other institutions and often with inadequate public consultation. Only in the late 1990's have national strategic decisions such as the Poverty Reduction Strategy Paper (IMF 2003) followed a bottom-up model, with significant public participation. "Policy", "plan", "programmes" and other terminologies such as "strategic plans", "strategy papers" and "master plans" are often used interchangeably.

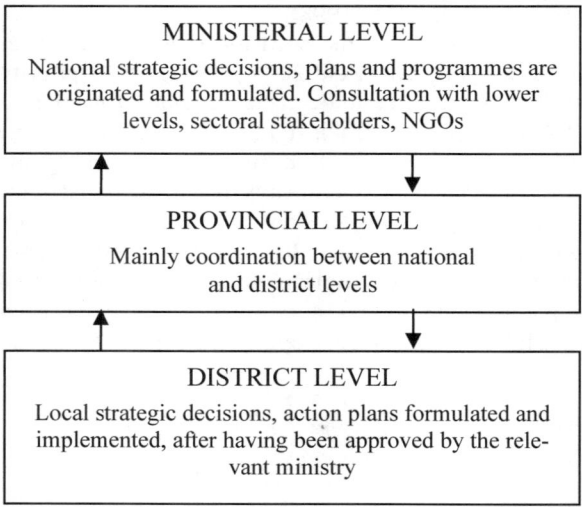

Fig. 20.1. Summarized process of formulating strategic development actions in Kenya (modified from The Coast Development Authority 2004)

National, regional and local development plans and programmes are normally originated by Heads of Departments in various ministries (national level) and the proposals submitted to a technical team within the ministry for scrutiny. The proposals are then sent to District Development Committees (local level), where they are scrutinized and approved before being sent back to the ministries (national level) for final approval, prioritization and allocation of funds (Fig. 20.1). Apart from conventional planning and control mechanisms, the author found a lack of regulated procedures to ensure that sustainability is integrated into the strategic development decisions.

In most of Kenya's national policy and sector documents the objectives, aims and targets are clearly stated. Terminologies such as "environmental protection" or "sustainable development" are often mentioned with criteria given, but without the relevant indicators that would assist monitoring and evaluation. In strategic decisions and development policy documents such as the Economic Recovery Strategy for Wealth and Employment Creation (ERSWEC) and the ruling party's manifesto, economic and social indicators are given but no environmental indicators are listed (NARC 2002; GOK 2003). It is notable that a link to sustainability objectives are also not clear, nor is the "sustainability" defined, contextualized and explaining within the Kenyan context. The five basic components and policy objectives of the Poverty Reduction Strategy Paper (IMF 2003) do not include the environment as one of their main principles and there is no regulatory framework for monitoring of environmental impacts pertinent to the implementation of the strategic decisions. Another example of inadequate sustainability criteria is The Coast Development Authority (CDA 2004) which stipulates that projects will be prioritized on "sustainability criteria", but these criteria provide a narrow and inadequate definition of sustainable development, with no measurable indicators attached. There is no regulatory mechanism to determine the sustainability criteria in such strategic decisions, allowing for sustainability to be defined variously depending upon the ideological opinions of the decision makers. This provides for errors of omission and commission in the strategic decisions-making process and also allows for inconsistent policies, thereby thwarting sustainable development which requires a systematic methodology (Thérivel et al. 1999).

20.3 History of Environmental Assessment of Strategic Decisions in Kenya

Tracing the history of SEA in Kenya reveals how SEA elements have been applied and potentially highlights entry points for an SEA framework. Until the enactment of the Environmental Management and Co-ordination Act of 1999, Kenyan laws did not have a legislative framework for project EIA, or *constitutional* rights to a clean and healthy environment. The development of national strategic policy-making tended to shift with dominant development paradigms, be they driven by donor agencies or not, with inadequate reflection on potential environmental implications and need for home-grown solutions (Wanjohi 1999). The enactment of the Environmental Management and Coordination Act of 1999 meant that there was a regulatory framework for environmental assessments in Kenya. Generally, the act:

- constitutionally recognized every Kenyan's right to clean and healthy environment
- gave every Kenyan legal basis (*locus standi*) to sue if there is reason to believe that there is environmental damage
- created and mandated key institutions such as the National Environment Management Authority (NEMA), to coordinate and manage environmental protection issues of the country
- provided guidelines for project EIA
- gave standards for compliance and set procedures for legal recompense and litigation
- gave rights to public access to environmental information transmitted to NEMA

Although the Environmental Management and Coordination Act of 1999 caters more for project EIA and is rather inadequate for SEA, one may argue that before the early 1990's environmental impacts of Kenya's strategic decisions were either:

- not at all considered
- done in order to comply with donor conditionality
- done on *ad hoc* basis without regulation or guidelines
- did not follow any pattern, methodology or procedural criteria
- inadequately monitored and enforced

A comprehensive regulatory mechanism to ensure that potentially adverse environmental impacts of Kenya's strategic decisions are appropriately considered did not exist. In reference to SEA's important aim of trickling down of strategic decisions from upper to lower levels (Thérivel et al. 1992), the authors did not find formal mechanisms to ascertain and guarantee that sustainability considerations in decision making trickle downstream. There is insufficient mechanism to ensure that environmental issues make the agenda at strategic decision making, meaning that sustainable development that balances social, economic and environmental issues have not been guaranteed. Even though Kenya has no SEA legislation, it is encouraging to note that the National Environmental Management Authority (NEMA) is already considering the development of an SEA framework (Mbegera 2004)

20.4 SEA Elements in Kenya and Tools to Implement them

Significant SEA elements have been variously implemented even though the label of "SEA" was not applied to them and they were not regulated or demanded by law. Table 20.1 shows eleven key SEA elements and a grading on how effectively they contributed to strategic decisions that have been made.

Table 20.1. Performance of SEA elements in strategic decisions since the 1990's summarized from policy and strategic decision documents in various sectors

SEA Element	Grading	Remark
Public participation	fair	NGOs active but local communities inadequately involved. Need for involvement during process of policy formulation at national, regional and sectoral levels
Cross-sectoral integration	fair	Various sectors, NGOs, stakeholders sometimes involved but still room for meaningful engagement. No legal framework for this process
Data and information availability	poor	Inadequate for all levels of spatial planning; data, information not easily accessible. No regulated policy on information management, access and sharing
Communication	fair	Slowly improving with govt agencies being more open and transparent. EMCA guarantees citizens right to environmental information, but govt rules a hindrance to accessing public information.
Agenda setting of environmental issues	fair	Often at behest of donor agencies, done at national level, but watered down in lower tiers. Needs great improvement. Donor driven, not home-grown
Clear criteria, objectives, targets and indicators	fair	Directional objectives and strategies are stated; but often poorly articulated, with inadequate criteria and benchmarks to measure performance. Even when criteria well documented, indicators are often weak or lacking. Research to develop easily measured and useful indicators needed.
Legislation, compliance, enforcement	fair	Exists for EIA, but not for SEA. Urgent need to overhaul old Acts; harmonize and streamline institutions and policies. EMCA still too young to evaluate
Institutional framework	fair	Urgent need to harmonize and streamline conflicting institutional interests. Need for enhanced technical and human capacity. Overlapping authority and mandate is a problem
Impacts that are global, transboundary, cumulative	fair	Cooperation with neighboring countries needs legislative mechanism for regulation. Example of recent decision to cooperate with Uganda and Tanzania in managing environmental issues in Lake Victoria seems well coordinated and implemented.
Monitoring	poor	Equipment and long-term data series is inadequate. Legislation framework for monitoring is urgently needed.
Tiering	fair	Various levels of planning and implementation exist, but effective vertical integration of sustainability concerns needs to be regulated, enforced and regularly evaluated. Need to avoid watering down of policy objectives at lower levels of decision making.

Good = effectively and satisfactorily performed; Fair = average performance, needs improvement; Poor = below average performance

The authors arrived at the grading of "poor", "fair" and "good" after scrutinizing their explicit use in the Kenyan national policy and development documents produced in the 1990's. Although no element was graded as "good", the situation is not futile as two elements (18%) scored "poor", while nine elements (82%) were graded "fair". Kenya can use SEA to improve the effectiveness of the elements by clarifying the procedures and also providing a methodological and regulatory framework for effectively applying them (Wood 2003; Partidário and Clark 2000; Sadler and Verheem 1996). A main tool to implement public participation in Kenya has been public meetings, workshops and written submissions. Community participatory tools such as Participatory Rural Appraisal (PRA) (Chambers 1992) and intersectoral multidisciplinary expert committees have also been commonly used to integrate various sectoral concerns during strategic decision-making. Unregulated by law, these tools identify and highlight the concerns and aspirations of local communities for strategic decision-making processes. EIA is regulated by law, and like PRA it enhances sustainability at project level but much less at higher strategic levels. The Environmental Management and Coordination Act is the only assurance that SEA elements will be considered in the strategic decision-making process. Other SEA elements are done out of mutual interests among stakeholders or out of donor conditionality, and without force of law. Various cross-sectoral actors have tended to use "Memoranda of Understanding" as a main basis for cooperation, for example WWF's cooperation with local communities in the management of vital biodiversity and natural resources (WWF 2004). The authors did not however find adequate quality control mechanisms in strategic decisions such as effective monitoring and external independent reviews, a weakness in methodology to achieve long term sustainability aims. Public participation in formulating strategic decisions has been increasing since the mid 1990's, though not yet fully realized (Kamori-Mbote 2000). Public participation in strategic decisions was often "adversarial" with Non-Governmental Organizations and sectoral representatives often lobbying and petitioning the government on environmental and sustainability issues. "Fit for purpose data" (see Chap. 48) is not readily available or accessible at all levels of decision-making (M'Mella and Masinde 2002). Under the Environmental and Management Coordination Act of 1999 a Kenyan citizen has a legal right to environmental information even though easier and simpler procedures of acquiring information by users should be implemented (Kamori-Mbote 2000). In their report on Environmental Information Circulation in Kenya M'Mella and Masinde (2002), listed the following as potential data and information limitations to SEA implementation:

- lack of coordination in data collection, especially between ministries and specialized institutions
- lack of quality control in data collection i.e. non-systematic, irregular data
- lack of skilled personnel, financial and technical resources
- limited computerization, to collect, interpret and report environmental data
- low level of standardization and non-compatibility of data sets and information systems

- inadequate involvement of NGOs and communities in the process of data collection especially at the local level

For information to be more accessible and useful to the public, the information should not only be provided in forms easily understandable to them, but also the time taken and unnecessary bureaucracy ought to be done away with.

20.5 Institutional Framework for SEA Elements

In 1974 a National Environment Service was established as lead agency to coordinate, promote, and oversee environmental activities (M'Mella and Masinde 2002), although little was achieved. Today there exist several environmental institutions, responsible for policy, research, implementation, standards, and coordination even though "there is a need to streamline mandates of existing institutions, particularly to reduce duplication and conflict of interests" (GOK 2003). For example the Environmental Management and Coordination Act does not clarify who the Public Complaints Committee or Environment Tribunal should report to (The East African 2003). Institutions dealing with strategic development decisions and environmental issues are coordinated at the national level by the Inter-ministerial Committee on Environment, chaired by the Permanent Secretary in the Ministry of Environment (GOK 2004). It is a forum for inter-sectoral integration and consultations. However, the National Environment Management Authority (NEMA) is the lead agency in coordinating environmental issues.

NEMA's main responsibilities include:

- coordination and management of environmental activities by other agencies
- monitoring and enforcing environmental standards and waste management
- advising government on environmental issues and policies etc

Other key institutions to support SEA elements are:

- National Environmental Council – deals with policy formulation and direction
- Public Complaints Committee – hears complaints from the public
- Environment Tribunal – resolves environmental disputes between parties
- Standards Review and Enforcement Committee – establishes, reviews and enforces environmental standards

Other key institutions potentially capable of contributing to national strategic decisions are:

- Kenya Institute of Public Policy Research and Analysis (KIPPRA) – does research and analysis on policies for long-term development objectives
- Institute of Policy Analysis and Research (IPAR) – does research on policies

In dealing with transboundary environmental issues, Kenya has not ratified the Espoo Convention (see Chap. 3), which is not yet open for ratification by non-

UNECE members (UNECE 2004). However, in adherence to its foreign policy principle of international cooperation (GOK 2004) the Kenyan Government has ratified most United Nations conventions and protocols, and further established key initiatives at policy and institutional levels to coordinate objectives of the protocols (M'Mella and Masinde 2002). These initiatives include:

- Development, review and harmonization of sectoral policies and strategic action plans for managing the environment and natural resources
- National Environmental Action Plan, 1994
- National Biodiversity Strategy and Action Plan, 1999
- Anti-Desertification Community Trust Fund
- Kenya National Cleaner Production Centre
- National Action Plan to Combat Desertification
- Poverty Reduction Strategy Paper
- National Population Policy for Sustainable Development
- National Development Plan (2002-2008)

From these initiatives one can surmise that an institutional environment potentially conducive to SEA is already being founded in the country. As Kenya shares significant biodiversity such as migrating wildlife and lakes with its neighbours, and aware of current impacts of transboundary genetic, atmospheric and hydrologic pollution, Kenya ought to use SEA to help it plan for long-term sustainable management of its development activities (Modak and Biswas 1999). SEA can provide relevant and timely information to aid the planners in making these strategic decisions (Wood 2003).

20.6 Need and Justification for SEA in Kenya

As a tool whose efficacy has been recognized by countries, major international organizations, regional economic communities and major financial institutions (Partidário and Clark 2000, Wood 2003, WCED 1987, UNECE 1992) this chapter argues that Kenya needs to adopt SEA as an effective framework for integrating environmental concerns into developmental strategic decisions. Over the last 10 years, Kenya's agricultural sector experienced declining productivity, among the reasons were according to GOK (2003) poor governance in agricultural institutions and lack of comprehensive legal framework to guide formulation of consistent policies. The lack of harmony between policies and legal frameworks (Omiti and Obunde 2002) can be significantly solved through an SEA methodology and framework that is embedded in the country's decision-making process (Dalkman et al. 2002). For example, in 2001, thirty percent of Kenya's indigenous forests comprising main water catchments were excised without an impact assessment and amidst massive public outcry (Namwaya 2003). In this case, an SEA framework would have supplemented project EIA by mandating procedural and substantive environmental considerations at earlier and higher levels of national decision-making. SEA would have predicted the adverse impacts of the excision and availed this information in a way that would have probably made EIA unnecessary, and also offered an alternative to the forest excisions.

Kenya's high incidence of HIV/AIDS, malaria, chronic crop failures, waste production and pollution, combined with urbanization, infrastructure development and other economic activities which imply intensive resource and energy use, are examples of sustainability problems that call for a tool to predict and highlight their potential adverse impacts and their implications. Balancing the needs of long term economic growth, environmental protection and society's welfare implies a need for a systematic and methodological tool that can predict potential adverse impacts and integrate them into the strategic decision-making process. The chapter argues that Kenya's unsustainable tourism industry and the problems mentioned earlier in this section can be solved by basing them on ecologically and socially sustainable paradigms (Brown 1998; Reid 2003) through SEA.

Mechanisms for integrating environmental considerations into Government

Macroeconomic frameworks and financial sector reforms are inadequate, requiring SEA to provide such mechanism (World Bank 2003; Wood 2003). Current unsustainable paradigm of globalization and its inequitable benefits to DCs like Kenya justify the use of SEA to predict and mitigate such adverse impacts well in advance. The government's recognition that Kenya's agricultural productivity is significantly dependent on forests and a healthy environment (GOK 2003), coupled with the statement that environmental problems in DCs are directly linked to unsustainable development (Munn 1979), justifies the suitability of SEA as a tool to foster sustainable development. SEA can be used to translate strategic sustainability priorities and criteria into measurable indicators (Partidário and Clark 2000). Given that the Government recognizes "the need to achieve broad macro and sectoral objectives and targets without compromising the health of the environment", SEA can contribute to realizing these aims by:

- Providing a regulated and methodological mechanism for integrating environmental considerations into national strategic decisions
- Facilitating and streamlining institutional and legal frameworks for delivering sustainable development.

Many sectoral policies were established long before the Environment Management and Coordination Act (EMCA) and therefore legislations and institutional mandates in conflict with EMCA must be overhauled. For instance "80% of the urban population remains dependent on charcoal as a key source of fuel, but charcoal production remains banned on public and trust lands yet it is not illegal to transport or market charcoal (GOK 2003). Such policy contradictions thwart sustainable production of charcoal as a strategic national source of energy. The country of South Africa and other Asian developing countries who share significant similarities with Kenya have successfully implemented SEA (Rossouw and Audouin 2000; Briffet et al. 2003) supporting the argument that Kenya's strategic decisions also stand to significantly benefit from SEA.

In summary, SEA can deliver the following benefits to Kenya:

- integrate, harmonize and aid domestic implementation of global treaties

- aid management of transboundary environmental issues
- mainstream environmental issues in strategic decision-making processes
- balance and integrate competing alternative strategic decisions on how to deal with pressing issues such as disease, desertification, climate change, poverty, gender imbalance and economic development actions, while not adversely impacting the environment
- clarify strategies, objectives, targets, criteria and indicators facilitating comparison and choice of alternatives and mitigation
- enhance and supplement project EIA
- aid legislative and institutional harmony, needed for delivering sustainable strategic decisions

20.7 How SEA will Supplement EIA and Physical Planning in Kenya

SEA will supplement and address project EIA limitations, because SEA deals with impacts not dealt with at project EIA (Wood 2003; Pritchard 1993). SEA is more proactive, considers a wider range of strategic alternatives, cumulative and synergistic impacts, potentially increasing EIA efficacy and reducing costs for EIA (Wood and Djeddour 1992). SEA provides a tiered impact assessment (Lee and Walsh 1992; Thérivel et al. 1992, Partidário and Clark 2000) providing a vital avenue for vertically integrating strategic actions and making them trickle down to lower levels systematically and methodologically. This will aid the harmonization and streamlining of environmental concerns in Kenya's strategic decisions resulting in efficient, effective implementation and monitoring, in addition to precipitating clear criteria and indicators at the various levels.

In a paper reviewing Physical Planning in Kenya, Omwenga (2001 p.5) argues that "inadequate spatial information has greatly contributed to the problem of ineffective and untimely preparation, implementation and management of physical development plans in Kenya [...] making planning preparation and implementation expensive". Therefore, SEA may prove invaluable in assessing the Physical Planning sector, guiding and highlighting opportunities for reform towards sustainability. For an effective SEA in Kenya, several key issues need to be addressed. Kenya should not inherit but adopt a home-grown SEA methodology and framework, suited to its priorities, its eco-environmental situation and embedded in the decision-making processes of the country. It will be crucial to establish relevant "fit for purpose" criteria and indicators that have been proven for scientific, technological, economic and socio-cultural suitability. The predictive techniques ought to drive data needs and must balance social, economic and environmental aspects based on:

- Impacts, not data availability
- Outcome, not output indicators

Kenya's SEA should develop a regulatory framework for community participation in strategic decision making, particularly in the management of natural resources. Where opinions of the public have been given, they must be documented and their use or non-use in the decision making process clearly documented. Kenya will need to establish a streamlined institutional and regulatory framework for comprehensive databases and information baselines, guided by data usefulness in strategic decision making and not by convenience of data collection. Considering that some policies have been weak and inadequate (Omiti and Obunde 2002; GOK 2003); and aware that their significant environmental impacts may have been a threat to sustainable development, Kenya may benefit most from an SEA framework that evaluates *all strategic decisions including policies*. To achieve efficiency in managing transboundary environmental issues; considering costs of developing SEA frameworks; aware of the similarities and contiguity among the East African countries; and aware that a regional Economic Community already exists, Kenya should consider adopting a regional SEA framework in concert with its neighbours. This should be more effective and more efficient in promoting sustainability than individual country SEA.

However, there are potential obstacles that could hinder the effective use of SEA in Kenya. A lack of "fit for purpose" data, inadequate financial, technical and human resources (GOK 2003; UNCHS 2001), political inertia and the misconception of what SEA is and can achieve will be a challenge that must be overcome. This is because SEA would be an innovative aspect of public policy formulation and may require time and experience (Partidário and Clark 2000) to implement effectively. Another obstacle is the lack of tailor-made SEA methodology and framework appropriate to the eco-environmental, socio-economic and juridico-political situation in a Developing Country like Kenya. The inadequate public participation and inadequate monitoring and evaluation due to poor databases, lack of monitoring policies and regulations, will also be an obstacle that needs to be addressed before an effective SEA is put in place.

20.8 Conclusions and Recommendations

While some SEA elements are already in place and several potential institutional arrangements may support SEA, a Kenyan SEA must be built on current success and not needlessly start from scratch. It should be tailor-made to serve the developmental priorities of the East African region, and not just the country, because transboundary strategic decisions may determine sustainable development in the long term. It will be effective only if a consistent and systematic approach is in place, is regulated and enforceable in law, and is supported by a political establishment that understands sustainability and SEA's role in achieving it. Legislative Acts written in the colonial times must be overhauled to reflect current realities and allow current policies to be operationalized maximally. EIA and SEA frameworks should not overlap, but be harmonized for efficient, effective and optimal operationalization. Kenya stands to benefit more from the option of an East African-wide SEA framework, and this should be a priority.

References[1]

Briffet C, Obbard JP, Mackee J (2003) Towards SEA for the developing nations of Asia. EIA Review 23:171-196
Brown F (1998) Tourism reassessed: blight or blessing? Butterworth Heinemann, Oxford
Buckley R (1998) Strategic environmental assessment. In: Porter L, Fittipaldi JJ (eds) Environmental methods review: Retooling Impact Assessment for the New Century. Army Environmental Policy Institute, Atlanta, pp 60-78
Chambers R (1992) Rural appraisal: rapid, relaxed and participatory. Institute of Development Studies Discussion Paper 311, Sussex
CDA – Coast Development Authority (2004) modified from The Management Plan of the Coast pg 105 posted on http://www.unep.org/eaf/Docs/keny/!15plann.pdf, last accessed 15.03.04
Dalkman H, Jiliberto R, Bongardt D (2002) Analytical strategic environmental impact assessment, developing a new approach of SEA. Internet address: www.taugroup.com, last access 10.02.04
Fischer TB (2002) Strategic Environmental Assessment: Transport and Land Use Planning. Earthscan, Publications Ltd, London
GOK – Government of Kenya (2003) Economic Recovery Strategy for Wealth and Employment Creation (ERSWEC) 2003-2007. Government Press, Nairobi
GOK – Government of Kenya (2004) Ministry of Foreign Affairs webpage. Internet address: http://www.kenya.go.ke/foreign.html, last accessed 12.03.04
IMF – International Monetary Fund (2003) IMF PRSP country report number 03/394 GOK, 12.09.2003. Internet address: www.imf.org, last accessed February 2004
Kamori-Mbote P (2000) Strategic planning and implementation of policies involved in environmental decision-making as they relate to Environmental Assessment in Kenya. IELRC working paper
Lee N, Walsh F (1992) Strategic environmental assessment: an overview. Project appraisal 7(3):126-136
M'Mella T, Masinde JM (2002) Environmental Issues in Kenya-National Environmental Secretariat: Environmental information circulation and monitoring system on the Internet. Regional Conference for African English Speaking Countries 22-26th September 2002, Accra Ghana
Mbegera M (2004) Director of Compliance, National Environmental Management Authority, personal communication by telephone
Modak P, Biswas AK (1999) Conducting Environmental Impact Assessment for Developing Countries. United Nations University Press
Munn RE (1979) EIA: principles and procedures. Scope 5. 2nd edn. John Wiley and Sons
Namwaya O (2003) Forest excisions in Kenya. The East African Standard, December 2003
NARC – National Rainbow Coalition (2002) Manifesto

---

[1] Note that legislation mentioned in the chapter is listed at the end of the handbook in the consolidated list of legislation (Appendix 2).

Omiti J, Obunde P (2002) Towards Linking Agriculture, Poverty and Policy in Kenya. Institute of Policy and Research (IPAR)

Omwenga M (2001) The Missing Link: Spatial Information Required in the Preparation And Implementation of Physical Development Plans in Kenya; paper for International Conference on Spatial Information for Sustainable Development, 2-5 October 2001, Nairobi

Partidário MR, Clark R (2000) Perspectives on Strategic Environmental Assessment. Lewis Publishers

Pritchard D (1993) Towards sustainability in the planning process: the role of EIA. Ecos 14(3-4):10-15

Reid DG (2003) Tourism, globalization and development: responsible tourism planning. Pluto press, London

Rossouw W, Audouin M (2000) Development of SEA in South Africa. Kluwer Academic Publishers, Dordrecht

Sadler B, Verheem R (1996) Strategic environmental assessment: assessment, challenges and future directions. MER series. The Hague, Ministry of Housing, spatial planning and environment (VROM)

The East African (2003) 18th November 2003, Nation Group

Thérivel R, Wilson E, Thompson S, Heaney D, Pritchard D (1992) Strategic environmental assessment. Earthscan, London

UNCHS – United Nations Centre for Human Settlements (2001) Land Information Service in Kenya. Nairobi, UNCHS (Habitat)

UNECE – United Nations Economic Commission for Europe (1992) Application of environmental impact assessment principles to policies, plans and programmes. Environmental series 5. UNECE, Geneva

UNECE – United Nations Economic Commission for Europe (2004) Ratifications to the UNECE Kiev SEA protocol 2003. Internet address: www.unece.org, last accessed 1.09.04

Wanjohi GJ (1999) Wajibu. A journal of social and religious concern 15(2):1140-1168

Wood C (2003) Environmental impact assessment: A comparative review 2nd edn. Pearson Prentice Hall, Harlow

Wood CM, Djeddour M (1992) Strategic environmental assessment: EA of policies, plans and programmes. Impact assessment bulletin, pp 3-22

World Bank (2003) World Bank group in Kenya, Kenya Consultative Group Meeting, Environment Donors Joint Statement. Internet address: http://www.worldbank.org/ke/cg2003/environ1ii.pdf, last accessed February 2004

WCED – World Commission on Environment and Development (1987) Our common future. Oxford University Press

WWF (2004) Lake Bogoria community conservation project. Internet address: http//www.panda.org/about_wwf/where_we_work/Africa/where/eastern_africa/Kenya/bogoria.cfm, last accessed 04.09.2004

21 SEA for Water Resource Management in Ghana

Eric Ofori

Brandenburg University of Technology (BTU), Cottbus, Germany

21.1 Introduction

Water resource development projects such as dams constitute a major direct and indirect cause of nature destruction and disruptions in the livelihoods of local communities worldwide. Although these projects have their positive and negative impacts, poor environmental assessments of the effects of such projects have led to the negative impacts far outweighing the positive impacts.

Ghana as a developing country has been grumbling over the negative impacts of such water resource development projects. Examples are the Volta and Kpong dams, which have impacted negatively on local communities and caused disruptions in the ecosystem of the Volta River. The irony of the situation is that Ghana now has to find sources of electricity for its increasing development needs. And hydropower seems to be the cheapest option as at now. However, the environmental impacts of large water resources development projects like dams seem to be a major concern of society now. This is one of the reasons why SEA and EIA are seen as tools for minimizing the impacts of such water resource development projects in Ghana and most developing countries.

This chapter takes a look at the impacts of large dam projects in Ghana and the suitability for SEA in water resources management in Ghana. The chapter first takes a look at the impacts of large dams in Ghana (Sect. 21.2), then some EA legislation and constraints in the country (Sect. 21.3) relevant to water resource management are mentioned, Sects. 21.4 and 25.5 looks at the issue of applying SEA and EIA to water resources management and the benefits such a sectorial SEA will bring Ghana, Finally suggestions are made on how such an SEA could be implemented taking Ghana's system of governance into consideration.

Implementing Strategic Environmental Assessment. Edited by Michael Schmidt, Elsa João and Eike Albrecht. © 2005 Springer-Verlag

21.2 Impacts of Water Resource Developments in Ghana

Ghana has had its fair share of impacts of water resource development projects. There are two major dams in Ghana, which are used for hydroelectric power generation. These dams are the Volta dam and the Kpong dams. The Volta dam has been the major supplier of electricity for Ghana. However, developments and improved access to this source of electricity has led to inadequacies in supply. As at now power from these two dams is unable to satisfy the country's demands and Ghana has now resorted to importing electricity to supplement it needs. However, hydropower energy has been identified as one of the best options for electricity production in Ghana (VRA 1995). As electricity demand is rising, new possibilities of hydropower production are been exploited. Already in 1995, 17 possible dam sites with further capacities of 1200 MW have been identified in Ghana (World River Review 1995).

Plans are advanced for the construction of a third hydropower dam on the Black Volta at the Bui gorge. This newly proposed Bui Dam has faced a lot of opposition both local and international because of the negative effects of the two dams mentioned above. All these hydropower projects involve the use of water resources and therefore adequate precaution needs to be taken by Ghanaians in protecting these water resources. This means that an integrated water resource management approach is needed to solve problems that may arise from such projects in the water resource sector in general as a compliment to project EIA. The ability of project EIA to solve some of the impacts of dams have been acknowledged but on a much broader scale, SEA is seen as the best option for Ghana in water resource management.

21.2.1 Impacts of the Volta and Kpong Dams

Table 21.1 shows some of the impacts of the two dam projects in Ghana. Lack of proper environmental assessment before the construction of these dams led to the many negative impacts outlined in table 1. Of current public and international interest is the abandoned Bui dam project, which is to be built on the Black Volta at Bui in Northern Ghana. This dam, which was abandoned, has now generated a lot of opposition both locally and internationally. This is because of its intended location in a national reserve and also its proposed effect on the last remaining species of hippopotamus in Ghana (Kalitsi 1995; Ghana wildlife Department 1991).

Table 21.1. Summary of impacts of the Volta and Kpong dams in Ghana (collected from Gordon 1999, Mill-Tettey 1989, Diaw and Schmid 1990, Smithsonian Institute 1974, Derban 1984)

Positive Impacts	Negative Impacts
Provision of hydroelectric power for development	Some communities living close to the dam have still not been connected to the national electricity grid.
Decrease in cases of river blindness infection (Onchocerchiasis), in local communities along the river because of the flooding of the breeding grounds of black fly (Simulium damnosum).	Volta dam caused the resettlement of approximately 80 000 people from 740 villages. Kpong Dam also displaced an additional 6 000 people (Mills-Tettey 1989, World River Review 1995). Some villagers resettled on lakeshores exposing themselves to various health impacts (VRA 1996b).
Opportunities for irrigated farming downstream	Increase in some water borne diseases such as bilharzias (Schistosomiasis), Sleeping Sickness (Trypanosomiasis) and malaria (type of malaria is endemic in Ghana and the mosquito vector breeds in the lake shores (Gordon 1999).
Enhanced fishing upstream: catch peaked at about 62 000 tonnes after impoundment of the Volta dam in 1969. Some have estimated a sustainable yield of about 40 000 metric tonnes/year (VRA 1996a).	Diminished fishing downstream
Increased access, better ways of transporting goods due to the artificial lake created; presently a ferry transports goods and people from the northern part of the country to the south.	Initially sedimentation was not found to be a problem; but now due to deforestation on the edges of the reservoir some sedimentation is being observed.
No data concerning positive impacts on flora and fauna was obtained.	Change in water quality in Volta Lake
	Clams and prawn populations have diminished (Smithsonian Institute 1974)
	Water quality is low in mineral nutrients except during the flood season
	Volta dam affected the coastal part of the Volta estuary, because of decrease in sediments and other important river minerals
	Proliferation of aquatic weeds upstream
	Most communities cut off from nearby big towns as a result of massive inundation (Diaw and Schmid 1990).
	The Kpong Hydroelectric development resulted in the increase in the incidence of bilharzias and hookworm (Derban 1984).

21.3 Environmental Assessment in Ghana

Environmental Assessment (EA) in Ghana is a little more advanced than most African countries. Since the inception of the Ghana Environmental Protection Agency (EPA) in 1995, impacts assessment studies have been improved in Ghana. There is no specific legislation or regulation of SEA in Ghana. What exists is an Act 490 spelling out the setting up of an agency responsible for the environment (i.e. EPA) and a regulation on EA in Ghana (see Sect. 21.3.1). Additionally there are specific guidelines on the conduction of EIA in the country. Sectorially, there are no specific legislation or policy on water resource management at present. However, there are guidelines and legislation, which will need to be taken into account in introducing SEA to the water resources sector. In Sect. 21.3.1 some of the institutions and legislations relevant to SEA and EIA, which could also be relevant to water resources management in Ghana are mentioned.

21.3.1 Institutional and Legislative Framework

Some of the institutions in the environmental sector which will be relevant for SEA in water resources management are: the Ghana EPA, Ministry of Science and Environment, National Development Planning Commission (GNDPC), Volta River Authority, Water Resources Commission, and the Energy Foundation. Others are the Energy Commission, the local government offices and NGOs relevant to the sector.

Ghana has a National environmental policy (NEP), which is part of the broader National Environmental Action Plan (NEAP) adopted in 1991. The main aim of this NEP is ensuring a quality of life for Ghanaians both present and future generations. It seeks to ensure reconciliation between economic development and natural resource conservation. This is one of the policies available in Ghana upon which SEA for the water resources sector could be based.

Some of the objectives of this policy are particularly geared towards and integrated approach to management of environmental resources at the 'higher' level. It calls for the "integration of environmental considerations into sectoral, structural and socio-economic planning at the national, regional district and grassroots levels" (EPA 2003).

The Environmental Protection Agency Act, 1994 (Act 490): This act created the EPA Ghana. This is the major agency in Ghana with the responsibility of advising on policy formulation, channel of communication between environmental bodies, issuance of permits, reviewing of EIA and environmental education and research

Environmental Assessment Regulations, 1999 (L.I. 1652): This is one of the legislations made under the E.P.A Act 1994 (Act 490). It is the backbone of EIA in Ghana detailing the procedures that must be followed in conducting environmental assessments. The Ghana EPA is responsible for its implementation. This regulation was recently amended to suit current trends in Environmental Assessments (EAs) in Ghana.

Energy Commission Act, 1997 (Act 541): The Energy Commission Act (Act 541) established an Energy Commission. It provides for the functions of the Commission relating to the regulation, management, development and utilization of energy resources in Ghana. Sections of this act details how environmental considerations should be taken in water resource development projects for energy purposes (hydro-dams).

Water Resources Act 522, 1996: This act resulted in the setting up of the Ghana Water Resources Commission (WRC). The commission is responsible for the regulation, management of the utilization of water resources and coordination of the relevant bodies in this sector. This commission also deals with issues of transboundary impacts of water resource projects.

Local Government Act, 1993 (Act 462): The Act 462 seeks to give a fresh legal meaning to government's commitment to the concept of decentralization. The SEA for water resource management will be more efficient if it is modeled around the Ghana decentralization scheme. This scheme ensures an adequate amount of public participation through local government and district assemblies in Ghana in issues of environmental and economic development.

A lot of programs have been implemented with the ultimate aim of protecting the environment. The Ghana Environmental Resource Management Program (GERMP), the Natural Resource Management Program (NRMP) and the Ghana Environmental Assessment Capacity development in Ghana Program (GEACaP) are examples of such donor-sponsored programs.

21.3.2 The Constraints to EA in Ghana

Despite all these environmental legislation and programmes, there exist a lot of constraints to EIA practice in Ghana. In a survey by Appiah Opoku (1999) various constraints to impact assessment studies were identified such as:

- Lack of organized baseline data
- Lack of environmental awareness in Ghana
- Lack of Local EA experts
- Institutional problems
- Financial constraints
- Lack of public involvement and therefore input
- Monitoring

21.4 Applying SEA and EIA to Water Resource Management in Ghana

According to McKinney (2003), sustainable development requires that water resources management simultaneously achieve the objectives of sustaining development for security and preserving the associated natural environment (McKinney 2003). Water resource management in Ghana got more attention in 1996 when the World Bank initiated the Ghana Water resource management study. This study provided a report titled "Ghana Water Resources Management – Challenges and Opportunities". The key findings of this study led to the Act 522 and the subsequent establishment of the Ghana WRC. Some of the findings are: myth of abundance of the resource in Ghana, inadequate data, absence of policy framework and public awareness (Ghana Government 1996; Ayibotele 1995). These key findings of this document have been considered in proposing the SEA for water resource sector in Ghana.

Large-scale water resources development projects have been a matter of concern to environmentalists, developers, and governmental agencies. This is the reason why adequate measures are been undertaken by governments in reducing the negative impacts of such projects. Recognizing the need for environmental management tools (e.g. SEA and EIA) in this regard is the key to successful curtailment of these impacts in countries such as Ghana. However, the issue of the proper implementation of SEA and EIA in Ghana's water resource sector is very tricky. Studies show that Ghana's EIA system is a bit more advanced in terms of legal and institutional framework (see Sect. 21.3). With regard to SEA, the country is still in its initial stages. Recognizing SEA as an integral part of the early planning process of water resources development projects (such as dams) in Ghana can help identify negative impacts way up the decision cycle. This is because of the ability of SEA and EIA to identify the positive and negative impacts of different design alternatives for a project like the Bui dam. The issue of sector specific SEA in the water resource sector in Ghana would ensure harmonization of environmental practices on various water resource development projects plan in the future, whiles at the same time providing guidelines for conducting environmental assessment studies in this sector.

In suggesting an SEA for Ghana's water resources management, various issues and environmental impacts of water resources development are raised. In Table 21.2 there is a look at how the various issues in SEA and EIA can be adapted to water resource management in Ghana. Key issues of consideration in proposing SEA for water resources management in Ghana are:

Legal and Institutional frameworks: Although EIA regulations do exist in Ghana (see Sect. 21.3.2), there are no legal or institutional frameworks concerning SEA. The only attempts at using SEA in Ghana have been on minimal and pilot basis. Example is the 1997 SEA on the Village Infrastructure Project in Ghana (Amoyaw-Osei 1997) and on the Ghana Poverty Reduction Scheme of 2002 (Nelson 2003). However, the various institutions (see Table 21.2) relevant for implementing SEA in Ghana are available and what is needed is the strengthening of these institutions based on the results on the study of EA constraints in Ghana (see Sect. 21.3.2). Legal frameworks for SEA in Ghana are also lacking but in this case some of the regulations on EA can be extended to cover SEA specifically for water resource management. As shown in Table 21.2, SEA targets a high level of the decision-making, and in this case the cabinet level of decision making apply to Ghana.

Alternatives to PPP: SEA could provide for consideration of a larger range of alternatives than is normally possible in project EIA. Therefore SEA is seen as a more effective means of weighing issues at the 'upper stream' so as to select the policies which has less effect on the environment. An important element of environmental assessment is the consideration of several alternatives. Assessment of alternatives is therefore to be adequately handled in making an environmental assessment effective. One important option to consider is the 'no-action' alternative. Despite the advantages of SEA in identifying alternatives to programs, it is equally difficult to achieve this feat. Therefore for a small country like Ghana, an ability to take on SEA would be seen as a major achievement for others in the sub region.

Background Data on Environment: Issues of environmental baseline data in Ghana are very inconsistent. There seems to be some form of data on the environment, however, these data has been dispersed among the various institutions responsible for the environment and the various NGOs operating in the environmental sector. Ghana lacks a central repository for accessing information on the environment. This is one of the constraints highlighted in the study by Appiah Opoku. However, such baseline data are very relevant for the conduction of any successful SEA and EIA.

Financial and Budgetary Restrictions: Often lack of political will has been identified as one of the constraints of EIA and SEA in the West African sub region, but this, when looked in detail, is due to financial restrictions. Most of the priorities of the Government are geared towards alleviating poverty and the creation of jobs. Therefore the little revenue generated is pushed more into these areas than for solving environmental problems. There is therefore an inadequate financial resource for SEA training and capacity building. However, EA requires money this is why in Table 21.2 suggestions are made concerning the allocation of project funds to EIA studies. SEA at the higher level could help implement this into law so that all developments in the water resource sector can follow this laid down regulation.

Table 21.2. Comparison between SEA and EIA – adapted to the water resource management in Ghana

	SEA (Policy stage)	EIA (Project based)
Subject	Ghana Water Resource Management Programme	Defined project titles on specific locations e.g. The Bui dam Project
Responsible Authorities	Ghana WRC, Ghana Planning Commission, Ministry of Science and Environment, Parliamentary sub-committee on environment	EPA Ghana, Local District Assemblies
Initiation Stage	High level of decision-making. In Ghana, at the cabinet level or before parliamentary approval	During project approval and funding stages
Alternatives	A range of options for incorporating SEA into a water management policy could be developed, including the 'reduction in demand' alternative (e.g. water saving schemes), which is equivalent to the do nothing alternative in EIA.	EIA alternative stage can also include the 'do nothing alternative. Identifying key issues and alternatives for specific water resource projects can help in developing a strategic level SEA which will intend solve project based impacts at the policy level.
Impact Prediction and Identification	All impacts on the environment and their mitigation measures as a result of the water resource program could be identified	Lists of alternatives are compared based on their impacts. In this case a matrix of impacts against projects could be used
Public Involvement	General incorporation of legal rules mandating detailed public involvement in projects likely to have significant environmental impacts. Special adaptation to Ghana's family system.	Public direct involvement in defined projects as a result of SEA legislation
Budget and Funding	Evaluation of cost and benefits including monetary values of the policy. In this case the water resource management policy should also have specific provisions (legal) allocating percentages of project funds to environmental assessment and mitigation	Adequate spending of project money on environmental impact studies and mitigation based on the SEA legislation
Monitoring	Monitoring and evaluation of the results of the water resource management program should be done as early as possible	Carrying out of follow up measures on specific water development projects and checking on proper implementation of monitoring programs

Donor Support and role of NGOs: SEA of Ghana's water resource sector needs the help of the donor community and NGOs in the environment. This is because financial restrictions have been the root cause of under-development in impact assessment studies in Ghana (see Sect. 21.3.2). The role of NGOs in Ghana has been limited to environmental education programs and helping in formulation and implementation of environmental policies. NGOs in Ghana are yet to be involved in environmental data gathering activities and research on the environment. Involvement of NGOs in environmental research would also help them in stating their case when it comes to issues of environmental protection and also protection of local rights and customs.

Issue of Capacity Building and Public Participation: Mandatory public participation in EAs and water resource development programmes has been lacking in Ghana (see Sect. 21.3.2). Although the steps of EIA do require this approach, there are no monitoring guidelines to ensure that this stage of the EIA process is fully enforced. This is one of the areas, which will be vital in developing SEA for Ghana's water resource sector. As shown in Table 21.2, legal guidelines could be incorporated into such programs through the use of SEA. This can complement the public participation step of EIA.

Capacity building has been identified as one of the areas, which can help in advocating for SEA (see Chap. 44). However, training workshops for personnel's and NGOs has mostly been directed at social problems in Ghana instead of environmental issues. Environmental and human settlement policies, strategies and programs of Ghana presently have been unable to ensure sustainable and active public participation at the societal level. This has been linked to the fact that foreign experts have prepared some of these policies with little emphasis on local values and customs (EPA 1997).

SEA can help recognize the need for adequate involvement of the public on policy level as well as on sector and project levels. However, achieving this feat is not always easy due to the fact that responsible establishments and other politicians do not always view public participation as advantageous. This has been so because of the discreet nature of Government policy in Ghana and some countries of the sub region.

21.5 Benefits of SEA for Water Resource Management in Ghana

The various impacts identified in the Volta and Kpong dam cases also puts up a strong argument to proposing an SEA for the water resource sector in Ghana. SEA could help in:

Complementing project EIA and improving the efficiency of the Ghana EPA: EIA practice in Ghana is restricted by certain constraints (see Sect. 21.3.2). These include institutional and legal weaknesses that exist in Ghana and the late stage at which EIA is applied to decision-making. SEA can therefore be used as a complement to project-level EIA to incorporate environmental considerations and alternatives directly into PPP design. Agenda 21 (UNCED), called for the undertaken of environmental assessment of proposed activities that are likely to have a significant adverse impact on the environment (UNCED 1992). SEA in Ghana will therefore be beneficial in eliminating some decisions at the policy level, even before EIA studies can commence, and thus saving money, and also reducing the number of project level EIA. This can serve as a means of making the Ghana EPA's work a bit easier and more efficient.

SEA will be able to identify Large Scale and Cumulative Impacts: SEA provides a better chance than project EIA to address cumulative and large-scale impacts. An SEA on water resource sector would also help identify early warning of large scale effects of water related development policies. Another advantage for Ghana is also that such an SEA would include cumulative impacts of small projects within the water resource sector, which might not pass for EIA. Small-scale impacts of successive dams (e.g. Volta and Kpong Dams) could be better identified and predicted at the policy stage by the use of SEA.

Issues of Large Scale Displacement: Large-scale water resource projects like dams are known to have adverse impacts on locals living along these water bodies. Improper prediction and mitigation measures can result in large displacement of people. An example is what is seen in the Volta and Kpong dam cases. Project level EIA can help in part to predict and mitigate these impacts to a certain extent, but these impacts and measures used in addressing them can be better solved at the policy level. This is one of the reasons why SEA is beneficial. Issues of conflict tend to evolve when people are being displaced or relocated and therefore environmental assessments have to be able to effectively resolve these conflicts. SEA would be able to give better regulations at the policy level for protecting the right of locals in cases of compensation and relocation in Ghana.

Prediction of impacts of water resource development: The Volta River in Ghana is presently harboring two large dams in Ghana and attempts are been made to put another dam at Bui in Ghana. Experience shows that multiple dams on a river significantly worsen the impact on ecosystems. Sediment entrapment can reach very high levels if a cascade of dams is developed. Typical impacts of dam construction have also been identified in the Volta and Kpong dam situation (see Table 21.1). All these negative impacts of dams outlined in Table 21.1, are as a result of the lack of proper environmental assessment during the construction of these two dams. This is one of the reasons why there is a need for Ghana to integrate SEA in its water resource management so as to prevent similar cases from happening especially with respect to the Bui dam. Integrating SEA in Ghana's water resource management could help in predicting and solving most of these impacts at an early stage in the decision-making process. This could be achieved in areas of weighing of alternatives and in broad based participatory approach. The various impacts identified in Table 21.1 could be managed with SEA. SEA could be used to respond to impacts of dams, example:

- In issues of water quality by helping predict before hand the changes that might occur in the water ecosystem before the plan is even implemented. In this way various alternatives of the plan could be considered including the 'no alternative' option and the best option taken.
- Downstream effects of damming a river can be very profound. These impacts can also be cumulative and sometimes move from ecological to economical. This is one of the reasons why the ability of SEA to predict even cumulative impacts could be advantageous in solving these problems before hand.
- Transboundary impacts can be a big problem when it comes to rivers transcending various countries. This is the case of the Volta River and therefore SEA can better help at the policy stage.
- In the issue of endangered species such as occurring in the case of the hippos at the Bui dam area can be solved well with SEA. Although project specific EIA can help predict these impacts and provide good and possible means of mitigating them, experience in Ghana have shown that the implementation of such project specific EIA recommendations can sometimes not be properly done.

21.6 Conclusions and Recommendations

SEA and EIA are tools required in environmental resources management of a country. The need for these tools becomes even greater if a country is or has already embarked on hydropower development of its water resources. The reason is that hydropower or dam development has very serious impacts on the environment both human and water and other natural resources. But because of a lack of development in most African countries these vital environmental management tools have been lacking or are improperly administered to various projects plans and programmes for development or well being of the people. In the end, the impacts of these very projects, which are supposed to improve the livelihood of the people, tend to make them worse off. It is therefore important that knowing the benefits of SEA to water resources management, the necessary action will be undertaken by the relevant authority in Ghana in order to implement an SEA in this sector.

For a developing country like Ghana in Sub-Saharan Africa, the situation is very much comparable to most developing countries. The lack of various policy level programs or the improper implementation of such programs because of lack of funds or lack of well-trained personnel (see Sect. 21.3.2) makes the very idea unappealing to the local populace. Although Ghana has made headways in policy formulations on the environment since the formation of the EPA, there exist a lot of lapses on the ground. An example is the lack of sector specific policies, which has left the various sectors including the water resources sector without adequate guidelines and/or policy upon which impact assessment is based. As a conclusion to this paper various areas have been identified that require further development:

Capacity building and Public Participation: Capacity building has been identified as a very important factor in integrated water resource management issues and also EIA and SEA studies (see also Chap. 44). Generating interest in Ghana's water resource development PPP would ensure sustainability. Various ways of capacity building in Ghana can be the training of staff of EPA, of NGOs in the environmental sector and of local environmental officers at the community level. The Ghana Center for Democratic Development has been training local NGOs and personnel in different areas of capacity building. This is very commendable but more training should be extended to environmental issues. In applying these processes therefore to the water sector, it is suggested that aside the legal and institutional requirement, adequate measures need to be put in place to ensure public participation. Public participation in SEA processes in Ghana would have to cover gender issues, public hearings and public review procedures following the traditional system of governance in Ghana. Informing citizens of their opportunities to participate in the system of governance beside the government is an important role of the NGOs.

Projects in the water resources sector would have to have a comprehensive EIA based public participation. The locals, who would be affected by the program, should be approached at the policy formulation stage through the use of SEA. Further reviews should also incorporate responses by the local communities through chiefs and the local assemblies as found in the Ghana's decentralization scheme. Various forms of information and data should be simplified and made available to the local communities through the chiefs and head of families ('Abusuapanin' – local name for head of family, based on Ghana's traditional extended family system). This means that the public would have the opportunity to be involved even at the project bidding stage, in selecting the company whose plan takes a more proactive approach towards the environment.

Public participation could be improved in the local communities where the impacts of water or energy projects could be felt by introduction of regional and community based energy/water-planning activities. These community-based forums could be in the form of local talks with the chiefs and people in affected communities. Various environmental flyers and souvenirs could be passed around while villagers are also given the opportunity to show how they would for example design a plan to protect river banks from encroachment by farmers. On the policy front, Ghana should try to make public participation mandatory in the preparation of the scoping report and in carrying out the full EIA and SEA study as outlined in the proceedings of the regional workshop on SEA in southern Africa (Tarr 2003).

SEA in Ghana and the Decentralization Scheme: Generally, SEA has proven relatively difficult to apply to policy, especially highest level of government decisions and actions. Taking this into consideration, an SEA on water resources management in Ghana should have a more flexible approach. One way that this can be done successfully in Ghana is to apply the SEA to draft legislation or policy. This may involve a flexible procedure outlining a brief documentation of environmental effects, which such a policy may bring. Other suggested forms of SEA for Ghana water resources sector could be applied on fundamentals of decision making such as PPP, pending cabinet approval.

Any form of SEA developed by the EPA for the water resources sector especially when dam construction is involved should follow the hallmarks of the Ghana local subsets system. The people who would be affected by the project should have a say as to which type of resettlement they envisage and whether in their opinion the project would be of benefit to them and the country. This would also prevent situations where locals resettle at dam and water resource projects sites.

The various barriers, which exist between the different Ghanaian Institutions, should be bridged by creating a better awareness of the main purpose of an SEA in the water resource sector. This could be achieved by organizing training workshops in a national and regional context. Such workshops should address officials (from all the ministries and agencies involved in the water resource use sector), experts, NGOs and non-experts. NGOs take on various roles in this regard, including education campaigns, assistance to government ministries in forming policy, legislation and regulations, independent assessment of environmental and water resource conditions in the country. Attempts should also be made during these workshops to develop training manuals, materials and framework highlighting the SEA processes with special adaptation to development of water resources in Ghana. Table 21.3 shows that objectives and targets in water resource management could be set and monitored using indicators.

Potential Data Sources for EIA/SEA in Ghana: In the quest to develop a water management program for Ghana using the environmental tool SEA, one factor, which comes across is access to data on the environment (see Sect. 21.3.2).

Experience have shown that in low and middle income countries like Ghana, existing environmental data may have accumulated in different places because of lack of a centralized data repository.

Table 21.3 Environmental objectives in Water Resources Management

	Environmental Objectives	Set Targets (Water Resources)	Potential Indicators
1	Sustainable Management of water resources in Ghana	Minimization of impacts of water resource projects Reduction of in pollution surface waters	Number of protected water bodies Reduced pollution in selected water bodies
2	Maintaining the ecological quality of surface freshwater	Protection of catchment areas of water resources in Ghana	Number of river catchments touched (number of farms located on river beds, catchments)
3	Education on discharges into freshwater systems in Ghana	Prevention of environmental damage from water resource development projects e.g. dams	Concentration of major ions like sulfate, nitrate in water bodies

Another possible cause may also be poor referencing of sources of available environmental data. This situation is also the case in Ghana. However, it should be clarified that these data may exist and in this case typical sources are national and local government offices, technical and academic institutions, NGOs, individual members of the public, libraries and other interest groups. One of the challenges of impact assessors working in Ghana is finding out who has what data. An attempt to improve the situation was made by the EPA with support form the Overseas Development Aid (ODA). This led to the creation of the Ghana Historical Data on Environment (HIDEN) system – a bibliographical database on environment and the Ghana INFOTERRA – Directory of Environmental Information Sources under the NEAP. But these databases are not very efficient now and attempts should be made to improve and expand the resource. In addition, there is a need for EPA Ghana to attempt to compile a register of all EIA and SEA case studies in Ghana for training and research purposes.

The issue of the use of Ghana's water bodies for the production of hydropower has its benefits and negative impacts (see Table 21.1). However, maximum benefit of using the countries water resources would be realized when proper integrated water management practices are put in place. PPP in the water resource sector would have to be formulated taking the environment into consideration. The issue of sustainable management of Ghana's water resources would have to involve the various stakeholders in the water sector.

Until now most communities displaced by the construction of the Volta dam are yet to get connected to the national electricity grid. Strategic planning in Ghana must take into account the medium to long-term impacts of new projects such as the Bui dam. This would help mitigate some of the negative impacts of massive projects constructed in Ghana using existing water resources. Such strategic predictions are made possible by the management tool SEA, therefore Ghana should be able to integrate this option into its water resources management programs. New and sustainable water management systems in Ghana would require the understanding of the various social and governmental policies that affect the country's water resources.

Ghana will gain a lot if this new environmental tool of SEA is implemented in the water resource sector. Along with the already existing EIA procedure in the country, the problems of water resource development and economic development could be well managed for the benefit of the citizens.

References[1]

Amoyaw-Osei Y (1997) Developing strategic EA in developing countries: the Ghana experience. Paper to 17th annual meeting of the IAIA, New Orleans, USA

---

[1] Note that legislation mentioned in the chapter is listed at the end of the handbook in the consolidated list of legislation (Appendix 2).

Ayibotele NB (1995) Institutional and legal aspects of water management in Ghana. In proceedings of the National workshop on water quality, sustainable development and Agenda 21, EPA, Accra, Ghana

Dalal-Clayton B, Sadler B. (2002) Strategic environmental assessment: a rapidly evolving approach, IIED, London

Derban LKA (1984) Health impacts of the Volta dam: perspectives on environmental impact assessment, D Reidel Publishing Company, pp 121-132

Environmental Protection Agency, Ghana (1997) Environmental review report on Ghana, EPA Publication, Accra, Ghana

Environmental Protection Agency, Ghana (2003) Ghana's climate change technology need and needs assessment report, UN Framework for climate change, EPA Publication Accra

Ghana wildlife Department (1991) Wildlife bulletin, Bui National Park animal attractions, Accra, Ghana

Gordon C (1999) Public health impacts of the Volta Lake. In: Gordon C, Ametekpor J (eds) The sustainable integrated development of the Volta basin in Ghana. Volta basin research project, University of Ghana, Accra, pp 50-62

Government of Ghana (GOG) (1996) Ghana's water resources – management challenges and opportunities, The Water Resources Management (WARM) Study, World Bank/GOG Paper

Kalitsi EAK (1995) Management of multipurpose reservoirs. The Volta experience, Workshop on reservoir management, environmental issues in the Senegal River basin, Dakar

McKinney DC (2003) Basin-scale Integrated water resources management in central Asia, Third world water forum presentation paper, Regional cooperation in shared water resources in central Asia, March, Kyoto
ww.ce.utexas.edu/prof/mckkinney/papers/ara/integratedWaterManagement,
last accessed 10.04.2004

Mills-Tettey R (1989) African resettlement housing: a revisit to the Volta and Kainji schemes. Habitat International 13(4):78

Nelson Peter (2003) Information on SEA of Ghana overty Reduction Programme (GPRP) and on the Sustainability Test. Land Use Consultants, Bristol, UK

Opoku-Appiah S (1999) Indigenous economic institutions and ecological knowledge: Ghanaian case study. The Environmentalist 19(3): 219-222

Schmid E, Diaw K (1990) Effects of Volta resettlement in Ghana – a re-appraisal after 25 years, Hamburg

Smithsonian Institute (1974) Environmental aspects of a large tropical reservoir; case Study of the Volta lake

Tarr P (2003) Regional workshop on SEA in southern Africa, New partnership for Africa's development NEPAD and Environmental Assessment, African Journal of Environmental Assessment and Management 7:1-13

UNCED (1992) Agenda 21, United Nations Conference on Environment and Development, United Nations, New York

VRA – Volta River Authority (1995) Pre-EIA Report, Bui Dam

VRA – Volta River Authority (1996a) Kpong Hydroelectric Project-Resettlement Programme

VRA – Volta River Authority (1996b) The Volta dam of Ghana, Akosombo, Ghana

World River Review (1995) International Rivers Network (IRN). Internert address: http://www.irn.org/pubs/wrr/9511wrr.html, last accessed on 06.10.2003

22 SEA in Ukraine

Vyacheslav Afanasyev

Department of Environmental Planning, Brandenburg Technical University (BTU), Cottbus, Germany

22.1 Introduction

This chapter deals with SEA in Ukraine. At present there is a lack of information regarding SEA within the Ukraine and the aim of this chapter is to fill this gap. This is important because the so called Newly Independent States (NIS) of Countries in Transition (CIT), where Ukraine and other former Soviet Union states belong, are usually reviewed all together (since the initial legal bases and conditions were very similar), despite the significant distinctive features of each country.

Sect. 22.2 starts by describing the political, socio-economic and environmental conditions, as well as the traditional decision-making and implementation processes, which had been historically practised in Ukraine while it was part of the Union of Soviet Socialist Republics (USSR). This information is useful for understanding the existing and possible obstacles that can block SEA, and for developing a set of recommendations regarding SEA implementation. Sect. 22.3 provides an up-to-date situation by analysing current conditions relevant to the SEA sector. The legal base and existing requirements are described in detail to provide the reader with a clear view of the present state of affairs. Sect. 22.4 offers an overview of some of the main hindrances in the way of effective application of SEA in Ukraine with regard to transparent and sound environmental strategic actions, and European Union integration. The possible levels of SEA development in Ukraine are discussed in Sect. 22.5 Finally, in Sect. 22.6 conclusions and recommendations to improve the implementation of the SEA procedure are presented.

22.2 Historical Background of SEA in Ukraine

Having gained its independence in 1991, Ukraine, as well as other former USSR republics, is still very much influenced – in terms of legislative procedures and regulations – by traditions of the Soviet system.

Implementing Strategic Environmental Assessment. Edited by Michael Schmidt, Elsa João and Eike Albrecht. © 2005 Springer-Verlag

The field of environmental assessment is no exception here. Therefore, there exists the need to briefly turn to the origin of the SEA and EIA in the Soviet Union.

It can be said that prototype EA procedures existed in the USSR since the 1970s. Some of them included SEA elements, which were basically represented by planning rules and regulations (such as standards and procedures for conducting site investigation and obtaining necessary permits); expert review procedures (carried out by the special expert committees of appropriate ministries and acting as a co-ordination and control mechanism, which addressed environmental aspects of planned activities); and the system of environmental planning called "Territorial Integrated Schemes of Nature Protection" (addressing environmental issues at a more general strategic level). The latter were entirely internal government procedures, closed and non-transparent to other parties than the state itself. The procedures foresaw no independent checks, offered no defined responsibilities of participants and were inevitably characterized by a high degree of subjectivity and the discretion of officials in charge (Cherp and Bonde 2000). What more, such sort of expertise seldom affected the existing goals of strategic plans and was incapable of bringing any significant changes to the overall system of socialistic planning (Patoka 2000).

In order to overcome these hindrances, in the mid-1980s, a system of State Environmental Expert Reviews was introduced. These aimed at expanding the range of environmental assessment procedures to the extent that would cover all environmentally significant activities at both project and strategic levels; making them fully independent of the developers, and more transparent and accountable; and ensuring that environmental assessment results are actually taken into account in decision-making by giving the "conclusions" of these reviews the status of legally binding directives (Cherp 2000a, 2000b). The innovations, however, did not result in any considerable improvements, and the situation remained unchangeable until the early 1990s, when the so-called OVOS - Assessment of Environmental Impacts ("Ozenka Vozdejstviya na Okruzhayushuyu Sredu") was implemented as a mandatory procedure for all the project developers. At the strategic level, responsibilities of the proponents stayed unaffected until the dissolution of the Soviet Union in 1991, when the newly-emerged independent states took over the control of their own legislation and then later came up with the improved regulatory measures, more appropriate for the existing situation at the national level.

22.3 The Existing Conditions of SEA Application in Ukraine

Presently, the two major regulatory systems covering the implementation of SEA are in place in Ukraine. The first one is Assessment of Environmental Impacts (OVOS) – a "survivor" of the old times, largely burdened by the Soviet non-innovative approach, and the second is the so-called system of Ecological Expertise (EE) – a set of more dynamic and up-to-date regulatory measures. The latter encompasses scientific research and practices of appropriate governmental agencies, ecological bodies and public communities, based on intersectorial ecological research, analysis and assessment of strategic, project and other materials, whose realization can potentially– or does at present – negatively affect the state of the environment and health of the citizens.

Table 22.1. Comparison of SEA, EIA and EE

	SEA	EIA	EE
Level of Application	Strategic actions (Policies, programs and plans level).	Physical projects.	Physical projects, strategic actions, environmental legislative acts.
Legal status	Not always obligatory.	Usually obligatory.	Obligatory, but has a lot of drawbacks.
Stage of decision making	As early as possible in strategic actions development process. Must precede EIA.	Usually conducted after a project design is well-advanced.	Before the project implementation, in strategic actions in the development phase.
Scope of analysis	Can be broad in time and space. Can consider cumulative, synergetic and indirect effects.	Geographically specific. Focus tends to be on direct physical effects of the project.	The terms of EE are defined by law (up to 45 days, max up to 120 days), concentrates on both strategic and project levels.
Consideration of alternatives	Can define whether an initiative should go forward, plus where and what type of projects should be implemented.	Focus is primarily on how to design a project to reduce adverse environmental effects.	Issues of sustainability are set as primary goals for strategic level, and minimization of impacts for project level.
Procedures	Procedures must be adapted to decision-making process within the relevant ministry. They include: scoping, stakeholders identification, participation, technical study on key issues, the SEA findings and conclusions.	Standard, government-wide procedures. Includes screening, scoping, mitigation, public participation, monitoring and report preparation steps.	Procedure is approved at ministerial level; it includes 3 phases: 1: check of required data for EE; 2: analytical analysis and evaluation of the object of EE; 3: preparation of the final report.
Public participation	Legally foreseen and implemented.	Legally foreseen and implemented.	Legally foreseen, but poorly organized due to bureaucratic hindrances leading to poor public access to information.

EE aims at preparing the conclusions as to the conformity of the planned or the running activities to the norms and requirements of the legislation on environmental protection, rational use and recreation of natural resources, and provision of ecological safety (Thérivel 1997).

Interrelations in the field of EE are regulated by the Law on Ecological Expertise, the Law on Environmental Protection, Decree About Listing Types of Activities and Objects, Considered to Have High Ecological Risk and other acts of Ukrainian legislation. Table 22.1 offers a functional comparison between SEA, EIA and EE (in the form the latter is presently carried out in Ukraine).

The main tasks of EE are:

1. assessment of ecological risk and safety of a planned or running activity;
2. organization of comprehensive scientifically-based assessment of objects of Ecological Expertise;
3. examination for compliance of objects of Ecological Expertise to ecological legislation requirements, sanitary code, building code and construction regulations;
4. assessment of impacts of objects of Ecological Expertise on environmental conditions, human health and natural resources;
5. evaluation of efficiency, completeness, substantiation and adequacy of environmental protection measures and human health;
6. preparation of objective, well-grounded, sound reports of Ecological Expertise.

The legal guiding principles of EE are:

1. guarantee of safe environment for human lives and health;
2. balance between ecological, economic, medical, biological and social interests and consideration of public opinion;
3. scientific grounding, independence, objectivity, all-inclusiveness, alternativity, prevention and publicity;
4. ecological safety, intersectorial and economic reasonability of realization of objects of Ecological Expertise, planned or running activities;
5. state regulation;
6. legality.

EE applies to the projects of legislative and other normative acts, as well as pre-project and project materials, documentation on the implementation of new technics and technology, substances and products whose implementation can result in the violation of ecological standards, cause negative influence on the environment and endanger the health of the people. After EE is carried out, the expert agencies are obliged to announce their conclusions through the mass media. In order to assess the public opinion, Ecological Expertise initiators hold public hearings or open sittings (EBRD 1994).

In the EE Law three types of EE in Ukraine are defined: state, public and others. EE decisions are taken into account as any other state expertise; they are legally binding and must be fulfilled. Public and other forms of EE are voluntary activities; their conclusions have the status of recommendation and could be taken into consideration by State Ecological Expertise (SEE) in the process of decision-making about further realization of the examined project.

SEE is obligatory for the following activities:

1. state investments programmes, projects of the development and placement the labour force; development of selected branches of national economy;
2. projects of general layouts of town planning, schemes of regional planning, schemes of general layouts of industrial objects, schemes of layout of plants in industrial zones and regions, and other strategic pre-planning documentation;
3. investments projects, technical-economical founding and calculations, project and working plans for construction of new and reconstruction and/or technical re-equipment of running industrial units, documentation for re-profiling, conservation and elimination of industrial or other national economy units, including military structures, which can cause negative impacts on the environment;
4. projects of law-making and other legislative acts that regulate relations in ecological (including nuclear) safety, environmental protection and use of natural resources and activities, which can negatively affect the environment and human lives and health;
5. documentation on implementation of new equipment, technologies, materials, which could create a potential danger to environment and human health.

According to the decisions of the Cabinet of the Ministers of Ukraine, the Government of the Autonomic Republic of Crimea, local Radas (Councils) of public Deputies or their executive committees, ecological situations, which took place in selected areas and regions, running projects and industry units including military complexes, which cause significant negative impacts on environment and humans health, can all be subjects of EE.

The procedure of EE foresees the following activities:

1. Preparatory phase: auditing of presence and adequacy of all the required materials and requisites for objects of EE and creation of ecological-expert committees in accordance to legislative requirements;
2. Main phase: analytical analysis of materials for EE, in selected cases – on-site monitoring and examination, and use of the obtained results for comparative analysis and evaluation of the level of ecological safety, adequacy and efficiency of the objects of EE;
3. Final phase: summarizing of experts' assessment of the obtained information and the consequences of activities of objects of EE, preparation of the final report and its presentation to the interested organizations and persons.

Public ecological expertise is carried out on the initiative of public organizations in any field of activity that demands ecological substantiation. It can be pursued simultaneously with SEE, by means of enrolling representatives of the public to expert commissions, and Public EE groups. Public participation in EE procedures can be realized through expressing ideas through the mass media, filing written remarks, proposals and recommendations, enrolling representatives of the public to expert commissions, and Ecological Expertise groups.

Other forms of ecological expertise can be executed on the initiative of the interested juridical and natural persons, on the contract basis, by specialized ecological-expert bodies and organizations.

Conclusions drawn by the SEE include brief description of the ongoing or planned activity, its impact on the state of environment, people's health, and the level of ecological risk of the measures aimed at neutralization of these possible influences. The conclusions of the SEE provide recommendations as to how the given activity can be carried out in such a way that it meets the standards of ecological safety, environmental protection and rational use and recreation of natural resources. Based on this analysis, a project or a strategic action is either accepted in its present state, or rejected and sent back to the proprietor for revision and reconsideration.

22.4 Obstacles for SEA Application in Ukraine

Despite the fact that Ukraine is one the 36 countries that had signed Kiev SEA Protocol during the fifth Ministerial Conference "Environment for Europe" held in Kiev on 21-23 of May 2003, it is difficult to review existing SEA capacities. SEA – as a newly introduced legal framework in the EU Member States for ecological assessment – has resulted in appreciable positive changes of the overall situation, but there still exist a number of impediments to progress in this field. The problem is that it is very difficult to talk about obstacles of SEA application due to the fact that the legal base for SEA procedure has not yet been developed in the national legislation system. However, some of these obstacles are quite evident even in the existing conditions.

1. After being influenced for a continuous period of time by the Soviet approach to environmental assessment, Ukraine now faces the challenge of stepping away from this long-lasting system and introduce something qualitatively new and appropriate for current conditions of the country's development. What we see now, though, is rather copying western countries' environmental policies and trying to impose them in the setting, which to a large degree differs from that of the Western Europe. What more, these newly introduced policies stubbornly continue to be based on old methodologies and standards. All this makes environmental assessment in Ukraine inconsistent and inappropriate for the existing conditions.
2. While most of the employees in the newly formed ecological NGOs are young, progressive specialists, whose philosophy and education have nothing to do with the Soviet system, the majority of the governmental decision-makers in the field of environment are representatives of the "old school" lacking flexibility and not infrequently hostile to innovations (Yaroshchuk 2003). This inevitably leads to the conflict of ideological interests, makes the government and the NGOs rivals rather than partners and creates a serious preclusion against the achievement of the ultimate common goal, which is cooperation in the field of environmental protection.

3. Lack of democratic tradition in the country's politics results in low level of public participation in decision-making, and though, as it has been mentioned in section 3, the necessity of such participation is acknowledged by the law on Environmental Expertise, it seldom works in practice and even if it does, the public is allowed to take part only at later stages of assessment, when it is difficult to affect the process of decision-making to any considerable degree.

In order to overcome these obstacles, this chapter argues that it is crucial that legislation suitable for SEA be developed by the law-makers. Strategic assessment should be introduced into Ukrainian system of environmental, social and economic regulations at different levels. This will contribute greatly to the effectiveness of public involvement and cooperation between the different parties that are in charge of the introduction of SEA process.

22.5 Possible Levels of SEA Development in Ukraine

There are four main levels at which SEA should be implemented in Ukraine. Each of them can perform specific functions that are both relevant to this very level and contribute to the overall success of cooperation (see Fig. 22.1). The following activities at the different levels could be introduced in order to develop and strengthen SEA in Ukraine:

The first level (the level of the Ministry of Environment and Nature Protection of Ukraine) should foresee the creation of the Institute of Environmental Policy; the elaboration of the legislative background and regulations for the SEA introduction; and the development of mechanism for SEA application in Ukrainian environmental policy.

The second level (the level of Local authorities) a special Department on the transition to sustainable development should be created and access to the materials on environmental policy should be provided.

The third level (Environmental NGOs) informational resources and services should be provided; round tables, training courses on sustainable development and SEA, and debates organized; and public participation activities undertaken and improved.

At the fourth level (Environmental Faculties of Ukrainian Universities) care should be taken that special courses on SEA in Ukraine and research in Western countries be elaborated; participation for students in projects connected with transition to sustainable development be provided; and courses for administrative employees of environmental branch aimed to raising their level of proficiency be organized.

Fig. 22.1. SEA Application in Ukraine at Different Levels

22.6 Conclusions and Recommendations

As can be seen, to introduce meaningful SEA provisions in Ukraine the institutional resistance to it should be overcome. In order to do this, the concept of SEA needs to be widely redefined as a tool for informing decision-makers, rather than a part of environmental permitting procedures. This is indeed a great challenge, because in Ukraine environmental assessment per se is regarded more as an addition to the process of issuing the final conclusions of the expertise (Cherp 2000c). Presently, for the decision-makers it is basically "yes" or "no" that counts, while the underpinning of the final decision largely remains an issue of secondary importance. Joint meetings, debates, round tables will in this case help shorten a gap between old-traditioned governmental policy-makers and the leaders of the new generation (Fischer 1999 and 2002).

Appropriate skills are required within government departments and agencies as well as the private sector (e.g. industry, environmental consulting companies) and the NGOs. There also exists the need for adequate capacity building (both human and financial) in these sectors.

Crucial as well is that Ukrainian environmental legislation be harmonized with the one of European Union. Since Ukraine strives to integrate into the EU, it would provide for more opportunities for the implementation of international projects and ease the bureaucratic protractions.

Public participation should be introduced at the early stages of environmental assessment. This will benefit both project developers and those carrying out the assessment, and enable the vast public to feel involved in the decision-making process at strategic level.

As NGOs are generally more flexible and motivated than state policy-makers, their role in promotion and implementation of SEA Directive should be enhanced. NGOs should be given more advisory power in the legislation development. This will allow to save much of the financial resources, the lack of which the government usually names as one of the most considerable constraints to the restructuring of the environmental sector (REC 1998).

For the acceleration of environmental assessment procedures, extra funding from the donor countries should be considered. In these cases, donors always require a thourough assessment of environmental impacts and thus facilitate the recepient countries to develop effective implementation solutions for such evaluation.

As it can be observed from the analysis offered in this chapter, Ukraine has a very high potential in terms of sustainable development and innovative approaches in the implementation of SEA. The country still has a lot of substantial institutional and procedual difficulties to ovecome, but still, it is noticeably moving forward, which is proved by the fact that the existing environmental conditions are improving steadily and that Ukraine has already joined a number of important international nature conservation, biodiversity protection and sustainable development conventions, the most recent of them being the new Kiev SEA Protocol, which was signed by 36 countries and the European Community.

References[1]

Cherp A (2000a) EIA in the Russian Federation. In: Lee N, George C (eds) Environmental Assessment in Developing and Transitional Countries. Wiley, Chichester

Cherp A (2000b) Environmental Impact Assessment in Belarus. In: Bellinger E et al. (eds) Environmental Assessment in Countries in Transition. CEU Press, Budapest

Cherp A (2000c) Integrating environmental appraisals of planned developments into decision-making in countries in transition. In: Lee N, Kirkpatrick C (eds) Sustainable Development and Integrated Appraisal in a Developing World. Edward Elgar, Cheltenham

---

[1] Note that legislation mentioned in the chapter is listed at the end of the handbook in the consolidated list of legislation (Appendix 2).

Cherp A, Bonde J (2000) Legal Acts on Environmental Assessment in Countries in Transition. Internet Address:
EBRD - European Bank for Reconstruction and Development (1994) Environmental Impact Assessment Legislation: Czech Republic, Estonia, Hungary, Latvia, Lithuania, Poland, Slovak Republic, Slovenia
Fischer TB (1999) Benefits from SEA application – a comparative review of North West England, Noord-Holland and EVR Brandenburg-Berlin, Environmental Impact Assessment Review 19:143-173
Fischer TB (2002) Strategic Environmental Assessment in Transport and Land use Planning. Earthscan, London,
http://personal.ceu.hu/departs/envsci/eianetwork/legislation.html, last accessed 25.04.2004
Patoka I (2000) EIA in Ukraine: History and Recent Developments. In: Bellinger E at al. (eds) Environmental Assessment in Countries in Transition. CEU Press, Budapest
REC - Regional Environmental Center for central and Eastern Europe and Directorate for the Protection of Nature and Environment of Croatia DPNE of Croatia (1998) Policy Recommendations on the Use of SEA in the CEE/NIS Region Fourth Ministerial Conference "Environment for Europe", ARH.CONF/BD.17
Thérivel R (1997) Strategic environmental assessment in Central Europe. Project Appraisal 12(3):151-60
Yaroshchuk Y (2003) SEA policy as an instrument for sustainable development achievement: issues and perspectives for Ukraine; 5th International Conference Business Styles and Sustainable Development, Kyiv, 2-6.4. 2003

23 Importance of SEA in China – The Case of Three Gorges Dam Project

Cynthia Huang and Jennifer Yang

Brandenburg Technical University (BTU), Cottbus, Germany

23.1 Introduction

This chapter is about the need for the implementation of SEA in China and how SEA could make a difference for cultural and archaeological conservation, and use the Three Gorges Project as an example to support the idea that SEA implementation has great significance for wise decision making in the field of cultural and archaeological conservation.

This paper is divided into two parts. The first part is about the implementation of SEA in China by introducing current environmental situation and policies, and the development of SEA in both the mainland of China and Hong Kong. The second part is about the importance of SEA for China through a Case study of Three Gorges Project. It illustrates the cultural and archaeological loss for the current ongoing hydroelectric project, and analyses the concept of SEA at the cultural and archaeological aspects and points out the great importance of SEA for decision makers at strategic level. In conclusion, a series of recommendations are proposed using SEA to improve the cultural and archeological conservation for this case study.

23.2 Environmental Situation and Policies in China

China has been experiencing a rapid economic development since the open-reform in the 1980s. But along with the rapid growth of economy, grave environmental problems have arisen, which have greatly threatened the Chinese people's living circumstances as well as the base of the economic development in China. Air and water pollution are the two most serious among the major environmental problems China has.

Implementing Strategic Environmental Assessment. Edited by Michael Schmidt, Elsa João and Eike Albrecht. © 2005 Springer-Verlag

Due to severe air pollution mainly caused by the combustion of coal, eight out of ten of the most polluted cities all over the world are in China, and some areas are seriously suffering from acid rain as well (China Daily 2002).

23.2.1 Environmental Challenges for China

According to the State Environmental Protection Administration (SEPA) of China, in most Chinese industrial cities, nearly half of the river systems are polluted and the airborne concentrations of dust and suspended particulate matter usually exceed environmental standards by several hundred percent (SEPA 2003). Only 37% of the surface water has met the national standards and less than 20% of urban wastewater is treated, while per-capita water resources is only one forth of the world's average (Zhang 2003). But in fact, China is not a country possessing an adequate amount of natural resources for more than 1.3 billion people to use randomly without any appropriate plan and necessary control.

Table 23.1 shows the rank of China's main natural resources among 144 countries. The table indicates that even China has relatively high ranks of natural resources in the world, the per-capita possession of natural resources are very low because of the high population. The per-capita possession of cultural resources is also insufficient. Wasting limited natural and cultural resources will certainly risk the life of the Chinese people and the existence of the society. Therefore, it is an extreme urgency for China to take great efforts to reduce the existing environmental pollution and ensure the development in a sustainable way.

Fortunately China now has realised the seriousness of the environmental pollution, the enormous stress on its environment, natural and cultural resources, and the importance of environmental protection for the sake of the sustainable development of China in terms of environment, economy and society. At the press conference, which was held by the State Council Information Office and the State Environmental Protection Agency, Zhu Jianqiu, vice-minister with SEPA promised that the Chinese Government would firmly fight against pollution and consistently improve the country's environment to make it possible to achieve sustainable development (China Daily of 1 June 2002).

China has attempted to control and mitigate its grim environmental problems largely through administrative procedures and efforts to increase public awareness.

Table 23.1. The Rank of China's Main Natural Resources (Beijing Evening News of April 11 1999)

Resource	Rank	Rank of Per-Capita Possession
Freshwater coverage	6	55
Land area	3	10
Cultivated land area	4	126
Mineral resources	3	80
Forest coverage	8	107

The environmental protection legislation is a basic and necessary means that can establish a national level standard for environmental protection and ensure the effective implementation of all the activities on environment without causing any great harm to the environment.

23.2.2 Chinese Environmental Policies

Since 1979, six environmental protection laws, nine resource conservation laws and twenty-eight pieces of environmental administrative regulation have been issued in China. And over seventy regulations have been issued by environmental protection bureau and over nine hundred regulations by local government. About four hundred national environmental standards and twenty-nine sector standards have been built up. All these laws and regulations have provided standards for the environmental protection in China, and the duties and responsibilities of all the Chinese people when dealing with environmental problems. For example, Art. 5 para 1 of the Law on Water Pollution Control from 1 October 2002 stipulates: "All units and individuals are duty-bound to protect water resources and have the right to supervise and inform on acts that pollute China's water resources." What's more, there are ten core environmental policies in China (Zhang 2003):

1. Three Synchronisations
2. Environmental Impact Assessment (EIA)
3. Pollution levy on discharges in exceeding of standards
4. The target responsibility system for environmental protection
5. The system for the quantitative examination of comprehensive improvement of urban environments
6. The pollution discharge permit system
7. The system of centralised pollution control
8. Time-limited treatment of pollution
9. Cleaner Production
10. Total Emission Control

The EIA Law 2003, one of the ten core environmental policies, was issued by SEPA in 2003. This law reveals the great attention China has paid to Environmental Assessment including EIA and SEA, which has been considered as an important tactic to achieve sustainable development. The EIA Law 2003 was passed and announced on 28 October in 2002 and executed on 1 September in 2003. This law could be seen as a milestone on the development of both China's environmental protection and the Environmental Impact Assessment. It will have tremendous influence on environmental protection, China's basic national policy, and the sustainable development strategy. It includes four chapters: Main Principles, EIA on Plans, EIA on Projects and Law Responsibilities.

The first Chapter states the reasons and aims for drawing up this law. It defines EIA as a method and system to analyze, predict and evaluate the possible environmental impacts arisen from the carrying out of plans and projects, expound countermeasures and means to avoid or mitigate negative environmental impacts, and monitor continuously. It suggests the scope of plans and projects for the application of EIA as well as the criterions for the application of EIA. Last but not the least, it declares the support of the Chinese government provides for the development and implementation of EIA.

Chapter two and three are the main body of this law. They stipulate the details of the implementation of EIA separately on plans and projects. For instance, it depicts the relevant institutions responsible for carrying out EIA on plans and projects. Without the report of the implementation of EIA, no plan or project can be approved. In Chapter four, it clearly proclaims that those people who are directly in charge of organising the implementation of EIA must undertake the law responsibilities in case if the assessment of EIA is not in line with the plans and projects.

EIA has been developing in China for more than twenty years reaching a relatively high level in both research and implementation aspects. The release and execution of the Law of the People's Republic of China on Environmental Impact Assessment marks a significant step in the development of EIA in China. This law supports the implementation of EIA and the quality of the implementation is also assured because of this. As the most important existing environmental assessment tool in China, the EIA law will absolutely be the base and reference of the legislation of SEA in the near future.

23.2.3 Development of SEA in China (Excluding Hong Kong)

However, SEA is still in the beginning period in China. On the contrary, EIA has become one of China's basic environmental management policies after more than twenty years development, and has played a very significant role in the control of new pollutions and the protection of ecological environment. It is estimated that 93.6% national and local construction projects has been evaluated with EIA in 2000 (SEPA 2003). But EIA can hardly affect the macroscopic decision-making process and resolve the accumulated and indirect influences coming up from the construction activities. In order to meet China's environmental protection policy-prevention as priority, SEA must be adopted to avoid grave decision-making fault and environmental damage. Enough positive and negative experiences in China and other countries have completely demonstrated that SEA is an efficient method to evade decision-making fault and involve environmental protection into comprehensive decision-making process.

The legislation of SEA has attracted China's attention from the local government to the central government. In the fifth meeting of the Standing Committee of Wuhan Municipal People's Congress, some representatives have even proposed that the founding of the SEA system has been of great urgency. In 2003, the National People's Congress of China appealed for the formulation of SEA law to establish the SEA system.

In the Mainland of China, a lot of projects have been assessed with EIA, but only limited policies, plans and programs have ever been evaluated with SEA. One good example out of the case studies of SEA application is SEA of Wastewater Reuse Policy: A Case Study From Tianjin in China (Xu et al. 2003). This article introduces the general situation of SEA development in China, recommends a schematic SEA process undertaken in China, and illustrates a case study for SEA aiming at Tianjin Wastewater Reuse Policy (TWRP) based on the suggested SEA process. The schematic SEA procedure advised in this case study adds more experiences to SEA implementation in China. This experience would be a good example and reference for the following SEA implementation in China.

23.2.4 Development of SEA in Hong Kong

In contrast, in Hong Kong, the Special Administrative District of China, the EIA Law has been officially executed since April 1998 which is five years earlier than in the Mainland of China. The project EIA system has been well established with more than 15 years of experience. Besides the well-established project-level EIA system, decision-making level SEA has been applied to major policies and planning strategies at the strategic and regional level since 1989. The highest level decision making body is the Executive Council which is responsible for examining the submitted information on environment implications. This is seen as a key basis for SEA implementation.

For example, the earliest is the Port and Airport Development Strategy in 1989 and the latest the Studies on Future Strategic Growth Areas – North Western New Territories and North Eastern New Territories in 1999. Three completed SEA reports, Territorial Development Strategy Review, Third Comprehensive Transport Study and The Second Railway Development Study, have been published as well. Moreover, an ongoing SEA study Hong Kong 2030 "Planning Vision and Strategy" has been carried out since July 2001, and will last until 2030. SEA has influenced the formulation and selection of strategic and regional development options. Due to the proper SEA input to the decision making process, various potential environmental destructions and problems were avoided. All these measures have made enormous contribution to the final achievement of maintaining Hong Kong's environmental sustainability (Hong Kong's Department of Environment 2003).

In a word, SEA has great significance for sustainability of the economic and environmental development in the mainland of China and Hong Kong. Especially in the mainland of China, where are full of abundant cultural and archeological heritage, SEA could be particularly helpful in the conservation of the cultural and archeological sites. The activities at the cultural and archeological heritage sites can affect the sites' environment and the economic development of the surrounding area and even the whole society, especially a country like China with high-speed developing economy and serious environmental problems. The protection of the sites to ensure its sustainable development becomes very important and some more strategic measures are needed to reach this final goal. SEA could be one of the choices to achieve the sustainability of the sites' development and the society. In the following section, a case study of Three Gorges Project will be demonstrated to prove that SEA could be of great importance in the decision making stage for the conservation of the cultural and archeological heritage in China.

23.3 The Case Study of Three Gorges Project

The Yangtze River is the third mightiest river in the world in terms of length and water flow. It is nearly more than 200 km from Fengjie to Yichang, which goes through the gorges of Qutang, Wu and Xiling. This is the so-called Yangtze Three Gorges. A dam to harness its unstoppable energy was first proposed in the 1920s, but no action was taken until April of 1992, when the National People's Congress approved construction of the Three Gorges Project. The Three Gorges Project is the largest water conservancy project in the world, which comprises a dam, two powerhouses and navigation facilities (see Fig. 23.2). The dam, a concrete gravity type, is situated in Sandouping of Yichang City, Hubei Province, China.

Box 23.1 gives a brief summary of the project. The Project consists of three major parts: the large dam across the Yangtze River, the hydroelectric power station houses and the navigation structures. The spillway section is located in the middle of the riverbed, the power station houses are placed at the toe of the dam on each side, the navigation structures are situated on the left bank (Yangtze Cruises Inc. 2003).

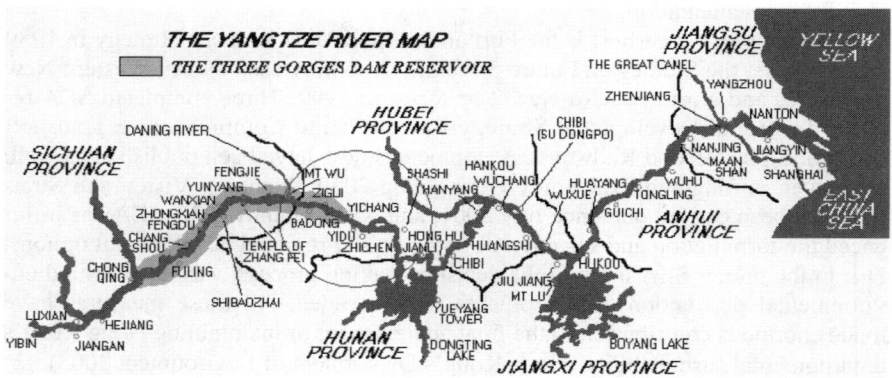

Fig. 23.1. Map of geographic location of the Three Gorges Dam Project (adapted from China Three Gorges Dam Cooperation 2002)

23.3.1 Features of Natural Landscape and Historic Relics

Natural Value: In general, the beauty of the Three Gorges lies in its majesty, perilousness, singularity and seclusion. This applies to every peak, shoal, cave and canyon in the Three Gorges. The mountains, water, springs, woods and caves compliment each others to achieve perfect harmony. The Three Gorges is one of the ten most famous scenery spots in China and highly praised as one of the world's famous tourism spots. Geologists regard that the Three Gorges has unique geological features that provide very important physical data for research.

Fig. 23.2. Layout of the Three Gorges Project (Yangtze Cruises Inc. 2003)

Box 23.1. Summary of the Three Gorges Project (China Daily Business Weekly of 3 Oct. 3 2000)	
Location	Sandouping, Yichang, Hubei province
Height	185 m
Length	3035 m
Expected investment	203.9 billion Chinese Yuan (US$24.65 billion)
Number of migrants	1.13 million
Installed power generation capacity	18.2 million kilowatts
Functions	Flood control, power generation, improved navigation
Construction timetable	1993-1997: The Yangtze River was diverted after four years in November 1997 1998-2003: The first batch of generators will begin to generate power in 2003 and a permanent ship lock is scheduled to open for navigation the same year. 2003-2009: The entire project is to be completed by 2009 when all 26 generators will be able to generate power.
Fund sources	The Three Gorges Dam Construction Fund Revenue from Gezhouba Power Plant Policy loans from the China Development Bank Loans from domestic and foreign commercial banks Corporate bonds

Cultural and Archaeological Value: Generally speaking, the various cultural relics with distinct characteristics scattered around the vast reservoir area include prehistoric cultural relics dating back to the Old Stone Age over two million years ago, cultural sites of successive ancient dynasties from the Xia Dynasty (21-16 Century BC) to the Qing Dynasty (1644-1911). It owns more than 60 archeological sites and ruins containing fossils and other evidence of extinct life forms and of human activity dating back to the Old Stone Age, and more than 80 archeological sites and ruins dating back to the New Stone Age. 470 sites and tombs dated from the Han dynasty (206 BC-220 AD) to the Six Dynasties period (386-589 AD) (Shao 2003). Six sites contain ancient inscriptions carved in stone that record dry-season water level, while ten other sites record the flood-season water levels. One of the best known of these sites is White Crane Ridge, which is an ancient indentation record of dry seasons in the history and may be the world's oldest hydraulic monitoring station.

Most importantly, the Three Gorges and surrounding area are one of the cradles of Chinese nationality and the center of the ancient Ba culture. Ba sites discovered here functioned as political, economic and cultural centers from Xia dynasty (21st – 16th century BC) to Qin dynasty (221 BC - 206 BC). There are more than 100 historic sites and tombs belonging to the Ba people who settled in the region about 4 000 years ago. A former curator at Beijing's National Museum of Chinese History describes the area as "the last and best place to study Ba culture (Kennedy 1999).

23.3.2 Impacts of the Three Gorges Project on the Natural Landscape

After completion of the Dam Project, the natural view will be influenced by the alteration in the water level, the river channel width and the flow velocity. Generally speaking, 5 % to 10 % of the Three Gorges will be inundated when the reservoir is built (Wu 1995). For instance, the gorge's feeling and the scenery of cliffs and canyons for the Kuimen and the other gorges nearby will be weakened. The whole Qutang Gorge and Wu Gorge, and the west part of the Riling Gorge will be affected by inundation and the east part of the Xiling Gorge will be affected by the outflow discharged from the reservoir. The rapids and the gorge's landscape in tributaries of the Yangtze River will be changed due to the rise of the water level, and the most beautiful scenery and cliffs of the "Little Three Gorges" on the Darting River will be heavily destroyed.

23.3.3 Impacts of Three Gorges Project on Cultural and Archaeological Relics

Chinese archaeologists and historians despair over the irreplaceable relics that will be sacrificed to flood control, electric power and navigation progress. In 1998, they drew up a list of the 1 282 most important sites including 441 sites of above ground cultural relics and 767 under earth cultural relics (Overhulse-King 2000).

The inundated relics date from New Stone Era to Qing Dynasty, including stone carvings made in the Eastern Han dynasty (25-220) and dozens of Buddhas and stone tablets carved with poems and prose dating back to Tang dynasty of the 7th to 10th centuries. Besides, hundreds of magnificent structures dating from the Ming and Qing dynasties – from the 14th to the early 20th centuries – were lost. Other important sites that will be inundated are the Buddhist monument at Single Pebble Village and the White Crane Backbone, which is one of the nation's key protected relics-in Fuling and some of the world's most important relics related to ancient transportation methods – the plank roads carved into the sides of the steep mountains.

One of the most important Ba sites is Lijiaba in Yunyang County, Sichuan. Lijiaba has been identified as one of the most important sites in the region and the major excavations were underway there during the 1998/99 archaeological season. This site has at least nine layers ranging in date from the Shang (ca. 2100-1700 B.C.) dynasty to the Qing dynasty (ca. 1800) (see Fig 23.3). The proper excavation and analysis of this site have to take many years under normal circumstances. However, the Culture Conservation of The Culture Conservation of the Three Gorges Project reservoir stops the excavation work. In addition, the Three Gorges Project will submerge 19 cities, 140 towns, more than 300 villages (Dai 2002).

Fig. 23.3. Findings in the ancient Tombs (Hubei TV 2004)

Thus about 1.13 million inhabitants in the dam area have to be relocated, and they will lose the roots of their culture. The important towns inundated are like Dachang (a 1 700-year-old town) and Fengdu – a city with two thousand years old culture.

23.3.4 Ongoing Culture Conservation

The waters of the Yangtze will rise more and more, and the new reservoir will reach its maximum depth by 2009. Facing the loss of natural and cultural relics and sites, the Chinese highest leaders address and support the preservation of these ancient relics and heritage. The State Council established a Three Gorges Project Construction Committee, to oversee a huge salvage operation, begun in June 2000, to record and preserve artifacts of immense historical importance. A fund of US $ 60 million out of the US $ five billion budget earmarked for resettlement was put toward the cultural rescue operation. (Huang 2002) The cultural rescue project aims to be completed by 2009 when the Three Gorge Dam is scheduled to go into operation and the whole area to be flooded. A feasible plan to conserve the most valuable ancient structures in the area was worked out. Based on the experts' ideas, some of them will be removed or rebuilt, some will have to be duplicated and the others will be put into museums.

In June 2003, plans were carried out to rescue Dachang, the 1,700-year-old town from flooding. To rescue and keep intact the town, a very bold and prudential project has begun to move the entire town, involving 38 classical residences and the integrated ancient walls. In order to keep the authenticity of the town, every brick, tile and beam was meticulously marked, torn down and shipped to the new location 5 km from the former town seat. The existing construction materials of these houses that remain useful have been treated against white ant infestation and rust, while substitutes have been purchased and used for some parts that were too decayed to be moved away. This project will cost over 30 million Yuan (US $ 3.61 million) (Xiao 2001).

Although many efforts from the government and social help have been put on the cultural salvage process in the Three Gorges area, the preservation of the stores of antiquities and the cultural relics have been unsuccessful. This might originate from lots of reasons such as shortage of funding, limited time schedule, insufficient research, theft, bureaucracy, etc. First of all, time is a big problem. The Three Gorges Project is supposed to end within the construction time schedule (1993-2009), however, archaeologists and historians believe that due to the great amount of the historical relics and archaeological sites in the Three Gorges area, the proper excavation and preservation operation will take 50 years to finish. Second, the rescue fund is only US $ 60 million, far from the fee of US $ 229 million, estimated by the archaeologists and historian. (Huang 2002) Although facing so many problems, to make way for the dam construction, archaeologists and historians have no choice but to save the most significant sites; however, conservation of these most important sites need more time, funds and research.

Take the White Crane Ridge for example, in order to protect this world's oldest hydraulic monitoring station, scientists have proposed two solutions. One is to make the models of White Crane Ridge Inscriptions, another is to build an underwater White Crane Ridge Museum. The latter is a pressure-free container, which is planned to build in the middle part of White Crane Ridge, where it contains 180 of inscriptions places. Due to the high cost and technique difficulty, the preservation work has ground to almost a complete halt.

Another threat now is theft. Thieves are robbing tombs and looting ancient sites, smuggling priceless objects out of the mainland to sell on the black market. It came as a real shock when a priceless bronze "money tree" unearthed from the Three Gorges area and dating back 2,000 years to the Han dynasty was sold for US $ 4 million in New York in 1996 (Overhulse-King 2000).

23.4 Three Gorges Project and SEA

23.4.1 Lessons Learnt from the Three Gorges Project

The greatest tragedy of the Three Gorges is the impact on culture and archaeology. Losing the irreplaceable cultural heritage and invaluable ancient treasures before they are properly surveyed and excavated, represents a major loss to the study of world civilisations in general, and to archeological work in China in particular. Looking back at the history of the project, the projects were planned some time ago – long before EIA or SEA. At the strategic stage, very few assessments were done to state goals for heritage and even fewer were made to the archaeological research and analysis. No sociologists, anthropologists or archeologists were invited to take part in the feasibility studies for the Three Gorges project. One reason for the poor results is that everything was done in a hurry without any regard to the archaeological and cultural heritage in parallel to other questions, i.e. the assessment focused on technical subjects such as the location of the future reservoir and the height of the dam.

In the 1990s, in the different stages of the project, national and international experts have been invited to propose a series of research and draw up EIA reports on evaluation and analysis on the potential ecological environmental problems resulting from the Three Gorges Project. Although the natural landscapes, archaeological and cultural heritages in the Three Gorges Area were regarded in the EIA report, and some endangered archaeological and cultural objects were dealt with, the structural or systematical levels of the archaeological and cultural heritage were rarely investigated. These are of strategic importance.

Because there is no assessment about Three Gorges Project impacts on cultural and archaeological heritage at a strategic level, it makes the cultural conservation later very difficult, especially for archaeological sites. Besides the already discovered relics, archaeologists believe that there must still be many potential unknown relics or heritage unearthed or some unexpected findings of archeology could be discovered in the reservoir area. Due to limited time and fund, almost all the archaeologists can do is to survey what will be lost, rather than actually excavate sites or undertake much in the way of salvage and preservation. On the other hand, after the dam construction, it is more difficult to pay more attention to the salvage operation. As a result, some relics have been handed over to thieves.

Therefore, conservation of cultural resources has strategic significance, and it is a prerequisite in the assessment of the project. It is necessary to point out cultural heritage of special interest very early and very clearly. A scientific assessment of archaeological and cultural heritage at a strategic level for a policy, plan and program is indispensable.

23.4.2 Archaeological and Cultural Heritage and SEA

It is inevitable to introduce the archaeological and cultural heritage in SEA. SEA should include such details for the purpose of assessing the significant direct and indirect effects of implementing the policy, plan or program on archaeological and cultural heritage at the strategic level. Where a policy, plan or program may have significant negative effects on culture and archaeology, it shall include a system for monitoring and mitigating such effect. Furthermore, it is also important to take the results of the assessment into account in future planning process.

It is probably easier to introduce cultural questions into the planning process at an earlier stage. The strategic questions concerning archaeological and cultural heritage might be:

- What are the influences on cultural or archeological structures?
- What elements will influence on ecological cultural or archeological systems?
- Are there risks of fragmentation, loss of quality in cultural or archeological areas?
- Will there be loss in unique cultural or archeological sites?
- Will there be loss of cultural or archeological interests?
- Will there be risk to lose customs and language?

23.4.3 Culture Conservation of the Three Gorges Project and SEA

If SEA was applied at the strategic level, the aims for SEA in the Three Gorges Project would contain different goals concerning sustainability. One would be a culturally sustainable dam construction system promotes archaeological and cultural development, support cultural and historical areas, takes care of regional characteristics and protects cultural and archeological sites. This means that the dam and the reservoir construction would be designed with regard for the surrounding natural landscape, archaeological and cultural heritage. These goals would be included in those considered even in strategic planning.

There would be a written report on the Assessment of Archaeological and Cultural and Heritage in SEA of the Three Gorges Project aiming to identify the cultural characteristics of archaeological and cultural heritage and cultural landscape on a structural or systematical levels, analyze the values and estimate the impacts on archaeological and cultural heritage, natural and cultural landscape of different levels, the closer area and the nearest environment, and describe what would happen to them and what that would mean for the future heritage. Probably after the SEA research, it would help the government to make a right decision. One alternative might be to launch a Three Gorges Archaeological and Cultural Heritage Survey and Conservation Project earlier before, as a precondition for the Three Gorges Dam Project; or the government might as early as the strategic level postponed the Three Gorges Dam Project to another 50 years to give sufficient time for the archaeological and cultural heritage survey and conservation at the Three Gorges area. Other alternatives might be the government redesign the dam and choose a new construction site perhaps in the upper reaches where there are less cultural heritages or reduce the dam scale. By this means part of the fund prepared for dam construction can be used first for the Archaeological and Cultural Heritage Survey and Conservation and the government still have sufficient time to get more funds for the dam construction in 50 years. Besides, the public participation and the international help are also very significant, thus a non-government international foundation 'The Three Gorges International Fund for Cultural Heritage' could be set up '. Imagine what will happen if this comes true. Here is a hypothesis:

Archaeological Survey: A comprehensive systemic accurate record should have been made for the entire archaeological heritage at the Three Gorges area since this had never done before. If those valuable Ba sites still existed, further and proper archeological investigation of this area would help all the archeologists to gain an understanding of the intriguing Ba culture, about which little is known at the present. Studying other archeological sites and tombs in the Three Gorges area would allow the archaeologists to figure out how the Chu culture (started the 11th century BC) and the Qin culture (began in 278 BC) developed and spread in the region.

Cultural Heritage Conservation: White Crane Ridge, the world's oldest hydraulic monitoring station would have been saved with the help of modern techniques. For example, the underwater museum for its conservation may be successfully designed if more efforts and fund are put on the feasibility study and scientific research, and more international technique aids are received. The stone carvings, dozens of Buddhas and stone tablets carved with poems and prose would have been preserved in the museum telling people the irreplaceable information about the region's history if scientific studies of a proper incision or rubbings are carried out. Hundreds of temples, houses and bridges, set against beautiful natural landscapes, dating from the Ming and Qing dynasties and providing a wealth of information about China's traditional cultures, would have been preserved and removed or rebuilt if more researches, time and money are allowed.

23.5 Conclusions and Recommendations

One of the main reasons leading to environmental pollution and cultural and archeological destruction is the decision-making fault. The environmental impacts and cultural destructions resulting from policies, plans and programs (PPP) are usually macroscopic, accumulated and long-term latent. If there is no consideration of environmental problems and cultural protection at decision-making stage, it will bring about disastrous consequences. The application of SEA can minimise the environmental and cultural devastations occurring from PPP.

From some specialists' point of views, sustainable development has developed into the direction and destination of the development of the whole human race society, as well as the action principle and guide of different countries governments for developing economy and promoting the society progress. SEA is seen as a tool to guarantee that environment and natural and cultural resources can be protected and used in a sustainable way in the economic development procedure of a society. It is a link between environment and development. It has made great efforts on promoting the environmental and cultural heritage protection in collaboration with the development of economy in different countries.

Thereby, SEA is one of the most efficient methods to balance the economic development and environmental and cultural heritage protection for attaining sustainability for the whole society. SEA has more significance for a fast developing country like China, which has plenty of cultural and archeological heritage and suffering from environmental pollutions at the same time. Take the Three Gorges Project for instance, it will bring many benefits in terms of electric power, flood control and navigation. On the other hand, it has caused very large negative impacts on the cultural and archaeological heritage.

It is therefore inevitable to seek for a SEA covering an assessment on archaeological and cultural heritage for decision-making. It is of great significance to promote the process of the implementation of the SEA in China. The current implementation of project EIA and the existing EIA Law are insufficient for the sustainable development of the society. On the basis of the current implementation of project EIA, the application of decision-making level SEA should have no delay in China.

References[1]

Beijing Evening News (April 11, 1999) Beijing Evening News Press, Beijing. Internet address: http://www.rrojasdatabank.org/chinnv02.htm, last accessed 12.05.2004
China Daily Business Weekly (October 3, 2000) Three Gorges Dam Project. Beijing
China Daily (June 1, 2002) China Daily Press, Beijing. Internet address:
http://www.chinadaily.com.cn/, last accessed 12.05.2004

---

[1] Note that legislation mentioned in the chapter is listed at the end of the handbook in the consolidated list of legislation (Appendix 2).

Dai Q (2002) Yangtze! Yangtze! Probe International's Three Gorges Campaign, Beijing

Environmental in China: China's Current Environmental Situation (2001) DC Consulting. Internet address: http://www.dckonsult.com/CurrentSituationinChina.htm, last accessed 15.05.2004

Hong Kong's Department of Environment (2003) Examples of Strategic Environmental Assessment (SEA) in Hong Kong. Internet address: shttp://www.epd.gov.hk/epd/english/environmentinhk/eia_planning/sea/ebook1.html, last accessed 26.08.2004

Hubei TV (2004) Display of Archaeological Relics at Three Gorges Area. Internet address: http://www.hbtv.com.cn/sanxiaxs/page/hq/zs.htm, last accessed 03.05.2004

Huang W (2002) The Countback of the Cultural Relics Preservation at the Three Gorges. Changjiang Daily

Overhulse-King J (2000) Tragedy in the Three Gorges. History of Ancient World (to ca. 499), Archaeological Magazine

Shao D (2003) Saving the Cultural Relics of the Three Gorges. China Daily of April 9

SEPA – State Environmental Protection Administration of China (2003). Internet address: http://www.zhb.gov.cn/eic/651333096108457984/index.shtml, last accessed 08/9/2004

Wu Y T (1995) Environmental Impact Statement for the Yangtze Three Gorges Project. China Science Press

Xiao X (2001) The Rescue of A 1700-year-old Town Towards to Entirely Remove. The Xin Wen Morning Newspaper

Xu H, Zhu T, Dai S G (2003) Strategic Environmental Assessment of Wastewater Reuse Policy: A Case Study From Tianjin in China. Journal of Environmental Assessment Policy and Management (JEAPM) 5(4):503-521

Yangtze Cruises Inc. (2003) Three Gorges Project / Three Gorges Dam

Zhang S (2001) FDI and Environment: Situation and Challenges in China. Peking University, China, May 8-9. Internet address: http://www.epe.be/fdi/fdidocuments/presentations/zsqfdi.pdf, last accessed 15.05.2004

Part V – Methodologies for SEA and Public Participation

Part V of the Handbook focuses in more detail on different methodologies for SEA and on the requirements and methods for public participation. The "environmental assessment" part of SEA cannot take place without appropriate methodologies. Chapter 24 describes a range of techniques that can be used in SEA. It begins with an analysis of what makes a good SEA tool, goes on to briefly review a series of commonly-used SEA tools, before discussing three SEA tools in more detail: impact matrices, Geographical Information Systems (GIS), and causal effect diagrams. Chapter 25 discusses other methodological approaches to SEA within a decision-making process context. The chapter outlines the differences between SEA and Project EIA, and identifies methodological tools that are particularly suitable to meet the requirements of SEA (as opposed to Project EIA).

Chapters 26 and 27 both deal with cumulative impact assessment. One of the key reasons for justifying the use of SEA is that it is supposed to be able to deal with cumulative impacts better than Project EIA. The SEA Directive requires an analysis of the likely significant impacts on the environment, including secondary, cumulative and synergistic impacts. Chapter 26 proposes a new approach to strategic level cumulative impact assessment focusing on the *resource* rather than the *action*. Chapter 27 evaluates how best to handle transboundary cumulative impacts in SEA, with particular reference to the experience in the Unites States.

The following three chapters address cultural issues and public participation - a crucial component of the SEA process. For example, according to the SEA Directive, the public needs to be given an opportunity to express their opinion on the *draft* of the plan or programme and respective environmental report (i.e. before the plan and programme is adopted). Moreover those opinions have to be taken into account in the final decision about the plan or programme. Chapter 28 discusses the elements that have to be considered within a cultural impact assessment procedure within SEA, and it proposes the use of indicators that use "cultural integrity" as a key criterion. Chapter 29 evaluates the requirements and the methods for Public Participation in SEA, and Chapter 30 discusses public participation for SEA within a transboundary context (using a pilot study for a cross border regional plan of the "Upper Lusatia – Lower Silesia" region - a zone where Germany, Poland and the Czech Republic intersect).

The last two chapters are more unusual in their methodological approaches. Chapter 31 suggests the development of quantitative SEA indicators using a thermodynamic approach. This chapter focuses on technology assessment and argues that developing sustainable technology is crucial. The chapter shows how quantitative information for SEA of technological options can be generated from a thermodynamic analysis. Chapter 32, on the other hand, proposes a structural and functional strategy analysis for SEA, based on causality diagrams. This last chapter focuses on two key aspects of SEA implementation: evaluating the quality of strategic action itself and discovering the (planned and non-planned) effects that may arise from the introduction of the strategic action.

24 Tools for SEA

Riki Thérivel[1] and Graham Wood[2]

1 Levett-Thérivel sustainability consultants, Oxford, England
2 School of the Built Environment, Oxford Brookes University, England

24.1 Introduction

Impact prediction, evaluation and description in strategic environmental assessment (SEA) cannot take place without some kind of framework and methodology: without "tools". This chapter describes a range of tools that have been used in SEA worldwide. It begins with an analysis of what makes a good SEA tool, goes on to briefly review a dozen commonly-used SEA tools, before discussing three of these in more detail: impact matrices, Geographical Information Systems (GIS), and causal effect diagrams. It concludes with a brief discussion of future trends in SEA tools.

24.2 What Makes a Good SEA Tool?

In project EIA, the more detailed and comprehensive the assessment is, the better. Mathematical modeling, risk assessment, scenario generation and sensitivity testing are all potentially useful EIA tools: they are "scientific" and expert-driven, aim for rigor and comprehensiveness, and require a detailed understanding of the environmental baseline.

But is this appropriate for SEA? A typical strategic action might cover thousands of hectares, lead to hundreds of projects, and last ten years or more, during which time a wide range of new technologies may or may not emerge. Yet a typical policy- or plan-maker might have only between 10 and 100 person-days in which to carry out the SEA. The planner may be able to predict the *types* of projects that the strategic action could lead to, but not *what* projects. In these circumstances – large area, little data, little time, and lots of uncertainty – one looks for tools that fulfill some key requirements.

Implementing Strategic Environmental Assessment. Edited by Michael Schmidt, Elsa João and Eike Albrecht. © 2005 Springer-Verlag

SEA tools *must*:

- fit into the decision-making timetable for the strategic action being assessed;
- help to improve the strategic action: make it more sustainable, more robust and easier to implement;
- be able to cope with various types of uncertainty; and
- identify and assess the key impacts of the strategic action and identify key changes needed.

Ideally SEA tools *should* also:

- take account of cumulative and indirect impacts, since a key benefit of SEA is that it has the potential to consider such impacts better than does project EIA;
- suggest alternatives and mitigation measures;
- allow alternatives to be compared;
- be robust and defensible; and
- be understandable by decision-makers, technical experts and the public.

What SEA tools *cannot* be is comprehensive and detailed.

24.3 SEA Tools

Tables 24.1 and 24.2 list a dozen tools that have been used in SEA worldwide and that fulfill these criteria. Table 24.1 and Fig. 24.1 summarize the key situations in which the tools could be used, though they should be seen as indicative rather than definitive. Table 24.2 summarizes the SEA stages at which the tools could be used. The rest of this section aims to give a 'taster' of what these tools can do, and provide a starting point for further reading. It is based heavily on Thérivel (2004).

Expert judgment involves one or more experts considering the relevant issue, possibly using formal approaches such as the Delphi technique. Expert judgment is quick, cheap, requires no specialist equipment, can lead to innovative solutions, copes with partial and unquantifiable information, and fosters innovation and information sharing. On the other hand, it has the potential for bias depending on the experts involved, and is perceived as unscientific.

Public participation: The public are informed, consulted or fully involved in identifying, predicting, assessing and mitigating the strategic action's impacts. This helps to ensure that the SEA is comprehensive and possibly makes the strategic action easier to implement. However it takes time and resources, can be dominated by specialist interests, and it is often difficult to interest the public in strategic level issues (Audit Commission 2000, Wilcox 1994).

Impact matrices show the sub-components of the strategic action (or alternatives to the strategic action) on one axis and environmental topics or objectives on the other axis. The resulting cells show the effect of the sub-components or alternatives on the environmental objectives.

Table 24.1. Key situations in which SEA tools could be used

Tool	tool works for...						tool copes with...			
	Policy level*	Program level*	Large area	Small area	Land use plan	Sectoral plan	Incomplete data	Uncertain data	Qualitative data	Few resources
Expert judgment	✓	✓	✓	✓	✓	✓	✓	✓	✓	✓
Public participation	?	✓	?	✓	✓	✓	✓	✓	✓	✓
Impact matrix	✓	✓	✓	✓	✓	✓	✓	✓	✓	✓
Quality of Life Assessment	✓	✓	✓	✓	✓	✓	✓	✓	✓	?
Maps /GIS	?	✓	✓	✓	✓	✓	?	?	?	?
Land unit partitioning anal.		✓	✓	✓	?	✓	✓			
Causal effect diagrams	✓	✓	?	✓	✓	✓	✓	✓	✓	✓
Modeling		✓	?	✓	?	✓	?	?		
Scenario/sensitivity anal.	?	✓	✓	✓	?	✓		?		
Multi-criteria analysis		✓	?	✓	?	✓		✓	✓	?
Risk assessment		✓	?	✓	✓	✓	?	✓	?	
Compatibility appraisal	✓	✓	✓	✓	✓	✓	✓	✓	✓	✓

* The tool's effectiveness at the plan level will be between that for policy and program level
Key: ✓ fully   ? partly   (blank) not

Impact matrices can be used to summarise and present the results of other analyses or as assessment tools in their own right. They are simple to use and transparent, but do not cope well with indirect or spatially based impacts. Sect. 24.4 gives more detail.

Quality of Life Assessment identifies what matters and why in an area. Its core idea is that the environment, economy and society provide a range of benefits for people, and that it is these benefits that need to be protected and/or enhanced. QoLA involves describing the relevant area; identifying the benefits/disbenefits that the area offers; analyzing how important each benefit/disbenefit is, to whom, why, whether there is enough of it, what (if anything) could substitute for the benefits, and what management implications this has for the benefit/disbenefit. This allows a list to be devised of things that any management of that area should achieve. QoLA acknowledges the complementary role of experts and local residents, and promotes enhancement of an area, but it is anthropocentric and can reduce the environment into its subcomponents (Countryside Agency et al. 2002).

Geographical information systems are a combination of a computerized cartography system that stores map data, and a database management system that stores attribute data. Links between map data and attribute data allow maps of the attribute data to be displayed, combined and analyzed with relative speed and ease. Longley et al. (2001) give a more detailed introduction to GIS, and Rodriguez-Bachiller and Wood (2001) explain the role of GIS in EIA. Sect. 24.5 gives more detail.

Table 24.2. SEA stages during which SEA tools could be used

Tool	Describe baseline	Identify impacts	Predict impacts	Evaluate impacts	Compare alternatives	Identify cumulative/ indirect impacts	Ensure coherence	Propose mitigation	Public participation	Monitoring
Expert judgment	✓	✓	✓	✓	✓	✓	✓	✓		✓
Public participation tools*	✓	✓	✓	✓	✓	?	✓	✓	✓	?
Impact matrix			✓	✓	✓	✓		✓	✓	✓
Quality of Life Assessment	?	✓	✓	✓	?			✓	✓	?
GIS	✓	✓	✓	✓	✓	✓		✓	✓	?
Land unit partitioning anal.			✓	✓	?	?		✓		
Causal effect diagrams	✓	✓	?	?	✓			✓	?	
Modeling		✓	✓	?	✓	?		✓		
Scenario/sensitivity anal.		✓	✓	✓	✓	?		✓		
Multi-criteria analysis				✓	✓			✓	?	
Risk assessment		✓	✓	✓	?			✓		
Compatibility appraisal							✓	✓	?	

Key: ✓ fully   ? partly   (blank) not
* e.g. visioning, Planning for Real, workshops

Land use partitioning analysis: Linear infrastructure cuts across land and fragments it: this affects nature conservation, tranquility, landscape etc. Land use partitioning aims to describe and assess this fragmentation by comparing the size and quality of areas of non-fragmentation before and after a programme of linear infrastructure construction. It deals with a topic that would otherwise be poorly considered and gives a good visual representation of impacts, but it requires GIS capability and much data, and its application is limited to only a few subjects. (European Environment Agency 1998).

Causal effect diagrams aim to identify the key cause-effect links which describe the pathway from initial action to ultimate environmental outcome. In doing so, they can also identify assumptions made in impact predictions, unintended consequences of the strategic action, and possible measures to ensure effective implementation. They are particularly useful for identifying cumulative impacts. Sect. 24.6 gives more detail.

Modeling involves making assumptions about future conditions with and without the strategic action, and calculating the resulting impacts. Models typically deal with quantifiable impacts: air pollution, noise, traffic, etc. Many models have evolved from EIA techniques and are computerized. Models are perceived to be objective, 'scientific' and rigorous.

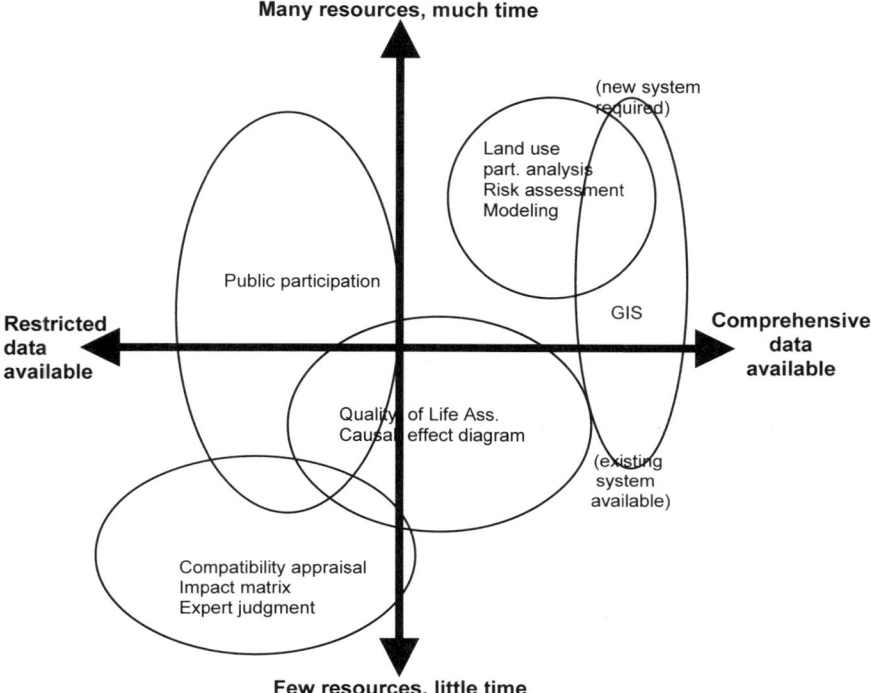

Fig. 24.1. Resource and data requirements of SEA tools

On the other hand, they are limited to impacts that can be quantified/modeled, can require large amounts of data, are often complex 'black boxes' that do not encourage public participation, and tend to promote project- rather than strategic-level thinking (Hyder 1999).

Scenario/sensitivity analysis: The impacts of a strategic action, or the relative benefits and disbenefits of different options often depend on variables outside the strategic action's control: for instance future noise levels in one authority could depend on whether an airport is built in a neighboring authority. The impact of a strategic action can be forecast and compared for different scenarios that describe a range of future conditions – sensitivity analysis – to test the strategic action's robustness to different possible futures. Sensitivity analysis aims to reflect uncertainties and provide ideas for reducing uncertainties, but can be time and resource intensive (Finnveden et al. 2003).

Multi-criteria analysis (also called multiple attribute analysis or multi-objective trade-off) analyses and compares how well different alternatives achieve different objectives, and helps to identify preferred alternatives. It involves identifying assessment criteria and alternatives; scoring how each alternative affects each criterion; assigning a weight (value of importance) to the impact; and aggregating the score and weight of each alternative. The scores and weightings are then multiplied and the results added up for each alternative. The alternative that scores most highly 'wins'. MCA is easy to understand, can be used in a variety of settings, acknowledges that society is composed of diverse stakeholders with different goals and values, and reflects the fact that some issues 'matter' more than others. However it can lead to very different results depending on who determines the weights and scores (Economics for the Environment 1999; Glasson et al. 2004).

Risk assessment estimates the risk that products and activities cause to human health and ecosystems. It involves identifying possible hazards, estimating their frequency, and identifying and analyzing their likely impacts. Risks – frequency times impacts – can be phrased as the probability of a specified event, eg 1 in 100 chance of a flood in area X in a given year; or as consequences, eg 10 homes flooded per year. Risk assessment can be used to compare alternatives on the basis of the risk that they cause, but only considers one aspect of the 'environment' (i.e. risk/safety), often extrapolates the risks at high dose levels of a pollutant to low dose levels with consequent uncertainties, and its results can vary greatly depending on the assumptions made (Economics for the Environment 1999).

Compatibility appraisal aims to ensure that the strategic action is internally coherent and consistent with other strategic actions. An *internal compatibility matrix* plots different components/statements of the strategic action on one axis and the same strategic actions on the other axis: matrix cells are filled in by asking 'is this statement compatible with that statement or not?'. An *external compatibility matrix* plots the strategic action against other relevant strategic actions: matrix cells are filled in by listing those statements of the strategic action that fulfill the requirements of the other strategic actions, or explaining how the evolving strategic action should take the requirements into account. Compatibility appraisal clarifies trade-offs and is easy to understand, but it is subjective (ODPM 2003).

Three of the most versatile of these tools are now discussed in more detail: impact matrices, GIS and causal effect diagrams. They are focused on because they are easy to understand, can take account of a wide range of data types (including incomplete or qualitative data), and can be used at various stages of SEA. They also represent a range of sophistication, different levels of resource use (see Fig. 24.1), and different strengths: one is particularly good for spatial analysis, and one for analyzing cumulative effects.

24.4 Impact Matrices

Impact matrices can be used as a basis for discussions – either between experts or with the public – about the impacts of a strategic action and possible mitigation measures. They can also be used as a presentational tool, to summarize the results of other more detailed studies about a strategic action's impacts.

Table 24.3 shows the simplest use of an impact matrix: to identify and mitigate the impacts of various sub-components of a strategic action. In such a case, a matrix would be drawn up with the strategic action's sub-components on one axis and various environmental/sustainability topics or objectives on the other axis. For each sub-component, a first question would be whether it is clearly written: if not, it might be possible to rewrite it to make it clearer. Each matrix cell would then be filled out, sub-component by sub-component, noting whether the sub-component:

- has a negative impact on the objective/topic (-): if so, the impact might be mitigated, for instance by rewriting the sub-component or adding a different sub-component.
- has a neutral (0) or positive impact (+): if so, it might be possible to rewrite the sub-component to make it even more positive.
- has an uncertain impact (?): if so, further information may need to be collected before the assessment can be completed, and the strategic action finalized.
- has an impact that depends on how the strategic action is implemented (I): if so, it may be possible to re-write the strategic action to ensure that it is implemented positively.

Obviously other symbols could be used instead, e.g. tick / cross, smiley/frowney face, red/amber/green, quantitative data from other studies. The cells could also be filled in with text which describes the type and magnitude of the impact, assumptions made during the assessment, etc.

Changes made to the strategic action as a result of this process would be documented, as at the final row of Table 24.3. These are the "mitigation measures" required by the SEA Directive. The process of filling in the matrix should focus on proposing appropriate mitigation measures.

Impact matrices can also be used to compare alternatives. In such a case, the different alternatives would be listed instead of the sub-components, and the final row of the matrix would explain what the preferred alternative is and why. For instance, based on the limited number of objectives in Table 24.3, alternative D looks more sustainable than alternative C.

Finally impact matrices can be used to identify the cumulative impacts of several sub-components of one strategic action, or several strategic actions.

Table 24.3. Example of partial impact matrix

Environmental/sustainability objective	strategic action sub-component or alternative			
	A	B	C	D
Conserve and enhance biodiversity	+	-	0	+
Promote the health of all residents	-	-	-	I
Maintain and enhance soil quality	0	+	0	+
Maintain and enhance the quality of ground and surface waters	+	I	-	+
...				
Proposed changes to the strategic action (sub-components):				

In such a case, for each environmental/sustainability topic, one would "read across" all of the sub-components' (or strategic actions') impacts on that topic to identify cumulative impacts. In Table 24.3, for instance, no one sub-component has a particularly negative impact, but all of them together have a cumulative negative impact on health. In summary, impact matrices are useful in SEA because they have a wide range of applications, are easy to understand, and help to identify mitigation measures.

24.5 Geographical Information Systems

Because GIS can spatially manifest many environmental problems, it has been much vaunted as a tool for EIA. However in practice the use of GIS in EIA has been limited for all but the largest and most expensive development proposals. Problems with data access and limited resources mean that developers have been reluctant to invest in GIS for one-off projects, and undertaking ongoing environmental simulation and decision support has not been a priority for other environmental organizations or regulators (Rodriguez-Bachiller 2004). The longer term nature of environmental planning and management at the strategic level should encourage a greater willingness to invest in and exploit GIS technology. Box 24.1 summarizes the advantages and disadvantages of GIS in SEA. The rest of this section outlines possible uses of GIS in SEA.

If an up-to-date GIS incorporating a spatially referenced database of relevant environmental information is available for the SEA, then at the most basic level the GIS can be used to generate *maps*, including overlay maps to illustrate the spatial coincidence of environmental features or 'themes'. Such maps could support the baseline assessment for the SEA and facilitate impact identification, hence serving to shape the SEA's scope. Basic spatial measurements (distance, area, length) and descriptive statistics relating to attributes contained in the database can all be quickly determined within a GIS, providing data that can be analyzed by other SEA tools such as multi-criteria analysis (MCA) or impact matrices.

Buffering and *clipping* provide the next level of analytical sophistication for SEA. Buffers delineate areas defined by a selected fixed distance around specified map features. In SEA, a buffer could be used as a proxy to represent a potential impact zone relating to development associated with the strategic action e.g. a zone of potential noise impact. Alternatively, the buffer might define a zone of constraint around sensitive locations such as areas of conservation value. Buffers can then be used to clip other map themes in the GIS: the clip acts as a 'cookie cutter', slicing out data contained in a second map e.g. a buffer used as a basis to represent noise effects might be used to clip a map of residential land use to identify all the homes potentially impacted upon, and (assuming the GIS database contains the relevant data) enabling an assessment of the socio-economic characteristics of the population affected. Again such functions could help to determine the scope of the SEA in terms of the spatial boundaries of the study and the issues that it should address, and could also provide data for comparing alternatives.

Box 24.1. Advantages and Disadvantages of GIS for SEA

Advantages:

- Helps to draw together multidisciplinary data to underpin and inform the SEA.
- The map making facility and basic analytical functions aid the description of the baseline environment and assist with scoping.
- Provides a powerful means for visualizing complex data, including "3D" visualization. This can provide an accessible means of presenting information for public participation and stakeholder involvement.
- Helps to identify constraints based on land use and designated areas.
- Can feed into other SEA methods such as MCA or impact matrices.
- Can identify cumulative impacts in respect of their spatial coincidence.
- Assists with the identification and evaluation of alternatives.
- Has powerful scenario simulation potential, particularly when combined with other complementary technologies.
- Provides a platform for drawing together and analyzing monitoring data once the strategic action has been implemented.

Disadvantages:

- Has significant data requirements. If the data is not readily available in a GIS compatible format, building up a useful GIS database is time consuming and requires specialist knowledge. This is particularly the case where GIS is combined with other modeling tools.
- An SEA would require staff that possesses knowledge of what analysis a GIS can provide and an ability to undertake at least some basic GIS analysis.
- Not all data available for an SEA will have a spatial expression, or it simply may not exist with a spatial coverage for the area under consideration.
- Where data sets of different scales are combined (e.g. when performing map algebra) the dataset with the smallest scale will determine the resolution that can be used, leading to a loss of information in larger scale datasets.
- Some map outputs can create the impression of 'sharp' boundaries which do not fully capture the more gradual or fuzzy boundaries between entities in reality e.g. soil maps, land cover.
- Does not readily capture the more dynamic aspects of a landscape e.g. the importance of migratory routes.
- Has a limited ability to identify impact interactions beyond simple spatial coincidence.
- The shape of some features, e.g. coastlines, may vary over time.
- Bringing together disparate datasets with multiple ownership and keeping them up-to-date so that they can be drawn on for a range of SEA (and subsequent project level EIA) requires institutional innovation and political will.
- Outputs can create an impression of greater certainty than is really possible. Any GIS output depends on the quality of the input data, and interpretation of the analysis must be informed by the context and limitations of these data.

GIS also provides a platform to develop more *complex models* that can support baseline assessment, scenario simulation and the identification and evaluation of alternatives in SEA. Many environmental processes or phenomena are spatially continuous in nature e.g. measurements of air quality could be made at any location. Clearly monitoring everywhere is impossible and measurements made at a limited number of locations must be relied on. Where there is a sufficient distribution of monitoring points over the study area, GIS can be used to *interpolate* between the values in order to generate a continuous surface or contour map of the parameter that might be used as part of the baseline assessment.

The ability of GIS to perform "*map algebra*" (also known as cartographic modeling) has perhaps the most widespread potential application in predicting impacts and identifying spatial alternatives in SEA. At the simplest level, map algebra can be used to identify areas that satisfy a series of predetermined rules or criteria using Boolean logic (and, or, not) which may be either inclusive (e.g. all areas within 2km of a rail link) or exclusionary (e.g. all areas that are more than 5km from a designated conservation area), or some combination of the two. By grading areas within map themes according to the degree to which they symbolize a constraint, this approach can be refined to produce a score which represents a gradient of relative attractiveness for potential implementation of the strategic action (see Box 24.2).

Box 24.2. SEA of UK Offshore Wind Energy Development

The UK government views offshore wind energy production as invaluable in meeting its pledge to supply 10% of electricity needs from renewable sources by 2010, rising to 20% by 2020 (DTI 2003). The government carried out SEA of the 'second round' of licensing offshore wind energy development, targeted within three strategic areas: the Thames Estuary, Greater Wash, and Liverpool Bay. The SEA involved a GIS driven spatial analysis, followed by a risk-based assessment of the likely effects of licensing, including cumulative impacts. The GIS analysis comprised an overlay mapping exercise to identify critical and relative constraints to windfarm development. Critical "maximum" constraints were:

- Existing and planned licensed areas for aggregate extraction and waste disposal;
- Oil and gas structures (including pipelines) and associated safety zones
- Sites of importance for cultural heritage (principally known ship wrecks)
- Cables (electricity, telecommunications etc) on the seabed
- Existing shipping and navigation lanes
- Areas proposed for windfarm development following the first round of licensing

"Relative" socio-economic and environmental constraints were:

- Shipping
- Fishing effort (in hours)
- Shellfish areas
- Habitats of marine conservation interest (designated / potentially designated)
- Fish spawning areas
- Fish nursery areas
- Seascape sensitivity

Box 24.2. (cont.)

Each relative constraint was rated using a score between 0 and 3. For example, fishing effort was rated as 0 where none occurred, 1 ("low constraint") for less than 500 hours per annum, up to 3 ("high constraint") for areas where more than 5000 hours of fishing occurred per annum. By combining the maps for each factor within a GIS it was possible to identify areas subject to several constraints to development (high score) and those with fewer constraints (lower score).

The GIS analysis was limited to factors for which sufficient baseline data existed and which were available for GIS mapping. For instance, whilst the initial intention had been to map ecological data such as species distribution and migration, "the availability, extent, format, resolution and quality characteristics of existing datasets have meant that for the greater part, this has not been feasible or practicable" (DTI 2003). To address this limitation, and in recognition of the fact that not all decision factors have a strictly spatial manifestation, a risk-based analysis of impacts associated with two proposed windfarm development scenarios (the 'most likely' and 'maximum credible' scenarios) was carried out. Impacts were quantified where possible, otherwise qualitative description was used based upon expert judgment and literature review.

The risk approach was expressed as a product of (i) the consequence of an impact (reflecting the interaction between an activity and a receptor) and (ii) the likelihood that the impact will occur: risk = consequence x likelihood. Again, a scoring system was used, this time based on different levels of consequence and likelihood of occurrence, resulting in overall impact significance scores of between 1 (minor consequence and unlikely to occur) through to 25 (serious consequences and certain to occur).

Where there were insufficient data and / or lack of knowledge to characterise the risk, the risk was described as "unknown". Outputs from the risk analysis included the identification of areas requiring further study and their relative priority; mitigation measures; and monitoring programmes for impacts considered to exhibit acceptable levels of uncertainty, but requiring further evidence to justify this assumption.

Similarly, where the GIS layers are graded according to the likely degree of impact, using map algebra a total impact score can be created, representing the cumulative spatial coincidence of impacts of different types. Where the resources and expertise are available, a full blown spatial version of MCA can be implemented (e.g. Carver 1991). This involves weighting the relative importance of the different criteria based on stakeholder perspectives, standardizing the data types represented in the map themes, and combining them to develop a map representing a relative 'index of attractiveness'. GIS allows the weighting of criteria to be altered with relative ease, and their spatial implications to be rapidly visualized.

Another widespread modeling capability of GIS is the ability to generate *Digital Terrain Models* (DTMs, also known as Digital Elevation Models, DEMs). These can be used to visualize topography, and as a basis for further maps describing slope and aspect. DTMs can be overlain with land use maps to create 'virtual' models of the landscape, and also provide the basis for predicting the area over which a proposal will be visible using the viewshed function common in many commercial GIS packages. For SEA involving potential development with a major visual effect, e.g. a programme of windfarm developments or electricity transmission lines, the DTM could be used to simulate the visual implications of various scenarios.

All of the examples outlined above involve modeling activities that take place entirely within the GIS environment. However, some of the most advanced recent developments in SEA-related modeling involve the use of *GIS coupled with other computer based tools*. At the most basic level, GIS can provide data to an external modeling package where further analysis is undertaken before importing the results back to the GIS for additional spatial analysis and mapping. For instance, a DTM may be used to provide terrain information to an air pollution dispersion model, outputs from which are then imported back to the GIS for interpolation and visualization of the predictions (often in the form of contour maps overlain on land use). Recently there has been a rapid growth in the number of commercially available software packages that perform this sequence of analysis in a seamless operating environment.

GIS can also be used in tandem with virtual reality modeling software e.g. World Construction Set. Whilst such impact modeling tools have their most obvious application in EIA, there is no reason why they cannot be used for scenario simulations of strategic actions. For example, Dolman et al. (2001) describe the use of GIS and virtual reality for visualizing "whole landscapes" under various landscape management scenarios designed to maximize amenity, environmental and biodiversity benefits.

Dynamic modeling of spatial systems using GIS can also be used in SEA, for instance for policy scenario simulation. The European Commission's (2004) project MOLAND (Monitoring land cover / land use dynamics) is an urban growth modeling tool based on the concept that the land use at a point (or "cell" in a raster GIS) is directly influenced by the neighboring land use. A series of "transition rules" are defined to quantify how a cell is affected by its neighbors with other land use functions. Using GIS spatial data inputs covering land use, suitability, zone maps (based on designations, protected areas etc), transport accessibility, and socio economic maps, MOLAND aims to model the implications of various policy changes over time e.g. the effect of altering transport networks or adjusting policy areas identified in zone maps.

Advances in *artificial intelligence* whereby the interaction of people, organizations, or ecological entities can be simulated to identify the 'global' effect of their interaction with their environment (so called agent based modeling) also hold promise as future GIS-SEA tools (e.g. Gimblet 2002). Finally, GIS can be used as an *integrating framework* for all of the stages of the SEA process (e.g. João and Fonseca 1996).

In summary, GIS has a range of uses in SEA, from the most basic mapping capabilities and description through to 'state of the art' scenario simulation involving a range of complementary technologies. Falling software costs, greater availability of digital data, and more widespread awareness of GIS (particularly as it has migrated to the desktop PC), all reduce the barriers to its adoption for SEA. The approach and level of GIS usage will reflect the resources and timeframe identified during SEA screening, and this in turn will largely reflect organizational and institutional arrangements.

24.6 Causal Effect Diagrams

Causal effect diagrams – also called network analyses, consequence analyses or causal chain analyses – aim to describe the web of relationships of natural systems. They involve, through expert judgment or possibly public participation, drawing the direct and indirect impacts of a strategic action as a network of boxes (inputs, outcomes) and arrows (interactions between them). Fig. 24.2, for instance, is a partial cause effect diagram showing the links between more money for buses (input) and air pollution (outcome), via two ways of spending the money: paying for buses to be converted to cleaner fuel and paying for new rural bus routes (obviously other ways of improving bus services also exist).

Causal effect diagrams can help to *identify constraints and assumptions* in impact predictions. Fig. 24.2 shows that paying for new rural bus routes will reduce air pollution only if people transfer some car journeys to bus journeys, which may not happen unless other measures are put in place (e.g. higher parking costs in town). Instead, paying for buses to be converted to cleaner fuel does not rely on other measures or assumptions. Fig. 24.2 suggests that, if improved air quality was the only outcome sought, then paying for buses to be converted to cleaner fuel would be more efficient and thus preferable than paying for new rural bus routes: the causal effect diagram can thus also help to *choose between alternatives*. On the other hand, new rural bus routes may have all kinds of additional outcomes, for instance improved access to services for people without use of a car (positive) or reduced exercise for people who switch from bike to bus (negative): these could also be added into Fig. 24.2. Causal effect diagrams are also one of the few tools that help to *identify cumulative impacts*. In Fig. 24.2, for instance, converting buses to cleaner fuels and fewer vehicle journeys in rural areas cumulatively help to reduce air pollution.

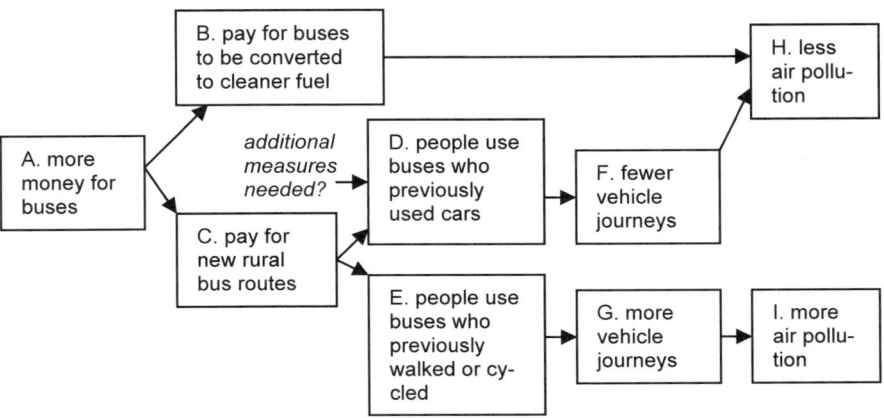

Fig. 24.2. Example of partial causal effect diagram

Causal effect diagrams are easy to understand and thus useful in public participation. They are transparent and not resource-intensive. They can help in scoping and as inputs to other SEA techniques such as modeling. However they can miss important impacts if not done well, do not deal with spatial impacts or impacts that vary over time, and the resulting diagram can become very complex.

24.7 Conclusions and Recommendations

In sum, a wide range of SEA tools exist, from simple 'expert judgment' to complex computer-based models. The best ones fulfill all the requirements of a good SEA tool: they can be carried out quickly, help to improve the strategic action, focus on key impacts, cope with uncertainty, take account of cumulative and indirect impacts, suggest and compare alternatives, are robust and are easily understandable. Several – for instance matrices and GIS – are a combination of analytical tool and presentational technique. Other SEA tools exist, normally variants of those discussed here, for instance overlay maps that do not use GIS, or vulnerability analyses which combine GIS and multi-criteria analysis.

Which tools to use in a given SEA depends on the type of strategic action, the time and resources available, the need (or not) for detail and rigor, and the availability of data. As a general rule, one should use the simplest possible tool consistent with the precautionary principle. It may be useful to start with simple tools, and move on to more detailed ones where necessary, for instance to "zoom in" on problem areas.

SEA is still in its infancy, and is often carried out by EIA experts. A key challenge will be to ensure that SEA remains strategic, and not a compilation of EIA-level detailed information which requires impossibly large quantities of detailed and comprehensive information. The right choice of tools will go a long way to achieving this, and to ensuring that SEA is timely, well-focused, and useful in decision-making.

References[1]

Audit Commission (2000). Listen up! Effective Community Consultation. Internet address: http://ww2.audit-commission.gov.uk/publications/pdf/mpeffect.pdf, last accessed 01.04.2004
Carver SJ (1991) Integrating multi-criteria evaluation with geographical information systems. International Journal of Geographical Information Systems 5(3):321-339
Countryside Agency, English Nature, English Heritage, Environment Agency (2002) Quality of Life Assessment. Internet address: http://www.qualityoflifecapital.org.uk, last accessed 14.04.2004

---

[1] Note that legislation mentioned in the chapter is listed at the end of the handbook in the consolidated list of legislation (Appendix 2).

Department for Trade and Industry (DTI) (2003) Offshore wind energy generation – Phase 1 proposals and environmental report, Prepared by BMT Cordah Ltd for DTI. Internet address: http://www.og.dti.gov.uk/offshore-wind-sea/process/, last access 15.04.2004

Dolman PM, Lovett A, O'Riordan T, Cobb D (2001) Designing whole landscapes. Landscape Research 26(4):305-335

Economics for the Environment (1999) Review of Technical Guidance on Environmental Appraisal, report to the DETR. Internet address: www.defra.gov.uk/environment/economics/rtgea, last accessed 15.04.2004

European Commission Institute for Environment and Sustainability (2004) MOLAND (Monitoring land cover / land use dynamics). Internet address: http://moland.jrc.it, last accessed 13.04.2004

European Environment Agency (1998) Spatial and Ecological Assessment of the TEN: Demonstration of Indicators and GIS Methods, Environmental Issues Series No 11, Copenhagen. Internet address: http://reports.eea.eu.int/GH-15-98-318-EN-C/en/seaoften.pdf, last accessed 15.04.2004

Finnveden G, Nilsson M, Johansson J, Persson A, Moberg A, Carlsson T (2003) Strategic Environmental Assessment methodologies: Applications within the energy sector. Environmental Impact Assessment Review 23: 91-123

Gimblet HR (2002) Integrating Geographic Information Systems and Agent Based Modelling Techniques for Simulating Social and Ecological Processes. Oxford University Press, New York

Glasson J, Thérivel R, Chadwick A (2004) Introduction to Environmental Impact Assessment, 3rd edn. UCL Press, London

Hyder (1999) Guidelines for the Assessment of Indirect and Cumulative Impacts as well as Impact Interactions. Report prepared for European Commission DG XI, Brussels

João E, Fonseca A (1996) The role of GIS in improving environmental assessment effectiveness: theory vs. practice. Impact Assessment 14(4):371-387

Longley PA, Goodchild MF, Maguire DJ, Rhind DW (2001) (eds) Geographic Information Systems and Science. Wiley, Chichester

ODPM – Office of the Deputy Prime Minister (2003) The Strategic Environmental Assessment Directive: Guidance for Planning Authorities, www.planning.odpm.gov.uk, last accessed 15.04.2004

Rodriguez-Bachiller A (2000) Geographical Information Systems and Expert Systems for Impact Assessment, Parts I and II. Journal of Environmental Assessment Policy and Management 2(3):369-448

Rodriguez-Bachiller A (2004) Expert Systems and Geographic Information Systems for Impact Assessment. Taylor and Francis, London

Rodriguez-Bachiller A, Wood G (2001) Geographical Information Systems (GIS) and EIA. In: Morris P, Thérivel R (eds) Methods of Environmental Impact Assessment 2nd edn. Spon, London

Thérivel R (2004) Strategic Environmental Assessment in Action. Earthscan, London

Wilcox D (1994) The Guide to Effective Participation, prepared for the Joseph Rowntree Foundation. Internet address: www.partnerhsips.org.uk/guide, last accessed 15.04.2004

25 Methodological Approaches to SEA within the Decision-Making Process

Beate Jessel

Institute for Geoecology, University of Potsdam, Germany

25.1 Introduction

SEA sets minimum procedural standards for an early comprehension of environmental issues in certain plans and programmes. Significant impacts on the environment shall be assessed and then considered in the *process* of working out and deciding over the plan or program. To be effective SEA should thus start as soon as possible and be fully integrated into the plan-making process with significant contributions at each decision making stage. The legal basis for this request is provided by Art. 4 para 1 of the SEA Directive laying down that SEA "shall be carried out *during the preparation* of a plan or programme and *before* its adoption or submission to the legislative procedure" (accentuation in italics by the author). Thus SEA shall become part of a decision-making process and methodological approaches have to be developed that suit the general aims of SEA as well as its procedural character.

Further, Art. 4 para 3 of the SEA Directive takes into account the fact that the assessment will often be carried out at different levels of the hierarchy and that it is therefore necessary to avoid duplication of work. Generally there is a tiered forward planning process which starts with the formulation of a policy or strategy at the upper level, is followed by a plan or programme at the second stage and finally results in a concrete project which then has to be submitted to an EIA (see Chaps. 1 and 6). Fig. 25.1 shows that such a tiered system can apply to different spatial and/or administrative levels (such as the national, regional and local level) but also to different kinds of plans and actions (such as land-use plans on different administrative levels, infrastructure plans or investment programmes). The different steps of assessing the environment within such a decision-making process have to be built upon each other in a coherent way and it has to be decided carefully at which process level certain matters are evaluated most appropriately. Further it has to be made sure that there is a working linkage providing that Project EIA can build up on the results of SEA.

Implementing Strategic Environmental Assessment. Edited by Michael Schmidt, Elsa João and Eike Albrecht. © 2005 Springer-Verlag

Considering these premises this chapter outlines the differences between SEA and Project EIA in procedural and in methodological aspects, and identifies methodological tools that in general are suitable to meet the resulting requirements of SEA. Some approaches that might be suitable for SEA and also demonstrate the integration of SEA and EIA into the decision-making process are exemplified:

- The identification of possible alternatives for dumping-ground sites,
- the identification and assessment of possible corridors and alignments for new roads and
- landscape plans which can serve as an information base for SEA.

The chapter discusses to what extent these approaches still meet the requirements of SEA and its necessary link to EIA and to what extent further methodological development is necessary.

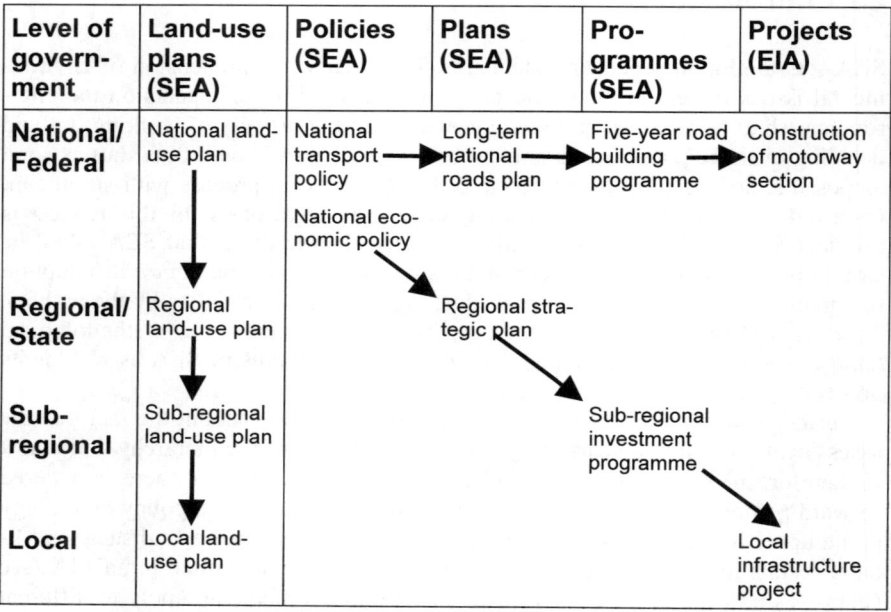

Fig. 25.1. Sequence of strategic and project-level actions and their correspondence to SEA and Project EIA (Wood 1995)

25.2 Procedural and Methodological Demands on SEA and EIA – a Comparison

So far in the literature the discussion on the implementation of SEA has been focused on procedural and legal aspects. Concerning the *methods* for prognosis and weighing of environmental impacts on this strategic level, there is yet a lack of experience in most European countries. For Sweden, for instance, Eriksson (2002 p.1) states that "[…] here is still a lack of common methodology for dealing with strategic issues" and that this therefore "[...] is a problem for an efficient application of SEA." Noblé (2002 p.4) also states very little consensus on the methodological principles of SEA (whereas most experts would agree on the overall concept of SEA) and gives a number of references to other authors who hold the same view.

Table 25.1. Comparison of selected aspects of SEA and Project-EIA (modified from Jessel and Tobias 2002)

	SEA (Plans and Programmes)	EIA (Projects)
Subject	Land Use or spatially relevant expert plans and programmes, which set the framework for development consent of future projects	Usually precisely defined projects on a certain site
Point of time	On the preliminary, programmatic level, accompanying often several (hierarchical) steps of the decision-making process	During approval procedure
Character of the instrument	Accompaniment of a decision-making process on a political and/or programmatic level -> emphasis on political and strategic questions	Comprehensive structuring of the decision process, systematised by environmental issues -> emphasis on technical questions
Procedural aspects:		
Scoping	Usually more complex and time-consuming because multiple activities, a larger area and/or a higher variety of alternatives are involved	Usually less complex, because the range of relevant activities and resulting impacts are more concise
Assessment of alternatives	Broader range of strategic alternatives (e.g. different traffic systems or waste treatment concepts) and spatial alternatives (e.g. site search over a wide area or identification of possible line alternatives)	Site and line alternatives with rather small modifications in site delineation or line course, often limited to issues of technical design

Table 25.1. (cont.)

Concernment and involvement of the public	The public is often concerned in a rather indirect way; resulting decisions with relevance for the society Aim: Broad public consent or sustainable compromise within different social groups Form of involvement: Use of special mediation and moderation techniques, visioning events and special media for spead of information, sometimes focused on relevant stakeholders	Individuals are directly concerned Aim: Project acceptance and project optimisation Form of involvement: Concise EIA-document, availability in public of the document, hearings, usually direct involvement of the public (open to everyone)
Criteria of assessment	For evaluation of plans and programmes assessment factors are legally determined only partly, criteria are often non-technical and qualitative	Basics are legally determined, partly very detailed and/or quantitative criteria are available
Change of the subject to assess during procedure	Likely to happen	Possible (but substantial modifications during approval procedure are unusual)

Content:

Impacts and impact area	Often wider range of impacts, larger impact area; not always essential that assessments have to be site-specific	Mostly smaller range of impacts, smaller and/or more clearly defined impact area; assessments have to be site-specific
Spatial und temporal reference	Partly not determined in detail, step-by-step realisation of plans or programs	Often known in detail
Database, methods	Aggregated environmental knowledge, mix of methods	Detailed environmental knowledge
Prognosis of impacts on environment	Rather undetailed, use of verbal and qualitative techniques (in particular scenario technique) Variety of future scenarios Prognosis has high uncertainty	In detail, with regard to rather exact interrelations, qualitative, as far as possible quantitative Possible future developments are rather manageable Uncertainty of prognosis exists
Secondary and interrelated effects, cumulative effects	Impacts on other fields, such as human ecology, social and economical aspects have to be considered	Mostly interrelations within natural sciences

Despite a similar basic structure, there are important differences between SEA and Project EIA in terms of procedural as well as methodological aspects. These are described in Table 25.1 and can be summarised and characterised by the following key points:

- The SEA technique needs to stress the management issues in a stronger way, as SEA should be active from the very beginning of the decision process. Methods for Project EIA are often adjusted for site-specific information and local impacts. They thus tend to be more detailed and comprehensive whereas SEA has to deal with strategic issues on a higher level of abstraction and to identify those key impacts of the strategic action that set the course for the following steps of the decision-making process (see also Chap. 24).
- A proper database as SEA demands a work focused on highly aggregated environmental data.
- Concerning the higher uncertainty of prognosis. The prognosis in the SEA-procedure will often not be able to make quantitative analyses but remain in the realm of qualitative statements (Finnveden et al. 2003; Hildén and Jalonen 2002). Mostly, the likely consequences of different alternatives will be presented as scenarios verbally described.
- SEA allows a more effective analysis of cumulative effects, whereas the individual case-by-case approach in project EIA makes it difficult to do so.

The compilation given in Table 25.1 primarily refers to SEA for plans and programmes, but many aspects are also true for SEA carried out to assess politics. Nevertheless for this field of action further aspects would have to be added, referring to the highly aggregated and even more abstract level on which political decisions usually are taken.

Despite the differences described in Table 25.1 it will also be necessary to establish a formal linkage between SEA and its related EIA to assure that SEA findings are carried over to the EIA level (Risse et al. 2003). The reasons that this linkage between EIA and SEA is not as easy as it seems can be found not only in the aspects mentioned in Table 25.1 but also in the fact that SEA is a relatively new approach whereas EIA is longer in practice with already well-established methods: SEA is still on the way to define its own status in the decision-making process and to get independent from the heritage of EIA.

Thus there actually is a big temptation to simply adopt EIA methods to SEA. But the differences between SEA and EIA that have been pointed out make clear that it is not possible to simply apply EIA methods to SEA and that a special methodological framework is needed (Chap. 24). Although we have known for several years (starting at least from 1996 when the draft of the SEA Directive was presented by the European Commission) that SEA will have to be introduced into the national planning systems, such a framework does not exist yet. Indeed, the lack of a structured framework that specifically addresses SEA requirements is often considered as one of the main constraints for the application of SEA (Noblé 2002).

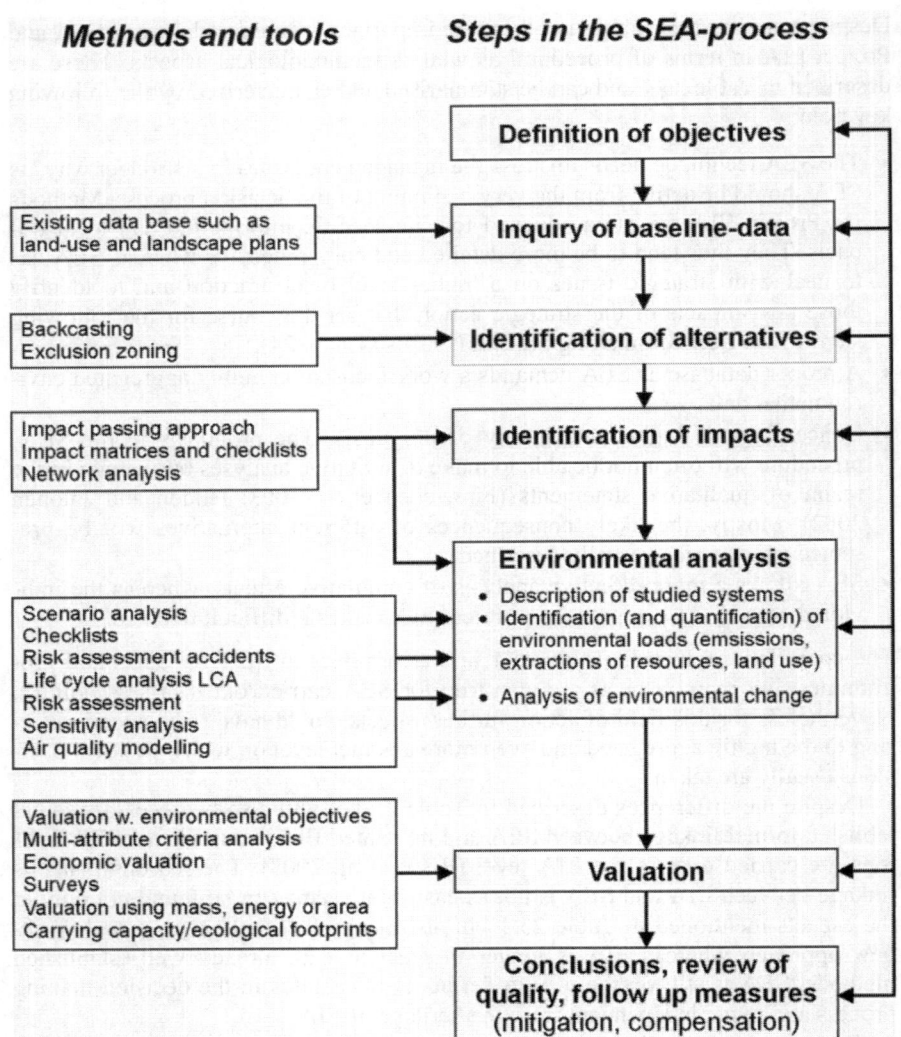

Fig. 25.2. Different tools suitable for different steps of the SEA-process (based on Finnveden et al. 2003 and Office of the Deputy Prime Minister 2003). The * denotes the tools that are dealt in this chapter. See also Chap. 24, which describes some of these approaches in detail

In an early study which supplied a survey of the practice of SEA in Great Britain, Thérivel (1998) stated that existing SEA were characterised by their qualitative approach, the relative brevity of the reports and low costs. Actually many of these results were approved by Short et al. (2004). In their survey they could further demonstrate that though being widely accepted within the administration current SEA practice thus exerts only little influence on plan content and objectives.

Sound and comprehensible methods are one important prerequisite to stimulate real change in the strategic alignment of a plan or programme. However, what is specific about methods for SEA? SEA methods should be more strategically oriented allowing the identification and comparison of alternatives, the systematic search for critical (negative and positive!) aspects, the identification and valuation of impacts on a highly aggregated level, the suggestion of mitigation strategies at an early stage and even the improvement of the strategic action (see Chap.1). As Fig. 25.2 shows they should also allow iterative feedbacks within the planning process going back to the earlier steps, probably even to the definition of the planning objectives, and challenging the results of these steps if this proves to be necessary. In this context Noblé (2000) points out that a common understanding of SEA which is primarily based on its strategic characteristics asks not only for the preferred option but probably also the preferred attainable ends rather than predicting the most likely outcomes of a predetermined type of action. The methods that will be suitable in each case also depend of the different steps within the process, and the function and aim of the particular SEA. Fig. 25.2 presents a summary of different tools that can be applied for different steps of the SEA process. It is also important to bear in mind what has been stated by other authors (e.g. Finnveden et al. 2003): Available analytical tools (such as life cycle analysis LCA, multi-attribute-analysis or economic valuation) primarily cover such types of environmental impacts related to pollutant emissions whereas on the other hand tools covering impacts on ecosystems and landscapes are more limited.

Despite the need for further methodological development, some approaches which can also support the decision-making process itself can be recommended principally for SEA. However, they have to be modified and incorporated in a consistent methodological framework. In the face of the current lack of SEA-methodologies this chapter now concentrates on methodologies for identifying landscape impacts and spacial impacts while also demonstrating the integration of SEA and EIA into the decision-making process. Examples of the methodologies that this chapter will evaluate are as follows:

- Exclusion zoning to determine site alternatives that are suitable for future projects by using suitability (positive) criteria and exclusion (negative) criteria;
- environmental risk analysis and risk assessment, that make it possible to deduce risk intensities and to identify spatial impacts using an overlay of spatial susceptibilities to certain impact intensities based on a coarse (i.e. a less detailed) database;
- making use of the area-wide information provided by land use or landscape plans that to a certain extent can be made available as a baseline information for SEA.

25.3 Identifying Possible Alternatives for Waste Dumping-Ground Sites

According to Art. 3 para 2a) of the SEA Directive plans and programmes which are prepared for waste management fall within the scope of SEA. In Germany, for example, according to the Waste Management Act, waste management plans have to provide information about the permitted technologies of waste management and about the areas that are suitable to deposit waste materials or to build waste management plants. Because they thereby set the framework for future development consent of projects listed in the Annexes I and II of the EIA Directive, an SEA has to be prepared (for details see Chap. 3). Within these waste management plans usually a systematic procedure for the delineation and assessment of sites for waste treatment is carried out. It can be taken as a concrete example that demonstrates how to start with strategic alternatives and to result in a proposal for a concrete project. Further on, the systematic identification of possible site alternatives for subsequent projects frequently also appears as a general task in SEA.

The procedure of identifying possible alternatives for waste dumping ground sites may consist of the steps shown in Fig. 25.3.

Phase 1: Within the SEA-process for waste management plans, strategic alternatives will have to be considered as a first step. For waste treatment in an administrative district this could mean the strategic decision whether residual waste should be disposed at a dumping ground or through waste incineration. On this level, scenarios for both alternatives will have to point out which environmental issues are affected and estimate the intensity of positive and negative impacts. Thereby special attention will have to be paid to mutual and cumulative effects. It also has to be pinpointed if impacts were displaced from one environmental sector to the other. A decision for waste incineration for example could mean significant air pollution but also entail the necessity of disposing the residues of the incineration process which will require some ground and affect soil and ground water to some extent. By drawing up scenarios and making use of aggregated parameters describing the emissions in each strategic alternative it will be necessary to balance the pro and con arguments for each alternative and to come to a decision which way or technology of waste management shall be taken.

Phase 2: In the next step, "taboo zones" in the considered area, e.g. an administrative district, in which a dumping ground shall be erected, are delineated. This process of identifying larger areas where a dumpsite may be located happens based on negative criteria, i.e. criteria which exclude building up dumpsites. In this phase, the assessment will focus on those criteria that are

- relevant for decision (key aspects) and
- appropriate for consideration on a less detailed scale without special or extensive data collection (e.g. by fieldwork). Prerequisite for such a strategy is an adequate database (i.e. geological and soil maps, maps of biotope types and conservation areas as they are provided by landscape plans (see Sect. 25.5).

25 Methodological Approaches to SEA 373

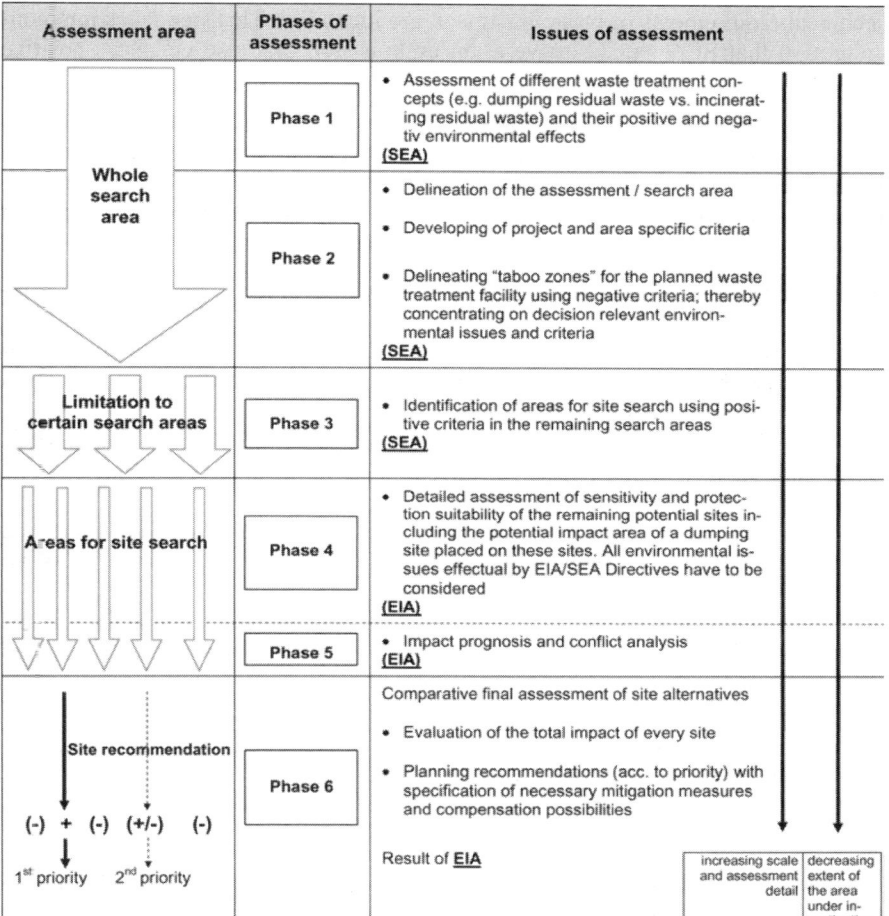

Fig. 25.3. Procedure for the delineation and assessment of sites for waste treatment

Geological layers and groundwater conditions play a decisive role in this context: Prevalent negative criteria for dumping grounds are for example areas with a high groundwater table, spas, areas with karst formations or drinking water catchment areas. Beyond this, areas with valuable populations like nature conservation areas or areas with a lot of valuable biotopes have to be considered. To reach an early coordination with other land-users, areas that are claimed for other purposes (e.g. settlements, infrastructure, agriculture) should be taken into consideration in this stage of the decision process, too.

Phase 3: The remaining areas – here they are called "search areas" – are further narrowed down using so-called suitability criteria (that is to say positive criteria): Potential sites, which meet the requirements of nature conservation as well as the technical requirements of waste treatment, are identified. This step takes into consideration that SEA should arrive at an optimisation that also enhances positive impacts.

Phase 4: For the areas found in the step above a detailed assessment of sensitivities and protection suitability is carried out then and mitigation and compensation measures are identified. Here, we are now on the level of EIA. For that purpose, necessary fieldwork and other surveys based on maps and existing data are carried out. Particularly, the following issues are important:

- Detailed assessment of geological profiles with detailed maps and drilling results;
- Survey of hydrological parameters (depth of groundwater table, velocity and directions of groundwater flow, quality of groundwater, quantitative and qualitative characterisation of hydraulic gradients);
- Delineation of soil conditions (soil types, chemical and physical characteristics of soils, such as cation exchange capacity, clay mineral and carbonate content, buffering capacity);
- Information on the climatic situation (macro-, meso-, microclimate, types and frequencies of meteorological conditions, air quality and the initial level of air pollution);
- Flora and fauna characteristics of the site search areas (elaboration of a detailed map with the different biotopes, data collection about protected and endangered species by detailed fauna and flora inquiries);
- existing traffic infrastructure, distance to residential areas;
- Assessment of visual landscape (e.g. visibility of the landfill body, its adaptation into the character of the surrounding landscape);
- Assessment of testimonials of cultural history (like cultural monuments, cultural and archaeological features).

The area under investigation must always be a possible site itself and the impact area of the waste treatment facility which must be evaluated as well.

Phase 5: Based on the Phase 4, the environmental impacts caused by every single site can be estimated and weighed. This analysis has to be extended to the different environmental issues of EIA. For dumping ground sites, the following impacts are especially relevant:

- Soil: Spatial extension of the facility; drift of contaminated waste material to adjacent areas; contamination of adjacent areas by surface-flow of leachate; impacts caused by dumping-ground-gases;
- Water: Contamination of ground- and surface water caused by leaching;
- Air: Disturbance of residents and surroundings through odours, dusts or germs; uncontrolled emission of dumping ground gases through defective gaskets, generation of inflammable or explosive mixtures caused by gas development;
- Human beings: Disturbance by noise, adverse impacts on recreation potentials;
- Animals and plants: Destruction or fragmentation of habitats;

- Visual landscape: Visual impression of the dumping site, interference with characteristic scales and proportions of a landscape.

In the assessment of environmental impacts the dumping ground site itself as well as traffic and other related facilities of the site (such as adjoining buildings) have to be regarded.

Phase 6: The result will be a comparative final assessment of site alternatives that takes into consideration the positive and negative impacts that have been investigated and leads into a concrete planning recommendation listing appropriate priorities, but also specifying mitigation and compensation measures.

This example refers to SEA as well as to EIA: In the field of waste disposal SEA comprises the assessment of different general waste treatment concepts and the delineation of principally suitable areas for sites (see phases 1-3 in Fig. 25.3). The following Project EIA (phases 4-6) specifies the assessment and comparison of the environmental impacts at the different sites in detail. Thus this example shall also demonstrate how SEA and EIA are in succession within the decision making process. Such procedures of exclusion zoning can easily be carried out by means of a GIS system and also be applied to the search for other projects such as wind power plants or industrial areas (further aspects of using GIS in SEA and its possible advantages and disadvantages are mentioned by Thérivel and Wood, see Chap. 24). However, what is still missing in this procedure (but usually should be a key point in any SEA applied to waste management) is a first and foremost step that considers what mechanisms should be put in place to minimize the production of waste.

25.4 Identification and Assessment of Possible Corridors and Alignments for New Roads

Like the procedure exemplified in Sect. 25.3 the decision process for linear projects, such as roads and railways is done in a sequence of steps which consider environmental issues. In Germany, for example, this process consists of the following main steps (see Fig. 25.4):

1. Consideration of environmental issues on the *level of determination of requirements*. If the analysis on this level comes to the conclusion, that a traffic infrastructure shall be erected between junction A and B, then
2. *spatial analysis* is carried out identifying, describing, and weighing the environmental issues as well as their interrelations for the area between the junction points A and B and identifying areas with different conflict density and conflict emphasis.

Based on the results of step 2 concrete route alternatives are developed. This work should be done in close cooperation of landscape planners and technical planners.

3. Finally, on the level of EIA, a detailed impact *prognosis and comparison of these alternative lines* is prepared.

In the case of Germany, for example, at present spatial analysis, the resulting determination of linear alternatives and the detailed assessment of these alternatives (e.g. for roads, railways or waterways) are altogether part of Project EIA whereas, for the determination of requirements as the first step, an SEA will have to be introduced (see also Chap. 7). Currently, there is a discussion among practitioners if the step of spatial analysis which, in Germany, is actually carried out as part of Project EIA using a procedure called "determination of alignment", should become part of the SEA (see also Fig. 25.4 which assigns both SEA and EIA to the step of spatial analysis). According to the philosophy of SEA, SEA should decide on strategic aspects and thus should be the stage where the approximate location of an alignment is determined. However many practitioners working on this field disapprove of this because they fear that SEA will lead to a simplification of the high level of quality already achieved in EIA studies. This example also accounts for the apprehension of methodological simplification in SEA, and reflects the deficits in establishing a methodological level within SEA that so far has failed to develop during the recent years.

In this section, the current methodology for assessing ecological risks within the German federal Transport Infrastructure Plan shall briefly be introduced (see Box 25.1). This Transport Infrastructure Plan is a national framework investment plan, which is adopted by the German Parliament and the federal cabinet. In this plan, possible traffic connections and parts of the traffic infrastructure net are subject to a cost-benefit-analysis including transport costs, expenditure of the preservation of transport infrastructure, traffic safety, accessibility, external costs and effort (which result for example from emissions and noise protection measures). Furthermore positive aspects like transport costs savings, accessibility improvement or impulses of infrastructure on regional employment are considered.

In addition to this the environmental risks that refer to a certain spatial context and cannot be considered in such a cost-benefit-analysis are assessed qualitatively. For this qualitative assessment the term "environmental risk assessment (ERA)" has been established. It refers to the circumstance that no exact impact analyses are carried out here but spatially attributable environmental risks are estimated. The main focus is on identifying serious environmental conflicts that result either from functional ecological interrelations that operate over a wide area or impacts on important regional or national protection areas. The ERA method that is briefly described in Box 25.1 and illustrated by Fig. 25.5 treats the different modes of transport in a comparable manner to ensure that the evaluation results obtained in any given case can be ranked in relation to other results. The approach used for road and rail projects on the one hand and the approach used for waterways on the other hand differ in details as a result of specific features: The status of planning and the spatially related and project-related documents available for federal waterways are, as a rule, more differentiated than for road and rail projects (Federal Ministry of Transport, Building and Housing 2003 p.48).

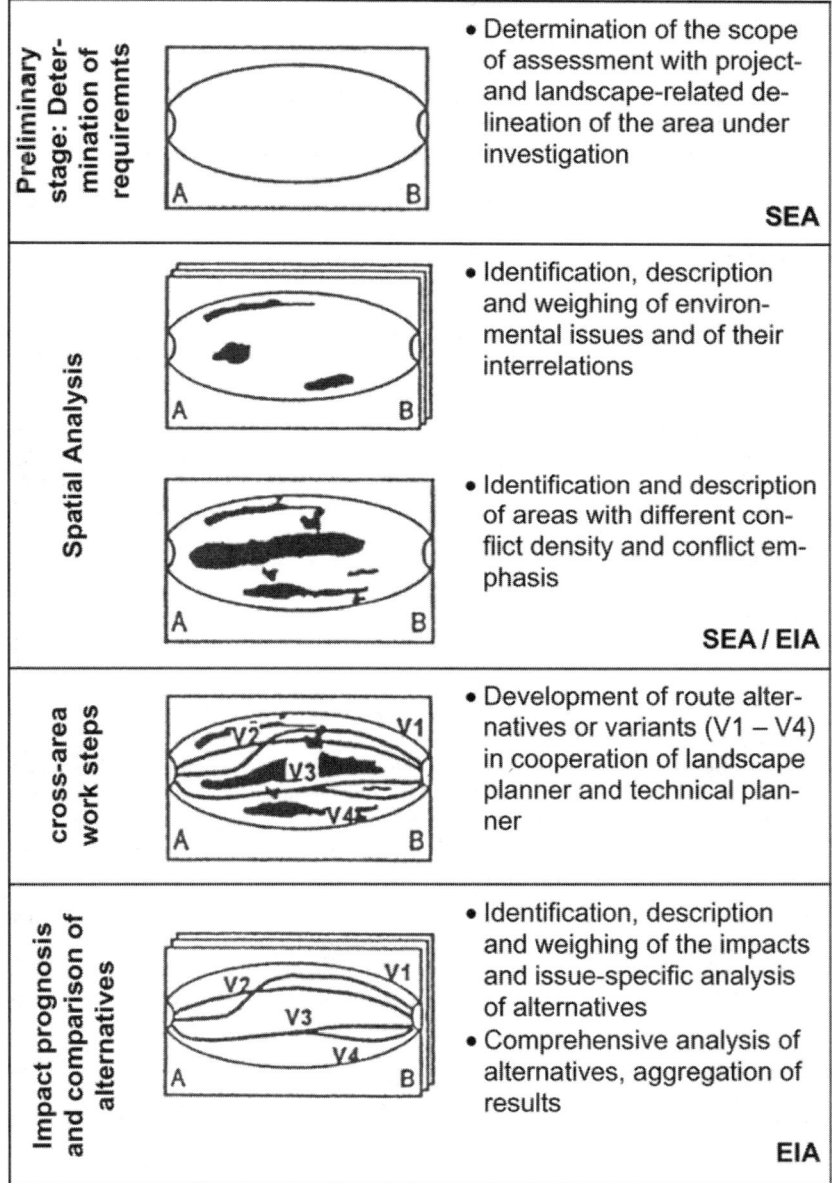

Fig. 25.4. Assessment of environmental issues in a multistage decision-making process (FGSV 2001)

> **Box 25.1.** Methodological approach within the German Federal Transport Infrastructure Plan (Federal Ministry of Transport, Building and Housing 2003)
>
> Environmental risk assessment within the German Federal Transport Infrastructure Plan referred to the assets human beings, fauna and flora, soil, water, air, the climate, landscape, cultural and other material assets. Furthermore, it was an important requirement that the traffic network should take account of existing conservation areas, in particular Natura 2000 sites. The current version of this plan was adopted by the federal cabinet in July 2003 and will be valid until 2015. The assessment comprised the following sequence of analysis and evaluation steps (see also Fig. 25.5):
>
> 1. *Spatial analysis and evaluation to determine the environment-related spatial resistance:* The land cover or land-use type, the status of protected areas, the relevant aims of regional planning policy and ecological features of nationwide and federal state-wide importance served as characteristics to determine the spatial resistance. The values of these characteristics were ordered ordinally (in categories such as low, medium, high, very high) according to their spatial sensitivity against traffic induced-impacts like fragmentation etc. The information provided in area-wide available landscape plans on the local and regional level was especially useful for the determination of the spatial sensitivity (see Sect. 25.5). Such landscape plans allowed, for instance, the identification of key points like areas with high density of valuable biotopes, sensitive soils or groundwater resources. Furthermore, they could be used to assess whether areas of the European network "Natura 2000" were affected.
>
> 2. *Environmental appraisal of the impacts of the different traffic projects:* For a first environmental appraisal of traffic projects on this level, besides type and dimension of projects, an estimation of the relevance of their environmental impacts (especially through required space, type and dimension of embankments and cuttings, barrier effects, noise and pollutant emissions) was crucial. According to these criteria the environmental impacts of transport infrastructure construction project types were assessed on a generalized level. The resulting intensity of impacts was spatially dependent; avoidance and mitigation measures were not taken into account yet.
>
> 3. *Determination of the environmental risk:* For this purpose, the single areas, which have been assessed for their spatial resistance and the intensity of the impacts acting on them by the traffic projects were linked with each other in a matrix of environmental risk. This procedure was a method that could work with highly aggregated, ordinal data and be done smoothly digitally using GIS.
>
> 4. *Final classification of the environmental risk:* The result that was achieved on the basis of digital area data was interpreted by experts and its plausibility was reviewed against the background of further spatial information available. Considering criteria as affected area shares, location in relation to Natura 2000 sites and barrier effects, five classes of environmental risk were created. Traffic projects that were classified as "5 - very high risk" or "4 - high risk" were not recommended for further consideration.

It is rather certain that the German Federal Transport Infrastructure Plan itself will be subject to SEA (cf. Hendler 2003). However, the environmental risk assessment that is usually carried out within this plan can be taken as a focal point to be extended to an SEA. This example demonstrates that, in the determination of requirements, the SEA analyses will be based to a lesser extent on concrete impact analyses and impact prognoses but will refer instead to some key indicators, such as affected area shares, location in relation to protection areas and barrier effects. The environmental risk assessment within the Federal Traffic Infrastructure Plan can be taken as a first device for SEA showing how to identify spatial impacts on a coarse database and/or on a less detailed scale.

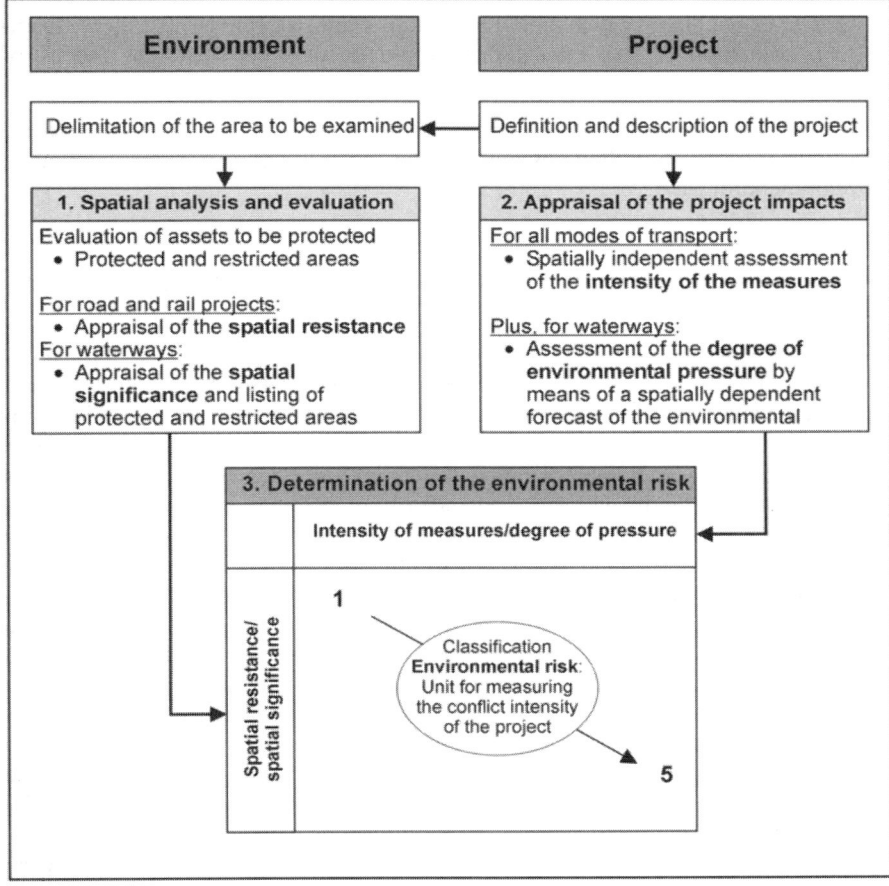

Fig. 25.5. Methodological framework of environmental risk assessment (Federal Ministry of Transport, Building and Housing 2003)

Although the methodology of the German Federal Transport Infrastructure Plan goes beyond the consideration of individual projects it still falls short of the SEA requirements:

- Impacts are still identified in relation to the different projects. These results are standing side by side whereas the examination of their interrelations and a point balance of the environmental impacts of the whole plan are still missing. Above all, a comprehensive examination considering the relationship between different modes of transport is still missing.
- Within the German Federal Transport Infrastructure Plan currently no assessment of alternatives is carried out. There is no consideration of different line alternatives but also – which would be an important strategic issue – no comparative evaluation of the effects that would occur when different modes of transport are chosen.

Thus the current practice of assessing environmental risk within the Federal Transport Infrastructure Plan obviously shows the need for further methodological development for SEA.

25.5 Landscape Planning as an Information Base for SEA

Art. 5 para 3 of the SEA Directive says that relevant information available on environmental effects of the plans and programmes that has been obtained from other sources may be used for providing the information necessary for the environmental report. Such information can for example be supplied by land use or landscape plans. The German planning system provides a good example insofar as the elaboration of landscape plans is mandatory in different scales: On the level of federal states, on a regional level of counties and administrative districts, and on a local level of communes and towns. The different kinds of landscape plans specify the aims and measures of nature preservation for the different administrative levels and thus provide information on different environmental issues that have also to be considered in SEA (like fauna, flora, soil, water, air, climatic factors and landscape). The state of these issues is depicted and valuated in an area-wide way. This entails for instance representations of the valued importance of different areas for the protection of species, biotopes, soil, water and visual landscape. In terms of species and biotopes assignments of areas to value categories criteria like the maturity and reconstructability of habitats, the species abundance, the existence of rare or endangered species and the functions of habitats for biotope connection can be found, as for soils for instance an identification of areas with the best or the worst agricultural and forestry production capacity, with especially rare or near-natural soil conditions or with high filter, buffer or transformation capacities of the soils.

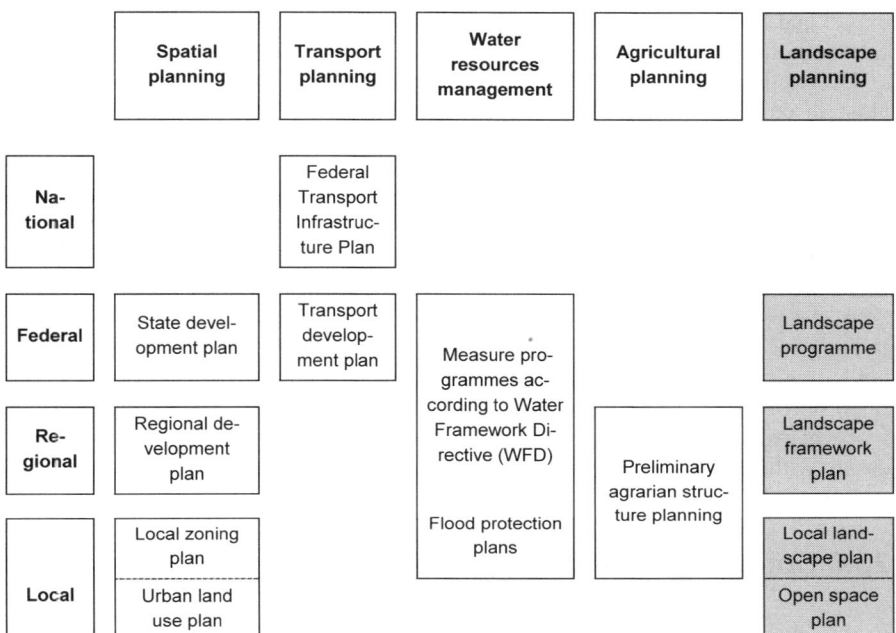

Fig. 25.6. Relations of the tiered levels of landscape plans to certain plans and programmes that have to be submitted to an SEA in Germany (Jessel at al. 2003)

The information provided by landscape plans can serve as a database for SEA, which set the frame for future planning areas (e.g. new residential areas) or linear projects (e.g. transport infrastructure). This support may be important insofar as in SEA often large areas and a wide range of impacts have to be considered, and on top of that, SEA-specific fieldwork is rarely done. In an investigation of a certain amount of SEA already carried out, Thérivel (1998) pointed out that only about one in five SEA-studies investigated described the baseline environment in their area. In the face of this lack of environmental baseline data existing landscape plans can be used for example to identify areas that have to be protected with highest priority (and thus should not be impacted) and areas with a high potential for conflicts. Landscape plans can also provide valuation criteria and thresholds to define the significance of impacts and to identify areas with a high ecological sensitivity in the landscape where significant effects are likely to happen. At the same time, with the help of landscape plans the impacts of different plans and programmes can be projected in the area under investigation to see their cumulative and synergistic effects in correlation to each other. Furthermore, different spatial scenarios can be based upon information from landscape plans. This makes clear, that the existence of area-wide and up-to-date landscape plans can provide a considerable support for working out an SEA (Jessel et al. 2003; see also Chap. 38) and can vitally reduce the resources needed for it. The tiered system of landscape planning corresponds with the different administrative levels and scales of other expert and land use plans (see Fig. 25.6) and thus can provide these plans with the information mentioned in this section.

Due to the overlap between SEA and landscape planning there has been an intense discussion in Germany if existing landscape plans, within the fields of land use, town and country planning, already fulfil the requirements of an SEA. However, the different tasks of both instruments (analysis of the environmental consequences by the SEA versus the conceptual approach related to planning by landscape planning) have to be considered. Nevertheless, the relation between landscape planning and SEA is a good example on how SEA can contribute to further development of national planning instruments: Thanks to the requirements of SEA, the process aspect of landscape planning can be improved and more topicality, as well as more frequent and exact updating, be stimulated. On the other hand landscape plans can provide SEA with data and assessment criteria related to the interests of nature preservation. Assuming that there is periodic updating, landscape plans can provide important information for the monitoring that is also necessary within SEA.

25.6 Conclusions and Recommendations

In this chapter, SEA is identified as a procedural tool providing information and structures for decision-making processes. It is important that the whole decision-making process starting from the strategic action and from drawing up planning objectives up to the implementation of a concrete project is accompanied by a continuous assessment of impacts. Within this process it is also essential to assure that the single steps are in succession in a coherent way and that the results of one step are incorporated in the following ones. The examples show that single devices for methodological support to SEA do exist (see also Chap. 24) – but a coherent methodological framework is still missing, i.e. a framework that above all deals with the *strategic* and *procedural* issues of SEA.

Nevertheless, methods are no end in themselves. They are not independent of the attitudes of their users and of people who take note of their results. Thus, apart from methodological discussions it seems above all necessary to create an awareness as to what SEA can contribute to enhance decisions. If this awareness exists in the administration, among decision-makers as well as in the public it will also stipulate the demand and promote the application of adequate methodologies. In fact, because even the most sophisticated methodologies for SEA will probably be of no use if the right spirit to apply them is missing (Thérivel 1998).

References[1]

Eriksson I-M (2002) Strategic Environmental Assessment in road- and transport system planning in Sweden. Unpublished, 10 pp

Finnveden G, Nilsson M, Johansson J, Persson A, Moberg A, Calrsson T (2003) Strategic environmental assessment methodologies – applications within the energy sector. Environmental Impact Assessment Review 23:91-123

Federal Ministry of Transport, Building and Housing (2003) Federal Transport Infrastructure Plan 2003. Basic features of the macroeconomic evaluation methodology, Berlin

Hendler R (2003) Zur Umweltprüfungspflichtigkeit des Bundesverkehrswegeplans (Requirements for environmental assessment of the Federal Traffic Infrastructure Plan). UVP-report 17(2):57-59

Hildén M, Jalonen P (2002) Key Issues in Strategic Environmental Impact Assessements – Finish Experiences. Extended and partly revised version of a paper submitted to Nordregio as "From idea to Practice – key issues in Finish SEA"

Jessel B, Tobias K (2002) Ökologisch orientierte Planung. Eine Einführung in Theorien, Daten und Methoden (Ecologically oriented planning – an introduction into theories, data and methodological tools). UTB Nr. 2280, Ulmer, Stuttgart

Jessel B, Müller-Pfannenstiel K, Rößling H (2003) Die künftige Stellung der Landschaftsplanung zur Strategischen Umweltprüfung (SUP) – Überlegungen zu den Möglichkeiten einer verfahrensmäßigen und inhaltlichen Verknüpfung (Future Relation of Landscape Planning to SEA – Consideration of the Possibilities of a connection of procedure and content). Naturschutz und Landschaftsplanung 35(11):332-338

FGSV – Forschungsgesellschaft für Straßen- und Verkehrswesen, Arbeitsgruppe Straßenentwurf (2001) (ed) MUVS – Merkblatt für die Umweltverträglichkeitsstudie in der Straßenplanung (Instructions on environmental impact studies in road planning)

Noble BF (2000) Strategic Environmental Assessment: What is it and what makes it strategic? Journal of Environmental Assessment Policy and Management 2(2):203-224

Noble BF (2002) The Canadian Experience with SEA and sustainability. Environmental Impact Assessment Review 22:3-16

ODPM – Office of the Deputy Prime Minister (2003) The SEA Directive: Guidance for Planning Authorities. Practical guidance on applying European Directive 2001/42/EC 'on the assessment of the effects of certain plans and programmes on the environment' to land use and spatial plans in England. ODPM, London, www.planning.odpm.gov.uk

Risse N, Crowley M, Vincke P, Waaub J-P (2003) Implementing the European SEA Directive: the Member States' margin of discretion. Environmental Impact Assessment Review 23:453-470

Short M, Jones C, Carter J, Baker M, Wood C (2004) Current Practice in the Strategic Environmental Assessment of Development Plans in England. Regional Studies 38(2):177-190

Thérivel R (1998) Strategic Environmental Assessment of Development Plans in Great Britain. Environmental Impact Assessment Review 18:39-57

Wood C (1995) Environmental Impact Assessment. A Comparative Review. Longman Scientific and Technical, Essex

---

[1] Note that legislation mentioned in the chapter is listed at the end of the handbook in the consolidated list of legislation (Appendix 2).

26 Strategic Level Cumulative Impact Assessment

Riki Thérivel

Levett-Thérivel sustainability consultants, Oxford, England

26.1 Introduction

One of the key reasons put forward for carrying out SEAs is that they are able to deal with cumulative impacts better than project EIAs can (Wood and Djeddour 1992; Thérivel et al. 1992). Cumulative impacts have traditionally been ignored or assessed poorly in EIAs, despite decades of regulations and some very good guidance (e.g. CEAA 1998, 1999; Court et al. 1994; USCEQ 1997). Those cumulative impact assessments that have been carried out have traditionally been very detailed analyses – described by Clark (2000) as "richer EIA" – of the impacts of a limited number of projects, and have been correspondingly resource-intensive.

The European SEA Directive requires an analysis of "*the likely significant impacts on the environment... These impacts should include secondary, cumulative, synergistic... impacts*" (Annex I footnote). At present the European Member States are still trying to figure out how to implement the main parts of the Directive, much less this small-font footnote near the end. The tendency would be to refer practitioners to the existing guidance on cumulative impact assessment (CIA), essentially suggesting that strategic-level CIA is "richer SEA".

However I would argue that a very different approach is needed. Practitioners are already concerned about the level of detail that the main Directive requirements should go into, and their resource requirements. CIA as "richer SEA" would simply add to this concern, causing cumulative impacts to be cheerfully ignored except under duress. Furthermore, given the wide range of activities and issues covered by the typical SEA, the SEA practitioner already struggles to identify key impacts and propose key mitigation measures: "richer SEA" would merely add trees, not help to distinguish the wood. Most importantly, the focus of CIA is arguably quite different from that of EIA and SEA. CIA focuses on the *resource* whilst EIA and SEA both focus on the *action* (either project or strategic level). Therefore a "richer SEA" would in fact give the wrong message because it would continue the focus on the action rather than the resource.

Implementing Strategic Environmental Assessment. Edited by Michael Schmidt, Elsa João and Eike Albrecht. © 2005 Springer-Verlag

This chapter suggests some approaches to CIA at the strategic level. It has evolved from work carried out in developing SEA guidance for English transport and land use plans. It is based on a working paper prepared by the author; discussions during a workshop involving 13 key government officials, academics and consultants; and subsequent development of the workshop findings into guidance.

This chapter gives definitions and explains how cumulative impacts occur; suggests some key principles of CIA; and explains how cumulative impacts can be identified, assessed and mitigated in SEA.

26.2 Definitions and Concepts

The aim of CIA is to identify, describe and evaluate cumulative (including synergistic) impacts, and enable them to be avoided, minimised or enhanced as appropriate. *Cumulative impacts* are the total impacts of multiple actions on a receptor, as shown in the example of Fig. 26.1. A *receptor* could be a geographical area, ecosystem, species, or section of the population. An *action* could be a plan or programme, a social trend, or an individual person's activities. It can have occurred in the past, or could occur in the future. Adverse cumulative impacts generally arise when impacts occur too often or too hard to allow the receptor to recover, for instance where fishing rates exceed the rate of fish replacement.

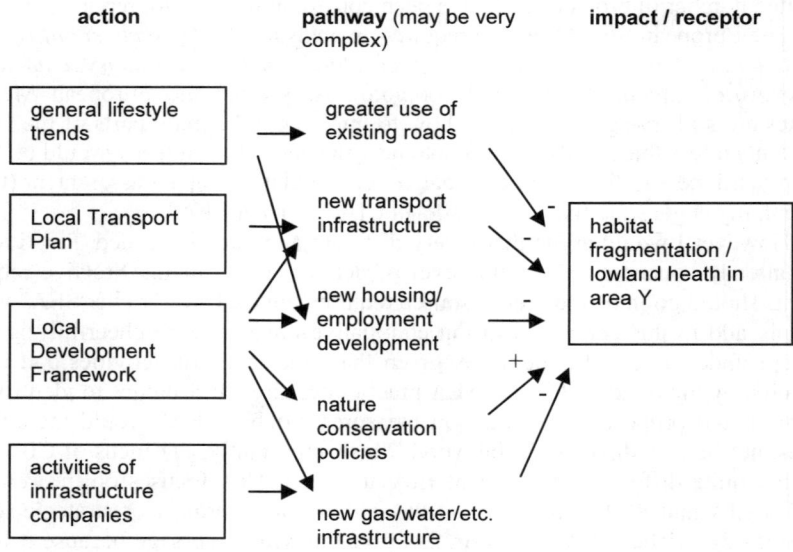

Fig. 26.1. Example of how a cumulative impact is caused

Examples of past cumulative impacts are habitat fragmentation, climate change, and decreasing rural services. Cumulative impacts can also be positive, for instance improved human health or access to open space. Table 26.1 shows examples of cumulative impacts linked to the topics listed in the SEA Directive. Multiple types of impacts can also affect the same receptor: for instance a community can be affected by noise and air pollution and severance from new development.

Cumulative impacts can be:

- Additive: the simple sum of all the impacts (e.g. new jobs in an area of high unemployment);
- Neutralising, where impacts counteract each other to reduce the overall impact (e.g. a development on the left river bank encroaches on the floodplain, but equivalent flood storage capacity is provided on the right bank);
- Synergistic, where impacts interact to produce a total impact greater than the sum of the individual impacts (e.g. a new footpath that links two existing footpaths, allowing more recreational opportunities than those provided by three separate footpaths). Negative synergistic impacts often happen as habitats, resources or human communities get close to capacity: for instance a wildlife habitat can become progressively fragmented with limited impacts on a particular species until the last fragmentation makes the areas too small to support the species at all.

Table 26.1. Examples of national level cumulative impacts linked to SEA Directive topics

Topic (SEA Directive Annex If)	Cumulative impacts (can be positive as well as negative)
Population	• community severance • inequalities in access to services • sections of population cumulatively affected by e.g. more development and associated traffic
Human health	• incidences of obesity, asthma, etc. • changes in crime levels • changes in accident levels
Biodiversity, fauna, flora	• fragmentation of habitats • changes in biodiversity • species extinction • loss of high quality agricultural land • soil erosion
Climatic factors	• climate change
Material assets	• rural diversification • changes in levels of, and efficiency of, resource use • changes in service provision (e.g. post offices, health care facilities)
Landscape	• changes in land use • changes in landscape character
Water	• eutrophication, acidification
Interrelation between factors	• loss of tranquillity

26.3 Principles of Cumulative Impact Assessment

Cumulative impacts are difficult to deal with on a project-by-project basis through environmental impact assessment, or indeed on a plan by plan basis. It is at the SEA level, and using a multi-agency approach, that they are most effectively addressed.

The focus of CIA is on receptors. CIA asks whether the total impacts on a given receptor of all actions, no matter who carries them out, form a significant impact. The test of significance of a cumulative impact is "can the receptor accommodate additional impacts?". An understanding of the thresholds/capacities of the receptors to deal with impacts is thus needed. For instance thresholds for assimilating greenhouse gas emissions are already being exceeded, whilst some landscapes might be able to accept considerably more development without a degradation in their character.

Cumulative impacts should be considered throughout the plan-making and SEA process, not at one stage. Finally, cumulative impacts on a given receptor are rarely aligned with political or administrative boundaries. CIA must use the relevant receptor boundaries: ecological boundaries for natural systems, socio-cultural boundaries for human communities (James et al. 2003).

26.4 Carrying out Cumulative Impact Assessment

Fig. 26.2 shows how CIA can be integrated with SEA and decision-making. Stages A to C are best carried out as part of the SEA scoping discussions early in plan-making, concurrently so that they inform each other. All stages are best carried out in discussions with the other key organizations that contribute to the impact.

Stage A. Identify cumulative problems and opportunities. As part of the SEA's identification of environmental or sustainability problems, cumulative impacts and/or receptors that are in decline and are near their threshold should be identified. The list of impacts from Table 26.1 can be supplemented with an analysis of regional and local level cumulative problems and opportunities, including:

- sections of population (e.g. groups particularly affected by accidents, or with poor access to services)
- geographical areas (e.g. non-tranquil areas, areas of poor accessibility)
- resources (e.g. air quality management areas, areas of soil contamination)
- ecosystems and species (e.g. heathland)

Many potential cumulative impacts may be identified. Consensus agreement will be needed about key impacts to focus on. The results of Stages B and C should assist in this.

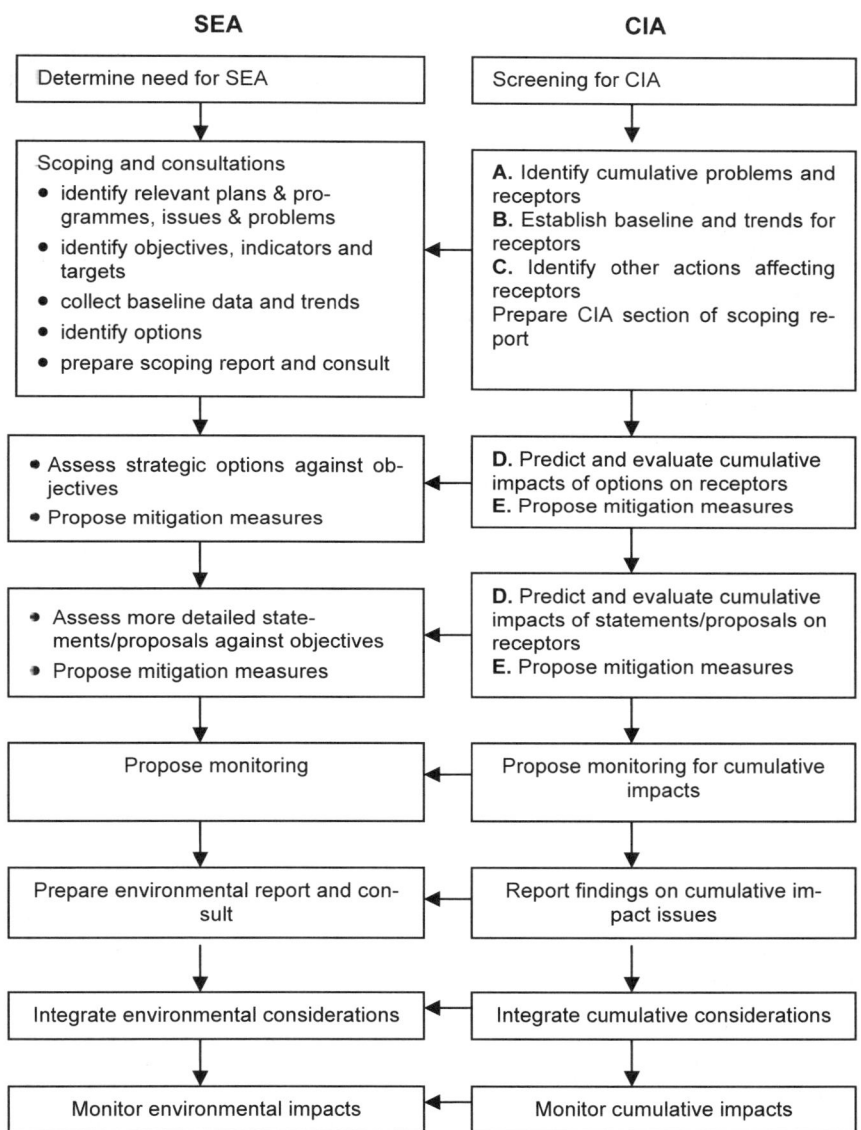

Fig. 26.2. Links between CIA, SEA and decision-making (Based on Cooper and Sheate 2004)

Note: Capital letters in bold refer to stages discussed in the text.

Stage B. Establish baseline and trends for receptors. An understanding of the trends and current status of the receptors without the proposed plan is needed to predict what their quality will be after the proposed plan is implemented. If a receptor is already degraded or recovering from a previous impact, then the impacts of new plans may be more serious. The time period and area over which different receptors/impacts are analysed will vary from receptor to receptor. For instance climate change issues will be international and span decades, whilst community severance is more local and potentially short-term.

Table 26.2 shows how the baseline could be documented. The focus of the baseline should be on outcomes (how the receptor will be affected) rather than inputs (what is affecting them or how the impacts are already being mitigated).

Table 26.2. Possible way of organising and presenting baseline data (example)

Impact / receptor	(Outcome) data	Comparators and targets	Trend	Problems
Habitat fragmentation / lowland heath in area Y	None available	None available	Getting worse (county ecologist opinion)	Local ecologist feels that this is a significant problem
Access to open space	82% of population lives within 400m of open space	Target 90%	Getting better: 79% in 1997, and likely to improve with development X (planned for 2007)	

Stage C. Identify other actions affecting receptors. This stage identifies, for each impact/receptor from A., the role of other actions in leading to or solving the problem. This could include a discussion of how the cumulative impact has arisen over time. Techniques for doing this include:

- Causal network analysis (e.g. Fig. 26.1);
- The last column of Table 26.3 can help to brainstorm plans that affect each receptor;
- Trend analysis where appropriate. This assesses the status of a receptor over time and projects this into the future using various assumptions: for instance future water use per household can be projected based on past trends in water use plus assumptions about the future use of water-saving or water-using technologies.

The analysis should consider the likelihood and timing/phasing of future actions: for instance whether one action depends on another one being completed first.

Table 26.3. Possible way of describing actions affecting receptors

Impact / receptor	Past actions	Possible future actions
Habitat fragmentation / lowland heath in area Y	Use of land for housing and transport infrastructure; agricultural intensification	Biodiversity Action Plan related habitat restoration (depends on funding). Increased house building (very significant)
Improved access to open space / residents in area X	No significant activities	Planning permission for new developments include requirements for open space provision. Improved footpath network (depends on negotiations with individual landowners)

Stage D. Predict and evaluate significant adverse cumulative impacts. Personally I feel that this stage, which is likely to be the longest and most costly in the CIA process, could be avoided if plan proponents were willing to mitigate all the impacts identified in Stage A to the fullest extent that is possible for them. After all, cumulative impacts are "cumulative" because they are already at the level of being a problem, so there is no need to prove that they require mitigation. However, authorities may prefer to know what level of cumulative impacts they are responsible for (in both senses of the word), and so may wish to predict and evaluate the impacts.

During the *prediction* stage – which identifies the *magnitude* of the impact, i.e. the difference between the with- and without-plan scenarios – the following points should be kept in mind:

- The type of plan and receptor will determine whether the prediction should be qualitative (e.g. "better/worse") or quantitative. Quantitative predictions may be more appropriate for changes in land use, accident levels etc; qualitative predictions for changes in habitat fragmentation, landscape character etc.; quantitative for more detailed programmes and qualitative for more strategic plans. Table 26.4 lists some techniques for identifying and predicting cumulative impacts.
- The predictions should include an indication of the level of, and reasons for, any uncertainties. Uncertainty increases at the higher planning levels because scales and issues are larger. The magnitude of cumulative impacts may depend on how the plan is implemented, e.g. whether new housing is located on a site of high or low biodiversity. Uncertainty should be documented, and measures to ensure that implementation minimises any negative impacts should be identified.
- Where appropriate assessment under the Habitats Directive is required, additional requirements apply: these are discussed in English Nature et al. (2004).

The *evaluation* stage – which determines the *significance* of the cumulative impact – should focus on testing the predicted impacts against the threshold/capacity of the receptor. Thresholds are the level of stress below which populations, ecosystem functions or quality of life can be sustained. This is different from the main

SEA process, which generally tests against environmental or sustainability objectives.

Table 26.4. Techniques for identifying and predicting cumulative impacts (based on Hyder 1999; James et al. 2003; Thérivel and Wood 2004; USCEQ 1997)

Method/description		Strengths	Weaknesses
Interviews etc.	Brainstorming sessions, interviews with experts etc. can help to identify cumulative impacts, provide data, analyse plan alternatives and components, and identify mitigation measures.	Flexible and can deal with subjective information.	Subjective and generally non-quantitative.
Matrices	A tabular format is used to organise and describe the interactions between actions and receptors. Table 26.5 shows an example.	Help to present and compare alternatives.	Do not address cause-impact relationships.
Causal chain anal.	Identifies and illustrates the cause-impact relationships that result in cumulative impacts. In doing so, it identifies assumptions made in impact predictions, unintended consequences of the strategic action, and possible measures to ensure effective implementation. Figure 26.1 shows an example.	Illustrates complex links and identifues cumulative impacts.	Can be cumbersome.
Modelling	Quantifies the cause-impact relationships leading to cumulative impacts. Modelling can take the form of mathematical equations describing cumulative processes, or expert systems that forecast the impacts of various scenarios.	Addresses cause-impact relationships and gives quantified results.	Needs much data, and extrapolation of data is still largely subjective.
Overlay mapping, GIS	A computerised cartography system that stores map data linked with a database management system that stores attribute data. Cumulative impacts can be displayed as superimposed (and possibly weighted) map layers. The vulnerability of receptors – how near they are to their thresholds – can also be mapped. CCW (2002) is an example.	Incorporate spatial data; help to set the boundary of the analysis; identify areas where impacts will be greatest.	Limited to impacts based on location. Digital data have cost, quality and scale issues.
Scenarios	The plan's impacts are described under different assumptions – scenarios – about future conditions outside the plan's control, e.g. under different economic growth scenarios.	Reflects uncertainties and suggests ways of reducing them.	Potentially time and resource intensive.
Extrapolation	The proposed plan's impacts are predicted based on data from similar existing plans.	Based on real data.	Choice of existing plans is crucial: context etc. may vary.

Table 26.5. Matrix showing cumulative impacts of several actions on receptors (cumulative impacts are read "down" as shown by the arrow)

Action	Impacts/receptors		
	Fragmentation of lowland heath in area Y	Climate change	Access to open space in area X
Improved rural bus services	0	0	++
Park & Ride at X	- on site of high biodiversity in area Y	? reduces city centre traffic but could increase traffic to P&R	0
Cycle network	0	+ helps switch away from car use	+
Cumulative impacts: 1 + 2 + 3	-	+	+

Thresholds can be identified based on existing guidance; existing targets such as Biodiversity Action Plan targets; and capacity studies. In cases of high uncertainty, it may be necessary to add a precautionary buffer to the targets.

Stage E. Propose mitigation measures. Mitigation measures should aim to first avoid loss or damage to the receptor and enhance it where possible; and then compensate for any unavoidable damage. This could include measures to achieve Biodiversity Action Plan targets; planning obligations regarding e.g. provision of open space or rural services (Conland and Rudd 2000; Piper 2001); and establishment of a "mitigation bank" which requires developers to compensate for loss or damage to a habitat by providing equivalent (or better) replacement habitat.

The level of mitigation that a given plan is responsible for is possibly the most contentious aspect of CIA. For most cumulative impacts, it will be impossible to precisely allocate responsibility among plan-makers regarding either the impacts they "cause" or what mitigation measures they should be responsible for. Arguably, today's plans should also be helping to remedy the cumulative impacts caused by past plans and by today's actions that are not covered by plans (for instance the obesity engendered by today's sedentary lifestyles). Some mitigation measures for a plan's cumulative impacts will only be capable of being delivered by parties other than the plan proponent, and should be discussed with those organisations. Vice-versa, the plan in question may be the best vehicle for delivering mitigation measures for another plan.

This is essentially a political, horse-trading stage which is informed by the findings of Stage D. If the plan is likely to adversely affect the integrity of a

Natura 2000 site, additional requirements apply: these are discussed in English Nature et al. (2004).

26.5 Conclusions and Recommendations

Strategic level CIA, if carried out as "richer SEA", would be impossibly complex and resource-intensive. It also would not address a key aspect of CIA, namely that it should focus on the resource, not the plan.

This chapter has suggested some simple techniques for identifying, predicting and evaluating cumulative impacts. They are not comprehensive, not rigorous, and for the most part not replicable. However they are quick enough to keep up with the plan-making timescale, and keep the emphasis on the receptor and on key impacts. Arguably they provide as good a basis as much more complex techniques for the (invariably political) discussions about who is responsible for cumulative impacts: those impacts that we and those who came before us have caused, but which, unless we deal with them, we will be passing down to our children.

Acknowledgements

This chapter is based heavily on guidance developed for the Department for Transport by TRL and Levett-Thérivel. I am very grateful to Helen Byron, Ric Eales, Chris Fry, Jon Grantham, Emma James, Roger Levett, Jeremy Owen, Maria Partidário, Jake Piper, Topsy Rudd, Bill Sheate, Stef Simmons, Nick Simon, Roger Smithson, Chris Smith and Paul Tomlinson for their help in writing this chapter and the guidance.

References[1]

CEAA – Canadian Environmental Assessment Agency (1999) Operational Policy Statement: Addressing Cumulative Environmental Impacts under the Canadian Environmental Assessment Act. Canadian Environmental Assessment Agency. Internet address: www.ceaa-acee.gc.ca/013/0002/cea_ops_e.htm, last accessed 21.5.2004

CEAA – Canadian Environmental Assessment Agency (1998) Cumulative Impacts Assessment Practitioners Guide. Canadian Environmental Assessment Agency. Internet address: www.ceaa-acee.gc.ca/013/0001/0004/index_e.htm, last accessed 21.5.2004

Conlan K, Rudd T (2000) Sustainable Estuarine Development? Cumulative Impact Study of the Humber. Journal of the Chartered Institution of Water and Environmental Management 14 (5):313-317

---

[1] Note that legislation mentioned in the chapter is listed at the end of the handbook in the consolidated list of legislation (Appendix 2).

Cooper LM, Sheate WR (2004) Integrating cumulative impacts assessment into UK strategic planning: implications of the European Union SEA Directive. Impact Assessment and Project Appraisal 22(5):5-16

CCW – Countryside Council for Wales (2002) Development of a methodology for the assessment of the impacts of marine activities using Liverpool Bay as a case study, CCW Contract Science Report No. 522. Internet address:
www.ccw.gov.uk/Images_Client/Reports/ ACF13C.pdf, last accessed 21.5.2004

Court JD, Wright CJ, Guthrie AC (1994) Assessment of Cumulative Impacts and Strategic Assessment in Environmental Impact Assessment. Commonwealth Environment Protection Agency, Barton, Australia

English Nature, Royal Society for the Protection of Birds, Environment Agency, Countryside Council for Wales (2004) SEA and Biodiversity: Guidance for Practitioners. EN, Peterborough

James E, Tomlinson P, McColl V, Fry C (2003) Final Report – Literature Review / Scoping Study on Cumulative Impacts Assessment and the Strategic Environmental Assessment Directive, report by TRL Ltd. for the Environment Agency. Internet address:
www.trl.co.uk/static/environment/cea_FinalReport.pdf, last accessed 21.5.2004

Hyder Consultants (1999): Study on the Assessment of Indirect and Cumulative Impacts as well as impact Interactions. Report for EC DG XI Environment, Nuclear Safety and Civil Protection. Internet address:
http://europa.eu.int/comm/environment/eia/eia-studies-and-reports/ guidel.pdf, last accessed 21.5.2004

Piper JM (2001) Assessing the Cumulative Impacts of Project Clusters: A Comparison of Process and Methods in Four UK Cases. Journal of Environmental Planning and Management 44(3):357-375

USCEQ – US Council on Environmental Quality (1997): Considering Cumulative Impacts under the NEPA. Internet address: http://ceq.eh.doe.gov/neap/nepanet.htm, last accessed 21.5.2004

27 Handling Transboundary Cumulative Impacts in SEA

Tyson Harty[1], Daniel Potts[2], Donald Potts[3] and Jehan El-Jourbagy[4]

1 Department of Zoology, Oregon State University, United States
2 Department of Ecology and Evolutionary Biology, University of Arizona, United States
3 Department of Forest Management, University of Montana, Missoula, United States
4 Attorney, Member of Oregon Bar and Environmental Advocate, United States

27.1 Introduction

The success of environmental policies such as SEA will ultimately be determined by their ability to promote a sustained reduction of environmental impacts due to human activities. Most importantly, in a world where environmental impacts are accumulating at ever-increasing rates, it is vital that policies such as SEA be especially successful at handling those impacts that cross national boundaries and are cumulative over time. Environmental impacts simply do not recognize political boundaries.

Indeed, it is the *accumulation* of the individually minor effects of multiple human actions over time that is primarily responsible for the continued observed degradation of natural systems, despite the many individual successes over the past few decades of governmental environmental infrastructures, such as the National Environmental Policy Act (NEPA) in the United States (US) (see Chap. 17), as well as other environmental policies throughout the world (CEQ 1997). For example, this potential for accumulation is illustrated by the San Francisco Bay estuary in California, where a single mile of the delta can be affected by the decisions of over 400 distinct federal, state, or local agencies (National Performance Review 1994). Cumulative environmental degradation is truly the result of a "tyranny of small decisions" (Odum 1982).

Sustainable development will be impossible if cumulative effects are not considered in environmental planning. Renewed interest in managing cumulative effects in environmental policy, however, offers hope that future environmental degradation may be limited.

Implementing Strategic Environmental Assessment. Edited by Michael Schmidt, Elsa João and Eike Albrecht. © 2005 Springer-Verlag

This chapter defines cumulative effects and describes their treatment using NEPA in the United States as the primary example due to its longevity over the past 30 years. Additionally, the United States with its multiple states provides an excellent lesson for the handling of transboundary cumulative effects across the different Member States of the European Union (EU) (see also Chap. 26).

This chapter compares and contrasts methods to identify potential cumulative effects associated with strategic actions and the potential for international trade-agreements to mitigate these effects across borders. The chapter describes the efforts of a coalition of universities and government agencies in addressing the cumulative effect of groundwater withdrawals on a watershed that spans the United States-Mexico border. In addition, an analysis of cumulative impacts associated with management of the United States National Park system, the first and most expansive park system in the world, is highlighted and its potential lessons for SEA in the EU are discussed. On a larger scale, the methods discussed here are used as the basis for recommendations for handling cumulative effects under SEA as it continues to develop in the EU.

27.2 Definition of Cumulative Effects

Implemented in 1969, NEPA requires every federal action to be preceded by either an Environmental Assessment (EA) or Environmental Impact Statement (EIS) addressing the direct, indirect, and cumulative effects on the environment. *Direct effects* are those caused by an action that occurs at the same time and place, such as the immediate loss of forest habitat following a timber harvest. *Indirect effects* are those that occur later in time or farther removed in distance, but are still reasonably foreseeable. In the case of timber harvest, indirect effects would include degradation of soil and water quality due to increased erosion. *Cumulative effects* are impacts on the environment that result from "the incremental impact of the action when added to other past, present, and reasonably foreseeable future actions regardless of what agency (federal or non-federal) or person undertakes such other actions" (40 CFR para 1508.7). Increased forest fragmentation and the associated impact on certain wildlife, as the result of multiple timber harvests in an area through time, is an example of a cumulative effect.

Despite their widespread and lasting impacts, cumulative effects are often underestimated relative to the direct and indirect effects of actions. Not until 27 years after its signing into law did the US Council on Environmental Quality's (CEQ) regulations, which have implemented the procedural provisions of NEPA over the past three decades, provide recommendations for assessing cumulative impacts in NEPA (i.e. 1997). In practice, cumulative effects have been the most challenging impacts to assess because of the difficulty in defining the spatial and temporal boundaries of the accumulation of environmental disturbances from multiple separate actions (CEQ 1997).

Definitions of spatial accumulation must take into consideration the impacts of other geographically nearby (or sometimes, not so nearby) actions that together have cumulative effects. Likewise, determination of temporal accumulation must assess not only the immediate impacts, but also the cumulative influence of past effects still present in the environment and any future effects for strategic actions yet unplanned. The inclusion of future effects in a cumulative effects analysis (CEA) can become difficult and controversial, since some future actions can seem remote or speculative. Generally, future actions can be disregarded if they are outside the spatial and temporal boundaries established for the CEA, if they affect resources that are not part of the CEA, or if they lack logical basis for inclusion in the CEA (Eccleston 2001).

Careful definition of spatial boundaries and temporal duration of impacts in a CEA is critical. If impacts are defined too broadly, then the CEA becomes unwieldy with too many influences to accurately and realistically assess. If defined too narrowly, then significant impacts beyond the immediate direct effects may be overlooked. In either case, decision-makers will be ill-informed about long-term consequences of their plans and programs (CEQ 1997).

Spatial and temporal accumulation of effects of a proposed strategic action occurs when the effects of existing and/or past actions have not fully dissipated. Accumulation can be *additive*, wherein effects in space or time are summed, or *synergistic*, wherein the total accumulation is greater than the sum of individual effects due to some interactive or amplifying relationship between the effects (Fig. 27.1). For example, hot water and nutrient discharges to a river can combine to cause algal blooms that deprive dissolved oxygen in a manner that is greater than the sum of each pollutant's effect.

27.3 Cumulative Effects Analysis in the US

To address the need for more thorough analyses of cumulative effects, the US CEQ issued a common framework to federal agencies for conducting CEAs in both EIS and EA (CEQ 1997) (see Chap. 17 for an explanation of EIS and EA in the US system). The goal of this framework was to incorporate CEA early in the planning process to improve decision-making. CEA can be integrated into several steps of traditional environmental impact assessment (EA or EIS): (1) scoping, (2) description of the affected environment, and (3) determination of environmental impacts. Additionally, the CEQ framework suggests that CEA be incorporated into the range of alternatives (including "no action") to the proposed strategic action, and that CEA be an iterative process throughout a strategic action, with re-evaluations of impacts as new information becomes available. A useful visualization of cumulative effects is the term "nibbling," coined by the Canadian Environmental Assessment Research Council to mean "the incremental erosion of a resource until there is a significant change or it is all used up" (Glasson et al. 1999). Scoping defines space and time boundaries for the effects themselves, rather than just the individual strategic action (CEQ 1997).

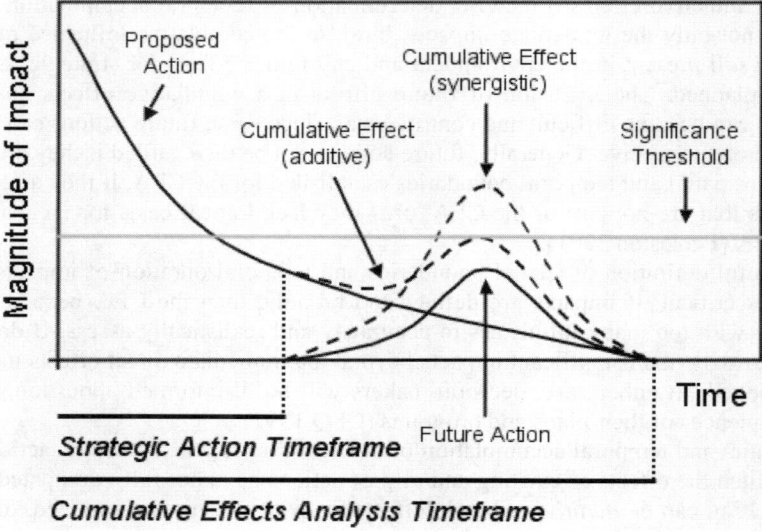

Fig. 27.1. Timeframes for strategic action-specific and cumulative effects analyses, illustrating both additive (i.e. summed) and synergetic effects (i.e. total accumulation is greater than the sum of individual effects) (Adapted from CEQ 1997)

Specific scoping for cumulative effects should determine whether the ecosystems, resources, and human communities directly or indirectly affected by the proposed strategic action have already been affected in the past or may be affected in the future by other agencies or the public. Most importantly, scoping can identify the need for interagency or transboundary cooperation regarding a proposed strategic action. At a minimum, adequate scoping should identify the following (adapted from Eccleston 2001):

1. Specific space and time boundaries for the effects of the proposed strategic action;
2. Expected impacts of the proposed strategic action within those spatial boundaries;
3. Impacts to the same area that have occurred or could be expected to occur from other past, proposed, and reasonably foreseeable future actions (federal, non-federal, or private);
4. Overall predicted impact to the area upon accumulation of individual impacts;
5. Potential expansion of impacts due to cumulative effects.

Information sources during the scoping process can include consultations with other agencies, comments from the public, expert opinion, new studies in support of the strategic action, planning activities, and knowledge and experience of specific analysts (CEQ 1997). Questionnaires, interviews, panels, and public meetings are all useful for organizing information; indeed, public participation and comment in the scoping stage is critical. There are multiple means by which to establish causal links between a proposed action and cumulative effects, each with strengths and weaknesses (Table 27.1). In practice, several of these methods, employed together, should yield the most thorough results. After this initial scoping procedure has been completed, the analysis process of the CEA can begin. Affected resources, ecosystems, and human communities should be characterized by their resilience to change and resistance to impacts. The direct, indirect, and cumulative effects should be gauged in relation to standard regulatory thresholds (e.g. governmental regulations and/or administrative standards such as air and water quality criteria) (CEQ 1997). Because cumulative effects assessment must consider past and other present actions, establishing *baseline conditions* for each resource, ecosystem, and human community must be taken (see Box 27.1).

Time-consuming and expensive, collecting baseline information on affected resources, ecosystems and human communities presents a challenge for most federal agencies. Especially important is looking beyond federal and agency-specific data sources to provide an *integrated assessment*. Such sources include individuals (e.g. landowners, residents, resource users), historical societies, universities (including libraries, museums, field stations), and private organizations.

Despite the importance of baseline conditions in determining the impact of cumulative effects, current efforts to establish a universal protocol for baseline assessment are minimal. In the US, the most comprehensive effort to catalog and monitor baseline ecological conditions has been the Environmental Monitoring and Assessment Program (EMAP) (see also the "Vital Signs" initiative of the US National Park Service in Box 27.1). EMAP was established in the late 1980s to "develop the scientific understanding for translating environmental monitoring data from multiple spatial and temporal scales into assessments of ecological condition and forecasts of the future risks to the sustainability of natural resources" (EPA EMAP 2002). Led by the US Environmental Protection Agency (EPA), EMAP is a multi-agency effort distinct from other federal environmental monitoring programs (CEQ 1997).

EMAP uses integrated biological indicators, such as aquatic macro-invertebrate fauna, to assess ecological condition, rather than traditional physical and chemical indicators. These indicators are measured within a statistically rigorous sampling design, using probability samples rather than site-specific subjective measures (CEQ 1997). EMAP also samples and monitors aquatic resources such as streams, lakes, and estuaries to maintain an ecosystem-oriented assessment. These attributes make EMAP more suited for CEAs, since regional ecological resources can be monitored and integrated with other data to determine more accurate and realistic baseline assessments for new proposed strategic actions (CEQ 1997).

Although EMAP has been successful at the regional level, it has not yet been implemented on a fully national scale in the US with regard to cumulative effects assessment. The goal of SEA efforts in the EU (e.g. via monitoring programs ongoing by the European Environment Agency) and elsewhere should be to establish an EMAP-like database infrastructure to create a dynamic source of baseline assessment data of affected resources, ecosystems and human communities on a continental scale. Such a database should use an overarching synthesis approach to realistically address and evaluate synergistic and interactive effects through metadata and trends analyses.

Ideally, EMAP and similar efforts could be incorporated with the World Conservation Monitoring Centre (WCMC) of the United Nations Environment Programme (UNEP) (http://www.unep-wcmc.org/). Established in 2000 as a world biodiversity information and assessment center, WCMC provides early warning of emerging environmental threats in forest, dryland, freshwater, and marine ecosystems.

Table 27.1. Methods for analyzing indirect, cumulative, and transboundary effects (modified from CEQ 1997)

Methods	Description	Strengths	Weaknesses
Questionnaires, Interviews	Useful for identifying region-specific cumulative effects issues as well as alternative actions	Flexible and potentially inexpensive	Subjective information; potentially biased interpretation
Public Meetings	A forum for community concerns and public input	Can build trust and consensus among public	Generate qualitative information
Checklists	Listing of potential effects associated with action	Systematic and concise	Cannot address interactions among effects or cause-effect relationships
Matrices	Tabular format to organize and quantify complex information	Comprehensive quantitative comparison of alternative actions	Do not address time and space issues or cause-effect relationships
Networks and System Diagrams	Graphical visualizations of multiple cause-effect relationships	Can assess indirect effects; aids in conceptualizing complex relationships	Do not address space and time; Requires prior knowledge of cause-effect relationships
Modeling, Trend Analysis	Mathematical or graphical representations of change through time	Aid in prediction of effects from multiple alternative actions	Data intensive, expensive, and technically rigorous
Geographic Information Systems (GIS) Mapping	Compilations of spatially explicit information to set boundaries on analyses	Can generate new insight in a CEA. Outstanding visual presentation	Data may be limited in some locations

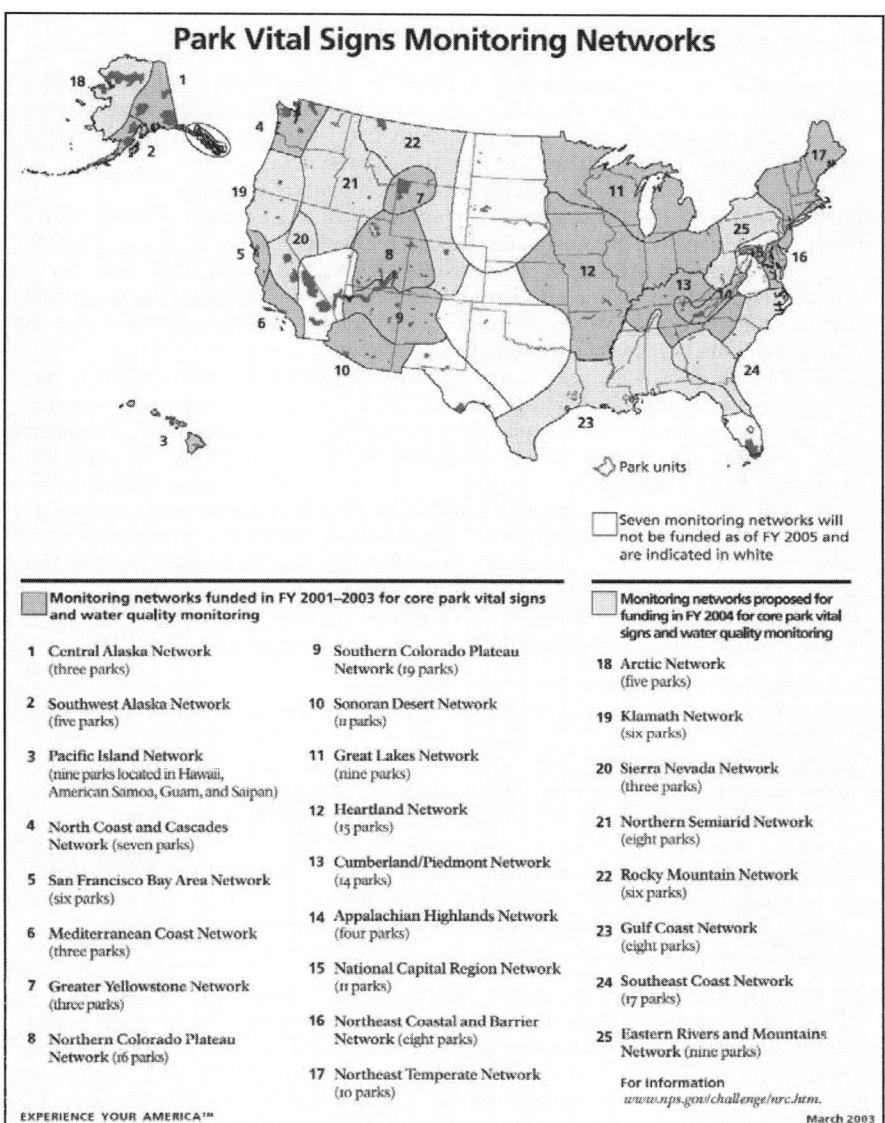

Fig. 27.2. Map of United States illustrating the "Vital Signs Monitoring Networks" of the National Park Service.

Established in 1999, 25 of the proposed 32 will be completed by the end of 2004. See Box 27.1 for more details. (Printed courtesy of US National Park Service)

Box 27.1. Cumulative Effects Monitoring of the US National Park System

Established in 1916, the US National Park Service (NPS) is responsible for over 32 million hectares of public land, preserving and protecting some of the world's most scenic and important natural resources in nearly 270 NPS units. Some of these units represent the last vestiges of once vast undisturbed ecosystems.

Many of the units are easily accessible to large numbers of people. In 2002, Yellowstone National Park hosted nearly 3 million visitors, Yosemite National Park, nearly 3.3 million. The NPS estimates that over 120,000 hectares in 195 of the park units have been disturbed by direct human activities, e.g. roads, trails, and campgrounds (NPS 2004a). Exotic plants, which reduce natural ecosystem diversity, infest over 1 million hectares. Moreover, the effects of indirect human activities—such as acid rain or global climate change—are largely unknown.

In 1999, the NPS announced a major effort to address these concerns and to improve the management of the natural resources under their care. This Natural Resource Challenge is essentially an action plan for Cumulative Effects Assessment. One of the basic programs identified in "The Challenge" is the Natural Resource Inventory and Monitoring Program (NRIMP).

NRIMP has a number of long-term goals, including: (1) to complete baseline inventories of basic biological and geophysical natural resources; and (2) to establish long-term monitoring programs to efficiently and effectively monitor ecosystem status and trends at various spatial scales (NPS 2004b). Cornerstone to this program are "Vital Signs," measurable early warning signals indicating changes that could impair long-term ecosystem health. Vital Signs Monitoring is a multi-step process (Jope 2001):

1. Identifying agents of change (e.g. visitors).
2. Identifying critical stressors that cause change (e.g. "trampling" or water use by visitors).
3. Determining responses in key ecological attributes. A stressor like "visitors' water use" could lead to a wide range of primary (direct) effects, like reduced streamflow, to secondary (indirect) effects, like loss of riparian vegetation and, in turn, loss of riparian bird habitat.
4. Identifying and ranking indicators. Pursuing the "visitors water use" example, measurable indicators of the effects of water withdrawal from an arid-land stream might include amphibian and macro-invertebrate population sizes and richness.
5. Establishing good protocols for sampling.

To facilitate collaboration, information exchange, and cost sharing, the roughly 270 NPS units are split into 32 monitoring "networks" linking parks that share similar geographic and ecological characteristics (Fig. 27.2). Each network designs a single, integrated program to monitor both physical and biological resources, such as air and water quality, threatened or endangered species, and exotic species. The list of Vital Signs will vary among the networks, reflecting the needs and resources of the component parks. The NPS will have begun implementation of Vital Signs monitoring programs in 25 monitoring networks by the end of 2004 and it is hoped that all 32 networks will be engaged by the end of 2005.

WCMC compiles biodiversity indicator research and makes extensive use of Geographic Information Systems (GIS) and other analytical technologies to visualize trends and identify new priorities for conservation action.

Following the assessment of baseline conditions in a CEA, the next step is to determine the environmental consequences of the proposed strategic action. The important cause-and-effect pathways of the proposed strategic action should be identified using system diagrams and models based on available data and theory. Then the likely magnitude and significance of cumulative effects can be analyzed. At this point, alternatives to the proposed action or to its particular components can be added to "avoid, minimize, or mitigate significant cumulative effects" (CEQ 1997). These alternatives (always including "no-action") can be ranked using matrices and compared with analyses of trends and impacts to ecosystems, economic, and social variables. The last and ongoing step to any CEA is the monitoring of cumulative effects following implementation of the proposed strategic action. Ongoing monitoring provides an iterative process that can identify the need for decreasing or increasing mitigation of cumulative effects, as well as provide new data for baseline determination in future strategic actions (CEQ 1997).

27.4 Transboundary Implications of Cumulative Effects

Transboundary environmental effects are of increasing concern in light of existing or planned international agreements deregulating trade (e.g. General Agreement on Tariffs and Trade [GATT], the North American Free Trade Agreement [NAFTA]) (see Box 27.2). Such trade-environment interactions involve a varied set of issues including pollution spillovers, social impacts, different standards of environmental regulation, and loss of sovereignty (Schramm 2000). Since no single country has responsibility for global cumulative impacts (e.g. global warming, atmospheric ozone depletion), there is considerable risk that such issues will receive inadequate attention. However, trade liberalization *can* have positive effects as well, such as an accelerated widespread conversion to cleaner energy technologies or universal environmental standards, if so implemented.

Even so, dealing with impacts to the unowned "global commons" (e.g. ocean resources, atmosphere) can be controversial. For example, in the absence of a multilateral environmental agreement, a recent GATT ruling favored the sovereignty of Mexico against the unilaterally imposed restrictions of its tuna imports by the US, saying that "a country may not use trade restrictions as a means of protecting environmental resources beyond its borders" (Schramm 2000). Nevertheless, when actions of one country contribute to environmental impacts of another, they should be analyzed as transboundary impacts. This is not always easy since questions of national sovereignty and access to data can limit both the desirability and practicality of such analyses. While no organization currently has the authority to prepare international environmental reports on agreements such as GATT, in 1999, the World Trade Organization (WTO) called for the creation of a World Environmental Organization parallel to the WTO that would establish a multilateral rules-based system for the environment (Schramm 2000).

Box 27.2. The Upper San Pedro Basin: Transboundary Cumulative Effects Mitigation in US-Mexico Borderlands

Semi-arid and arid regions cover approximately 30% of Earth's terrestrial surface and are characterized by the limiting role of water in ecological processes and human activities (Noy-Meir 1973). Worldwide, human activities threaten to permanently alter semi-arid and arid landscapes by rapid urbanization, ground water mining, overgrazing, and fire suppression (Swetnam and Betancourt 1998).

The Upper San Pedro Basin (USPB) is a borderland region covering 7610 km^2 in southeastern Arizona, USA and 1800 km^2 in Sonora, Mexico. A tributary of the Gila River, the San Pedro River flows north through the USPB and is the last permanently flowing, unimpounded stream in the region. Located between the Sonoran and Chihuahuan deserts, the river and its attendant riparian forest provide habitat for a rich assemblage of terrestrial and aquatic fauna including 250 species of migratory birds. A unique biological and cultural resource, the US federal government designated 60 km of the river north of the Mexican border as the San Pedro River National Riparian Conservation Area (SPRNCA) in 1988. SPRNCA is managed by the Bureau of Land Management, U.S. Department of Interior.

San Pedro River stream flow and riparian forest condition is closely linked to groundwater levels in the USPB (Goodrich et al. 2000; Stromberg et al. 1996). Conservation of San Pedro River biological resources is complicated by groundwater pumping for mining, ranching and municipal uses on both sides of the U.S.-Mexican border (Pool and Coes 1999). A groundwater dispute in the USPB provided a test of international environmental law under the North American Free Trade Agreement (CEC 1999).

Recognizing the need for a better understanding of the requirements of the biological and human communities in the USPB, the National Science Foundation began a science and technology center named Sustainability of semi-Arid Hydrology Riparian Areas (SAHRA) at the University of Arizona in 2001 (www.sahra.arizona.edu). Composed of researchers from eleven universities and five federal agencies in the southwestern US, SAHRA promotes sustainable management of water resources in semi-arid regions through interdisciplinary research, public outreach, and the education of water-users and managers.

To provide resource managers and public-policy decision makers with better information on the cumulative effects of groundwater pumping in the USPB, SAHRA is developing predictive watershed and regional-scale hydrologic models which incorporate ecosystem function (e.g. riparian vegetation evapotranspiration) climatic variability and projected human population growth to predict the effect of various proposed water resource actions on the biological communities supported by the San Pedro River.

To integrate physical and biological modeling of water resources with public-policy development, SAHRA hydrologists are collaborating with sociologists and economists in investigations of non-market valuation of water resources, alternative conservation strategies and the development watershed councils in both Mexico and United States. In addition, SAHRA promotes water resources education and collaboration with non-governmental organizations in integrated watershed management on both sides of the border (SAHRA Annual Report 2003).

Mitigation of transboundary cumulative effects is complicated by differences in national standards and enforcement of environmental law. In the case of US-

Mexico borderlands (see Box 27.2), where vast differences in governance exist, mitigation of cumulative effects associated with groundwater withdrawals is being pursued at a variety of scales. At the highest levels of government, the issue is being addressed through a multilateral trade agreement. At the local level, community involvement and exchange in the form of watershed councils is building trust between stakeholders and helping communities on both side of the border to identify common concerns.

27.5 Conclusions and Recommendations

The EU has a unique opportunity to set a new standard in the successful management of transboundary cumulative effects both within and across national borders of its member states. The evolving CEA process under NEPA in the US can provide a starting point for SEA managers and policy makers in the EU. Vital Signs monitoring in the U.S. national park system (Box 27.1) and research and public-policy oriented partnerships such as SAHRA (Box 27.2) could provide useful examples for future efforts to address cumulative environmental impacts in the EU.

The EU should make every effort to link individual national environmental policies under an overarching common structure, using examples of continental and worldwide monitoring and cumulative effects databases, such as EMAP in the US and the WCMC at the United Nations level. Only by incorporating a fundamental commitment to understanding and considering *cumulative effects* in every step of environmental assessment can the true environmental costs of human actions be measured and the necessary alternatives be introduced to avoid or minimize such costs. If this commitment by policy makers and the public is not realized, then the accumulation of human impacts will continue, and sustainability will never be achieved.

References[1]

CEC – Commission for Environmental Cooperation (1999) Ribbon of life, an agenda for preserving transboundary migratory bird habitat on the Upper San Pedro River. Montreal, Canada. Internet Address: http://www.cec.org, last accessed 15.01.2004

CEQ – Council of Environmental Quality (1997) Considering cumulative effects under the National Environmental Policy Act. Executive Office of the President, United States, January 1997. Internet address: http://ceq.eh.doe.gov/nepa/ccenepa/ccenepa.htm, last accessed 15.01.2004

Eccleston CH (2001) Effective Environmental Assessments: How to Manage and Prepare NEPA EAs. Boca Raton, Lewis Publishers, Florida

---

[1] Note that legislation mentioned in the chapter is listed at the end of the handbook in the consolidated list of legislation (Appendix 2).

EPA EMAP (2002) Research Strategy: Environmental Monitoring and Assessment Program. Document No. PA 620/R-02/002. U.S. Environmental Protection Agency. July 2002. Internet address: http://www.epa.gov/docs/emap/html/pubs/docs/resdocs/EMAP_Research_Strategy.pdf last accessed 15.01.2004

Glasson J, Thérivel R, Chadwick A (1999) Introduction to Environmental Impact Assessment: Principles and Procedures, Process, Practice, and Prospects, 2nd edn. UCL Press, London

Goodrich DC et al. (2000) Preface paper to the Semi-Arid Land-Surface-Atmosphere (SALSA) Program special issue. Agricultural and Forest Meteorology 105(1-3):3-20

Jope KL (2001) An approach to identifying "vital signs" of ecosystem health. In: Harmon D (ed) Crossing Boundaries in Park Management: Proceedings of the 11th Conference on Research and Resource Management in Parks and on Public Lands. The George Wright Society, Hancock, Michigan, pp 399-406

National Performance Review (1994) Creating a Government that Works Better and Costs Less. Washington, D.C.

Noy-Meir I (1973) Desert ecosystems: environment and producers. Annual Review of Ecology and Systematics 4:25-51

NPS (2004a) Inventory and Monitoring of Park Natural Resources: Discovering and Protecting America's Natural Heritage. US National Park Service. Internet Address: http://www.nature.nps.gov/protectingrestoring/index.htm, last accessed 15.01.2004

NPS (2004b) Protecting and Restoring: Overview. US National Park Service. Internet Address: http://www.nature.nps/protectingrestoring/IM/inventoryandmonitoring.htm, last accessed 15.01.2004

Odum WE (1982) Environmental degradation and the tyranny of small decisions. Bioscience 33:728-729

Pool DL, Coes AL (1999) Hydrogeologic Investigations of the Sierra Vista Subwatershed of the Upper San Pedro Basin, Cochise County, Southeast Arizona. U.S. Geological Survey Water Resources Investigation Report 99-4197. Tucson, AZ

SAHRA (2003) Science and Technology Center for Sustainability of semi-Arid Hydrology and Riparian Areas (SAHRA), 4th Annual Report to the National Science Foundation, 8/1/2002-8/31/-2003. Tucson, AZ: Dept. of Hydrology and Water Resources, University of Arizona, 2003

Schramm WE (2000) Evaluating trade agreements for environmental impacts: a review and analysis. In: Partidário MR, Clar R (eds) Perspectives on Strategic Environmental Assessment. Lewis Publishers, Boca Raton Florida

Stromberg JC et al. (1996) Effects of groundwater decline on riparian vegetation of semi-arid regions: the San Pedro, Arizona. Ecological Applications 6(1):113-131

Swetnam TW, JL Betancourt (1998) Mesoscale disturbance and ecological response to decadal climatic variability in the American Southwest. Journal of Climate 11:3128-3147

28 Cultural Integrity as a Criterion of SEA

Engelberth Soto-Estrada[1], Rina Aguirre-Saldivar[2] and Shafi Noor Islam[3]

1, 3 Environmental and Resource Management, Brandenburg University of Technology (BTU), Germany
2 Postgraduate Department of Environmental Engineering, National University of Mexico

28.1 Introduction

Environmental evaluation procedures like SEA have become a key component in development planning and decision making. With the aim to anticipate and mitigate its possible effects, the necessity for a better understanding of the environmental consequences created by the implementation of policies, plans and strategic actions has been recognized. The knowledge of environmental mechanisms has been developed gradually, through wide participation of stakeholders and the consideration of new evaluation criteria such as cultural impacts.

Due to the diversity that characterizes the cultural environment, opposed by definition to the natural environment (Vandenberghe 2003), the consideration of this element and the evaluation of its impacts have not been well integrated into the decision making process including SEA until present (Thérivel and Partidário 2002). Although some agencies have developed guidelines for the cultural impact assessment, noticeable differences persist in the way in which these impacts have to be evaluated. Until present, there exists no recognized and internationally approved declaration to define the aspects that must be included into a cultural impact assessment (UNEP 2003).

In this sense, the objective of this chapter is to generate a discussion with respect to the elements that could be considered within a cultural impact assessment procedure, to propose the use of evaluation indicators to define the concept of cultural integrity as a valuation criterion. In this chapter the problematic of identifying the cultural environment is mentioned, the inclusion of this aspect within the procedure is analyzed, the concept of culture and its possible indicators is discussed.

Implementing Strategic Environmental Assessment. Edited by Michael Schmidt, Elsa João and Eike Albrecht. © 2005 Springer-Verlag

Further, the construction of an evaluation methodology is outlined and finally, some recommendations are identified for a future implementation.

28.2 The Impact on Cultural Integrity

The diversity of modern activities is causing impacts on the cultural environment; a common example is the use of culture as a "trade good" for the tourism industry. The host community is generally the weak part of the relationship host-guest; taking any influence from the guest side in order to fulfil its requirements (UNEP 2003). The impacts arise when changes in the cultural values and customs take place and its integrity affected. These changes frequently take place in the community structures, family relationships, traditions, ceremonies and morality. As it happens frequently when different cultures converge, the socioeconomic impacts are ambiguous. The same impacts (evaluated objectively) are seen as beneficial for some groups and perceived as negative for others. According to the United Nations Environment Programme (UNEP 2003), the cultural impact caused by the tourist industry can be divided into the following influence areas:

Change or loss of values and identity: characterised by the "commodification" when culture is sold as a product and its characteristics must be adapted to the tourist requirements, the "standardization", the process to satisfy tourists' desires for family facilities in an unfamiliar environment (well-known fast-food restaurants or hotel chains), "loss of authenticity and staged authenticity", originated by the adaptation of the traditions to the tourists' taste, and the "change of productive patterns", that is the adjustment made by the local craftsmen to cover the new demand.

Cultural clashes: promoted by means of the convergence of cultures, the cultural clashes can arise through "economic inequality" due to consumption patterns and lifestyles of visitors that tend to influence the locals by developing a sort of copying behaviour, "annoyance produced by the behaviour of the visitors" who fail to respect (out of ignorance or carelessness) local customs and moral values, or by "job level friction", since normally the best positions are taken by the non-locals.

Physical influences causing social stress: with the development of the industry "resource use conflicts" can appear generating a competition between tourism and local populations for the preference over the use of services like water or energy, and "conflicts with traditional land-uses", in which the local communities are frequently the losers due to strong economic pressures.

Ethical issues: aspects as "crime generation", the practice of "child labour" and "prostitution and sex tourism". Although these aspects can not always be attributed directly to the tourism, there is a direct relation with the development of tourist facilities.

All these factors press the culture of the host community, causing changes in the values and local identity, that the "sacred" may not be respected when perceived as goods for trade (Simpson 2002 p.2; UNEP 2003). The result can be the overexploitation of the "social carrying capacity" (limits of acceptable change in the social system inside or around the destination) and "cultural carrying capacity" (limits of acceptable change in the culture of the host population) of the local community that derives in frictions between the haves and have-nots, between the visitors and the locals. In the following lines, some relevant aspects of the consideration of cultural aspects within SEA are mentioned.

28.3 Consideration of Cultural Elements within SEA

SEA includes strategic actions with repercussions on the environment or those that are already considered by other regulation instruments of the European Community. The Directive defines in its third Article, "scope", those sectors for which the realisation of an SEA is obligatory, including the tourist industry. Additionally, it is mentioned that the environmental report will have to consider the

> "likely significant effects on the environment, including on issues such as biodiversity, population, human health, fauna, flora, soil, water, air, climatic factors, material assets, cultural heritage including architectural and archaeological heritage, landscape and the interrelationship between the above factors" (Art. 5, Annex I f of the SEA Directive).

It is considered that the SEA legislation reflects an advance in the inclusion of cultural aspect, when taking its analysis when the strategic decisions are made. However, a procedure for the assessment and monitoring of impacts on culture does not exist yet. In order to consider this issue, some cultural elements are proposed to be used as evaluating indicators in order to define the effects of strategic actions on cultural integrity of a community.

The use of evaluation indicators supposes advantages for the accomplishment of the cultural impact assessment at SEA level, since it could facilitate the obtaining and analysis of information, the prediction of impacts, the definition of mitigation measures and the establishment of a monitoring procedure. As an example of the use of possible indicators, the cultural element "language" can be mentioned. The integrity of this element in a community can be determined by the use of statistical data in a suitable time period.

The State of Quintana Roo, on the Caribbean Sea, is the greatest receiver of foreign tourism of México and its territory contains an outstanding Maya heritage. In the last 20 years the population of Quintana Roo has increased up to 6% annually, whereas the indigenous communities have grown with 3%. This does not mean a reduction in the birth rate, but the loss of the use of traditional languages within these communities, since these statistics are based on the use of language (INI 2003).

In order to define a list of suitable indicators, the first step should be to delineate the elements that compose the term culture, the second to establish a mechanism for the selection of indicators for its evaluation.

28.4 The Concept of Culture

The term culture is used in at least three different senses: philosophical, anthropological and common. Culture in the broadest sense refers to everything that was created by humans and is socially transmitted and reproduced. Culture also represents a symbolic expression, the emanation of the "soul" of a community that differentiates that community from others and determines its "whole way of life, from birth to grave, from morning to night and even in sleep" (Eliot 1948 p.31 cited in Vandenberghe 2003 p.462).

According to the United Nations Educational, Scientific and Cultural Organization (UNESCO 2003), culture is regarded as the set of distinctive spiritual, material, intellectual and emotional features of a society or a social group and that it encompasses, in addition to art and literature, lifestyles, ways of living together, value systems, traditions and beliefs. Culture is the link in the chain that connects humans and makes the development of individuals possible. Additionally, culture has been defined as "the integrated pattern of human knowledge, belief, and behaviour, language, ideas, customs, taboos, codes, institutions, tools, techniques, works of arts, rituals, ceremonies and other related components" (Encyclopaedia Britannica 1991 p.784).

According to the previous definitions the elements that compose culture can be recognised in three different components, being "knowledge", "behaviours" and "products"; these three components stress the general integrated functions of culture (see Fig. 28.1).

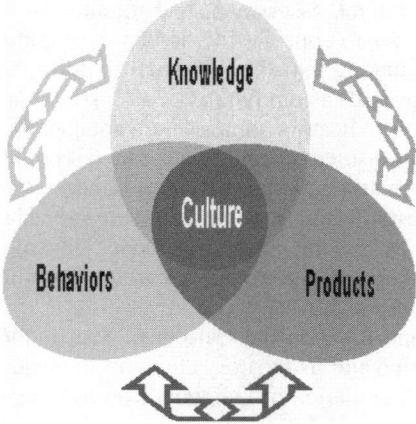

Fig. 28.1. Components of Culture

Although these components provide an approach to the methodological definition of a cultural element, some limitations subsist since it can not be easily identified by containing only one of the three components. Let us analyse the inclusion of some cultural elements. In this case, an investigation of the concept "food" is proposed, since it presents an essential characteristic of any human community. Food (raw material) according to the philosophical definition, represents a product created by humans. Therefore it can be identified in the category of "products". In the same way, the procedures used to prepare this food (process from raw material to products) can be transmitted and repeated, thus it is possible to locate this concept as "knowledge" and furthermore the culinary habits represent a part of the symbolic expression of the respective community and can be located within "behaviour".

Considering this, it is important to mention that this chapter does not try to catalogue – in an exhaustive way – the elements included in the concept of culture, but to provide ideas for the handling of "characteristic features", in form of indicators within the SEA procedure.

The elements that compose culture could also be divided through the consideration of cultural levels (see Fig. 28.2). In this system, the "cosmovision" represents the central level of cultural expression; it expresses the deep-rooted values maintained by a group of people. The "cultural values" differentiate one culture from another; according to Rokeach (1973), a "value" derives from a relatively permanent belief in which, in a particular situation, a specific way of conduct is personally or socially preferable. The "customs" are usually local and reflect the preferences for the type of food, dress, dance etc. Finally, the last level is composed by the "personal preferences".

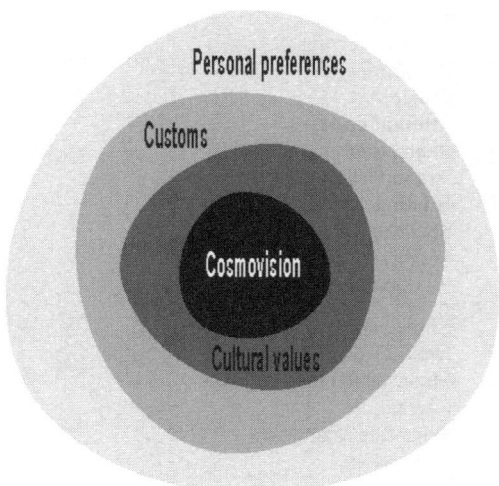

Fig. 28.2. Cultural Levels

By means of the classification, represented in Fig. 28.2, Table 28.1 has a list which is divided according to cultural level.

Table 28.1. Determination of Cultural Elements According to Cultural Level

Cultural level	Cultural element
Cosmovision	Religion
	Politics
	Economy
	Knowledge
Cultural values	Rites and beliefs
	Stories, myths and legends
	Systems of communication
	Language
	Writing
	Alphabet
	Literature
	Symbolic sign: oral and physical
	Transmission of experience
	Education
	Architecture
	Magic
	The subjective belief of a common origin: Identity, Ethnic
Customs	Food habits
	Dress
	Festivities
	Music and song
	Performing arts
	Dance
	Cultural goods
	Recreation
	Art and Calligraphy
	Hunting
	Social structure
	Social organisation
	Productive methods
Personal preferences	Choice and decision
	Dress
	Music and song
	Recreation
	Sports and games
	Food habits

When observing the elements in Table 28.1 some questions arise:

- What is the importance of each element?
- Which elements must be analysed within SEA?
- Which are the elements that determine the changes in a community?

Opinions are divided between the relation culture-development. A school of thoughts supports that culture determines how humans interact with nature, the environment and the universe. It is clear that ultimately all forms of development, including human, are determined by culture (Encyclopaedia Britannica 1991). Another school considers that this definition of culture is broad and argues that it must be restricted to the ways of living together. It must be noted that culture is not static, but dynamic and continually changing.

Different anthropological theories have tried to explain the factors that determine cultural changes. For "structuralism", for example, the important part is to know the social structure and the individual thoughts that are manifested by writing, myths and legends; therefore, the structure of the human language is equivalent to the society (Levi-Strauss 1963). On the contrary, "neo-evolutionism" considers social development processes; the point of departure is the intensification of agriculture, which leads to private property, then to specialisation, arising social stratification of classes (the "haves" and "have-nots"), a political centralisation and so on; cultural development is essentially the fight of mankind against environment in order to get the resources to sustain existence and to preserve the species (White 1959).

Different cultural groups modify the natural environment in a characteristic way creating unique regions. The geographic study of human-environment relationships is known as "cultural ecology" (Rubenstein 2003). The cultural ecology supposes that the factors that determine the cultural change are multiple. It is affirmed that the environment (the enviro-cultural adaptations) is an important factor of cultural change, since a deep change on it influences the amount and distribution of resources and therefore the amount of population, mainly in societies that have a less developed system of technology and that are less developed (Bohannan and Glazer 1988). The human is considered to be an organism that works within its physical environment and culture as a "catalyst", an adapter to this environment. However, as it is considered that a specific environment originates a specific society with its specific culture; also culture changes the society and originates the environment.

Other investigators indicate the necessity to study the "hidden tensions" of a culture, which motivate the communities to experience cultural changes (Theory of the conflicts), as well as the "mental factors" as determinants of the cultural differences throughout the world (Goodenough 2002), or "the symbolic factors" in order to find out the values of a culture (Turner 1975 cited in Herrero 2002). In general, it is affirmed that the importance of any cultural element is based on two conditions:

- it must serve to maintain the social system (the necessities of the group); and
- it must contribute to the psycho-biological necessities of the members.

As it is observed, the analysis of the concept of culture and its breakdown into elements, considering its diversity and subjectivity, is not a simple task. However, it is considered that the identification of cultural elements and their hierarchical structure should be carried out according to the scale at different cultural levels (see Fig. 28.2 and Table 28.1), remembering that the changes in the "cosmovision" will affect the "cultural values", the "customs" and the "personal preferences". It can be assumed that the more an indicator fits within the scale of cultural levels, the greater will be its potential effect on cultural change.

In this way, the selection of cultural elements and their hierarchical discrimination would suppose the consideration of the theoretical concepts previously mentioned, and the elements that would be used as evaluation indicators for the cultural impact assessment will be properly defined in the existing culture theories.

28.5 The Cultural Impact Assessment

The cultural and natural environments are linked together. The cultural (social) environment behaves at the same time as a receiving system of the alterations produced on the cultural environment and as a generator of the modifications produced on it.

The concept of "cultural impact" alludes to the consequences perceived by a human community due to the implementation of any strategic action or project that causes significant changes in its conception of world, values, customs or personal preferences. The need to outline a set of principles and guidelines for assessing the cultural impact is based on the demand for diminishing cultural damage, especially in those countries with weak standards for the preservation of their patrimony and heritage.

Taking into account possible or detected cultural impacts, some governments have begun to formulate policies and to establish competent agencies for the protection of the cultural patrimony. To mention some examples, the Environmental Council of the State of Hawaii (ECSH 1997) proposes the accomplishment of the cultural evaluation procedure based on consultation with experts, meetings with communities, interviews, compilation of oral histories, and finally an analysis of the aspects of subsistence, commerce, residence, agriculture, recreation and religious and spiritual customs in a final report.

In another example, the Mexican Environmental Agency (SEMARNAT) suggests the inclusion of cognitive aspects, values and collective norms, beliefs and signs, considering information of the influence zone of the project (land uses, sites of importance for the community), the historical patrimony (existing historical-artistic and archaeological monuments) and the level of acceptance of the project in the community (SEMARNAT 2002). The analysis procedure is proposed to be based on the evaluation of available statistical data, considering a reference period of at least 30 years.

Moreover some Districts of New Zealand (e.g. Waimakariri or Clutha) carried out cultural impact assessments with the aim of determining how the local communities may be affected by wastewater discharges. These studies were elaborated on the basis of literature review, interviews and site visits with the overall goal to analyse the impacts of these aboriginal groups upon spiritual values and their productive vocation (CDC 2003; WDC 2003).

Although these proposals are focused on the valuation of projects subject to environmental impact assessment (EIA), it is supposed that they are useful when carrying out the evaluation of cultural impacts at PPP level, considering that SEA:

- may includes a large scale, involving multiple activities and communities;
- may has to cope with limited information where, for instance, environmental data collected in different countries are incompatible, limited or come from evaluations with different detail levels (Thérivel and Partidário 2002);
- may is subject to the application of diverse legislations and policies;
- may has to face uncertainty about future environmental, economic and social conditions, as well as uncertainty about consequences that could derive from the application of the strategic action;
- may has to contemplate a great variety of impacts, including the additive effects of many small projects or management schemes for which SEA is not required, synergistic impacts, where the impact of several projects exceed the sum of their individual impacts, and global impacts such as biodiversity decrease and greenhouse gases emissions (Wood 1995 cited in Thérivel and Partidário 2002).

28.6 Cultural Impact Assessment Procedure

Considering the procedure defined in the SEA Directive and in its document of implementation (EC 2003), the stages that are considered within a cultural impact assessment are (see Fig. 28.3) – considering during the whole process the public participation and the assessment of the stractegic actions' alternatives:

- *Considering PPP objectives and targets*, including the time-scale defined for its attainment, since the greater the time-scale, the greater the uncertainty associated within the prediction of the future impacts.
- *Identifying alternative PPP,* allows the decision maker to determine which strategic action is the best option.
- *Scoping*, considering the PPP's scale, since different "scales" of PPP (e.g., regional, local) should address different types of impacts. For instance, an international-level SEA should focus primarily on global issues, while a local SEA should emphasize community aspects.
- *Establishing cultural indicators*. These must be defined from a list of pre-established indicators that can be evaluated, considering four main aspects: cosmovision, cultural values, customs and personal preferences.
- *Describing the baseline environment* to identify cultural indicators which the PPP expected impacts can be appraised. It also provides information for the monitoring stage of predicted impacts.
- *Predicting possible impacts* from selected indicators, with the use of different techniques: scenarios, checklists, sensitivity analyses, etc., or considering the combination of several of these methods.

- *Determining the significance of impacts*, which can be based on the cultural integrity criterion.
- *Considering mitigation measures*, which is the mayor advantage of SEA over EIA since it allows the consideration of a wider range of mitigation measures in a more appropriate stage of decision-making.
- *Monitoring.* The selected indicators can be used for monitoring of predicted impacts.

It is proposed a set of cultural elements that could be used as indicators within the evaluation process that can be selected from a list of pre-established indicators as a part of the results of this investigation. The selection will have to be made considering the importance that the cultural element has for a particular community (SEMARNAT 2002), in addition taking into account, as it was mentioned before, its location within the scale of cultural levels (see Fig. 28.2).

Additionally, an evaluation system for cultural impacts will be established to allow the determination of the significance of cultural integrity considering the rules for the evaluation of synergistic and cumulative impacts in order to generate prospective scenarios.

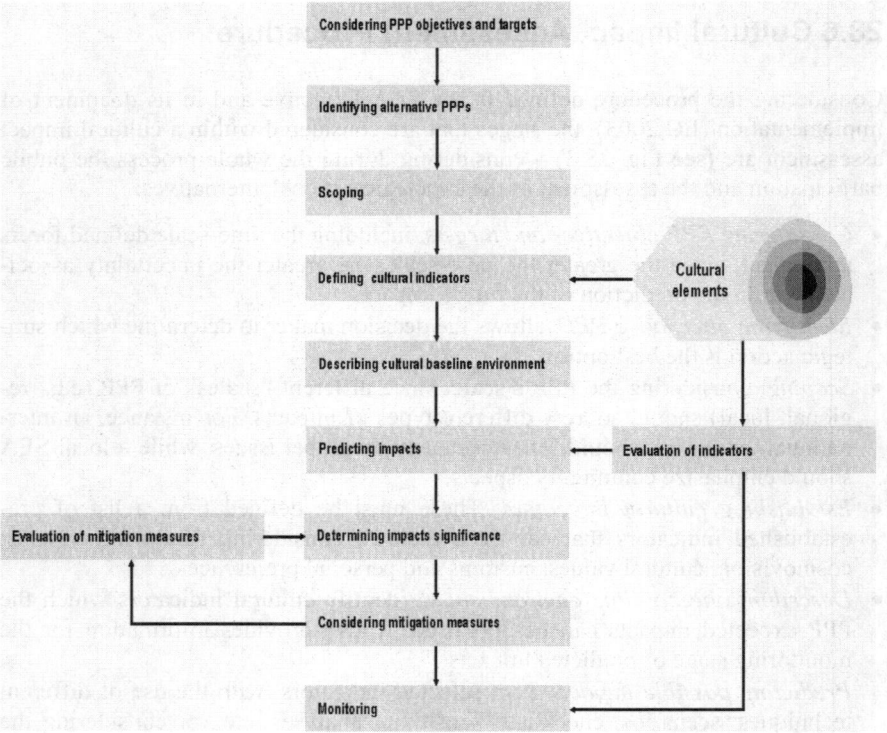

Fig. 28.3. Cultural Impact Assessment Procedure

28.7 Conclusions and Recommendations

SEA can be considered to represent an advance in the conservation of cultural integrity, when defining the analysis of projects of sectors that can press it as compulsory, specifying the inclusion of the cultural heritage as an evaluation element and undertaking its valuation when strategic decision are made. Therefore it is advisable to evaluate cultural impacts on a well defined basis.

The elements that compose culture can be divided into four cultural levels: cosmovision, cultural values, customs and personal preferences. Finally, the following research elements are suggested:

- The list of cultural elements, from which the evaluation indicators can be selected, have to be developed.
- The mechanism of assigning values to these indicators, which lead to the evaluation of the significance of cultural integrity.
- The establishment of rules to obtain synergistic and cumulative impacts, which allow the generation of prospective scenarios also need to develop.

References[1]

Bohannan P, Glazer M (1988) High Points in Anthropology. McGraw-Hill, Inc., New York

CDC (2003) Cultural Impact Assessment on the Wastewater Discharges from the Clutha District Council Operated Treatments Plants. Clutha District Council. Internet address: http://www.cluthadc.govt.nz/, last accessed 20.02.2004

ECSH (1997) Guidelines for Assessing Cultural Impacts. Environmental Council of the State of Hawaii, Office of Environmental Quality Control. Internet address: http://www.state.hi.us/health/oeqc/guidance/cultural.htm, last accessed 1.2.2004

EC – European Commission (2003) Guidance on the Implementation of Directive 2001/42/EC on the assessment of the effects of certain plans and programmes on the environment, Luxembourg

Eliot TS (1948) Notes towards a Definition of Culture. Faber & Faber, London. Cited in: Vandenberghe F (2003) The Nature of Culture. Towards a Realist Phenomenology of Material, Animal and Human Nature. Journal for the Theory of Social Behaviour 33 (4):461-475

Encyclopaedia Britannica (1991) The New Encyclopaedia Britannica 3, London

Goodenough O (2002) Information Replication in Culture: Three Modes for the Transmission of Culture Elements through Observed Action, in Imitation in Animals and Artefacts. M.I.T. Press, USA

Herrero J (2002) Curso básico de antropología social y cultura (Basic course of social and cultural antheopology). Universidad Ricardo Palma, Lima, Perú. Internet address: http://www.sil.org/capacitar/antro/cursoantro.htm, last accessed 20.01.2004

---

[1] Note that legislation mentioned in the chapter is listed at the end of the handbook in the consolidated list of legislation (Appendix 2).

INI (2003) Diagnostico de los Pueblos Indígenas de Quintana Roo (Analysis of the indigenous villages of Quintana Roo). Instituto Nacional Indigenista. Internet address: http://www.ini.gob.mx, last accessed 15.9.2004

Lévi-Strauss C (1963) The Structural Study of Myth, in Structural Anthropology. Basic Books, New York

Rubenstein J M (2003) The cultural landscape an introduction to human geography. Pearson Education Inc., Oxford

SEMARNAT (2002) Guía para la presentación de la manifestación de impacto ambiental del sector Turístico (Guidelines for presenting environmental impacts in the tourism sector). Modalidad: particular. Secretaría de Medio Ambiente y Recursos Naturales, Dirección General de Impacto y Riesgo Ambiental

Simpson J (2002) Tourism, Cultural Integrity and Environmental Conflict. Gawler High School, South Australia. Australian Geography Teachers' Association National Conference

Thérivel R, Partidário MR (2002) The Practice of Strategic Environmental Assessment. Earthscan, London

Turner V (1975) Symbolic Studies. Annual Review of Anthropology. Cited in Herrero (2002)

UNEP (2003) Socio-Cultural Impacts of Tourism. United Nations Environment Programme, About Sustainable Tourism. Internet address:
http://www.uneptie.org/pc/tourism/sust-tourism/social.htm, Accessed 10.10.2003

UNESCO (2003) Universal Declaration on Cultural Diversity. United Nations Educational, Scientific and Cultural Organization. Internet address:
www.unesco.org/confgen/press_rel/021101_clt_diversity.shtml, Accessed 10.10.2003

WDC (2003) Cultural Impact Assessment, Eastern District Sewerage Project. Waimakariri District Council. Prepared by Ecological Services. Internet address:
http://www.waimakariri.govt.nz/, last accessed 20.2.2004

White L (1959) The Evolution of Culture: The Development of Civilization to the fall of Rome." McGraw-Hill, New York

Wood C (1995) Environmental Impact Assessment: A Comparative Review. Longman, Harlow. Cited in Thérivel and Partidário (2002)

29 Requirements and Methods for Public Participation in SEA

Stefan Heiland

Leibniz Institute of Ecological and Regional Development (IOER), Germany

29.1 Introduction

Up to now the discussion concerning SEA neglected questions on consultation and public participation (both terms are used here as synonyms) in SEA. This applies both to the German situation (where the author's main expertise lies) and to the international "state of the art". A GEOBASE search, on all the references that mentioned SEA between 1997 and August 2004, returned 133 records. Only 14 abstracts included the term "participation" and only four references seem to have put a strong focus on that issue (Kornov 1997; Kravchenko 2002; Mathiesen 2003; McCracken and Jones 2003).

This is most surprising considering the importance that public participation will most probably have for environmentally relevant strategic actions in the future. This chapter therefore wants to contribute to the further discussion about requirements and methods of public participation in SEA. It pays special attention to the requirements that have to be fulfilled by public participation within the SEA. Therefore it investigates different questions in the following sections.

- What are the chances of public participation, which factors may exacerbate its realization? (Sect. 29.2)
- What does "public" mean? Which actors must therefore get the possibility to participate? (Sect. 29.3)
- What does participation mean? Which influence do the stakeholders have on the decisions of the planning agencies? (Sect. 29.4)
- Which methods exist to fulfil the requirements on public participation and which experiences have already been made? (Sect. 29.5)

The chapter shows that an effective public participation offers different opportunities, e.g. to improve and qualify the plan or programme, to identify and mitigate conflicts in an early stage, to improve the transparency and comprehensibility of the planning process and its results as well as the public acceptance of the plan or programme. To utilize these chances, it is helpful or even necessary to use methods that have especially been used for informal public participation – for example in Local Agenda 21 Initiatives or within urban development processes.

29.2 Opportunities and Obstacles of Public Participation in SEA

Public participation within SEA offers a variety of opportunities for planning processes, which, however, can be jeopardized by different objections and obstacles (Danielzyk et al. 2003; Grotefels and Uebbing 2003; Heiland 2003a; Sheate et al. 2001a). The following opportunities can be named:

- Public participation enhances the transparency of decision-making processes by offering the opportunity to an effective and broad participation in planning processes,
- Public participation enhances the completeness, validity and reliability of the relevant information,
- Public participation adds to the appropriate assessment of all concerned interests,
- Public participation helps to identify and mitigate conflicts in an early stage of the planning process,
- Public participation allows to take into account the needs of the concerned population and the environmental concerns in the subsequent specific projects better then up to now,
- Public participation facilitates a better understanding between planning agencies, public authorities, NGOs and citizens,
- Public participation leads to an higher legal certainty for the planning agencies and to a simpler and faster procedure in the following planning steps as well, and, therefore
- Public participation generally leads to a better result as regards to content and to a better acceptance of the strategic action by the (concerned) population.

These opportunities are juxtaposed to apprehensions and objections which can lead to an insufficient support of public participation by relevant actors, for example planning agencies, planners and authorities. In consequence the chances of public participation may not be utilised and the possible quality of the results may not be approached. The following apprehensions are frequently mentioned (Heiland 2003a):

- higher work load,
- higher costs,

- higher requirements concerning often unfamiliar informal methods of participation,
- higher need of adjustment and modification of the strategic action while public participation takes place,
- delay of the planning process.

It is necessary to bear in mind those apprehensions in order to mitigate the subsequent obstacles of public participation in SEA.

29.3 What Does "Public" Mean?

The constitutive regulations to public participation in SEA are to be found in Art. 6 SEA Directive (cf. Chap. 30). The SEA Directive makes a difference between the public in general which has the right of access to the draft plan or programme on the one hand (Art. 6 para 1) and – as a subset of the public in general – the public which has the right to express its opinion on the draft plan or programme and on the accompanying environmental report on the other hand (Art. 6 para 2). It is up to the member states "to identify the public for the purposes of paragraph 2, including the public affected or likely to be affected by, or having an interest in, the decision-making subject to this Directive, including relevant non-governmental organisations" (Art. 6 para 4). As it may be difficult to differ between "affected" and "not affected" or even "having an interest in" and "not having an interest in" (Grotefels and Uebbing 2003), in Germany Federal Building Code and the Federal Spatial Planning Act do not make such a difference. Therefore it can be assumed that public participation (in Germany) means "participation of everyone".

Besides that, the question is discussed, whether it is necessary to enlist all members of the public in general or whether it is sufficient to enlist only representatives, such as associations or NGOs. The supporters of "representative participation" refer to the high expenditure of "general participation", whereas the supporters of this solution emphasize that representative participation does not fulfil the demand of the SEA Directive (Danielzyk et al. 2003; Grotefels and Uebbing 2003). Also the quality of the required information and the result of the SEA may not be as good as it could be, if only a few selected actors have the opportunity to participate. According to Sheate et al. (2001a p.3) "widespread involvement of stakeholders, policy makers and the wider public is crucial for successful SEA". However, participation in different loads of intensity and in different forms is needed in accordance to different administrative and spatial levels (see Sect. 29.5.1).

29.4 What Does Participation Mean?

29.4.1 Different Ways to Understand Participation

Participation can be, and usually is, interpreted in different ways – from merely giving information over allowing taking part in decision-making processes up to a joint elaboration of solutions. Above that, participation can refer to the basic question whether something should be done at or just on details of a plan and its fulfilment (Heiland 2003a). Fig. 29.1 gives an overview about different levels of participation, concerning the involvement of the participants in the decision. In a different way Danielzyk et al. (2003; following Bischoff et al. 1996) distinguish between three functions of public participation:

1. *Information* includes public relations without any possibility for the public to give statements or to take influence on the decision-making process.
2. *Participation* offers the opportunity to the public to express its opinions in an active manner. (In this case participation is defined as one function of 'public participation' which implies a difference between "participation in a narrow sense" and "participation in a broader sense". This shows obviously the ambiguous content of the term participation.)
3. *Cooperation* refers to decision-making processes between equal partners and includes the possibility for jointly developed solutions.

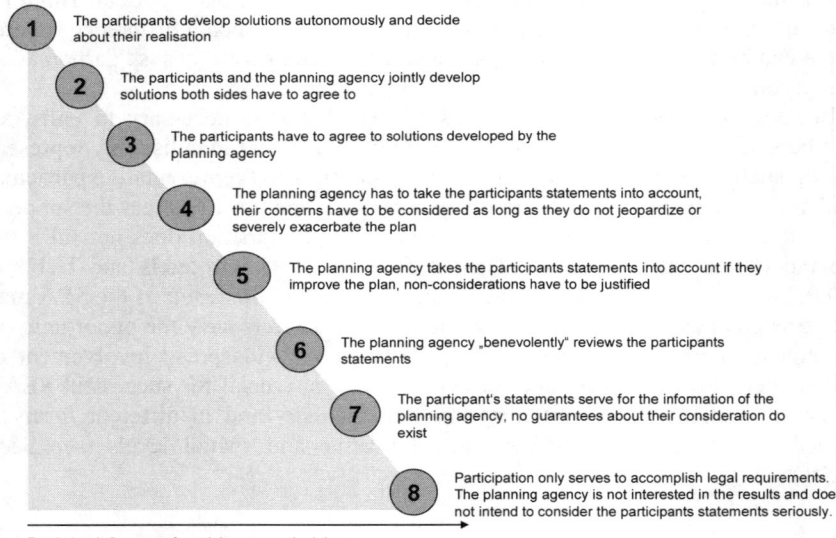

Fig. 29.1. Levels of participation (Heiland 2003a)

Planning agencies and public authorities should be well aware of these different meanings of the term participation and assume that therefore the expectations on the participation process may deeply differ between (parts of) the public and themselves – irrespective of the legal position. It is crucial that the level of the participation and its meaning for the strategic action is made clear to all participants in advance. Otherwise two risks may arise: 1) the risk of frustration about an (assumed) "alibi-participation", in which the opinions of the participants do not have a real influence on the strategic action and therefore 2) the related risk of low acceptance of the participation process and the strategic action as well. Both risks may lead to the so-called "consultation fatigue".

How is participation defined by the SEA Directive? According to Art. 6 para 2 the public "shall be given an early and effective opportunity within appropriate time frames to express their opinion on the draft plan or programme and the accompanying environmental report". The expressed opinions "shall be taken into account during the preparation of the plan or programme and before its adoption or submission to the legislative procedure" (Art. 8) and the public has to be informed about how the opinions have been taken into account and about the reasons for choosing the plan or programme as adopted (Art. 9, para 1). This means that neither SEA in general nor the consultation process in particular offer a right or a guarantee to take textual influence on the decision. "SEA is a supporting tool for the decision-making process, not more, but not less (Sheate et al. 2001b p.15).

29.4.2 Procedural Requirements for Public Participation

The SEA Directive does not make a statement concerning the concrete procedure of the consultations, it just names requirements on the procedure mentioned above (Sect. 29.4.1). Therefore the member states have the opportunity to design the participation process in very different ways. A simple compliance or a narrow interpretation of the SEA Directive may not be enough to utilise all chances of public participation. In addition to legal concerns, the social context in which participation is embedded has also to be taken into account. This context decides about the acceptance, the effectiveness and the achievement of the objectives of strategic action, SEA and consultation.

Public participation always meets certain social expectations concerning the process and the results. If the expressions of opinions do not affect the strategic action at all, the public will get the impression, that participation is only an "alibi-participation", which is not worthwhile. This again may lead to poor acceptance of the consultations and of the strategic action as well. For this reason it is necessary to take into account that during the last decades many, mostly informal, participation methods have been developed and applied and that many citizens demand the opportunity to participate (even if they do not always join in).

Participation, understood as mere information of the public and as dutiful ticking off the expressions of opinion does therefore not accomplish the tasks of the SEA Directive. Above that it does not take advantage of the chances of public participation within SEA. „Successful SEA is an active, participatory and educational process for all parties, in that stakeholders are able to influence the decision-maker, and the decision-maker is able to raise awareness of the strategic dimensions of the policy, plan or programme" (Sheate et al. 2001a p.4). Which requirements have thus to be fulfilled to ensure an early and effective participation within appropriate time frames?

Requirements concerning publication and dissemination of the plan or programme and the environmental report are:

- Utilisation of all helpful means of dissemination, including public media and internet
- Reasonable accessibility to and opening hours of the offices, where the information can be studied
- Easy and cheap availability of the documents, for example by internet-download
- Comprehensibility and visually pleasing layout of the documents; this refers especially, but not exclusively, to the non-technical summary of the environmental report
- The means used for publication and dissemination must not lead to an exclusion of certain persons or groups from participation. For example, the dissemination via Internet is not sufficient as long as not everybody has access to this media (see EC 2003 p.39)

Time and duration of participation also have to fulfil certain demands:

- Consultations must take place at a time, where no irreversible decisions have been taken yet. Therefore it might be useful to enlist the public even before the draft plan or programme has been prepared, so that the draft does not have to be altered profoundly during or after the participation process (Grotefels and Uebbing 2003). A very early consultation can also be required, because informal preparations for the strategic action itself might sometimes take place long before the official beginning of the procedure – and therefore some important preliminary decisions might be taken in advance (see Sheate et al. 2001b p.14).
- The period of time to express one's opinion must be long enough to allow the public to become familiar even with abstract strategic actions. "Different time frames may be appropriate for different types of plan or programme but care should be taken to allow sufficient time for opinions to be properly developed and formulated on lengthy, complex, contentious or far-reaching plans or programmes" (EC 2003 p.36)

Even if the SEA Directive does not regulate those requirements in detail, planning agencies may take advantage of paying regard to them. Otherwise the public might accuse the agency of not having been informed well enough – apart from legal considerations. Only if the public *feels* to be satisfactorily informed the opportunities of public participation can be seized. Bearing this in mind, a broad and pro-active consultation process should be in the interest of every planning agency.

29.4.3 The "Participation Paradox"

Even if participation seems to be very helpful or even crucial within the SEA – does the public (individuals, associations, NGOs and so on) really want to participate? This could be doubted for some good reasons, especially if strategic actions are concerned. Strategic actions on this level of decision are usually very abstract, far away of "everyday life" and their consequences for one's own needs may be difficult to be estimated. Therefore the interest to participate will diminish with the increasing abstractness of plans or programmes.

On the other hand, the range for possible decisions usually is much vaster at the level of strategic actions than on the level of specific projects, because political commitments and the so called inherent necessities are usually much stronger on project-level. For this reason participation within SEA could offer the public an opportunity to get more influence on the design of subsequent plans and projects.

Both aspects together lead to a phenomenon that could be called "participation paradox": The more influence the public could take on the subsequent design of plans and projects, the less it avails itself of the opportunity to participate. But with the diminishing influence in subsequent and more concrete planning steps the wish to participate increases. Experiences with different forms of public participation confirm this fact. Regarding this, the apprehension that pro-active and informal participation methods and requirements would set off an "avalanche of participation" seems to be unfounded (see Grotefels and Uebbing 2003, case studies in Sheate et al. 2001b).

29.5 Methods of Public Participation within SEA – Possibilities and Experiences

29.5.1 The Variety of Methods

The SEA Directive does not include details concerning forms and methods of consultations. "There are many different methods and techniques for public consultation. These range through seeking written comments on draft proposals, public hearings, steering groups, focus groups, advisory committees or interviews. It will be important to select the most appropriate form of consultation for any given plan or programme" (EC 2003 p.39). Therefore it is possible to apply not only formal but also informal methods that have been developed for example in urban development or Local Agenda processes. The internet will play a vital, but only additional, role too – be it only to inform the public or to feature the opportunity to express opinions, for example by online-forms (see Danielzyk et al. 2003; EC 2003 p.39).

Table 29.1 offers an overview about potential methods and their suitability for the three different levels and intensities of participation mentioned in Sect. 29.4.1 (information, participation, cooperation). Of course different methods can or must be combined.

Table 29.1. Possible methods of public participation and their suitability for information, participation, cooperation (+ suitable, ++ very suitable)

	Information	Participation	Cooperation
Press	++		
Other print-media (e.g. flyer)	++	+	
Internet	++	++	
Presentations / discussion	++		
Exhibition	++		
Public display	++		
Explanation	+	++	
Open council	++	++	
Working groups / Workshops		++	++
Mediation		++	++
Round Table		++	++
Future search		++	++
Agenda Conference		++	++
Open Space		++	++

For information about most of the mentioned methods see Bischoff et al. (1996), for Open Space see Owen (1997), for Agenda Conference see Heiland (2003b)

For example presentations will usually be part of an open council, exhibitions may be part of an open space. Above that, it has to be considered that participation and cooperation include or call for intensive information.

The (combination of) methods and the procedure of the consultation have to be chosen depending on the spatial and administrative level of the strategic action. A local land-use plan will permit and as well require other methods than a national programme. In smaller municipalities the whole population can be invited to attend events personally, such as Open Councils, Workshops etc. On higher administrative levels where usually more people and actors are concerned it will only be possible to invite selected persons for personal statements. This should be well chosen representatives of all relevant social groups, while the participation of interested individuals has to be ensured by opportunities for written statements. It has to be guaranteed that both forms of participation are taken into account equally. On every spatial and administrative level general participation and representative participation will play a complementary role – with different emphasis and different methods.

Attention must be paid to the fact that informal methods might not be a priori accepted by planning agencies. Higher costs and expenditures may be arguments against those methods. It can be replied that the advantages they offer, such as better results and higher acceptance of the strategic action, mitigated resistance against subsequent projects and therefore less time delay will probably outweigh possible disadvantages. However, doubts have to be taken seriously, especially if they refer to financial and personal problems of public authorities.

29.5.2 Case Studies

Case studies about and experiences from consultations in SEA are still rare due to the facts, that only a few SEA have been carried out (and documented) yet and that even less have taken special regard of public participation. The following examples are taken from Sheate et al. (2001b). They were accomplished voluntarily and before the SEA Directive had been adopted. Nonetheless they might give valuable advices and stimulations for public participation within SEA in the future. For some further case studies from Austria see Chap. 42.

The study "SEA and Integration of the Environment into Strategic Decision-Making" (Sheate et al. 2001b) includes 19 international case studies on national, regional and local level. Only 12 of these 19 examples are called SEA by the study itself, only in 7 out of 12 cases public participation took place and final conclusions could be withdrawn, in one case only associations were enlisted. Therefore only 6 significant case studies remain, where the public had (at least theoretically) the possibility to participate.

The case study "Canada – Framework of SEA for Trade Negotiations" (Sheate et al. 2001b pp.22-29) had some severe weaknesses concerning public participation. According to the authors, the time frame of 45 days offered not enough time for a thorough review of the complex topic, the notification of the participation process in the Canada Gazette and in the planning agencies web site was inadequate and during the negotiations was given no possibility for public participation at all.

The integrated SEA of the Danish Report on National Planning (Sheate et al. 2001b pp.30-38) offered two opportunities for public involvement: first during a public pre-consultation period before the SEA took place, secondly during a public hearing before the scope of the environmental assessment was fixed. To both dates the required information was only given via the homepage and the newsletter of the National Spatial Planning Department. Mainly counties, municipalities and organisations but very few individuals made comments and most comments "did not relate to the environmental effects of the plan. This lack of attention may be explained by the fact that the evaluations of environmental effects indicated only directions for the environment rather than more concrete predictions of impacts" (Sheate et al. 2001b p.37). According to the study public participation seems to have been the weakest part of the SEA. Obviously it is not enough to offer the opportunity to participate if the chosen methods are not sufficient. Little individual participation is not amazing if information is only given via two media which are hardly recognized by "usual" citizens.

Within the SEA of the Wind Power Plan for the Spanish region Castilla y León (Sheate et al. 2001b pp.122-130) public participation was envisioned and theoretically possible, but the opportunity could hardly be seized out of several reasons: 1) the time frame for public review (30 or 45 working days, depending on different provinces) was too short due to the length of the document and a missing non-technical summary; 2) the notification for the consultations was made only through the Official Gazette of Castilla y León; 3) the availability of the plan was limited, no copies were made and consultation was only possible in the developer's offices; 4) the plan was divided into 9 provincial plans, the integral document was only available in the capital cities of the provinces. This example too shows that effective public participation is impossible unless the necessary preconditions are guaranteed.

The SEA to the revision of the land-use plan of the municipality of Weiz, Austria, (Sheate et al. 2001b pp.7-14) was integrated into the procedure for revising the land-use plan according to existing law. On this account every inhabitant of Weiz could comment on the drafted plan and the environmental report within eight weeks in written form, all comments had to be taken into account by the competent authority, the result and its reasons had to be published. The following methods were used: formal meetings within the municipality, scoping meetings, round tables and a public hearing. The responsible project group "conducted a short and easy-to-understand explanation paper and published it in the local official newspaper including a simplified map of the drafted land-use plan and a nontechnical summary of the environmental statement. Nevertheless, the public felt that the SEA was somewhat abstract" (Sheate et al. 2001b p.13), what made participation more difficult. From an environmental point of view it can be criticised that the SEA started after the revision of the land-use plan. On this account important preliminary decisions had already been taken, so that the City Council did not choose the most environmentally sustainable option. On the other hand, the SEA raised the environmental awareness of the local actors and caused no time delay in revising the land-use plan. Sheate et al. (2001b p.10) summarize that the project "showed that the quality of the communication structure is one of the crucial issues for successful SEA".

A similar case is the SEA of the revision of the land-use plan with integrated landscape plan of the city of Erlangen, Germany (Sheate et al. 2001b pp.56-63). Here again the procedure took place within the legal framework of the German Federal Building Code. The mechanisms of communication and participation were clearly defined. In addition to formal procedures, informal measures were used, for example discussions within the eight City district councils, special meetings with farmers, Open Councils for all citizens or a symposium with neighbour municipalities. Similar to the Weiz case study the integration of SEA in existing legal procedures strengthened the integration of environmental issues into local decision-making.

29.6 Conclusions and Recommendations

As the SEA Directive does not include any detailed specifications about the procedure of public consultation, a relatively vast range of different methods, forms and intensities remains. Public consultation should be used in order to ensure effective public participation that could contribute to better results of the strategic action, to an early revelation and solution of conflicts, to transparency and to a higher acceptance of the strategic action and the planning agency as well. To seize these opportunities it is necessary to give feedback to the public how its participation affected the decision. Also methods and techniques should be used that have been applied mainly in informal participation processes, for example in urban development or within Local Agenda 21. Despite higher expenditures that may be necessary for those methods in the beginning, experience shows that the neglect of participation requirements can lead to severe consequences afterwards: less acceptance of the plan, higher resistance and opposition against it, time delay and the necessity of complex, extensive and expensive changes of the plan or programme.

References

Bischoff A, Selle K, Sinning H (1996) Informieren, Beteiligen, Kooperieren. Kommunikation in Planungsprozessen. Eine Übersicht zu Formen, Verfahren, Methoden und Techniken (Information, participation, cooperation. Communication in planning processes. A survey of forms, procedures, methods and techniques). Dortmunder Vertrieb für Bau- und Planungsliteratur, Dortmund

Danielzyk R, Knieling J, Hanebeck K, Reitzig F (2003) Öffentlichkeitsbeteiligung bei Programmen und Plänen der Raumordnung (Public participation on programmes and plans for land use planning). Ressortforschungsvorhaben BMVBW. Bundesamt für Bauwesen und Raumordnung (BBR), Bonn

EC – Commission of the European Union (2003) Guidance on the Implementation of Directive 2001/42/EC on the assessment of the effects of certain plans on the environment. Brussels

Grotefels S, Uebbing C (2003) Öffentlichkeitsbeteiligung in der Raumordnung. Anforderungen der Richtlinie über die Prüfung der Umweltauswirkungen bestimmter Pläne und Programme vom 27.06.2001 (Plan-UP-RL) an die Aufstellung von Raumordnungsplänen (Public participation in land use planning. Requirements of the SEA Directive from 27.06.2001 on the compilation of land use plans). NuR 8:460-468

Heiland S (2003a) Die Öffentlichkeits- und Behördenbeteiligung in der SUP - Chancen, Nutzen, Erfordernisse (Participation of the authorities and the public in SEA – opportunites, benefits, requirements). UVP-report 2003, Sonderheft zum UVP-Kongress 2002: 93-96

Heiland S (2003b) Informieren, bewerten, planen, handeln: Agenda-Konferenz – ein Verfahren der Bürgerbeteiligung (Inform, evaluate, plan, act: Agenda-Conference – a method of public participation). UVP-report 17:180-183

Kornov L (1997) Strategic environmental assessment: sustainability and democratization. In: European Environment 7(6):175-180

Kravchenko S (2002) Effective public participation in the preparation of policies and legislation. Environmental Policy and Law 32(5):204-208

Mathiesen AS (2003) Public participation in decision-making and access to justice in EC environmental law: The case of certain plans and programmes. In: European Environmental Law Review 12(2):36-51

McCracken R, Jones G (2003) The Aarhus Convention. Journal of Planning and Environment Law (7):802-811

Owen H (1997) Open Space Technology, A User's Guide. San Francisco, 2. Aufl.

Sheate W, Dagg S, Richardson J, Wolmarans P, Aschemann R, Palerm J, Steen U (2001a): SEA and Integration of the Environment into Strategic Decision-Making. Final Report.

Executive Summary. European Commission. Contract No. B4-3040/99/136634/MAR /B4, London 2001. Internet address: http://europa.eu.int/comm/environment/eia/sea-support.htm, last accessed 26.03.2004

Sheate W, Dagg S, Richardson J, Wolmarans P, Aschemann R, Palerm J, Steen U (2001b) SEA and Integration of the Environment into Strategic Decision-Making. Final Report. Volume 3. Case Studies. European Commission. Contract No. B4-3040/99/136634/M AR/B4. London 2001. Internet address: http://europa.eu.int/comm/environment/eia/sea-support.htm, last accessed 26.03.2004

30 Public Participation for SEA in a Transboundary Context

Harry Meyer-Steinbrenner

Department of Questions in Principle of the Environment, Agriculture and Forestry, International Cooperation, EU; Saxon State Ministry of the Environment and Agriculture, Germany

30.1 Introduction

Public participation and transboundary issues play a vital role in the SEA procedure. In all EU Member States, including the ten accession states, the SEA Directive came into force in 2001. On the other hand, some EU Member States such as Germany have not fully adopted the Directive yet.

Germany intends to amend the EIA Act by adding rules and SEA procedural steps such as public participation, transborder consultation, content of an environmental report and monitoring. Considering the federal structure of Germany all 16 federal states (*Laender*) will set up their own rules and regulations which will recognise specific requirements of the regions. Not only the EIA Act contains legal provisions on SEA but also other acts contain SEA rules (see Chap. 7). The most important provision of this kind is the Federal Spatial Planning Act, which rules the requirements of SEA on regional planning procedures.

Despite the lack of practical experience with the SEA Directive some practical experience with national SEA-procedures could be obtained before the SEA-Directive was published. Experience with SEA before the EC Directive was published have been documented for example by Dusik at al. (2001) for Central and Eastern Europe, the Newly Independent States of the Former Soviet Union (NIS) by Cherp (2001), Norway by Swensen (2001), Czech by Vaclaviková and Jendrike (see Chap. 13), Poland by Maćkowiak-Pandera and Jessel (see Chap. 14), Rzeszot (2001), Bulgaria by Grigorova and Metodieva (2001), and Slovenia by Kontic and Marega (2001).

Implementing Strategic Environmental Assessment. Edited by Michael Schmidt, Elsa João and Eike Albrecht. © 2005 Springer-Verlag

Tackling all types of plans, both existing and proposed planning systems from local to regional level in England, the Office of the Deputy Prime Minister (2003) has published SEA guidelines for planning authorities. This publication provides guidance on environmental assessment of English land use and spatial planning in accordance with the European SEA Directive. It also makes recommendations on how public participation should be carried out. A recent comprehensive overview of European SEA case studies has been published by ICON (2001a). Outside Europe, a good example of public participation for SEA in spatial planning is the SEA Guideline Document in South Africa published by the Department of Environmental Affairs and Tourism (CSIR 2000). This document reports on six case studies of different land use types such as the Cape Town 2004 Olympic Bid and the huge industrial site of Somchem Krantzkop, Wellington.

All these studies show that public participation in some countries played a vital decision making role in spatial planning during the entire SEA process. Nevertheless, the SEA Directive strengthens this instrument mainly through the involvement of NGOs, citizens and experts in a broader way.

In Germany, an official way of public participation is well tested within the regional planning process, especially in terms of consultation with authorities. However, such consultation often took place only when a new proposal of a regional plan was elaborated (Hoppenstedt 2003).

31.2 Public Participation in the SEA Directive

As defined in Art. 6 on "Consultations" and Art. 7 on "Transboundary consultations" of the SEA Directive, the responsible authorities have to set up public participation and consultation procedures for the SEA as early as possible. Art. 6 requires that:

> "the draft plan or programme and the environmental report [...] shall be made available to the authorities [...] and the public."

Para 2 states

> "The authorities [...] and the public [...] shall be given early and effective opportunity within appropriate time frames to express their opinion on the draft plan or programme and the accompanying environmental report before the adoption of the plan or [...] its submission to the legislative procedure."

Art.7 requires transboundary consultations

> "Where a Member State considers that the implementation of a plan or programme being prepared in relation to its territory is likely to have significant effects on the environment in another Member State, or where a Member State likely to be significantly affected [...]" It also states that "[...] the Member State in whose territory the plan or programme is being prepared shall, before its adoption or submission to the legislative procedure, forward a copy of the draft plan or programme and the relevant environmental report to the other Member State."

The SEA Directive does not require full consultation of the public until the Environmental Report is finished. In support of issues and option's report, the responsible authority may find it useful to publish environmental and sustainability information in conjunction with scoping and alternatives development. This therefore helps to facilitate discussion earlier in the SEA process. Feedback from the public may also provide more information or highlight new issues for the Environmental Report or sustainability appraisal (Office of the Deputy Prime Minister 2003).

30.3 Methodological Aspects

In Germany, responsibility concerning Regional Development Planning is not at the federal level but at the Ministry of Interior in all sixteen different *Laender*. Environmental Assessment however is a concern to the local Environmental authorities. Regarding SEA in regional development planning, it is assumed that the Ministries of Environment organise and perform public participation and consultation. Ultimately SEA in regional planning can only take place in close cooperation with the Ministries of Environment, Ministries of Interior and the responsible planning agencies such as regional planning associations or communities.

Effective public participation needs transparent and well explained decisions. This enhances people's understanding and adds value to the final decisions made at the planning process. It gives the public insight into the complicated planning process and probable long-term environmental problems. It promotes timely disclosure of relevant information to participants in the environmental planning and decision-making process.

The organization of public participation in regional planning according to the SEA Directive is not carried out in practice yet, due to lack of certain procedures. However, what is relevant to the public in a particular procedure is still open. Furthermore, the questions how to involve the public, how to provide access to information and the content of this information stay open as well. These questions however can only be clarified by practicing SEA in regional planning procedures.

It is presupposed that the development and application of appropriate methods to engage responsible parties like stakeholders or the public at strategic level could be difficult, nonetheless it is essential. Particular effort is required to identify the "affected public". NGOs may be able to act as a proxy for the general public, but not in all cases. It may be necessary to establish an organized and/or qualified public participation for the purpose (ICON 2001b). The Aarhus Convention gives an orientation for minimum requirements (see Chap. 3).

Objects of public participation and consultation procedures are both key issues of the draft plan and the Environmental Report. Both documents must be available at the same time, as an integral part of the consultation process, and the relationship between both documents should be clearly indicated.

According to Annex I of the SEA Directive the report should document the following issues:

- The baseline situation and the description of any problems
- The foreseeable effects of the draft plan and how they were evaluated
- How preferred options were chosen
- How environmental considerations were deemed necessary in the plan
- How mitigation measures have been incorporated into plan
- The proposed monitoring arrangements

In cases where plans go through different successful consultation exercises, the draft plan should be kept under review. If this results to changes in predictions and evaluated effects, new information should be made available in the same way as before.

30.4 Public Participation for SEA in a Transboundary Context

It is much more complicated to organize public participation in a transboundary context according to Art. 7 of the SEA Directive. In this case, a close cooperation with the concerned environment and planning authorities of the affected countries is essential. Public participation should be organized as a common task. Obviously different legislation and proceedings of the EU States considerably influence the course of public participation procedures. Unfortunately however, very little experience has been acquired with regard to public consultation in transboundary context. Therefore, only the execution of multi-national SEA projects could deliver the needed experience about procedures and methods.

Despite lack of experience, it is obvious that public participation in a transboundary context plays a prominent role in cross border spatial planning. It improves relations between people and countries, and prevents transboundary environmental conflicts. It also develops civil societies and democracy among countries in the EU region.

30.5 Pilot Project

In order to gain the necessary practice with the application of the SEA Directive in cross border regional planning, the East German State of Saxony, also known as the Free State of Saxony, scheduled a pilot project for revision of the regional plan "Upper Lusatia – Lower Silesia". This is the place where Germany, Poland and the Czech Republic meet (Fig. 30.1). The pilot project is two years long and will run from June 2004 to June 2006. The goal of the project is to find common procedures and proceedings, as well as technical methods of SEA in a trans-boundary context. Questions of extended public participation and cross border organization of strategic monitoring are also a key focus of the pilot project.

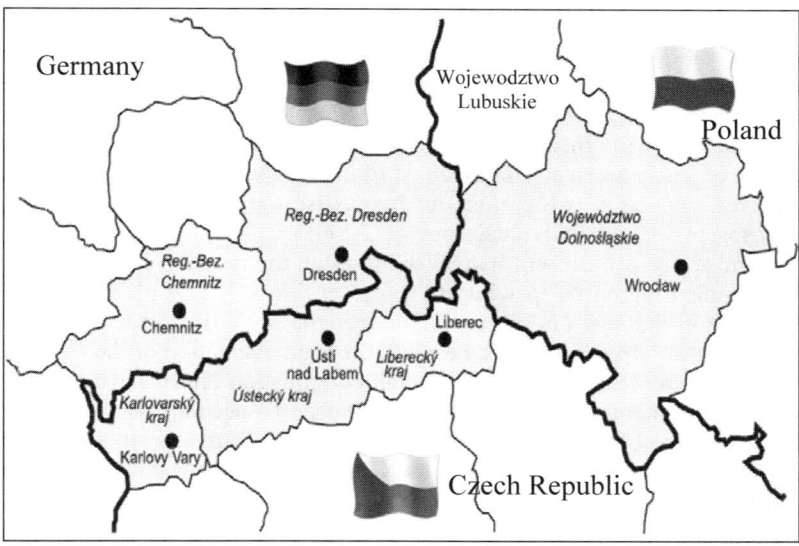

Fig. 30.1. Trilateral border region of Germany (Saxony), the Czech Republic and Poland

Note: The project area comprises the Regierungsbezirk (Region) Dresden, the Polish regions Wojewodztwo (Region) Dolnoslaskie and the southern parts of the Wojewodztwo Lubuskie, the Czech krajs (Regions) Liberecky and Ustecky

The extended participation covers consultations in neighbouring regions as well as cross border States. In Saxony, SEA of regional planning needs cross border participation with EU accession States such as Poland and the Czech Republic. Considering the fact that all three states have different legislation, culture, historical preconditions, planning requirements and public awareness, much effort in consultation and mediation is needed to achieve the above stated goal (compare the situation in Germany, see Chaps. 7 and 29, in the Czech Republic, see Chap. 13 and in Poland, see Chap. 14).

One of the most difficult problems is the language barrier between the three regions. Authorities often do not communicate properly due to language problems. The risk of misunderstanding and misinterpretation of planning results is high. The three states overcome this problem through regular regional consultations between the responsible Ministries and regional technical authorities. This also promotes acceptance of SEA in this region.

A special problem in focus is how to involve and manage the public. This requires strategic involvement of public institutions, schools, universities, press, churches, NGOs and so on. The communication chains between these stakeholders are developed differently. The project will show how an optimal public management can be achieved.

The expectations of all three partner regions are high concerning the project results. Responsible authorities therefore need clear results and recommendations about this issue. Saxony, which has four of the five planning regions along its international border with Poland and the Czech Republic, has high expectations about the project results and recommendations. The pilot project through consideration of existing research activities and projects shall develop solutions and give recommendations to the textual form of the SEA, cross border participation and coordination of environmental assessment. According to the EU requirements, coordination of environmental assessment is of high priority and relevance. This can be achieved with SEA, which aims to harmonise the consideration of environmental aspects between individual states (EC 2003).

The aim of this project will be realised through the revision of a regional plan for the planning region "Upper Lusatia – Lower Silesia". This planning region is the *only* one in the new *Laender*, which experiences a cross border participation liability with *two other* national states with the implementation of the SEA procedure. Thus, on the one hand especially high coordination efforts have to be made, but on the other hand benefits are to be expected. To accomplish the aim of the project before the SEA Guideline comes into force and to develop recommendations for SEA in Saxony, Poland and the Czech Republic, the following working steps are planned:

- Through close cooperation with the responsible authorities of the three partner states, an assessment concept on SEA in the planning region "Upper Lusatia – Lower Silesia" shall be developed with its practical applicability tested. As already mentioned, of special importance is the reconcilement of cross border environmental impacts in regional planning and the organization of environmental monitoring.
- Through cooperation of Polish and Czech actors the extent of project results' transferability to Polish and the Czech Republic regional planning shall be analysed.
- After a compatibility assessment of technical, legal regulations and instrumentation of SEA in Saxony, specific requirements of a partner country could be identified.
- The last step is the creation of a transnational SEA practice proposal to reconcile SEA in regional planning. This concept contributes to the implementation of the described requirement of the EU for a harmonisation of environmental interests.

In order to ensure a close cooperation between Saxon, Czech and Polish project partners, regular communications, working meetings as well as international workshops are planned. Further developed instructions and regulations of SEA in the partner states within the 25 EU Member States, will be published. This shall promote a discussion on cooperation in a cross border transnational context.

Table 30.1 gives overview of the work schedule in the pilot project. This has to be agreed upon at the trilateral kick off meeting on 1 October 2004.

Table 30.1. Work packages and work steps

Number of Work step	Tasks
Work package 1: Analysis of regional planning (contents/aims), derivation of conditions for SEA (running time till 10.2004)	
1.1	Analysis of regional planning
1.2	Elaboration of an indicator/criteria set for SEA
1.3	Consultation on contents of SEA and indicators/criteria with Saxon, Polish and Czech partners
Work package 2: Data investigation / elaboration of a catalogue of required environmental data in Saxony, Poland and the Czech Republic (running time till 01.2005)	
2.1	Fixing of themes and details of environmental assessment
2.2	Data collection in Saxony, Poland and the Czech Republic
2.3	Documentation of necessary existing SEA data
Work package 3: Elaboration / testing of designated methods/contents for SEA referring to Saxon regional planning (running time till 09.2005)	
3.1	Development and testing of the designated methods/contents for SEA of the Regional Plan "Upper Lusatia-Lower Silesia"
3.2	Discussion of results of the first work step with Saxon, Polish and Czech project partners
	Workshop on 08.2005
Work package 4: Data investigation/elaboration of a catalogue on necessary environmental data in Saxony, Poland and the Czech Republic (running time till 10.2005)	
4.1	Discussion on possibilities of transborder cooperation/participation of authorities, stakeholders, NGOs and the public
4.2	Documentation of a common concept for transborder participation
4.3	Testing of the designated participation concept for the Region "Upper Lusatia – Lower Silesia"/utilization together with Saxon, Polish and Czech project partners
Work package 5: Environmental monitoring (running time till 03.2006)	
5.1	Analysis of necessary environmental monitoring
5.2	Comparison of existing environmental monitoring and necessary environmental monitoring according to the SEA Directive
5.3	Discussion and designation of a common proposal for transborder monitoring

Table 30.1. (cont.)

Work package 6: Comparability of contents/methods of SEA in Saxony with SEA in Poland and the Czech Republic (running time till 04.2006)

6.1	Consultation on contents of SEA for regional planning in Saxony, Poland and the Czech Republic, discussion on results of the tests
6.2	Analysis/documentation of possibilities and limits of a common approach

Work package 7: Data investigation/elaboration of a catalogue of needed environmental data in Saxony, Poland and the Czech Republic (running time till 06.2006)

7.1	Analysis of further international approaches referring SEA in regional planning
7.2	Discussion and elaboration of similar and specific approaches to SEA procedures in all three partner states Saxony, Poland and the Czech Republic

Final workshop (06.2006)

The project with the title "Strategic environmental assessment for regional planning – development of a transnational assessment and practice concept for Saxony, Poland and the Czech Republic" is financed by the EU-INTERREG IIIA Program. Project leader is the Leibnitz Institut für ökologische Raumentwicklung e.V. (IÖR), Dresden. Project partners are the Regionaler Planungsverband Oberlausitz-Niederschlesien, Bautzen (Regional Planning Authority Upper Lusatia-Lower Silesia, Bautzen) as well as the Brandenburg Technical University (BTU) in Cottbus. The project is supervised by the Saxon State Ministry of Environment and Agriculture (SMUL) in close cooperation with the Saxon State Ministry of Interior (SMI).

30.6 Conclusions and Recommendations

The SEA Directive requires the responsible authorities to set up public participation during the entire assessment procedure of a plan or programme as early as possible. Most of the EU Member States already have more or less extensive consultation procedures for regional planning but they only take place after the new proposal of a regional plan has been accomplished. It is supposed that the SEA Directive will improve participation procedures and lead to better understanding of planning steps and goals. Public participation plays a prominent political role in cases where planning in transboundary context is elaborated. It therefore helps to prevent environmental and political conflicts among nations.

To gain the required practice in the application and implementation of this new instrument, the Free State of Saxony is carrying out an SEA-pilot project for the

Regional Plan "Upper Lusatia-Lower Silesia" in the trilateral region of Saxony, Poland and the Czech Republic, which runs from June 2004 utill June 2006. Questions concerning transborder consultation and public participation, application of strategic environmental indicators, set up of transborder monitoring, and the management of SEA within the legal and procedural frame of regional planning will be worked out.

References[1]

Cherp A (2001) Strategic Environmental Assessment in the Newly Independent States of the Former Soviet Union. In: The Regional Environmental Center for Central and Eastern Europe (ed) Proc. of Int. Workshop on Public Participation and Health Aspects in Strategic Environmental Assessment. Szentendre, Hungary, pp 57-69

Department of Environmental Affairs and Tourism (CSIR) (2000) Strategic Environmental Assessment in South Africa. Guideline Document. Pretoria, South Africa

Dusik J, Sadler B, Mikulic N (2001) Developments in Strategic Environmental Assessment in Central and Eastern Europe. In: The Regional Environmental Center for Central and Eastern Europe (ed) Proc. of Int. Workshop on Public Participation and Health Aspects in Strategic Environmental Assessment. Szentendre, Hungary, pp 49-56

EC – Commission of the European Union (2003) Guidance on the Implementation of Directive 2001/42/EC on the assessment of the effects of certain plans and programs on the environment. Brussels

Grigorova V and Metodieva J (2001) Strategic Environmental Assessment of Varna Municipality Development Plan. In: The Regional Environmental Center for Central and Eastern Europe (ed) Proc. of Int. Workshop on Public Participation and Health Aspects in Strategic Environmental Assessment. Szentendre, Hungary, pp 81-85

Hoppenstedt A (2003) Strategic Environmental Assessment (SEA) in Germany-Case Study Regional Planning. Internet address:
http://www.sazp.sk/eia/zbornik/html/english/12.htm, last accessed 08.08.2004

Imperial College Consultants Ltd (ICON) (2001a) SEA and integration of the environment into strategic decision-making. Vol. 3: Case studies. London

Imperial College Consultants Ltd. (ICON) (2001b) SEA and the integration of the environment into strategic decision-making. Executive summary. London

Institute of Environmental Management and Assessment (IEMA) (2002) Perspectives: Guidelines on participation in environmental decision-making. IEMA, London

Kontic B, Marega M (2001) Strategic Environmental Assessment in Slovenia: Summary of Methodological Topics. In: The Regional Environmental Center for Central and Eastern Europe (ed) Proc. of Int. Workshop on Public Participation and Health Aspects in Strategic Environmental Assessment. Szentendre, Hungary, pp 127-134

Office of the Deputy Prime Minister (2003) The Strategic Environmental Assessment Directive: Guidance for Planning Authorities. London

Rzeszot U (2001) Strategic Environmental Assessment of Regional Land-use Planning: Lessons from Poland. In: The Regional Environmental Center for Central and Eastern

---

[1] Note that legislation mentioned in the chapter is listed at the end of the book in the consolidated list of legislation.

Europe (ed) Proc. of Int. Workshop on Public Participation and Health Aspects in Strategic Environmental Assessment 77-80. Szentendre, Hungary, pp 77-80

Swensen I (2001) The Application of Strategic Environmental Elements in Land-use Planning in Norway. Small Steps to Improvement. In: The Regional Environmental Center for Central and Eastern Europe (ed) Proc. of Int. Workshop on Public Participation and Health Aspects in Strategic Environmental Assessment Szentendre, Hungary, pp 71-76

31 Developing Quantitative SEA Indicators Using a Thermodynamic Approach

Jo Dewulf and Herman Van Langenhove

Environmental and Clean Technology Research Group ENVOC, Ghent University, Belgium

31.1 Introduction

A wide number of people, institutions and organizations have proposed a definition of 'Sustainability'. One widely accepted definition is from the Brundlandt Report, stating that sustainable development is a development which fulfils the needs of present generations without endangering the possibilities of fulfillment of the needs of future generations (World Commission on Environment and Development 1987). From this definition, it is clear that sustainable development can only be accomplished if it is approached in a multidisciplinary way. The definition gives rise to questions in philosophic, economic, political, social and natural sciences. Elkington (1997) translated the sustainability issue into a concept with a social, economic and environmental bottom line. From this perspective, there is no doubt that natural sciences are highly important. They do not only study the environmental system, but they are also the basis of technology that manufactures the goods for society, starting from resources delivered by the natural environment. Technology is indispensable in a sustainable development since it provides the goods to fulfill the needs, now and in the future. However, technology simultaneously endangers the possibilities of current and future generations. The threat can arise from direct threats, such as acute danger or toxicity. On a longer time scale, the interaction of technology with the natural environment has to be taken into account. It extracts material and energy resources out of the ecosystem depleting technological potential for future generations. It also emits waste products into the natural environment that can interfere with natural ecological mechanisms, resulting in reduced resource production capacity. The pressure on the environment originates from our industrial society that requires a tremendous flow of materials.

Implementing Strategic Environmental Assessment. Edited by Michael Schmidt, Elsa João and Eike Albrecht. © 2005 Springer-Verlag

SEA allows one to take the right decisions with regard to impacts of our modern society on the environment (Office of the Deputy Prime Minister 2003; TRL ltd and Collingwood Environment Planning 2002). There is no doubt that SEA can be of value if one strives to develop sustainable technology. SEA in general, and in case of technology particularly, is well served by making use of a lot of information being embedded in indicators (Thérivel and Partidário 1996). SEA should make use of indicators, either qualitative or quantitative, that take into account social and economic issues, next to the physical chemical realm, the latter including environmental aspects. This chapter focuses on technology assessment: developing sustainable technology is crucial when one aims at sustainable technology. It shows how thermodynamics can contribute to developing quantitative indicators being of value in SEA. To do so, it starts with an analysis of the interactions of technology with the environment (Sect. 31.2), followed by a brief explanation of thermodynamics that can be used (Sect. 31.3). Sect. 31.4 shows how quantitative information for SEA of technological options can be generated from thermodynamic analysis.

31.2 Sustainability and Technology

Technology is a key issue in our modern society affecting sustainability in general and the environment in particular. Therefore, SEA of technology is essential if SEA wants to contribute to a better and more sustainable society. Today, technologists become more and more aware of the environmental impact of technology. They target at environmental impact reduction as is illustrated in Sect. 31.2.1 and 31.2.2. Some of them make also use of metrics to assess the environmental impact (Sect. 31.2.3).

31.2.1 Reduction of the Direct Environmental Impact of Technology

In order to reduce the environmental impact of technology, cleaner and clean technology came up in the nineties (Clift 1995, 1997; Clift and Longley 1995). Internationally, Cleaner Production is defined by the United Nations Environment Programme (UNEP) as a conceptual and procedural approach to production that demands that all phases of the life cycle of a product or of a process should be addressed with the objective of prevention or minimisation of short- and long-term risks to human health and to the environment. Going beyond Cleaner Production, Clean Technology has been defined as a means of providing a human benefit which, overall, uses less resources and causes less environmental damage than alternative means with which it is economically competitive. This latter definition concentrates not on products or processes, but on the benefit which the product or process provides, matching the fulfilment of needs in the Brundlandt definition of sustainability.

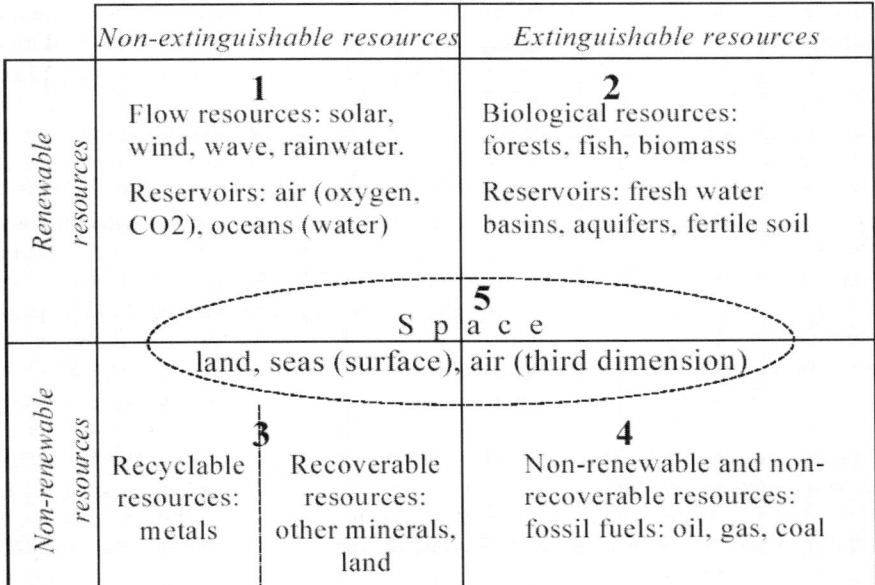

Fig. 31.1. Natural resources: classification in extinguishable and non-extinguishable resources, renewable and non-renewable resources, and recyclable, recoverable and non-recoverable resources (European Commission 2002)

The clean technology strategy is stronger in terms of sustainability because of several reasons. It can give insight being of value for SEA of technological options. By considering the whole production process with all possible environmental interactions, it does not only look to the effects of all emissions but it takes also into account the role of resources to be extracted out of the natural environment that are required to start up the production. The attention for resource intake is growing from an environmental sustainability point of view because of their potential depletion and because of the fact that they are the precursors of emissions that are generated later on. Resources can be divided in several types, as is illustrated in Fig. 31.1. By taking into account the type of resources in designing overall production chains, clean technology can also result in a reduction of the economic cost since it eliminates expensive end-of-pipe technology.

31.2.2 Technology within the Industrial Metabolism

Cleaner and clean production strategies are based on a life cycle approach, i.e. reduction of the environmental impact of the generation and disposal of a product to provide a service by looking to the interface between the production and consumption chain and the natural environment. However, our current industrial society is a complex network of material flows. Product generation, usage and disposal have not only an interface with the natural environment, but also with the industrial metabolism. Indeed, products can be partially based on waste materials instead of virgin natural resources; the product after its end of life may be the resource for the generation of the same product (recycling) or another product (conversion). This means that the environmental pressure from a product is not only the result of direct resource consumption and waste emission immediately related to its life cycle, but also of its embedment in the industrial metabolism. SEA for technology can be well aided by the industrial metabolism concept. This approach is the major idea in the industrial ecology theory (Graedel and Allenby 1996; Lowe et al. 1997). Here, the minimization of the indirect environmental impact of the generation, usage and disposal of a product is achieved by an adequate integration in the industrial system. There is no longer a distinction between resource and waste: in principle all materials have the two characteristics simultaneously. Materials are kept as long as possible in the industrial metabolism. Linear flow through material systems should be converted as much as possible in cyclic systems. The first principle of industrial ecology has to be mentioned here: individual firms have to be connected into industrial ecosystems. Whereas the principles of industrial ecology are quite clear, the assessment of the 'cyclic character' and metabolic efficiency of industrial systems is more difficult, especially from a quantitative point of view.

31.2.3 Metrics for Technology Assessment

The principles and concepts to come to an appropriate strategy to develop sustainable technology are outlined above. SEA of technology can make use of these principles, particularly if it can make use of qualitative and/or quantitative indicators that reflect to what extend these principles and concepts can be brought into practice. Indeed from the one-liner "you cannot control what you don't measure", development of quantitative assessment indicators is essential for an effective environmental strategy and policy.

The quantitative assessment of the sustainability of different technological options is not straightforward. Based on the cradle to grave principle, currently Life Cycle Assessment (LCA) may be one tool of choice. LCA consists of a goal and scope definition, an inventory analysis, an impact assessment and an interpretation (ISO 1997).

Currently, new indicators focusing on the overall industrial metabolism show up. One popular and easy-to-communicate indicator is the Ecological Footprint (EF). Ecological Footprinting was developed in the 1990s by the academics Mathis Wackernagel and William Rees in Canada (Wackernagel and Rees 1995; Chambers et al. 2001).The Ecological Footprint of a population (i.e. an entire industrial metabolism) is the area of land and water ecosystems needed to provide the resources and assimilate the waste of the population being studied. Since the area of land owned or controlled by the population is usually a finite and identifiable quantity, it can be compared to the EF. This juxtaposition of the actual land available and the EF and the difference between the two, termed "the ecological deficit", serves to show the dependence of the population of ecosystem situated outside of the political area. Several shortcomings of the concept have been mentioned, such as its lack of transparency, the sole focus on carbon dioxide with respect to emitted gases and the hypothetical translation of the carbon dioxide emissions into a required area to absorb them (Ecotec 2001).

Other indicators that are able to cover a whole industrial metabolism are Total Material Requirement (TMR) and Material Flow Analysis (MFA) (Bringezu and Schütz 2001; Spangenberg et al. 1999). Here, the mass flows as the result of our industrial society are quantified. The unit of quantification is mass in kg. The choice of this unit makes the concept easy to understand; however it is simultaneously the weakness of the concept: reduction of the environmental pressure of our material flows or dematerialisation is more than weighting kilograms (van der Voet et al. 2003). For example, the potential of 1 kg of sand as a resource is different from 1 kg of natural gas; also the environmental impact of their waste products is different.

From the examples above, it is obvious that adequate tools for SEA are not easily available. It is even questionable if one single indicator or even one single set of indicators can do the job for all cases where SEA is requested. According to Levett (1998), we should take a modest 'fitness-for-purpose' approach in developing indicators, i.e. using different indicator sets for different purposes, rather than straining to produce a single definitive set of sustainable development indicators. In this sense, a set of indicators for SEA particularly for technological options might be targeted, provided that one starts with a consistent basis that takes into account the physical chemical realm both in the industrial metabolism and in the natural environment.

In the next paragraphs, we show a limited set of quantitative indicators that can be helpful for SEA of technology. The limited set of indicators is based on thermodynamics that consider the physical chemical realm in the natural environment and the industrial metabolism that is the result of the implementation of different technological options.

31.3 Thermodynamics – its Basic Laws and the Exergy Concept

Both our industrial society and the natural environment are subjected to the basic laws of thermodynamics. Thermodynamics is a discipline in natural sciences that studies the transformation of energy and materials through processes. The basic laws of thermodynamics are rather abstract and were for a long time hard to apply in practice. In the late twentieth century, applications in engineering are found, exemplified by the handbooks of Çengel and Boles (1994) and Bejan (1997). Also applications in the study of natural ecosystems are reported (Jorgensen and Müller 2000). Nowadays thermodynamics is one of the basic courses in numerous curricula, e.g. in physics, biology, chemistry, engineering.

Two laws, the so-called first and second law, are the basis of thermodynamics. The first law states that, in general terms, energy cannot be created or disappear, but only be transformed. All industrial and natural processes only transform energy and materials, without loss of energy if one makes a balance between inputs into and outputs out of the considered processes. Inputs and outputs are all kinds of energy and material: kinetic energy (e.g. wind energy), radiation energy (e.g. solar radiation), potential energy (e.g. hydropower), renewable resources (biomass), non-renewable resources (fossil resources), waste materials etc. The second law of thermodynamics states that all processes generate entropy, reflecting a quality loss of the input energy. The exergy concept, sometimes called availability, is the thermodynamic tool which expresses the quality of energy: it is the amount of energy which can be used to produce work with respect to an environmental equilibrium state, the so-called dead state. In other words, exergy is the useful energy that can be derived out of materials. Particularly this second law gets increasing attention (Minkel 2002).

The irreversibility of each natural and technological process and the inherent degradation of the quality of energy are quantified by the loss of exergy. This degradation is illustrated in Fig. 31.2. Starting from inputs with relatively high exergy content, a process degrades the exergy content. The exergy approach provides a quantitative base with the same unit (Joule) for all types of energy carriers and materials. The exergy concept has the capability to serve as a basis to quantify the sustainability of technology, taking into account all processes that are initiated by this technology: the technological process itself processes in the natural environment and those in the industrial metabolism.

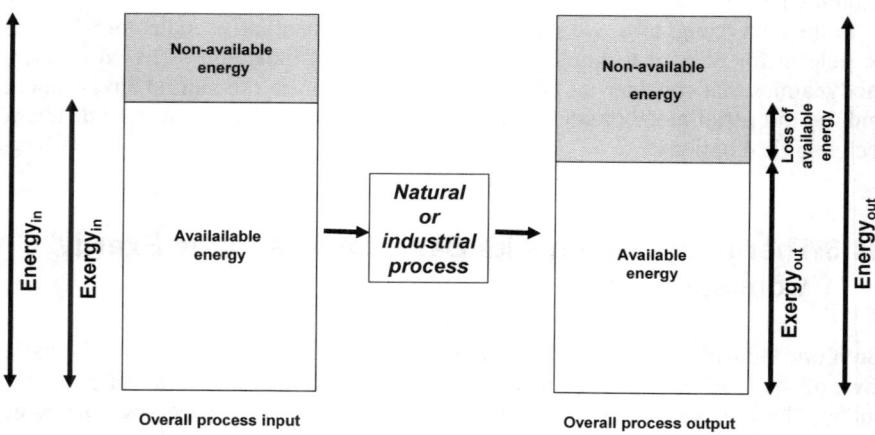

Fig. 31.2. Analysis of a process with the basic laws of thermodynamics.

Note: The first law states that all energy going into the process is equal to the energy leaving the process (energyin = energyout); the second law states that the available energy or exergy leaving the process is always lower than the available energy or exergy brought to the process (exergyin > exergyout) due to entropy generation

31.4 Quantitative Information for the SEA of Technology from the Thermodynamic Approach

Currently, the thermodynamic approach to assess the environmental sustainability of technological options is intensively explored. A thorough consideration of all physical chemical processes related to a technological option to produce a product to accomplish a service is illustrated in Fig. 31.3. This scheme is essential if one aims at well established indicators to assess the physical chemical realm in the SEA. Three interacting major sections can be found: the studied technology and consumption chain to accomplish the service demanded by society (middle), processes initiated in the ecosystem due to necessary natural resource extraction and generated emissions (left), and processes in the industrial metabolism interacting with the production and consumption chain (right).

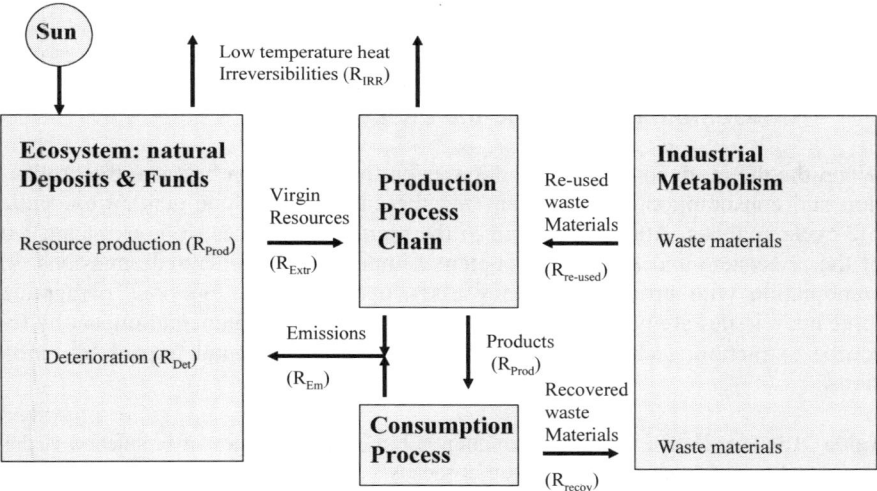

Fig. 31.3. Exchanges between the production and consumption processes, the ecosystem and the industrial metabolism.

Note: Exchanges: R_{Extr} (extraction rate of resources out of the ecosystem (J/s)), $R_{Re-used}$ (consumption rate of re-used waste materials (J/s)), R_{Em} (emission rate of both the production and consumption chain (J/s)), R_{Recov} (rate of recovery of waste materials (J/s)), R_P (rate of delivery of products (J/s)); Exergy losses: R_{IRR} (rate of irreversibility generation (J/s)); Changes in the ecosphere: R_{Prod} (rate of resource production (J/s)), R_{Det} (rate of deterioration of the ecosphere due to emissions (J/s))

31.4.1 Focus on the Production and Consumption Process Chain

In the centre of Fig. 31.3, the production and consumption chain is represented. For a long time, this has been the main focus of applications of thermodynamics, i.e. assessing the thermodynamic performance of technology as exemplified in the handbooks "The exergy method of thermal plant analysis" of Kotas (1985) and "Exergy analysis of thermal, chemical and metallurgical processes" of Szargut et al. (1988). Whole production and consumption chain analysis has resulted in the so-called cumulative exergy consumption (CExC) indicator. The cumulative exergy consumption is the total amount of useful energy or exergy that has to be invested to deliver the desired product. Table 31.1 gives examples for a series of typical products. The table does not only give the required inputs by CExC figures, but also the overall efficiency of the production chain through the ratio of exergy embodied in the product over the cumulative exergy consumption. Information in Table 31.1 shows the depletion of environmental resources induced by product generation, an issue that can not be overlooked in SEA of technology.

31.4.2 Focus on the Interaction between the Production and Consumption Chain and the Ecosystem

When the thermodynamic analysis focuses on the interaction between the production and consumption process chain and the environment, one gets an exergetic life cycle analysis. Attention is paid to the required resources to be extracted out of the ecosystem and to the environmental impact of the generated emissions. In combination with process efficiency assessment as in the previous paragraph, three main issues are now covered by the assessment: environmental impact by resource extraction, technological efficiency and environmental impact by emissions.

Table 31.1. Cumulative exergy consumption (CExC) and efficiency of production chains for a set of typical products in our industrial society

Product	CExC (MJ/kg)	Efficiency (%)	Reference
Copper	147.4	1.4	Szargut et al. 1988
Zinc	198.9	2.6	Szargut et al. 1988
Aluminium	250.2	13.2	Szargut et al. 1988
Methanol	73.1	30.7	Szargut et al. 1988
Acetylene gas	236	20.7	Szargut et al. 1988
Propane gas	61.6	79.3	Szargut et al. 1988
Concrete	1.1	0	Dewulf et al. 2001
Paper	59.9	27.5	Szargut et al. 1988
Glass	21.1	0.8	Szargut et al. 1988
Polyethylene plastics	86.0	54.1	Dewulf et al. 2001
Polystyrene plastics	91.9	45.7	Dewulf et al. 2001
Electricity	4.17*	24.0	Szargut et al. 1988

* Unit is MJ cumulative exergy consumption per MJ electricity

Translation of these three factors in quantitative sustainability indicators has been done in several ways. Dewulf et al. (2000) assessed the resource intake by the introduction of a renewability parameter α. Investigating the sustainability of technological processes requires the consideration of the nature of resources that are taken in (see also Fig. 31.1). Resources can be divided into two categories: the renewables, i.e. those which are generated in the natural environment at a rate at least as high as the consumption rate and the non-renewables with higher industrial consumption than natural production rates. All energy carriers and materials brought into the production process have to be traced back to the natural resources they require. The coefficient α reflects the fraction of renewable resources out of the total set of resources that are extracted from the natural environment. Next to this renewability parameter, the authors combined process efficiency and abatement efficiency, defined in a similar way as Cornelissen (1997), in one overall efficiency indicator η. This indicator reflects technological performance and potential environmental impact of emissions. They combined these two environmental sustainability indicators α and η into one sustainability indicator S, being equal to $0.5*(\alpha+\eta)$. In Table 31.2, the sustainability analysis is presented for ethanol that is currently produced both from fossil resources via hydration of ethylene and from biomass through fermentation. Additionally, they studied a hypothetical route based on photovoltaic generation of electricity, hydrolysis and hydrogen reaction with carbon dioxide to produce ethanol. The results show that this latter hypothetical route and the renewable based route are the strategic options of choice.

The major difficulty in these environmental exergetic life cycle analyses is the treatment of negative effects embodied in the generated waste. The approaches presented in this paragraph so far, calculate the exergy required for the abatement of emissions prior to their release into the ecosphere. The abatement in practice depends on limits imposed by authorities. These imposed emission limits are not necessary non-effect limits. Dewulf and Van Langenhove (2002a) considered the emissions as they are released in practice and presented a thermodynamic analysis of their environmental effects. The deterioration process, i.e. exposure of the ecosystem and human population to these emissions, ending up in deteriorated environmental systems and lost human lives, is also subjected to the second law of thermodynamics.

Table 31.2. Ethanol production: renewability coefficient α, overall efficiency η and sustainability indicator S (Dewulf et al. 2000)

Production pathway	α	η	S
Synthesis from fossil resources	0.0002	0.365	0.183
Biomass/Fermentation	0.998	0.00694	0.502
Synthesis from Hydrogen /CO_2, with solar driven H_2 production	0.911	0.0645	0.488

Note: Indicator α represents the usage of renewable resources in the production and consumption chain; indicator η the overall efficiency, including the emission abatement requirements. S is the overall sustainability indicator. All three indicators are scaled between 0 (non-sustainable) and 1 (full sustainable)

Table 31.3. Loss of exergy (MJ) due to the production and disposal after use (best case) of 1 kg of synthetic organic polymer (Dewulf and Van Langenhove 2002a)

Polymer	$(\partial Ex_{eco})_{resources}$	$(\partial Ex_{eco})_{exposure}$	$(\partial E_{soc})_{exposure}$	ΔEx_{total}
Polyethylene (LDPE)	86	286	0.2	372
Polypropylene (PP)	85	178	0.2	263
Polyethylene terephtalate (PET)	87	688	0.4	775
General Purpose Polystyrene (GPPS)	92	399	0.3	491
Poly vinylchloride (PVC)	67	3000	0.8	3068

Note: $(\partial Ex_{eco})_{resources}$: loss of exergy in the ecosystem due to resource extraction; $(\partial Ex_{eco})_{exposure}$: loss of exergy in the ecosystem due to exposure to emissions; $(\partial Ex_{soc})_{exposure}$: loss of exergy in the society due to loss of human lives due to exposure to emissions; ΔEx_{total}: overall loss of exergy

The analysis allowed them to come up with one overall exergy loss indicator composed out of three fractions, as illustrated for plastics in Table 31.3. Two losses of exergy are situated in the environment, i.e. due to resources extraction and due to environmental degradation caused by exposure to emissions. The third loss is damage to the human population due to exposure to the emissions.

In conclusion, the thermodynamics based indicators in Tables 31.2 and 31.3 can be of value in SEA of technological options, as illustrated for the production options of ethanol and for the selection of plastics.

31.4.3 Focus on the Interaction between Production and Consumption Chain and the Industrial Metabolism

Since the 1980s, second law or exergy analyses of whole industrial metabolisms were reported. Frequently studied metabolisms are national economies with natural resource intake on one hand and products delivered on the other hand. This is illustrated with the example of Japan in 1985, showing the conversion of exergy embodied in the resource base into exergy embodied in products made available to the individuals to fulfill their needs. It shows that our modern industrial society largely relies on non-renewable and exhaustible resources.

Recently, Ayres et al. (2003) presented a historical overview of the nature of the USA resource intake for the twentieth century, illustrating the growing share of fossil fuels as exergy to drive the US economy.

From these studies, one gets important information on the magnitude and nature of the natural resource intake of overall industrial metabolisms. For more particular products and services, Dewulf et al. (2002b) developed an industrial metabolism assessment for different solid waste products. Cardboard, newspaper and plastics production and waste disposal options were studied, taking into account the consequences they generate for other products in the industrial metabolism. Indeed, if for example solid waste products are incinerated with heat and electricity production, they save virgin resource intake for the energy production. Simultaneously, the incineration option implicitly demands a continued virgin resource intake for the manufacture of solid products. Recycling however results in savings on this latter virgin resource intake but results in continued virgin resource requirements for heat and electricity production.

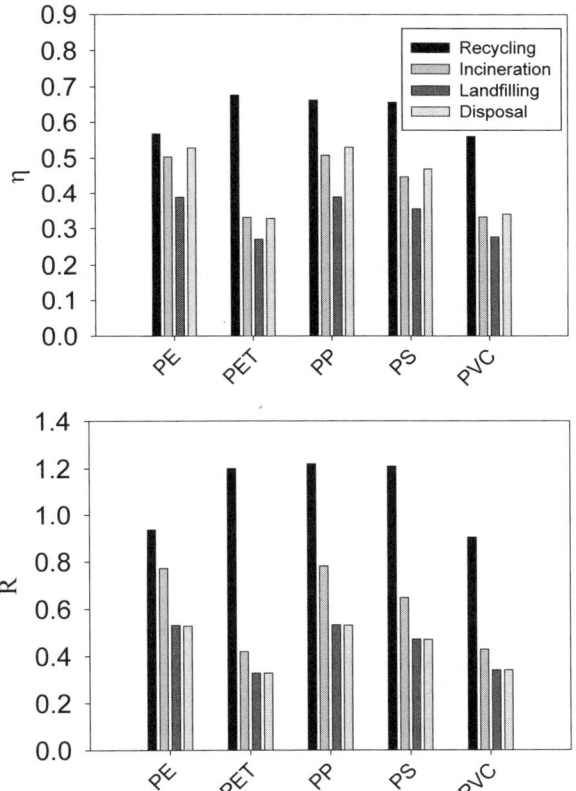

Fig. 31.4. Process efficiencies η and metabolic performance ratios R for different industrial metabolisms of plastics products poly ethylene (PE), poly ethylene terephtalate(PET), poly propylene (PP), poly styrene (PS) and poly vinyl chloride (PVC).

Note: Metabolic options are recycling, incineration with heat and electricity production, landfilling with methane gas recovery for heat and electricity generation, and disposal (i.e. throw away) of the plastics waste materials (Dewulf and Van Langenhove 2002b)

Fig. 31.4 shows that for an accurate assessment the overall metabolic performance is to be considered rather than the product technology performance solely. In the poly ethylene (PE) example, heat, electricity and PE are produced from virgin resources and PE waste is thrown away (disposal). Process analysis shows that this option is more efficient than landfilling with methane gas recovery for energy purposes and incineration with energy production. However, due to the savings that are realized for the overall metabolism through landfilling and in particular with incineration, the quantitative analysis shows that the throw away metabolism has the weakest sustainability. The quantitative information as illustrated in Figure 6 can contribute to SEA of waste disposal options of all types of materials.

31.4.4 Interaction of the Production and Process Chain with the Ecosystem and the Industrial Metabolism

Recently, Dewulf et al. (2004) proposed a set of five universal environmental sustainability indicators for the assessment of products and production pathways, taking into account the three interacting sections presented in Fig. 31.3: the technology and consumption chain (middle), processes within the ecosystem (left) and processes in the industrial metabolism (right). The indicators are based on the second law of thermodynamics and scaled between 0 and 1. The parameters reflect input of previously used materials in the industrial metabolism to generate the new product (ρ), recoverability of the generated materials after their application (σ), the efficiency of the product generation (η), the renewability of the virgin resources used in the production and consumption chain (α), and the (un)toxicity of emissions generated during the production and consumption stage (τ). Production chains totally based on virgin resources result in zero scores on parameter ρ, whereas technologies delivering products out of (mainly) waste materials, such as recycling, have scores near 1. Products designed for recycling (DFR: design for recycling; DFD: design for disassembly) have high scores on parameter σ, typically durable goods. Fuels, used to generate kinetic or thermal energy in heating or transport applications ending up in gaseous emissions, will have scores near zero. The performance of the production and consumption route assessed by the parameter η is assessed in a similar way as in cumulative exergy consumption studies. The parameter α quantifies the renewability of the consumed virgin materials in a same way as defined in the second last paragraph. Finally, the toxicity of emissions is quantified by parameter τ: it is based on the calculation of the deterioration rate in the environment and in the human population as a result of the exposure to the emissions initiated by the production and consumption process.

The applicability is illustrated for different production pathways of alcohols for the manufacture of surfactants. These alcohols can be from petrochemical or oleochemical origin. In brief, alcohols from petrochemical resources require (1) crude oil and natural gas production, (2) crude oil distillation, desalting and hydrotreating, n-paraffin production from crude oil distillates, (3) natural gas processing, (4) crude oil and natural gas based ethylene production, (5) n-olefin production, and (6) alcohol production. Oleochemical alcohols require two subsequent production stages. In a first stage, vegetable oils, such as palm oil, are produced through agriculture and extraction, requiring renewable inputs and a number of non-renewables for production and application of pesticides and nutrients, and for agricultural and extraction equipment. In a second stage, the crude vegetable oil has to be refined and converted into the final alcohol through triglyceride splitting and hydrogenation, or through methyl ester production and hydrogenation followed by hydrolysis.

A thorough thermodynamic analysis results in a comprehensive environmental sustainability assessment as shown in Fig. 31.5. For the petrochemical pathway, the sustainability in terms of renewability, recovered waste and re-use of materials is "nil", so that the sustainability indicators above zero are only owing to a quite high efficiency of 0.44 and an untoxicity score of 0.09. Indeed, in this production pathway, there is no use of renewables or recovered waste. The final alcohol product is input material for detergents, where the final use stage (e.g. households) currently allows no re-use. The toxicity studied is limited to the alcohol as a final product. Incorporation of the alcohol in detergents as poly alcohol ethoxylates should be taken into account when indicators are investigated for the final end-use product, i.e. detergents. The sustainability picture of palm oil based alcohol is quite different. The main difference is a better use of renewables (α=0.998) and lower emission of toxics (τ=0.304). However, efficiency is quite low when compared to petrochemical based alcohols, due to a weak photosynthesis efficiency (η=0.004). The items of recovered waste and re-use of materials for oleochemical based alcohols are similar to the petrochemical based ones. In conclusion the five issues in Fig. 31.5 being relevant to SEA of technology are quantified by a thermodynamic aproach.

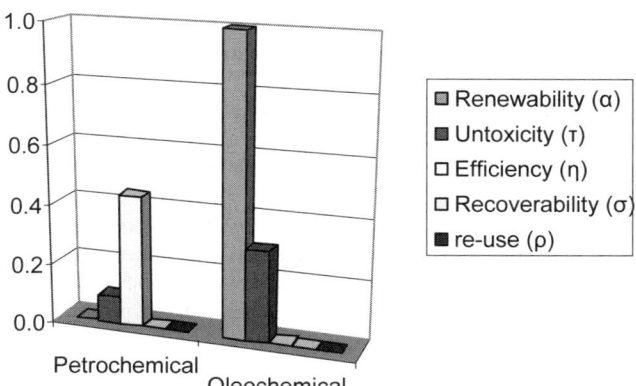

Fig. 31.5. Scores of petrochemical and oleochemical based alcohols on the developed set of 5 physical-chemical sustainability indicators.

Note: Scores are scaled between 0 (non-sustainable) and 1 (full sustainable) (Dewulf and Van Langenhove 2004)

31.5 Conclusions and Recommendations

Aiming at implementation of SEA at a high level, is dependent on the way information is gathered, whether it is in the economic, social or physical-chemical field, the latter including environmental effects. Where information can be quantified, SEA should make use of it. This chapter has shown that quantitative information for the physical chemical effects of technology development can be generated through thermodynamics. Based on a strong thermodynamic basis, it is possible to present quantitative indicators that reflect important aspects in SEA of technology: usage of renewable resources, waste emission, re-use of materials, destination of products at the end of life and process efficiency. Due to its generic character, thermodynamic analysis enables the development of indicators through a quantitative assessment of processes within the developed technology itself, but also processes that are linked and initiated by the technology, taking place both in the environment and in the industrial metabolism.

SEA is served by a thorough physical chemical analysis. Thermodynamics can do so. However, this thermodynamic approach should be integrated in SEA frameworks. There, it should offer objective information on physical chemical effects of decisions made or to be made, next to indicators that reflect social and economic implications.

References[1]

Ayres RU, Ayres LW, Warr B (2003) Exergy, power and work in the US economy, 1900–1998. Energy 28:219–273

Bejan A (1997) Advanced Engineering Thermodynamics, 2nd edn. John Wiley and Sons, Toronto

Bringezu S, Schütz H (2001) Total material requirement of the European Union. Technical report no. 55, EEA, Copenhagen

Çengel YA, Boles M.A. (1994) Thermodynamics, an engineering approach, 2nd edn. McGraw-Hill Inc., New York

Chambers N, Simmons C, Wackernagel M (2001) Sharing Nature's Interest : Ecological Footprints as an Indicator of Sustainability. Earthscan Publications Ltd., London

Clift R (1995) Clean technology – an introduction. J. Chem. Tech. Biotech 62:321-326

Clift R (1997) Clean technology – the idea and the practice. J. Chem. Tech. Biotech 68:347-350

Clift R, Longley AJ (1995) Introduction to clean technology. In: Kirkwood RC, Longley AJ (eds) Clean technology and the environment. Blackie, London pp 174-198

Cornelissen RL (1997) Thermodynamics and sustainable development. PhD thesis, Universiteit Twente. Cornelissen, Enschede

---

[1] Note that legislation mentioned in the chapter is listed at the end of the handbook in the consolidated list of legislation (Appendix 2).

Dewulf J, Van Langenhove H, Mulder J, van den Berg MMD, van der Kooi HJ, de Swaan Arons J (2000) Illustrations towards quantifying the sustainability of technology. Green Chem. 2:108-114

Dewulf J, Van Langenhove H, Dirckx J (2001) Exergy analysis in the assessment of the sustainability of waste gas treatment systems. Sci. Tot. Environ 273:41-52

Dewulf J, Van Langenhove H (2002a) Assessment of the sustainability of technology by means of a thermodynamically based life cycle analysis. Environ. Sci. Pollut. Res. 9:267-273

Dewulf J, Van Langenhove H (2002b) Quantitative assessment of solid waste treatment systems in the industrial ecology perspective by exergy analysis. Environ. Sci. Technol 36:1130-1135

Dewulf J, Van Langenhove H (2004) Integrating industrial ecology principles into a set of environmental sustainability indicators for technology assessment. Resources, Conservation and Recycling: in press

Ecotec (2001) Ecological Footprinting. In: Chambers G (ed) European Parliament - Directorate General for Research, the STOA Programme. Paper PE 297.571/Fin.St.

Elkington J (1997) Cannibals With Forks: The Triple Bottom Line of 21st Century Business. Capstone, Oxford

European Commission (2002) Towards a European strategy for the sustainable use of natural resources. European Commission, Directorate-General Environment, Report of the Stakeholders Meeting, April 10, 2002

Graedel TE, Allenby BR (1996) Design for Environment. Prentice Hall, New Jersey

ISO (1997) Environmental management – Life cycle assessment – Principles and framework, ISO14040

Jorgensen SE, Müller F (2000) Handbook of ecosystem theories and management. CRC Press-Lewis Publishers, Boca Raton

Kotas TJ (1985) The exergy method of thermal plant analysis. Butterwoods, London

Levett R (1998) Sustainability indicators - integrating quality of life and environmental protection. J. R. Statist. Soc. A 161:291-302

Lowe EA, Warren JL, Moran SR (1997) Discovering industrial ecology – An executive briefing and sourcebook. Battelle Press, Columbus

Minkel JR (2002) The meaning of life. New Scientist, 5 October 2002:30-33

Office of the Deputy Prime Minister (2003) The SEA Directive: Guidance for Planning Authorities. Practical guidance on applying European Directive 2001/42/EC 'on the assessment of the effects of certain plans and programmes on the environment' to land use and spatial plans in England. London: ODPM. Internet address: http://www.odpm.gov.uk/stellent/groups/odpm_planning/documents/page/odpm_plan_025198.pdf, last accessed 21/09/2004

Spangenberg JH, Hinterberger F, Moll S, Schütz H (1999) Material flow analysis, TMR and the mips-concept: a contribution to the development of indicators for measuring changes in consumption and production patterns. Int. J. Sustain. Development 2:491-505

Szargut J, Morris DR, Steward FR (1988) Exergy analysis of thermal, chemical and metallurgical processes. Hemisphere Publ. Corp., New York/Springer Verlag, Berlin

Thérivel R, Partidário MR (1996) The Practice of Strategic Environmental Assessment. Earthscan Publications Ltd, London

TRL Ltd and Collingwood Environmental Planning (2002) Analysis of Baseline Data Requirements for the SEA Directive – Final Report. TRL Ltd. Internet address:

http://www.trl.co.uk/static/environment/SWRA%20SEA%20Report%2011.pdf, last accessed 21/09/2004

Van der Voet E, van Oers L, Kikolic I (2003) Dematerialisation: not just a matter of weight. CML report nr. 160. Centre of Environmental Science, Leiden

Wackernagel M, Rees W (1995) Our Ecological Footprint: Reducing Human Impact on the Earth. New Society Publ, Gabriola Island

Wall G (2002) Conditions and tools in the design of energy conversion and management systems of a sustainable society. En. Convers. Manage 43:1235-1248

World Commission on Environment and Development (1987) Our Common Future. Oxford University Press, New York

32 A Structural and Functional Strategy Analysis for SEA

Anastássios Perdicoúlis

University of Trás-os-Montes e Alto Douro, Portugal

32.1 Introduction

This chapter focuses on two aspects of SEA implementation: the quality of strategies and discovering their likely effects. A structural and functional analysis based on causality diagrams checks, on one hand, for quality characteristics of the strategy, and determines whether strategies are prepared to achieve their objectives; on the other hand, it explores ways in which both planned and non-planned effects may arise from strategies. The method is mostly to be used in a qualitative mode, and provides a high transparency in strategy- and decision-making; this transparency and simplicity make it suitable for community participation. This analysis has much more value when accompanying the preparation of strategies, rather than serving merely as a groundwork test for already completed strategies.

In the interest of sustainable development, SEA has been conceived to be more than a formal pre-implementation test for strategies. SEA is supposed to participate and help in the *creation* of development strategies. This function is novel compared to classic EIA frameworks and practice, and appears clearly stated in milestone SEA guidance documents, such as in the SEA Directive (Art. 4 para 1). However, producing development strategies in conjunction with SEA is a challenge for SEA practitioners and planners alike, particularly when they are working in close collaboration (Stoeglehner 2004; Thérivel and Minas 2002; Smith and Sheate 2001).

For the time being, SEA keeps a clear focus on the effects of certain strategies on society, economy, and nature, i.e. in terms of changes of the quality of the environment. Aspects of quality of the strategies themselves, such as consistency and clarity, are normally beyond SEA's scope. In other words, SEA is interested in the results that strategies are likely to produce, but not in the quality of the strategies themselves that are submitted for assessment.

Implementing Strategic Environmental Assessment. Edited by Michael Schmidt, Elsa João and Eike Albrecht. © 2005 Springer-Verlag

The quality of the strategies themselves is likely to influence the SEA process (e.g. by resolving uncertainties or inconsistencies) and the outcomes of the strategy (e.g. by identifying the right issues for intervention). With this in mind, this chapter focuses on two key aims: a) to check for the quality of strategies, and b) identify how effects arise from these strategies.

This chapter proposes a *single* method in order to achieve these two aims. The chapter starts by setting key definitions in Sect. 32.2, explains the protocol of the strategy analysis, illustrated by a concrete example in Sect. 32.3, discusses the contributions, usability, and the special feature of the method (i.e. causality diagrams) in Sect. 32.4, and closes with conclusions and recommendations.

32.2 Background and Definitions

Several common groups of methods are used in SEA and similar processes, including matrices, modeling, checklists, expert opinion, etc. (EC 1999). The most appropriate method that manages to satisfy the double objective set for this work can be sought in the Network and Systems Analysis methodology. In fact, the common advantages of this group of methods are explicit mechanisms of cause and effect, and the use of diagrams to understand the effects of strategies. The common difficulties are the absence of spatial and temporal scale, and the potential risk of creating diagrams which are too large (EC 1999 pp.39-44).

The method of choice from the identified group of Network and Systems Analysis is a structural and functional analysis based on causality diagrams, used to verify strategy quality and also to perform an analysis on the potential effects of strategies. This method extends the greater "SEA Toolkit", along with the other methods presented in this book.

Before proceeding with the presentation of the strategy analysis, it is necessary to make two clarifications: (a) what is a "good" strategy in a functional perspective as explained in the introduction, and (b) what is the terminology to be used for strategies, planning, etc.. These definitions are necessary in each SEA implementation, and the following two lists serve merely as indications or suggestions.

To evaluate development strategies for their internal consistency, focus, direction, etc., a set of criteria needs to be developed. The following set of criteria has been developed with functionality in mind, i.e. how the proposed actions are leading to the objectives:

1. *Logical:* The actions of the strategy lead to the set objectives, and all tiers or levels of aggregation are respected
2. *Holistic:* The strategy takes objectively into account and integrates well with other relevant proposals and activities in the geographic vicinity (referring to the past, present, or future); clearly takes stakeholders into account
3. *Coherent:* The strategy does not involve conflicts or duplication of actions, internal or external
4. *Efficient:* The strategy reaches its objectives with direct and coordinated or tiered actions, not through long-winded and fragile schemes

5. *Unambiguous:* The strategy explains clearly the starting points (conditions), direction (vision), and goals (exactly what is expected to be reached)
6. *Illustrative:* The strategy is presented with different, appropriate media: text, geographic maps, charts, diagrams, photographs, etc.

Although not difficult, terms used in planning often vary in meaning or use. To ensure efficient communication within the community, the elements of the community development strategy must be identified. This can be done with a key, such as the one in the following list:[1]

1. *Priority* is a broad issue of importance to the community; it does not indicate neither "what the community wants to achieve", nor "what the community wants to do towards it"
2. *Aim* is "what the community wants to achieve", expressed in a nutshell, at the highest tier, i.e. the most aggregated level of information; this is a common synonym of "objective"
3. *Goal* is also an expression of "what the community wants to achieve", but in a lower tier than aim or objective; in most cases this is accepted in qualitative terms; it also functions as the motive or the "reason why" for the actions (to follow); typical identifiers: reduce, maximise
4. *Target* is also "what the community wants to achieve", usually in quantitative terms; targets are usually as detailed as indicators, so they may not function as motives for the actions; rather, they are comparable to conditions, and thus permit the evaluation of the strategy's performance; typical identifiers: X% in Y years
5. *Impact* is an non-planned outcome or effect of the strategy (i.e. a "side effect" of the strategy)
6. *Guideline* is "what the community intends to do", expressed in a condensed manner, i.e. high tier or level of aggregation of information; this is the highest strategic level for the purposes of this chapter; guidelines direct for the more detailed action; typical identifiers: promote, encourage
7. *Action* is also "what the community intends to do", but at a more detailed tier or strategic level; actions could be directly implementable projects, or just one aggregation level above them; typical identifiers: monitor, provide
8. *Condition* is an observation on the state of the community; for the purposes of this chapter, condition is taken in its dis-aggregated form "indicators", equivalent to that of the target; information at this low-aggregation level is usually directly measurable

This is a generic definition key, and variations in the use of the above terms is to be expected in different cultural contexts. Perhaps using a key like this as a reference is more important than having the false sense that these terms are "commonly understood" within, or among communities.

---

[1] Various synonyms may be used for these definitions, according to language or cultural contexts. Also, not all levels of hierarchy presented here are obligatory to exist in the planning practice of the various communities.

32.3 The Method of Strategy Analysis

The method of strategy analysis in this chapter starts by identifying and laying out the elements of the strategy, proceeds by linking them together in cause-and-effect relationships, and ends by determining how the strategy works, what is missing, whether the links are logical, whether its objectives are satisfied, what are the non-planned effects, etc. This strategy analysis method aims

- to provide a means to evaluate the strategy in a profound and organized manner – for instance, characterising it as coherent, logical, efficient, etc. – and thus improve the strategy, and
- to seek the likely effects of the strategy, whether planned or non-planned.

Under the analysis heading, there are two parts: a structural and a functional analysis, as explained in the next two paragraphs.

As soon as a strategy is subject to an analysis, it is necessary to explore its contents and determine what types of "elements" make it up, such as actions, goals, and guidelines. This is a structural perspective, so the corresponding analysis is generically called "structural analysis". The names of the categories identified in the structural analysis are expected to vary among strategies, but should be consistent within the same strategy.

To see "how good, or fit for the purpose" a community development strategy is, is essentially to look for "how the strategy will most likely work on the ground". This is a functional perspective, so the analysis that corresponds to this end is generically called "functional analysis". Furthermore, since the particular functional analysis used in this chapter employs cause-and-effect, or causality diagrams, it can be specifically called "causality analysis".

The analysis presented in this chapter is meant to be qualitative, as the complexity of the community development strategies is generally quite high. For procedural purposes, the steps of the analysis are described below:

1. *Identify and classify* the elements of the strategy into objectives, actions, etc. This can be done in the list format of the strategy presentation. It might be convenient to assign different colors, boxes, or letter styles to each type of elements, e.g. red and italic for objectives, black for actions, etc. This is where to discover whether all the necessary types of elements for the strategy are present, in a first look.
2. *Lay out* first all the objectives, and then all the conceived actions that would lead to the former. This gives the "elements" of the strategy in a basic diagrammatic form, and the layout may be arranged in any way that seems organized, illustrative, or generally convenient.
3. *Link* the elements of the strategy in the way they were originally conceived, so that they reveal how one action leads to another action, and finally to an objective (and, eventually, to non-planned effects). This is the step where the "functionality" of the strategy is laid out. These causal diagrams can be created with pen and paper, or by using any computer software for diagramming, drawing, or flowcharting (Sterman 2000 pp.137-190; Glasson et al. 1994 p.117). At this step it is also possible to verify the true functions of the elements, comparing to

the classification made in the previous step. For instance, what had been classified as an "action" may be really a "general guidance" because it contains or suggests more immediate actions.

4. *Simulate* the strategy: this is where the functionality of the strategy is tested. The basic simulation works best with the diagrammatic form. A starting point is defined, followed by the actions that should normally lead to the objectives. If all the pathways make sense and nothing is missing, then it is to be concluded that the community development strategy is finished and ready for evaluation. If not, then the procedure indicates to the next step.
5. *Enter* the workshop: this is where missing strategy elements are added, links between elements are corrected, the use of elements is optimised, and pathways are corrected or made more efficient (e.g. shorter, more direct and robust). This final step may produce a completely new strategy.

The method is illustrated with an example from a local authority development strategy. Box 32.1 presents a single sector: housing.

Fig. 32.1 shows the housing sector of the local authority development strategy, presented in a diagrammatic form (causality diagram) by direct translation from its list form (Box 32.1). The observed relationships between the strategy elements for the housing sector (Box 32.1 and Fig. 32.1) are presented in Fig. 32.2. The diagonal dashed line separates causes, or "what to do" (top left) from effects, or "what to achieve" (bottom right), while the arrows denote the causality between the strategy elements. For instance, guidelines appear to shape the goals and targets, while conditions appear to shape the guidelines and actions.

Box 32.1. Original list form of the housing sector of a local authority strategy

- *Priority*: Affordable and safe housing
- *Aim:* To meet the priority housing needs of the community, provide safe homes in a safe environment and help key workers get into the housing market
- *How?*

1. *guideline* Ensure sufficient affordable and safe homes to *condition* meet the priority needs of the community
2. *target* Maximise the number of affordable homes, *target* increase the number of smaller units and *target* balance the mix of housing, to take into account *condition* the differing ages and incomes of people in the district
3. *goal* No more than six families in bed and breakfast accommodation
4. *goal* No one on the housing waiting list for longer than a year
5. *action* Source adequate funding to maintain and enhance affordable housing and *action* look at co-ordinating housing across district boundaries
6. *action* Investigate alternative ways to help people get into the local housing market
7. *goal* Raise the standard of safety in existing housing and *guideline* promote safe housing especially with regard to fire, crime, location and building structures
8. *goal* Increase the independence and safety of ageing and less mobile people in the district by *guideline* addressing their housing needs more adequately

---

However, the causality logic in Fig. 32.2 reveals some unexpected facts: in the original strategy there is no relationship between guidelines and actions, or goals and aims. These two couples of strategy elements, being tiers with hierarchical relations, were expected to have certain links as explained in the second half of Sect. 32.2.

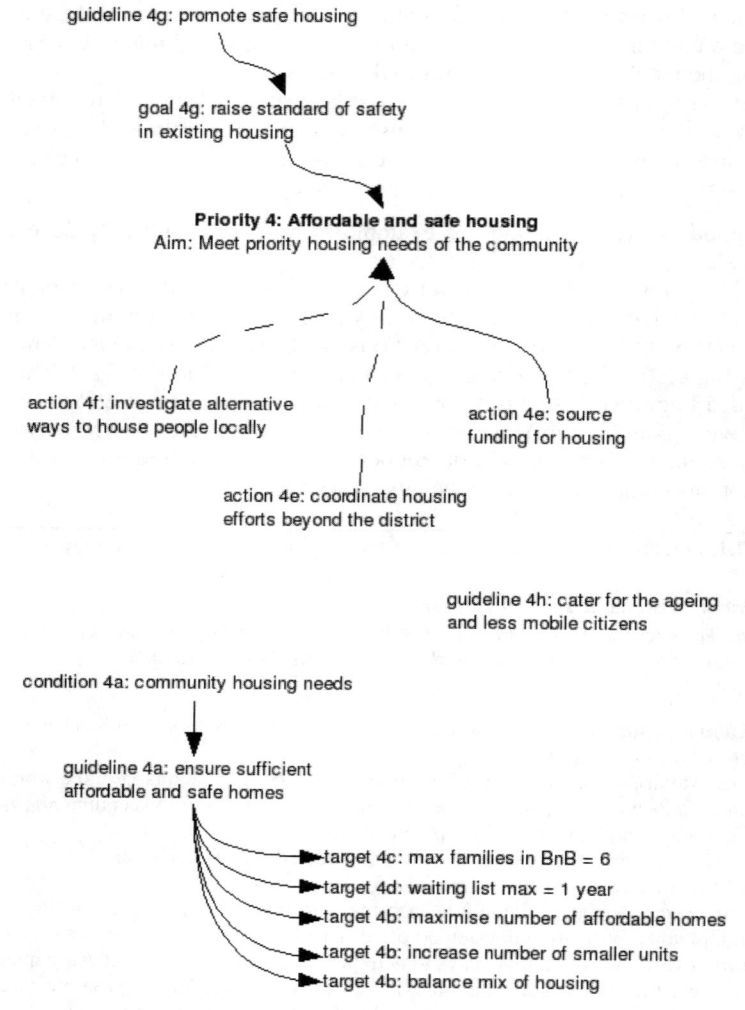

Fig. 32.1. Structural and functional analysis: causality diagram of the housing sector of a local authority development strategy

32 A Structural and Functional Strategy Analysis for SEA   465

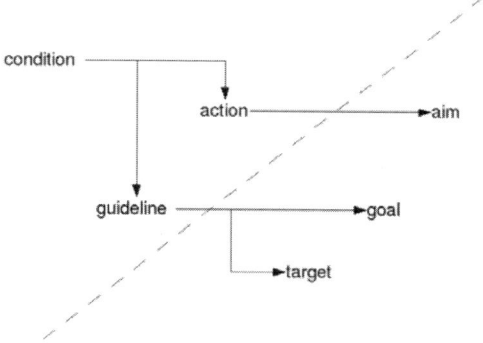

Fig. 32.2. Causality logic of the housing sector strategy in its original formulation

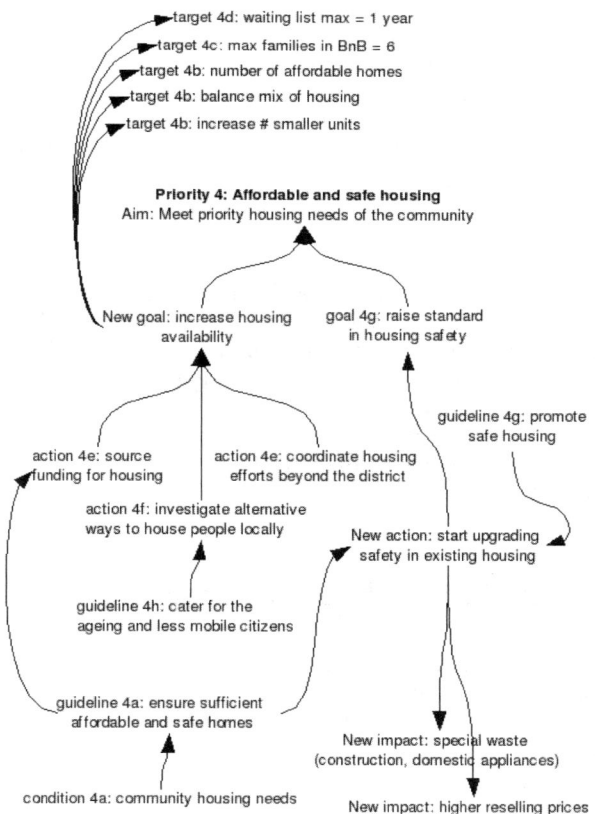

Fig. 32.3. The same housing sector after structural and functional modifications

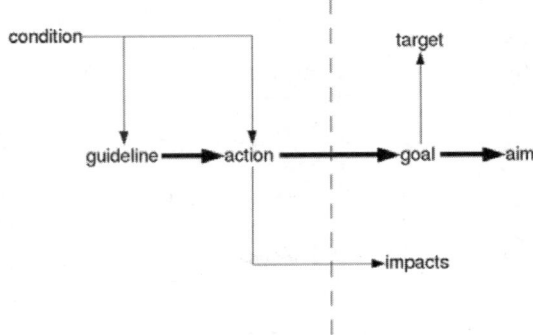

Fig. 32.4. Causality logic of the same housing sector in the modified strategy

Fig. 32.3 presents a modified version of the housing sector of the same strategy, after a structural and functional analysis with the proposed method. Structural changes are those that introduce or delete elements in the strategy or the system of intervention; functional changes are those that modify the relations between these elements. For instance, the new strategy features structural changes such as two non-planned effects, marked as "impacts", and a new goal that addresses the existing five targets (top of Fig. 32.3). Functional changes include modifications of causality sequences, such as guideline 4a not addressing the five targets directly, but going through a new action 4e and a new goal (left side of Fig. 32.3).

The causality logic of the modified strategy in the particular housing sector of the example (Fig. 32.4) respects the hierarchical links between guidelines and actions as well as between goals and aims, exactly as explained in the second half of Sect. 32.2. These four elements, two on the left and two on the right of the dashed vertical line in Fig. 32.4, constitute the main cause-and-effect spine of the causality logic, marked with a thicker arrow line (Fig. 32.4). In addition, the modified causality logic contemplates non-planned effects, marked as "impacts" (Figs. 32.3 and 32.4), which are side-effects of planned actions, not contemplated in the original version of the strategy.

32.4 Contribution of the Strategy Analysis

This section summarises and criticises the contributions of the proposed method for strategy analysis. The main issue is the improvement of the strategy, followed by the usability of the method, and finally the special feature of the method: the causality diagrams.

32.4.1 Strategy Improvement

The contributions of the structural and functional analysis in the improvement of the development strategy in the particular example can be summarised as follows:

1. *identification* of the possible pathways in which the strategy actions could lead to the objectives or goals, including correction the causal links between strategy elements
2. *identification and categorisation* of each strategy element, e.g. action, goal, guideline – thus avoiding conceptual and functional confusions
3. *indication* what other elements could be taken into consideration to turn the strategy more realistic, e.g. stakeholders' wishes and actions, other plans and activities, external to the sector or geographic area
4. *encouraging* to look for effects, how they appear, and how they can be controlled (encouraged or discouraged), i.e. at which intervention points
5. *visibility* for the (re-) elaboration of the the strategy, i.e. new goals, new actions, re-directed interactions, etc.

Additional benefits of the method, but not illustrated in this example, include the following:

1. *seeking* which action elements serve more than one objective, i.e. to indicate and incentivate re-use of strategy elements for different purposes
2. *seeking* what types of feedback loops that could be encountered, i.e. positive or negative (reinforcing or counterbalancing)
3. *indicating* what delays are expected to be encountered – which is important for the intermediate results of the strategy

32.4.2 Usability

The structural and functional analysis presented in this chapter can be used as a method for elucidating strategies submitted to SEA, with the intent to explore whether all the important elements and interactions are present, and whether the cause-effect relationships appear to be likely. The causality diagram can be used as a technique for the analysis-exploration-forecasting of all effects, including cumulative and indirect effects, as well as interactions among effects when the situations become complex.

The criteria proposed in this chapter to assess the quality of strategies could be used as a preliminary "quality screening" for the strategies submitted to SEA. This screening would have the advantage to raise awareness that not all strategies are of the same quality, for example, some are clearer and some more efficient than others. In a more pro-active manner, the proposed structural and functional analysis could be used to assist SEA in the construction of good strategies, particularly in aspects such as showing logical, consistent, and orderly links of the strategy elements.

32.4.3 Causality Diagrams

The causality diagrams are the protagonists in the analysis. Conceptually, they are related to mind maps and causal loop diagrams (Sterman 2000; Glasson et al. 1994). Causality diagrams carry a high importance in terms of scientific explanation (Okasha 2002), and yet can be quite accessible to anyone without specific preparation. In practice, planners and stakeholders have the freedom to draw their own diagrams in any way they find useful and comprehensible, because their rules are very simple and are defined clearly in the beginning of the diagramming.

Causality diagrams are not any more demanding or unusual than geographic maps. The following list explains their similarities:

1. they occupy large sheets of paper, or big computer screens
2. they display information that would be very difficult to visualize or examine in text form
3. anyone can "read" and "draw" them somewhat intuitively, but this capacity improves significantly with a little training
4. when ambition takes over, one may get flooded with information, and get confused with the representational complexity; the art in both cases is to display the essential for the purpose
5. they are capable of treating small and large "objects", i.e. geographic areas in the case of maps, and strategies in the case of causal diagrams
6. their verification or validation is more or less empirical, and it is done by checking the elements and relations in both the maps and causal diagrams against the facts or knowledge

So, the utility of the causal diagrams, as indicated in this example, justifies the introduction of yet another instrument in SEA or planning. Causal diagrams stand well on their own, as well as with other techniques (e.g. lists, matrices), and seem to have aided much in the evaluation, presentation, and re-structuring or improvement of the community development strategy in the example.

32.5 Conclusions and Recommendations

This chapter presented a method for the structural and functional analysis of strategies for SEA purposes. In a given example, the method examined the internal consistency and focus of a strategy, and also sought and displayed transparently how its most significant planned and non-planned outcomes would arise. The method also permitted the accompaniment of strategy making, in a way that SEA might be seeking to implement in the near future. The method was based on causality diagrams, which have an accessible interface and a learning curve close to that of geographic mapping – although causality diagrams treat causal rather than spatial relations. This particular simplicity of the method makes it suitable for communication with the whole community, without requiring any previous special training.

The chapter presented and illustrated the method using a certain choice of planning terminology. It is recommended that this terminology be modified and adapted to the cultural context of different community applications. Also, it is recommended that the method be used principally in a qualitative mode, with low investments in time and materials; this would make the method more practical and worthwhile to implement. Finally, it is recommended that the method be applied as early as possible in the conception and design of a strategy.

Acknowledgements

This work has been developed at Oxford Brookes University, with funding from the Masters in Environmental Technology of the University of Trás-os-Montes e Alto Douro, Portugal. Special thanks to Professor Riki Thérivel of Oxford Brookes University for her guidance in the development of this work.

References

Bryson JM (1995) Strategic Planning for Public and Nonprofit Organizations: a Guide to Strengthening and Sustaining Organizational Achievement. Jossey-Bass, San Francisco

Carmona M (2003) An International Perspective on Measuring Quality in Planning. Built Environment 29(4): 281-287

Cooper LM, Sheate WR (2004) Integrating Cumulative Effects Assessment into UK Strategic Planning: Implications of the European Union SEA Directive. Impact Assessment and Project Appraisal 22(5):5-16

EC – European Commission (1999) Guidelines for the Assessment of Indirect and Cumulative Impacts as well as Impact Interactions. Luxemburg: Office for Official Publications of the European Communities

Glasson J, Thérivel R, Chadwick A (1994) Introduction to Environmental Impact Assessment. UCL Press, London

Güell F, Miguel J (1997) Planificación Estratégica de Ciudades. Gustavo Gili, Barcelona

Kelly ED, Becker B (2000) Community Planning: An Introduction to the Comprehensive Plan. Island Press, Washington DC

Okasha S (2002) Philosophy of Science. Oxford University Press, Oxford

Smith SP, Sheate WR (2001) Sustainability Appraisal of English Regional Plans: Incorporating the Requirements of the EU Strategic Environmental Assessment Directive. Impact Assessment and Project Appraisal 19(4):263-276

Sterman JD (2000) Business Dynamics. Irwin McGraw-Hill, Boston

Stoeglehner G (2004) Integrating Strategic Environmental Assessment into Community Development Plans – a Case Study from Austria. European Environment 14(2):58-72

Thérivel R, Partidário MR (1996) The Practice of Strategic Environmental Assessment. Earthscan, London

Thérivel R, Minas P (2002) Ensuring Effective Sustainability Appraisal. Impact Assessment and Project Appraisal 20(2):81-91

Part VI – SEA for Abiotic and Biotic Resources

Part VI focuses on the use of SEA for particular resources. For example, the SEA Directive requires the assessment of the effects of strategic actions on the environment, including issues such as fauna, flora, biodiversity, population, human health, soil, air, water, climatic factors, material assets, cultural heritage, landscape and the interrelationship between them all. Chapter 33 discusses the use of SEA in protecting soil resources, which, due to the nature of soil, deals with soil, water, air, climatic factors, landscape and the interrelationship between these factors.

The following two chapters address water-related resources. Chapter 34 evaluates how best to implement SEA, based on existing EIA experience in water resources. The chapter proposes that it is possible to use lessons learned from Project EIA to improve the implementation of SEA. The chapter suggests that for the successful implementation of SEA it is necessary to have appropriate environmental data (driven by a well-designed environmental monitoring programme), the development of a system to monitor the environmental effectiveness of new strategic actions, and a rigorous peer review system for all major SEA reports. Chapter 35 explores the contribution that SEA could make towards sustainable planning and management of water resources and, in particular, evaluates the links between SEA and the Water Framework Directive.

The last two chapters of Part VI deal with the important issue of biodiversity. Chapter 36 evaluates how best to assess biodiversity in SEA. In this chapter, different approaches to the definition and measurement of biodiversity in a SEA context are discussed. Additionally, the major approaches to the evaluation of biodiversity are checked for their relevance for deriving assessment criteria in SEA. In a complementary way, Chapter 37 analyses different biodiversity programmes related to SEA at global, European and national levels and how they can influence biodiversity protection and management. The major driving forces affecting biodiversity at different spatial scales are described, and related problems of data availability at different scales are discussed.

33 Soil Resources and SEA

Robert Mayer

Department of Urban and Landscape Planning, University of Kassel, Germany

33.1 Introduction

This chapter discusses how SEA may contribute to the protection of soils and help to make best soil use in a modern society in which a multitude of interests – public as well as private – and landuse types come together on the same physical surface. It is shown – taking the present situation in Germany as an example – what soil protection means in this context, which methods for a successful implementation are available, and which data are required to make SEA an effective planning instrument.

Soil is a non-renewable resource forming the interface between hydrosphere, atmosphere and lithosphere. It is inhabited by organisms and is the base for human life by regulating natural material and energy cycles. Soil not only controls many of these processes but is, vice versa, subject to influences from human activities as well as from the other environmental components, biotic or abiotic. This makes soils that are never static but are contiuously changing in structure and in the dynamics of internal processes. Since most visible changes are rather slow compared to human life, soils may keep traces from historic events over centuries which qualifies some variants as archives of natural and cultural history.

So, exposed to impacts from many sides, at the same time controlling other environmental and life processes, planning of soil use or activities for soil protection must necessarily be preventative and strategic because causality in ecological functions is often poorly understood, and restoration of disturbed or degraded soils is technically often impossible, or very expensive.

In recognition of this, the European Commission forwarded, in April 2002, to the parliament a communication 'Towards a Thematic Strategy for Soil Protection' (COM(2002) 179 – 2002/2172(COS)).

Implementing Strategic Environmental Assessment. Edited by Michael Schmidt, Elsa João and Eike Albrecht. © 2005 Springer-Verlag

The Committee on the Environment, Public Health and Consumer Policy, to which the communication was referred, adopted the motion for a resolution which

- calls on the Commission to ensure that greater account is taken of soil protection in the SEA Directive;
- calls on Member States for EIA and SEA Directives to be applied when implementing the urban and regional thematic strategy.

The SEA Directive requires the assessment of the effects of strategic actions on the environment, including on issues such as soil, water, air, climatic factors, landscape, and the interrelationship between these factors (Art. 5 para 1; Annex I f). Art. 5 demands that "environmental protection objectives" be put into operable terms at international, Community or Member State level. Viewed from the standpoint of law, according to Sommer (2003) this includes all goals and measures aiming at securing or improving the environment. Primarily it applies to standards defined for environmental media, or measures suitable to reach the standard. The decision whether the expected effects from a PP are relevant in respect to quality standards and norms can only be made in relation to the individual case.

Consequently soil has to be analyzed and described for the individual case, on the territory in question not only with regard to its structural and physico-chemical status but also with regard to soil functions, i.e. the multitude of soil based and related processes relating soils with the other environmental media (atmosphere, hydrosphere, lithosphere, biosphere). Not only is the status quo functionality to be considered but also the presumed changes under expected impacts following from strategic actions.

The potential for SEA as an instrument for environmental protection must be seen in making existing environmental legislation more effective. While the aims of SEA are clearly expressed, the methods by which these aims could be reached are not available in a complete, standardized form ready to be used for all sets of impacts as well as local soil and land use conditions. Also SEA does not provide any scales for the evaluation of effects. Sect. 33.2 therefore takes a look at existing environmental legislation in Germany and on the European level in which soils form an essential objective. This legislation is investigated (1) for methods appropriate to reveal effects and (2) for scales of evaluation allowing a judgement upon relevance and magnitude of effects.

The sections to follow give a short review of the scientific base (Sect. 33.3) and the methods (Sect. 33.4) which modern soil science provides to relate environmental impacts (from PP) with effects upon soils and soil functions. Sect. 33.5 tries to judge strength and weakness of the methods at present and takes a look at the availability of data, GIS-based soil surveys included, which are required for their application. Sect. 33.6 stresses the advantage of a sound data base commonly used for different purposes in environmental planning. Sect. 33.7 gives conclusions and recommendations.

It seems quite clear that the task is not accomplished once a methodology, together with a sound data, base is made available to avoid negative effects of PP upon soils. Full implementation of the SEA directive requires similar efforts for all environmental media in question. The environmental report must reflect the interaction between various media and reveal synergetic effects from all impacts. It must finally be accepted in its judgement by all individuals and groups as guidance for local and regional planning.

33.2 Links with Environmental and Planning Legislation in Germany

In Germany protection of soils is explicitly or implicitly included in number of sectoral and cross-sectoral planning instruments as part of environmental legislation. The guiding idea of SEA lies in the linkage between environmental media with the focus upon the effects. Planning in Germany uses the term 'Schutzgüter', values and benefits to be protected. Together with human health, environmental media are protected by law. Soil as environmental medium in the crosspoint of numerous interacting processes is, in some aspects and functions or in total, protected by several competing regulations. The evaluation scales and procedural requirements of SEA have to be integrated into existing legal, institutional and instrumental frames at different hierarchical levels. In order to do this we will take a look upon the procedures by which environmental legislation is implemented in practical planning in Germany and investigate the possibility for SEA to use a similar approach and eventually apply the same methodology (see also Chap. 7).

33.2.1 EIA Act

The EIA Directive was followed at the German national level by the EIA Act which came into force in 1990. Similar to EIA, the SEA Directive is not an independent planning instrument. Experiences in the Member States in implementing the EIA Directive, as amended by Directive 97/11/EC, are included in a report under http://europa.eu.int/comm/environment/eia/report_en.pdf (last update 08.08.2003). The report addresses the procedural base including screening criteria and minimum information requirements.

Looking at the history of the SEA Directive we find that essentially the procedural steps have been adopted from the existing EIA system, with adaptation to the requirements of the planning level (Feldmann 2000). Both procedures are focussing upon environmental effects and there are parallels in the procedural steps. But there are two important aspects which bring about major differences between EIA and SEA with consequences for any methodical approach as well as in its implementation and control of effects. SEA is located in an earlier phase of planning, evaluation of effects upon impact must therefore remain on a much more general level. The second aspect is the scope of the projects: SEA covers larger areas, with consequences for the parameters to be investigated, survey methods and number of alternatives. The relation between scale and methods will be considered in Sect. 33.4.

33.2.2 Nature Conservation Act

The German federal Nature Conservation Act is not very specific in soil protection which is explicitly addressed in a very general manner (Art. 2.3: *Soils must be kept in a way as to sustain their functionality in natural cycling. [...] Soil erosion is to be avoided [...]*), not taking into regard local or regional variability of soils. But there are two regulations in the act which give access to a more effective way of soil protection in planning.

The first is to be seen in Art. 13-15 which requires plans at different scales (community, region, federal state level) as environmental reports for the whole of the territory (*Landschaftspläne, Landschaftsrahmenpläne, Landschaftsprogramme*), in which the environmental situation is to be evaluated, and measures for protection and development are to be described. When preparing these plans, the specific qualities and sensitivities of soils have to be taken into regard on each scale, and evaluation has to obey the general goals of the law. The second regulation is put down in Art. 19-21 concerning "interventions in the natural surroundings" (*Eingriffsregelung*). It comes into effect under defined conditions when a surface, in private or public ownership, is searching for authorization to make changes in land use. In the plans to be submitted to the authorities the presumed effects, resulting from land use changes, have to be described and evaluated according to the principles of the law, taking into regard the soil conditions and sensitivities in the impacted areas.

In the whole one can see that the regulations arising from federal Nature Conservation Act have goals similar to those of SEA and have, likewise, to cope with very different scales from local to state level. It will be seen later what use can be made of the methodology with regard to soil effects established until now in this context.

33.2.3 Soil Protection Act

The function for anthropogenic use potential linked to the natural process of soil development is the primary goal of the German federal Soil Protection Act. This means that it is not in the scope of this law to prevent soils from degradation as long as this is to be considered as a natural process. Only negative changes by anthropogenic impact are relevant in view of the act and restoration of hazardous waste is included explicitly as a special form of soil degradation from the past. In contrast to this, SEA covers both positive and negative changes from impacts.

But do positive changes from impact really exist with regard to soils? Can soils be improved by impacts resulting from PP? We can easily figure out a situation in which this could be the case, but it is necessary then to precisely relate cause and effects in such a situation. Taking as an example the emission of atmospheric pollutants like nitrogen which may be enhanced through implementation of a PP. Nitrogen could then, via deposition to soil and transfer into plants, *improve* the nutrient and productivity function of the soil. But at the same time, other functions (with respect to water, natural biota) may be affected in an opposite direction, and the environmental report would have to list and evaluate *all* of the impacts. It can perhaps be stated, in the context of this example that the intention of SEA lies rather in the protection of environmental media than in their improvement beyond natural capacities. The Soil Protection Act comes only into vigour if standardized methods and quality standards are set with regard to hazardous waste, for polluted old industrial sites, and for urban land use. These quality standards can not readily be adopted in the context of SEA, but they can serve as a frame, together with the general principles of soil protection expressed by the act, for soil-related evaluation of impacts in the environmental report. Methods for assessment and survey of soil-related parameters can be used as well.

33.2.4 Further Legislation

German federal Building Code and Regional Planning Act gain their significance for soil protection from setting of priorities in land use. Areas of highly vulnerable soils or for the purpose of erosion control may be put under protection. In this context the Soil Protection Act and the Nature Conservation Act come into effect, with the issues described above, by contributing to the decision in urban land-use planning (*Bauleitplanung*). Immission Control Act and, on the European level, IPPC Directive, Hazardous Waste Directive and Nitrates Directive and are dealing with quite specific impacts upon soils. The procedural issues in respect to these regulations and the methodology are not directly relevant in the context considered here. But soil-related quality standards and norms may serve as a frame for the scaling of evaluation criteria in the judgement of environmental effects.

From this short review of existing environmental acts, directives and regulations we conclude that they are all aiming, like SEA, at avoiding negative impacts upon soils from different sources: industries, housing, land use practices, etc. Therefore it is scale, size of area considered and the multitude of impacts to be taken into account simultaneously which characterizes SEA. Common feature for all regulations, including SEA, is that soil shall not only be protected as a piece of nature, as a value of its own, but rather in its ecological function with respect to water, turnover of matter, and in its capacity to support human life and the life of plants and animals. Consequently the first interest is put upon methods relating impacts with effects upon soil functions. There is no set of standardized routine methods available for all regulations, and in view of the great variety of impacts and soil types reacting very differently we cannot really expect such a methodology. The specific requirements of SEA in this regard is looked upon next.

33.3 Impacts and Effects – Adequate Methods for Soil Evaluation

As has been shown above, in the context of SEA soil protection primarily means protection of soil functionality, or prevention from negative, undesired effects caused by anthropogenic impact. This is, obviously, a goal different from preserving the structure or integrity of soil components. Apart from their economic value, soils have physical and chemical properties. Linked to these properties are functions in natural processes as well as in its reaction upon anthropogenic impacts. Why is it that soils are put under the protection by law and – in our context – by SEA? The reason is that man and the society is interested in soil functions. The interests may arise and be motivated from very different sources. In a systematic view upon environmental and planning legislation, Mengel (2004 pp.25ff) has distinguished materialistic-rational values from inherent non-materialistic values attached to objects or types of objects. In relation to soil as an environmental medium, the view upon the typology of soils is clearly dominating in the context of SEA compared to soil protection of defined pedons. Scales for evaluation are usually determined by rational (utility) considerations. Looking upon the problem from the planning side, we first account for impairment of functions, negative effects, as likely consequence from the implementation of strategic actions. Scope and degree of impairment are to be evaluated, and only after having made this judgement the soil properties underlying these functions come into view, and measures to counteract negative effects may be chosen or suggested.

Soil is a heterogeneous, multi-phase system. Therefore reaction upon external impact is difficult to analyze, to describe and to predict precisely in terms of space and time. Yet our knowledge is sufficient for successfully planning agricultural and silvicultural crops. A problem arises when the scale becomes large and measures to be taken for soil protection remain unspecific in localization. If we have a distinct, small-scale soil pattern in a given area, adequate measures to counteract or to prevent negative effects will similarly show a small-scaled pattern. It could also happen that data on actual soil conditions or future impacts are not available in the same scale. In this case predictions become quite ambiguous. Consequently, only very generalized recommendations for soil protection instead of precisely defined action plans can be given for plans on a high hierarchical level, as e.g. territories of countries, federal states or watersheds of larger rivers. The situation is similar in other environmental sectors like in pollution control, flood control or water protection. But most environmental problems touch theses sectors at the same time, and what is good for the soil is very often good for other media as well. If the soil scientist gives the advice that for soil protection the emission of toxic substances to the atmosphere should be reduced in order to prevent negative effects upon soil biota, this is in agreement with goals of the emission control, protection of groundwater and nature conservation.

Summarizing this, SEA widens the possibilities for results from scientific research to be brought into the debate of environment-effective plans and projects from the very beginning. The formal procedure of the assessment enforces the scientific community to explain its knowledge of general validity in the context of a particular plan and project, and to explain and discuss it with politicians and in public. For the scientists it opens the chance to interfere with legislation, thus shaping and improving policies in a good sense.

The SEA Directive does not provide or require a fixed catalogue of methods to be applied, neither are the criteria, parameters and indicators specified, which should be used to analyze and to describe the actual state and future development of the soil, nor is this done for the form in which the environmental report is to be put down. The SEA Directive provides only the frame with regard to the procedure, methods and essential contents, and the members of the Community have to fill the gaps, if possible by making use of existing methods and instruments. In this case, compatibility and standardization of methods is required.

The environmental report, its scientific arguments and the underlying data base have to be in such a form that they can be understood and judged for plausibility by non-scientists, public and politicians. Each patch of soil is naturally endowed with a number of functions, usually called soil functions, which can be fulfilled at various degrees from individual soils or soil types. Usually in environmental legislation these functions are addressed in a very general way as shown in Box 33.1.

Since soil is a dynamic system and variable in time, its functions are variable as well. Soils can be considered under aspects of functionality and of potentiality (Fürst et al. 1992). For practical purposes each function (resp. *potential* function) must be divided into a number of sub-functions as, e.g., cycling of water, of plant nutrients, of chemicals, of organic matter as part of the "support of life – function". The names used to describe functions or sub-functions are in fact descriptive of the evaluation criteria. Thus having made the problem operable, soil scientists are able to specify the physical and chemical soil properties, or *soil parameters*, which are responsible for the function.

Box 33.1. Soil functions

Natural functions
- Maintenance of cycles of matter and energy
- Function as medium for temporary storage and buffering of transfers
- Function as medium for transformation of matter

Function for anthropogenic use
- Supply of raw materials
- Space for living and settlements
- Site for use by agriculture and silviculture
- Site for industry, public use, traffic etc.

Function to achieve natural and cultural history

each soil type is witnessing, just by its existence, the changing environmental conditions at each specific site, thus giving record of
- the natural history
- the cultural history (soil use by man)

of this site

Thus having made the problem operable, soil scientists are able to specify the physical and chemical soil properties, or *soil parameters*, which are responsible for the function. Usually these parameters can be analyzed and measured, and sometimes are technically manipulated to improve soil use for a specfic function (agricultural crops, groundwater renewal and protection, etc.). They can be used for evaluation with respect to function.

Tables 33.1 and 33.2 contain a list of typical impacts often encountered in PP, together with the main effects to be expected and with functional and key parameters for the effects. Functional parameters are indicating the function, their change upon impact can be observed or measured directly, but can perhaps not be explained in terms of a physical or chemical model.

Table 33.1. Impact by mechanical stress and precipitation, effects on soils, and related soil properties

Impacts	Direct effects >> consecutive effects	Functional parameters and indicators	Key parameters for the function in question
Effects related with water and physical soil properties			
sealing or clogging of soil surface	intake of water reduced >> surface runoff increases >> erosion >> water storage reduced	infiltration capacity of soil surface	soil texture organic matter content
vegetation cover of soil disturbed or destroyed	disturbance or loss of soil aggregates in the surface layer > surface runoff increases >> erosion >> water storage reduced	root development and density at soil surface aggregate stability in soil surface layer	soil texture, clay content organic matter content
pressure upon soil surface by machines, vehicules, in settlements, in agriculture and forestry	reduction of pore space in upper soil horizons > water storage capacity reduced >> anaerobic conditions in root horizon >> mineralization of organic matter disturbed >> soil vegetation damage	aggregate stability in soil surface layer, water storage capacity in upper soil layers	soil texture, clay content organic matter content, growth conditions for soil vegetation and edaphon

Table 33.2. Impacts by chemicals, effects on soils, and related soil properties

Effects related with chemicals and soil chemical properties

emission and deposition of acids via atmosphere (rain, snow, fog)	soil acidification > mobilization of toxic substances, especially heavy metals (from soil minerals or anthropogenic) > loss of nutrients (basicity) >> pollution of ground and surface water >> toxic effects in plants and soil animals >>reduced plant growth and crop quality	acid neutralizing capacity soil acidity (pH) increase of dissolved metals in soil solution acidification of seepage water	base saturation carbonate content stores of silicates in mineral soil horizons
emission and depositon of persistent toxic substances via atmosphere or by direct application (dust, refuse, sewage sludge, fertilizers and pesticides, chemicals etc.)	accumulation of toxic substances in soil, unexpected mobilization due to various causes >damage to soil organisms and vegetation >>health risks, esp. for children >> pollution of ground and surface water	adsorption capacity of soil for toxic substances degree of saturation of adsorbing soil matrix with toxic substances (pre-loading)	soil texture, clay content organic matter content, base saturation, carbonate content
emission and depositon of degradable toxic organics via atmosphere or by direct application (pesticides, industrial emissions, chemicals, sewage sludge etc.)	accumulation, usually transitory, of toxic substances in soil, unexpected mobilization due to various causes >damage to soil organisms and vegetation >>health risks, esp. for children >> pollution of ground and surface water	adsorption capacity for toxic substances and microbial activty in soil degree of saturation of adsorbing soil matrix with toxic substances (pre-loading)	soil texture, clay content, organic matter content, pH, base saturation, carbonate content, microbial diversity in organic soil horizons

Key parameters are considered to be causal for the effect, based upon knowledge of the process chain initialized by the impact. Usually they can, too, be observed or measured quantitatively. Both types of parameters are crucial for soil evaluation and for the statement of relevance and sustainability of effects in the project area or some subunit of it. Tables 33.1 and 33.2 contain only the most common impacts from human activities and the most relevant of effects and parameters. It includes only natural functions inherent in various degrees to soils without any technical treatment. Enhancement of this functionality would require technical measures or counteractions, the cost of which would have to enter in an economic evaluation. From an economic point of view, in the course of the SEA process this automatism makes sure that external effects from impact are not neglected, and privately owned soil remains protected by property rights according to its vulnerability, in the interest of the public. In this context a multi-annual project of the Statistical Office of the European Commission (EUROSTAT) has to be mentioned. TEPI stands for "Towards Environmental Pressure Indicators for the EU" (http://www.e-m-a-i-l.nu/tepi/default.htm).

The objectives of the project are to calculate and present priority pressure indicators in each environmental policy field, for all EU Member States and to present, where possible, the contributions of the economic sectors to the environmental pressure (TEPI 2004). The indicators, as defined, could point at environmental impacts when a PP is likely to increase environmental pressure in this specific respect, perhaps with effects relevant for SEA. Several indicators defined by TEPI are soil-related like, e.g., emission of nutrients in agriculture, waste incinerated, consumption of toxic chemicals, discharges of heavy metals, etc.

Viewing the list of impacts and effects in Tables 33.1 and 33.2, two questions arise if we want to make SEA a successful instrument in environmental protection:

1. which methods are available to relate impact on one hand with effects on the other, and are parameters given in the list a sufficient base for applying the method, or are additional parameters necessary?
2. are information and data available for the area touched by the PP, and is the spatial resolution adequate to the requirement of SEA?

The following examples are taken from Germany, but applicability of methods is not restricted to a specific country. In fact, similar methods are used in many other countries, which, once again, stresses the necessity for standardization of methods and quality indicators on the European level.

33.4 Methods Relating Soil Properties with Functions

SEA starts with the determination of potential impacts from the PP upon environmental values including soil. The information and data for the area concerned by the PP must specify spatial and time scale, but also intensity of the effects likely to result from impact, for only *relevant* effects are mandatory for an assessment. What is the scientific base for a sound prognosis of effects? In Annex 1 the Directive demands that *"the environmental report prepared [...] shall include the information that may reasonably be required taking into account current knowledge and methods of assessment"*. This means in practical terms that the soil in the project area must be evaluated according to its sensitivity towards external impacts. What is the methodology that practical soil science and soil survey can offer in order to respond to these demands? The soil scientist will first look for parameters which can be measured and results be surveyed and laid down in a map, giving the degree of sensitivity of a given soil towards external impacts. Such impacts could be, e.g., mechanical disturbance, input of acids or toxic substances, wind, water, thermal impacts etc. All of these impacts may be absorbed or buffered by the soil, or they may alter, disturb or deteriorate it. Soil science has developed numerous of methods relating all sorts of impacts with effects in soils. The strength and intensity of the effects will certainly depend upon the intensity of the impact, but it will also depend upon soil inherent parameters. The way in which soil reacts can be called *sensitivity*. With regard to practical application it is important that the sensitivity of soils can be evaluated quantitatively, i.e. that a classification of sensitivity parameters allows a ranking. Ideally we would hope that in the project area soil survey provides a map showing the spatial distribution of soil types classified or ranked in sensitivity classes.

Looking for adequate methods, the intention of SEA must be kept in mind. The environmental report must contain relevant information on environmental effects, especially here with regard to soils, taking into account the objectives and the geographical scope of the PP. The PP are to be conducted in a way that negative effects are minimized. The methods can only be applied when input parameters representative of the soils in the area considered for the PP are available and accessible. Usually this will be a soil map. The methods must yield a surface pattern depicting the sensitivity of soils upon impacts at individual sites, and in this they should fit into a simple classification system. A local or regional survey should offer a map reflecting the pattern of soil sensitivity in the area subject to the PP.

Soils can be evaluated with regard to different aspects. One aspect is to look at key parameters, the term *key* indicating that this parameter is responsible for a soil function (see Table 33.3). These parameters stand for a soil-inherent property which is directly related with a soil function. As an example, the carbonate content of mineral soil horizons is the cause for their buffering capacity. It prevents or counteracts acidification of soil and seepage water being caused by acid rain or similar impacts. Carbonate content can therefore be used to rank soils according to their sensitivity towards acid impact.

Another aspect of evaluation can be to measure a function directly, without looking at structural and chemical base of the function. In Tables 33.1-33.5 these are called functional parameters because they are the result of a number of soil-inherent properties which must not necessarily be known. As an example, the structural stability of a soil horizon can be directly measured. A low stability would indicate high sensitivity towards impact of water droplets or mechanical pressure, with various negative consecutive effects like increase in surface runoff, erosion etc. Again this parameter can be used to rank soils according to their sensitivity.

A large number of methods is available for the evaluation of soils for planning, both under protection and sensitivity aspects. A very useful guidance has been elaborated by the German Federal Soil Association (Bundesverband Boden 2001). The focus is put on precaution-oriented evaluation of soils in urban land use planning. An excellent review is also put forward in the final report from Planungsgruppe Ökologie + Umwelt (2003) where 64 methods are cited for the evaluation of natural and archivary soil functions with reference to data availability and adaption to different planning scales. With these methods soil can be evaluated with regard to a specific function, as defined in Sect. 33.3 (Box 33.1). Usually this aptitude can be ranked, i.e. the method gives information about how good the function in question is supported by the patch of soil analyzed. In addition to this, 18 methods for non-chemical soil risks are reviewed, i.e. risk for the integrity of the soil profile with regard to structural soil properties (erosion, compaction).

Many of the methods cited are going back to the methods bank, a concise assemblage of evaluation methods edited by Müller (1997).

Table 33.3. Selected methods for risk assessment related to soil functions (adapted from Müller 1997)

Risks for integrity of natural soil functions

risk assessment for defined soil functions	key and functional parameters	recommended scale for application
susceptibility for erosion by water 6.4	soil texture, slope	1:10 000 – 1:500 000 high resolution methods available depending from availability of site data
susceptibility for erosion by wind 6.5	soil texture, stone content, soil type, humus content, bulk density, groundwater level*	1:10 000 – 1:500 000 high resolution methods available depending from availability of site data
susceptibility for compaction and loss of pore space 6.7	soil texture, stone content, soil type, humus content, bulk density, carbonate content, groundwater level*,	all scales
susceptibility for surface clogging 6.19	soil texture, clay content	all scales

Note: The figures in the left column indicate the method of risk assessment described in (Müller 1997). The parameters denominate soil properties by which the aptitude of a defined soil is evaluated to comply with the specific function, * indicates non-soil parameters to be included in the evaluation (e.g. topographic position, meteorological data, land use type).

Table 33.4. Selected methods for risk assessment related to soil functions (adapted from Müller 1997)

Risks for soil functions related with water and site hydrology

risk for soil functions	key and functional parameters	recommended scale
supply of water for plant growth (water availability) 6.1	soil texture, stone content, soil type, bulk density, humus content, land use type*	all scales
regulation of seepage flow towards groundwater 6.2	soil texture, stone content, soil type, humus content, slope*, rainfall data*	all scales

Risks for soil functions related with chemicals

risk for soil functions	key and functional parameters	recommended scale
retention of toxic organic substances 6.8	soil texture, stone content, soil type, pH, humus content, bulk density, carbonate content, type of organic*, groundwater level*, land use type*, meteorological data*, slope*	all scales
retention of heavy metals 6.9	soil texture, stone content, soil type, pH humus content, bulk density, carbonate content, metal type*, groundwater level*, land use type*, meteorological data*, slope*	all scales
susceptibility for soil acidification under forestry use 6.18	soil texture, clay content, humus content, pH, potential cation exchange capacity, bulk density, soil type, bedrock type, exposition, meteorological data*, forest type*	1:5 000 – 1:200 000
susceptibility for transfer of nitrate towards groundwater 6.16	soil texture, stone content, soil type, humus content, bulk density, field capacity, meteorological data*, slope and exposition*, land use type*	all scales
susceptibility of agricultural surfaces for transfer of nitrate during winter period 6.10	soil texture, stone content, soil type, humus content, bulk density, carbonate content, groundwater level*, meteorological data*, surface inflow and runoff*, slope*	1:5 000

Note: see note of Table 33.3

Tables 33.3-33.5 give an overview of available methods for the evaluation of soils with regard to the complete set of natural functions/potentials, or with regard to specific functions as listed in Box 33.1. The list is far from exhaustive. Since the degree of standardization in this field is very low and the conditions under which methods are applied very variable and, finally, the accuracy requirements are quite soft in view of the overall uncertainty of planning (see Sect. 33.5), it is not difficult for a soil expert to develop an ad-hoc method for evaluation based on best available data. This is frequently done with good success, yielding better results for planning decisions than a poor guess without any field data. As an example we may look at a method developed by Arnold and Vorderbrügge (1996) to evaluate soils for their potential of crop productivity and for the development of natural biotopes. Input data include soil parameters like texture and root accessibility, field capacity and soil pH. As a result the soil can be classified and evaluated with regard to nature conservation requirements and land use types.

Table 33.5. Selected methods for risk assessment related to soil functions (adapted from Müller 1997)

Soil functions related to agricultural and silvicultural use or natural vegetation

risk assessment for defined soil functions	key and functional parameters	recommended scale for application
site dependent growth potential 6.11	soil texture, stone content, soil type, humus content, bulk density, meteorological data	1:50 000 – 1:500 000
site dependent soil fertility 6.12	soil texture, stone content, soil type, humus content, potential cation exchange capacity, pH, base saturation, bulk density, root depth*	1:5 000 – 1:200 000
supply of basicity (liming requirement) 6.12	soil texture, stone content, humus content, bulk density, pH	1:5 000
water supply conditions (bodenkundliche Feuchtestufen) 6.3	soil texture, stone content, soil type, humus content, groundwater level*, bulk density	all scales

Note: see note of Table 33.3

33.5 Strength and Weakness of Methodologies, Data Requirements

Deficiencies in current SEA practice are very often, as far as soils are concerned, the lack of area-specific data. Either there are no data at all, or they are not available in necessary scale or resolution. In addition to this, the inhomogeneity aggravates the problem. For example, in Germany the soil survey is until today the responsibility of the federal states (for details see Chap. 7), each of which has its own tradition and for a long time even has used its own methods of survey, standards and presentation in maps. If we go beyond the national level, the discrepancy in this respect is even greater.

But it must be stated that there have been in recent years considerable efforts in standardizing methods and defining norms. With regard to the methodology, the method bank cited in Sect. 33.4 is already the result of inter-state and international cooperation in developing methods.

Compared to the problems of availability of soil data and evaluation methods, accuracy of the results seems to be of minor importance. As long as there is good base in area specific soil data, the accuracy requirements in the evaluation of effects are not excessive. The large area SEA are usually covering leads to a high level of uncertainty with respect to future environmental and economic conditions; and hence to uncertainty about the developments and impacts from PP. In this situation highly precise prognoses built on uncertain grounds make no sense. To give an example: if we want to predict soil acidification, as a result of environmental impact from a PP, on a surface of some 1 000 km^2 with a soil pattern ranging from Podsol to Rendzina, than a five step scale in the evaluation of susceptibility towards acidification is sufficient.

The methods listed in Tables 33.4-33.6 are marked by the numbers as used in Müller 1997. Some of these methods require additional procedures to estimate basic parameters from field data (e.g. the effective bulk density, utilizable field capacity, water permeability of saturated soil). These supporting methods are also included in the methods bank.

Some of the key and functional parameters in Tables 33.4-33.6 (second column) represent an essential input to several evaluation methods (first column). Soil texture for example is responsible for and related with many soil processes, because it determines the form and size of the soil pores, the size of reactive surface and its chemical characteristics. Grain size distribution (texture) is easy to be measured and most soil maps give information about soil texture in high resolution. It has been mentioned in the previous section that the applicability of the methods in Tables 33.4-33.6 depends upon the scale for which evaluation of effects can be derived. In each single case the question will raise whether data are available for the area considered in the form of soil maps. Box 33.2 reflects the present situation in Germany, according to a report from Planungsgruppe Ökologie + Umwelt (2003).

> **Box 33.2.** Coverage of soil maps at different scales in Germany, providing data for soil sensivity to effects
>
> - survey level (1:100 000 to 1:200 000)
>
> Soil maps available for >40 % of the territory, percentage rapidly increasing, 1:200 000 (in the western federal states), or 1:100.000 (in the eastern federal states)
>
> - intermediate level (1:25 000 to 1:50 000)
>
> the mean coverage of the territory is around 30 % , percentage of coverage in digital form rapidly increasing
>
> - scales $\geq$ 1:10 000
>
> On forest areas almost complete coverage by *Forstliche Standortkarten* ($\geq$ 1:10 000) which include information on soils.
> On all areas under agricultural use complete coverage by land evaluation map (*Bodenschätzungskarte*) usually in the scale $\geq$ 1:5 000.
> Both types of maps are usually not available in digitalized form

The soil survey published by authorities from each of the federal states of Germany as "*Amtliche Bodenkarte* 1:25 000" is increasingly made available in digitalized form. The soil map from Münden, edited by the federal state of Hesse, may serve as an example. The report to the map (Emmerich 1997) contains several thematic evaluation maps: exchange capacity, base saturation, field capacity for soil water, risk for nitrate loss to groundwater, accumulation capacity for pollutants. These parameters are among the key and functional parameters listed in Tables 33.4-33.6, ready to be used in evaluation procedures.

This is just one example for the way how authorities and institutions producing soil maps provide inherant information for planning purposes. It shows on the other hand the potential for soil evaluation when digitalized soil data are incorporated in a Geographical Information System (GIS).

In this context the soil information systems should be mentioned, which have been established by several authorities during recent years, usually by making use of a GIS system (Scholles 1999). On the national level, the Federal Environmental Agency has taken responsibility for survey and evaluation of information on soil status, pollution, impacts, input and loss of materials under different aspects (bBIS 2004). On the level of federal states, there are similar programmes under way. The link between practical application of soil information as needed for example by SEA, and scientific research in landscape ecology is treated under theoretical aspects and with regard to case studies in Tenhunen et al. (2001).

The degree in depth of information and the parameters to be included in the environmental report is not fixed in advance. Usually the authorities responsible for environmental protection have to define and select these in each individual case. But participation of more authorities could make sense, and the rules for the procedure in this respect are still under discussion (Pietzcker and Fiedler 2002). Therefore as long as there are basic soil data available for the project area, as for instance texture and soil type represented in a soil map, the soil expert will be able to point out the best method to make valid predictions on likely effects and risks.

In the discussion planners are often expressing their doubts whether the inclusion of highly aggregated soil information for planning decisions concerning large surfaces, as it would be the case for most PP in SEA, is feasible. This is in fact a problem and the solution for it is not trivial. But the case studies presented in recent literature (see for instance Riedel and Lange 2001; Tenhunen et al. 2001; Kaule 2002; Jessel and Tobias 2002) show that combination of geo-statistical and remote sensing methods with GIS provides a number of tools for prediction, evaluation and avoidance of negative environmental effects on a scientific base. The gap between soil research, survey techniques and planning practice becomes smaller by user-friendly presentation of soil information and evaluation methods. For example, the soil information system BODIS developed in Schleswig-Holstein (BODIS 2004) started in 2002 to make information of the land evaluation map in scales $\geq$ 1:5 000 available until 2006, online and in digital form. Finally on the European level the CORINE (Coordination of Information on the Environment) deserves recognition as a programme providing spatial environmental data including maps suitable as information base for SEA.

33.6 Synergies from the Use of a Common Environmental Data Base

In order to avoid duplication of assessment, the Directive allows that relevant information available on environmental effects of PP obtained at other levels of decision-making or through other Community legislation may be used for providing information for the environmental report (Art. 5 para 2 of the SEA Directive). Synergetic effects from different cross-sectional and specific planning instruments can make plans more effective in the results for each of it. Usually the implementation of all measures, duties and regulations based on environmental law are, to a certain degree, opposed to the interests of individuals; enforcement by authorities is often provoking resistance. Enhancement in the acceptance of PP in public opinion can certainly be improved through the SEA process by leading to an early participation of the public and thorough information about background and limitation of decisions. Soil protection is a goal for several environmental regulations and acts, the information on which each of them must base its decision, can be and should be the same when it covers the same area. Collecting and analyzing environmental data is time consuming and expensive. Therefore positive synergies from the use of a common environmental data base may be expected. The environmental report as required by SEA gives adequate information on quality, status and values of the environmental media in a form that can be understood from non-experts. In a research investigation launched by the German Federal Agency for Nature Conservation, Scholles et al. (2003) have shown that existing planning instruments and potentials can in fact be linked in order to achieve better results.

Table 33.6. Soil protective goals in German planning

soil related objectives and goals of the plan	concepts and methods for enforcement	data and information required for evaluation and future development
Cross-sectoral planning on various hierarchical levels (federal, state, region, community)		
safeguarding natural cycles and soil functionality; restoration of degraded; protection from soil degradation, sealing and loss	declaration of areas for protection of highly vulnerable soils with regard to integrity and soil functions; erosion control; conservation of soils; setting of priorities in land use formal definition of best practice in land use	soil maps in appropriate scales;
Nature conservation planning (federal and state according to Nature Conservation Act)		
protection of environmental values as defined by law => safeguarding natural cycles and soil functionality; restoration if degraded; protection from soil degradation, sealing and loss on respective planning level and scale: protection of local requirements are declared and delineated in the plan and are transformed into a develoment concept for the community territory	"interventions in the natural environments" (Eingriffsregelung)	Environmental Report with an inventory of environmental values (Landschaftsplan)
Sectoral planning with soil protective effects (agriculture and forestry, transportation)		
soil protection and restoration or melioration if degraded – protection of environmental values as defined by federal Nature Conservation Act sustainable forest use; soil protection as defined by BWaldG organization and planning of roads >> and in this context >> minimize soil loss and sealing; optimize soil protection and restoration, protection of environmental values as defined by federal Nature Conservation Act	formal definition of best practice in forestry; approval of plan	soil maps; land evaluation maps;

Scholles et al. (2003) remind of the basic function of landscape planning in Germany for the integration of environmental protection requirements into the existing system of spatial planning. This intention is in accordance with the goals of SEA. The authors show that the requirements for landscape planning in Germany can satisfy those for the environmental report (SEA) to a large extent when properly prepared and applied according to existing regulations. It can in total be used for SEA as base for information and evaluation. Yet German landscape planning does not set goals for environmental protection in general, but explicitly for the environmental values as defined by the federal Nature Conservation Act. Alternatives for a project/plan are not in its scope. The requirement for the environmental report to be summarized and explained for the non-scientific public is not mandatory in German landscape planning. In spite of these differences the efficiency of the existing planning system will certainly be improved when SEA is implemented according to its intentions.

Table 33.6 shows a list of soil protective goals implicitly or explicitly set in various acts and regulations of German planning legislation, adapted from *Planungsgruppe Ökologie + Umwelt* (2003, Annex I, p.A5 ff). The regulations listed are very different in their intentions and goals. While most of them explicitly include environmental protection there are others which have in focus quite different objectives, but are legally forced to take into account environmental issues. The federal and state transportation planning system can be taken as an example. Although having the quality of an administrative order and not that of a legal instrument, the federal transportation plan complies with the definition of PP for which SEA is obligatory. Individual segments of a federal road is subject to the regulation for interventions in the natural surroundings (*Eingriffsregelung*), according to federal Nature Conservation Act. By this approach one could expect to find a routing alignment with smallest environmental impact. But the question wether SEA has to be applied at all, and what the methodology would be adequate in this case, is not yet settled and is the object of various research programmes (see for example Hendler 2003; Wende et al. 2003). If the result will be that SEA is to be applied soil maps of satisfactory quality are required. A similar situation exists in the waste and recycling as well as in the water sector (Kraetzschmer 2003).

33.7 Conclusions and Recommendations

The large number of planning regulations, in which environmental protection is at least one, if not the only goal, shows that there is a broad field for SEA to assure proper and timely coordination of planning activities. The methodology is very difficult to formalize just because plans and programmes (PP) are of very different nature, and the environmental law to be applied depends upon just this nature. For example, according to existing legislation in Germany a transportation plan for the federal highway system is covered by a procedural framework completely different from planning for powerplants or an energy wind park. In transboundary PP the situation is even worse.

Consequently the methodology must be systemized according to the likely effects upon soil, caused by impacts from all relevant sources. The system of effects must be crossed with the presumed impacts from a specific PP. At the crosspoints we will have to develop measures to protect soils from impacts or to find countermeasures suitable to avoid negative effects. This procedure can only be successful if environmental effects are predicted on a sound data base specific for the region/surface considered for the PP.

Following the arguments of Bachmann and Thoenes (2000), soil protection as a general environmental goal can only be achieved on such a base, available at an early stage of planning. Until today soil protection is suffering not so much from a lack in knowledge and information but rather from deficiencies in the planning procedures to make use of this knowledge and to implement it in decisions. SEA puts a procedural frame to such efforts when large surfaces are presumably or potentially affected by impacts from a PP, provided that there is a consistent link between type of soil, soil value with respect to environmental standards, and a realistic estimate of the magnitude of impacts. While setting of quality goals and standards is a matter of society and politicians advised by soil experts, the linkage between soil on one side, in its distinct local form and shape, and impacts and effects on the other side is a field open for scientific research. Poor knowledge in this respect produces poor results in soil protection. The examples presented above were chosen in order to testify, that research in this field has already produced considerable progress good enough to be tested in the planning practice.

References[1]

Ad-hoc-AG Boden der Geol. Landesämter und d. BGR (1994) Bodenkundliche Kartieranleitung. 4. Auflage, Hannover (in Kommission: E. Schweizerbart'sche Verlagsbuchhandlung Nägele und Obermiller, Stuttgart)

Arnold H, Vorderbrügge Th (1996) Beiträge des Bodenschutzes zum Naturschutz – am Beispiel von thematischen Bodenschutzkarten zum Produktions- und Biotopentwicklungspotential. Jb. Naturschutz in Hessen 1:67-70

Bachmann G, Thoenes H-W (2000) Wege zum vorsorgenden Bodenschutz, Erich Schmidt Verlag, Berlin

bBIS (2004) Bundesweites Bodeninformationssystem des Umweltbundesamtes, web: http://www.umweltbundesamt.de/altlast/web1/steckbriefe/steckb07.htm, last accessed 04.05.2004

BODIS (2004) Umweltbericht des Landes Schleswig-Holstein – Bodeninformationssystem. Internet address: http://www.umwelt.schleswig-holstein.de/?23367, last accessed 04.05.2004

---

[1] Note that legislation mentioned in the chapter is listed at the end of the handbook in the consolidated list of legislation (Appendix 2).

Bundesverband Boden (ed) (2001) Bodenschutz in der Bauleitplanung – Vorsorgeorientierte Bewertung, BvB Materialien vol. 6. Erich Schmidt, Berlin
Emmerich K-H (1997) Erläuterungen zur Bodenkarte von Hessen 1:25 000, Blatt Nr. 4523 Münden, Wiesbaden
Feldmann L (2000) Strategische Umweltprüfung (SUP) - Zwei Drittel des Weges zur EG-Richtlinie geschafft. UVP-report 2:109-110
Fürst D, Kiemstedt H, Gustedt E, Ratzbor G, Scholles F (1992) Umweltqualitätsziele für die ökologsiche Planung. UBA-Texte 34(93) Berlin
Hendler R (2003) Zur Umweltprüfungspflichtigkeit des Bundesverkehrswegeplans. UVP-report 17(2):57-59
Jessel B, Tobias K (2002) Ökologisch orientierte Planung (ecologically oriented planning). Ulmer, Stuttgart
Kaule G (2002) Umweltplanung, Ulmer, Stuttgart
Kraetzschmer D (2003) Umweltprüfung für Pläne und Programme des Abfall- und des Wasserrechts. UVP-report 17(2):64-67.
Mengel A (2004) Naturschutz, Landnutzung und Grundeigentum. Frankfurter Schriften zum Umweltrecht 34, Nomos Verlagsgesellschaft Baden-Baden
Müller U (1997) Auswertungsmethoden im Bodenschutz - Dokumentation zur Methodenbank des Niedersächsischen Bodeninformationssystems (NIBIS), Hannover, Loseblattsammlung
Pietzcker J, Fiedler Ch (2002) Die Umsetzung der Plan-UVP-Richtlinie im Bauplanungsrecht. UVP-report 3:83-93
Planungsgruppe Ökologie + Umwelt (2003) Zusammenfassung und Strukturierung von relevanten Methoden und Verfahren zur Klassifikation und Bewertung von Bodenfunktionen für Planungs- und Zulassungsverfahren mit dem Ziel der Vergleichbarkeit, Bund-/Länder-Arbeitsgemeinschaft Bodenschutz LABO, July 2003, Hannover
Riedel W, Lange H (2001) Landschaftsplanung, G. Fischer, Berlin, Heidelberg
Scholles F (1999) GIS-Einsatz in der UVP – Eine Einführung. UVP-report 4:197
Scholles F, Haaren von C, Myrzik A, Ott S, Wilke T, Winkelbrandt A, Wulfert K (2003) Strategische Umweltprüfung und Landschaftsplanung (SEA and landscape planning). UVP-report 17(2):76-82
Sommer K (2003) Umweltschutzziele in der Strategischen Umweltprüfung. UVP-report 17(2):82-84
Tenhunen JD, Lenz R, Hantschel R (eds.) (2001) Ecosystem Approaches to Landscape Management in Central Europe. Springer (Ecological Studies 147), Berlin
TEPI (2004) Towards Environmental Pressure Indicators for the EU, Statistical Office of the European Commission (EUROSTAT).
web: http://www.e-m-a-i-l.nu/tepi/default.htm, last accessed 13/02/04
Wende W, Gassner E, Günnewig D, Köppel J, Langenheld A, Kerber N, Peters W, Röthke P (2003) Anforderungen der SUP-Richtlinie an Bundesverkehrswegeplanung und Verkehrsentwicklungsplanung der Länder. UVP-report 17(2):60-63

34 Towards the Implementation of SEA – Learning from EIA for Water Resources

Damian Lawler

School of Geography, Earth and Environmental Sciences, The University of Birmingham, UK

34.1 Introduction

The main purpose of SEA is to facilitate early and systematic consideration of potential environmental impacts in strategic decision-making' (Finneveden et al. 2003 p.92). In short, it is to take a longer-term and more holistic view of environmental effects and sustainability than is possible for the project-based foci of EIA. It was made official in the EU in July 2001 under the SEA Directive, with a margin of discretion allowed for member states to propose their own procedures, integration timescales and legislative frameworks, although 21 July 2004 was the published application date (Risse et al. 2003). SEA deals with plans, policies and programmes, although the SEA Directive focuses on plans and programmes. Of additional relevance to water resources is the Water Framework Directive (see also Chap. 35) which place responsibilities on governments, environmental protection agencies and regulators to ensure an appropriate quality and quantity of water for a nation's human population and its ecosystems. This has led to many recent papers from the EU dealing with sustainable watershed management (e.g. Hedo and Bina 1999; Tsakiris 2002). This normally includes some kind of Environmental Assessment (EA) process.

This chapter proposes that we can use lessons learned from project EIA to improve the implementation of SEA. The principle of what is termed here 'EIA-SEA experience linkage' is evident in EA studies. For example, Noble (2004) implicitly took this approach in the case of the role of assessment panels in SEA, and Finnveden (2003) comments explicitly on the useful transfer of EIA *principles* to SEA policy development, although the applicability of EIA *methods* may be less straightforward. The chapter is timely, given the lack of current methodological guidance for SEA components, and it is hoped that it can contribute to the development of SEA principles, approaches, methods and tools that will undoubtedly take shape over the next ten years.

Implementing Strategic Environmental Assessment. Edited by Michael Schmidt, Elsa João and Eike Albrecht. © 2005 Springer-Verlag

The chapter proposes, in particular, three key needs for the successful implementation of SEA: (1) appropriate environmental data quality and quantity driven by a well-designed and resourced environmental monitoring programme; (2) the development of a system of what is called here 'Concurrent Programme Appraisal' (CPA) – the SEA equivalent to 'post-project appraisal' in EIA, to monitor the environmental effectiveness of new plans, programmes and policies; and (c) a rigorous Peer Review System for all major SEA reports.

Although the concepts proposed are applicable to many nations, examples to demonstrate the need for change in EA thinking are drawn mainly from the Former Soviet Union (FSU) republics. Few hydrological or environmental studies have been published for these nations, however, yet they represent very important foci because they are characterised by scarce water resources, a legacy of environmental pollution, highly-variable hydrological systems, significant climate changes and complex water resource development issues. Also, many EIA have been completed in the FSU recently, particularly in relation to development of their vast oil and gas reserves. It is argued here that the *key* threat to the successful introduction of SEA in the FSU is the collapse of environmental monitoring networks, and the consequent cessation in environmental data capture vital for monitoring the success of plans, policies and programmes. The chapter first summarises the existing methodology and context for water resource project EIA studies within the energy sector (Sects. 34.2-34.4), then identifies some of the challenges in applying EIA procedures (Sects. 34.5-34.7), and finally proposes a number of solutions to these challenges within a revised *Water Impact Assessment* (WIA) module incorporated within project EIA, and *Strategic Water Assessment* (SWA) methodologies integrated within SEA (Sects. 34.8-34.9).

34.2 Water in the SEA and EIA Processes

The survey by Cherp (2002a) revealed that, in the FSU, there is little SEA, few methodological provisions for SEA, and limited mandatory EIA legislation. Instead, SER (*State Environmental Review*) and OVOS (Assessment of Environmental Impacts) is used for projects, and SER-type approaches also for strategic issues (Cherp 2001a; 2001b; Cherp and Lee 1997). Since 1996, SER has sometimes been used in a dual approach alongside a 'western' style, modern EIA methodology introduced by overseas companies and consortia, at the World Bank's encouragement. Clearly, it is desirable to take longer-term strategic views of the energy and water sector. Cherp (2001a) found that there was resistance to SEA in the FSU generally, because it is seen as intruding on State's right to control strategic issues, unlike EIA and individual project developers. SER can be applied to strategic issues (Cherp 2001b), though with little legislative enforcement it may generate modest results. No systematic evaluation of the patchy take-up of SEA in the FSU has yet been undertaken (Cherp 2001b).

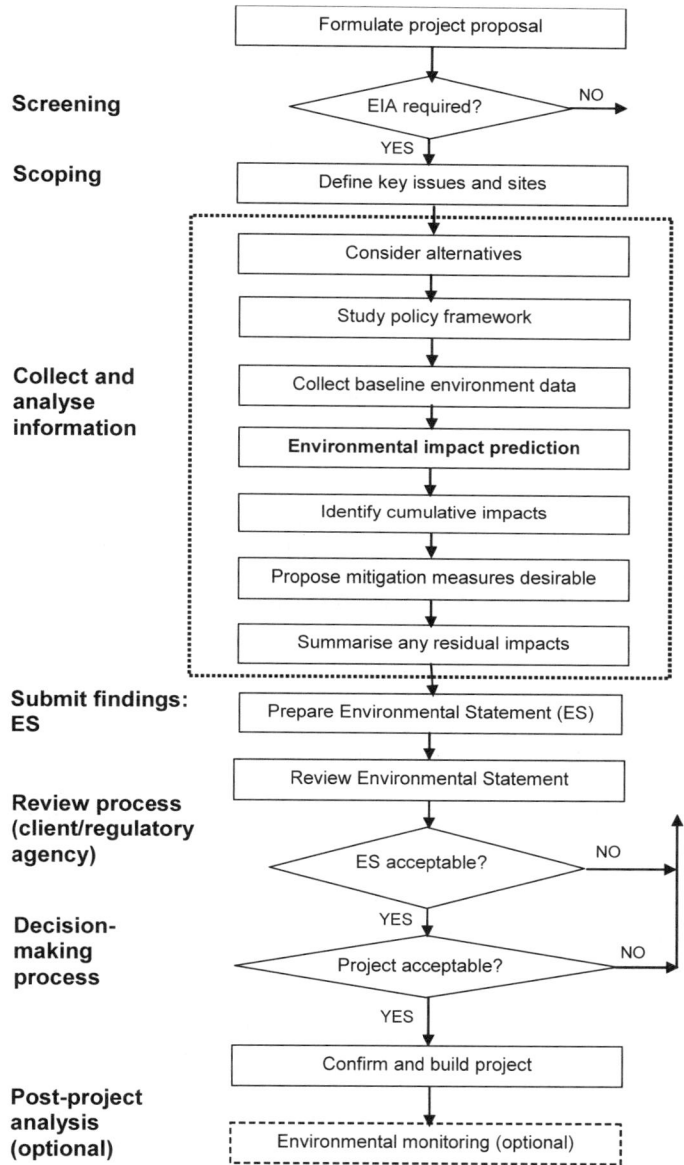

Fig. 34.1. Main stages in a typical modern EIA process (after Glasson et al. 1999)

It is timely and instructive, therefore, to examine how EIA currently operates in the FSU, identify the strengths and limitations of these approaches, suggest some solutions to some of the difficulties, and finally examine how the lessons learned from EIA application can help to ease and strengthen the introduction of SEA. A typical EIA process is shown in Fig. 34.1 EIA includes a prediction of the likely environmental effects that a specific development would be expected to produce (Glasson et al. 1999), the results of which are presented in Environmental Statements (ES) (Fig. 34.1). ES should also discuss: a description of the development project; alternatives considered; policy framework; baseline environmental conditions; environmental impact assessment of the development; cumulative impacts; mitigation measures recommended to reduce undesirable impacts to an acceptable levels (e.g. Johnson et al. 1999); and a summary of any residual impacts remaining after mitigation techniques are adopted (e.g. Fig. 34.1).

34.3 Water and Energy Resources of Azerbaijan

Azerbaijan is a Newly Independent State (NIS) of the FSU, located between Russia, Iran, Georgia and Armenia. It lies in the semi-arid zone: mean annual precipitation varies from approximately 400 mm per year in the west to just 100 mm per year at the Caspian Sea. Azerbaijan provides a very useful test of the effectiveness of the hydrological components of EIA procedures because numerous EIAs have been completed here since 1996, and is a good example of an FSU republic with important energy resources to develop. Also, Azerbaijan lies in the semi-arid zone characterised by scarce, valuable water resources (Wolfson and Daniell 1995), key geopolitical significance, a transitional economy (Bradshaw 1997), and numerous oil and gas development projects, many with western involvement. The rivers are sensitive to development disturbance because, owing to their high mountain origin, rivers in Azerbaijan are high-energy systems and are already transporting large loads of sediment which are liable to bind with any introduced contaminants in a hazardous manner. It has a history of environmental pollution.

When water and energy environments meet, however, conflicts can arise. There has been very substantial development of Azeri oil and gas resources since the mid-1990s (Efendiyeva 2000; Saiko 2001; Lawler 2002), with more planned. Oil is extracted from the Caspian Sea to Sangachal terminal, near the capital, Baku, in eastern Azerbaijan. It is then piped through northern Azerbaijan to the Black Sea via Russia via the Northern Route Export Pipeline (NREP). A second line runs from Sangachal via the Western Route Export Pipeline (WREP) across central Azerbaijan into Georgia, to the Supsa oil terminal on the Black Sea. A third pipeline (SCP) runs alongside the WREP, and a fourth development, a major oil pipeline (BTC), is currently under construction (2004) to link Azerbaijan with Turkey through Georgia. The World Bank funds some of the development. Loans are conditional, however, upon full EIA being carried out to western standards (Lawler 2002).

34.4 Environmental Impacts Summary

34.4.1 Pipeline Impacts on Water Resources

A short summary of environmental impacts of pipeline construction and operation on scarce and vulnerable surface and groundwater resources will suffice here. This will demonstrate the range and complexity of impacts that had to be assessed using existing EIA methodologies, including some less obvious issues. Construction across rivers, for example, can lead to sediment pollution, while pipeline leakage can result in hydrocarbon contamination of water bodies, habitats and downstream receptors such as reservoirs, wetlands and groundwater bodies. Pipeline river crossings need to be assessed carefully and comprehensively: see, for example, the framework established by Lawler et al. (1996a). A further impact can arise during "hydrotesting". This is the initial checking of pipeline integrity for leaks, before oil is transmitted, by pumping high-pressure water through the system. This places great demand on water supply in a semi-arid environment. Also, in some places, however, as in Azerbaijan, routes can include old, refurbished pipelines which contained stale crude oil. Clearly, hydrotesting then produces a large volume of contaminated water which needs safe disposal. As part of the EIA, therefore, a method statement has to be agreed and published for the acceptable disposal of hydrotest water. Other statements also need to be produce for the safe disposal of effluents generated during the construction or operation process, including waste water produced at temporary construction sites and permanent terminal and pump stations.

34.4.2 'Reverse' Impacts

Contemporary EIA, however, should also consider the 'reverse' impacts of the environment on the development. For pipelines, these include the impacts of bank erosion and channel instability on pipeline failure at river crossings (e.g. Lawler et al. 1996a). Unprotected pipelines can be undermined, abraded or corroded by river flows and coarse sediment transport which in turn, can cause further environmental damage. Assessment of flood and mudflow risk to facilities built on river or coastal floodplains is also included, as is the risk of earthquake damage in seismically-active environments.

34.5 Challenges for EIA Procedures in the FSU

EIA carried out in Azerbaijan for overseas consortia have been detailed and modern in approach, and several strengths are identifiable. For example, EIA improves on previous environmental policies, some of which have been seen by some as deficient (e.g. Cherp 2001b; Micklin 1988; Saiko 2001). EIA has also stimulated development of links and research between FSU and western scientists. New environmental technologies and approaches have also been introduced, and various long-term environmental needs have been identified. Each key stage in Figure 1 is afforded a chapter in its own right in ES. This includes the 'problem area' of cumulative impacts which are specifically addressed in later pipeline ES: cumulative impacts represent a neglected topic, and are often cited as one of the reasons for developing SEA (e.g. Cherp 2001b). ES include non-technical summaries, and public meetings are held in the regions and the capital, as is good practice (and also enshrined as a general principle within the Water Framework Directive).

However, the following challenges for EIA in the FSU can also be identified:

1. 'Multidisciplinary', not 'interdisciplinary', approaches used
2. Limited project scoping
3. Collapse of environmental monitoring networks
4. Data availability and quality uncertainties
5. Few published data available
6. Infrastructural, logistical and (transitional) economic problems
7. International difficulties, e.g. related to geopolitical issues with surrounding regions and republics and language or culture barriers

In particular, much environmental science is carried out in a *multidisciplinary*, rather than an *interdisciplinary*, way (O'Riordan 2000). Such compartmentalisation can occur within EIA, where logistics make it difficult for related specialists (e.g. hydrologists, hydrogeologists, freshwater ecologists and land contamination specialists) to get together to study common issues or sites. Second, a restricted scoping process can result if insufficient involvement of the environmental specialists occurs. This may exclude very important sites, impacts, hazards or interdisciplinary issues, constraining the EA. However, a key problem affecting EA worldwide is the appropriateness of the *environmental data* available for the various stages of Fig. 34.1. This issue forms the main focus of Sects. 34.6 and 34.7.

34.6 Water Resource Data Appropriateness: General Limitations for EA

The limitations of water data are numerous, and relate to the following issues:

34.6.1 Dynamic, Episodic and Seasonal Processes, and Data Representativeness

Environmental systems generally are much more dynamic than is commonly appreciated, especially by non-specialists who cannot visit sites at the *key* times of day or year. In particular, high mountain systems, including Azerbaijan and Georgia, incorporate mobile braided rivers, eroding river banks (Lawler et al. 1997a; 1997b), high sediment fluxes (Lawler et al. 1996b; 2003), mudslides, mud volcanoes and active seismicity. High temperatures and wind speeds are also common. Fig. 34.2 shows that mountain river flows are especially sea*sonal,* with classic snowmelt seasons starting around April (e.g. Lawler et al. 2003). Seasonality itself can vary over short distances and through time (e.g. Walsh and Lawler 1981). Strong seasonality can make it difficult for non-specialists to appreciate the true power of hydrological systems. For example, visits to rivers in winter when at low flow can suggest that rivers are insignificant and quiescent (see Fig. 34.2). This can lead to the dangerous and erroneous assumptions that the rivers (a) are unimportant as water resources, and (b) pose no threat to developments. Field observations at the *key periods* are essential, therefore. For fluvial systems in Azerbaijan and Georgia, therefore, as for most northern hemisphere mountain rivers, this should include the active periods in the March-July snowmelt season (e.g. Fig. 34.2).

Note how, if field observations are confined to the winter period (e.g. Fig. 34.2), the major flows and events of the year are entirely missed, leading to a substantial underestimate of discharge rates, flow energy, sediment loads and resource significance.

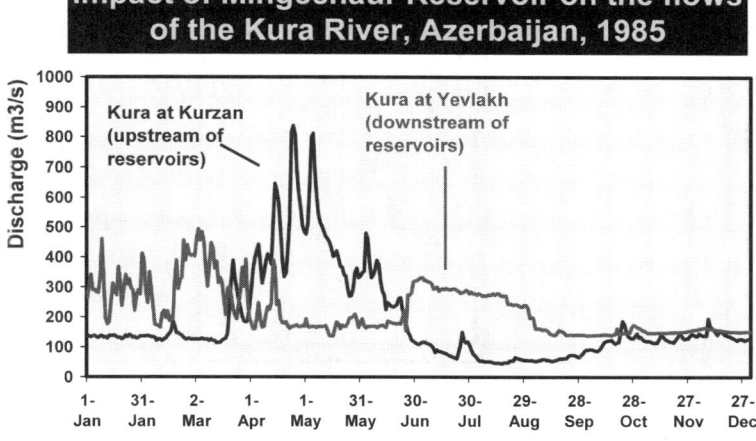

Fig. 34.2. Typical annual flow regime of the River Kura in Azerbaijan, the major transboundary river of the Caucasus region. Field visits made in November, for example, are unlikely to be representative of the extreme flows or water quality values which arise, and will miss completely the main peak in river discharge between March and June, even below the reservoirs

The seasonality problem is exacerbated by the *episodic* nature of many hydrological and geomorphological processes too (e.g. flash floods and 'first-flush' pollution events). Just as the correct *spatial* scale should be used for EIA studies (João 2002), so the correct *temporal* resolution should be adopted if at all possible. For example, very large (order of magnitude) changes in river water quality are almost universally common during storm events. A typical example, for the urban River Tame in Birmingham, UK, demonstrates this (Fig. 34.3).

The data in Fig. 34.3 are at 15-minute resolution. However, for example, note how if a hypothetical 2-week sampling methodology had been adopted instead, it would have missed completely the crucial water quality variations at diurnal and storm-event timescales, including:

1. all the high-flow events;
2. the associated key peaks in water turbidity (an order of magnitude higher than background levels)
3. the associated sudden decreases in electrical conductivity
4. the crucial ammonia spikes which are toxic to fish life
5. the diurnal variations in river temperature

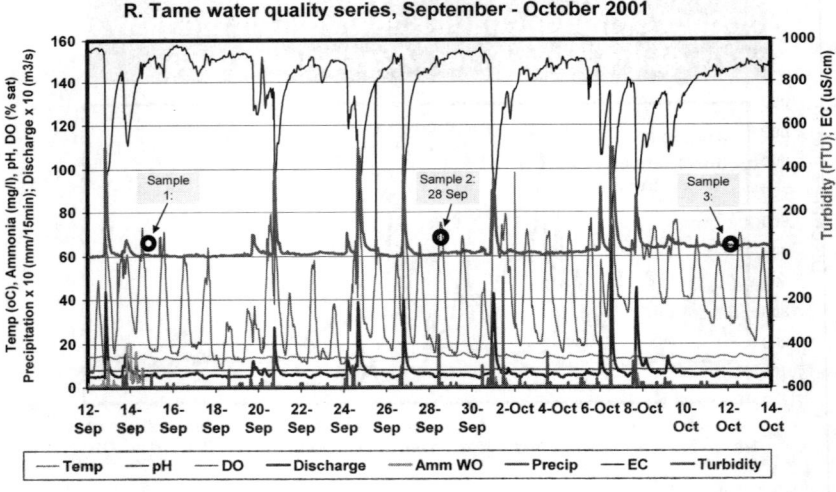

Fig. 34.3. Water quality time series for September-October 2001, recorded automatically at 15-minute resolution for the headwater River Tame monitoring station at James Bridge, West Midlands, UK. Because of the fast, transient and very substantial variations in flows and water quality responses in relation to precipitation inputs, the hypothetical 2-week sampling scheme indicated by black circles would entirely miss the key features of water pollution response here (Temp = temperature; DO = dissolved oxygen; Amm WO = Ammonia concentrations at Water Orton further downstream; Precip = Precipitation; EC = Electrical Conductivity)

6. the diurnal variations in vital dissolved oxygen (DO) which, instead, have been recorded *only* at unrepresentatively high levels by the normal manual daytime sampling: DO is temporarily enhanced here during daylight hours because of aquatic plant photosynthesis. Note how DO actually drops each night to very low levels ($< 20\%$). However, such daytime measurements would yield a dangerously and falsely optimistic impression of river health. This could result in the granting of planning permission for a project or plan on the assumption that the 'healthy' fluvial ecosystem was able to withstand a DO reduction. The same 'misdiagnosed dynamics' arguments can be advanced for air quality too. Such episodicity and transience is superimposed on 'normal' seasonal changes, and creates additional problems of data representativeness for EIA and SEA. This makes continuous monitoring essential to detect the peaks and troughs of water quality change (a) to help understand the processes, and (b) to ensure *continuous* compliance with international standards.

34.6.2 Poorly-Defined Environmental Changes

At the longer timescale of gradual environmental change, data adequacy is also questionable. EIA still tend to focus on *baseline* environments, and neglect evaluation of *recent* and *future* environmental changes (e.g. Krenke and Kravchenko 1996) predicted over the project design life. Azerbaijan and Georgia have experienced significant recent environmental changes (e.g. Hadiyev 1996), but these do not tend to be well-defined because limited research has been completed. What is uncertain also is the degree of homogeneity of hydrological time series (e.g. Craddock 1979) or the spatial coherence of environmental changes. Very recent changes are especially poorly documented (see Sect. 34.7).

34.6.3 Data Quality Issues

Environmental data must be recorded for the right variables, in the right places, at the right times, with an appropriate resolution, accuracy and precision. This problem is not restricted to FSU nations, of course, and it has influenced the Water Framework Directive thinking in the EU. However, in the FSU, it is very difficult to obtain metadata (data about data) to allow judgments to be made on dataset appropriateness. The collection techniques are often uncertain, and the nature, source and magnitude of errors is generally unknown, as are accuracy and precision and homogeneity of records. Water quality measurements are generally sparse, so that environmental baseline conditions are not defined well. Threshold and cumulative impacts are also difficult to delineate. The high-resolution data ideally needed, as in Fig. 34.3, are simply not available. On-site data collection is difficult for local and visiting scientists because of logistics, and great reliance has to be placed on manual infrequent measurements at a few sites only. For example, gauging stations are often far from pipeline crossings. Averages for most variables are uncertain but extremes are virtually undefinable for many variables relevant to EIA or SEA.

34.7 Specific Environmental Data Issues in the FSU

Four further difficulties make EIA work specifically in the FSU problematic, because they exacerbate the problems noted above, and collectively are likely to impede considerably the development of SEA approaches in these NIS:

Collapse of Environmental Monitoring Networks in 1990s: A virtual collapse of environmental monitoring networks occurred in the early 1990s, following the political and economic changes spreading through the FSU republics in the Post-Soviet period (e.g. Bradshaw 1997). Very limited *contemporary* hydrological data are available. Encouragingly, some moves have now started to re-initiate environmental monitoring. Many long datasets terminate in 1991.

Logistical problems in the FSU: Infrastructural, resource, energy and political problems in the FSU can also hinder EIA field survey work, especially travel and access to remote sites, water sample processing and data analysis. Poorly resourced laboratories can understandably limit the number and reliability of analyses carried out, and may make it difficult to demonstrate compliance with international protocols and standards.

Geopolitical tensions in region: Such tensions also hinder access to certain (border) sites, and, at longer timescales, understandably diverts energy and resources away from environmental assessment and protection. Some environmental tensions may be eased by adoption of the Espoo Convention on transboundary effects (Nazari 2003).

Data QA, storage, retrieval and access issues: There are difficulties with data access on occasions, and Quality Assurance (QA) procedures, data storage facilities, transparency, and public access to environmental datasets is not as well developed as in western nations (e.g. under the Water Framework Directive for EU countries). The signing of the Arhus Convention by Azerbaijan was potentially a significant step forward for EA in that country. However, Bektashi and Cherp (2002) identify several areas where the full provisions of the Convention are still not being met in Azerbaijan, and Cherp (2001b) comments on the relative secrecy with which data are held in the Newly Independent States generally.

34.8 A Revised Water Impact Assessment (WIA) Procedure

To meet these challenges, and given the vital importance of water resources, it is argued here that a robust *Water Impact Assessment* (WIA) procedure be formally established within EIA, and extended to SEA. WIA integrates a fresh generation of assessment protocols, many of which can be adapted for SEA purposes in a SWA (see Sects. 34.9 and 34.10), including:

1. *Integrative and holistic hydrological assessments.* EIA protocols should encourage interdisciplinary studies of common hydrological themes by specialists working in tandem to produce *holistic* water resource assessments, including interface issues, an approach encouraged by the Water Framework Directive.

2. *Investigation of recent and future hydrological changes.* Future EIA protocols should require incorporation of recent and future environmental change, including hydrological changes, at least over the development design life.

3. *A new 'pre-scoping' phase.* The introduction of 'pre-scoping' phase for WIAs, in which greater *specialist* input by hydrologists is secured and fed into the main scoping stage often carried out by non-specialists, would help to ensure that key water resource issues or sites are not overlooked in the EIA.

4. *Post-project appraisal.* EIA includes a prediction of likely environmental consequences (Fig. 34.1). EIA legislation, therefore, should incorporate the *requirement* to monitor *actual impacts* of a development in the form of 'post-project appraisal' and the results published. *Actual* impacts may then be related to those *predicted* in the EIA, as a gauge of EIA effectiveness and comprehensiveness, or the relative predictability of environmental systems. These will form a bank of reference studies for the benefit of *future* EIA, hydrological studies or water resource schemes.

5. *Peer review of ES.* New EIA legislation should insist on external peer review of Environmental Statements by specialists in the fields represented, including hydrology. Peer review is now standard for scientific work, and ES should now be embraced to promote quality assurance. Careful thought must be given to its introduction, as ES may contain commercially or politically sensitive material. Reviews could be published to inform all interested parties and help to identify strengths and gaps in *existing* ES, and suggest areas for EIA revision, before planning permission is granted.

34.9 SEA into the Future: Learning from Project EIA

Fig. 34.4 shows a view of SEA processes and methods from Finnveden et al. (2003). Clearly, the EIA-SEA experience linkages discussed above – with respect particularly to the all-pervading issue of environmental data appropriateness, will be applicable to most stages in an SEA. They are perhaps most relevant, though, to the 'environmental analysis', 'valuation' and 'quality review' steps in the Finnveden et al. (2003) process (Fig. 34.4), and to several steps in the SEA methodology of Hedo and Bina (1999), including 'defining the reference framework', 'choosing indicators and prediction techniques, 'impact prediction and evaluation of plans or programmes' and 'devising a monitoring plan'.

An important overarching theme is that any likely undesirable impacts of a development are detected and remedied *early*, *either* under an SEA procedure or a project EIA. There is a danger that some projects could slip between the two if linkages are weak (Risse et al. 2003). Making both SEA and EIA rigorous and complementary, and ensuring connectivity between the two, should reduce this risk. Above all, as Fischer (2003) argues, we should avoid the premature abandonment of EIA approaches. EIA and SEA should be mutually supportive and complementary, not alternatives, each with its specific spatial scales and methods. To this end, and developing from the recommendations in section 8, this chapter proposes in particular three key needs for the successful implementation of SEA: (1) appropriate environmental data quality and quantity, driven by a well-designed, managed and resourced environmental monitoring programme; (2) the development of a system of 'Concurrent Programme Appraisal' (CPA) – newly proposed here as the SEA equivalent to 'post-project appraisal' in EIA, to monitor the environmental effectiveness of new plans, programmes and policies; and (c) a rigorous Peer Review System for all SEA reports.

Note that proposal (2) suggests that monitoring be specifically *regular, contemporary and ongoing with the plan itself*: it is therefore contrary to the suggestion of Arts (1998, cited in Risse et al. 2003) that monitoring be delayed till after the first plan has run its full course, and simply relegated to a part of the planning cycle, rather than to any environmental cycle or response lag. The clear danger here, of course, is that environmental damage – possibly irreversible damage – is not detected till too late for remedial action. For example, if a transport plan proposes an increase in winter road de-icing dressings or a switch from salt to urea, this could increase ammonia concentrations or electrical conductivity levels in local and regional watercourses, with possibly deleterious implications for aquatic ecosystems (Lawler et al. 2004). Catching this problem early in the life of a plan, though, before populations have collapsed irrecoverably, may well allow remedial action and changes to the plan to be effected without incurring excessive financial or environmental cost. Indeed, early monitoring of plans is fully supported in Article 10 of the SEA Directive. Proposal (2) above, therefore, is likely to be both environmentally sustainable and cost-effective – an excellent combination!

This will clearly place a certain demand on resources. However, such is the overriding importance of environmental health and the threats to this from certain PPP and future environmental change, any attempts to cut back on environmental monitoring programmes should be resisted. Evidence abounds for the need for monitoring at the appropriate frequency: a selection only (e.g. Figs. 34.2 and 34.3) is presented here to demonstrate the highly variable nature of many water quality indicators which needs to be documented. After all, it is only through these efforts we can actually check that activated PPP are having negligible environmental effects. The problems faced in the Hedo and Bina (1999 p.264) study are illuminating here: in northern Spain, SEA of water resource plans (hydrological and irrigation plans) were not able to predict water quality impacts as robustly as conservation impacts, because of "a lack of baseline data (e.g. regarding groundwater pollution, which is particularly relevant to agriculture)". Simple checklist techniques had to be used for water quality instead.

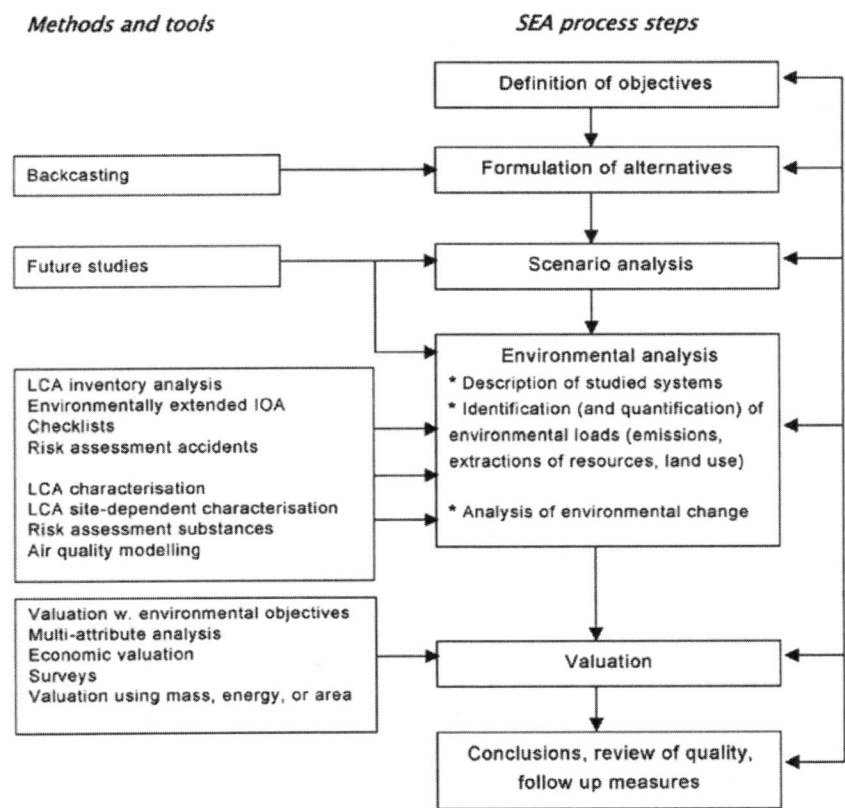

Fig. 34.4. The stages of the SEA process, and appropriate methods and tools (from Finnveden et al. 2003)

Modelling can be very useful, of course, especially for scenario-evaluation, sensitivity analysis and forecasting, but it is no substitute for accurate, precise and appropriately-resolved data on the key environmental quantities. In particular, a rigorous temporally *continuous* environmental monitoring network for water (and air) is required to detect with appropriate sensitivity and robustness the change to water quality and quantity which is fundamental to life. This would also allow EU nations to demonstrate that water quality targets have been met under the Water Framework Directive. A positive solution to this would be for the EU to earmark funds specifically for qualifying member states to develop or upgrade their environmental monitoring networks to appropriate international standards, so that all objectives of relevant Directives, including the Water Framework Directive and the SEA Directive, can be met, and shown to have been met, including environmental compliance.

Recently, from 2002, 'western' EIA in the FSU have become Environmental and *Social* Impact Assessments, and include predictions of project impacts on human health, employment and economy. Perhaps we need now to look to develop SEA in the same way, and it is proposed here that a Strategic Environmental and *Social* Assessment (SESA) – to embrace impacts on society more overtly – could provide a useful, parallel contribution. Indeed, Hedo and Bina (1999) already suggest social impacts included within SEA for plans in Castilla y Leon.

34.10 Conclusions and Recommendations

Noble (2004 p.4) argues that "SEA remain less than satisfactory and [...] the challenge is to consider how practitioners can ensure the quality of strategic decisions". In this chapter, some of the solutions advanced from parallel environmental and EA studies elsewhere, aim to contribute to more robust SEA protocols for the future. These should be applicable to a wider range of environmental components than water alone, and beyond the FSU nations used as examples here. Resources are needed, however. Of the improvements advanced, this chapter argues that three key developments are especially important: (a) a full re-establishment of environmental monitoring networks to generate the appropriate data for EA, including some continuous data at key sites for key variables; (b) a Concurrent Programme Appraisal (CPA) procedure to drive ongoing monitoring of the effectiveness of new plans, programmes and policies in SEA; and (c) a rigorous, external, assessment panel system to review the major SEA reports in relation to internationally-established performance criteria, and recognized societal constraints in less economically-developed nations. This proposal parallels calls elsewhere (e.g. Lawler 2002) for a compulsory Peer Review system of project-based EIA procedures and ES.

Bektashi and Cherp (2002) have argued that the success of Environmental Assessment processes in Azerbaijan depends on whether the present systems can be improved. Western-style EA is in its infancy in the FSU, but has gained in popularity since 1996, and rapid improvements are being made. It has been the main aim of this chapter to demonstrate some of the limitations of the present EA systems, and to present ways of addressing these challenges in SEA processes.

Current EIA protocols have strengths and limitations, which are compounded in the FSU. These relate to: highly dynamic and variable processes; paucity of data; restrictive scoping; limited interdisciplinary approaches; and logistical and economic constraints. To improve the water resource components of future EIA, it is argued that a new, formal *Water Impact Assessment* (WIA) module be developed to include all relevant water bodies. WIA specifically includes: holistic water resource assessments; evaluation of recent and future environmental changes; a specialist pre-scoping stage; post-project appraisal with key monitoring; and external peer review of Environmental Statements. The adoption of these should significantly improve future EIA worldwide, and especially in the FSU, although "effective EA reform should be sensitive to specific needs of the transitional societies and sensitive to their unique societal context" (Cherp 2001b p.335).

Furthermore, these proposed EIA improvements could be developed to drive refined SEA protocols, collectively termed here 'Strategic Water Assessment' (SWA), to deliver long-term strategic planning of water resources within a strictly defined framework. Both would include the translatable components from WIA described above. The resultant broadening of scope and sharpening of focus should encourage a strengthening of methodologies over time, to enhance the water assessment, and other, components of EIA and SEA. This author is in agreement with Fischer (2003) that an SEA procedure lacking a strongly directed, well-defined structured framework, could well be counter-productive. A well-designed structure, with strongly defined rather than arbitrary components, should simplify budgeting. The key advantage, though, is that it will help to ensure that the SEA does not overlook key stages, issues, type-sites, processes, time-lags, interrelationships and cognate disciplines, the important Peer Review process and con*current*, not retrospective, monitoring of the environmental effects of plans and programmes.

Acknowledgements

I am very grateful to the UK Environment Agency for the data of Fig. 34.3.

References[1]

Bektashi B, Cherp A (2002) Evolution and current state of environmental assessment in Azerbaijan. Impact Assessment and Project Appraisal 20(4):255-263
Bradshaw MJ (ed) (1997) Geography and Transition in the Post-Soviet Republics. Wiley, Chichester
Cherp A (2001a) SEA in the Newly Independent States of the former Soviet Union. Szentendre, Budapest
Cherp A (2001b) EA legislation and practice in Central and Eastern Europe and the former USSR: a comparative analysis. Environmental Impact Assessment Review 21:335-361
Cherp O, Lee N (1997) Evolution of SER and Ovos in the Soviet Union and Russia (1985-1996). Environmental Impact Assessment Review 17:177-204
Craddock JM (1979) Methods of comparing annual rainfall records for climatic purposes. Weather 34:332-346
Efendiyeva IM (2000) Ecological problems of oil exploitation in the Caspian SEA area, Journal of Petroleum Science and Engineering 28 (4):227-231

---

[1] Note that legislation mentioned in the chapter is listed at the end of the handbook in the consolidated list of legislation (Appendix 2).

Finnveden G, Nilsson M, Johansson J, Persson A, Morberg A, Carlsson T (2003) Strategic environmental assessment methodologies - applications within the energy sector. Environmental Impact Assessment Review 23:91-123

Fischer TB (2003) Strategic environmental assessment in post-modern times. Environment Impact Assessment Review 23:155-170

Glasson J, Therivel R, Chadwick A (1999) Introduction to Environmental Impact Assessment. UCL Press (2nd edition)

Hadiyev YJ (1996) Calculation of fluctuations of the atmospheric circulation in estimation of future changes of climate in the Trans-Caucasus (in Russian with English and Azeri summary), Azerbaijan Academy of Sciences, Baku

Hedo D, Bina O (1999) Strategic environmental assessment of hydrological and irrigation plans in Castilla Y Leon, Spain. Environment Impact Assessment Review 19:259-273

João E (2002) How scale affects environmental impact assessment. Environmental Impact Assessment Review 22:289-310

Johnson KD, Martin CD, Davis TG (1999) Treatment of wastewater effluent from a natural gas compressor station, Water Science and Technology 40(3):51-56

Krenke AN, Kravchenko GN (1996) Impact of future climate change on glacier runoff and the possibilities for artificially increasing melt water runoff in the Aral SEA basin. In: Jones JA, Changmin L, Ming-Ko W, Hsiang-Te K (eds) Regional Hydrological Response to Climate Change, Kluwer, Dordrecht, pp 259-267

Lawler DM (1995) Turbidimetry and nephelometry. In: Townshend A (ed) Encyclopedia of Analytical Science. Academic Press vol. 9, pp 5289-5297

Lawler DM (2002) Safeguarding water resources in the Former Soviet Union in an era of transition: establishing Water Impact Assessment with the EIA process. In: Proceedings of the 5th International Conference on Water Resources Management in the Era of Transition, 4-8 September 2002, Athens, European Water Resources Association Publication, p 10345

Lawler DM (2004) Turbidimetry. In: Townshend A (ed) Encyclopedia of Analytical Science, 2nd edition. Academic Press (in press)

Lawler DM, Sljivic S, Caplat M (1996a) Assessing the environmental impact of the Birmingham Airport Link pipeline. In: Gerrard AJ, Slater TR (eds) Managing a Conurbation: Birmingham and its Region, British Association for the Advancement of Science Meeting. Brewin Books, Studley Warwickshire, 75-89

Lawler DM, Björnsson H, Dolan M (1996b) Impact of subglacial geothermal activity on meltwater quality in the Jökulsá á Sólheimasandi system, southern Iceland. Hydrological Processes 10:557-578

Lawler DM, CouperthwaitevJ, Bull LJ, Harris NM (1997a) Bank erosion events and processes in the Upper Severn basin, Hydrology and Earth System Sciences 1:523-534

Lawler DM, Thorne CR, Hooke JM (1997b) Bank erosion and instability. In: Thorne CR, Hey RD, Newson MD (eds) Applied Fluvial Geomorphology for River Engineering and Management. John Wiley, pp 137-172

Lawler DM, McGregor GR, Phillips ID (2003) Influence of atmospheric circulation changes and regional climate variability on river flow and suspended sediment fluxes in southern Iceland. Hydrological Processes 17:3195 – 3223

Lawler DM, Petts GE, Foster ID, Harper S (2004) Turbidity dynamics and hysteresis patterns during spring storm events in an urban headwater system: the Upper Tame, West Midlands, UK. Science of the Total Environment, in review

Micklin PP (1988) Desiccation of the Aral Sea: a water management disaster in the Soviet Union. Science 241:1170-1175

Nazari MM (2003) The transboundary EIA convention in the context of private sector operations co-financed by the International Financial Institution: two case studies from Azerbaijan and Turkmenistan. Environmental Impact Assessment Review 23:441-451

Noble BF (2003) Strategic environmental assessment quality assurance: evaluating and improving the consistency of judgements in assessment panels. Environmental Impact Assessment Review 24:3-25

O'Riordan T (2000) Environmental Science for Environmental Management, Prentice Hall

Risse N, Crowley M, Vincke P, Waaube J-P (2003) Implementing the European SEA Directive: the Member States' margin of discretion. Environmental Impact Assessment Review 23:453-470

Saiko T (2001) Environmental Crises. Pearson Education, Harlow

Tsakiris G (2002) From single purpose planning to sustainable watershed management, European Water Resources Association Bulletin 1:1-2

Wolfson Z, Daniell Z (1995) Azerbaijan. In: Pryde PR (ed.) Environmental Resources and Constraints in the Former Soviet Republics. Westview Press, Colorado, 235-250Press, Colorado, pp 235-250

35 Links between the Water Framework Directive and SEA

Natalia Gullón

Ministry of Environment, Spain

35.1 Introduction

We are facing a key moment, in which the efforts to implement both the SEA Directive and the Water Framework Directive coincide, together with a special consciousness on hydraulic resources, after the International Year of Freshwater 2003, and in a world context of water crisis. The purpose of the Water Framework Directive is "to establish a framework for the protection of inland surface waters, transitional waters, coastal waters and groundwater" (Art. 1 of the Water Framework Directive). The Water Framework Directive requires, among others, the preparation of river basin management plans and to develop and implement programmes of measures that will address significant pressures and impacts. What are the links between the Water Framework Directive and the SEA Directive? Do they overlap? Do river basin management plans require an SEA? This chapter explores the contribution that SEA could make towards a sustainable planning and management of water resources, with particular reference to the Water Framework Directive. The chapter starts by explaining how sustainability is nowadays a key concept in water policy, and later explains the strengths and weaknesses of carrying out SEA in the water sector. The central part of the chapter describes the links between the Water Framework Directive and the SEA Directive, finalizing with some conclusions and recommendations.

35.2 From a "Hydrological Policy" to a Sustainable "Water Policy"

Over the last decades, the priorities of our society have been changing and environmental consciousness has gained importance.

Implementing Strategic Environmental Assessment. Edited by Michael Schmidt, Elsa João and Eike Albrecht. © 2005 Springer-Verlag

Similarly, this is also a fact in the water sector: A growing importance is given to the protection of the natural resources. But nowadays, we are going a little further, from "environmental policies" to "sustainability policies". The Amsterdam Treaty incorporates the statement that "environmental protection requirements must be integrated into the definition and implementation of Community policies and activities, in particular with a view to promoting sustainable development" (Art. 6 EC Treaty 1997). There is clearly a need to integrate the three dimensions of sustainability into the process of decision making: environment, social equity and economic growth. In the same way, the traditional *hydraulic policy* is extending towards a more global concept of *water policy*, in which the idea of sustainable development has to be incorporated. The Spanish Department of Environment (2000) underlines that "It is a question of searching a harmonization and complementarity of interests, maintaining the balance between economic growth and the limits and needs of the environment, in order to achieve not only the higher outputs, but the well-being of citizens in the medium and long term".

The focus is not only nature, but environment as a broad concept, including quality of life. In this context, it is clear that SEA can contribute to an integrated assessment of sustainability of decisions. However, as Thérivel and Partidário (1996) state, it is necessary to move SEA practice from a purely rationalist and technocratic decision-making approach and limited opportunities for public-participation, to one which reflects "collaborative consensus-making" approaches" to planning.

SEA should be not only an ex-post instrument evaluating an already approved policy, plan or program and proposing retrospective actions, but should be a procedure to apply during the planning process, strengthening the positive and minimizing the negative environmental effects of the strategic action. In short, SEA should optimise environmentally the policy, plan or programme.

35.3 Strengths of SEA in the Water Sector

Within the water sector, SEA of decision-making is crucial, not only due to the own nature of the resource, but also because of the peculiar characteristics of hydraulic projects. Water is basic for life. It is present in every ecosystem, and water quality is a reflection of the quality of the entire natural environment. According to the Water Framework Directive, "Water is not a commercial product like any other but, rather, a heritage which must be protected, defended and treated as such".

Water resource is finite and vulnerable, and this is the reason why the purpose of the Water Framework Directive is "to establish a framework for the protection of inland surface waters, transitional waters, coastal waters and groundwater". Moreover, water has a clear influence on all other sectors, as it constitutes the key for development, in particular to generate and maintain economic resources through agriculture, fishery, industry, transport, tourism and generation of energy. As a consequence, water availability is a key factor shaping the development of areas. Water resource issues are a high-priority concern in relation to a state's economy and land use planning.

In addition, over the last decade there has been a considerable advance in the understanding of water, not only as an economic value, but also in its social, environmental, cultural and religious dimension (UN 2003). The book *"Water in Spain"* expresses that "The water resource is not only a natural resource and an economic good, but it plays an environmental role: inputs to the economy, waste receptor, aesthetic enjoyment or spiritual satisfaction of people" (Spanish Department of Environment 2000).

We all make use of water, and this gives special importance to public participation, as it was expressed in one of the basic principles of the International Conference on Water and Environment, held in Dublin in 1992: "Water development and management should be based on a participatory approach, involving users, planners, and policymakers at all levels". The other three *Dublin principles* are: *Freshwater* is a finite and vulnerable resource, essential to sustain life, development, and the environment; *Women* play a central part in providing, managing, and safeguarding water; *Water* has an economic value in all its competing uses and should be recognized as an economic good.

Water provision is always a controversial, strategic and complex problem. Water is a scarce and strategic resource, subject to a constant demand, and therefore a sustainable management, preservation and assignation of the resource is essential for life, reconciling the demands for uses of different nature and social function. At the same time, this is not an easy task, because traditionally all issues related to water have always had an institutional character, and water has been an element subjugated to a strong regulation (Spanish Department of Environment, 2000). The governance of the water sector faces uncertainties, changing situations and conflictive requirements, resulting from the multiple interests linked to water. "Water should be regarded as a finite resource having an economic value with significant social and economic implications regarding the importance of meeting basic needs." (Agenda 21, Chapter 18 UNCED 1992*)*.

On the other hand, due to the distinctive features of projects related to the hydraulic resource, Environmental Impact Assessments (EIA) are normally limited to a tool for the correction of impacts. Project level EIA have proved to be not enough, as they are conducted at a later stage of the decision-making process and usually arrive too late, after the great objectives have been set, the major alternatives have already been selected and the main strategic decisions have been taken in previous stages, during the preparation of policies, plans and programmes (for instance, the decision to build or not a dam, and its location).

Given that environmental assessments are preventive and proactive instruments, its efficiency is higher when applied at early stages of the decision making chain, and when used with more global approach, when they are integrated in the planning processes previous to the projects. This is especially remarkable in the water sector. In addition, water related projects have important indirect effects, as for example the development of agriculture or urban use. All those effects have to be assessed, and SEA could help to do it.

As well, SEA can detect cumulative impacts of different projects: the economic, environmental, social and political impacts can be acceptable when considering an isolated project, but unacceptable when taking into account both the direct and indirect effects of projects, programs, plans and policies acting in synergy. SEA provides the framework for the articulation of individual projects into complementary design, implementation and management, in a way that is coherent and respectful with the ecological, social, political and economic conditions (Arce and Gullón 2000).

SEA increase the possibility of analysing and proposing alternative solutions and incorporating sustainability criteria throughout the planning process, as they carry the principles of *sustainability* down from policies to individual projects (Arce and Gullón 2000).

A crucial SEA alternative to be taken into account is "obviating development", which in the water sector could be materialized by considering alternatives based on reducing the need for water-related developments. This means introducing planning objectives of demand reduction, and not only increasing the offer. For example, if people save water then not so many dams will need to be built.

35.4 Weaknesses of SEA in the Water Sector

The main obstacle for SEA is the complexity of the planning process, especially in hydrological planning. Decisions are interdependent and they can be subject to later modifications. Decisions taken are of different types and present different degrees of detail. They may concern from general objectives of water policy to specific options to meet the demand. It is very difficult to identify the precise moment in which decisions are taken. It is therefore necessary to systematize the process of decision making and to identify the significant moments.

Moreover, it is also complex to assess environmental effects at a strategic level, as there are different scales and different levels of detail (João 2002; Osterkamp 1995; Sposito 1998), there is a lack of uniformity of the data which makes it difficult to know what to assess and how, as well as which indicators to use. The Guidance Document No 8 for the implementation of the Water Framework Directive expresses that "Not only the area where the activities will be implemented should be considered, but the whole area where their impact may be felt" (EC 2002 pp.2.7).

Water resource is usually subject to conflicting demands and incompatible interests, which make it difficult to fulfill in a satisfactory way all the different needs. The participation of all the agents involved is crucial, in order to reach the most suitable consensus; but at the same time this is not easy at all.

The level of uncertainties is very high: hydrological planning is setting objectives and strategies that will be detailed later on in projects and actions. How to assess something that is continuously changing, and that has a different degree of definition along time? Hydrological planning has environmental effects that are simpler to predict: for example, when programming dams, canals or water transfers. But when setting priorities for water use, for example, the effects are much less precise and hardly predictable.

Hydrological planning is normally flexible and continuous, and it develops every so often, depending of the needs or problems arising. It is not closed, as the selection of one alterative may in turn have different possible forms of materializing. Assessing the effects of PP means assessing the projects which are brought to fruition, as the plan is being carried out. Therefore, the effects of a PP depend on the type of subsequent proceedings. The great sum of indirect effects makes it even more difficult to carry an SEA of hydrological planning. For example, water availability may determine the development of agriculture.

35.5 SEA Directive and the Water Framework Directive

We are facing a key moment, in which both the SEA Directive and the Water Framework Directive are being implemented. Both directives are complementary and, in a way, synergistic. The main objective of both Directives (Art 1 of the Water Framework Directive and Art 1 of the SEA Directive) is the protection of the environment and natural resources (see Table 35.1). Although according to the European Commission Guidance for the implementation of the SEA Directive "it is not possible to state categorically whether or not the River Basin Management Plan and the Program of Measures are within the scope of the SEA Directive" (EC 2003 p.48).

Table 35.1. Comparison between the objectives of the Water Framework Directive and the SEA Directive

SEA DIRECTIVE (Art. 1)	Water Framework Directive (Art. 1)
The objective of this Directive is to provide for a high level of protection of the environment and to contribute to the integration of environmental considerations into the preparation and adoption of plans and programmes with a view to promoting sustainable development, by ensuring that, in accordance with this Directive, an environmental assessment is carried out of certain plans and programmes which are likely to have significant effects on the environment.	The purpose of this Directive is to establish a framework for the protection of inland surface waters, transitional waters, coastal waters and groundwater which: (a) prevents further deterioration (b) promotes sustainable water use (c) aims at enhanced protection and improvement of the aquatic environment (d) ensures the progressive reduction of pollution groundwater and prevents its further pollution, and (e) contributes to mitigating the effects of floods and droughts.

Overall, the Water Framework Directive aims at achieving good water status for all waters by 2015, and it is centred in the concept of integration as key to the management of water protection within the river basin district: Integration of environmental objectives; Integration of all water resources; Integration of all water uses, functions and values into a common policy; Integration of disciplines, analyses and expertise; Integration of all significant management and ecological aspects relevant to sustainable river basin planning including those which are beyond the scope of the Water Framework Directive such as flood protection and prevention; Integration of a wide range of measures, including pricing and economic and financial instruments, in a common management approach for achieving the environmental objectives of the Directive. Integration of stakeholders and the civil society in decision making, Integration of different decision-making levels that influence water resources and water status (local, regional or national).

Both Directives require similar environmental assessments, as can be seen in Table 35.2. In this table, both the exact contents of the Water Framework Directive and the SEA Directive are compared, as well and the specific requirements for the River Basin Management Plans, always from the point of view of environmental assessment. The Water Framework Directive gives more importance to biophysical and chemical issues, and also asks for an economic analysis, whereas SEA Directive requires the assessment of impacts concerning not only of biodiversity, fauna, flora, soil, water, air, climatic factors and material assets, but also population, human health, landscape and cultural heritage, including architectural and archaeological heritage.

Table 35.2. Comparison between the Water Framework Directive and the SEA Directive

	SEA Directive	Water Framework Directive	River Basin Management Plans
	Art. 5 and Annex 1: *Preparation of an environmental report*	Art. 5 and Annexes II and III	Annex VII: River Basin Management Plans
Assessment	(a) contents, objectives and relationship with other relevant pp;		
	(b) state of the environment;		
	(c) environmental characteristics of areas likely to be affected;	Art. 5.1: characteristics of river basin district.	1. Characteristics of the river basin district (RBD).
	(d) environmental problems;		3. Identification and mapping of protected areas.
	(e) environmental protection objectives;		5. Environmental objectives for surface waters, groundwaters and protected areas.

Table 35.2. (cont.)

Assessment	(f) likely significant effects on environment: biodiversity, population, human health, fauna, flora, soil, water, air, climatic factors, material assets, cultural heritage, landscape;	Art. 5: Review of the impact of human activity on the status of surface waters and on groundwater. Economic analysis of water use.	2. Significant pressures and impact of human activity on water. 6. Summary of the economic analysis of water use.
	(g) measures to prevent, reduce and as offset these effects;	Art. 11: Programme of measures	7. Summary of programmes of measures.
	(h) reasons for selecting the alternatives, description of the assessment;		
	(i) measures concerning monitoring;	Art. 8: Monitoring of surface and groundwater status and protected areas.	4. Map of monitoring networks, and of the results of the monitoring programmes.
	(j) non-technical summary.		8. Register and summary of other detailed programmes and management plans for RBD sub-basins, sectors, issues or water types;
Public participation	Art. 5 and 6: public participation. Art. 7: Transboundary consultation.	Art. 14: Involvement of all interested parties in the implementation of this Directive, in particular in the production, review and updating of the RBMPs.	9, 10 and 11: Information and consultation measures, results and consequent changes to the plan. List of competent authorities. Contact points and procedures for documentation and information.
Decision-making	Art. 8. Art. 9: information on the decision.	Art. 14. Art. 15: Reporting.	9. Summary of the public information and consultation measures taken, their results and the changes to the plan made as a consequence.
Monitoring	Art. 10	Art. 11	

In relation to public participation, the requirements of Water Framework Directive for public involvement go beyond those of the SEA Directive (Table 35.3), establishing specific timescales. The Water Framework Directive gives great importance to public participation. Recital 14 stated that "The success of this Directive relies on close cooperation and coherent action at Community, Member State and local level as well as on information, consultation and involvement of the public, including users." There is a specific Guidance Document (No. 8) on public participation, active involvement, consultation and access to information and background documents, explaining extensively who should be involved, how and when, the scope, scale and timing of public participation as well as useful techniques and examples. This document expresses that "It follows from Art. 14 of the Directive that active involvement should be encouraged at all scales where activities take place to implement the Directive" (EC 2002 pp.2.7).

Table 35.3. Public participation in the Water Framework Directive and the SEA Directive

SEA Directive	Water Framework Directive
The *authorities* shall be consulted when deciding on the *scope and level of detail* of the information which must be included in the environmental report. (Art. 5 para 4) The *authorities and the public* shall be given an early and effective opportunity within appropriate time frames to express their opinion on the *draft plan or programme and the accompanying environmental report* before the adoption of the plan or programme or its submission to the legislative procedure. (Art. 6) Other Member State, when it considers that the implementation of the plan or programme being prepared is likely to have significant effects on its environment, or where a Member State likely to be significantly affected so requests. (Art. 7: *Transboundary consultations*)	Recitals 14 and 46 Art. 14: *Public information and consultation* 1. Member States shall encourage the active involvement of all interested parties in the implementation of this Directive, in particular in the *production, review and updating of the River Basin Management Plans*. Member States shall ensure that, for each River Basin District (RBD), they publish and make available for comments to the *public*, including *users*: (a) a timetable and work programme for the production of the plan, including a statement of the consultation measures to be taken, at least three years before the beginning of the period to which the plan refers; (b) an interim overview of the significant water management issues identified in the river basin, at least two years before the beginning of the period to which the plan refers; (c) draft copies of RBMP, at least one year before the beginning of the period to which the plan refers. On request, access shall be given to background documents and information used for the development of the draft RBMP. 2. Member States shall allow at least *six months* to comment in writing on those documents in order to allow active involvement and consultation. 3. Paragraphs 1 and 2 shall apply equally to updated RBMPs.

Programmes of measures are defined in River Basin Management Plans developed for each river basin district (RBD). Both the River Basin Management Plan and the Programme of Measures are prepared and adopted by an authority (Art. 2 a) para 1 of the SEA Directive), they are required by the respective Member States' provisions to transpose the Water Framework Directive (Art. 2 a) para 2 of the SEA Directive), they fall within the scope of the SEA Directive (water management, Art. 3 para 2 a) of the SEA Directive) and are likely to have significant environmental effects (Art. 3 para 1 of the SEA Directive) but it should be studied, on a case by case basis, whether they set the Framework for future development consent of projects listed in Annexes I and II of the EIA Directive (Art. 3 para 2 a) of the SEA Directive), or (in view of the likely effect on sites, Art. 3 para 2 b) of the SEA Directive, have been determined to require an assessment pursuant to Art. 6 or 7 of the Habitats Directive and the respective national provisions.

As it has been demonstrated, both Directives are complementary and synergistic. Therefore, it would not be very difficult or resource-consuming to carry out the SEA, as a separate but integrated procedure in relation to the Water Framework Directive. This environmental assessment would help to make better River Basin Management Plans.

35.6 Conclusions and Recommendations

SEA may constitute a powerful tool to help in decision making, in accordance with the modern society demands, as it includes sustainability criteria in the processes of planning. A change in mentality and values is needed, in order to take into consideration environmental objectives from the beginning of the hydrological planning process, and this will contribute to better planning and decision making processes.

The SEA Directive will enforce that change, extending and anticipating environmental assessment to more general stages, previous to the elaboration of projects, and establishing an administrative procedure for their evaluation. It is possible to integrate hydrological planning and SEA, and both the Directive on SEA and the Water Framework Directive are complementary.

The process of decision making in the water sector should be systematized as far as possible, identifying the significant moments in which decisions are taken, in order to ensure that environmental concerns are properly taken into consideration, together with social, technical and economic factors.

The objective is that the planning body incorporates environmental criteria in planning and selects the best solution, so that the assessing body only verifies that the procedure has been followed properly, the results of the decisions are not evaluated. All the Member States are now facing the challenge to incorporate SEA Directive into their national legislations, and this in principle may imply a substantial change in the process of design and approval of plans and programmes, and at the same time an opportunity to review environmental assessment and the role that each administration plays in it. All the EU countries are therefore facing a especially interesting moment, in which coincide the efforts to apply the Water Framework Directive and the SEA Directive, together with a special consciousness on all subjects related to the water resource. This stage presents an opportunity for reflection, action and awareness of a sustainable planning, management and protection of hydraulic resources. In this context, River Basin Authorities may play a key role, showing their willingness to deal with the exigencies and necessities of modern society.

References[1]

Arce R, Gullón N (2000) The Application of Strategic Environmental Assessment to Sustainability Assessment of Infrastructure Development. Environmental Impact Assessment Review 20(3):393-402

EC – European Commission (2002) Guidance on public participation in relation to the Water Framework Directive. WFD CIRCA

EC – European Commission (2003) Guidance for the implementation of Directive 2001/42/EC on the assessment of the effects of certain plans and programmes on the environment, Brussels

João E (2002) How scale affects environmental impact assessment. Environmental Impact Assessment Review 22(4):287-306

ODPM (2003) The SEA Directive: Guidance for Planning Authorities. Practical guidance on applying European Directive 001/42/EC 'on the assessment of the effects of certain plans and programmes on the environment' to land use and spatial plans in England

Osterkamp W (ed) (1995) Effects of scale on interpretation and management of sediment and water quality. IAHS Publication No. 226 International Association of Hydrological Sciences

Partidário MR (1999) Strategic Environmental Assessment – Principles and potential. In: Petts J (ed) Handbook on Environmental Impact Assessment, Blackwell, London

Spanish Department of Environment (Ministerio de Medio Ambiente) (2000), Libro Blanco del Agua en España (Spanish White Paper on Water). Ministerio de Medio Ambiente de España, Madrid

Sposito G (ed) (1998) Scale dependence and scale invariance in hydrology. Cambridge University Press, Cambridge

Thérivel R, Partidário MR (eds) (1996) The practice of Strategic Environmental Assessment. Earthscan Publications Ltd, London

UNCED (1992) Agenda 21 – Rio Declarration on Environment and Development

World Commission on Environment and Development (1987) Report "Our Common Future", Oxford University Press

---

[1] Note that legislation mentioned in the chapter is listed at the end of the handbook in the consolidated list of legislation (Appendix 2).

36 Assessing Biodiversity in SEA

Udo Bröring and Gerhard Wiegleb

Department of General Ecology, Brandenburg University of Technology (BTU), Cottbus, Germany

36.1 Introduction

Biodiversity refers to number and difference of living objects over a wide range of observational levels, in particular genes, species and ecosystems (Art. 2 of the Convention on Biodiversity). Biodiversity management is regarded as a part of an integrated sustainability approach. The science of biodiversity is facing six major difficulties: defining biodiversity, measuring biodiversity, giving a causal explanation of biodiversity, assessing the function of biodiversity, evaluating biodiversity, and making decisions for biodiversity (Wiegleb 2003; 2004d). Definition and measurement of biodiversity are basic ecological tasks, mostly in the context of explanation and prediction of biodiversity patterns and dynamics. Evaluation and decision-making are characteristic for environmental planning, management and policy (Jessel 1996). Common to both aspects is the necessity of understanding the causal and functional relationships of biodiversity.

Unfortunately the connection between the domains of ecology and planning is often negatively affected by different use of terminology and lacking distinction between factual and evaluative statement on both sides (Peters 1997). In ecology measurement of biodiversity is based on counting species richness or calculating a diversity index. Measurement of biodiversity in planning and decision-making has to take into account the language in which planning and management objectives are formulated. Both species and habitat oriented parameters are used for the definition of planning and management objectives and subsequent distance-to-target measurement. So far no generally accepted framework for biodiversity evaluation has been developed. A pluralistic approach based on a discursive development of guiding principles for regional and global biodiversity protection seems to be adequate.

SEA can be taken as an extension of EIA beyond single impacts and projects (Richtlinie 2001; Secretariat of CBD 2002). The ultimate aim of SEA is to anticipate all possible environmental impacts of comprehensive strategic actions long before single projects are implemented or single measures are applied.

Implementing Strategic Environmental Assessment. Edited by Michael Schmidt, Elsa João and Eike Albrecht. © 2005 Springer-Verlag

Biodiversity is affected by many strategic actions on various political levels (worldwide, Europe, EU, country, federal state, municipal; see Fürst and Scholles 1998; Byron 2000; Byron et al. 2000; Scholles et al. 2003). In this chapter different approaches to the definition and measurement of biodiversity in an SEA context are discussed. Additionally, the major approaches to the evaluation of biodiversity are checked for their relevance for deriving assessment criteria in SEA.

36.2 Definition of Biodiversity

Biodiversity is often equated with "number of species" or "species richness". However, biodiversity is more than that. Biodiversity can be observed at the genetic level, the level of the species, and the level of the habitats and ecosystems (Convention on Biodiversity). There is an intricate hierarchy of biological entities, on which diversity can be observed. Hierarchy on one level, for example the population genetic level, can be the prerequisite for hierarchy on another level such as the worldwide multitude of species. From a biological viewpoint, "biodiversity" refers to differences within and between biological objects on all biological levels of observation (Wood 1997; Wiegleb 2003, 2004a). Biodiversity science thus deals with number (α-diversity) and difference (β-diversity) of biological objects (Gaston 1996). Therefore, biodiversity science is more than systematics (Groombridge and Jenkins 2002). Natural history of biodiversity provides us with the description and "mapping" of genes, species, and ecosystems, while ecology of biodiversity provides us with the explanation and prediction of biodiversity and its functions in ecological systems on various scales (Begon et al. 1996).

Additionally, biodiversity has an economic, a social and an ecological dimension (Preamble of the Convention on Biodiversity). Fig. 36.1 shows the three columns of sustainability and their interfaces. Any assessment is not carried out by scientists. Scientists collect and systematise preferences held by individuals (WBGU 1999), the final decision is one by decision makers:

- As "Homo oeconomicus", judging according to money or other utilities.
- As "Homo sociologicus", judging according to social values such as acceptance, justice, equity etc.
- As "Homo oecologicus", judging according to ecologically oriented currencies such as genes, species, habitats etc.

Thus, the social dimension of biodiversity goes beyond nature conservation. Conservation of biodiversity is mostly dealing with the maintenance, sustainable use and restoration of a desired degree of biodiversity in ecological systems. Overall socioeconomics of biodiversity is additionally dealing with all legal, ethical, social and economic aspects of biodiversity management (Costanza and Folke 1997, Wiegleb 2004b, 2004d). The most important interface between the columns of sustainability are the references states. Reference states denote states of the environment which are ecologically possible, socially accepted and economically feasible (O'Riordan and Klee-Stollmann 2003).

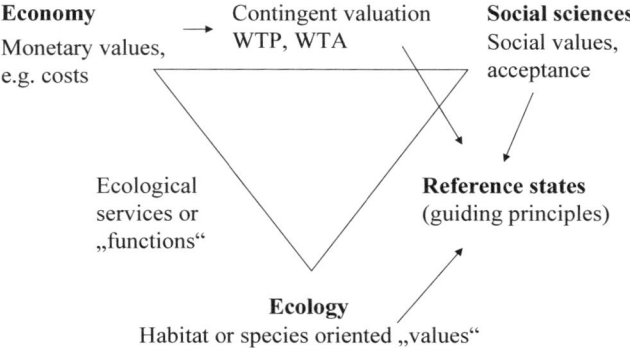

Fig. 36.1. The sustainability triangle (WTP = willingness to pay, WTA = Willingness to accept)

SEA can be regarded a promising instrument for setting reference states ("environmental objectives", Sommer 2003) for a future sustainable development of society, economy, and landscape.

36.3 Measurability of Biodiversity in Theory and Practice

Biodiversity is used in ecology both as "dependent" and "independent" variable (Brown et al. 2001). Both the study of the "determinants of diversity" (e.g. time, area, disturbance, productivity, spatial heterogeneity, competition etc.), and the study of the "functional importance of diversity" (for ecosystem stability, production, material cycling, evolutionary progress etc.) require quantification of diversity. Additionally, quantification is a prerequisite for evaluation in any planning context, including SEA.

36.3.1 Measurement in an Ecological Research Context

Quantification of biodiversity is possible on different hierarchical observational levels. On each level typification eventually leads to distinguish different biotic entities, such as landscape (landscape types), ecosystem (ecosystem types), community (community or habitat types), population (species, functional group, guild, trophic level), individual (sex, age class) and gene (phenotype, genotype). Fig. 36.2 shows that there are four different approaches to quantification of biodiversity (Wiegleb 2003).

- In case of α- or within-diversity single entities are counted (e.g. species or interactions) and summed up to additive indices. Alternatively, at least two entities (e.g. species and individuals) are combined to a relational index.

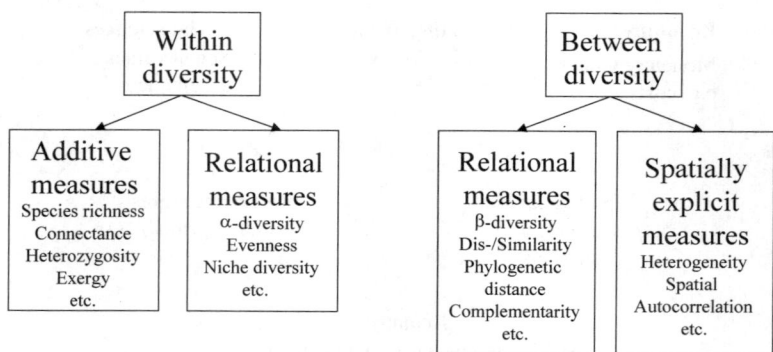

Fig. 36.2. Theoretical measures of biodiversity (adapted from Wiegleb 2003)

- In case of β- or between diversity relational indices are also widespread. Most of them are not spatially explicit and refer to similarity between ecosystems or dissimilarity between species (Gaston 1996). Additionally, spatial explicit indices can be used in particular in the study of landscape heterogeneity.

36.3.2 Measurement in a Planning Context

For hypothesis testing in ecology, biodiversity is based on one of the above indices, such as the Shannon-Wiener index. For practical reasons biodiversity conservation or management goals are not formulated in terms of such indices. Ecological language can be translated into planning language by using indicators (e.g. correlates or surrogates, Duelli and Obrist 1998; Simberloff 1999). Correlates are numerical variables with an empirically proven correlations to a diversity variable of interest (e.g. plant diversity indicating total biodiversity). Surrogates are analogues of which a relation to biodiversity can be assumed based on experience knowledge (e.g. the occurrence of a flagship species indicating ecosystem health).

Direct measurement approaches to biodiversity are summarised in Table 36.1, focussing on measurement of species diversity. Depending on the kind of information available species presence, population size of selected species, total species number or species number of selected groups, the whole species spectrum, or the number of rare and threatened species can be considered. On a larger spatial scale also the contribution of a species to large scale diversity counts. All parameters can be used to define a goal of biodiversity protection, management or restoration. Afterwards, the distance-to-target measurement can be carried out.

Indirect measurement approaches to biodiversity are listed in Table 36.2. Number and size of protected areas, presence of desired ecological processes, degree of naturalness, ecotope structure, habitat structure or state, rarity of habitats, and overall habitat spectrum can be considered on the local level. Again, the contribution of a habitat to large scale habitat pattern and to network function can be taken into account separately (Durka et al. 1997). As in the preceding case parameters can be used to define a goal of biodiversity protection, management or restoration and subsequent distance-to-target measurement.

Table 36.1. Direct methods of biodiversity measurement

Criterion	Measurement
Species presence	Presence-absence
Population size	Effective population size, extinction probability
Species number	Total species number, diversity index, "species deficit" (relative to a reference state), "species surplus" (relative to species-area relation)
Species spectrum	Habitat specific species, character species
Rarity or threat	Number and degree of rare and endangered species, a rarity index
Contribution of a species to large scale diversity	Iterative, rule-based algorithms based on complementarity or other optimization procedures

Table 36.2. Indirect methods of biodiversity measurement

Criterion	Measurement
Number and size of protected areas	Absolute number and size of areas, number above a threshold level
Ecological processes	Observation, combined with ecological knowledge
Degree of naturalness	As "distance" (e.g. "time for development") to Potential Vegetation (if a correlation to diversity is proved)
Ecotope structure	Classification of abiotic conditions such as moisture, nutrients, relief etc. (ranking of ecotopes)
Habitat structure or state	Ocurrence of spatial niches, food plants etc. (ranking of habitats)
Rarity of habitats	Number, degree, a rarity index
Habitat spectrum	Combination, optimum arrangement, irreplaceability or representativity
Contribution to network function	Spatial arrangement, shape, stepping stone or corridor function, accessability etc.

Biodiversity measurement in an SEA context is to a large extent dependent of the available data. Because of the limited direct measurability of biodiversity (see Table 36.1) in many cases it will have to rely on sum indicators or surrogates as listed in Table 36.2 (see Chap. 37). This is not much different from other fields of spatial planning (Fischer-Kowalski et al. 1993; OECD 1998).

36.4 The Value of Biodiversity

The Preamble of the Convention on Biodiversity lists "intrinsic, ecological, genetic, social, economic, scientific, educational, cultural, recreational and aesthetic values" of biological diversity. This list is insufficient and needs specification (see Australian Government 2003; IUCN 2004). Four different value approaches are presented here, which are amply discussed in the current literature (Wiegleb 2004b, 2004c).

36.4.1 A Philosophical Classification of Nature Values

A philosophical classification of nature values which likewise apply to biodiversity values is shown in Table 36.3. The value categories (instrumental value, eudaemonic value, moral value, and "absolute intrinsic value") were regarded as exhaustive by Krebs (1996a). Instrumental values ("use values" in economic language) comprise both current and future uses and direct and indirect use. Instrumental values are well-founded in philosophical reasoning. They are created by exchange or replacement by money. The concept of "eudaemonic values" (called "eudaemonic intrinsic values" by Krebs 1996a) comprises a number of different value categories. Aesthetic values are frequently treated under this argument. Both "emotional" or "sentimental" values (for example the emotional binding to one´s home area) and religious or transcendental values are included. All of these values relate to the "good life" of the individual. Moral values (called "moral intrinsic values" by Krebs 1996a) refer to moral obligations of man to other beings. Moral values of humans are well-founded in occidental philosophical traditions. However, possible extension of moral intuitions to other living beings is strongly controversial.

Pathocentric, teleologic (biocentric) and holistic (ecocentric) arguments have been proposed (Krebs 1996b). While pathocentric approaches are not applicable to compound entities such as species and ecosystems, biocentric and physiocentric positions tend to provoke unsolvable dilemmas. They do not include any rule for dealing with conflicting aims and values, and would therefore, make decision-making in planning procedures almost impossible. Excluding these viewpoint does not necessarily mean that the project of "environmental ethics" has failed. Only the crude viewpoints are flawed disregarding such as the influence of different ecological perspectives on the outcome of an ethical judgement (see Swart et al. 2001). As all European nature conservation laws (including the SEA Directive) are based on a moderate or axiological anthropocentrism (Schneider 2000), extension to biocentric or ecocentric reasoning has to be rejected as unnecessary.

Values that cannot be classified into the preceding groups are called "absolute intrinsic values" by Krebs (1996a). Among others, "diversity" is regarded as belonging to this group. Despite the fact that biodiversity values are often used in the current conservation discussion, they are not well rooted in western philosophical traditions.

Table 36.3. A classification of values of nature (based on Krebs 1996a)

Value category	Value arguments
Instrumental values	Values related to direct and indirect use, ranging from basic needs (food) to landscape ecological functions
Eudaemonic values	Values related to the "well-being of human soul", including e.g. aesthetic, emotional and religious values
Moral values	Values related to moral obligations of man to other beings (including "responsibility", "care" and "commitment")
Absolute "intrinsic values"	Values which are only derived from properties of the protected object itself (e.g. diversity, individuality)

According to Krebs's viewpoint neither instrumental values, eudaemonic values nor moral values can be connected to the diversity aspect of nature. However, Krebs (1996a) herself raised some doubts whether "absolute intrinsic value" are a meaningfull concept. The more comprehensive account of ecological ethics in Krebs (1996b) does not mention this category.

36.4.2 The Concept of Total Economic Value

Reading the recent biodiversity literature is supporting the impression that valuation of biodiversity has become a major domain of economics. Economic valuation of biodiversity deals with the identification of economic benefits, their monetarization and their calculation for specific situations. Economic valuation is not independent of philosophical value theory, even though it has developed a specific nomenclature. The intersections between both approaches are outlined in Schneider (2000). Including considerable modifications, a widely used classification of benefits (see e.g. Lerch 1996; Fergus 1997; WBGU 1999; Faucheux and Noel 2001) distinguishes six components of Total Economic Value (TEV, see Table 36.4). In its original form, the concept assumes the existence of additional "intrinsic values" ("intangibles"). Intangibles are excluded a priori from the approach because of an assumed nonmeasurability in monetary terms. However, sometimes intrinsic values are included under the concept of existence value.

The central distinction is made between use values and non-use values. Use values are either divided into direct and indirect values, or consumptive and non-consumptive values. They form the basis of welfare theory. For quantifying economic use values various direct and indirect methods are available in ecological economics (see Garrod and Willis 1999; WBGU 1999; Faucheux and Noel 2001).

"Symbol values" comprising spiritual or cultural use values are less well accessible by economic evaluation methods. On the other hand, they are responsible for differences in values-in-use and values-in-exchange in goods such as stamps, autographs of stars etc. and apparently apply to biodiversity, too. Option values (including "quasi-option" values, see Becker-Soest 1998) typically denote current ignorance of potential value or utility.

Table 36.4. Total economic value of nature

Value category	Specification
Direct use values	Outputs directly consumable
- Economic use values	Consumptive: source of fish, game, timber, fuel wood etc.
	Non-consumptive: recreation, ecotourism, education, science etc.
- Symbol value	Spiritual attitude, cultural identification
Indirect use values	Functional benefits and services
- Ecological function value	Flood control, climate, photosynthesis, nutrient cycles, waste assimilation etc.
- Option value	Source of future drugs, genes for plant breeding, complements for new technology, substitutes for depleted resources, unexplored amenites of conserved habitats
Non-use values	Noninstrumental values
- Existence value	Satisfaction that a good is there, beauty of a species or ecosystem, i.e. nature experience or knowledge of continued existence
- Bequest value	Environmental legacy, taking the altruistic perspective, including intergenerational equity and equity between economies

Existence and bequest value are difficult to distinguish among each other and from "intrinsic value" (Elsasser 1996) and even more resistant against quantification.

Symbol, option, existence and bequest value reflect a high degree of uncertainty in measuring the total economic value of an environmental object or project (Schneider 2000). Insufficient quantification may lead to an "evaluation gap" (Lerch 1996) based on insufficient methodology[1]. Therefore, the limits between monetarizable and nonmonetarizable values deserve more detailed accounts in the future. Nevertheless, TEV is a usefull concept for structuring the biodiversity value discussion (Nunes and van den Bergh 2001). It indicates that despite meth-

---

[1] On the other hand, Schneider (2000) even assumed an "evaluation surplus" caused by the unwanted influence of "commitment". A systematics treatment of biases leading to inexact estimation of economic value in contigent valuation methods is found in Elsasser (1996).

odological problems (Faucheux and Noel 2001) many environmental values can be monetarised.

36.4.3 The Concept of Ecosystem Services

As already introduced in Figure 36.1, ecosystem services refer to another interface between ecology and society (Costanza and Folke 1997). This approach was recently elaborated by MEA (2002). According to this approach services can be "provisory, supporting, or cultural".

One problem of this approach is that ecosystem services need not necessarily relate to diversity measured in terms of species richness. Provisory services comprise direct instrumental use, for food, raw materials, and pharmacy incl. bionics. Cultural services comprise instrumental and non-instrumental use for recreation, education, science, nature experience, amenity etc. list The insurance values can be added to this (Turner et al. 1993) for a risk averse society. In Western societies "risk aversion" can be regarded as a part of their culture.

Supporting services include indirect services, buffer, filter and regulation functions such as water purification, pollination, climate regulation etc. Obviously the concept of ecosystem services has both regional and global importance. On the regional level ecosystem services and species richness can be summarised as natural capital to which property rights may be assigned (e.g. to a national state). Therefore incentives for protection may be derived easily (Lerch 1996). The protection of biotic goods and resources has become a norm in all civilised countries. Both environmental protection and nature conservation are instrumentalistic, resource-oriented approaches based on prudence and rationality.

On the other hand it is unknown whether local diversity is an indicator of global diversity and whether it can be used for monitoring global changes. In this case, diversity and related concepts (complementarity, rarity etc.) would become very important as operational indicators for a desirable state of nature. If local diversity is determined by global diversity this would be the starting point of interesting scientific questions, on global vs. local determinants of biotic communities, expectations about deviation of diversity from random distributions etc. If local diversity is a prerequisite of global diversity things would be different. Being a prerequisite implies that global diversity is not only a collective property or the sum of all local diversities, but an emergent property of the biosphere resulting from interactions of local effects. So far it is unknown which of these alternatives are true. If local diversity or any pattern of local diversities (with diversity cold and hot spots) would be a prerequisite of global diversity and global diversity a prerequisite of global persistence, a good reason for protecting and restoring local diversity would arise. In this case, the protection of ecological processes (see above) could be replaced by something feasible, namely local and regional biodiversity management. However, it is also unknown how local causality of diversity maintenance is propagated to the global scale. This was amply shown by various papers in Ricklefs and Schluter (1993).

36.4.4 Ecologically informed Values of Biodiversity

In Table 36.5 a preliminary list of properties of ecological systems is presented, which can be regarded as "ecologically oriented" or "ecologically informed" justifications for preserving and restoring biodiversity. While being unrelated to the mainstream of philosophical and economic writing they are widespread in current ecological and nature conservation literature (Wulf 2001). Unfortunately, ecologists often introduce these scientific concepts as "values" (e.g. Primack 1993; Usher and Erz 1994), even though in the first instance they are based on factual information (e.g. species number, diversity indices, phylogenetic distance). Thus, evaluation is either implicit or subjective. Natural History Museum (2003) is more careful in pointing out the necessity of making choices among the various currencies of diversity offered by scientiest. It should be elucidated how these naturalistic values can be integrated into a consistent value system. The ecological properties listed in Table 36.5 are regarded as causally independent of each other.

Richness can be calculated for all systems that are decomposable into countable units. The α-diversity is ultimately a measure of the syntactic information stored in a system. The way how the information content of living systems can be measured will be discussed separately (Wiegleb 2004a). As biodiversity is unevenly distributed in space, β-diversity, dissimilarity, heterogeneity and complementarity represent an independent idea of diversity. They represent context dependent information measures. "Context" comprises both spatial relations and ecological or phylogenetic similarity (Natural History Museum 2003; Vane-Wright et al. 1991).

Table 36.5. Domains of ecologically informed values of biodiversity

Domain	Specification
Richness	α-diversity, a hot spot; total information stored in composition, structure or function of biota, ultimately a complexity measure
Complementarity	β-diversity, similarity, redundancy dissimilarity (with respect to phylogenetic and ecological information), patchiness, heterogeneity
Rarity	Endemism (small area), low number of populations, small population size
Threat	Vulnerability, low numbers (population vulnerability), narrow habitat range (habitat suitability)
Irreplaceability	Distinctiveness, uniqueness, individuality, irreproducibility, reversibility, resilience, part of minimum structure
Representativity	Typicalness, "Eigenart"
Area size	Size, shape of area (= effective size), distance, connectivity (species-area relation, fragmentation, patch-matrix)
Naturalness	Wilderness, undisturbed ecological processes, authenticity

Irreplaceability has several different connotations, first of individuality and uniqueness (having a slightly moral connotation similar to rarity), second of reversibility of processes or irreproducibility, and third of being an essential part of a larger system (representing a minimum structure). Representativeness has a slightly eudaemonic connotation (typicalness or even peculiarity), while still being quantifiable (based on frequency of a character). If it is strictly dealt with the scientific and measurable part of the concepts a definition of a family of concepts logically and information theoretically related to each other and to diversity is possible (see Faith and Walker 1996; Ferrier et al. 2000)[2].

Rarity and threat cannot be regarded as value arguments per se. First, frequent species serve as basis for most food chains. Second, it is not the aim of conservation to have as many rare species as possible. However, the protection of rare or endangered species may be a strategy or a guiding principle to maintain diversity. Spellerberg (1992) named threat as a main priority criterion (besides utility and distinctiveness!). Rarity and threat have a moral connotation (based on the fear of loss). Rarity can be measured in information theoretical terms, while threat or endangerment can only be estimated based on probability statements. A multicriterial approach is combining the measurement and protection of diversity, rarity and threat in so-called "hotspots" (Reid 1998). Area size cannot be a specific value argument on its own right but a spatial or structural surrogate for both divers and natural areas (Wiegleb 2004b). All the properties mentioned are mostly based on empirical scientific constructs (diversity indices, iterative selection algorithms etc.). The way how such constructs (the syntactic information) can be transferred into value (pragmatic information) has to be elucidated (Wiegleb 2004b).

In case of naturalness the relation to diversity may be doubted. Genetically manipulated organisms, invasive plants and animals, domestic plants and animals incl. their breeds and cultivars enhance diversity without being "natural" in the common sense of the word (Mc Isaac and Brün 1999). Natural areas need not be more diverse than disturbed ones. Most probably, naturalness belongs to a second group of ecologically informed values, to which also wilderness and undisturbed ecological processes belong (Blumrich et al. 1998). A strictly biodiversity-oriented argumentation should overcome the dichotomy between natural and disturbed ecosystems in ecology and between natural and cultural landscapes in nature conservation.

36.5 Conclusions and Recommendations

Biodiversity as defined and outlined above is directly affected by strategic actions given the following circumstances:

- A program or plan is completely devoted to biodiversity protection (e.g. a species or ecosystem protection program).

---

[2] Scientifically informed appreciation of nature goes beyond simply biophilia or other crude naturalistic arguments.

- A cross-cutting planning instrument is addressing biodiversity issues (e.g. a landscape program or land-development plan containing a nature conservation chapter).

Biodiversity is additionally indirectly affected in the following cases:

- A program or plan is aiming at the protection of air, soil and water (e.g. an anti-desertification program). Usually beneficial side-effects can be expected.
- A program or plan is aiming at the regulation of primary land use (agriculture, forestry and fisheries). Usually mitigation of negative side-effects can be assumed. While biodiversity programs in a strict sense usually address the levels of species and ecosystems (habitats, biotopes) the genetic level has to be taken into account in case of the release of genetically modified organisms.
- A program or plan has obvious environmental relevance (ecotourism development, traffic planning etc.).

Biodiversity is neither static nor given. Both measurement and evaluation of biodiversity depends of social conventions and agreements. None of the values approaches presented justifies a value of biodiversity per se. It is known that global diversity easily recovered after each event of mass extinction in the past. The value of biodiversity is best agreed among people by means of a participatory discussion of environmental goals. This was called the discoursive approach (O'Hara 1996; Wiegleb 1997). It should be asked: Who is allowed to take part in the process of aim development and specification? Scientists, planners, land users, NGO conservationists, the nature conservation administration and other possible groups (e.g. "wise users") may be involved in such a process. Secondly, it should be asked: Which is the optimal procedure to achieve a consensus or compromise in case of conflicting aims? Conflicting aims in nature conservation are based on conflicting interests of different social groups or individuals. Two different approaches to develop aims are being widely used in planning (Blumrich et al. 1998; Vorwald and Wiegleb 1996):

- Expert models. Experts decide which factual and normative information to include in the planning process, whom to ask, which planning steps to take etc.
- Open planning (also called participatory method, discoursive planning). In this case guiding principles are modified by scientific input and discussion in the course of an interactive process. "Scientific input" includes habitat-oriented inventarization of biota, abiotic conditions, site history and remote sensing. Ideally, there is a constant discussion of all groups involved. Each participant contributes to the process with her or his knowledge and worldviews in order to find a generally acceptable and accepted plan of action.

Thus, open planning can transfer individual preferences or value judgments into intersubjectively acceptable goals beyond being simply arbitrary conventions. Discourse is one of the possible procedures (besides markets, institutional power, and legislation; Becker-Soest 1998) for aggregating individual preferences to socially accepted goals. This has to taken into account also in an SEA context (Scholles et al. 2003). It is crucial that biodiversity considerations are appropriately addressed in SEA and recent published guidance (see English Nature 2004) is an important step in that direction. SEA can only provide the crash barriers for evaluating biodiversity. In practice, each single case has to be evaluated based on the available information.

References[3]

Australian Government, Department of the Environment and Heritage (2003) Biodiversity and its Value. Biodiversity Series, Paper No 1. Internet address: http://www.deh.gov.au/biodiversity/publications/series/paper1/, last accessed 20.09.2004

Becker-Soest D (1998) Institutionelle Vielfalt zur Begrenzung von Unsicherheit. Ansatzpunkte zur Bewahrung biologischer Vielfalt in einer liberalen Wettbewerbsgesellschaft (Institutional diversity can limit uncertainty. Approaches to the maintenance of biological diversity in a liberal competition society). Metropolis, Marburg

Begon M, Harper, JL, Townsend CR (1996) Ecology: Individuals, Populations and Communities. 3rd ed. Cambridge Univ. Press, Cambridge

Blumrich H, Bröring U, Felinks B, Fromm H, Mrzljak J, Schulz F, Vorwald J, Wiegleb G (1998) Naturschutz in der Bergbaufolgelandschaft – Leitbildentwicklung (Nature conservation in former brown coal mining areas – development of guiding principles). Studien und Tagungsberichte 17:1-44

Brown JH, Ernest SK, Parody JM, Haskell JP (2001) Regulation of diversity: maintenance of species richness in changing environments. Oecologia 126:321-332

Byron HJ (2000) Biodiversity Impact – Biodiversity and Environmental Impact Assessment: A Good Practice Guide for Road Schemes. The RSPB, WW-UK, English Nature and the Wildlife Trusts, Sandy

Byron HJ, Treweek, JR, Sheate, WR, Thomson, S (2000) Road developments in the UK: an analysis of ecological assessment in Environmental Impact Statements produced between 1993 and 1997. Journal of Environmental Planning and Management 43:71-77

Costanza R, Folke C (1997) Valuing ecosystem services with efficiency, fairness, and sustainability as goals. In: Daily GC (ed) Nature's Services – Societal Dependence on Natural Ecosystems. Island Press, Washington DC, pp 49-68

Duelli P, Obrist M (1998) In search of the best correlates for local biodiversity in cultivated areas. Biodiversity and conservation 7:297-309

Durka W, Altmoss M, Henle K (1997) Naturschutz in Bergbaufolgelandschaften des Südraumes Leipzig unter besonderer Berücksichtigung spontaner Sukzession (Nature conservation in former brown coal mining areas south of Leipzig with special reference to spontaneous succession). UFZ-Bericht 22/1997:1-209

English Nature, Royal Society for the Protection of Birds, Environment Agency and Countryside Council for Wales (2004) Strategic Environmental Assessment and Biodiversity: Guidance for Practitioners. English Nature, Peterborough

Elsasser P (1996) Der Erholungswert des Waldes. Monetäre Bewertung der Erholungsleistung ausgewählter Wälder in Deutschland (The recreational value of forests). Sauerländer's, Frankfurt/M

---

[3] Note that legislation mentioned in the chapter is listed at the end of the handbook in the consolidated list of legislation (Appendix 2).

Faith DP, Walker PA (1996) How do indicator groups provide information about the relative biodiversity of different sets of areas?: on hotspots, complementarity and pattern-based approaches. Biodiversity Letters 3:18-25

Faucheux S, Noel JF (2001) Ökonomie natürlicher Ressourcen und der Umwelt (Economics of natural resources and the environment. Metropolis, Marburg

Fergus D (1997) Citizens' Forum on Public Policy: Monetization of Environmental Impacts of Roads. Internet address: http://www.geocities.com/davefergus/Transportation/4CHAP4.htm, last accessed 20.09.2004

Ferrier S, Pressey RL, Barrett TW (2000) A new predictor of the irreplaceability of areas for achieving a conservation goal, its application to real-world planning, and a research agenda for further refinement. Biological Conservation 93:303-325

Fischer-Kowalski M, Haberl H, Player H, Steurer A., Zangerl-Weisz H (1993) Das System verursacherbezogener Umweltindikatoren (The system of causal-agent oriented environmental indicators). Schriftenreihe des IÖW 64/93:1-73

Fürst D, Scholles F (1998) Plan-UVP und Verfahrensmanagement am Beispiel der Regionalplanung (Plan EIA and project management in regional planning). Internet address: http://www.laum.uni-hannover.de/ilr/publ/sup/sup.html, last accessed 20.09.2004

Garrod G, Willis KG (1999) Economic Valuation of the Environment. Methods and Case Studies. E. Elgar, Cheltenham

Gaston KJ (ed) (1996) Biodiversity: a Biology of Numbers and Difference. Blackwell, Oxford

Groombridge B, Jenkins MD (2002) World Atlas of Biodiversity. Earth's Living Resources in the 21st Century. Univ. of California Press. - The most important maps are also available under http://www.unep-wcmc.org, last accessed 20.09.2004

IUCN (2004) Monitoring and Evaluation. Internet address: http://www.iucn.org, last accessed 20.9.04

Jessel B (1996) Leitbilder und Wertungsfragen in der Naturschutz- und Umweltplanung (Guiding principles and value questions in nature conservation and environmental planning). Naturschutz und Landschaftsplanung 28:211-216

Krebs A (1996a) "Ich würde gerne aus dem Hause tretend ein paar Bäume sehen". Philosophische Überlegungen zum Eigenwert der Natur (While leaving my home I would like to see trees. Philosophical considerations concerning the value of nature). In: Nutzinger HG (ed) Naturschutz-Ethik-Ökonomie: Theoretische Grundlagen und praktische Konsequenzen. Metropolis Marburg, pp 31-48

Krebs A (1996b) Ökologische Ethik I. Grundlagen und Grundbegriffe (Ecological ethics I. Foundations and basic concepts). In: Nida-Rümelin J (ed) Angewandte Ethik. Die Bereichsethiken und ihre theoretisch Fundierung. Ein Handbuch, pp 346-385

Lerch A (1996) Verfügungsrechte und biologische Vielfalt (Property rights and biological diversity). Metropolis, Marburg

McIsaac GF, Brün M (1999) Natural environment and human culture: defining terms and understanding worldviews. Journal of Environmental Quality 28:1-10

Millenium Ecosystem Assessment (MEA) (2004) News updates. Internet address: http://www.millenniumassessment.org/en/products.chapters.aspx, last accessed 20.09.2004

Natural History Museum (2003) Biodiversity and Worldmap. Biodiversity: Measuring the Variety of Nature and Selecting Priority Areas for Conservation. Internet address: http://www.nhm.ac.uk/science/projects/worldmap/index.html, last accessed 20.9.04

Nunes PA, van den Bergh JC (2001) Economic valuation of biodiversity: sense or nonsense? Ecological Economics 39:203-222

O'Hara S (1996) Discursive ethics in ecosystem valuation and environmental policy. Ecological Economics 16:95-107
OECD – Organisation for Economic Co-operation and Development (1998) Towards Sustainable Development. Environmental indicators. Paris
O'Riordan T, Stoll-Kleemann S (eds) (2003) Biodiversity, Sustainability and Human Communities. Protecting beyond the Protected. Cambridge Univ. Press, Cambridge
Peters W (1997) Zur Theorie der Modellierung von Natur und Umwelt (On the theory of modelling nature and the environment). Dissertation TU Berlin, Berlin
Primack RB (1993) Essentials of Conservation Biology. Sinauer, Sunderland Massachusetts
Reid WV (1998) Biodiversity hotspots. TREE 13:275-280
Ricklefs RE, Schluter D (eds) (1993) Species Diversity in Ecological Communities. Historical and Geographical Perspectives. University of Chicago Press, Chicago
Schneider J (2000) Die ökonomische Bewertung von Umweltprojekten. Zur Kritik an einer umfassenden Umweltbewertung mit Hilfe der Kontingenten Evaluierungsmethode (Economic evaluation of environmental projects. On the critique of an exhaustive environmental valuation by means of contingent envaluation method). Physica, Berlin
Scholles F, v Haaren C, Myrzik A, Ott S, Wilke T, Winkelbrandt A, Wulfert K (2003) Strategische Umweltprüfung und Landschaftsplanung (SEA and landscape planning). UVP-report 17:76-82
Secretariat of the Convention on Biological Diversity (2002) Vorläufige Leitlinie für die Einbeziehung von Biodiversitätsaspekten in die Gesetzgebung und/oder das Verfahren von Umweltverträglichkeitsprüfung und Strategischer Umweltprüfung. Beschluss VI/7 A) der Vertragparteien des Übereinkommens über die biologische Vielfalt auf ihrem sechsten Treffen. Den Haag, 7.-19. April 2002
Simberloff D (1999) The role of science in the preservation of forest biodiversity. Forest Ecology and Management 1115:101-111.
Sommer K (2003) Umweltschutzziele in der Strategischen Umweltprüfung (Objectives of environmental protection in SEA). UVP-report 17: 82-84.
Spellerberg IF (1992) Evaluation and Assessment for Conservation. Chapman and Hall, London
Swart JA, Van der Windt HJ, Keulartz J (2001) Valuation of nature in conservation and restoration. Restoration ecology 9:230-238
Turner RK, Pearce D, Bateman I (1993) Environmental Economics. An Elementary Introduction. Johns Hopkins University Press, Baltimore
Usher MB., Erz W (eds) (1994) Erfassen und Bewerten im Naturschutz (Original title: Wildlife Conservation Evaluation). Quelle and Meyer, Heidelberg
Vane-Wright RI, Humphries CJ, Williams PH (1991) What to protect? Systematics and the agony of choice. Biological Conservation 55:235-254
Vorwald J, Wiegleb G (1996) Anforderungen an Leitbilder für die Entwicklung von Bewertungsverfahren im Naturschutz (Requirements for guiding principles in the development of assessment procedures in nature conservation). Aktuelle Reihe BTU Cottbus 8/96:38-49
WBGU – Wissenschaftlicher Beirat der Bundesregierung Globale Umweltveränderungen (1999) Welt im Wandel: Umwelt und Ethik. Sondergutachten 1999 (Nature and ethics in a changing world). Metropolis, Marburg. Internet address of the English version: http://www.wbgu.de/wbgu_sn1999_voll_engl.html, last accessed 20.09.2004
Wiegleb G (1997) Leitbildmethode und naturschutzfachliche Bewertung (The method of guiding principles' development and nature conservation assessment). Zeitschrift für Ökologie und Naturschutz 6:43-62

Wiegleb G (2003) Was sollten wir über Biodiversität wissen? Aspekte einer angewandten Biodiversitätsforschung (What should we know about biodiversity? Aspects of an apllied biodiversty research). In: Weimann J, Hoffmann A, Hoffmann S (eds) Messung und Bewertung von Biodiversität: Mission impossible? Metropolis, Marburg, pp 151-178

Wiegleb G (2004a) The measurement of biodiversity in ecology. Forum. Internet address: http://www.tu-cottbus.de/BTU/Fak4/AllgOeko/, last accessed 20.09.2004

Wiegleb G (2004b) The value of biodiversity. Forum. Internet address: http://www.tu-cottbus.de/BTU/Fak4/AllgOeko/, last accessed 20.09.2004

Wiegleb G (2004c) Ecologically informed values of biodiversity for conservation and restoration. Forum. Internet address: http://www.tu-cottbus.de/BTU/Fak4/AllgOeko/, last accessed 20.09.2004

Wiegleb G (2004d) The measurement of biodiversity for nature conservation. Forum. Internet address: http://www.tu-cottbus.de/BTU/Fak4/AllgOeko/, last accessed 20.09.2004

Wood P (1997) Biodiversity as a source of biological resources. A new look at biodiversity values. Environmental Values 6:251-268

Wulf A (2001) Die Eignung landschaftsökologischer Bewertungskriteien für die raumbezogene Umweltplanung (Suitability of landscape ecological criteria for spatial environmental planning). Libri, Norderstedt

37 Biodiversity Programmes on Global, European and National Levels Related to SEA

Gerhard Wiegleb and Udo Bröring

Department of General Ecology, Brandenburg University of Technology (BTU), Germany

37.1 Introduction

Different driving forces account for biodiversity on a global and regional level. Only knowledge about the interactions between driving forces and human action guarantees successful implementation of SEA. On the global scale biodiversity is influenced by many strategic actions, however, technical assessment is difficult as the causal chains are not well-known. On the regional level, biodiversity is amenable to SEA assessment in various ways. Biodiversity in individual EU countries affected by world wide, European, national, and Federal state programmes (such in the case of Germany). All of these programmes are derived from or obliged to the Convention on Biodiversity. Decision-making in SEA is based both on economic and ecologically-oriented evaluation. Economic evaluation results can be taken as an input to Cost-benefit Analysis or Multicriteria Analysis. Ecologically oriented evaluation methods can be used as a separate evaluation approach or blended with economic evaluation in a decision-supporting methods. An example of a combined approach comparing the results of Multicriteria Analysis and Hasse-Diagramm Technique is presented. The example is taken from application of the Water Framework Directive to stream restoration.

Recently, biodiversity has become a major concern in both ecological sciences and natural resource management (Groombridge and Jenkins 2002; Wiegleb 2003). Several instruments have been advanced in order to reach the aims of future biodiversity protection and preservation (Groombridge and Jenkins 2002): improving taxonomic knowledge and capacity, preserving agricultural genetic resources, defining and inventorizing ecosystems, including NGOs in decision-making processes, applying the precautionary pinciple in legislation, implementing systematic conservation planning, and negotiating multilateral agreements for long-term and larges scale success.

Implementing Strategic Environmental Assessment. Edited by Michael Schmidt, Elsa João and Eike Albrecht. © 2005 Springer-Verlag

The instrument of SEA is clearly devoted to the precautionary principle (Kleinschmidt and Wagner 1998; Secretariat of the Convention on Biological Diversity 2002). In this chapter the major driving forces of biodiversity on different spatial scales are described. Different programmes on the world wide, European, and national level are analyzed with respect to their potential interference with biodiversity protection and management. The problem of data availability is briefly addressed. Finally, some methodological issues concerning the procedure of decision-making in SEA are presented.

37.2 Determinants or Driving Forces of Biodiversity

Conflicting theories as to causal and functional relations of biodiversity exist. Despite scientific disagreement about details (see Begon et al. 1996), Fig. 37.1 shows shows a noncontroversial summary of the determinants of species diversity. The ultimate source of biodiversity is biological evolution. The speciation-extinction dynamics is to a large extent regulated the water-energy balance of the earth (Hawkins 2004). Therefore global climate change has a major influence on biodiversity. The water-energy balance does also influence the global water-heat and productivity gradients. Favorable abiotic and biotic resource conditions allow the survival of more individuals, and thus the survival of more species.

On a local scale, additional factors of biodiversity regulation become effective (Fig. 37.2). Besides historic effects which had an influence on the species pool in the past (e.g. the glaciations in Central Europe) proximate effects such as immigration from the species pool, area size, stress, productivity, and disturbance become effective (Brown et al. 2001; Hubbell 2001). Favourable conditions again lead to more individuals on the local scale and, thus, to more species. Additionally, biotic interactions in the local community such as competition, predation, and mutualism may regulate an equilibrium number of species.

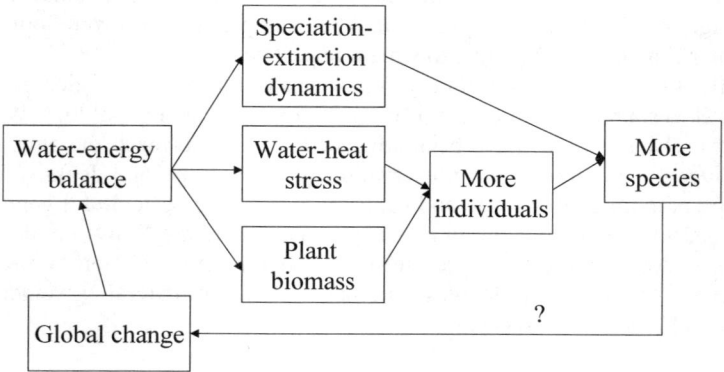

Fig. 37.1. Determinants of species diversity on a world-wide scale (adapted from Hawkins 2004)

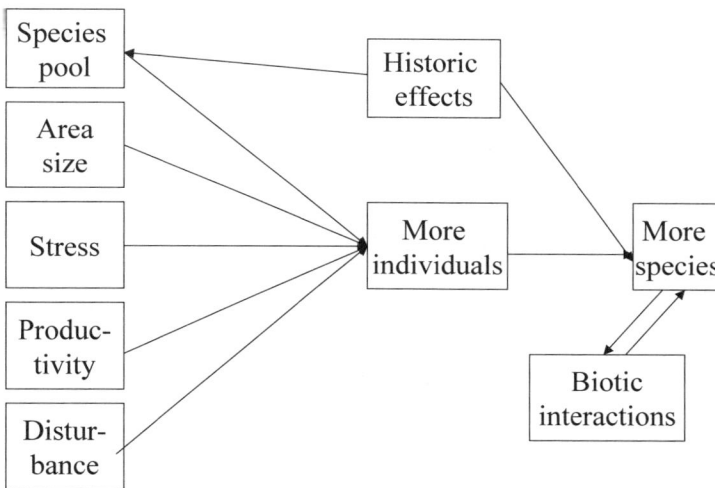

Fig. 37.2. Determinants of biodiversity on a local or regional scale

Management and planning measures on the local and regional level directly interfere into the causal network. Spatial planning will inevitably influence area size of favorable habitats, stress and disturbance intensity as well as productivity of habitats and landscapes. Therefore SEA can be used to avoid, minimise or mitigate any negative influence on the driving forces of the generation and maintenance of biodiversity.

37.3 Biodiversity in SEA

European Community Biodiversity Clearing-house Mechanism (BfN 2004) lists relevant sources for biodiversity assessment. All of them refer to the Convention on Biodiversity: Convention on Biodiversity at Global Level, Convention on Biodiversity at Pan-European And Regional Level, Convention on Biodiversity at European Community Level, Convention on Biodiversity at National Level, other biodiversity related conventions.

In this chapter, programmes of different levels, that is worldwide programmes, EU-wide programmes, nationwide programmes for of EU countries, and programmes on the *Federal* state level within Germany are reviewed. Generally those programmes can be divided in two groups, i.e. applied and monitoring programmes for planning and management, and research programmes.

37.3.1 Worldwide Programmes

The Convention on Biodiversity was signed in Rio de Janeiro in 1992. The main aims are the conservation of biological diversity, the sustainable use of its components, and the fair and equitable sharing of the benefits arising out of the utilization of genetic resources.

Worldwide programmes and notifications of the Convention on Biodiversity underpin these aims. The Convention on Biodiversity has five thematic work programmes: Marine and coastal biodiversity, Agricultural biodiversity, Forest biodiversity, Biodiversity of inland waters, and Biodiversity of dry and sub-humid lands.

In addition United Nations Environmental Programme (UNEP) and the associated World Conservation Monitoring Centre have various instruments such as "Key cross-cutting issues of relevance" and "thematic areas". Key cross-cutting issues relevant to all thematic areas are: Biosafety; access to genetic resources; traditional knowledge, innovations and practices; intellectual property rights; indicators; taxonomy; public education and awareness; incentives; and alien species.

Other conventions and associated programmes focus on different aspects of biodiversity. Numerous programmes related to habitats, species, regions, protected areas etc. are found under UNEP (2004a), to name a few:

- Convention on Wetlands of International importance especially Waterfowl Habitat (Ramsar Convention 1971)
- Convention concerning the Protection of the World Cultural and Natural Heritage (World Heritage Convention 1972)
- Convention on International Trade in Endangered Species (CITES 1973)
- Convention on the Conservation of Migratory Species (CMS or Bonn convention 1979)
- Cartagena Protocol on Biosafety (2000)
- Man and Biosphere programme (MAB 1972) established by the UNESCO. Scientific research is carried out in order to understand the relations between man and biosphere.

From a methodological point of view changes in global biodiversity can only be assessed by means of the scenario methods (Anders et al. 2004). Scenarios describe possible futures by highlighting the implications of different policy decisions (Groombridge and Jenkins 2002). The Global Environmental Outlook approach (GEO-3, ten Brink 2000) includes the following scenarios "market first", "policy first", "security first", and "sustainability first". In none of the scenarios first priority is given to biodiversity. Biodiversity protection is the casual outcome of the scenario. The negative side effects of the policy first and security first scenario are obvious. While "policy first" might eroded the constitutional site from inside, "security first" might lead to prevailing of international inequality and other sources of conflict. Most disputed is the "market first" scenario. It is not clear whether it will inevitably lead to a strong decrease in habitat quality and quantity. Assignment of property rights to biodiversity may as well have a positive effect on biodiversity (Lerch 1996). It might be a future task for SEA to investigate the consequences of such scenarios for selected areas, ecosystem types, or economy types in more detail.

37.3.2 Habitats Directive and EU Wide Programmes

The aims of EC environmental policy are described by Knopp (2003) based on relevant EU legislation:

- to preserve and protect the environment and improve its quality;
- to protect human health;
- to use natural resources in a prudent and sensible manner;
- to support international measures to overcome regional and global environmental problems.

Main programme concerning biodiversity protection in Europe is the Habitats Directive. A complete overview of EU biodiversity policy is given in O'Riordan and Klee-Stollmann (2003).

A related programme is the Wild Birds Directive. This Directive establishes a general system of protection for all species of naturally occurring birds in the wild state in the European territory of the Member States. This general system prohibits:

- deliberate killing or capture.
- deliberate destruction of, or damage to, their nests and eggs or removal of their nests.
- taking their eggs in the wild and keeping these eggs even if empty.
- deliberate disturbance of these birds particularly during the breeding and rearing period and keeping birds belonging to species which may not be hunted or captured.

The "Special Area of Conservation" (SAC) of the Habitats Directive and the "Special Protected Area" (SPA) are integrated in the NATURA 2000 system of protected sites. Further plans and programmes are (some examples):

- Bern Convention approved in the EC with the decision 82/72/EEC

The Council of Europe Convention on the Conservation of European Wildlife and Natural Habitats was adopted on September 1979 in Bern (Switzerland) and came into force on 1 June 1982. The aims of the Convention are "to conserve wild flora and fauna and their natural habitats, especially those species and habitats whose conservation requires the co-operation of several States, and to promote such co-operation. Particular emphasis is given to endangered and vulnerable species, including endangered and vulnerable migratory species." Protected species and prohibited means and methods of killing, capture and other forms of exploitation are listed in four appendices.

- EU action plan to boost Environmental Technologies for innovation, growth and sustainable development (ETAP 2004)

The Commission adopted the Environmental Technologies Action Plan in 2004. ETAP was prepared in common by the Commission services for Research and for Environment, with the co-operation of other services for specific actions. "It is an ambitious plan to further environmental technologies within the EU and globally. It seeks to exploit their potential to improve both the environment and competitiveness, thus contributing to growth and possibly creating jobs. It sets out a number of actions that the Commission will take and some that other stakeholders, such as industry and national and regional governments should undertake for the plan to be successful." (ETAP 2004)

- Commission acts to protect endangered animals and plants in the EU's mountain regions

A list of 959 nature sites in mountain regions of the European Union was approved. The approval signifies the protection of these areas and their endangered animal and plant species. The protection of these species is scientifically considered to be of European importance, and efforts are necessary to ensure biodiversity and the conservation of natural fauna and flora in these mountain regions. It is also a major step forward in establishing Natura 2000, the network of protected sites in the EU.

- DIVERSITAS

DIVERSITAS is an international global environmental change research programme. General tasks are to promote integrative biodiversity science, linking biological, ecological and social disciplines in an effort to produce socially relevant new knowledge, and to provide the scientific basis for an understanding of biodiversity loss, and to draw out the implications for the policies for conservation and sustainable use of biodiversity.

- European programmes derived from the Bonn Convention

The following agreements can be regarded as European contributions to implementation of the Bonn Convention: Agreement on the Conservation of African-Eurasian Migratory Waterbirds (AEWA 1999); Agreement on the Conservation of Small Cetaceans of the Baltic and North Seas (ASCOBANS 1994); Agreement on the Conservation of the Black Seas, Mediterranean and Contiguous Atlantic Area (ACCOBAMS 1995), and Agreement on the Conservation of Populations of European Bats (EUROBATS 1995).

37.3.3 Nation Wide Programmes with Special Reference to German Legislation

German legislation concerning biodiversity is laid down in the EIA Act and the Federal Nature Conservation Act. Protected goods according to the EIA Act are:

Air, soil, water, animals and plants, landscape and interrelations. The biodiversity aspects is restricted to wild living plants and animals. Protected goods according to Federal Nature Conservation Act are: The efficiency of nature's economy („household of nature"), the long-term usability of natural goods, wild-living plants and animals in their natural environment, and the diversity, peculiarity and beauty of nature and landscape. Under this law biodiversity covers both natural goods and their usability.

National programmes defining more detailed approaches to biodiversity are advanced by the Federal Agency for Nature Conservation (BfN). Several Research and Development-Projects (*F+E-Vorhaben*) in the framework of the environmental research plan of the state government: UFOPLAN 2004 (*Umweltforschungsplan*) and Testing and Generating Projects (*E+E-Vorhaben*) as well as nature protection projects (*Naturschutzgroßprojekts*) are funded (see BFN 2004). National programmes in Germany have not yet undergone any SEA.

37.3.4 Federal State Programmes in Germany with Special Reference to Brandenburg

In Brandenburg, biodiversity protection is regulated in the framework of the Brandenburg Nature Conservation Act of May 2004. Environmental policy in the Federal State of Brandenburg is following a defined set of aims:

- To perceive nature conservation as a societal task,
- To build up habitat networks,
- To safeguard conservation areas,
- To "ecologise" land and resource use,
- To enhance acceptance of nature and landscape protection,
- To modernise nature conservation law.

Efforts are closely related to the NATURA 2000 network of protected sites (see above). Programmes actually exist as to species protection and management of large protected areas (*Großschutzgebiete*) (see MUNR Brandenburg 2004). Further examples from Germany are:

- SEA for Spatial Plan in the Region Danube-Iller

An integrated SEA was carried out by an ecological risk analysis as to the identification of preferential sites for buildings and infrastructure (*Bauleitplanung*). Biodiversity is represented here only as "habitat types" (*Biotope*). As a result a map is presented with zones of restriction poor and restriction free sites with long and short distances to the railway (see Bundesamt für Bauwesen und Raumordnung 2004).

- SEA in the development planning in North-Rhine-Westphalia

Three cases of application of SEA were distinguished in regional planning which differ in their substantial and methodological requirements: (1) regional planning decisions on projects which later will require EIA at the authorisation stage (e.g., waste disposal installations, quarrying), (2) regional planning decisions on spatial areas without direct link to projects (e.g., areas for settlement), (3) general planning objectives which do not directly refer to projects but affect spatial development and the environment of the planning area. Furthermore, a distinction has to be drawn between the amendment of an existing plan and the development of a new plan. The later involves a great number of separate decisions (Auge and Wagner 2004).

37.4 Available Data for Biodiversity Assessment

Any SEA is dependent of the availability of data. In some cases, new data might not be collected because of the lack of money and the lack of time. In those circumstances, existing data have to be taken into account. Such data exist on various spatial scales and various degrees of precision. A typical data source the World Atlas of Biodiversity (Groombrige and Jenkins 2002). It is underpinned by other data bases provided by IUCN (2004); Natural History Museum (2003) and UNEP (2004b). The European community has its own data bases, for example on biogeographical regions. Some typical examples:

Distribution maps of species are available for many species, in particular birds, mammals, and higher plants for many countries of the world. Recently great progress has been made in the inventory of species.

- Summary maps of species numbers and density (Groombridge and Jenkins 2002, maps 4.7 and 5.3; including information about biodiversity hotspots, Reid 1998, being often differentiated according to taxonomic groups).
- Vegetations maps, based on ground mapping, or combined approaches of mapping, satellite imagery and simulation models (Groombridge and Jenkins 2002, maps 5.2 and 5.8).
- Thematical maps on human impact, its type and intensity (Groombridge and Jenkins 2002, e.g. maps 4.2 and 4.4).
- Satellite imagery is becoming more popular with ongoing technological progress. Satellites images can provide data about land use types, vegetation cover, or disturbed landscapes, all kind of indirect information helping to interpret more detailed species data.
- Numerous data bases and meta data bases on biodiversity information can be found on the INTERNET (Berendsohn et al. 2004; Biodiversity Information Network 2004).

On the regional and national level regular Environmental Reporting is a major source of information. Reporting is a national and international obligation based on laws and agreements. However, monitoring and reporting can never replace the assessment in an SEA procedure itself. It is only a necessary prerequisite.

37.5 Decision-Aid in Biodiversity Protection

Evaluation is a prerequisite for decision-making (Bastian and Schreiber 1999, Barrow 1999). Evaluation methods have to be established that are able to measure the absolute or relative value of the good of interest. In this respect SEA does not differ form any other planning technique.

37.5.1 Evaluation Methods of Environmental Economics

Tab. 37.1 shows an overview of economic methods for evaluating the Total Economic Value of biodiversity. The value types are derived from Chap. 36, while the methods are taken from Faucheux and Noel (2001) and Nunes and van den Berghen (2001). As has been pointed out by many authors, the table is highly theoretical. Economic methods usually do not measure the biodiversity aspect of a good but the resource quality only. Biodiversity in the strict sense is a non-marketable good. Therefore any biodiversity measurement in monetary terms will necessarily remain incomplete and has to be complemented by other methods (Wiegleb 2004a, 2004b, 2004c, see also below).

All economic methods listed try to aggregate values and preferences held by individuals (Fergus 1977, Garrod and Willis 1991; WBGU 1999). Results can be the basis of a Cost-benefit Analysis (CBA). In a Cost-benefit Analysis all utility functions are substitutable and can be monetarised by one of the preceding methods. Simple algebraic will yield a satisfying solution given the assumptions are correct. CBA is impossible, if people have lexicographic preferences (Schneider 2000). CBA usually considers only one environmental project in order to calculate the marginal value of an environmental improvement or deterioration. Cost-efficiency Analysis (CEA) compares two or more projects in order to find out which one can reach the same effect with lowest costs. Thus CEA may be more appropriate in an SEA context, as like EIA, SEA is usually more interested in the comparison of alternative scenarios, than in the calculation of the marginal value of a single rare species.

If CBA or CEA are not applicable, alternative methods are available. Multi-criteria Analysis can be used if at least one utility function cannot be monetarised. Aggregation of values is dependent on weighting of criteria and of scaling (Rauschmeyer 2001, Strassert 1995).

Table 37.1. Methodological approach to monetary biodiversity valuation

Direct measurement			⇔	Indirect measurement	
Contingent Valuation Method	Travel cost method	Hedonic price method	Market prices	Input-output analysis
Nonuse values, option values	Non-consumptive direct use values	Real estate prices, wages	Costs, opportunity costs	Local economic activity

Multi-criteria Analysis is the most widespread evaluation instrument of EIA, and may therefore become the preferred method in SEA, too. In Risk Analysis (RA) utility functions are unknown or uncertain. For biodiversity assessment still no generally applicable methodology is available.

37.5.2 Ecologically-Oriented Evaluation

Ecologically oriented evaluation always has to complement economic evaluation. There are more than 2000 ecologically-oriented evaluation methods of biotic goods and biodiversity (Spellerberg 1992; Usher and Erz 1990; Wiegleb 1997). Most of them are based either on species occurrence (overall diversity or selected target species) or on habitat characters (surrogates, Bröring and Wiegleb, 2004 this issue). All of them depend on clear-cut definition of a reference state (Wiegleb 1997; SER 2004). They can be aggregated by Multicriteria Analysis or analyzed by alternative methods (e.g. Hasse Diagram Technique, Brüggemann and Fromm 2001).

Ecologically oriented evaluation methods can be executed separately from economic methods. The main problem of integrating biodiversity assessment into SEA is that there are no generally recognised standards or reference states of biodiversity. In many cases the informational basis is lacking for deriving biodiversity values from general assumptions or preferences. Therefore, status quo analyses of the environment is not possible to a reasonable extent and evaluation of biodiversity based on calculations of the differences between a given state and desired reference states is not possible. On the other hand it is impossible to give operational evaluations without a calculation of relations between status quo and such reference states.

37.6 A Hypothetical Example

A hypothetical SEA of a restoration or river basin management plan of a lowland stream is taken as an example. The plan is assessed in the general framework of the Water Framework Directive. The example shows that the Water Framework Directive itself cannot be the object of an SEA. Its aims are too general needing specification and regionalisation in any case. However, the restoration or river basin management plan can be subjected to SEA.

Four hypothetical measures, which according the directive are regarded as beneficial for the environment, are analyzed with respect to their effect on four biodiversity indicators. The measures are water quality improvement, channel restoration, implementation of marginal strips alongside the river course, and integrated drainage area management (Wiegleb 1991). Reappearance of migrating fish species, suppression of invasive plant species, creation of representative bank habitats, and increase of regional habitat diversity are regarded as relevant biodiversity indicators.

Table 37.2. Effectiveness assessment of measures related to the Water Framework Directive (3 = very effective, 2 = effective, 1 = less effective, 0 = uneffective)

Biodiversity indicators Programme measures	Reappearance of migrating fish species	Suppression of invasive plant species	Creation of representative bank habitats	Increase of regional habitat diversity	Total
Water quality improvement	2	0	0	0	2
Channel restoration	3	2	1	2	8
Implementation of marginal strips	0	3	3	3	9
Integrated area management	1	1	2	1	5

Table 37.2 shows an assessment of the assumed effectiveness of the measures. Neither "water quality improvement" nor "integrated drainage area management" have the highest positive impact (3) on any of the biodiversity indicators. "Channel restoration" will foster reappearance of migrating fish species.

"Implementation of marginal strips" will likewise be favourable for suppression of invasive plants, creation of representative bank habitats, and increase of regional habitat diversity. Both measures show the highest total values. For multicriteria decision aid both the added effectiveness values are compared with the costs and social acceptance estimated in an ordinal scale (Table 37.3). Channel restoration and marginal strips are not only most effective, but also the cheapest ways to reach the goals. On the other hand they are not very popular among the population. Water quality improvement still has the highest prestige. Integrated management may be preferred by experts, but as it is provoking interest conflicts it might be the least popular measure among the population.

Table 37.3. Multicriteria decision-aid for assumed programme effects (*informed estimations/guesses)

	Effectiveness for biodiversity	Costs*	Acceptance*	Total
Water quality improvement	2	1	5	8
Channel restoration	8	3	3	14
Implementation of marginal strips	9	4	3	16
Integrated area management	5	2	1	8

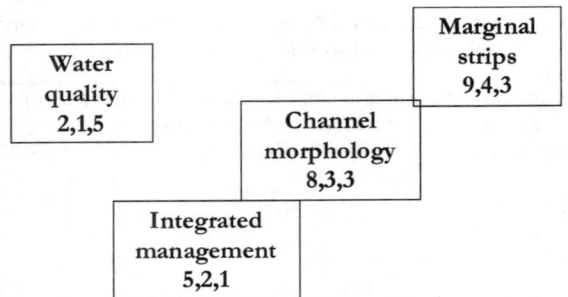

Effectivity, costs, and acceptance are ordered according to the ≥ rule.

Fig. 37.3. Hasse Diagram ordination of the evaluation results

Another evaluation approach based on Hasse-Diagram Technique (Brüggemann and Fromm 2001; Wiegleb and Herr 1999) is shown in Fig. 37.3. It is shown that, based on the criteria used, marginal strips are superior to channel morphology improvement which is itself superior to integrated management. Water quality improvement cannot be compared to the other measure, at least not on the basis of the information available. Therefore any decision in favour of water quality improvement has to be regarded as a political decision which is not based on biodiversity assessment.

37.7 Conclusions and Recommendations

There are many programmes around which claim positive effects on biodiversity. As long as causal links between driving forces and biodiversity is unexplained, and the data base for concrete evaluations is insufficient, satisfying results of SEA can not be expected. SEA methodology relating to biodiversity is largely undeveloped. There are however a few publications dealing with this topic (e.g. TRL Ltd. and Collingwood Environmental Planning 2002; English Nature et al. 2004). A standardised catalogue of biodiversity indicators does not yet exist. It is assumed that biodiversity indicators of the future will not refer to singe species occurrence but to aggregate indices such as species density of well known taxon groups or indirect indicators (surrogates) such as habitat classifications (Bröring and Wiegleb 2004). A standardised methodology of evaluation of and decision-making for biodiversity does not exist. This is due to the fact that ecologically-oriented evaluation methods are too much devoted to nature conservation traditions instead of adopting the wider approach of resource protection. On the other hand biodiversity is not amenable to economic valuation, therefore methodological progress is required.

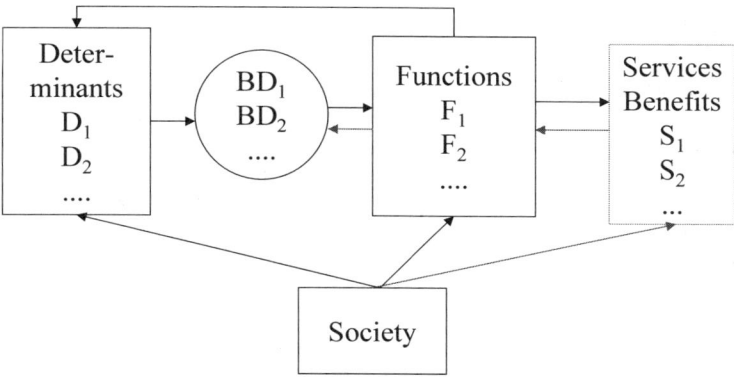

Fig. 37.4. Societal interest in biodiversity functions (adapted from Wiegleb 2003)

This chapter therefore recommends the following: before implementation, the compatibility of SEA with existing legislation, in particular EIA legislation, has to be checked (Fürst and Scholles 2003). The development of competing methodologies has to be avoided. The development of a methodology beyond simple monitoring and reporting schemes has to be fostered. Monitoring and reporting cannot replace responsible decision-making. The development of meaningful large scale biodiversity indicators has to be improved. Indicators systems such as OECD (1998) do not meet the requirements of SEA. Additionally, sets of regional indicators have to be compiled. Modern decision-making techniques such as the scenario technique and Hasse Diagram technique have to be integrated in decision-making procedures.

For this purpose it is recommended to proceed according to the approach outlined in Fig. 37.4. Selection of biodiversity indicators BD1,2 should not rely on existing data. Existing data usually have been collected at different times with different methodology, intensity, and purpose. Two alternatives exist: either indicators are strictly based on a scientific approach (left hand side of the diagramm). The full network of determinants (D1,2) and functions (F1,2) of biodiversity has to be analysed in order to finally select the right indicator. One attempt to reach scientifically sound indicators was presented by Müller (1996). The other alternative (right hand side of the diagram) indicates a kind of shortcut: The society has to define the services and benefits (S1,2) of biodiversity in which it is interested in (see for example Millenium Ecosystem Assessment 2002). In that case, only that subset of biodiversity has to be investigated and assessed that is responsible for the desired services. Indicators for desired services may be found much easier. They can also be integrated into methodological recommendations for SEA in an understandable language. Input of scientific knowledge into planning is necessary, but science must not dominate decison-making (O'Riordan and Stoll-Kleemann 2003). This will furthermore enhance credibility of the SEA approach.

References[1]

Anders K, Prochnow A, Schlauderer R, Wiegleb G (2004) Szenario-Methode als Instrument der Naturschutzplanung im Offenland (The scenario-method as an instrument in environmental planning in open landscapes). In: Anders K, Mrzljak J, Wiegleb G, Wallschläger D (eds) Handbuch Offenlandmanagement. Am Beispiel ehemaliger und in Nutzung befindlicher Truppenübungsplätze. Springer, Heidelberg, in Press

Auge J, Wagner D (2004) Case Study "Strategic Environmental Assessment in Regional Planning". Internet address: http://www.uvp.de/welcome.html?http://www.uvp.de/aktuell/BauleitUvp.html, last accessed 20.09.2004

Barrow CJ (1999) Environmental Management. Principles and Practice. Routledge, London

Bastian O, Schreiber KF (1999) Analyse und ökologische Bewertung der Landschaft (Analysis and assessment of landscapes), 2nd ed. Spektrum, Heidelberg

Begon M, Harper JL, Townsend CR (1996) Ecology: Individuals, Populations and Communities, 3rd ed. Cambridge Univ. Press, Cambridge

Berendsohn W, Häuser C, Lampe KH (2004) Biodiversitätsinformatik in Deutschland: Bestandsaufnahme und Perspektiven (Biodiversity informatics in Germany. Inventarisation and perspectives). Bonner Zoologische Monographien 45. Also available under http://www.bgbm.fu-berlin.de/BioDivInf/Docs/BinD/33.htm, last accessed 20.09.2004

BFN (2004) F+E-Vorhaben, E+E-Vorhaben, Naturschutzgroßprojekte, Verbändeförderung, (Large-scale nature conservation projects). UFOPLAN. Internet address: http://www.bfn.de/, last accessed 20.09.2004

BfN – Bundesamt für Naturschutz (2004) Informationsplattform (Information platform). Internet address: http://www.biodiv-chm.de/konvention/F1052472490/HTML, last accessed 20.09.2004

Brown JH, Ernest SK, Parody JM, Haskell, JP (2001) Regulation of diversity: maintenance of species richness in changing environments. Oecologia 126:321-332

Brüggemann R, Fromm O (2001) Partielle Ordnungen: Möglichkeiten und Grenzen am Beispiel der Biodiversität (Partially ordered sets. potentials and limits). Jahrbuch Ökologische Ökonomik 2 (Ökonomische Naturbewertung), pp 181-200

Bundesamt für Bauwesen und Raumordnung (2004). Internet address: http://www.urban21.de/raumordnung/moro/fallbeispiele/donau_iller.htm, last accessed 20.09.2004

English Nature, Royal Society for the Protection of Birds, Environment Agency and Countryside Council for Wales (2004) Strategic Environmental Assessment and Biodiversity: Guidance for Practitioners. English Nature, Peterborough

ETAP – Environmetal Technologies Action Plan (2004). Internet address: http://europa.eu.int/comm/environment/index_en.htm, last accessed 20.09.2004

Faucheux S, Noel JF (2001) Ökonomie natürlicher Ressourcen und der Umwelt (The economy of natural resouces an the environment). Metropolis, Marburg

Fergus D (1997) Citizens' Forum on Public Policy: Monetization of environmental impacts of roads. Internet addres: http://www.geocities.com/davefergus/Transportation/4CHAP4.htm, last accessed

---

[1] Note that legislation mentioned in the chapter is listed at the end of the handbook in the consolidated list of legislation (Appendix 2).

20.09.2004

Fürst D, Scholles F (2003) Plan-UVP und Verfahrensmanagement am Beispiel der Regionalplanung (Plang EIA and project management in regional planning). Internet address: http://www.laum.uni-hannover.de/ilr/publ/sup/sup.html, last accessed 20.09.2004

Garrod G, Willis KG (1999) Economic Valuation of the Environment. Methods and case studies. E. Elgar, Cheltenham

Groombridge B, Jenkins MD (2002) World Atlas of Biodiversity. Earth's living Resources in the 21st Century. University of California Press. - The most important maps are also available under http://www.unep-wcmc.org, last accessed 20.09.04

Hawkins BA (2004) Are we making progress toward understanding the global diversity gradients? Basic and Applied Ecology 5:1-3

Hubbell SP (2001) The Unified Neutral Theory of Biodiversity and Biogeography. Princeton Univ. Press, Princeton

IUCN (2004) Monitoring and Evaluation. Internet address: http://www.iucn.org, last accessed 20.09.2004

Kleinschmidt V, Wagner D (1998) Strategic Environmental Assessment in Europe: Fourth European Workshop on Environmental Impact Assessment. Kluwer Academic Publishers, Dordrecht

Knopp L (2003) Instruments of European law and questions concerning their implementation at national level as exemplified by water law, immission control and waste law. In: Schmidt M, Knopp L (2003) Reform in CEE-countries with regard to European enlargement. Institution building and public administration reform in the environmental sector. Springer, Berlin, pp 3-14.

Lerch A (1996) Verfügungsrechte und biologische Vielfalt (Property rights and biological diversity). Metropolis, Marburg

MLNR Brandenburg (2004). Internet address: http://www.mlur.brandenburg.de/n/b_n.htm, last accessed 20.09.2004

Müller F (1996) Ableitung von integrativen Indikatoren zur Bewertung von Ökosystem-Zuständen für die Umweltökonomische Gesamtrechnung (Deriving integrative indicators to assess the status of ecosystems in a Total Economic Balance). Studie für das Statistische Bundesamt Wiesbaden, Universität Kiel, pp 1-130

Natural History Museum (2003) Biodiversity and Worldmap. Biodiversity: Measuring the Variety of Nature and Selecting Priority Areas for Conservation. Internet address: www.nhm.ac.uk/science/projects/worldmap/index.html, last accessed 20.09.2004

Nunes PALD, van den Bergh JCJM (2001) Economic valuation of biodiversity: sense or nonsense? Ecological Economics 39:203-222.

OECD – Organisation for economic co-operation and development (1998) Towards Sustainable Development. Environmental indicators. Paris

O'Riordan T, Stoll-Kleemann S (eds) (2003) Biodiversity, Sustainability and Human Communities. Protecting beyond the Protected. Cambridge Univ. Press, Cambridge

Rauschmeyer F (2001) Entscheidungshilfen im Umweltbereich. Von der mono-kriteriellen zur multi-kriteriellen Analyse (Decision-aid in the environmental sector. From mono-criteria to multi-criteria analysis). Jahrbuch Ökologische Ökonomik 2 (Ökomische Naturbewertung):221-242

Reid WV (1998) Biodiversity hotspots. TREE 13:275-280.

Schneider, J (2000) Die ökonomische Bewertung von Umweltprojekten. Zur Kritik an einer umfassenden Umweltbewertung mit Hilfe der Kontingenten Evaluierungsmethode (Economic evaluation of environmental projects. On the critique of an exhaustive environmental valuation by means of the contingent envaluation method). Physica, Berlin

Secretariat of the Convention on Biological Diversity (2002) Vorläufige Leitlinie für die Einbeziehung von Biodiversitätsaspekten in die Gesetzgebung und/oder das Verfahren von Umweltverträglichkeitsprüfung und Strategischer Umweltprüfung. Beschluss VI/7 A) der Vertragparteien des Übereinkommens über die biologische Vielfalt auf ihrem sechsten Treffen. Den Haag, 7.-19. April 2002

SER (2004) A Primer of Ecological Restoration. Internet address: www.ser.org, last accessed 20.09.2004

Spellerberg IF (1992) Evaluation and Assessment for Conservation. Chapman and Hall, London

Strassert G (1995) Das Abwägungsproblem bei multikrieriellen Entscheidungen – Grundlagen und Lösungsansätze unter besonderer Berücksichtigung der Regionalplanung (The problem of consideration in multi-criteria decisions – foundations and solutions with spezial reference to regional planning). Peter Lang, Franfurt/M

Sten Brink B (2000) Biodiversity Indicators for the OECD. Environmental Outlook and Strategy, a Feasibilty Study. RIVM report 402001014, Bilthoven

TRL Ltd and Colligwood Environmental Planning (2002) Analysis of Baseline Requirements for the SEA Directive - Final report. Internet address:
http://www.trl.co.uk/static/environment/SWRA, last accessed 20.9.04

UNEP (2004a) Conventions. Internet address:
http://www.unep.ch/conventions/geclist.htm#biodiv, last accessed 20.09.2004

UNEP (2004b) Forests. Internet address:
http://www.unep-wcmc.org/forest/datasets_maps.htm, last accessed 20.9.04

Usher MB, Erz W (eds) (1994) Erfassen und Bewerten im Naturschutz (Wildlife Conservation Evaluation). Quelle and Meyer, Heidelberg

WBGU – Wissenschaftlicher Beirat der Bundesregierung Globale Umweltveränderungen (1999) Welt im Wandel: Umwelt und Ethik. Sondergutachten 1999 (Nature and ethics in a changing world). Metropolis, Marburg (English version:
http://www.wbgu.de/wbgu_sn1999_voll_engl.html, last accessed 20.9.04)

Wiegleb G, Herr W (1999) Bewertung im Rahmen der Ems-UVS 1993 (Assessment in the framework of the EMS EIA 1993). In: Wiegleb G, Bröring U (eds) Implementation naturschutzfachlicher Bewertungsverfahren in Verwaltungshandeln. ln. Aktuelle Reihe BTU Cottbus 5/99: 119-129

Wiegleb G (1991) Die wissenschaftlichen Grundlagen von Fließgewässer-Renaturierungskonzepten (The scientific foundations of restoration concepts of running waters). Verhandlungen der Gesellschaft für Ökologie 19/3:7-15

Wiegleb G (1997) Leitbildmethode und naturschutzfachliche Bewertung (The method of guiding principles' development and nature conservation assessment). Zeitschrift für Ökologie und Naturschutz 6:43-62

Wiegleb G (2003) Was sollten wir über Biodiversität wissen? Aspekte einer angewandten Biodiversitätsforschung (What should we know about biodiversity? Aspects of an applied biodiversity research). In: Weimann J, Hoffmann A, Hoffmann S (eds) Messung und Bewertung von Biodiversität: Mission impossible? Metropolis, Marburg, pp 151-178

Wiegleb G (2004a) Ecologically informed values of biodiversity for conservation and restoration. Forum. Internet address: http://www.tu-cottbus.de/BTU/Fak4/AllgOeko/, last accessed 20.09.2004

Wiegleb G (2004d) The measurement of biodiversity for nature conservation. Forum. Internet address: http://www.tu-cottbus.de/BTU/Fak4/AllgOeko/, last accessed 20.09.2004

Wiegleb G (2004c) The value of biodiversity. Forum
http://www.tu-cottbus.de/BTU/Fak4/AllgOeko/, last accessed 20.09.2004

Part VII – Implementing SEA in Spatial and Sector Planning

This Part focuses on the implementation of SEA for particular sectors. The SEA Directive, for example, is quite broad and applies to plans or programmes in eleven sectors: agriculture, forestry, fisheries, energy, industry, transport, waste, water, tourism, telecommunications and land use planning. Chapter 38 starts by discussing the link and the overlap between SEA and landscape planning. The chapter argues that the potential integration between these two instruments could produce mutual benefits and enhance SEA's acceptability by municipal, regional and sectoral planning institutions. Chapter 39 does a similar type of evaluation but in relation to urban planning.

Chapter 40 discusses SEA in transport planning in Germany. In Europe, a large proportion of SEA methodology has been developed for the transport sector. In Germany, in particular, several national and state-level plans and programmes have been subject to a type of environmental assessment since the early nineties and this is why this chapter concentrates on the early development of SEA methodologies in the German transport sector. Chapter 41 discusses how SEA might be able to contribute to the improvement of agricultural land use methods. The chapter argues that the SEA Directive's exclusion of certain agricultural plans and programmes is a mistake, which weakens the implementation of SEA in the European Union.

Chapter 42 evaluates the SEA process for waste management using the example of the Viennese Waste Management Plan. The chapter addresses issues of methodologies and data used in waste-management SEA. The chapter also describes an innovative SEA model for pro-active stakeholder participation called "the SEA Round Table". Finally, Chapter 43 addresses the use of SEA associated with the mining industry, with particular regard to mining operations in Turkey.

38 Co-ordination of SEA and Landscape Planning

Frank Scholles[1] and Christina von Haaren[2]

1 Institute of Regional Planning and Regional Science (ILR), University of Hanover, Germany
2 Institute of Landscape Planning and Nature Conservation (ILN), University of Hanover, Germany

38.1 Introduction

In Germany, SEA for plans and programmes is confronted with an existing legal instrument of environmental precaution, landscape planning. (The SEA Directive does not distinguish plans from programmes and at least in German practice they cannot be clearly distinguished from each other. Therefore in this article the term "plans" means "plans and programmes".) Installed with the Federal Nature Conservation Act in 1976, landscape planning is now an established instrument, although there is criticism about its heterogeneity in the 16 German *Laender* and lack of implementation (e.g. Kiemstedt et al. 1990).

Since one of the tasks of German landscape planning is to assess the existing and anticipated status of nature and landscapes, including any resultant conflicts, the relationship between this approach and future SEA has to be investigated. Maybe SEA can also learn from this established approach. Because a legally based landscape planning has not (yet) been developed to the same extend in other EU Member States, the investigation is focused on Germany.

After a short description of the situation in Germany (Sect. 38.3), we identify intersections of SEA and landscape planning (Sect. 38.4), regarding procedure, stages, contents, and methods. We propose a model for the integration of both instruments. But at first, it is necessary to briefly clarify the intention of the SEA Directive (Sect. 38.2), which is still frequently misinterpreted in Germany.

Implementing Strategic Environmental Assessment. Edited by Michael Schmidt, Elsa João and Eike Albrecht. © 2005 Springer-Verlag

38.2 The Intention of the SEA Directive

The intention of the SEA Directive is not to carry out hundreds of small project EIA at high expenditure, thus controlling and, if necessary, sanctioning a plan as carried out for the Federal Transport Infrastructure Plan (cf. Hoppenstedt 1999). Nor it is, as stated by Schmidt-Eichstaedt (2004 p.92), to restrict the assessment to those plan contents that clearly set the framework for certain projects or that influence certain protected areas. Typically, proponents of this approach call the instrument "Plan-EIA". The aims of the SEA Directive are rather:

- to early integrate environmental considerations and to promote sustainable development,
- to accompany and procedurally assess the planning process, thus stimulating a self-reflective control,
- to assess strategic contents such as aims and objectives or claimed demands with straightforward methods,
- to publicly discuss plans,
- to increase decision-making transparency,
- to contribute to an efficient planning process by avoiding duplication or competition of assessment and by allowing for the coordination of existing instruments (like landscape planning).

In addition, the transposition into national law should lead to more understandable regulations for practitioners who are only occasionally involved in SEA. Most of the plans that will be subject to SEA must be drawn up or continued only once within ten or more years. So, unlike project developers who have experience with carrying out an EIA for their projects, most planners are not accustomed to carrying out environmental assessments.

38.3 The Situation in Germany

The complicated process of transposing the SEA Directive into German law is described in Chap. 7. This chapter therefore focuses on the relation to landscape planning.

In 2002, a fundamental amendment of landscape planning regulations was made in the Federal Nature Conservation Act. The changes in the law include the obligation to draw up local landscape plans as a basis for land use planning and all action by the authorities concerning the environment. The Federal Nature Conservation Act is just a framework or "skeleton" law. This is why its regulations must be transposed by each of the 16 *Laender* into their respective nature conservation act to take effect. This transposition offers the opportunity to shape and to supplement the coordination with SEA that has already been laid out by the "skeleton" law.

To do this successfully, some crucial questions about the future relation of the two instruments must be answered: Can landscape planning fulfil the requirements of the SEA Directive or can SEA replace landscape planning? Or will the two instruments interfere with each other? Which similar tasks should be integrated in order to avoid duplication of assessment?

In Germany, fortunately, there is no need to develop SEA from scratch. The existing SEA approaches in Germany have been cited as models in EU projects that were carried out to justify the SEA Directive (Dom 1997; Wagner 2000). Nevertheless, opponents claim that for juridical, administrative, economic and systematic reasons (cf. overview in Fürst and Scholles 1998; Haaren et al. 2004) implementation of SEA is impossible or at least too costly and thus inadequate for the majority of German plans. Other critics have claimed that many German plans do not require an SEA because they already consider the environmental factors in an appropriate way. Some manage to take both positions at the same time (e.g. Bavarian motion in the representative organ for the *Laender* on Federal level in Deutscher Bundesrat 1997)

38.4 The Intersection of SEA and Landscape Planning

SEA and landscape planning have four aspects in common: procedures, stages, contents concerning environmental factors, and methods.

38.4.1 Procedures

To investigate potential procedural overlaps between SEA and landscape planning, we must briefly describe the German Planning System, focusing on the relationship between spatial/zoning planning and landscape planning. Fig. 38.1 shows that spatial and zoning planning in Germany consist of a hierarchical system with four levels: the *Laender*, the regional, and the municipal levels as well as the level that covers parts of a municipality. Federal Government has abandoned formal planning and now provides only an informal orientation framework (BMBau 1993). In most of the *Laender*, landscape planning has been implemented on the same four levels. Consequently, spatial/zoning and landscape planning are usually drawn up on the same planning levels, often by the same authorities and in close relationship to each other (cf. Fig. 38.1, for more information cf. Kiemstedt et al. 1998; Beckmann 2001).

Landscape planning with blanket coverage and coincidence at all levels of spatial and zoning planning provides for the acquisition and integration of the interests of nature conservation and landscape management required by SEA in a tiered way. Thus, crucial modules of the environmental report are likely to already exist through landscape planning. Data acquisition is unlikely to be a prominent problem for SEA, because most required data will have been collected by landscape planning anyway.

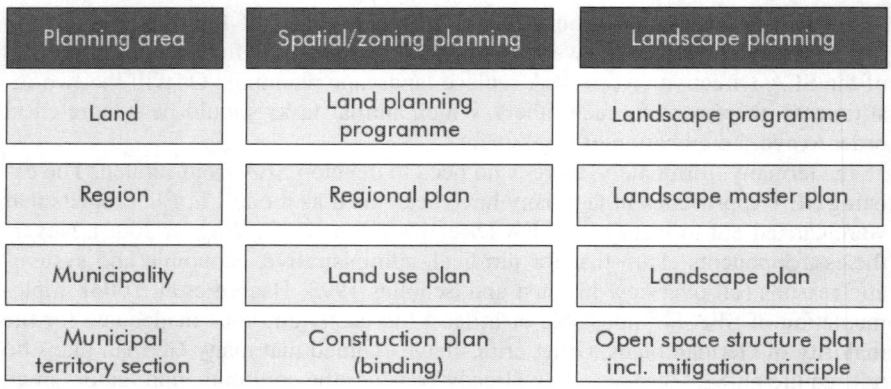

Fig. 38.1. The hierarchical systems of spatial/zoning and landscape planning in Germany (changed from Kiemstedt et al. 1998 p.12)

SEA will have the task to document that and how the integration is achieved. Procedural relations to SEA for sectoral plans are less strong – due to a lack in coordination of planning periods and levels. But sectoral plan SEA should consult landscape planning for basic data and evaluation.

38.4.2 Stages

The environmental report, as it is laid down in Art. 5 and Annex 1 of the SEA Directive, is at the heart of the SEA because the contents are compiled here. The right side of Fig. 38.2 shows the sequential steps of an environmental report according to Art. 5 and Annex 1. Since the order of the indents in Annex 1 clearly does not introduce a sequence, the sequence of the steps in Fig. 38.2 has been produced to facilitate the comparison between SEA and landscape planning. Because some of the terms used in the SEA Directive are not common in German methodology and planning practice, they had to be interpreted and transposed into conventional planning terms in a second step. This has also been done to better elucidate the intersections.

Concerning the relation of spatial/zoning planning and landscape planning, the amended Federal Nature Conservation Act sets out in Art. 14 para 1: "[...] Such plans shall contain information about:

1. the existing and anticipated status of nature and landscapes
2. the concretised aims and principles of nature conservation and landscape management
3. an assessment of the existing and anticipated status of nature and landscapes on the basis of these aims and principles, including any resultant conflicts
4. the requirements and measures a) to avoid, reduce or eliminate impairments to nature and landscapes [...]

When drafting landscape plans, allowance shall be made for their usability in regional plans and zoning plans." It then adds in Art. 14 para 2: "Planning and administrative procedures must consider the content of landscape plans. In particular, the content of landscape plans must be consulted when evaluating environmental compatibility [...]. Where a decision does not make allowance for the contents of landscape plans, justification must be given." The cited legal basis for future landscape planning in Germany combined with existing guidelines of the *Laender* for recent landscape plans, helps derive steps for drafting a landscape plan which have relevance for SEA, esp. for the environmental report (Fig. 38.2, left side). The comparison clearly demonstrates that future landscape planning can support most of the steps of the environmental report. The amended act has not redefined landscape planning from scratch instead it was based upon previously implemented approaches. Therefore, it seemed to be useful to analyse good landscape planning practice in the *Laender*. This analysis shows that prediction of conflicts in a landscape plan and assessment of environmental impacts of a zoning plan in a way we would expect future SEA to assess it, is good practice of landscape planning in most of the *Laender*, as illustrated by examples from Rhineland-Palatinate (Fig. 38.3), Lower Saxony (Fig. 38.4) or Saxony (cf. Chap. 39).

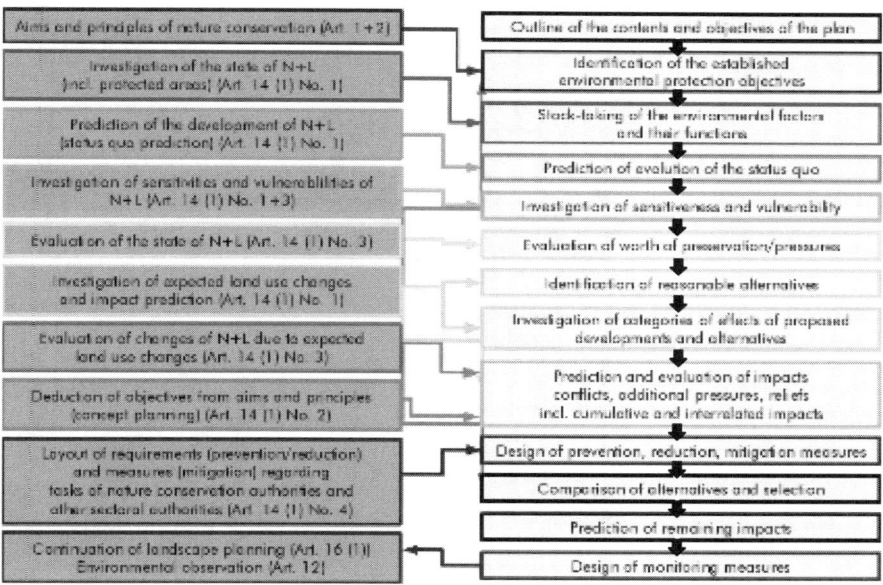

Fig. 38.2. Sequential steps of an environmental report in conventional planning terms (right side) and contributing steps of landscape planning (left side)

Nature and landscapes appraisal of development proposals

Acceptable structural completion

Just acceptable proposal, open space structure arrangements for construction plan

Questionable proposal, environmental compatibilty to be proven in construction plan

Inacceptable proposal, abandonment or alternative site required

Restriction of development required

Green structures within existing built up areas particularly required

Nature and landscapes appraisal of transport proposals

Just acceptable proposal, landscape planning accompanying plan required

Questionable proposal, environmental impact assessment required

Inacceptable proposal, recommendation to abandon

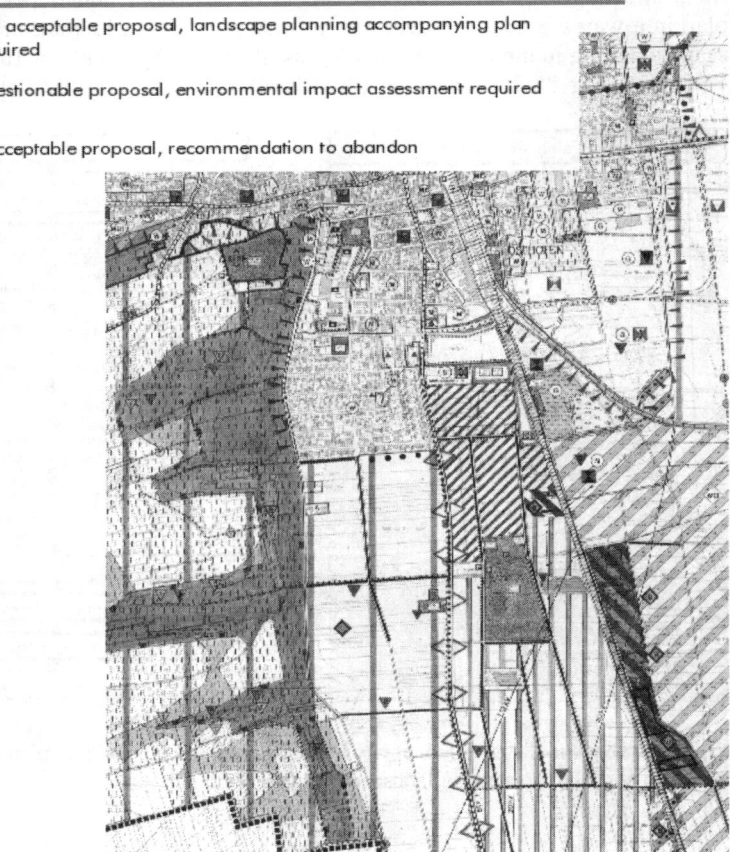

Fig. 38.3. Appraisal of conflicts in a landscape plan in Rhineland-Palatinate (taken from Miess and Miess 1992, legend translated)

New land for housing, mixed use, industry and a special site for a centre for waste treatment (proposals for developments taken from the land use plan, City of Oldenburg, Urban Planning Office)

 Proposals for development with severe conflicts between land use plan and landscape plan

 Proposals for development with conflicts between land use plan and landscape plan

 Proposals for development with limited conflicts between land use plan and landscape plan*

* At least basic conflicts will result that do not regularly arise before the development

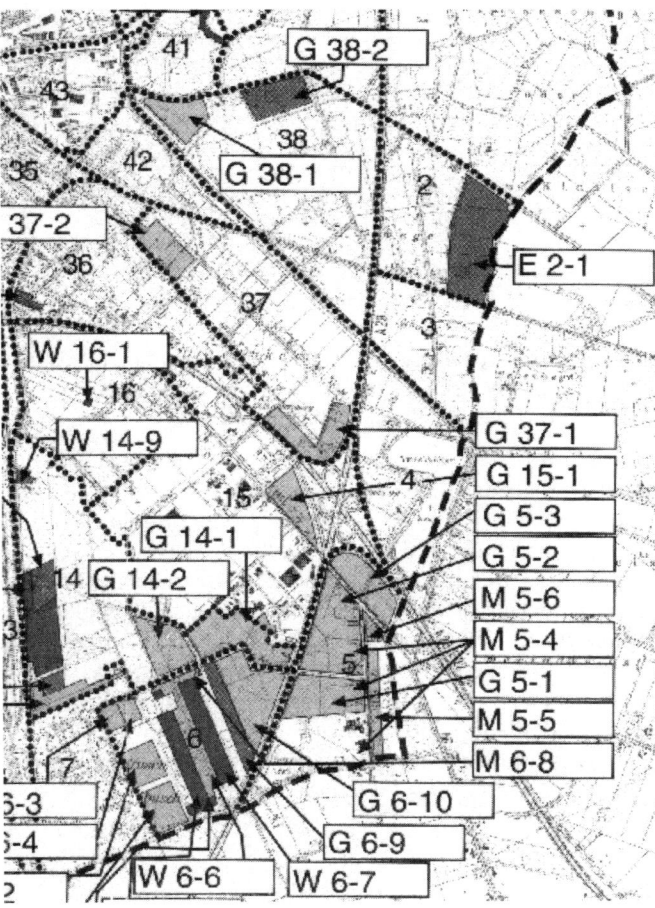

Fig. 38.4. Prediction of conflicts in a landscape plan in Lower Saxony (taken from Stadt Oldenburg 1996, legend translated)

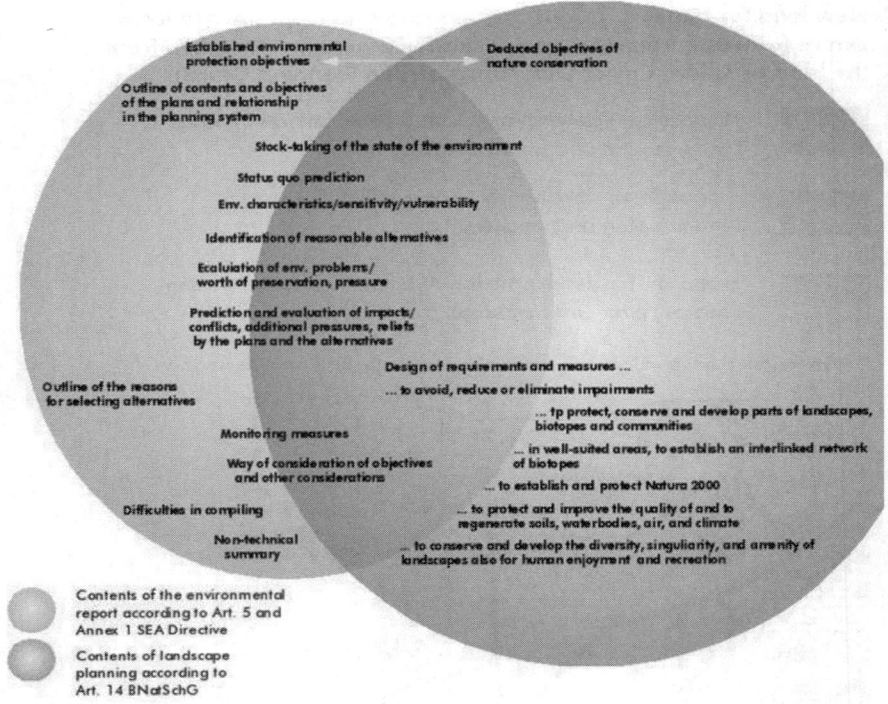

Fig. 38.5. Intersection of the contents of the environmental report according to Art. 5 and Annex 1 of the SEA Directive and contents of landscape planning according to Art. 14 Federal Nature Conservation Law (from Scholles et al. 2003, translated)

Consequently, besides the procedural intersections, considerable intersections of content can be ascertained (cf. Fig. 38.5, the figure shows the principle of the intersection of the content, it is not supposed to indicate the dimension of this intersection). Landscape planning is able to contribute to central contents of the environmental report. However, present landscape planning cannot contribute to every step of the environmental report. Vice versa, not all contents of landscape planning can be used for the environmental report. Consequently, no SEA, however qualified, will be able to replace the function of landscape planning because the function and brief of an assessment instrument do not coincide with those of a conceptional and planning instrument. However, the two instruments have a great deal in common, which forms a good basis for good practice.

38.4.3 Environmental Factors

Fig. 38.6 shows the intersection of environmental factors mentioned in Annex 1 of the SEA Directive and their interrelationships, overlaid with the natural factors landscape planning has to deal with as laid down in Art. 1 and 2 of the Federal Nature Conservation Act.

The natural factors form a subset of the environmental factors as defined in Annex 1 which means SEA must include all natural factors. Landscape planning does not deal with human health and population except for recreation, it usually covers selected issues of chemical loads of air, water and soil or as a rather general survey. It does not address those parts of the cultural heritage and those material assets that are not related to landscapes. Therefore, it can be said that landscape planning provides evidence for all the environmental factors in accordance with its legal brief.

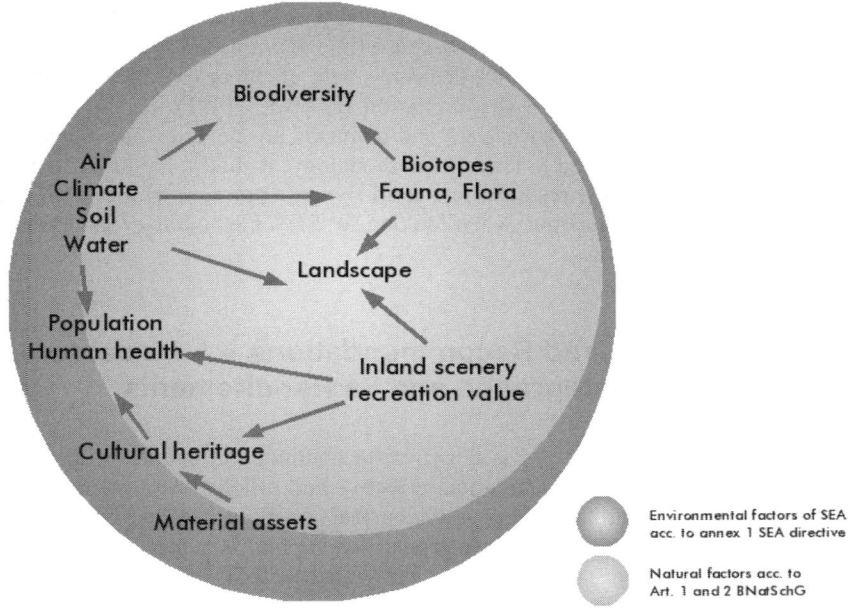

Fig. 38.6. Environmental vs. natural factors (from Scholles et al. 2003, translated)

Landscape planning is the only precautional instrument that addresses the interrelationships between factors and cumulative impacts in a spatial context (Scholles et al. 2003). This, however, is done in a different extent and based upon the aims and principles of nature conservation and landscape management. Further information from authorities competent for sectoral environmental laws is indispensable. Since landscape planning does not treat all factors, shifts of impairments are possible in theory. Thus, an SEA of landscape planning is conceivable.

38.4.4 Methods

The interception between landscape planning and SEA with respect to the applied methods is further elaborated by Jessel in this book: Landscape planning methodology can support SEA methodology. Inventory and appraisal of ecological and social landscape functions (Haaren 2004; Langer et al. 1985), Spatial Sensitivity Assessment based on overlay technique (e.g. Schemel 1985; Scholles 2001a), Ecological Risk Analysis, an ordinally scaled derivate of the utility analysis (Bachfischer 1978; Scholles 1997, 2001b), Environmental Risk Assessment (Hoppenstedt 1999), or Ecological Balancing (Kanning 2001) are suitable methods for assessing plans; whereas the original utility analysis and cost benefit analysis have proven to be inappropriate. The former are good practice in Germany but hardly noticed in the international discussion because little is published in English. There is no lack of appropriate methods for the assessment of site or route alternatives in Germany instead a problem of dissemination (cf. Dom 1997). Straightforward methods to assess strategic alternatives and demands are nevertheless scarce. On the basis of the existing good practice in methodology, it should not be difficult to adapt to the new task. This result differs form Bunge's opinion, who did not take landscape planning into account, when looking for SEA methodology in Germany (see Chap. 7).

38.5 Conclusions and Recommendations – A Model for Procedural Integration and its Requirements

If procedural landscape planning is drawn up or continued in parallel with spatial and zoning planning, it will facilitate an effective and efficient SEA, which is integrated into both planning processes. Given that, as discussed by Federal Government, a zoning plan (i.e. land use or construction plan) is subject to an SEA as well as a landscape plan and given that the landscape plan makes considerable contributions to the environmental report for the zoning plan, then the procedure can be integrated as shown in the sketch in Fig. 38.7. The stages: scoping, consultation of authorities and the public, information of authorities and the public and especially monitoring can be integrated for both planning processes. In drafting the landscape plan and accompanying the drafting of the zoning plan, the landscape planners can provide for the draft of the environmental report both for the zoning plan and the landscape plan. Thus, we are able to use existing good practice and, therefore, handle SEA in an efficient way. If landscape planning itself should become subject to SEA – as proposed by the Federal Ministry for the Environment – the procedure could be very similar (cf. Fig. 38.7).

In order to be able to benefit from this model, some prerequisites should be met by forthcoming legislation. The intersections with SEA that have been laid out in the Federal Nature Conservation Act should be shaped and supplemented in the 16 *Laender* acts. This means:

- Landscape planning is required on all levels of spatial and zoning planning.
- A narrow temporal co-ordination with continuation intervals of 10 to 15 years must be achieved to simultaneously draw up or continue a landscape plan and a spatial/zoning plan.
- Prediction and evaluation of probable impacts of plans and land use changes on the environment (not only nature and landscapes) should be set out explicitly in the legislation of every Land as a task of landscape planning.
- Consultation of the public must be incorporated in landscape planning (be that according to the SEA Directive or to the Public Participation Directive).
- The contents of landscape plans must be considered when evaluating environmental impacts, justification must be given if not following these contents.
- Monitoring of landscape plans must be implemented.

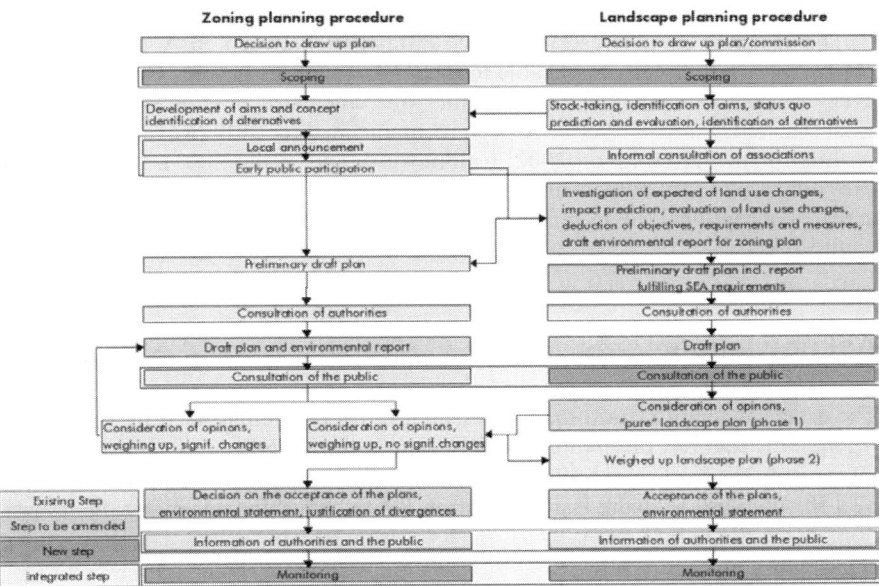

Fig. 38.7. A model for an integrated procedure of zoning and landscape planning with SEA

There are further legal recommendations to the forthcoming SEA Act and amendment of the Construction Act:

- The EIA/SEA Act should reference section 2 of the Federal Nature Conservation Act
- It should also authorize the integration of SEA requirements into landscape planning according to *Laender* nature conservation acts (to "toughen up" landscape planning for SEA).
- Synchronized planning procedures for both planning systems should be introduced in the Construction Act and the spatial planning acts of the *Laender*.

If *Laender* legislation were amended in the described way, it would form a suitable legal framework for an effective use of the potential of landscape planning and help introduce SEA in an efficient way in Germany. The described coordination and integration of the instruments would produce mutual benefits and enhance their acceptability in municipal, regional and sectoral planning institutions.

The findings of the research on co-ordination between SEA and landscape planning are not directly applicable to other EU-countries besides Germany. However, in some European countries similar instruments of environmental planning exist (Haaren et al. 2000) or have been recommended recently (Royal Commission on Environmental Pollution 2002; Cullingworth and Nadin 2001). In these countries our results may help to find specific approaches to a systematic integration of active, conceptional environmental planning and the reactive assessment instrument SEA.

Acknowledgements

The article is mainly based upon a research project commissioned by the Federal Agency for Nature Conservation with resources from the Federal Ministry for Environment, Nature Conservation and Nuclear Safety under UFOPLAN FKZ 802 82 130. We wish to thank Bartlett Warren-Kretzschmar for reviewing the translation.

References[1]

Bachfischer R (1978) Die ökologische Risikoanalyse (The ecological risk analysis), Dissertation, TU München

Beckmann P, Fürst D, Scholles F (2001) Das System der räumlichen Planung in Deutschland (The spatial planning system in Germany). In: Fürst D, Scholles F (eds) Handbuch Theorien + Methoden der Raum- und Umweltplanung, Handbücher zum Umweltschutz 4. Dortmund, pp 36-53

---

[1] Note that legislation mentioned in the chapter is listed at the end of the handbook in the consolidated list of legislation (Appendix 2).

Bundesministerium für Raumordnung, Bauwesen und Städtebau (ed) (1993) Raumordnungspolitischer Orientierungsrahmen: Leitbilder für die räumliche Entwicklung der Bundesrepublik Deutschland (Strategies for the spatial development of the Federal Republic of Germany). Bonn-Bad Godesberg

Cullingworth B, Nadin V (2001) Town and Country Planning in the UK. Publisher, London

Deutscher Bundesrat (1997) Beschluss des Bundesrates Vorschlag für eine Richtlinie des Rates über die Prüfung der Umweltauswirkungen bestimmter Pläne und Programme (Bundesrat decision proposal for a directive on the assessment of the effects of certain plans and programmes on the environment). Bundesratsdrucksache 7093/ 97, Berlin

Dom A (1997) SEA Developments in the European Union. In: UVP-Förderverein (ed) UVP in der Bundesverkehrswegeplanung. Die Bedeutung der Plan-/Programm-UVP zur Sicherung einer umwelt- und sozialverträglichen Mobilität. UVP-Spezial 14:62-66

Fürst D, Scholles F (1998) Plan-UVP und Verfahrensmanagement am Beispiel der Regionalplanung (Plan-EIA and management by procedure. The example of regional planning). Internet address:
http://www.laum.uni-hannover.de/ilr/publ/sup/sup.html, last access 08.06.2004

Haaren C v (ed.) (2004) Landschaftsplanung (Landscape planning). Publisher, Stuttgart

Haaren C v, Kügelgen B v, Waaren-Kretzschmar B (2000) Landscape Planning in Europe – Conference report; printed by the Ministry of Enviroment Lower Saxony

Haaren C v, Scholles F, Ott S, Myrzik A, Wulfert K (2003) Strategische Umweltprüfung und Landschaftsplanung (Strategic environmental assessment and landscape planning). Forschungsprojekt des Bundesamts für Naturschutz, Hannover

Hoppenstedt A (1999) Die Umweltrisikoeinschätzung (URE) von Straßenbauprojekten im BVWP (1992) (The environmental risk evaluation of road projects in the FTIP (1992). In: Buchwald K, Engelhardt W (eds) Verkehr und Umwelt. Umweltbeiträge zur Verkehrsplanung, Umweltschutz - Grundlagen und Praxis 16/II. Bonn, pp 179-191

Kanning H (2001) Umweltbilanzen - Instrumente einer zukunftsfähigen Regionalplanung? (Environmental balances – instruments for a sustainable regional planning?). UVP-Spezial 17

Kiemstedt H, Haaren von C, Mönnecke M, Ott S (1998) Landscape Planning - Contents and Procedures, edited by the Federal Ministry of the Environment, Nature Conservation, and Nuclear Safety, Bonn

Kiemstedt H, Wirz S and Ahlswede H (1990) Gutachten "Effektivierung der Landschaftsplanung" (Study "Making landscape planning more effictive"), Berlin (UBA-Texte 11/90)

Langer H, Haaren C v, Hoppenstedt A (1985) Ökologische Landschaftsfunktionen als Planungsgrundlage - Ein Verfahrensansatz zur räumlichen Erfassung (Ecological landscape functions as a basis for planning – a procedural approach for their spatial identification). Landschaft + Stadt 17(1):1-9

Miess B, Miess M (1992) Landschaftsplanung Osthofen. Beitrag zum Flächennutzungsplan der Stadt Osthofen, Landkreis Alzey-Worms (Landscape planning Osthofen. Contribution to the land use plan of the city of Osthofen, county of Alzey-Worms), Bonn

Royal Commission in Environmental Pollution (2002) Environmental Planning report. Internet address: http://www.rcep.org.uk/epreport.htm, last accessed 27.09.2004

Schemel H-J (1985) Die Umweltverträglichkeitsprüfung (UVP) von Großprojekten (The environmental impact assessment (EIA) of large-scale projects), Beiträge zur Umweltgestaltung A 9, Regensburg

Schmidt-Eichstaedt G (2004) Die Richtlinie zur strategischen Umweltprüfung aus kommunaler Sicht (The Directive on strategic environmental assessment from the local point of view). In: Hendler R, Marburger P, Reinhardt M, Schröder M (eds) Die strategische Umweltprüfung (sog. Plan-UVP) als neues Instrument des Umweltrechts. Erich Schmidt, Berlin, pp 81-98

Scholles F (1997) Abschätzen, Einschätzen und Bewerten in der UVP. Weiterentwicklung der Ökologischen Risikoanalyse vor dem Hintergrund der neueren Rechtslage und des Einsatzes rechnergestützter Werkzeuge (Estimating, assessing and evaluating in German EIA. Development of ecological risk analysis on the background of the recent laws and the application of software tools). UVP-Spezial 13

Scholles F (2001a) Die Raumempfindlichkeitsuntersuchung (Spatial sensitiveness analysis). In: Fürst D, Scholles F (eds) Handbuch Theorien + Methoden der Raum- und Umweltplanung, Handbücher zum Umweltschutz 4. Dortmund, pp 247-252

Scholles F (2001b) Die Ökologische Risikoanalyse und ihre Weiterentwicklung (The ecological risk analysis and its developments). In: Fürst D, Scholles F (eds) Handbuch Theorien + Methoden der Raum- und Umweltplanung, Handbücher zum Umweltschutz 4. Dortmund, pp 252-267

Scholles F, Haaren von C, Myrzik A, Ott S, Wilke T, Winkelbrandt A, Wulfert K (2003) Strategische Umweltprüfung und Landschaftsplanung (Strategic environmental assessment and landscape planning). UVP-report 17(2):76-82

Stadt Oldenburg (1996) Landschaftsplan Oldenburg (Landscape plan Oldenburg). Oldenburg

Wagner D (2000) Ansätze für die Umweltverträglichkeitsprüfung (UVP) für Programme und Pläne in der Bundesrepublik Deutschland (Approaches of environmental impact assessment (EIA) for programmes and plans in the Federal Republic of Germany). Raumforschung und Raumordnung 58

39 Urban Planning and SEA

Markus Reinke

Leibniz Institute of Ecological and Regional Development (IOER), Dresden, Germany

39.1 Introduction

This chapter describes the aims and contents of urban planning and landscape planning in Germany which have a strong link to SEA. The strong linkage results from the objective of landscape planning to create an ecological basis for urban planning. This link ensures an improved consideration of the requirements of environmental protection in addition to SEA, which from now on has to be carried out for all urban plans as required by the SEA Directive. Due to the existing obligation to ensure a consideration of environmental aspects (gathered during the combination of urban planning and landscape planning), in Germany already a planning system with similar objectives to the ones in the SEA Directive exists.

The chapter gives a brief overview of the German urban planning system (Sect. 39.2), the liabilities to consider environmental protection in land use planning according to the German Building Code (Sect. 39.3) and contents and quality of German landscape planning as an ecological basis for urban planning and SEA (Sect. 39.4). The chapter concludes with a presentation of new liabilities derived from the SEA Directive in comparison to already existing contents of German landscape planning (Sect. 39.5). Further suggestions are made on how to accomplish new requirements of the SEA Directive such as the consideration of cumulative environmental effects (Sect. 39.6). The acceptance of new aspects of the SEA Directive by local authorities shall be presented finally in Sect. 39.7.

Implementing Strategic Environmental Assessment. Edited by Michael Schmidt, Elsa João and Eike Albrecht. © 2005 Springer-Verlag

39.2 Existing Liabilities due to the Federal Building Code

The German urban planning shall ensure a sustainable urban development by preparing urban area developments in communities for the next 15 years. This is divided into two levels (see Fig. 39.1). At the level of preparatory land use planning (scale 1:10 000) decisions are made on the dimension and location of e.g. future development zones or by-passes. At a more specific level, the so called legally binding land use planning, preliminary planning decisions, e.g. appointments of the constructional tightness, are improved or the number of storeys is determined (Braam 1999).

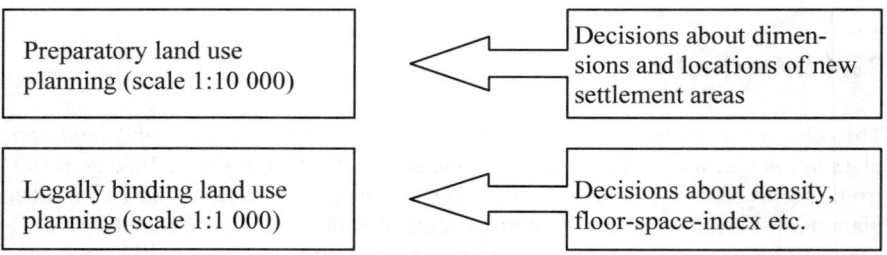

Fig. 39.1. Levels of German urban planning

SEA has to be carried out for preparatory land use planning as well as for legally binding land use planning (Building Code after amended by the Act to Adapt the Building Code to EU Directives). Especially, at the level of preparatory land use planning, SEA is of a particular importance. At this level, dimensions of new construction areas, the relation of internal and external development and alternative sites are estimated. These present basic decisions for land use consumption and environmental compatibility of land use.

It had been mandatory for German urban planning to consider environmentally relevant issues far before the SEA Directive came into force. Art. 1 of the Building Code describes the objectives of land-use plans as follows:

> "Land-use plans shall […] contribute to securing a more human environment and to protecting and developing the basic conditions for natural life. In the preparation of land use plans, attention must be paid to the following:
> […] the requirements of environmental protection pursuant to section 1a and through the use of renewable energy sources, nature protection and the preservation of the countryside [landscape management], with particular reference to the ecological balance in nature, water, air, soil […] and climate".

In addition to the purpose of nature conservation, a sustainable urban development shall ensure an economic and socially equitable utilisation of land (Battis et al. 2002). Therefore the aims of nature conservation are part of a catalogue of requirements which has to be thoroughly weighed by the local authority during the creation of urban land use plans.

Another important statement in Art. 1a of the Building Code ("consideration of environmental concerns") with relevance for urban planning is:

> "Land shall be used sparingly and with due consideration; the extent to which it is sealed by development shall be kept to a minimum.
>
> In the course of the weighing process [...] the following matters shall be considered:
>
> 1. the content of landscape and other plans, particularly those produced under water, waste and pollution control legislation".

Landscape planning, as laid down in the Federal Nature Conservation Act, is given relevance in the Building Code by considering environmental interests in the context of urban planning. Additionally, the application of SEA has the potential to contribute to a high level of environmental protection during the formation of plans and programmes.

39.3 Relation of Landscape Planning and Urban Planning

German landscape planning is a special planning for nature conservation. Landscape planning includes a description and assessment of natural resources: plants, animals, visual quality of landscape, water, soil and climate. Furthermore, it delivers a concept for protection, maintenance and development of these natural resources. Landscape planning can be an essential information basis for urban planning in order to fulfill commitments to the Building Code through consideration of environmental interests. Landscape planning also has the function of providing a measure for estimating the environmental tolerance (compatibility) for urban planning (Art. 14 para 2 of the Federal Nature Conservation Act). Considering the contents of landscape planning, nature conservation makes up a considerable part of a preparatory land use plan. In some conservation legislations of the German *laender* landscape planning is classified as an "ecological basis for urban planning" (see for instance Art. 7 of the Saxony nature protection legislation).

In the past, landscape planning contributed successfully to an improved consideration of environmental aspects in urban planning. A research study in 100 Saxon municipalities from 1993 to 2002 showed that the consideration of environmental aspects in the preparatory land use plans was improved, in the case a qualified landscape planning could be used (Fig. 39.2). Thus 60 % of the investigated preparatory land use maps were of satisfying or good quality from an environmental perspective with regard to the choice of new settlement areas (Reinke 2001, 2003).

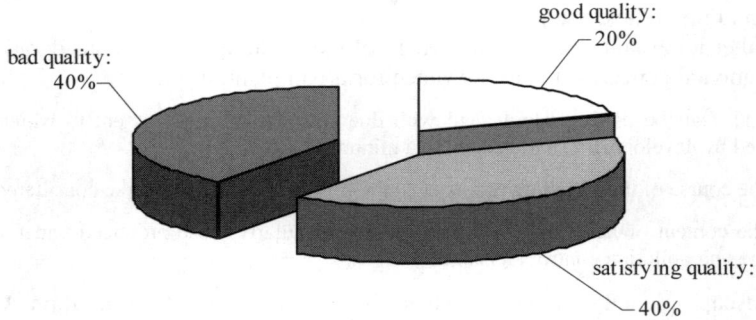

Fig. 39.2. Quality of preparatory land use plans after consideration of environmental aspects during the selection of new settlement areas (with Landscape plans as a basis; Reinke 2001)

These selected new settlement areas were mostly areas which were not of value for the conservation of natural resources, although the environmental aspects in the choice of new settlement areas (with landscape plans as a basis; Reinke 2001) natural resources soil, water and climate had been considered tendentiously less valuable than plant or animals and the visual quality of landscape. In contrast to the advantages which result from high-qualified landscape plans for urban planning (strength) Fig. 39.3 shows the quality of preparatory land use plans under consideration of environmental importance by site decisions. This is influenced by missing, inadequate or strongly obsolete landscape plans. Opposite to the 60 % of satisfying and good preparatory land use plans improved by qualified landscape plans (Fig. 39.2), Fig. 39.3 shows 60 % of the preparatory land use plans being of bad quality.

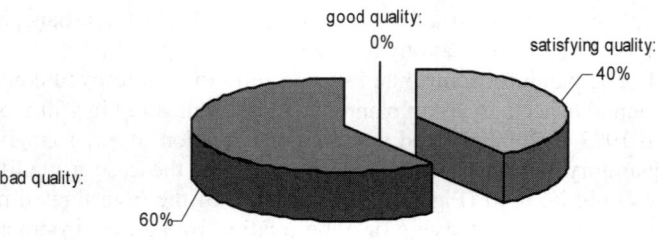

Fig. 39.3. Quality of Preparatory land use plans after consideration of environmental aspects during the selection of new settlement areas (without landscape plans as a basis; Reinke 2001)

It could be concluded, that qualified landscape plans result in an environmentally "safe" choice of sites for settlement enlargements. Landscape plans improve the consideration of environmental issues in urban planning but landscape planning does not evaluate preparatory land use plans in a comprehensive way. It does not assess the extension of new settlement areas or the ratio between internal and external developments. It rather designates areas which are of value for conservation and sensitive against impairments. Recent landscape plans often include an assessment of sites of new planned settlement areas but do not assess the extension of new settlement areas as mentioned above.

German landscape planning is an important basis for SEA, especially concerning the description of nature. But at present it does not cover all the contents required by the SEA Directive (Haaren et al. 2000). Landscape planning does not contain:

- data on "human health" or "culture and real assets"
- public consultations as defined in Art. 6 of the SEA Directive
- a far-reaching assessment of alternatives in urban planning as is demanded by the SEA guidelines of the EC (2003). Landscape planning contains a basic assessment of alternatives of sites of planned settlement areas but does not query the necessity of settlement enlargement or the settlement structure (often with emphasis on land consumption through one-family-houses).

39.4 Analysis of SEA Liabilities Exceeding Previous Requirements

Comparing the contents of the SEA Directive with the environmental statements of the federal Building Code, the following deductions could be made: The description and the assessment of the natural resources "plants / animals, visual quality of landscape / recovery, soil, water, climate / air, human health and culture / real assets" are necessary for SEA. These requirements are mainly achieved with data from German landscape planning and their consideration in urban planning. Only the aspects human health and culture / real assets are not dealt with in landscape planning. SEA requires an early "integration of environmental considerations into the preparation and adoption of plans and programmes" (Art. 1 of the SEA Directive), "clearly documented" (Art. 5 and 9 of the SEA Directive) and "extensive environmental assessment" (Annex I of the SEA Directive) of plans and programmes. In addition to the aspects on "human health" and "culture and real assets" the timing, the complexity and the documentation of an environmental assessment are the essential differences between SEA and the current urban planning procedure.

Table 39.1. Contents of landscape planning in comparison to SEA

Landscape planning	SEA
Plants and animals	Plants and animals
Countryside	Countryside
Soil	Soil
Water	Water
Climate	Climate
	Human health
	Culture and real assets
	Public consultations
Assessment of different sites	*More extensive alternative examination*

"Timing of the assessment": Until present environmental assessment took place rather late in the German urban planning process. An environmental assessment and a comparison with alternative sites did not take place *before* well-defined suggestions of sites for settlement enlargement were made. In this way, former concepts, first approaches and targets were not examined from an environmental perspective *at all*. Now the SEA Directive requires an early assessment of environmental effects, first approaches and objectives will have to be assessed much earlier. Table 39.2 shows the established steps of the German urban planning procedure in comparison with these possible steps of SEA.

Table 39.2. Established steps in the German urban planning compared to possible steps of SEA

Sequences of work in preparatory land-use planning	Sequences of work in SEA
definition of aims to settlement development	Scoping: designation of framework of research
adoption of preparation; public announcement by town council	
letting of contract	added research: provision of environmental data (natural resources as the SEA Directive demands)
analysis of data; acquirement of development forecast (e.g. population growth), assessment of aims	first environmental assessment: assessment of impairments, discussion of strategic alternatives (forecast, aims)
completion of preliminary design	first documentation: reasons for assessed alternatives / kind of assessment / consideration

Table 39.1. (cont.)

advanced public participation	Consultations with public and non-profit-organisations on SEA
analysis of objections and support representations; preliminary design with explanatory report	second environmental assessment: strategic alternatives and assessment of site alternatives*
adoption of preliminary design (town council)	second documentation: reasons for assessed alternatives/ kind of assessment/ consideration*
agreement on planning proposals with neighbouring communities; participation of other administration	Control 1: nature conservation administration assesses the weight of environmental consideration
public inspection of the design; public participation; following assessment of objections and supporting representations, if necessary revised version of design	Consultations with public and non-profit-organisations about SEA
completion of design for Preparatory land-use plan, acceptance through town council	Completion of the environmental report; contents: 1. assessment of environmental effects on the plan, description and arguments / reasons for assessed alternatives 2. documentation of considered objections and support representation on environment matters 3. reasons/arguments for decisions 4. designation/statement of environmental monitoring
assessment and approval of preparatory land-use plan through planning approval administration (social and economic considerations...)	Control 2: assessment if the SEA liabilities are fulfilled

*Usually the point of time for environmental assessment in landscape planning

Table 39.2 illustrates that the assessment of environmental effects carried out earlier in landscape planning was hardly extended in its contents. Thus additional steps or changes have to recognise the chronological running of the environmental assessment procedure. SEA enforces a change from a single assessment to a continuous environmental assessment. SEA shall accompany the entire planning process of preparatory land use planning in order to enforce better documentation of environmental considerations (Haaren et al. 2000).

39.5 Suggestions for Contents of SEA and the Assessment of Alternatives

After an inquiry by the Federal Agency for Nature Conservation landscape planning is a significant ecological basis for urban planning (BfN 1998); first and foremost in its relation to a sustainable selection of sites for settlement enlargements. About 60 % of interviewed landscape planning and urban planning experts confirm a complete or satisfying environmental consideration during the selection of sites for settlement enlargement in cases where qualified landscape plans exist. On the contrary 70 % of these experts estimate that environmental effects caused by the dimension of settlement enlargement were not assessed (BfN 2004).

Agreement was reached that environmental effects caused by the intensity of land consumption and the consumption of the resource "area/landscape" for settlements are not sufficiently assessed up to date. In comparison, environmental effects of alternative sites are generally better assessed. Consequently, the following central question arises: Is the environmental assessment of alternative sites sufficient enough to satisfy the requirements of SEA according to the SEA Directive?

To be able to answer this question, the guidance for the implementation of the SEA Directive of the EC is used as a basis for advice (EC 2003).

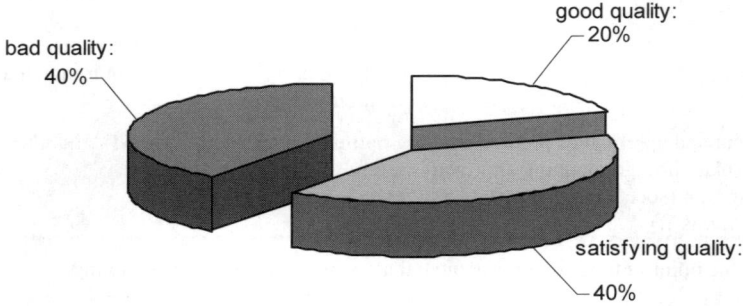

Fig. 39.4. Assessment of sites in preparatory land use planning (BfN 2004)

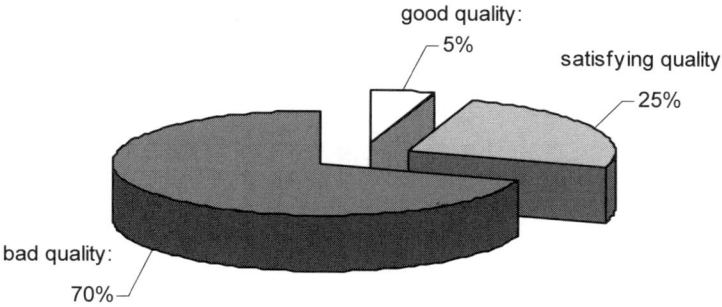

Fig. 39.5. Assessment of the dimension of urbanisation in preparatory land-use planning (BfN 2004)

One objective of this guidance is to encourage a uniform conversion of the SEA Directive in all EU Member States. The following statements are included:

"In the text of the directive it is not explained what a "rational alternative of a plan or programme" is. The most important criteria for the agreement about possible rational alternatives should be the aims and the geographical application area of a plan or programme. […] In practice mostly different alternatives are checked within one plan […].

Possible alternatives of objectives in preparatory land use plans or regional plan reports are for example: areas which are selected for special activities or purposes used in a different way so that other areas will be selected for the original purpose [consequently liability to an assessment of alternative sites!].

For plans and programs which are valid for a long time/period […] different scenarios of development are a possibility to assess the alternatives and their environmental effects." (EC 2003 p.30)

The liability to assess "*secondary, cumulative, synergistic and some other effects*" on the environment (Annex 1) and the detection of alternative scenarios suggest that in SEA of preparatory land use planning alternative sites for settlements and also the density of buildings must be assessed (see Fig. 39.6).

With the consideration of one specific site *only* – as it is common in urban planning and landscape planning – cumulative environmental effects can not be described or assessed. Furthermore, SEA has to make an enquiry if connected habitats or water regime economy of an area will be considerably impaired due to a high density of settlement areas, infrastructure etc. (Cooper and Sheate 2002).

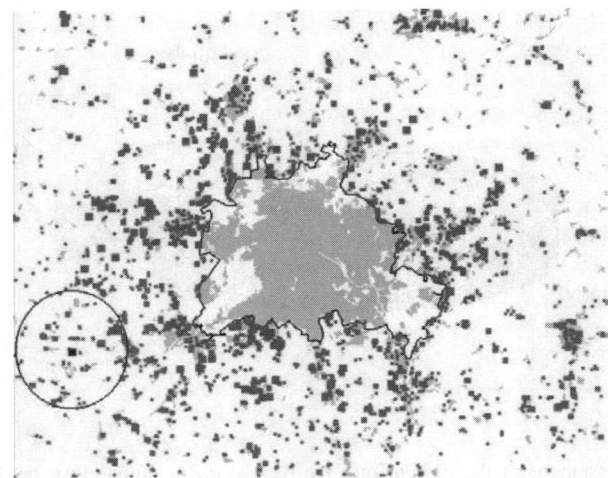

Fig. 39.6. A very high density of development zones in the surrounding countryside of Berlin (changed from Siedentop 2002)

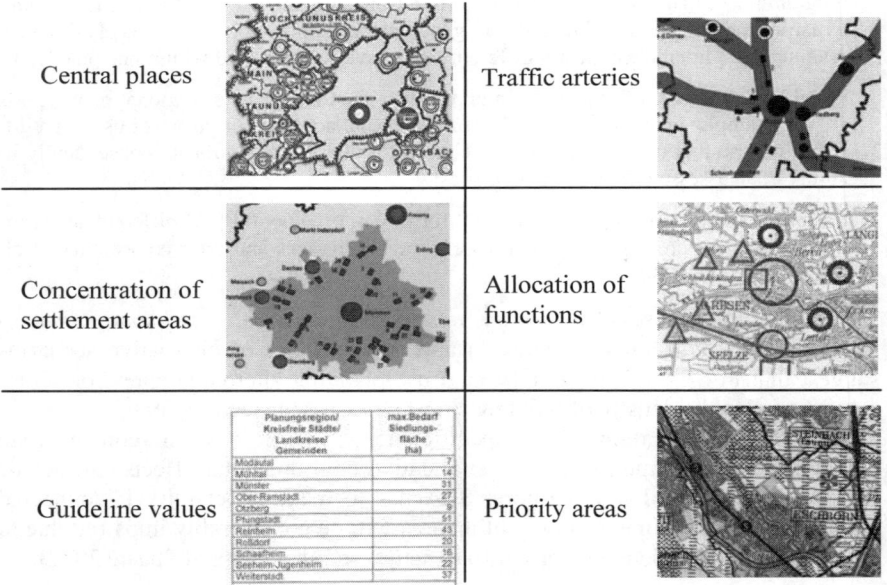

Fig. 39.7. Assessment of settlement structure, e.g. traffic arteries

Therefore SEA for urban planning should assess environmental effects of selected land uses and strategic alternatives at early planning phase (Bonde and Cherp 2000). The following illustration shows some strategic assessment aspects of regional planning. For instance it has to be examined which environmental effects settlement areas or traffic arteries have (e.g. effects on connected habitats, effects of "over-development" on the land, commuting and its emissions) or which environmental effects function-allocations in specified areas will have (e.g. raw material exploitation, recreation area). he challenge is that it is scarcely possible to assess these aspects with detailed environmental data of a specific site. In fact, more abstract assessments (through urban sprawl/land-bisection caused by traffic amounts, population density and others) are needed (Lell 2003). Presently a gap exists between the detailed information of a site (which can be used for the environmental assessment of different sites) and the need for more abstract indicators to assess the environmental effects, e.g. the settlement structure or the traffic arteries. trategic assessments are of value for preparatory land use planning. From an environmental perspective it can be estimated:

- if the potentials in cities (infill or "brownfield" development, areas of dereliction) are used with the aim to save the external landscape
- if the selected dimension of settlement enlargement is adequate and sustainable
- if highly concentrated urban structural shapes can contribute to a sparingly settlement development

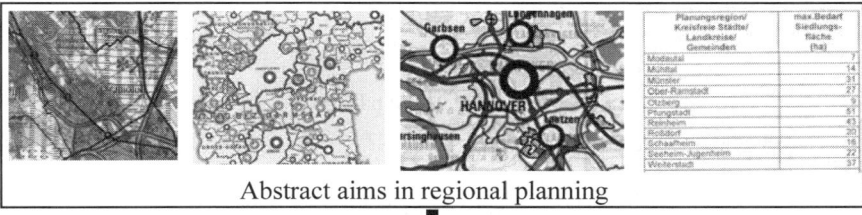

Abstract aims in regional planning

Environmental assessment – how to do it?

Detailed environmental information about the site

Fig. 39.8. Conflict between abstract assessment of impairments and detailed information of the site

To what extent and within which scope strategic assessment can describe the environmental effects for example of a preparatory land use plan will depend on qualified contents (allocation of simple methods for indicator sets for this assessment) and on the acceptance of the local authority. The local authorities (as the assessing instance) choose the rational alternatives in regard to the criteria if they are intelligible for their political committees. Moreover the availability of practical methods for an assessment influences their capability to decide.

39.6 The Question of Acceptability

The preceding reflections lead to the question, if the new instrument of SEA will be accepted by the community, planning offices and urban panning authorities or not. At the moment it seems that in general scepticism in Germany is increasing with the introduction of new planning and assessment instruments such as SEA. An inquiry by the commission of the German Federal Agency for Nature Conservation (BfN) shows a more detailed situation.

Two-thirds of the communities and top communal associations estimate that there is a need to reduce the utilisation of open space for settlement areas in a quantitative (mass reduction) and in a qualitative dimension (better consideration of environmental aspects in the location of settlement areas) with the help of SEA. A similar opinion exists among most planning offices and urban planning authorities which work in the field of urban planning.

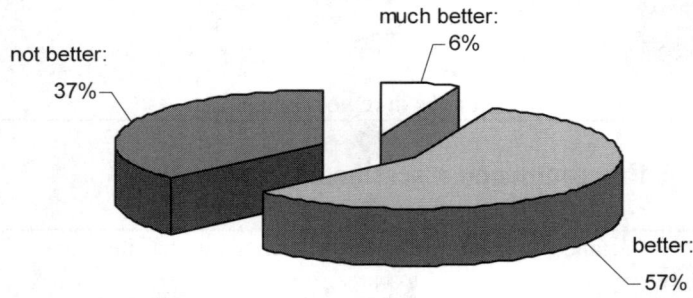

Fig. 39.9. Parts of communities which estimate a chance for a better environmental precaution out of the SEA (in regard to urbanisation; BfN 2004)

The inquiry made by the German Federal Agency for Nature Conservation (BfN 2004) emphasises that assessment of alternatives including the dimension of planned settlement areas will be accepted by the greater majority of the local authorities. Future acceptance will mainly depend on how extra effort required for carrying out SEA in urban planning could be limited through the utilisation of approved and established procedural approaches by the federal landscape planning (assessment of natural resources).

39.7 Conclusions and Recommendations

In the past, German landscape planning qualified urban planning in considering requirements of environmental protection. With the SEA Directive having a similar goal – to ensure better consideration of environmental aspects in the planning process – it became possible to adopt contents and methods for SEA from landscape planning. Furthermore SEA gives the planners the chance to avoid formerly made mistakes, which led to poor examples of landscape plans, e.g. caused by not sufficient data collections of environmental resources.

Nevertheless the SEA Directive requires a more far-reaching assessment than the one provided by German landscape planning at present. Especially the two aspects of "human health" and "culture and real assets" are not dealt with in landscape planning.

Due to these reasons, there is a need to develop new methods for specific aspects of environmental assessment and to combine these with the approved methods for other aspects from German landscape planning. As a result there is a need to develop a widely accepted and effective approach for SEA.

References[1]

Battis U, Krautzberger M, Löhr RP (2002) Baugesetzbuch (Building Code Commentary), 8th edn. C.H. Beck, München

Bonde J, Cherp A (2000) Quality review package for strategic environmental assessments of land-use plans. Impact Assessment and Project Appraisal 18:99-110

Braam W (1999) Stadtplanung – Aufgaben, Planungsmethodik, Rechtsgrundlagen (Urban planning – tasks, planning methods, laws), 3rd edn. Werner-Verlag, Düsseldorf

BfN – Bundesamt für Naturschutz (1998) Berücksichtigung von Naturschutz und Landschaftspflege in der Flächennutzungsplanung – Ergebnisse aus dem F+E-Vorhaben 80806011 (Consideration of nature conservation and landscape care in the preparatory

---

[1] Note that legislation mentioned in the chapter is listed at the end of the handbook in the consolidated list of legislation (Appendix 2).

land-use planning – Results out of research-project 80806011). Angewandte Landschaftsökologie 17. Bonn-Bad Godesberg

BfN – Bundesamt für Naturschutz (2004) Flächeninanspruchnahme – naturschutzpolitische Strategien, Instrumente und Maßnahmen; Status Quo Analyse – Ergebnisse aus dem F+E-Vorhaben 80382010 (Area requirements – nature protection strategies, instruments and measures, status quo analysis - Results out of research-project 80382010). Bonn-Bad Godesberg

Cooper LM, Sheate WR (2002) Cumulative Effects Assessment – a review of UK Environmental Impact Statements. Environmental Impact Assessment Review 22(1):5-16

EC – Commission of the European Union (2003) Guidance on implementation of the Directive 2001/42/EC on the assessment of the effects of certain plans and programs on the environment. Brussels

Haaren C v., Hoppenstedt A, Scholles F, Werk K, Runge K, Winkelbrandt A (2000) Landschaftsplanung und Strategische Umweltprüfung (SUP) (Landscape planning and Strategic Environmental Assessment (SEA)). EIA-report 1:44-29

Lell O (2003) Strategische Umweltprüfung- Initialstatement (Strategic Environmental Assessment – initial statement). UVP-Report, UVP- Kongress 12.- 14.06.2002, Sonderheft (17):11-16

Reinke M (2001) Qualität der kommunalen Landschaftsplanung und ihre Berücksichtigung in der Flächennutzungsplanung im Freistaat Sachsen (Quality of landscape plans and their consideration in the preparatory land-use planning in the free state of Saxony). LOGOS-Verlag, Berlin

Reinke M (2003) Einfluss der Landschaftsplanung auf den Vollzug der Eingriffsregelung in der Flächennutzungsplanung – Eine Erhebung am Beispiel Sachsen (Influence of landscape planning on the implementation of the German mitigation act in the preparatory land-use planning – an inquiry in Saxony). Naturschutz und Landschaftsplanung 35(2):46-49

Siedentop S (2002) Kumulative Wirkungen in der Umweltverträglichkeitsprüfung (Consideration of cumulative effects in the environmental impact assessment) Dortmunder Beiträge zur Raumplanung 108

40 SEA in Transport Planning in Germany

Wolfgang Stein[1], Jürgen Gerlach[2] and Paul Tomlinson[3]

1 State Enterprise for Roads in the Federal State North Rhine-Westphalia, Germany
2 University of Wuppertal, Germany
3 Centre for Sustainability at TRL Limited, England

40.1 Introduction

The SEA Directive was adopted by the commission over three years ago. The members of the EU had three years to implement the Directive but because of this short time there is still a lack of experience in applying SEA in Europe overall. A large proportion of the existing methodology has been developed in the transport sector. In Germany several national and state level plans and programmes have been subject to a type of environmental assessment since the early nineties. This chapter provides an overview of the early development of SEA methodologies in German transport planning.

In Germany, road constructors were the first to develop a methodology for EIA in the eighties. This was done voluntarily before the German EIA Act had come into force (FGSV 1990) because the road-planning process was accelerated by EIA. Thanks to Project EIA, road-constructors could identify the main constraints to a new road project at an early stage and thus have ample time to address the problem. Road constructors believed that the consultations with authorities and the public would increase the acceptance of the proposed project.

With SEA things are similar. Many early experiences with SEA across Europe come from the transport sector (see Chap. 7; ECMT 2000 and 2004; EU 1994; Fischer 1999, 2000, 2002; Tomlinson 2002, 2003, 2004) In Germany there are several studies on the federal, state and regional level (e.g. EU 1994; FMT 2003; Stein 2000), but still none of them fulfil all requirements of the directive. These studies have been made in spite of having no national SEA Act. The German transport authorities and politicians know that they have to consider environmental issues in all their decisions, because the pressure imposed by the public and the Green Party in Parliament.

Implementing Strategic Environmental Assessment. Edited by Michael Schmidt, Elsa João and Eike Albrecht. © 2005 Springer-Verlag

In this chapter the results of the recent discussions between the experts in the German Research Association for Road and Transport (FGSV 2004) and within the German road authority are presented along with several recommendations based on references listed below and the experiences of the authors. The chapter follows the structure of the SEA process and includes screening (Sect. 40.2), scoping (Sect. 40.3), identifying and describing the environmental effects (Sect. 40.4), assessing alternatives (Sect. 40.5), decision making process (Sect. 40.6) and monitoring (Sect. 40.7). Consultation of authorities and the public are only mentioned generally as there is a lack of experience linked to environmental plannings on this topic in Germany (see Chap. 7).

40.2 Screening

The first question is: Which transport sector plans and programmes will need SEA in Germany (see Chap. 7)? At the national level, land-use plans concerning transport planning need SEA, while at a regional level, regional planning is also dealing with transport infrastructure. These plans are legally required to be subject to SEA. At the state level it is not yet clear for plans and programmes SEA will be carried out. However, out of the 16 German *Laender*, three states already have experiences in environmental assessment for their transport infrastructure plans: North Rhine-Westphalia, Baden-Wuerttemberg and Brandenburg. Finally, at the federal level The Federal Transport Infrastructure Plan (FTIP) will be subject to SEA. Special rules will be laid down by the Federal Minister for Transport (see Chap. 7).

It should be mentioned that there is a very unusual decision making process in traffic and transport planning in Germany. The FTIP for example has the task to choose between about 2 000 infrastructure measures – mostly new roads or railways. Even on the state level a transport plan has to decide about 500 projects. Generally the task of transport plans in Germany is to decide, which projects with which financial budget should be legally fixed. Determining the approved alignment for projects is done in the second step. The German transport plans are a novelty in Europe. Some other European countries start transport planning on the corridor or study area level, rather than from a bottom-up list of projects. As the methods described in this article are based on the German planning situation so they are not necessarily transferable to other European planning systems.

At the moment it seems that there will be some plans where an SEA will be obligatory, such as the federal transport plan at national level. But there will be others like the municipal transport plan, which are no legal plans and not captured by the Directive so that the SEA is not liable. However, the question is not "is it legally obligated to make an SEA?" but "is it clever to do that?" So it would be useful to focus on the benefits of SEA.

Germany has an increasing traffic volume, especially in North Rhine-Westphalia – over 2% per year over the last ten years resulting in increasing noise, emissions, fragmentation of landscape and so on despite improved mitigation measures. This background is important for the following benefits of SEA in transport planning:

- Application of the principles of sustainable development by enabling strategies to avoid and reduce environmental disadvantages of the transport system while increasing the benefits. This proactive approach can provide an inproved framework for project delivery minimising the need for reactive costly bolt-on mitigation measures.
- Enables consideration of cumulative effects and those not addressed by individual projects.
- Enhancing the quality of the planning process by more timely and relevant information about all the consequences of plans, programmes and projects. This may also save resources from poorly designed projects.
- Providing for a better informed public debate on environmental issues to potentially ease the burden during project delivery.

To realize these benefits the introduction of SEA to all transport plans in a hierarchical manner is recommended even where SEA is not mandatory. These benefits are not just seen by the environmental community as the economists within the UN ECE have also called for SEA to be implemented across Europe (Kiev SEA Protocol). When an economic body such as the UN ECE demands SEA, the advantages for the developers must be assumed to be greater than the burden.

The motives to establish the Directives for EIA and SEA were similar: Mainly the economists demanded a coherent environmental assessment all over Europe, not the stakeholders that care about the protection of the environment.

40.3 Scoping

40.3.1 What has to be done in SEA?

While the SEA Directive seeks to incorporate environmental issues into plan making, this is delivered through a series of procedural requirements a key one being the need to publish an Environmental Report. The contents of the Environmental Report are described in Appendix I of the SEA Directive with Box 40.1 providing a draft of a structure for an Environmental Report for the transport sector taken from FGSV 2004. This document also provides a draft for a product specification as a rule for steps which has to be done in an SEA. SEA must be integrated into the whole planning process rather than an add-on towards the end if it is to be effective and the procedural requirements of the Directive make this add-on role less likely.

While planning the early stages of the plan-making activities consideration should also be given to the requirements of SEA. As a result the SEA should help identify and assess:

- new projects,

- alternative projects and designs,
- phasing and priorities among projects,
- means to avoid, minimise, mitigate the adverse effects and maximize the beneficial consequences

The Environmental Report should also provide a basis for informed public debate, for the decision making process (see Sect. 40.6) and subsequent EIA.

Box 40.1. Proposal for a structure of an environmental report for an SEA in the transport sector (FGSV 2004, in parentheses requirements of the SEA Directive, Annex I)

1. Description of the traffic plan/programme (a)
    1.1. Objectives of the plan/programme
    1.2. Interrelationships to other traffic plans/programmes
    1.3. Implementation procedure
    1.4. Conditions for traffic prognosis and reasons for choosing the scenario

2. Scoping
    2.1. Avoiding duplication of assessment
    2.2. Methods used
    2.3. Planning area
    2.4. Examination period
    2.5. Data sources
    2.6. Forecast horizon
    2.7. Scenarios

3. Sustainable development and environmental protection objectives (e)
    3.1. Established objectives
    3.2. Other objectives
    3.3. Reasons for excluding other objectives
    3.4. Indicators
    3.5. Environmental quality standards

4. State of the environment without implementation of the plan/programme (b,c,d)
    4.1. Environmental assessment of the current planning area
    4.2. Likely evolution of the environment until the forecast horizon based on the chosen scenario
    4.3. Relevant environmental problems, in particular those relating to any areas of a particular environmental importance
    4.4. Assessment of environment using objectives above (see 3.1-3.5)

5. Environmental effects of the current traffic network (d)
    5.1 Current network and likely evolution until the time of prognosis (status-quo-network)
    5.2 Environmental effects of the status-quo-network, current and for the forecast horizon
    5.3 Assessment using objectives above (deficiency analysis)
    5.4 Recommendations to reduce the environmental effects of the status-quo-network

Box 40.1. (cont.)

6. Environmental effects of the projects (f,g,h)
   6.1. Description of the projects including their interrelationships
   6.2. Reasons for the projects from traffic's view
   6.3. Statement on how the recommendations (see 5.4) have been taken into account
   6.4. Description of and reasons for selecting the alternatives (system-, site- and design-alternatives)
   6.5. Measures to avoid, reduce and offset adverse effects
   6.6. Negative environmental effects (environmental impact) including secondary, cumulative, synergistic, short, medium and long-term, permanent and temporary effects
   6.7. Positive environmental effects (environmental relief, see 6.6)
   6.8. Assessment of the projects and their alternatives using objectives above (see 3.)
   6.9. Recommendations for integrating the projects into the plan/programme from environment's view

7. Environmental effects of the new traffic network including the projects (f, g, g)
   7.1. Description of the new network
   7.2. Statement on how the recommendations (see 6.9) have been taken into account
   7.3. Negative environmental effects (environmental impact)
   7.4. Positive environmental effects (environmental relief)
   7.5. Assessment of the new network using objectives above (see 3.)
   7.6. Recommendations for the new network from environment's view

8. Recommendations for decision making process
   8.1. Legal requirements and scope of discretion
   8.2. Conflicts between environmental and other interests
   8.3. Recommendations for decision making process
   8.4. Recommendations for traffic policy

9. Monitoring (i)
   9.1. Overview (objectives, indicators, local and temporal requirements, responsibilities, data sources, remedial actions in case of unforeseen adverse effects)
   9.2. Measures to monitor the network and the conditions of the chosen scenario
   9.3. Measures to monitor the projects in the following planning stages
   9.4. Measures to monitor the projects after realisation

10. Difficulties in particular technical deficiencies or lack of know-how (h)

11. Non-technical summary (j)

40.3.2 How to Decide which Projects Should be Assessed?

SEA deals with plans and programmes setting the framework for projects that may require an EIA in the later stages of the planning process as Annex I or Annex II projects or where Natura 2000 sites may be affected. So there must be a decision on how to create these projects and how to make them part of the plan.

Transport planning in Germany is based on a bottom-up approach with lower planning tiers proposing projects for inclusion within the next high tier's plan. Hence states make proposals to the federal transport plan, while the states examine the proposals of the communities. This procedure has the benefit that valid information about the need of new infrastructure projects is collected, but it also has one disadvantage:

There might be proposals which are important for the community (or its mayor or his party) but not for the state, they would wrongly burden the state's budget. Other projects may be of value at a state level. To overcome this disadvantage a systematic study of the existing transport network should be undertaken to examine how to make best use of the existing infrastructure and to consider the need for new transport projects. The multi-modal studies undertaken in the UK provide examples of this approach. Through such studies new projects could be identified and assessed on a comparable basis, thereby improving the efficiency of resource allocation.

40.3.3 How to Avoid Duplication of Assessment?

In German road planning there are three steps in the planning process: The first is the infrastructure plan, the second is the determination of alignment and the last is the approval procedure (see Chap.7). In the last two steps there are EIA-procedures, in the first SEA should be carried out. So the next question is: How to avoid duplication of assessment between these tiers? To answer this question the following differences between SEA and EIA should be recognized:

- SEA tend to cover larger geographic areas than EIA.
- Given the larger study area the focus is upon using existing data rather than commencing new surveys.
- In SEA the focus is on the entire infrastructure network and the budget for the projects, in EIA only one project is assessed.
- At the project level localized and specific effects are examined, while within the SEA, network level effects and generalized local effects are examined where they can be meaningfully recognized at this early stage.
- Beyond alignment and design alternatives considered within EIA, SEA introduces other alternatives, i.e. alternative modes (see description in Sect. 40.5).
- SEA considers cumulative effects which are often overlooked at project level.

40.4 Identifying and Describing the Environmental Effects

40.4.1 What are the Significant Effects of a Transport Plan?

One of the reasons for creating SEA was that plans and programmes have other effects on the environment than projects. The whole is more than the sum of all parts and so the type and scale of the effects considered in SEA differ from those of an EIA, for example:

- *Global climate change:* There is increasing confidence in a correlation between CO_2-emissions and natural catastrophes like storms and floods. Even in Germany, insurance-corporations complain about increasing payments for damages which are consequences of global climate change (Münchener Rück 2003). Also fauna and flora is affected: Until 2050 a million of species will be lost (Nature 2004). So it is the objective of the national government (BREG 2002) to reduce all CO_2-emissions in Germany by 25% until 2005 and by 30% until 2010. These objectives may be achievable in industry and energy-sector. However, 20% of the CO_2-emissions are a result of road traffic and this has not been reduced in recent years.
- *Fragmentation:* One of the most serious problems for several species is fragmentation of their habitats, which may result in the remaining area falling below the minimum area needed to support the species. New transport infrastructure can contribute to this effect. The government's objective is that until 2020 at least 10% of the whole area in Germany should have ecological restrictions in order to build up a coherent habitat network.
- *Sealing and compacting soil:* About 7% of the country is urbanized with a daily increase of about 105 hectares. Continuation of this trend will lead to the entire country being urbanised by end of the century. The national government has the objective to reduce the daily increase to 30 hectares until 2020 (BREG 2002).

Beyond such network's effects SEA also should address recognizable project level effects. In this way it can inform the scoping process for the project EIA leading to a more targeted and efficient EIA. SEA will also aid the avoidance and more effective mitigation of adverse environmental impacts at a project level and help ensure that appropriate project budgets are set.

40.4.2 How to Assess the Environmental Effects?

When the effects are identified they also need to be assessed. Which environmental quality standards can be used? The SEA Directive requires a description of "the environmental protection objectives, established at international, Community or Member State level, which are relevant to the plan or programme and the way those objectives and any environmental considerations have been taken into account during its preparation". Among the "established" objectives are UN Conventions, EC Directives, acts and regulations. However, these are not sufficient for the assessment as objectives are needed for all environmental effects in order to judge the importance of the forecast effect. Hence it is necessary for objectives to be established that relate to the locality and the type of plan. These may include objectives not yet established, for example drafts of established objectives or recommendations of commissions of the government as well as locally derived objectives as a result of public consultation.

40.5 Assessing Alternatives

Art. 5 of the SEA Directive requires the identification, description and assessment of reasonable alternatives. To answer the question which types of alternatives should be examined, the environmental objectives of an SEA for transport plans should be reviewed to identify ways of:

- Avoiding or reducing significant adverse effects on the environment
- Achieving environmental objectives and improving environmental quality according to the principles of sustainable development.
- Avoiding traffic and delivering a change to modes with less impact.
- Improving the planning process by consideration of environmental effects at the earliest stage.
- Saving time and money by avoiding mistakes in the following planning stages.

Consideration of alternatives also helps respond to the following questions on environmental objectives often raised during the political discussions regardless of whether a strategic assessment is undertaken:

a) Which areas would be affected by significant environmental effects and is it possible to avoid these areas? This question leads to location-alternatives.
b) Is there a chance to avoid adverse effects by changing the project design? This question leads to technical alternatives.
c) Would measures to deliver modal change, satisfy the demand for transport with lower environmental effects? This question leads to the assessment of alternative modes.
d) Will the planned network fit the environmental objectives? If the transport plan doesn't achieve these objectives perhaps other opportunities to reduce environmental impacts should be considered, for example changing the conditions within the traffic forecasting scenarios, such as assumptions for the proportions on different modes, by altering land use patterns or fiscal rules or by opportunities for car-sharing. Such policy alternatives may rest with higher level planning jurisdictions and hence some liaison will be required as to their practicability. These topics are beyond SEA and lead to the policy level.

Also the zero option is not a real alternative in SEA because it does not help to achieve the goals of the transport plan. Nevertheless it should be described to make clear the positive environmental effects of the plan.

40.6 Decision-Making Process

How can it be ensured that environmental issues are considered in the decision making process? The following points can help but carry no guarantee:

- Integrate SEA in the planning process (see Sect. 40.3.1) and do not try to undertake sectoral planning.
- Find objectives and standards in all sectors and not only in the environmental sector and measure all together.
- Give clear recommendations of project priorities from the environmental point of view. These will be weighed with economic development, traffic, safety, spatial order or urban development.
- Avoid complexity in the environmental report and focus on relevant issues.
- Create comprehensible methods for balancing interests.
- Give several present informations in different stages.

The last two points need further explanation. What means "comprehensible methods for balancing interests"? In each plan or programme there is an interest-balancing process in which environmental interests are balanced against others, for example safety, spatial order or urban development. Fig. 40.1 shows a matrix with the environmental risks on one side and the benefits for other interests on the other side where after assessment all projects can be enlisted. Each point in the matrix stands for one project. There is a region in the left upper corner of the matrix, where the decision is very clear: These projects will be part of the plan, because they have low risks and high benefits. On the other hand there is a region in the right bottom corner with high risks and low benefits, Such projects will rarely be accepted as a part of the plan. In the rest of the matrix, the decision will be unclear. It is the trade-offs in this decision space have to be presented to the politicians.

In order to enlighten this decision space, some methodological requirements should be fulfilled. Firstly an early definition of categories of results is needed, for example "Yes", "No" or "Yes under a condition", such as prove the effectiveness of mitigation measures in later planning stages. Secondly advantages and disadvantages of the projects, especially concerning environment, should be described. A presentation as shown in Fig. 40.2 might be better than a multi-criteria-analysis or a cost-benefits-analysis.

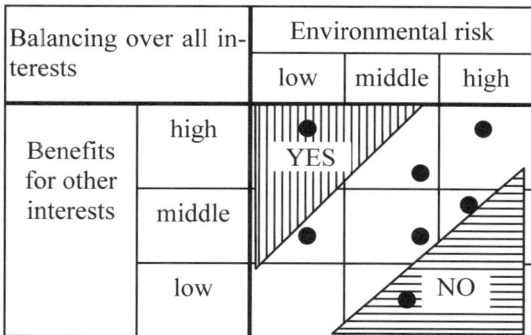

Fig. 40.1. Balancing interests (Stein 2003)

In this example the project's effects to all important interests are described in a three class assessment, divided in positive and negative effects. The relationship between the advantages and disadvantages can be seen, as well as across the different interests. In the example shown in Fig. 40.2 environmental advantages are greater than the disadvantages and over all interests advantages are in majority. This type of presentation leaves the weighing decisions to the politicians. It also shows the scope for the political decision and does not give the illusion that the decision has already been made with the spurious accuracy of numerical methods. No specific guidance can be given on how the results should be presented to the public or politicians. Instead the approach should be designed to reflect the local circumstances of the plan. Whatever approach is adopted, it is important that the mechanisms for public involvement should be designed at an early stage of the plan-making process. As known from the author's experience the results of the SEA should be presented in several stages of information throughout the SEA process. For example the discussion of the Environmental Report for the Road Infrastructure Plan North Rhine-Westphalia (Stein et al. 2000) within the road authority and in Parliament was based on the following information:

- List with an overview over all projects,
- list for each project with all results,
- list for each project with all environmental assessments,
- Environmental Report (about 5 sheets of text, 8 maps) for each project and photos and videos of the critical projects.

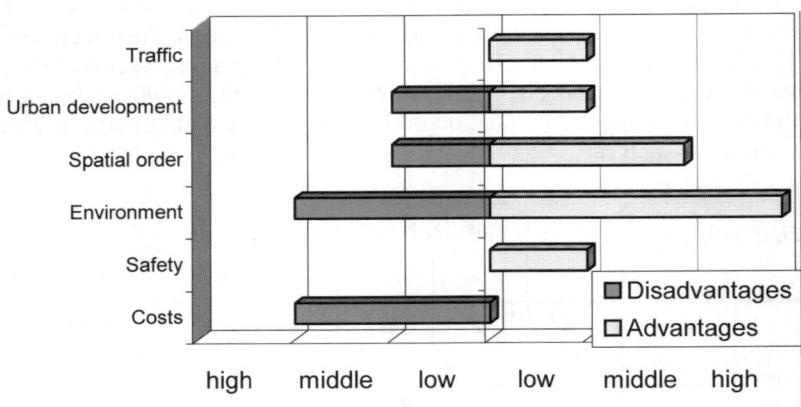

Fig. 40.2. Advantages and Disadvantages of a project (Stein 1997)

40.7 Monitoring

There are no experiences in monitoring transport plans in Germany and there are several difficulties:

- Complexity of environment makes the selection of suitable indicators a demanding task.
- Lag time between the action and the impact on the environment. It will be difficult to prove correlation between parts of the plan and their effects on environment in the absence of control stations.
- The large number of projects implemented over a long time period makes the task even more difficult.

There must be monitoring of the network effects as well as the projects' effects in order to clarify whether the environmental objectives are achieved or not. In addition monitoring should also observe whether the assumptions of traffic development are fulfilled.

40.8 Conclusions and Recommendations

The decision-making process in traffic planning in Germany is unlike most other European countries. The aggregation of many measures within one plan makes the assessment and selection of projects difficult. However, by considering the essential differences between SEA and EIA, duplication of assessment should be avoided. The main aspects which will be considered in SEA are global climate change, fragmentation, and urbanization, thereby reducing the complexity of the Environmental Report. Objectives and standards have to be established in all sectors and not only in the environmental sector. The main task in all countries is to integrate SEA in the whole planning process and to avoid doing sectoral planning.

References[1]

EU (1994) Road Program Nordrhein-Westfalen (Germany). In: SEA Existing Methodology. EU-Commission, DHV Environment and Infrastructure BV

BREG – Bundesregierung (2002) Perspektiven für Deutschland – Unsere Strategie für eine nachhaltige Entwicklung (Prospects for Germany – our strategy for sustainable development) Beitrag der Bundesrepublik Deutschland zum Weltgipfel für Nachhaltige Entwicklung im Aug./Sept. 2002 in Johannesburg. Internet address: http://www.bmu.de/files/nachhaltigkeit_strategie.pdf, last accessed 10.03.2004

---

[1] Note that legislation mentioned in the chapter is listed at the end of the handbook in the consolidated list of legislation (Appendix 2).

ECMT (2000) Strategic Environmental Assessment for Transport, European Conference of Ministers of Transport, Paris
ECMT (2004) Assessment and Decision Making for Sustainable Transport, European Conference of Ministers of Transport, Paris
Fischer TB (1999) Benefits from SEA application - a comparative review of North West England, Noord-Holland and EVR Brandenburg-Berlin, EIA Review 19:143-173
Fischer TB (2000) Die strategische Umweltprüfung in England, den Niederlanden und Deutschland (SEA in England, Netherlands and Germany). UVP-Report 4:221-225
Fischer TB (2001) The practice of environmental assessment for transport and land use policies, plans and programmes, Impact Assessment and Project Appraisal 19(1):41-51
Fischer TB, Seaton K (2002) Strategic environmental assessment – effective planning instrument or lost concept? Planning Practice and Research 17(1):31-44
Fischer TB, Siemoneit D (2002) Die Strategische Umweltprüfung – das Beispiel des Regionalplans Lausitz-Spreewald in Brandenburg (SEA for the Regional Plan Lausitz-Spreewald). UVP-Report 2001 (5):253-258
FGSV – Forschungsgesellschaft für Straßen und Verkehrswesen e.V. (1990): Merkblatt zur Umweltverträglichkeitsstudie (MUVS)
FGSV – Forschungsgesellschaft für Straßen und Verkehrswesen e.V. (2003) Strategische Umweltprüfung von Plänen und Programmen, Referate und Ergebnisse eine Workshops am 25. November 2002 in Wuppertal
FGSV – Forschungsgesellschaft für Straßen und Verkehrswesen e.V. (2004) Merkblatt zur Strategischen Umweltprüfung, Entwurf März 2004
FMT – Federal Ministry of Transport (2003): Federal Transport Infrastructure Plan 2003: Basic features of macroeconomic evaluation methodology. Internet address: http://www.bmvbw.de/Anlage13389/Federal-Transport-Infrastructure-Plan-2003-Basic-features-of-macroeconomic-evaluation-methodology.pdf, last accessed 21.09.2004
Gerlach J et al. (2002) Umweltziele in der Strategischen Umweltprüfung von Plänen und Programmen im Verkehrssektor, In: Straßenverkehrstechnik 10(2002):549-565
Münchener Rück (2003). Internet address: http://www.diw.de/deutsch/publikationen/wochenberichte/docs/02-35-2.html#HDR0, last accessed 16.01.2003
Nature (2003): Feeling the heat: Climate change and biodiversity loss. Nature 427 (145). Also http://www.nature.com/nature/links/040108/040108-1.html, last accessed 05.03.2004
Stein W (1997) Verkehrsplanung und Umwelt in den Bundesländern. In: UVP in der Bundesverkehrswegeplanung. UVP Spezial 14. UVP-Förderverein
Stein W, Smeets P, Wolff F (2000) Stand und methodische Weiterentwicklung der Plan-UVP des Landesstraßenbedarfsplans NRW. UVP-Report 2:84-89
Stein W (2003) Strategische Umweltprüfung in der Verkehrsplanung in Deutschland. In: Internationales Verkehrswesen 9:405-410
Tomlinson P (2004) The Evolution of Strategic Environmental Assessment, Integrated Assessment and Decision-Making in the United Kingdom, in ECMT, Assessment and Decision Making for Sustainable Transport, European Conference of Ministers of Transport, Paris
Tomlinson P, Fry C (2003) Strategic Environmental Assessment and its Relationship to Transportation Projects, in Transport Projects, Programmes and Policies, Ashgate Publishing Limited

Tomlinson P (2000) New Approaches to the Environmental Assessment of Multi-Modal Transport Schemes, paper presented at the Institute of Highways and Transportation, Alan Brant Workshop, Leamington Spa, April 2000

Tomlinson P (2002) Streamlining Strategic and Project Environmental Appraisal Practice, Paper presented to Landor Conference Transport Options Appraisal and Scheme Delivery, 27 June 2002, London

Tomlinson P (2003): South West Area Multi-Modal Study (SWARMMS), in Ministry of the Environment, Government of Japan, Effective SEA System and Case Studies, Tokyo, Japan

41 SEA for Agricultural Programmes in the EU

Michael Schmidt, Harry Storch and Hendrike Helbron

Department of Environmental Planning, Brandenburg Technical University (BTU), Cottbus, Germany

41.1 Introduction

The aim of this chapter is to discuss the implementation of SEA for the agricultural sector with special emphasis on the programme level, where the framework is set for project EIA at a lower tier. Although agriculture is one of the sections within the scope of the SEA Directive (see Chap. 2), most discussions and studies about the implementation of SEA have concentrated on land use and transport planning (see Chap. 40). It is therefore important to redress this imbalance by investigating agriculture. As formulated in Art. 1, the SEA Directive aims to provide for a high level of protection of the environment during the preparation and adoption of plans and programmes. The main focus of this chapter is therefore to reflect on how SEA could contribute to agricultural land use becoming more environment-friendly in the future.

The chapter starts by evaluating the current situation concerning EU agricultural policy and agricultural land use. It also evaluates the environmental impacts that agriculture can cause. As a precondition to understanding the characteristics of the agrarian sector and its specific relationship to SEA, Sect. 41.1 highlights how land use is influenced by the requirements from policy to programme level.

Sect. 41.2 gives a detailed overview of the surprising exclusion of the dominant part of the agricultural plans and programmes from the scope of the SEA Directive. The German situation will be used as an example at programme level. Not only does Germany have a wide planning system, but it also represents a typical highly industrialised agricultural country in the EU with comprehensive development planning for rural areas. Out of the numerous plans and programmes that exist in the agricultural sector in Germany, the German preliminary agrarian structure planning (gAEP) presents the *only* programme that would call for SEA. The consequences of the exclusion of other important plans and programmes are stressed with the example of the German agricultural planning system.

Implementing Strategic Environmental Assessment. Edited by Michael Schmidt, Elsa João and Eike Albrecht. © 2005 Springer-Verlag

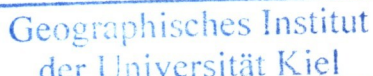

Section 41.3 explores certain pre-conditions that will have to be met before SEA could play an important role in European agriculture and therefore would become part of the preparation and adoption of gAEP. Especially the latest proposed Common Agricultural Policy reform (CAP reform 2003) from Luxemburg, which will have to be implemented by the EU Member States in 2005, nourishes high expectations concerning a more environment-friendly agricultural production. According to the SEA Directive, SEA will only be a tool to assess site-related objectives in agriculture. Therefore Sect. 41.3 analyses which objectives currently exist and how the proposed forthcoming CAP reform will likely change the situation.

The chapter closes with some conclusions in Sect. 41.4 on how SEA might be able to contribute to the improvement of agricultural land use methods and with recommendations which ensure that specific preliminary conditions are in place. The exclusion from the SEA Directive of most of the plans and programmes of such an important sector as agriculture is seen as a mistake, which weakens the implementation of SEA in the EU. However, the possible introduction of the CAP reform in 2005 will hopefully contribute to SEA to really being fully implemented in the agricultural sector.

41.2 EU Agricultural Policy 1992-2005

The aim of this section is to give a background as to how agriculture is influenced by the EU agricultural policy and impacts. It is necessary to understand the current situation in the EU, in order to be able to analyse how SEA could improve the situation in the long term. The EU policy influences national policy and therefore affects the objectives of agricultural programmes such as in the case of Germany. The benefits of SEA for the agricultural sector can only be understood if the relationship between agriculture and the environment becomes clear. Two elements of agricultural policy of the EU are key:

1. Common Agricultural Policy (CAP), which deals since 1962 with price support directly to the farmer and
2. Restructuring of agricultural life that includes set-aside schemes, extensification, farm diversification, input control grants relating to nitrogenous compounds, pesticides and herbicides, organic farming and early retirement schemes and which deals with regional payments to farmers since 1998 (Hawke and Kovaleva 1998).

According to Kaletta et al. (2001), the EU political framework had initially food provisions assurance as a goal and not environmental protection. Hawke and Kovaleva (1998 p.1) affirmed: "Environmental protection and enhancement was not an original objective of the common Agricultural Policy". Although the EU was created in 1957, environment was only first mentioned in the Directive on Farming in Mountain, Hill and Less-Favoured Areas in 1975. The overproduction of food problem (Kaletta et al. 2001) provoked restructuring in the use of agricultural resources and lead to the reform of CAP in 1992. This reform contained so-called "accompanying measures" including an agri-environmental scheme (AEP) (Council regulation 2078/92). These measures had the objective to support farmers, who participated voluntarily, financially and not to save endangered species (Hergersberg 2001). No efficient systematic monitoring and evaluation of the environmental effects of the AEP – i.e. if the money was well spent – were done. The reason was claimed to be a lack of methods (EC 1998b).

In July 1997 the European Commission launched Agenda 2000 (EC 1997). One of the reasons was a funding deficit that had to be faced in relation to the EU-enlargement, e.g. unaffordable payments in the livestock sector (Schmidt and Storch 2004). Among the aims of the rural development policy were the protection of the environment and rural heritage and the promotion of the multifunctional role of agriculture (EC 1997).

Since 1998 separate policies provide the foundation for agri-environmental matters within the Union. Besides the simultaneous demand to conform production to global market conditions as well as ensure farmers' welfare an additional objective was the "maintenance of biotic and abiotic landscape characteristics through *sustainable* land use" (Kaletta et al. 2001 p.288). *Sustainable agricultural production* must protect and enhance particular farmland landscapes, their wildlife habitats and historic features, support lively and diverse village-communities, guarantee social well-being, long-term benefits, ecological safety and high quality of products, promote sensitive innovative farming techniques that meet environmental, health and quality standards and improve opportunities for public enjoyment (EC 2004). The second objective was influenced by agriculture being a declining industry, which had to "give up and back out of the land on marginal areas with naturally bad conditions" (Braun 1995 p.100).

In 2004 it is still the case that payments made directly to farmers for certain amounts of produced goods play the major role in agricultural policy. This fact influences the present agricultural land use in the EU with two different trends:

1. Persistent specialisation and intensification of arable and livestock farming systems on profitable areas, leading to major losses of biodiversity and abandonment of traditionally grazed areas.
2. Extensification and land set aside of less profitable areas with marginal yields. Land abandonment, undergrazing and lack of capital to maintain or improve farm infrastructure are creating, besides positive effects on the environment, new environmental pressures (EEA 2003 p.43).

Preliminary conditions for the implementation of the SEA Directive are site-specific objectives, which did not exist in the history of the EU agricultural policy and will not be introduced before 2005. The next section will clarify this situation by revealing the present dependencies of CAP and agricultural land use in the EU and in Germany.

41.3 Influence of CAP on Agricultural Land Use

This section highlights to which extent the Common Agricultural Policy (CAP) influences agricultural land use. For the implementation of SEA it is essential to know if, or to which extent, CAP influences specific activities on the individual site and thus causes negative and positive effects on the environment. The authors' intention is to make clear, that European agriculture has got important spatial effects. As the objectives of CAP have changed in the last decades, from self-supporting agriculture over avoidance of over-production to sustainable and organic farming, the countryside changed accordingly (BMVEL 2003). Land use plays a major role in the context of agricultural policy and environmental effects. Agriculture covered 44 % of the territory of the EU (EU-15) in 1997 making it one of the largest percentages of the area of the EU (Caradec et al. 2004).

For instance the German farmers are, with nearly 50 % agricultural use of the whole territory, the main land users and landowners with property rights (Statistisches Bundesamt 2004). These dimensions concerning land use must be explicitly noticed in the decision-making process if SEA is mandatory for agricultural plans or programmes at all. This is especially important considering that farmers can change their cultivation, i.e. land use, on a large area within a very short time period.

Agricultural activities are directly and indirectly affected by CAP. Until 2004, with product related funding, CAP brought about considerable changes in land use. The agricultural policies "induced farmers to farm the land more intensively by making it more profitable for them to bring previously marginal land into production and to apply greater amounts of 'non-labour inputs' such as fertilisers and pesticides to each hectare of land already in production" (Hawke and Kovaleva 1998 p.28). Land use changes such as conversion from lower to higher intensity, abandonment or extensification cause positive as well as negative effects on the environment. Examples of negative effects are: contamination of surface, groundwater's and soils from the effects of fertilisers and pesticides, a continued decline in soil productivity from excessive soil erosion and nutrient run-off losses, hazards to human health, destruction of ecosystems and landscape degradation (see Fig. 41.1). An example of a convincing positive effect is: maintenance of valuable landscape and habitat types through certain agricultural land use practices and thus survival of many habitats and species only due to agricultural policies and measures (Wascher 2000). The CAP support development of EU agriculture led to environmental damage on a large scale. Although the type and degree of ecological degradation in different areas depended on the applied agricultural production method and the regional natural conditions (Kaletta et al. 2001).

The goal of sustainable land use was, before the introduction of agri-environmental programmes (AEP) with regional reference to the individual site, often not realized as reported by Kaletta et al. (2001), due to subsidy programs that *inadequately* addressed the unique local and regional landscape and ecosystem characteristics as well as the negotiation terms of farmers.

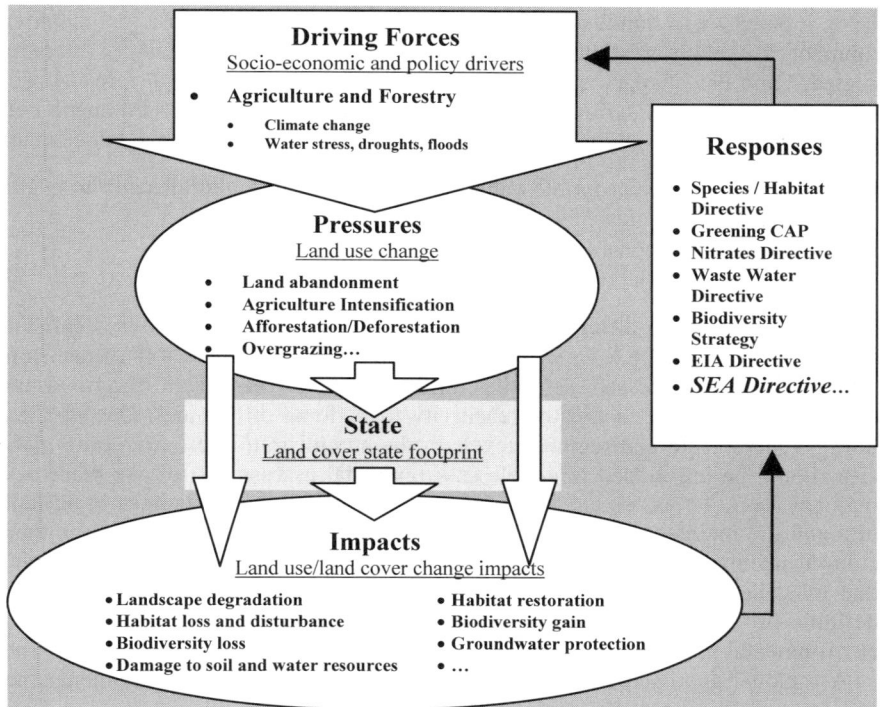

Fig. 41.1. Driving Forces Pressures State Impacts Responses (DPSIR)-framework in relation to agricultural land use change and responses including such as the SEA Directive (modified from EEA 2001)

To assess the environmental impact, the reference situation prior to the land use change has to be considered. Fig. 41.1 shows the variety of possible sources of pressures on land use change in agriculture, which depend on political and socio-economic driving forces. Agricultural driving forces are large scale effects of human activities related to the agricultural sector. For example Wascher (2000 p.28) identified those as being "the action of establishing agricultural production systems on the land (cultivation, conversion, changing geomorphology, buildings, facilities and infrastructure)", "periodic land use changes a part of the long-term agricultural management" or "on-field farm management (ploughing, mowing, livestock rearing, irrigation, application of agro-chemicals, harvesting etc.)." The subsequent impacts on the environment are therefore changing with constraints and benefits originating from top and "leaking down". At the same time impacts influence the responses such as the introduction of the SEA Directive, and these responses affect decisions made on the higher level (bottom up).

Buckwell et al. (1997) confirmed, that there exists a greater knowledge on the relation between environmental effects and land uses, whilst the relation between the agricultural policy and the land uses is less understood. Thus we need to develop a better understanding of the relationship between agriculture and the environment, as wealth creation and environmental quality are increasingly interconnected. Land use changes cause positive or negative environmental impacts. The relationships between agricultural policy, land use changes and environmental effects cannot be accurately forecasted. The following deficiencies have to be faced:

- A general lack of appropriate methodologies in the agricultural sector to evaluate these dependencies
- Constraints when considering different levels of scale for SEA
- Problems when addressing planning for the agricultural sector

Table 41.1 shows the different levels of scale for environmental assessment. The adoption and use of SEA varies with regard to the levels and sectors of decision-making that are addressed, as Sadler argued in 1998. The design of agricultural policies should reflect a greater sensitivity to regional differences. Decentralisation to federal state and regional levels in the administration of agricultural policies should be introduced to enable environmental assessment to take place at a regional level. To assess environmental impacts of structural changes in agriculture and the maintenance of agricultural landscapes, a regional approach to agricultural planning is required. It will be necessary to define the regional accuracy that must be present to meet the environmental targets. This means we need a definition of areas which will be subject to the implementation of specific agri-environmental measures.

A regional approach should be accompanied by the following advantages according to the European Commission (1998b):

- coherence with other sectoral policies to prevent application of competing measures
- appropriate implementation of regulatory environmental standards

Table 41.1. Different levels of scale for environmental assessment (EC 1998a)

	SEA (plans/programmes)	EIA (projects)
Data	mixture between descriptive and quantified	mainly quantified
Objectives; scope of impacts	National, regional and local	mainly local
Alternatives	More efficient use of existing infrastructure, fiscal measures, spatial balance of location etc.	location, technical variants, design etc.
Methods for impact prediction	simple (often based on matrices and use of expert judgement) with high level of uncertainty	complex (and usually based on quantified data)
Outputs	broad-brush	detailed

- clear definition of environmental aims and priorities
- measurable agri-environmental targets

The essence of the regional approach is that clear observation of the range of needs and potentials can lead to appropriate agri-environmental measures and targets. SEA could potentially be an important instrument to support agricultural planning at regional level, which is examined in the next section.

41.4 SEA for Agricultural Plans and Programmes

As already discussed in Chap. 2, SEA will only affect plans and programmes at Member State level and below. EC legislation and programmes as well as policies, which are too general to be subject to environmental assessment (e. g. they imply no specific spatial component such as common agricultural policy mechanisms (see Table 41.1), are excluded. Having adopted the definitions of Plan and Programme from Sadler and Verheem (1996) for the agricultural context in Table 41.2, it can be stated that these two tiers, suitable for implementing SEA, do exist in the agricultural planning of the EU Member States. Nevertheless the SEA Directive excludes important agricultural plans and programmes in advance, which shall be stressed in this section. It focuses on all agricultural plans and programmes with direct or indirect impacts on the individual area under the conditions shown in Fig. 41.2.

Table 41.2. Definitions of plan and programme and their link to agriculture (modified from Sadler and Verheem 1996)

SEA tier	Definition	Example from agricultural sector
Plan	A purposeful, forward looking strategy or design, often with coordinated priorities, options and measures, which elaborates and implements policy.	Extensification programme
Programme	A coherent, organized agenda or schedule of commitments, proposals instruments and/or activities that elaborates and implements policy. Programs make plans more specific by including a time schedule for specific activities.	Preliminary Agrarian Structure Planning

41.4.1 Agricultural Plan or Programme under the Scope of the SEA Directive

An analysis of the contents of the SEA Directive with special emphasis on agriculture shall reveal the legal framework for potential plans and programmes that require to be assessed by SEA. As presented in Chaps. 2 or 3 the SEA Directive states:

> "Subject to the SEA Directive are all plans and programmes, including those co-financed by the EC, as well as any modifications to them which are subject to
> - preparation and/or adoption by an authority at national, regional or local level or
> - which are prepared by an authority for adoption, through a legislative procedure by Parliament or Government, and
> - which are required by legislative, regulatory or administrative provisions (SEA Directive, Art. 2 a))".

Further SEA shall be carried out for all plans and programmes,

> "which are prepared for agriculture [...] and set the framework for future development consent of projects listed in Annexes I and II to the EIA Directive, the "positive list" or which, in view of the likely effect on sites, have been determined to require an assessment pursuant to Art. 6 or 7 of the Habitats Directive (Art. 3 para 2 a) and b) of the SEA Directive)".

In relation to agriculture, what is important to emphasise are plans and programme not subject to the SEA Directive:

> "[...] (agricultural) financial or budget plans and programmes or [...] plans and programmes co-financed under the current respective programming periods[1] for Council Regulations (EC) No 1260/992 on Structural Funds and No 1257/993 on support for rural development from the European Agricultural Guidance and Guarantee Fund (EAGGF)" (Art. 3 para 8 and 9 of the SEA Directive).

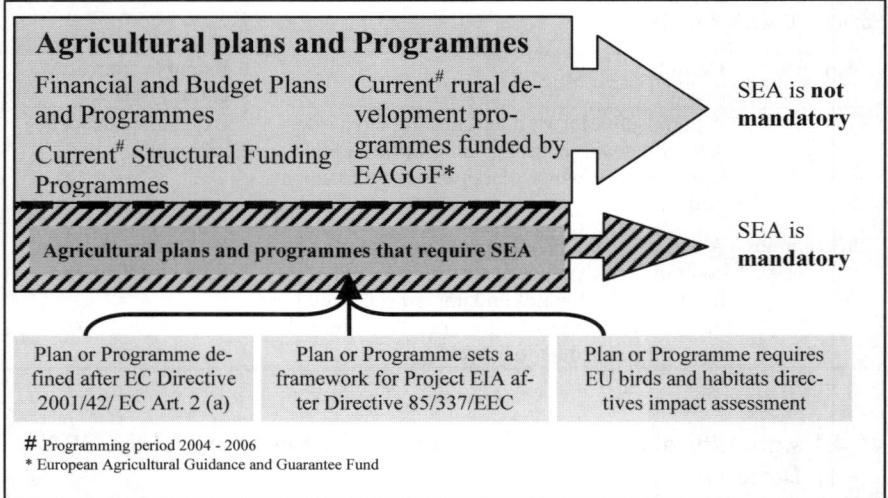

Fig. 41.2. Scope of the SEA Directive for agriculture

---

[1] 2004-2006

This means that according to the SEA Directive most of the existing EU agricultural plans and programmes are excluded from SEA. This is illustrated in Fig. 41.2. Considering the financing of these excluded agricultural programmes plus taking into account some of the most important EU programmes like LEADER+, INTERREG, LIFE programme, the Extensification Payment Scheme (EPS), land set aside and AEP; in Germany for example these covered 35 % for the improvement of rural structures, 31 % for the improvement of production and marketing structures and 21 % for sustainable land use in 2002. The major investment sum went into the promotion of individual farms as enterprises (BMVEL 2003) and therefore not into site-specific land use changes. Thus the influence of agricultural planning with relevance to the impact on land use, i.e. grown crops and dependant use of fertilizer, is financially and legally restricted.

41.4.2 Projects that Require EIA and their Influence on SEA for Agricultural Plans and Programmes

An agricultural plan or programme that sets a framework for Project EIA according to the EIA Directive has to be assessed by SEA. Therefore the lower tier of agricultural projects will be investigated in order to find out which higher plans and programmes are generally relevant for the implementation of SEA. Further the type of projects, their dimensions and effects on the site shall be understood. Agriculture does not feature in the Annex I projects of the EIA Directive where environmental assessment is a mandatory requirement. Agricultural projects are listed in Annex II and are subject to discretionary assessment where it is considered that a given project will have *"significant effects"* on the environment; such as in the case of:

(a) the restructuring of rural land holdings
(b) the use of uncultivated land or semi-natural areas for intensive agricultural purposes
(c) Water-management projects for agriculture
(d) Initial afforestation where this may lead to adverse ecological changes and land reclamation for the purposes of conversion to another type of land use
(e) Poultry-rearing installations
(f) Pig-rearing installations

EIA as the legal requirement for certain projects likely to have significant effects on the environment is therefore mandatory in the EU for all above-mentioned projects with a specific dimension. According to Annex I of the EIA Directive it is mandatory for installations for the intensive rearing of poultry or pigs with more than 85 000 places for broilers, 60 000 places for hens; 3 000 places for production pigs (over 30 kg); or 900 places for sows.[2] An example for a mandatory project EIA are the setting up and operation of an installation for intensive breeding

---

[2] For further information about the relation between EIA Directive and the IPPC Directive concerning livestock production see Grimm (2001).

of hens with 42 000 or more number of places, or 2 000 or more rearing pigs of a minimum of 30 kg living weight. Further for instance the changes of rural structure through land consolidation or specific water economic projects are such EIA-projects.

Lee and Walsh (1992 p.265) argued that "one of the deficiencies of environmental assessment is that the majority of agricultural activity falls outside project level assessment" because it is classified as *"non-project" action*. Non-project action in the good farming practice of 2004 includes for example the cultivation of maize, which generally contributes to a higher degree to groundwater nitrification than the cultivation of cereals. Only the land use change from a former site not cultivated by farmers to an agriculturally used site is a project for which EIA is mandatory.

Concluding, potential agricultural plans and programmes that require SEA would create the framework for rather punctual industrial projects that have to be assessed by EIA or land consolidations on a larger scale. They must not be financial or budget plans or programmes, but should have an impact on the area. The next section presents a German programme that matches those conditions.

41.5 Suitability of the German Preliminary Agrarian Structure Planning (gAEP) for the implementation of SEA

The German Preliminary Agrarian Structure Planning[3] – here with the acronym gAEP instead of AEP to differentiate from the EU-wide agri-environmental programme – presents an example of an agricultural programme at a higher level than the project EIA level. The goal of this section is to analyse if potential links for SEA can be found, which outweigh the likely exclusion of the dominant land user agricultural planning as regards to SEA.

The gAEP is an informal not legally binding sector planning according to Art. 1 para 2 of Federal Act on the Improvement of the Agrarian Sector. Therefore participation is voluntary for the farmers, as long as the gAEP does not set the framework for subsequent project EIA. gAEP becomes formal, *only* if it delivers targets or statements for the selection of the sites of projects have to be assessed by EIA. According to Borchard et al. (1994 p.227) gAEP shall be used as an "instrument for the comprehensive development of rural areas". This requires an extensive spectrum of objectives in combination with the consideration of site-specific potentials. The three main tasks of gAEP are:

- To develop a site-specific vision as a basis for the abolition of agrarian structure deficiencies, for the implementation of agrarian structure objectives and programmes and for the realisation of non-agrarian projects.
- To investigate and define the need for measures in the respective areas after having carried out an analysis of the potentials and constraints of an area.

---

[3] Agrarstrukturelle Entwicklungsplanung.

- To set up a concept for the development of rural areas that can deliver priorities for land use and for periods of time and contribute to the harmonisation of land use conflicts (Borchard et al. 1994 p.227).

The gAEP is:

- Subject to preparation and/or adoption by the Federal boards for land consolidation at regional or local level, through a legislative procedure by Parliament
- Required by legislative, regulatory or administrative provisions, if these set the framework for subsequent project EIA
- gAEP is a sector planning making concrete plans related to the area, changing rural structures
- an authority is responsible for procedure and deliverance of certain reports; applications and registration for funding is involved
- precautionary measures have to be undertaken; it is mandatory to prepare and carry out the EIA procedure

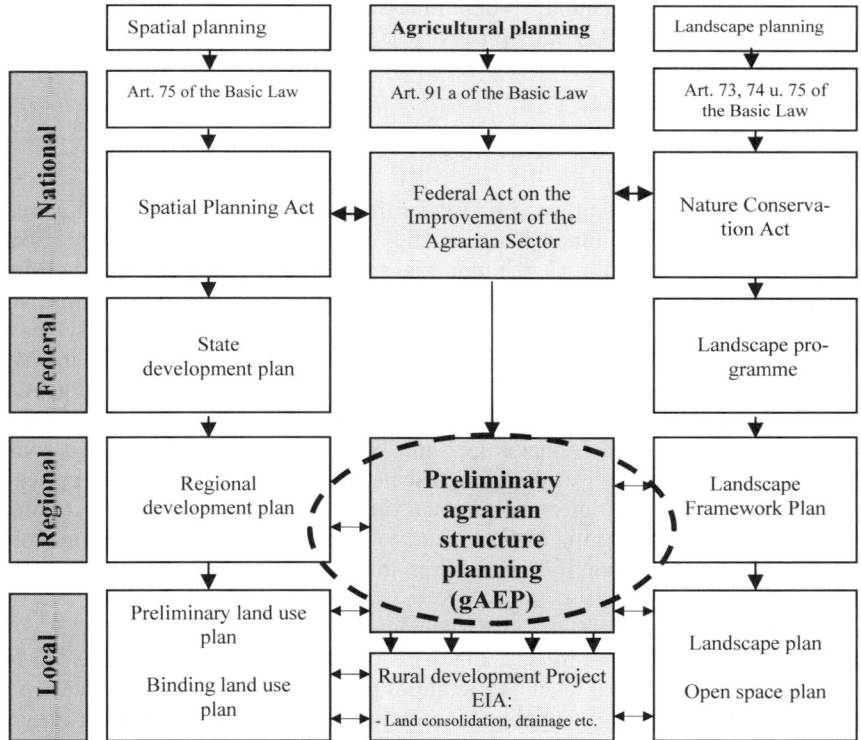

Fig. 41.3. Integration of gAEP into the German planning system showing the ideal SEA level relevant for SEA (note that funding policies, programmes and plans are not represented) (changed from SMUL 2000)

The German preliminary agrarian structure planning is a basis for implementing agri-structural requirements and programmes and thus it is an "important decision-making instrument" for an integrated and sustainable development in rural areas (Kohl 1995 p.228). It aims to promote an integrated land development and prepare the implementation of related projects, which can be also projects that require EIA. The main function is to nominate structural deficits, e.g. a poor road network, make recommendations for different forms of land use, e.g. poultry farms, and propose the location and dimension of projects for which EIA is mandatory. A further task of the gAEP is to develop specific *development "visions"* for rural areas and to suggest measures and concepts to achieve these (MLUR Brandenburg 2004). Fig. 41.3 presents the integration of the gAEP into the German planning system. As can be seen in Fig. 41.3 there exist no relevant agricultural programmes at the federal level. However, it can be argued that the regional level, i. e. the location of gAEP, is the best level in scale for carrying out SEA.

Fig. 41.3 shows that planning is happening in the German agricultural sector. There is a strong link between landscape planning, spatial planning and agricultural planning. Spatial planning programmes have to integrate agricultural concerns into their objectives. Although gAEP is widely not considered as planning with effects on the land use it functions as such and therefore requires SEA.

41.5.1 Suitability of gAEP with Regard to Scale Issues

In order to assess the suitability of gAEP for the implementation of SEA it is necessary to sum up the mentioned characteristics of German agriculture and to check the scale at which agricultural planning takes place. Agricultural practices often contradict ecological and aesthetic land management objectives and therefore agriculture is a sector known to have or likely to cause environmental effects. These impacts grew with European agriculture becoming largely industrialized, intensificated and specialised at least until the formation of Agenda 2000. The potential impacts on the environment will last a very long time as do landscape changes through deforestation and drainage. Lee and Walsh (1992) pointed out that some of the agricultural activity was on a micro scale involving numerous small agricultural holdings all with minor environmental impacts. The cumulative effect could be seriously damaging for the environment, as where increased fertiliser or pesticide use for example from 100 such farms impacts on one river catchment (see further Köppel et al. 2004).

It is mandatory for the German parliaments to implement SEA in gAEP if it sets a framework for EIA-projects. Fig. 41.4 clarifies the scale level affected by gAEP at regional and local level. Agricultural planning therefore is apparently not only settled at project level, but above it and therefore this chapter strongly argues that SEA is suitable for gAEP. At this level it is possible to obtain the data needed to carry out SEA and to operate agricultural planning in a sustainable way.

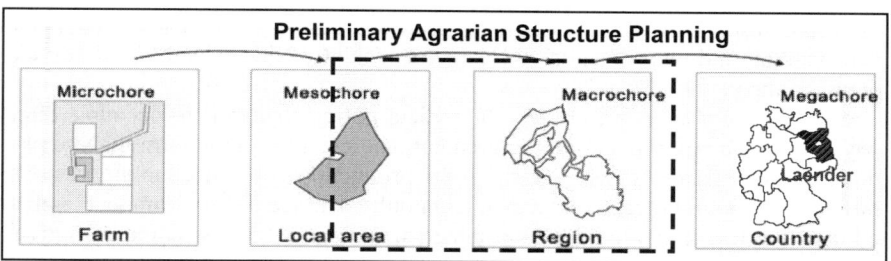

Fig. 41.4. Potential ideal level of gAEP at SEA level for the environmentally appropriate operation of agricultural planning above the EIA-project level (changed from Caspersen and Stenstrup 1997)

The preventive nature of SEA would be of great benefit for a sustainable rural development on a regional level. Sustainable agricultural land use could further be promoted by SEA through optimised agro-industrial production methods after the assessment of impacts. Nevertheless, SEA for gAEP would accompany the benefit of a premature assessment on a higher level when selecting the sites for EIA-projects. However SEA does not function as a tool to influence agricultural activities. As we realized in the past, product-related subsidies cannot contribute to our common goal of sustainable living. The most recent CAP reform, which will hopefully be introduced in 2005, contains a new way of site-specific funding.

40.5.2 Effects of the most recent CAP reform on gAEP and SEA

This section shall evaluate how SEA could make the current situation different in the future, as long as two conditions are in place. The first condition for SEA to be implemented for agricultural planning is that spatial planning exists. It was stressed in the previous analysis that this is the case. The second condition is that the most recent CAP Reform comes into force in 2005 as planned.

The current CAP (until 2004) had damaging effects on the physical environment in rural areas, because high support prices had encouraged over-production, which in turn has led to over grazing of pastures and excessive application of fertilisers and pesticides. The results are the pollution of surface and groundwater and the destruction of cultural landscapes. An important reason for recent reforms was the EU enlargement (Schmidt and Storch 2004). The establishment of AEPs has done little to offset these effects caused by the CAP production support, because agri-environment measures are voluntary and the ubiquitous and unconditional support for agricultural production by the CAP support is financially more attractive.

The fundamental reform of the CAP (implemented in European law by Council Regulation 1782/2003) will completely change the way the EU supports its agricultural sector. Instead of having to produce particular products to obtain subsidy, under the reformed CAP, farmers will have more freedom to decide what crops and livestock to produce. In addition to this farmers are also required to undertake important responsibilities towards the protection of the environment, animal health and welfare, and public health in return for receipt of this subsidy. The key elements of the new CAP for 2005 are:

Single Farm Income Payment: The radical policy change is to decouple direct payments from any production requirement, and create a single farm income payment for EU farmer's independent from production. This decoupling removes production specific incentives, which potentially damage the environment, and facilitates the integration of the environmental measures into the common agricultural policy (CEC 2002 p.19).

Table 41.3. Legally binding Statutory Management Requirements from CAP Reform (EC) 1782/2003 (Cross compliance) (CEC 2002)

Ref No	EC Directive / Regulation	Cross compliance Requirements
1	Wild Birds Directive	Art. 3 requires Member States to take action to secure or re-establish habitats for all naturally occurring wild birds. Art. 4 para 1,2 and 4 requires Member States to take special protection measures for certain species of bird, including the establishment of Special Protection Areas (SPAs). Art. 5 prohibits the significant disturbance of wild birds, the damage and removal of their nests. Art. 7 permits hunting of wild birds subject to conditions. Art. 8 prohibits certain means of killing wild birds.
2	Groundwater Directive	Art. 4 and 5: Farmers should comply with relevant legislation and codes of good practice for the protection of groundwater against pollution.
3	Sewage sludge Directive	Art. 3: Use of sewage sludge treated in accordance with the Directive: Requirements to prevent contamination caused e.g. by heavy metals (observation of specified harvesting intervals etc.)
4	Nitrates Directive	Art. 4 and 5: Farmers should comply with the mandatory legal measures to protect water against pollution caused by nitrates.
5	Habitats Directive	Art 6 requires Special Areas of Conservation to be designated for protected habitats and species. Art 13 requires prohibition of destroying, cutting or uprooting of protected plant species. Art 15 requires prohibition of certain methods of killing or taking wild species. Art 22 b) requires regulation of introduction of non-native species where prejudicial to native wildlife.
9	Plant Protection Products Regulations	Farmers are required to comply in full with Plant Protection Products Regulations (e.g. sufficient 'buffer zones' to water courses).

Avoidance of abandonment of agricultural production: Limited combined elements may be maintained to avoid abandonment of agricultural production. If farmers could decide not to produce, i.e. cultivate parts of their land, this would lead to an abandonment of land in less favoured areas. The definition of "good agricultural practice and environmental condition" is targeted at the risks of land abandonment (cross compliance conditions).

Cross compliance conditions according to Fischler (2003): This means that in the future a farmer receiving direct payments will be required to respect approximately 18 legislative provisions of "Statutory Management Requirements" (Annex III of Regulation (EC) 1782/2003) as well as maintaining the land in "good agricultural and environmental condition", *even if the farmer is not producing a crop.* Otherwise farmers will forfeit part of their farm income payment. Table 41.3 presents only those legislative provisions that are linked to environmental standards. All conditions are already – in 2004 – legally binding to farmers.

The Regulation of "good agricultural practice and environmental condition" is mainly targeted at the risks of land abandonment (Council of the European Communities 2003 p.5) and Member States are permitted to define minimum requirements within a European framework (Table 41.4). The definition of those standards has to be fulfilled on a national or federal level. Land abandonment is of direct concern for sustainable land use and thus creates a link to SEA, which could contribute to the prevention of negative effects of land abandonment.

Some of the responses in Fig. 41.1 show parallels to the new standards from the CAP reform from Luxemburg in Table 41.4. For instance the percentage of the farm land used as pastures must not change and hedgerows have to be maintained without additional funding from AEPs. The requirements for cross compliance will obtain a "baseline" level of environmental protection, and incentives (rural development programmes) should be only available to help farmers to take further action.

Modulation: A reduction in direct payments for larger farms to finance the new rural development policy, new measures to promote the environment, quality and animal welfare, this should be a re-orientation of support into environmentally sensitive farming. Under the new rural development policy, the coverage of less favoured areas has been extended to include areas where environmental restrictions apply. Farmers who choose voluntarily to participate in these new rural development schemes may be required to observe additional cross compliance standards as a pre-condition of their participation.

SEA and the CAP reform from Luxemburg would find a common focal point in their site-related targets. Cross compliance will guarantee certain environment-friendly measures with direct effects on the area and at the same time SEA would contribute to more environmentally-sensitive alternatives of cultivation. With the new CAP reform more specific site-related objectives in EU agriculture are expected.

Table 41.4. Good Agricultural and Environmental Condition Framework as defined in Annex IV of the Council Regulation 1782/2003

Issue	Target	Standards
Soil erosion	Protect soil through appropriate measures	Minimum soil cover Minimum land management reflecting site-specific conditions Retain terraces
Soil organic matter	Maintain soil organic matter levels through appropriate practices	Standards for crop rotations where applicable Arable stubble management
Soil structure	Maintain soil structure through appropriate measures	Appropriate machinery use
Minimum level of maintenance	Ensure a minimum level of maintenance and avoid the deterioration of habitats	Minimum livestock stocking rates or/and appropriate regimes Protection of permanent pasture Retention of landscape features Avoiding the encroachment of unwanted vegetation on agricultural land

41.6 Conclusions and Recommendations

This chapter has argued that in order for the SEA Directive to play an important role in the agricultural sector in the EU two key conditions are needed: first agricultural planning has to deliver objectives with effects on the individual area and second the new CAP Reform from Luxembourg has to be implemented by 2005. The chapter has emphasised the German situation, but this would be equally relevant for most other EU Member States, as they are exposed to quite similar situations as regard to agricultural planning. Until the present (2004), agriculture in most EU-countries takes place without long term and appropriate planning with effects on the land use. Although targets of national plans and programmes, like gAEP, cause impacts on the land use, they do not define specific objectives based on the characteristics of individual sites. Therefore agricultural land use could *only* be made more environment-friendly by SEA through improvement of the agro-industrial production methods and an optimised selection of sites. If the CAP reform of Luxemburg will come into force in 2005 as planned, the EU will face a new era in agricultural planning. Farmers will finally receive funding dependant on the dimensions and characteristics of their agricultural area and no longer on their farms' economic productivity. Even sanctions would be possible in the future and therefore SEA could play an important role for the prevention of wrong decisions at a higher level.

Considering the case study of Germany used in this chapter, the gAEP will hopefully become more environmentally relevant with the new CAP reform in 2005. It would then be able to formulate regional objectives and targets; it would not stay as informal or as voluntary as it is at present. Instead, the gAEP would have to respect more environmental requirements and targets such as an efficient monitoring. Environmental targets would have to be formulated for gAEP in a close cooperation with Landscape planning (see Fig. 41.3 and Chap. 38) in order to reach the newly introduced environmental standards of the EC in the agricultural sector.

Concerning agricultural *activities,* i.e. set-aside or extensification, gAEP is not the best instrument to assess how environment-friendly they are, as environmental impacts on the land caused indirectly by EC-funding programmes are recognised much too late. Moreover a large part of agricultural activities is classified as "non-project activity" (e.g. the use of fertilizers and pesticides), and large structural projects that require EIA such as large-scale land consolidations do not occur so often any more. With a radical reform of CAP at EU level, which makes the Member States responsible to define environmental standards on a national or federal level, long term planning with objectives for the individual site will become necessary even for projects and cross compliance activities on the farm level. One important task will be the definition of "visions" for the agricultural development of European regions. For instance the maintenance of the region of Extremadura in Spain requires a vision which differs to a great extent from that one of the Spree forest in Brandenburg, Germany. Cross compliance links direct funding to the compliance of certain environmental quality standards defined after the 2003 "good agricultural and environmental practice" (see Table 41.3). The definition of those standards has to be fulfilled at a national or federal level. This chapter argues that the following opportunities and potentials of SEA could be expected after having been implemented into gAEP, when the most recent CAP reform will have come into force in 2005:

- As soon as statements with spatial relation to specific sites are made, SEA will gain importance.
- Planning is necessary for the purpose of protecting specific sensitive areas.
- SEA enhances the chance of a do-nothing option, because not as many resources have been invested into the concrete planning, as for instance in the land consolidation procedure.
- To make use of the preventive characteristic of SEA, a common planning strategy or cooperation together with targets from landscape planning and other sector planning, e.g. forestry could be formulated on a higher level from the perspective of agriculture.

In relation to the case study of Germany the following recommendations can be made: one of the most important tasks for the gAEP will be the protection of agricultural land use on marginal sites and of extensification in order to maintain the cultural landscapes of the EU. As the implementation of the CAP reform of 2003 is still in development, it is not yet quite clear, how it will look and influence our countryside. The Member States responsible task will be the integration into the national planning system. Nevertheless it is assumed that at least one year of practice could give useful results for a first evaluation of positive effects on the environment. This means that before 2006, or more likely 2007, no evaluations will be carried out. Considering the development of CAP in the future, SEA will become important, when more plans or programmes with objectives for the individual site exist. The role and implementation of SEA in EU agriculture could be discussed in more detail, when the basis of more site-specific targets will be established. If agricultural planning adopts requirements from the recent CAP reform in 2005, further advantages of SEA would include:

1. The consideration of cumulative impacts (see Chaps. 26 and 27), e.g. from many mass breeding farms in one region, which could affect air and water, e.g. by contaminating rivers through chemicals used. Due to the very complex relationship between agricultural policy or land use and its effects on the environment, both direct and indirect effects of a proposal could be examined.
2. Proposal of a large range of alternative land cultivation and mitigation measures e.g. to avoid sensitive sites at a regional level.
3. More open and transparent planning processes through public participation and consultation.
4. An additional information pool for farmers to raise acceptance and increase knowledge about alternative production methods in specific areas.
5. Long-term monitoring and review of how sustainable strategies are developed in agriculture and a positive criticism of society and its organisation.

If the mentioned pre-conditions are met in the future, not only the German agricultural programme gAEP but also further EU agricultural plans and programmes will have to be assessed by SEA in order to improve the environmental situation in the EU agriculture. SEA is considered to be a suitable and important tool for the agriculture sector and should therefore be implemented for programmes like gAEP. Farmers have a vital role to play in the conservation of biodiversity and traditional landscape and thus achieve the goal of sustainable farming. Their willingness to cooperate with environmental planners in the long term has to be supported by a good knowledge about the benefits of SEA. Although a large part of agricultural plans and programmes is excluded from the SEA Directive, spatial planning exists, which explicitly deserves attention and should not be neglected. SEA would be a suitable instrument to predict and assess site-specific impacts of plans and programmes and thus to move towards sustainable agricultural land use, if the CAP reform comes into force, as described earlier. Thus agricultural land use in the EU could move towards sustainability with the help of SEA. Certainly it will need a long time to tackle the root of environmental problems: the way resources are used in society overall.

References[4]

BMVEL (2003) Ernährungs- und agrarpolitischer Bericht 2003 der Bundesregierung (Report of the German Government on food and agricultural policy)

Borchard K, Kötter T, Brassel T (1994) Agrarstrukturelle Vorplanung: Vorschläge zur inhaltlichen und konzeptionellen Neugestaltung eines Instruments zur Entwicklung ländlicher Räume (Preliminary Agrarian Structure Planning: Proposals for contents and conceptual reorganization of the instrument fort he development of rural areas). In: Schriftenreihe des Bundesministeriums für Ernährung, Landwirtschaft und Forsten. Reihe B: Heft 81: FlurbereinigungLandwirtschaftsverlag, Münster, pp 226-228

Braun J, (1995) Flächendeckende Umstellung der Landwirtschaft auf ökologischen Landbau als Alternative zur EU – Agrarreform dargestellt am Beispiel Baden-Württembergs Agrarwirtschaft (Transformation of agriculture to organic farming on the entire area as alternative to the EU agrarian reform with the case study of agricultural economy in Baden Wuerttemberg). Agrarwirtschaft – Zeitschrift für Betriebswirtschaft, Marktforschung und Agrarpolitik Sonderheft 145

Caradec Y, Lucas S, Vidal C (Eurostat) (2004) Agrarlandschaften: Mehr als die Hälfte der europäischen Fläche wird landwirtschaftlich genutzt, Internet address: http://europa.eu.int/comm/agriculture/envir/report/de/terr_de/report.htm, last accessed 09.09.2004

Caspersen OH, Stenstrup J (1997) GIS-Based Methods for Rural Land Use Analysis Based on Integrated Agricultural Control Systems IACS In: Proceedings of the International Conference on Geo-Information for Sustainable Land Management (SLM): Enschede, 17 - 21 August 1997.

CEC – Commission of the European Communities (2002), Communication from the Commission to the Council and the European Parliament. Mid-Term Review of the Common Agricultural Policy, COM(2002)394, Brussels: CEC

Council of the European Union (2003) CAP Reform – Presidency Compromise (in agreement with the Commission), Note from Presidency to Delegations, 30 June, 10961/03, Brussels

EC – European Commission (2004) Towards a more sustainable agricultural production. Internet address: http://europa.eu.int/comm/agriculture/foodqual/sustain_en.htm, last accessed 09.09.2004

EC – European Commission (1997) 'Agenda 2000', Vol 1, 'For a stronger and wider Union', COM (97) 2000 Final

EC – European Commission (1998a) A Handbook on Environmental Assessment of Regional Development Plans and EU Structural Funds Programmes, August 1998, Part 2: Structural Funds and Environmental Assessment

EC – European Commission (1998b) State of application of regulation (EEC) no. 2078/92: Evaluation of Agri-environment Programmes, Commission working document – DG VI, Brussels

EEA – European Environment Agency (2001) Towards spatial and territorial indicators using land cover data. Technical report No. 59, Copenhagen

---

[4] Note that legislation mentioned in the chapter is listed at the end of the handbook in the consolidated list of legislation (Appendix 2).

EEA – European Environment Agency (2003) Europe's environment: the third assessment, Environmental assessment report No. 10, Copenhagen

Fischler F (2003) CAP Reform, Committee on Agriculture and Rural Development, European Parliament. Speech 03/356, 9 July 2003, Brussels

Grimm E (2001) Environmental legislation in the European Union to reduce emissions from livestock production. Paper presented at XXVI Annual Meeting of the Chilean Society for Animal Production (SOCHIPA), International Symposium on Animal Production and Environmental Issues, Session II: Animal Nutrition and Environmental Issues: Reduction of Emissions from Animal Production – Legislation and Environmental Technologies in the USA and the European Union

Hawke N, Kovaleva N (1998) Agri-environmental law and policy. Cavendish publishing, London

Hergersberg P (2001) Agrarprogramme nutzlos für den Umweltschutz? Holländische Forscher: Geförderte Maßnahmen sichern nicht die Artenvielfalt (Agricultural programmes useless for the Environmental protection?). Pressemitteilung vom 12.11.2001, http://www.deutsche-landwirte.de/110201d.htm, last accessed 20.09.2004

Kaletta T, Helming K, Kächele H, Khorkov A, Müller K, Philipp H-J (2001) Sustainable land use: An Interdisciplinary Demonstration Project in northeast Germany. In: Stott DE, Mohtar RH, Steinhardt GC(eds) Sustaining the Global Farm. Selected papers from the 10th International Soil Conservation Organization Meeting held May 24-29, 1999 at Purdue University and the USDA-ARS National Soil Erosion Research Laboratory pp 288-292

Kohl A (1995) Die Agrarstrukturelle Vorplanung – ein Instrument zur Entwicklung ländlicher Räume? (Preliminary Agrarian Structure Planning – An instrument for the development of rural areas?). Zeitschrift für Kulturtechnik und Landentwicklung 36: 27-229

Köppel J, Peters W, Wende W (2004) Eingriffsregelung Umweltverträglichkeitsprüfung FFH-Verträglichkeitsprüfung. Ulmer, Stuttgart

Lee N, Walsh F (1992) Strategic Environmental Assessment: an Overview. Project Appraisal 7(3), cited in Hawke and Kovaleva 1998

MLUR Brandenburg (2003) Fördermaßnahme Agrarstrukturelle Entwicklungsplanung (AEP) (Promotion measure, prelminiary agrarian structure planning (gAEP), Internet address: http://www.brandenburg.de/land/mlur/politik/foerder/rl_aep.htm, last accessed 21.11.2003

Sadler B, Verheem R (1996) Strategic Environmental Assessment: Status, Challenges and Future Directions, Ministry of Housing, Spatial Planning and the Environment, Publication number 54, The Hague, The Netherlands

Sadler B (1998) Strategic Environmental Assessment: Institutional Arrangements, Practical Experience and Future Directions, Institute of Environmental Assessment, International Workshop on SEA, Organised by the Japan Environment Agency Tokyo, November 26-27, 1998

Schmidt M, Storch H (2004) Transborder regions and administrative boundaries: institution building on the basis of common environmental values and its role in the enlargement of the EU. In: Schmidt M, Knopp L (eds.) Reform in CEE-Countries with Regard to European Enlargement – Institution Building and Public Administration reform in the Environmental Sector. Springer, Heidelberg, pp 99-112

Statistisches Bundesamt (2004) Bodenflächen nach Art der tatsächlichen Nutzung in Deutschland (German land use areas according to their actual use). Internet address: http://www.destatis.de/basis/d/umw/ugrtab7.htm, last accessed 20.09.04

Stott DE, Mohtar RH, Steinhardt GC (eds) (1999) Sustaining the Global Farm. Selected papers from the 10th International Soil Conservation Organization Meeting held May 24-29, 1999 at Purdue University and the USDA-ARS National Soil Erosion Research Laboratory

SMUL (2000) AEP im Freistaat Sachsen (Preliminary Agrarian Structure Planning in Saxony). In: Borchard K, Kötter T, Brassel T (1994) Agrarstrukturelle Vorplanung. Schriftenreihe des BML, Reihe B: Flurbereinigung, Heft 81, p 8

Wascher DM (ed) (2000) Agri-enviromental indicators for sustainable agriculture in Europe. European Centre for Nature Conservation (ECNC), Tilburg

42 SEA of Waste Management Plans – An Austrian Case Study

Kerstin Arbter

Arbter – SEA consulting and research, Austria

42.1 Introduction

This chapter is about waste management SEA. Waste management is one of the traditional SEA sectors with more than ten years of experience. In the Netherlands waste management SEA dates back to the late eighties and early nineties, for example, the SEA for the Dutch Ten-Year Programme on Waste Management or the SEA for the 3rd Provincial Waste Management Plan of Gelderland. In Finland, an SEA for the Finnish National Waste Management Plan was carried out in the mid nineties (European Commission 1997; Verheem 1998) and in Austria the first pilot-SEA started in 1999. Up to 2004, two waste management SEA had been carried out in Austria, the SEA for the Viennese Waste Management Plan and the SEA for the Waste Management Plan of the Province of Salzburg (Arbter 2004).

To start with, this chapter introduces the main fields of application of SEA in the waste management sector, distinguishing between the level of needs, technologies (methods) and capacities and the level of location. Then the SEA for the Viennese Waste Management Plan is described as a case study. Furthermore, after addressing the issues of methodologies and data used in waste management SEA, it moves on to the topics of process design and public participation. In this connection an SEA model for pro-active stakeholder participation called "SEA Round Table" is described. The chapter ends up with some general remarks and conclusions for applying SEA to waste management plans.

42.2 Fields of Application

SEA in the waste management sector can be applied at two levels:

- Level of needs, technologies (methods) and capacities
- Level of locations

At the level of needs, technologies (methods) and capacities fundamental questions of waste management can be answered, for example:

- *If* and *why* a project or measure is needed. For example, if a city or a province needs new waste treatment facilities or if growing waste volumes could be stopped by reducing waste generation and recycling measures
- *What technologies* or *methods* for waste treatment or which waste avoidance and recycling measures, including awareness programmes educating the public to recycle more, would be best for the specific situation
- *What capacities* of waste treatment facilities would be needed

At the level of locations, the question of sites can be answered, such as:

- *Where* a new waste treatment facility should be built.

The next level is the design level, where the question of *how* new treatment facilities should be constructed is answered. This is already also the project level, where EIA can be used within the licensing process (see Thérivel 2004; Verheem 1998).

42.3 The SEA for the Viennese Waste Management Plan

Vienna had experienced growing volumes of waste, increasing requirements from landfill legislation, and bottlenecks in its waste treatment facilities for several years. Confronted with these problems, the Environmental Commission of Vienna, a kind of "environmental ombudsman", launched an SEA in order to provide a waste management plan that would resolve these problems by 2010. The waste management authority decided to engage with a wide range of stakeholders. The *SEA Round Table* was developed as a model fostering pro-active participation of stakeholders (see Sect. 42.5). This process started in 1999 and was completed in 2001 (Arbter 2003).

From the beginning, environmental, economic and social aspects were taken into account. This went beyond the requirements of the SEA Directive, which mainly focuses on environmental effects. But nevertheless, this was a central requirement in the Commission's brief for the study, because the Viennese Waste Management Plan should comprise these three dimensions of sustainable development. This SEA was applied at the level of needs, technologies (methods) and capacities. The fundamental issues to be resolved were (Arbter 2003):

- How can you get to the root of the waste problem? Which waste minimisation and recycling options must be implemented to solve the problems?
- Are additional waste treatment facilities needed in Vienna to cope with the waste generated before 2010?
- Which treatment technologies are best suited to the specific circumstances in the city?

- Which waste treatment facilities should be chosen? How could the capacity of Vienna's existing facilities be optimised? What treatment capacities are needed for the newly built facilities?

Some important results of the SEA were (Arbter 2003):

- Waste avoidance and material recycling: The city of Vienna should step up its efforts to improve the quality and reduce the quantity of waste. The goal is to significantly decrease the growing quantities of wastes needed to be treated and their pollutant concentrations. An annual budget of five million Euros should be spent on further waste avoidance measures.
- Fermentation plant: the ground needs to be prepared for a biogas facility that is capable of processing 25 000 tonnes of fermentable waste.
- New waste incineration plant: Vienna should build a new waste incinerator with 450 000 tonnes annual capacity. One older incinerator shall be closed down. The existing sorting and processing plant shall cover the extra capacity which is needed in case of a breakdown.
- A monitoring group shall be established, whose task will be to keep an eye on the successful implementation of the Viennese Waste Management Plan.

The issues of the locations of new waste management facilities were explicitly excluded from the discussions within the SEA process. In December 2001, the Vienna City Council made its decision concerning the waste management plan – it followed the recommendations of the SEA. Subsequent to the SEA, an assessment process to find the best suitable location for the recommended new treatment facilities was initiated. Following the results, the licensing processes of the new waste incineration plant and the new fermentation plant commenced, implementing the recommendations of the SEA. For the time being, these two Project EIA are carried out without remarkable delays or problems, because all the fundamental questions regarding need, technologies (methods) and capacities had already been answered within the SEA. In addition, a strategy group for waste avoidance was founded, managing the annual budget of five million Euros for waste avoidance measures.

42.4 Methodologies and Data

To assess the likely significant effects of waste management plans on the environment, either quantitative or qualitative methods or a combination of both can be used – as in all other SEA sectors. The choice depends highly on the availability of data. Only if good quality data is available, an assessment could be carried out using energy or waste removal chain flowcharts with appropriate simulation models or life cycle analyses, which produce quantitative results concerning, for example, the amounts of emissions, energy saving or energy use and costs.

The SEA for the Ten Years Waste Management Programme of the Netherlands 1995-2005 used life cycle analyses. This is a standardised assessment method in which the total environmental impact of a certain activity is determined. The complete life cycle of a good or service, from the input of basic materials, through product manufacturing, use of the product or service to processing of residual materials, is included. The impact is then re-calculated into a so-called "environmental profile" of the activity by use of "classification-factors" (Commission of Environmental Impact Assessment of the Netherlands 1994). For the SEA for the Viennese Waste Management Plan a chain flow simulation model for energy and waste removal was used. It was a highly sophisticated mathematical model which produced quantitative data on emissions, waste amounts, energy production and use of space.

As convincing as these kinds of methods may seem, in many cases at the strategic level, it remains doubtable if impacts are really seriously quantifiable. At least, the Austrian experiences show that the mathematical simulation models do not always produce reliable results. Data gaps which very often occur at strategic planning levels hamper serious modulation. Sophisticated models can of course produce data, but often they only pretend to be accurate. Too detailed impact predictions can also reduce the intelligibility of the assessment and draws attention to unessential questions, instead of focusing interest to the most important issues (Saarikoski 1997).

Therefore simpler, qualitative evaluations are often recommended for waste management SEA (Arbter 2001; Commission of Environmental Impact Assessment of the Netherlands 1991). This means that environmental effects are for example, described verbally (possibilities and threats of alternatives) or estimated on ordinal scales between + + (very good) to - - (very bad), always accompanied with a written explanation of the evaluation. The use of these methods in Finland showed, that even if exact data is missing, it is possible to make credible and transparent impact predictions when appropriate methods are used (Saarikoski 1997). Only transparent results will be accepted by the public and by decision-makers – which in turn is a prerequisite for SEA to influence decision making. Examples of necessary data for waste management SEA are:

- Size and composition of the current waste flows per category and the expected development including data on prevention and reuse or recycling of waste
- Specific emission and energy data of the different processing and disposal techniques
- At location level: data on the environment, nature and landscape protection, logistics of supply and removal (transport activities)

Possible indicators for impact prediction are:

- Percentage of reuse or recycling
- Volume and method of dumping
- Emissions of heavy metals (Hg, Cd)
- Emissions of polycyclic aromatic hydrocarbons, dioxins, SO_2, NO_x, CO_2 and CH_4
- Odour
- Net energy production

- Costs
- Flexibility
- Space occupation
- Acceptance of measures by the public

In any case, the assessment should be focused on a small number of reliable and illustrative indicators which can be understood easily and which really can support decision making. A large volume of produced data could cause confusion and hide the main results of the SEA. Of course, indirect effects should also be taken into account in the assessment. For the Viennese Waste Management Plan the substitution of single oil or coal heaters by district heating produced in waste incineration plants was an important issue. It was assumed, that emissions by domestic fuel could be reduced, if more households would have access to district heating.

One of the most important facts to enhance the credibility of the SEA results is that all assumptions and data sources must be clearly documented. Gaps in knowledge and information must be stated clearly and their significance for the decision on the strategic action should be indicated. Of course at the strategic level, factors of uncertainties are significantly greater than at the Project EIA level.

When choosing the alternatives to be assessed, the whole chain of the waste production and removal should be taken into account. In addition, waste minimisation, reuse and recycling measures must be part of comprehensive alternatives if questions of need, technology (methods) and capacities are to be answered. A simple comparison of the effects of differing waste treatment techniques is not sufficient for waste management SEA at this level.

42.5 Process Design and Public Participation

The design of the SEA process is at least as important as the choice of appropriate assessment methods. The Austrian experiences show that two issues of process design highly influence the efficiency of SEA:

- The integration of the planning process and the SEA process to one common, well interlinked process
- The pro-active participation of the stakeholders concerned and information and consultation with the general public

42.5.1 Integration of Planning Process and SEA Process

The integration of the planning process and the SEA process to one common process means, that the elements of both processes are fully interlinked. The SEA starts with the first ideas of the strategic action, and *before* the latter has been drafted completely. The stages and elements of both processes are already linked together (see Fig. 42.1). This full integration offers three main advantages:

- Planning and SEA can influence each other continuously. Environmental aspects can be taken into account step by step at each stage of the process.
- If SEA only starts when the strategic action has already been drafted, there is the threat that in the meanwhile important decisions were taken, that can not be influenced by SEA any more, even if negative environmental effects are foreseeable. In this case, also a strong identification with the strategic action by the planning authority could lead to strong resistance if SEA results would make changes of the strategic action necessary and the role of SEA could be undermined. A fully integrated SEA reduces these threats.
- Furthermore, in a fully integrated SEA defined planning and environmental goals can be taken into account more effectively. If environmental goals only appear after finally drafting the strategic action, it is extremely difficult to integrate environmental standards with hindsight.

If planning and SEA are carried out as separate processes, this could also cause problems for public participation, especially if the SEA results and the public comments do not adequately feedback into the planning process (Saarikoski 1997).

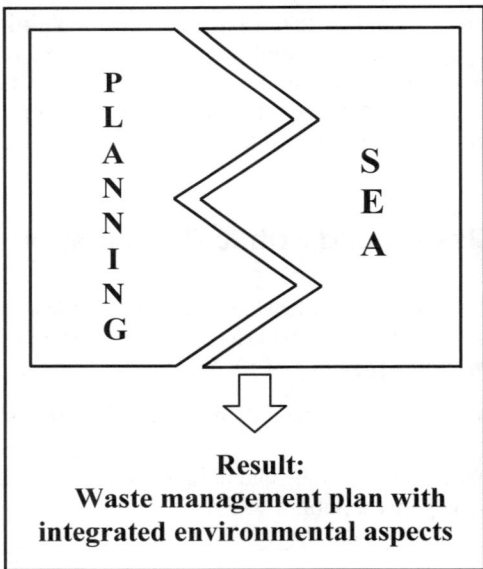

Fig. 42.1. Integration of planning and SEA process to one common process (adapted from Arbter 2004, with kind permission of Neuer Wissenschaftlicher Verlag, Vienna Graz)

42.5.2 Pro-active Participation – Viennese SEA Round Table

As the example of the SEA for the Viennese Waste Management Plan shows, pro-active participation is a crucial factor for efficient SEA, especially for highly controversial "hot topics" with large public interest. At strategic planning levels, values and policies of the stakeholders concerned influence the SEA process a lot. To tackle this, the different points of view need to be reconciled in a structured process. A mere technical analysis of environmental impacts is not enough for successful SEA. Furthermore; pro-active participation can foster the acceptance of the waste management plan, even if unpopular measures have to be taken, like the construction of waste incinerators or raising of waste management fees.

Therefore a special SEA model was created for the SEA of the Viennese Waste Management Plan: the *SEA Round Table* (see Fig. 42.2). The SEA was carried out by an SEA team, it consisted of members of: the relevant authority for waste management and environmental authorities, external waste management experts, and representatives of different environmental NGOs. The NGOs took part in the process as representatives of the public. They were termed "the qualified public". The team worked together from the very beginning and was responsible for the planning and the SEA results. They came to a consensus in nearly all aspects on the best solution for the Viennese Waste Management Plan (Arbter 2003).

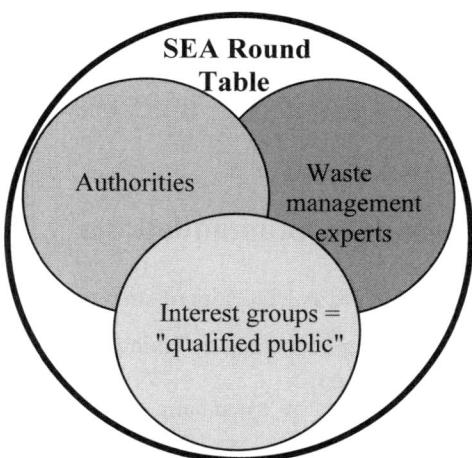

Fig. 42.2. SEA team of the SEA Round Table model, consisting of representatives of the authorities concerned, of external waste management experts and representatives of the qualified public (interest groups concerned) (adapted from Arbter 2004, with kind permission of Neuer Wissenschaftlicher Verlag, Vienna Graz)

The strengths of the SEA Round Table are (Arbter 2004):

- Members of the SEA-team can pro-actively influence the SEA process and its results and they can voice their points of view effectively and at the right time
- Comments can be taken into account from the very beginning, when changes can be made without time loss
- The chance to reconcile the interests at stake during the SEA process and coming to broadly supported planning solutions with integrated environmental aspects
- Higher identification with the results and more acceptance
- Better chances to implement the recommended measures without delays and resistance
- Building up a base of confidence within the SEA team and network effects for the implementation of the strategic action

Similar approaches of consensus driven SEA processes in the waste management sector are known from Finland (Saarikoski 1997). Beside this pro-active participation model for stakeholders at the round table, the general public should be informed continuously and they should be given the opportunity to comment on interim and final results during the SEA process. This form of broader participation was missing at the SEA for the Viennese Waste Management Plan, but it was tested with another Viennese SEA, the SEA for urban and transport development in Vienna's North-East. There, several public forums were held and an SEA homepage (www.wien.at/stadtentwicklung/supernow; in German language) was provided. The experiences were convincing and a general recommendation for a combination of (1) pro-active stakeholder participation at the round table and of (2) information and consultation opportunities for the general public can be given also for waste management SEA.

42.6 Conclusions and Recommendations

In summary, waste management SEA have a lot of strengths and advantages:

- Guarding of strategic planning decisions: by taking environmental aspects into account as early as possible
- Optimising of decision making: by systematic evaluation of advantages and disadvantages of several alternatives
- Fostering sustainable solutions: by taking into account environmental, social and economic effects equally weighted
- Fostering unbiased discussions: by providing serious and well-founded information for decision making
- More acceptance and faster implementation of the strategic action: by transparent and well-supported decisions through public participation
- Democratising planning and decision making: by public participation

- Releasing Project EIA: by answering questions concerning the need, technologies (methods), capacities and locations in advance within the SEA process at strategic planning level

What might be unique regarding waste management SEA is, that SEA can foster a broad and comprehensive understanding of waste management, not only focusing on different waste treatment facilities, but also taking waste avoidance, reuse and recycling measures into account for strategic planning. SEA can also contribute to a serious, unbiased discussion about waste management measures, even if they are unpopular, i.e. new waste treatment facilities in larger cities or raised waste management fees.

At the first glance, waste management SEA might also have disadvantages because they can be time consuming and they cost money, especially if they build on sophisticated simulation models. But regarding the whole chain from the first waste management policy to the implementation of single measures, the opposite effect can be noticed. In the course of the processes time and even money can be saved if serious SEA are carried out, because time delays and costs for legal proceedings can be avoided, especially if acceptance for single measures is created through pro-active participation. The Viennese example already shows that the EIA for the recommended waste incineration plant – of course a really unpopular measure – could be handled without serious problems. One of the reasons for this was that waste avoidance measures were also taken with appropriate resources.

Summing up, if waste management SEA do not get lost in complicated simulation models without appropriate data, they are convincing in fostering high quality planning and better results for the environment.

References[1]

Arbter K (2001) Wissenschaftliche Begleitstudie zur SUP zum Wiener Abfallwirtschaftsplan – Endbericht (Scientific accompanying study to the Strategic Environmental Assessment of the Viennese Waste Management Plan). Bundesministerium für Land- und Forstwirtschaft, Umwelt und Wasserwirtschaft, Wien

Arbter K (2003) Mediated SEA: the Viennese experience. The environmentalist 15:19-22

Arbter K (2004) SUP – Strategische Umweltprüfung für die Planungspraxis von morgen (SEA – Strategic Environmental Assessment for tomorrow's planning practice). Neuer Wissenschaftlicher Verlag, Vienna Graz

Commission of Environmental Impact Assessment of the Netherlands (1991) Advisory guidelines for the environmental impact statement Ten Years Programme Waste of the Waste Management Council 1992-2002. Utrecht

Commission of Environmental Impact Assessment of the Netherlands (1994) Advisory guidelines for the environmental impact statement Ten Years Programme Waste of the Waste Management Council 1995-2005. Utrecht

---

[1] Note that legislation mentioned in the chapter is listed at the end of the handbook in the consolidated list of legislation (Appendix 2).

European Commission (1997) Case Studies on Strategic Environmental Assessment, Final Report, Volume 1. Brussels
Saarikoski H (1997) Environmental Impact Assessment in Strategic Waste Management Planning. Publications of the Finnish Environmental Institute 164, English summary
Thérivel R (2004) Strategic Environmental Assessment in Action. Earthscan, London
Verheem R (1998) EA in Dutch Waste Management: from Policy to Project. Short background paper for the October 1998 Austrian EU SEA Workshop

43 Mining Industry and SEA – An Example in Turkey

Evren Yaylaci

Brandenburg Technical University (BTU), Cottbus, Germany

43.1 Introduction

This chapter examines possible advantages of SEA for the mining sector and argues that project EIA is insufficient *on its own* for both environmental and social impact consideration and protection for the mining sector. Additionally, this chapter attempts to provide a preliminary evaluation of how SEA could improve the qualities of environmental and social impact assessment for mining industry in developing countries, such as Turkey.

The mining sector is important for Turkey due to two reasons. The first one is that electricity production is highly dependent on coal-burning in Turkey. For example, the electricity produced from coal and lignite increased from 44.2% in 1993 to 47.6% in 2000 (SSI 2004). The second reason is that Turkey exports coal, chromite, boron and copper on a worldwide basis (Nationmaster 2004).

In addition to these, the reason why Turkey has been chosen as an example is that Turkey demonstrates the general characteristics of a developing country. For instance, lack of environmental consideration in general and coal and lignite mining to supply the increased electricity demand, are similar to other developing countries world wide (IIASA and WEC 1998; UIC 2002). This situation has, in recent years, put enormous pressure on relatively clean and undisturbed environment of these countries. At the same time, untouched mineral resources in rural areas of developing countries are becoming economic for extraction, which could cause environmental and social damages if the social and environmental assessments would not be practiced well. In addition, Turkey has to improve the present legislation and administration considering environmental issues in order to fulfil the requirements for joining the European Union (EU).

This chapter has four sections. In Sect. 43.2, basic information about mining, such as the main phases and the characteristics of mining operations, some of the major environmental, social and economic impacts related to mining industry and the importance of SEA for mining sector are discussed.

Implementing Strategic Environmental Assessment. Edited by Michael Schmidt, Elsa João and Eike Albrecht. © 2005 Springer-Verlag

Sect. 43.3 briefly addresses possible challenges of SEA implementation and application in developing countries. The last section, Sect. 43.4, covers the conclusions and recommendations.

43.2 SEA and the Mining Industry

This section aims to discuss why a strategic action level environmental and social assessment of the mining industry investments is important and how SEA could help decision-makers in this regard. This section starts with a brief discussion about the mining industry while the following two sections cover SEA's importance and potential benefits for the mining industry on environmental, social and economic issues.

43.2.1 The Mining Industry

The mining industry is one of the most important pillars of industrialisation and high standard of living in our society. This is not only because it supplies the raw materials for the industry but also the amount of mining products used for energy production is still significantly high. For instance, Turkey produced 47.6% of its electricity in 2000 (SSI 2004) and approximately 40% of world electricity was produced in 2001 (IEA 2004) from coal and lignite, which shows the strong connection between the mining and the energy sector in general. For this reason, energy and mining projects are considered and discussed together in Sect. 43.2.2.

Unfortunately, the mining activities have several impacts on the environment, society and culture whilst providing invaluable raw materials to our standard of living. Since keeping the life quality at least at the same level is not possible without raw materials and consequently mining industry, what must be done is to minimise its negative impacts on environment, society and culture. At the same time, since the input of mining industry is non-renewable resources, making it a more sustainable enterprise must be another aim.

After World War II, the demand for non-fuel minerals sharply increased because the world population doubled between 1959 and 1990. The production of the six major base metals (aluminium, copper, nickel, tin, lead and zinc) increased more than eightfold (Azcue 1999). During the five decades since World War II, as a result of this increasing demand in mining industry products, it caused a sharp supply effort in regions rich of natural resources, especially in developing countries.

On the contrary, the environmental protection and sustainable development conscience arose over the last three decades throughout the industrial world and relatively later, in developing countries. As a result the much slower development of environmental protection conscience in the public and environmental protection legislations in governmental policies, during 1970s and mid-1980s mining activities resulted in very serious environmental problems in and around mining areas.

Mining is a set of operations aimed at the exploration, discovery, and extraction of minerals of economic value. This procedure is controlled by market conditions that respond to the law of supply and demand (Anon 1998). The different principal phases of mining activities that can be summarised as:

1. Exploration
2. Project development
3. Mine operation-extraction
4. Transport and storage
5. Processing
6. Mine closure and reclamation (restoration)

Each of these phases comprises of a sequence of processes and these processes can also have different impacts. For example, project development and mine operation have dumps as waste and also they can cause land-degradation, ecosystem disruption, slope failures and acid mine drainage. Processing activities can cause environmental hazards like dust, emissions, tailings, leachate ponds slags as waste and ecosystem disruption, acid mine drainage, chemical leakages, toxic dusts and compounds of carbon/sulphur/nitrogen metal particulates. Transport and storage can cause land-degradation, ecosystem disruption, dust and chemical leakages (Auty and Mikesell 1998).

43.2.2 SEA's Importance for the Mining Industry

With developments in technology and with modern practices, many of the negative effects of the mining operations, mentioned in Sect. 43.2.1, can be reduced or avoided at the project level. Nevertheless, project EIA is insufficient on its own for covering and preventing some major environmental impacts of mining, such as cumulative and transboundary effects of mining operation. In addition to these, the capability of EIA to develop strategic actions, such as national energy and mining policy and programmes, is limited because EIA considers a specific project and it evaluates the project locally. On the contrary, for example, to develop a national level energy policy needs consideration of alternative ways of energy production, i.e. thermo-, hydro-power and renewable energy, for short, intermediate and long term. At the same time, environmental, economic and social impact evaluation of these alternatives in local, regional and national level must be carried out.

In this respect, although EIA has given significant acceleration about the environmental consideration in the mining industry, it does not cover all expectations to contribute to the overall goal of sustainable development, which is minimising the negative impacts of mining and energy industry. At the same time alternatives to maintain of current production rates should be developed and investigated. Therefore, to be able to maximise the benefits of environmental assessments for the mining industry and to consider and compare the alternatives of the mining operations is very important. For instance, renewable energy instead of coal-burning, recycling rather than preliminary raw material extraction could also eliminate the possibility of poorly develop projects' negative impacts on environment and society. Therefore, higher level consideration of mining operations, which can be strategic action level, could be helpful.

As Partidário and Clark (2002 p.3) discuss, "decision-makers increasingly believe that Strategic Environmental Assessment has the capacity to influence the environment and sustainability nature of strategic decisions". The main reasons of why SEA can be of potential benefit for the mining and energy industries and its potential advantages will be highlighted in the following sections.

Environmental Issues: Early Consideration of Possible Environmental Impacts

Mineral extraction and mineral processing create a number of environmental problems at the both local and national level. All these potential hazards can appear because of the different principal phases of mining activities, which are discussed in Sect. 43.2.1.

The negative impacts of the mining activity on environment in different stages should be considered separately. This *consideration* must be started at an *early stage of mining activity* in order to avoid serious future pollution. In addition to this, the operation can be more economically effective, if this consideration starts early stages because once pollution appears it needs a very difficult process in terms of time and money to overcome these problems. Especially an early consideration of cumulative impacts is not possible on project level because as Nierynck (2000 p.4) points out "while efforts have been made to address cumulative impacts issues by extending the scope of project-level EIA, in general, the time and space boundaries at this level are inappropriate scale of the problems."

In this respect, as Nierynck (2000 p.4) discusses, "SEA can serve as an *'early warning system'* by allowing the early identification and management of cumulative impacts, e.g. the accumulation of impacts from multiple projects, directly or indirectly influenced by a policy, plan or programme".

Environmental Issues: Long Term Monitoring

In addition to the importance of an early consideration of potential negative impacts of mining, the *long term monitoring* after the closure of a mining site is very important for environmental protection because of the actual environmental damage caused by solid mine wastes deposited in the tract (Eggert 1994). When water and oxygen interact and are oxidised with solid base metal, precious metal, uranium, diamond and coal mines' wastes, where the sulphide minerals are common, acid mine drainage is created (Ersan et al. 2003). For this reason, long term monitoring is necessary to ensure that closure and reclamation activities are carried out completely and successfully.

Although the consideration of potential impacts and long term monitoring stages are included in EIA, especially in developing countries the ability to enforce laws and monitoring performance is relatively lacking. Even in developed countries, due to limited resources and technical expertise, many mines are not frequently inspected. Therefore, as Russo (1999 p.363) points out "one does not have to search very hard to find an energy project that does not have all of the required environmental and social mitigation measures promised in the EIA". In this regard, Table 43.1 highlights the lack of monitoring and enforcement with the example of the United States and some developing countries. (Data for Turkey is not available.) As it is seen in Table 43.1, for instance, in Venezuela one inspector is available for inspecting 133 mine sites and this ratio is very close or even higher in some states of USA, such as one inspector for 191 mine sites in Utah, USA. However, the advantages of the most developed countries are the ability of preparing well planed projects, due to experience and financial strength, having strong laws and regulations in use.

Another reason for the insufficient long term monitoring ability at the project level, in addition to lack of technical expertise, insufficient budget and weakness of the legislative background in developing countries, ministries and other governmental agencies have generally poor intuitional capacity in performing the day-to-day activities that ensure a healthy mining and energy sector. Asking to the same organisations to implement environmental and social protection programmes and monitor them successfully is for a long time very problematic (Russo 1999). Therefore, a long-term view and strengthening the long term monitoring capacity with strategic actions can help the project EIA for the mining projects. Another important possible gain of SEA in terms of long term monitoring can be that the long-term view of SEA could serve as a standard for all mining projects. As a result, the long term monitoring activity is followed more efficiently and the quality of the monitoring activities for all the projects would be same with limited budget, equipment and technical expertise.

Table 43.1. Capacity to Monitor and Enforce Laws (extended from Miranda et al. 1998, cited in Miranda et al. 2003)

Country or State	Number of Inspectors	Number of Mines	Ratio of Inspectors to Mines
Zimbabwe	4	300	1:75
Venezuela	3	400*	1:133
Arizona, USA	13	538	1:41
Colorado, USA	15	1944	1:129
Idaho, USA	6	65	1:11
Montana, USA	20	1100**	1:55
Nevada, USA	13	225	1:17
New Mexico, USA	7	185	1:26
Utah, USA	4	766***	1:191

*Includes exploratory Concessions
**Includes 1000 "notice mines" (5 acres or less)
***Includes 334 "notice mines" and 326 mining exploration projects

Environmental Issues: Cumulative Impacts of Mining Operation

Another important reason why higher level environmental consideration is important for mining investment is that project-specific environmental assessment tries to prevent possible damage by placing specific limits on the amount of harmful emissions and by assessing fees on emissions above a certain level (Auty and Mikesell 1998). In other words the 'polluter pays' approach is widely used to prevent possible damage. Nevertheless, in case there are more than one mining site's and/or industry's complex or there is a combination of these situated in a specific region, and although all the individual harmful emissions of the operations are lower than allowed limit, the cumulative emission (cumulative environmental damage) may be higher than the regulated limits. In such a case, the cumulative negative impacts of these operations can not be foreseen on project level because "cumulative impacts may be experienced after long delays and at a distance from their causes" (Barrow 2003 p.113).

For these reasons, with a higher level of environmental assessment, cumulative effects and cumulative negative impacts of mining operation may be foreseen. As an example, Turkey has one of the biggest marble reserves in the world and about 90% of the country's total quarries are located in Western Turkey, mainly on the Marmara Island, in the Marmara Sea and in the province of Afyon. Approximately 65% of Turkey's total marble production is supplied from this region. Marble is generally produced in small-scale and in unsophisticated quarries. The number of operating marble quarries is estimated to be more than 1 000 (Yildiz 2003). Consequently, approximately 650 marble quarries are working in the western part of Turkey and considerable amount of them are located in the same province, named Afyon.

Although, all of these quarries are considered on a project level environmental assessment and some of them have EIA (for quarries having higher than 5 000 m^3/year capacity EIA is mandatory, MoEF 2004), the cumulative environmental impacts of these workings in the region are unknown. For example, even all these workings' possible harmful emissions, i.e. dust, can be controlled individually; the air quality of the region may be lower than the allowed limit. Additionally, the future projects in the region will contribute cumulative air quality problems. One possible solution for this situation is to not to allow new operations in the region in order to be able to reduce and control the cumulative impacts, but this is not an applicable solution for mineral exporter developing countries because generally economic profits are considered before environmental protection.

However, with strategic actions, at least additional environmental problems of future operations can be predicted more realistically and some precautions may be improved. Secondly, to the cumulative impacts of future projects in a region also contributes the need for increasing infrastructure for operations, processing and transportation. At the same time, urbanisation sourced from new job opportunities will cause new environmental impacts in the region.

Therefore, for such complex and interlocking situations, making quantitative predictions and evaluations could be more realistic to see the possible cumulative future impacts and to improve alternative ways to overcome the problem (Barrow 2003). However, quantitative predictions, regional planning and development plans, are expensive and time consuming activities. Also they are not the responsibility of the action-leading agent, so project based environmental assessment does not consider all these impacts and all the possible ways of preventing them.

In the short term, although it may not be a direct solution of cumulative environmental problems, SEA can deal with many of these difficulties with its ability of "looking from the top".

Socio-economic Issues: Social Impacts of Mining Operation

As it is mentioned in Sect. 43.2.1, mining and energy industry investments cause several environmental problems. These environmental impacts affect the social life and the investments change the social-economic conditions in the region deeply. Therefore, all of these must be considered by not only the action-leading agent but also by the competent authority. However, it is often the case with a project-based environmental approach, e.g. in the EIA report, that *social impacts* are not considered in detail (Ortolano and Shepherd 1995) or even if they are considered, the final decision may not be objective because the decision is made based on the EIA report that is presented by the action-leading agent. In addition, social impact assessment (SIA) could be useful with EIA for consideration of social impacts during the decision-making process. Unfortunately, SIA is not used commonly and its possible benefits have not been understood yet also by mining companies (Joyce and MacFarlane 2001; Ortolano and Shepherd 1995). In addition two different assessment evaluation means interdisciplinary work of different governmental offices, NGOs and scientists, which reduces efficiency and increases the evaluation time because of lack of communication skills in developing countries (see Sect. 43.3).

Secondly, during mining, generally the investment in a region causes high rate of immigration. The reason is not only the main investment itself, but also the auxiliary industries that cause economical and social improvements. As a consequence, all or most of the economic activities depend on the mining operation. Additionally, in developing countries, to work at a mining company becomes more attractive than being a farmer in terms of income. This causes transformation of the local economic activities and is exacerbated by next generations during the lifespan of the mining operation. Therefore, local economic activities can change dramatically in the mining operation areas.

However, when the main investment ends, the region may be faced with both social and economic problems, e.g. high unemployment, reverse immigration and lack of production ability of non-mining related products, such as, agricultural products and other industrial products due to the loss of local economic activity skills during generations. As an example, the town of Maden case can be given. Maden is an administrative district of Elazig Province in Turkey. The settlement in the region is very old, due to the rich copper reserves. Therefore, the copper mining is the main source of economic activity during Maden's lifespan (GOoEP 2004). This 'traditional' economic activity changed in mid 1990's because, the mining operation had been shrunk and the main economic activity had become insufficient to employ the new generations in the region. As a consequence the new generations started to leave the town in order to find jobs in the big cities, because during the mining operation no other industry had been invested in the town.

Therefore, if such a project had been implemented with a long term planning, possible long term social problems could have been prevented by the development of programmes and plans. At the same time, the transformation of the town to a self-sufficient livelihood could be more effective and faster. This is because, with project EIA, the action-leading agent can not be only responsible to improve and implement development programmes and plans for possible future unemployment, caused from changes in the traditional economic activities in the region. Additionally, to develop new working areas for the locals after projects end also can not be covered with project EIA by the action-leading agent. Another important reason is that to predict the possible future social impacts of an investment is the decision-makers responsibility, therefore with SEA's early warning and wide information sharing possibility (cross-and inter-departmental communication) can serve as a social impact prediction and evaluation system for decision-makers.

Socio-economic Issues: Economic Reasons

Thirdly, in addition to environmental and social impacts of mining activities, there are several economy-based reasons, which support the need of strategic action level deliberation of mining sector. Market conditions, economical and political conditions in a country are very important for investors, because mining is highly cyclical and capital-intensive industry, with a long lead time between initial investment and commercial production (NRCAN 2004). This means investors want to see political and also economical stability before making the investment.

For this reason, developing countries should improve long-term programmes that give guarantee on no change of conditions during investment especially for high capital intensive investments to make the country safe for international investors. As an example, after the change of the related mining law in 1985, 17 international gold companies started to search gold deposits in Turkey but due to some political problems and public rejection based on poorly developed environmental reports of mining projects, in 2000 only three of them have started to invest. Other 14 companies left Turkey after having decided that conditions were not suitable for investment (General Directorate of mineral Research and Exploration of Turkish Republic 2004). As it can be seen here, SEA does not only serve as an environmental consideration and protection mechanism but also economy and development instrument for developing countries.

Socio-economic Issues: Public Participation

Last but not least, lack of public participation in decision-making process of mining and also energy investments is another important subject. Insufficient public participation can lead to problems between the public and the decision authorities and as a result, projects may be postponed or scraped. This means loss of time and money for both decision authorities and investors.

Importance of public participation is widely accepted, because as Zillman et al. (2002) discuss, the understanding *public concerns* in advance can strengthen the credibility of the policy, plan, programme or projects' decision in several ways, such as, decision makers may need *to build a consensus* among different or opposing interests. In addition to these, public involvement can help develop the credibility and trust in the decision-making process that is needed for such a consensus, and over time, a visible commitment to understanding and responding to public concerns can help *build a sense of public trust* and credibility in the decisions of the department or agency.

As it is known, public participation is a part of EIA procedure and it must be followed by the action-leading agent to complete the EIA procedure. It aims basically to gather economic, environmental and social information from the public and to inform the public about potential actions or alternatives, and the potential consequences of these actions (FEARO 1988 cited in Ortolano and Shepherd 1995). But the problematic point of public participation in the EIA procedure is that it often occurs too late to attain fully its aims because "the proposed project has already been conceived, and thus key decisions on project size and location will have already been made" (Ortolano and Shepherd 1995 p.20). Therefore, on project level it is not realistic to expect that the project is altered completely to satisfy the public, which may put extra costs or need complete changes in the way of operation. For example, to change the way of exploitation or processing, to prepare better mine closer and reclamation plans. To make this clear with an example, Bergama-Ovacik Gold Mine Project in Turkey can be discussed.

Ovacik Mining Site is situated next to the village of the same name, 12 kilometres west of Bergama (ancient Pergamon) in western Turkey. The site is close to many archaeological centres – Troy, Assos, Ephesus, and Pergamon, which attracts thousands of tourists every year, within easy travel distance (Newmont 2002). The mining site covers some part of the village and lies on the highly productive agricultural area (AoTAEC 2001).

The gold reserve was discovered in 1989 in the site. A feasibility study was conducted in 1992 and the EIA report of the Ovacik Gold Mine was approved by the Turkish Ministry of Environment in 1993. Finally, the Normandy Mining's US $45 million project was approved in late March 1996 and the construction of the plant finished in 1997 (Burton 1996, AoTAEC 2001). The project covers an open pit and an underground mine in the same site.

The problems that lead to the public rejection about Ovacik Mine, can be summarised very shortly: the processing of the mine runoff by cyanide leaching and the disagreement between the land owners and the company. As a result of public rejection, the operations had to be interrupted from time to time and therefore the company lost not only money but also the public trust, due to its lack of enthusiasm for public participation and failing to develop consensus during project preparation. The problem is not sourced directly from insufficiency of EIA alone, but developing country specific problems as well (see Sect. 43.3). The following conclusions can be drawn from this case study:

- Although during the EIA process the project was introduced to the local people, it seems the possible actions, such as use of cyanide and their consequences were not clearly explained. This can be inferred because the public rejection started after the preparation of an EIA report when the use of cyanide started to

be talked in public by scientists and NGOs. Therefore, this example shows how crucial it is to consider public opinion at proposed project level, especially with subjects like cyanide use or mining in a national park.
- Some other international companies are planning new gold mining operations in Turkey and if these companies decide to start operations, public participation is necessary because the number, variety and interests of stakeholders are different and due to the lack of public consensus about cyanide use and possible profits of gold mining for Turkey, new projects could also face public rejections at local and national levels.
- In addition, other economic activities in the region, such as agriculture and tourism, must be considered and the economic and social benefits of mining at the short, medium and long term must be compared by decision-makers. Additionally, the evaluation of these must be carried out quantitatively on the higher level to avoid speculations about the decision and also different interest groups' manipulations that may disturb the sense of public trust. In this respect, as Wiseman points out (2000 p.157) "environmental assessment tools, which can address issues at a level above the project-specific, are needed".
- Finally, gold reserves and possible economic profits of them had been discussed in different platforms to influence the public opinion but these discussions resulted in not more than speculations. Therefore, a platform is needed to discuss costs and the benefits of the extraction of the minerals and to prevent mismanagement of the non-renewable resources in Turkey. As it is seen in this example, project level is not enough as such a platform for developing a public consensus and preventing the mismanagement of the mineral resources.

43.3 Challenges of SEA in Developing Countries

In Sect. 43.2, possible advantages and strengths of SEA for the mining industry in Turkey, and possibly in other developing countries, were discussed. In this section the reasons, why it is not easy to make SEA a useful tool for reducing the environmental, economic and social problems derive from mining operations and maximizing benefits of mineral extraction, are discussed.

Due to the concept of SEA, successful application highly depends on several factors and these factors derived from or are affected by social, economic and political conditions in a country. Possible barrier factors for successful SEA application for mining operations in developing countries can be summarised as:

1. Organisation and coordination problems
2. Corruption
3. Political instability
4. Technical problems
5. Financial problems

Systematic and continual analyses of environmental impacts (see definitions given by Thérivel et al 1992; Sadler and Verheem 1996, cited in Nierynck 2000) and consideration of alternatives, for instance coal mining for thermo-electric, dam construction for hydro-electric, natural gas and renewable energy applications, are the key elements of SEA. In this respect, to be able to manage the systematic and continual analyses for early consideration and long term monitoring of mining sites and evaluation of alternatives, good organisation, coordination and communication skills are needed in the governmental hierarchy, between different ministries, governmental agencies, NGOs, scientists and public. On the contrary, in most developing countries, the organisation, coordination and communication are highly problematic both in the governmental hierarchy and between the government public relations.

Corruption maybe one of the key reasons of mismanagement of the natural resources and it reduces the effectiveness of the legislations in use. There are several issues contributing to corruption and some of them are lack of transparency, lack of access to information, lack of access to justice and lack of public participation in decision-making. When we consider the objectives of the SEA, such as, public participation for improving the quality of the strategic action, making the decision-making process democratic and transparent, it is seen that some of these objectives can not be generally processed as it should be due to corruption.

Maybe the one of the most important characteristics of SEA is its long term approach. However, to see a long term political stability, which affects the feasibility of strategic actions, is not generally possible in developing countries (Barrow 2003). It is clear enough that without political strength to apply well prepared and organised SEA and strategic actions is not possible. In addition to these, even the strategic action and SEA is well prepared for the mining sector, if the legislation background does not support the strategic action then the feasibility of SEA decreases. For these reasons, political instability causes poor quality environmental laws and regulations, which can lower the feasibility of SEA.

Technical problems cover mainly the lack of technical expertise and database, lack of monitoring and controlling. Actually, when we consider the monitoring and controlling, maybe the most important criteria is political will. Additionally, the ability, capacity and quality of the monitoring and controlling depend on the qualified technical and/or scientific expertise. However, in Turkey for example, generally because of the organisation and communication problems, the limited technical expertise and equipment cannot be used efficiently. In addition to this, some NGOs have scientific and technical experts due to the lack of public participation they cannot be a part of strategic actions. Another point is that databases, e.g. air and water quality, bio- and cultural diversity, in developing countries are very limited and generally it is hard and burdensome to fulfil the clumsy bureaucratic procedures to get these data. This reduces the effective evaluation and reduces the chances to give feasible decisions about mining operations.

Last but not least, finance can be a problem or quality lowering barrier for SEA in developing countries. Although SEA's short time frame, low cost and de-emphasis of final report is an advantage (Brown 2000), due to the economical instability and economy crises, developing countries face fast and unexpected economic and social changes. These changes mean losing time and effectiveness of the SEA.

43.4 Conclusions and Recommendations

Since the introduction of EIA, some 35 years ago, the world has changed dramatically. On the one hand, the overall goal of sustainable development presents a new vision, with its long term consideration and higher level evaluation of all actions and their consequences on environment, economy and society. On the other hand, although EIA has gained very big acceleration of environmental protection at the project level, without higher level evaluation the *project-to-project approach of EIA does not guarantee environmental quality* because as Wood (1988) and Barrow (2003) discuss EIA is not sufficient above project level for environmental protection. For these reasons, in addition to EIA, for higher level strategic actions, SEA can be used successfully to avoid or reduce environmental and social negative impacts of mining investments with evaluation of alternatives, cumulative effects and that both economic and environmental benefits can be derived from the application of SEA (Nierynck 2000).

In addition to these, the possible advantages and strength of SEA for mining sector can be summarised as:

- the consideration of alternatives especially before energy investments is very important, because to make changes after investment of a mining activity or construction of a dam has started cause irreversible negative impacts on environment and society and cause loss of huge amount of money. Therefore, SEA's to encourage examination of alternatives in all levels (programme, plan and project level) approach (Partidário and Clark 2000) is very important.

- a strategic action level environmental assessment would provide a chance to alleviate or abandon environmentally unsound concepts before they were turned into projects (Ortolano and Shepherd 1995) that could mean time and money saving for both potential action-leading agent and the competent authority.

- in the short term, SEA could increase the environmental consideration much faster as it has been in developing countries, especially for contributing to the controlling and monitoring activities in the mining areas. Additionally, providing a minimum environmental quality with the existing regulations could be possible with well develop strategic actions. As a result of this, the environmental consideration in the mining industry could be integrated for full mine life cycle, "from cradle to grave".

- in the long term, better understanding of sustainable development, optimum resource utilisation and implementation of direct mineral-derived wealth into other activities in affected communities, such as small business, renewable and local sustainable local utility development, health and protection of environmental quality, so that the country is not left worse off than if the mineral wealth had never been generated (Epps 1996).

The possible disadvantages and weaknesses of SEA for the mining sector can be concluded as;

- the main possible weakness of SEA could base on the lack of experience and lack of experts for the mining sector in Turkey
- the interactive approach, a level of cross- and inter-departmental communication and cooperation abilities are generally very low in Turkey that could reduce the power and strength of SEA application
- in general as Ortolano and Shepherd (1995 p.17) discuss "programme and policy decisions often evolve over time, making it difficult to identify what constitutes the program"
- finally, Ortolano and Shepherd (1995 p.17) believe that "the scope of a programme could be hard to define, both spatially and temporally, and this makes assessing impacts even more uncertain than usual"

Finally, the recommendations for further studies about SEA structure and specific sectoral application of strategic action assessments in Turkey can be summarised as:

- Due to the very limited scientific study about SEA and different sectoral applications of SEA, it would take some time to develop well planed strategic actions in Turkey. Therefore, further studies about SEA and pilot sectoral application of SEA should be carried out.
- Brown (2000 p.133) discusses "objective of the SEA must be much more than the production of an SEA report". In this respect, as it is very common for EIA process in Turkey, project EIA is often seen as a way of justifying a development, not reducing problems (Barrow 2003). Therefore the mentality of putting the main effort on the preparation the paper documentation must change in order that high beneficial strategic action level assessment can be developed.

References

Anon (1998) Fundamentals of Mineral Law and Policy, Natural Resources and Sustainable Development, Mineral Resources Forum education documents. Internet address: www.natural-resources.org/minerals/education/docs/Mineral%20Law%20&%20Policy-Unit1.pdf, last accessed 16.08.2004

Arzcue JM (1999) Environmental Impacts of Mining Activities: Emphasis on Mitigation and Remedial Measures. Springer, Berlin

AoTAEC – Association of Turkish Architects and Engineers Chambers (2001) Chambers of Chemistry-Environment-Geology- Metallurgy Engineers' Critique Report of Bergama-Ovacik Gold Mine TUBITAK-YDABCAG Specialists Commission's Report, Ankara

Auty RM, Mikesell RF (1998) Sustainable Development in Mineral Economies. Clarendon Press, Oxford

Barrow CJ (2003) Environmental and Social Impact Assessment: An Introduction. Arnold London, Oxford University Press Inc, New York

Brown AL (2000) SEA Experience in Development Assistance Using the Environmental Overview In: Partidário MR, Clark R (eds) Perspectives on Strategic Environmental

Assessment. Lewis Publishers Boca Raton, London New York Washington D.C., pp 131-141

Burton B (1996) Normandy's Turkish Foray, Mining Monitor Vol. 1 No 2. Internet address: http://www.minesandcommunities.org/Company/bergama1.htm, last accessed 16.08.2004

Eggert RG (1994) Mining and the Environment: International Perspectives on Public Policy, Resources for the Future. Washington D.C.

Epps JM (1996) Environmental Management in Mining: an International Perspective of an increasing Global Industry, Journal of the South African Institute of Mining and Metallurgy 2:67-70

Ersan HY, Dagdelen K, Rozgonyi TG (2003) Environmental Issues and Eco-Based Mine Planning. In: Ozbayoglu G (Ed) Proceeding of the 18th International Mining Congress and Exhibition of Turkey (IMCET). The Chember of Mining Engineers of Turkey Press, Ankara, pp 11-17

Gavorner's Office of Elazig Province (GOoEP). Internet address: www.elezig.gov.tr, last accessed 6.09.2004

General Directorate of Mineral Research and Exploration (MTA) of Turkish Republic. Internet addres: http://www.mta.gov.tr/forum/tral-ov.asp, last accessed 16.08.2004

International Energy Agency (IEA) (2004) Energy Statistics: Evaluation of Electricity Generation by Fuel from 1971 to 2001. Internet address:
http://www.iea.org/statist/index.htm, last accessed 14.09.2004

International Institute for Applied Systems Analysis (IIASA) and World energy Council (WEC) (1998), Global Energy Perspectives, Computer Based Learning Unit,University of Leeds. Internet address:
http://www.iiasa.ac.at/Research/ECS/docs/book_st/wecintro.html, last accessed 14.09.2004

Joyce SA, MacFarlane M (2001) Social Impact Assessment in the Mining Industry: Current Situation and Future Directions, Mining, Minerals and Sustainable Development of International Institute for Environment and Development Report No: 46, IIED and World Business Council for Sustainable Development (WBCSD)

Ministry of Environment and Forest of Turkish Republic (MoEF) (2004) Environmental Impact Assessment Legislative Procedure, Ankara, Turkey. Internet address: http://www.cedgm.gov.tr/cedyonetmeligi.htm, last accessed 16.08.2004

Miranda M, Burris P, Bincang JF, Shearman P, Briones JO, Viña AL, Menard S (2003) Mining and Critical Ecosystems: Mapping the Risks Report, Appendix 2, World Resource Institute. Internet page: http://materials.wri.org/pubs_pdf.cfm?PubID=3874, last accessed 16.08.2004

Nationmaster (2004) Internet address: www.nationmaster.com, last accessed 17.05.2004

NRCAN – Natural Resources Canada (2004) Internet address:
http://www.nrcan.gc.ca/miningtax/inv_2.htm, last accessed 16.08.2004

Newmont's internal newsletter (October 2002), The Gold Standard, Ovacik Enjoys Success, Despite Challenges. Internet address:
http://www.newmont.com/en/ourbusiness/operations/other/ovacik/minestory.asp, last accessed 13.12.2003

Nierynck E (2000) Strategic Environmental Assessment, Capacity Building for Environmental Management in Vietnam, Vietnam Information for Science and Technology Advanced (VISTA). National Centre for Scientific and Technological Information Documentation. Internet address:

www.vista.gov.vn/VistaEnglish/VistaWeb/learn/Env.%20Planning/texts%20in%20Eglish/P1Chap6.pdf, last accessed 13.09.2004

Ortolano L, Shepherd A (1995) Environmental Impact Assessment. In: Vanclay F, Bronstein DA (eds) (1996) Environmental and Social Impact Assessment. John Wiley & Sons Ltd, Chichester, pp 3-31

Partidário MR, Clark R (2000) Perspectives on Strategic Environmental Assessment. Lewis Publishers Boca Raton, London New York Washington D.C.

Russo T (1999) Environmental Impact Assessment for Energy Projects In: Petts J (ed) Handbook of EIA Vol: 2, Environmental Impact Assessment in Practice: Impact and Limitations. Blackwell Science Ltd., Oxford

State Statistics Institute (SSI) of Turkish Republic (2004) Internet address: www.die.gov.tr/istTablolar.htm. last accessed 06.09.2004

Uranium Information Centre (UIC) Ltd, (July 2002) World Energy Needs and Nuclear Power, Nuclear Issues Briefing Paper 11. Internet address: www.uic.com.au/nip11.htm, last accessed 14.09.2004

Wiseman K (2000) Environmental Assessment and Planning in South Africa: The SEA Connection. In: Partidário MR, Clark R (eds) Perspectives on Strategic Environmental Assessment. Lewis Publishers Boca Raton, London New York Washington D.C.

Wood C (1988) EIA in Plan Making. In: Wathern P (ed) Environmental Impact Assessment: Theory and Practice. London: Unwin Hyman, pp 98-114

Yildiz N (2003) Mining Sector in Turkey, Ankara. Internet address: www.maden.org.tr/yeni3/english/mining_sector_in_turkey/miningsectorinturkey%202003.pdf, last accessed 16.08.2004

Zillman DN, Lucas AR, Pring G (2002) Human Rights in Natural Resource Development: Public Participation in the Sustainable Development of Mining and Energy Resources, Oxford University Press, Oxford

Part VIII – Conclusions

This final Part concludes the Handbook by discussing issues related to best practice and the future of SEA. Chapter 44 addresses capacity building and SEA. The chapter discusses the issue of capacity-building needs for more rigorous implementation of SEA. Capacity building aims to engage existing human, technical and institutional resources in the most appropriate and effective way. The chapter evaluates the policy, institutional and technical capacity-building needs for good-practice SEA.

Chapter 45 presents a critique of SEA from the point of view of the German industry, where SEA is seen as not being of additional value over and above Project EIA. In contrast, Chapter 46 discusses best practice in relation to the use of SEA for energy and industry in the UK. The chapter examines why SEA might be attractive to industry and how SEA can be applied within a private sector organisation. The chapter uses a regional transmission network SEA case study undertaken by a privatised electrical utility. The case study is used to evaluate how SEA frameworks can be established inside a private sector organisation and the practical issues that have to be addressed by corporate decision-makers. Finally, chapter 47 concludes the Handbook with an 'SEA outlook' by discussing SEA's future challenges and possibilities.

44 Capacity-Building and SEA

Maria do Rosário Partidário

Universidade Nova de Lisboa, Portugal

44.1 Introduction

Many interpretations of what SEA is and can deliver appear to currently exist around the world. Previous chapters provide a fair display of forms and approaches to SEA as currently used in different countries and jurisdictions. Three main reasons could be suggested to explain why all these multiple forms appear to exist. The first is related to the certainly different cultural interpretations of what strategic decision-making is about. The second may be the stage of incipiency, or immaturity, of the concept of SEA as a decision-support tool, given the early learning stage about its role and capacity to inform and facilitate strategic decision-making. Finally the third reason for the existence of multiple interpretations of what SEA is about may be a sign of its inappropriate utilization as an assessment tool. Not only it is often suggested as a promising panacea that will resolve the insufficiency of Project EIA but there are actually cases where SEA has been inadequately used, replacing EIA. As a consequence this runs the risk of making SEA a non-specific, and perhaps even a useless, tool in the near future.

This chapter discusses the issue of capacity-building needs for wise and more rigorous implementation of SEA. Capacity-building is at the vanguard of all sound and balanced development processes that engage existing human, technical and institutional resources in the most appropriate and effective way. While the "learning by doing" principle is important and must be encouraged, it also needs to be backed up by consistent drivers that, based on actual results, stimulate innovative actions and wise practices. This also applies to SEA.

The starting point for these reflective notes is based on the author's experience in running professional training courses on SEA over the past six years in national and international contexts, with sectoral and multi-sectoral focus. This experience involved the training of, and interaction with, about 500 different professionals from various parts of the world. The chapter will elaborate on the policy, institutional and technical capacity-building needs for good practice SEA.

Implementing Strategic Environmental Assessment. Edited by Michael Schmidt, Elsa João and Eike Albrecht. © 2005 Springer-Verlag

It starts by exploring the meaning of capacity-building for SEA in Sect. 44.2, discusses the priorities for SEA as a strategic impact assessment tool in Sect. 44.3, thus leading to an elaboration of needs and key drivers in SEA capacity-building in Sect. 44.4. General guidance for SEA training, in the form of rules-of-thumb, is offered in Sect. 44.5, before Sect. 44.6 concludes with recommendations on the training needs for SEA as a trigger for enlarged institutional and professional capacities.

44.2 Meaning of Capacity-Building for SEA

Building capacity is the process of creating conditions for effective and efficient performance and improved capabilities to do something. Many definitions of capacity-building can be found in the literature. Box 44.1 shows a comprehensive definition for capacity-building provided by the Human Strategies for Human Rights in their website (HSHR 2004). A range of definitions on capacity-building are available at the Adept website (ADEPT 2004).

When developing an approach to building capacity for SEA it is first necessary to make very clear "what is" that SEA is expected to perform and deliver. Considering the multiple forms of SEA that exist around the world, as previously mentioned, it is critical to clarify first what is the purpose, or objectives, of SEA, what is the role that SEA is expected to perform, and what is the format that SEA will adopt. And subsequently identify the factors and activities for which capacity needs to be built.

Considering first the purpose of SEA and what is that SEA is expected to perform and deliver, it would be a waste of time to attempt to unite the various concepts and practices put forward in this book into a coherent and meaningful standard approach to SEA.

Box 44.1. Human Strategies for Human Rights Definition of Capacity-Building (HSHR 2004)

Capacity building is a continuing process that enables the sustainability of a professional, relevant and legitimate organization that works to fulfill its stated mission. It is not just training but rather a combination of factors/activities focused on the improvement of an organization's performance in relation to its mission, working environment and practical resources.

To varying degrees, it includes the investment in property and equipment, training, information and communication strategies, personnel, and private and public sector relationships. The primary goal is to increase an organization's effectiveness in a changing social, economic and political environment. This is achieved by improving internal management structures, working processes and procedures, as well as strengthening partnerships among various actors in the development market process.

Organizational capacity building is about strengthening institutions that deliver services or products to those outside of the organization. An organization with strong capacity should be able to survive the departure of its Founder or Senior Executive.

SEA is, and will always be, different from country to country. And most probably it will be as different as the underlying decision-making systems to which it will apply (Bina 2003; Partidário 1996, 2000). But this is not the same of saying that all possible forms of SEA will be equally good in addressing the following four fundamental aims of SEA:

- informing decision-making in a timely manner;
- enabling the integrated consideration of all relevant issues in decision-making;
- improving the quality of decisions;
- making overall development processes easier and more sustainable.

In 2002 the International Association for Impact Assessment (IAIA) formally adopted SEA Performance Criteria (see Box 44.2). The performance criteria were developed throughout three years of discussions in IAIA annual conferences (between the 1998 IAIA Conference in Glasgow and the 2000 IAIA Conference in Hong-Kong), a discussion process led by Rob Verheem (an expert in EIA and SEA in the Netherlands and worldwide). Its purpose was to establish, on international grounds, what would be meant by good quality SEA process in view of enhancing the credibility of strategic decisions. The background assumptions were that a good quality SEA process should inform planners, decision makers and affected public on the sustainability of strategic decisions, facilitate the search for the best alternative and ensure a democratic decision making process. The performance criteria were tested by different authors in specific national contexts (for example see Fischer 2002 and ELARD 2004) and can be consistently indicated as a reference on how SEA should be performing as an impact assessment tool. For example ELARD (2004) used the IAIA criteria as the basis of a comparative framework to assess different possible models of SEA for land-use planning in Lebanon, and engaged different institutional views in that process.

In this chapter the performance criteria help to define what is internationally benchmarked as proper SEA, even though the criteria in Box 44.2 are likely to adopt different meanings and understandings in different contexts. Very often concepts such that of integration or that of sustainability raise the most diverse interpretations. As Sect. 44.3.2 further explores, words can be used alike but sometimes express very different meanings. This is also the case where cultural differences in decision-making can turn the standardization of such concepts totally irrelevant in situations where the shortage of evidence can be detrimental to the establishment of more detailed international ruling.

However in order to put some more flesh into the meaning of SEA as considered in this chapter, and therefore into a meaningful focus on capacity-building needs for SEA, Sect. 44.3 attempts to articulate key priorities to develop SEA into a strategic impact assessment (Strategic IA) sense. In other words, an SEA that is designed to address and work with strategies, rather than with solutions or outcomes, and that uses impact assessment, rather than planning, methodological approaches (Partidário 2004).

Box 44.2. IAIA SEA Performance Criteria (IAIA 2002)	
Integrated	• ensures an appropriate environmental assessment of all strategic decisions relevant for the achievement of sustainable development • addresses the interrelationships of biophysical, social and economic aspects • is tiered to policies in relevant sectors and (transboundary) regions and, where appropriate, to project EIA and decision making
Sustainability-led	• facilitates identification of development options and alternative proposals that are more sustainable[1]
Focused	• provides sufficient, reliable and usable information for development planning and decision making • concentrates on key issues of sustainable development • is customised to the characteristics of the decision making process • is cost and time effective
Accountable	• is the responsibility of the leading agencies for the strategic decision to be taken • is carried out with professionalism, rigor fairness, impartiality and balance • is subject to independent checks and verification • documents and justifies how sustainability issues were taken into account in decision making
Participative	• informs and involves interested and affected publics and government bodies throughout the decision making process • explicitly addresses their inputs and concerns in documentation and decision making • has clear, easily understood information requirements and ensures sufficient access to all relevant information
Iterative	• ensures availability of the assessment results early enough to influence the decision making process and inspire future planning • provides sufficient information on the actual impacts of implementing a strategic decision to judge whether this decision should be amended and to provide a basis for future decisions

What is critical is that addressing strategies requires strategic minds, and being strategic means:

- Look forward and beyond, in wider time and spatial contexts and scales;
- Adopt and keep such wide perspective, helped by objectives and targets;
- Ensure a logic, structured and flexible approach;

---

[1] i.e. that contribute to the overall sustainable development strategy as laid down in Rio 1992 and defined in the specific policies or values of a country.

- Find the right path (in a route plan) that enables a short time action towards the wide perspective;
- Re-align the route whenever needed by changing contexts.

44.3 Current Key Priorities for SEA as a Strategic Impact Assessment Approach

The review of current SEA practice stimulates ideas on many priorities for action to improve current problems that SEA is facing, and also to enhance the opportunities already identified. Box 44.3 offers a snapshot of good common sense issues that SEA, as a strategic impact assessment approach, should be addressing widely.

> **Box 44.3.** Strategic impact assessment should be about...
>
> - Setting minds for strategic perspectives
> - Working together
> - Using windows of opportunity
> - Feeding information that is really needed
> - Communicating
> - Avoiding conflicts (e.g. inter-generational, inter-sectoral, geographic)

Keeping this level of thinking, three priorities, that try to address these issues, appear therefore to be the most important at the present time:

1. Improve SEA focus on decisions and SEA relationship with decision-making
2. Improve communication skills and mechanisms in SEA
3. Make SEA attractive and increase win-win opportunities

44.3.1 Improve SEA Focus on Decisions and the Relationship of SEA with Decision-Making

The first priority is to make SEA *more focused on decisions* (causes) rather than on impacts (consequences) and *improve the relationship between SEA and decision-making* (Thissen 2000). SEA has to act as a facilitator in a process-driven rather than product-driven approach. Having as a key SEA requirement the production of a, more or less, fat report is perhaps not the most advisable way to approach strategic dimensions.

Strategic decisions can not afford to wait too long for very complete reporting. It is often said that information is only important if you can use it, but useless if not available. Therefore the purpose has to be to provide a decision-maker with the right information, at the right time, in a timely, short, concise and focused format, that avoids complexity and makes the information usable (Partidário 2000).

Strategic decision-making is often non-structured, made of small, incremental and iterative decisions taken at, sometimes unexpected, decision moments (Clark 2001). The key is to try to understand how the decision process functions, what are the main essential decision moments, what is that the decision-maker needs to know to help him or her carry on, and address their issue in their timing, when they need it. Such decision moments in strategic decision processes, or "checking points" (Partidário 1996), are decision windows for SEA, the right time for checking (externally or internally) the information needs of strategic processes or the information that should be fed in, because it is relevant, precise and useful for better decision-making.

44.3.2 Improve Communication Skills and Mechanisms in SEA

The second priority is *communication*. Terminology creates perspectives and expectations. Often the same term has different meanings and different terms mean the same thing. Current tensions with respect to understanding the purposes of multiple impact assessment tools, including SEA in its various forms, derive mostly from communication issues. For example SEA and sustainability assessment can be argued to be not much different, when broadly viewed through the perspective of strategic assessment mainly focused on the quality of development decisions (Bina 2003; Partidário 2004). However, as the multiple advocates of sustainability assessment argue (Devuyst 1999; Gibson 2001; George 2001; Lee and Kirkpatrick 2001; Noble 2002; Jenkins et al. 2003), there is a major difference in emphasis and scope of assessment between the SEA and sustainability assessment, and even perhaps between different sustainability assessment approaches (Pope et al. 2004). The same tensions in concepts and contents appear to exist in relation to the meaning of programme and planning and the rather random way in which countries use these terms, with programme being used either before, or following planning, thus implying very different concepts and strategic meanings.

The unclearness in speeches, or writings, for example through the misuse of different terms or in the absence of clarification of its associated meaning, is likely to generate tensions and consequent conflicts, and these can be responsible for important strategic impacts (such as the overlapping, or otherwise unresolved clearance regarding institutional responsibilities on the management of certain issues because of badly used terminology and respective definitions).

There is also a need to communicate within, and across, different scales and spheres of action (from policy analysis to project management), professional practices, disciplines and respective terminologies. As policy-makers, planners or impact assessment professionals it is necessary to think and act horizontally if we want decision-makers to act the same way. Efforts should be placed on communicating what is relevant, and leave behind what is not a priority. This may require social and political thesaurus that will translate problems into opportunities, values into costs, deep fundamentalism into pragmatic objectivity.

For example, the use of the strategic decision-makers' terminology should be part of the strengthening relationship and articulation of SEA with the decision-making process. There is a need to work better on communication capacities, to learn how to translate values into technical terms and how to develop common areas of understanding, spoken and written, with diverse disciplines and sectoral interests (Vicente and Partidário 2002).

44.3.3 Make SEA Attractive and Increase Win-Win Opportunities

The third priority is to make SEA more attractive and to make the best *marketing for strategic impact assessment*. It is crucial for the success of SEA as a strategic impact assessment tool to convince potential users, particularly high-level governmental senior officials in sectoral decision-making, or private strategic business decision-makers, of the benefits and added value that SEA can bring into decision processes.

Good communication and timely inputs of relevant information are actually powerful marketing tools in this respect. However, there is more to say about making SEA appealing. No high-level business decision-maker, or politician, will spend one minute with an issue or tool that does not bring benefits and advantages. At the end of the day, all comes down to costs and benefits, but not necessarily expressed in cashflow. For example saving time in licensing processes is always a benefit. Producing studies that can be truly helpful from a strategic management perspective can be a benefit. Producing fewer studies to get the same information is also a benefit!

SEA needs to be made more appealing. SEA needs to turn out as an incentive rather than as an obstacle. Problems must be turned into opportunities. Actors and promoters of SEA must get a quick understanding of what are priorities in the decision world, instead of keeping the focus on impact assessment priorities and spending the whole time searching for the right solutions, that will be irrelevant by the time they reach decision-makers when the decision moment is gone. Tools do not have priorities, decisions have.

44.4 What are the Needs and the Key Drivers in SEA Capacity-Building?

SEA has a distinctive role to play in relation to other tools such as Project EIA, cumulative impact assessment, policy analysis or planning. Despite its original roots as an impact assessment tool, SEA has a major role to play in creating and facilitating new thinking in decision-making: a strategic and an integrative thinking. Strategic thinking and integration are major players in support of a new kind of development that must be sustainable.

The priorities suggested in Sect. 44.3 require appropriate professional skills, adequate institutional settings and a platform of pragmatic, yet feasible, rules that encourage effective use for needed outcomes. Again there is no simple answer, the forms and shapes for capacity-building in SEA are multiple, depending on intended targets and cultural specificities.

Development is about realities, expectations, priorities and choice. These altogether make up the context in which SEA operates. These also make the essence of decision-making. Increased capacities for SEA are therefore inherently linked to improvement in decision-frameworks, including:

- policy design
- political engagement
- leadership (motivation, prioritization, governance)
- institutional architecture

Policies are, or should be, masters of development. In strategic approaches policies are the strings that operate the mechanisms. Strategically, the *design of policies* become a sensitive dimension in a tiered development system that is expressed through plans, programmes and projects and where explicitly stated policies are absent or ill-defined. The absence of policies can leave medium to long-term development without direction!

Decision-makers, particular at higer-level management systems, are the actual string operators. If their thinking is not in accordance with this new thinking, any effort made at later stages will be academic except to minimize or reduce impacts of previously taken decisions. Even in the absence of explicitely stated policies major decisions encapsulate political intentions. *Political engagement* and commitment provides the master energy.

Leadership provides direction, ensuring motivation, long-term perspectives, coherent and transparent priorities and rules of governance. It is inherently linked to political commitment and engagement, it involves strong attitudes and determined actions towards development. This is how SEA can have an impact on options made for future development. But SEA can only do that if designed to fit the context in which it operates, understanding and acting over social processes and political options, requiring technical analysis just to the extent that it reduces uncertainty and allows greater scientific confidence.

Box 44.3 argued that SEA should be about "working together" and "avoiding inter-sectoral or geographical conflicts". Vincente and Partidário (2002) and Genter (2004) provide theoretical arguments for the importance of communication in strategic processes, particularly in articulating the technical and the political speeches. These are social and political processes that run across institutional borders, fundamentally dependent on the *institutional architecture* and engineering that characterizes the context of decision-making, therefore critical to the success of SEA.

These four ingredients can be argued to define a capacity for better decision-making frameworks, and for better SEA, but which depend however from the fulfilment of certain conditions that determine the context for decision-making. The following may not be the only needed conditions, but are used here to describe critical political, technical and financial factors in enabling the right political context, and capacities, for strategic decision-making:

- Institutional cross-sectoral relationships, internally and externally. In other words, it requires cooperation and dialogues across domains and levels of responsibility, as much as across sectors. This is in order to enable cross sharing of views and priorities, both within the same and across institutions;
- Communication, new forms of participation schemes, terminologies, dialogues, and other mechanisms that enable constructive interaction, the establishment of perspectives, the raising of strategic thinking on one hand and the pragmatic agreement on feasible actions on the other;
- Resources – human and financial resources – the basic ingredients of any intended action, whether immediate or tiered in time;
- Technical and scientific capabilities, the know-how or the capacity to find solutions, perhaps not necessarily engineering solutions but often political solutions for polarized and apparently unsolvable conflicts.

Stimulated by the trade and impact assessment debate George et al. (2001) discussed political (institutional and legal), technical and financial components in capacity-building for trade impact assessment. Box 44.4 shows a summary of recommendations towards the set up of capacity-building programmes. Whatever the arrangement made and its classification, key issues tend to remain basically the same.

Box 44.4. Recommendations for Capacity-building programmes in Trade and Impact Assessment (George et al. 2001)

Political (Institutional and Legal) components:

- Legal basis or integration into the policy development process
- Integrated decision-making
- Gain ministers' support
- Coordination with other authorities
- Public involvement
- Improve procedural functions (refining screening and scoping mechanisms, enable access to reports, monitoring and corrective action)
- Leadership for increasing motivation

Technical components:

- Creating guidelines (process, sectoral, impact-specific)
- Creating and strengthening review system and ensure access to and quality of reports
- Establishing technical expertise
- Choice of appropriate methods
- Appropriate mitigation actions

Financial and human resources components:

- Funding the responsible (environmental) agency
- Improve cost-effectiveness
- Recruiting and training staff at responsible agency and building competent consultancies

Based on the needs for improved decision-making frameworks and the components of capacity-building suggested by George et al. (2001), Box 44.5 suggests essential drivers in the process of creating capacity for SEA. These include policy, institutional and technical drivers, but also financial and human resources, key issues in SEA that have been advanced by different authors on various occasions (Partidário 1996; Sadler and Verheem 1996). Such key issues must be adopted, put in place as part of capacity-building approach for SEA. Interestingly each of these issues have a specific role to play, whether providing the policy strings and the leadership, the institutional architecture and the power behind actions, or the technical needs and tools.

What is common to all these issues is that they make part of a system that can be described as a very general decision framework, the context in which SEA and strategic decision-making are likely to become more effective. Difficulties in implementation are however variable and are strongly related to the socio-cultural, economic and political characteristics of the decision-making contexts. Sect. 44.5 follows by addressing the particular component of training for SEA capacity-building. All of the issues identified in Box 44.5 are determined, or operated, by people, the human resources, critical factors in the effectiveness of any system.

Box 44.5. Key Drivers in SEA capacity-building

Policy drivers:
- policy framework for sustainability, environment, development
- policy interactions, priorities, policy tools to operate the essential policies
- decision-making structure, whether formal and informal, but focused, open and effective, marked by political engagement and inter-sectoral cooperation
- leadership in mastering processes of involvement, of dialogues, of prioritization, of values management and consensus-building to get choices that make up decisions.

Institutional drivers:
- institutional framework, relationships, interactions, mechanisms
- accountibilities, essential in decisions quality control
- power relationship - sharing and strengthening power

Technical drivers:
- cultural mind-set
- typical tools (planning, evaluation, decision)
- needs assessment, both in substance and in process
- guidance
- communication capacities
- participation tools, dialogues
- consensus-building traditional knowledge – a pool of essential rules-of-thumb and good common sense

Financial and human drivers:
- availability of financial resources
- cost-effectiveness practices
- expertise and competence of human resources

While capacity-building goes far beyond training, the fact is that trained resources are the key pieces in the mechanism that can make the success of any system, any decision framework, or otherwise its inoperation, and eventually its collapse.

44.5 Training for SEA Capacity-Building

Training, in particular professional training, has grown to be a major area of activity in the last decades, clearly competing with operational deliveries in project activity as far as consultancy is concerned. The demand is enormous, stimulated by several factors:

- the emergence of new tools that have widespread application;
- a fast development of new information;
- increasing job opportunities for professionals in countries at all levels of development;
- the need to know more, to have more information – the power of being informed;
- the importance of organizational learning processes.

The definition given on capacity-building at the opening of this chapter clearly states that capacity-building is more than training. While not filling up capacity-building needs, training has been so far an indispensable activity and an instrument for capacity-building. It clearly remains a starting point, out of which the key tools and directions are provided for further initiatives.

Whether speaking about professional skills, institutional needs, or political capacities, training is at the forefront of all essential instruments. No matter if it is shaped through massive and standard training or otherwise dedicated and specialized training, delivered as international, national or local training or even through in-house facilitation of practical hands-on training.

Whether offered by competent national and international training institutions or requested by regional, national or local authorities that have to ensure deliveries, training requires particular skills. Guidance and training manuals on all subjects have been produced over the years by those that offer training (universities, multi-financial and United Nations organizations and even national or local governments) and often by those that request training, as a form of expressing what form of training they wish to receive.

SEA training is widely spread, often following the experience gained with years of training on EIA, particularly as EIA was the entrance door used by SEA to get in the world of professional practice. However increasingly training requirements for SEA must follow less standardized models and seek to meet the increasingly acknowldged and distinctive nature of SEA, as discussed in the beginning of this paper.

Box 44.6 offers basic recommendations for setting up and running SEA training. These are simple rules-of-thumb that result from a personal experience of several years of practice with running SEA training in national and international contexts, for sectoral and multi-sectoral purposes, having trained, and interacted, with about 500 different professionals, in various parts of the world, facing and learning with their inherent different needs and realities.

> **Box 44.6.** Rules-of-thumb for running SEA training
>
> A. Start by understanding:
>
> 1. the decision-making system in place
> 2. who are your target groups
> 3. what are participants backgrounds
> 4. what are participants expectations with the training
> 5. which are the institutional aims (particularly when there is a contracting institution)
>   => Design the course accordingly to meet expectations and within the time available (half a day or five days?)
>
> B. Target group:
>
> => Training on SEA is aimed at helping decision-making more than for technical analysis, which can be basically carried out with planning or EIA techniques and methods. So:
>
> 1. Target for senior level policy and planning professionals
> 2. If possible aim at top or senior management, in which case do not spend more than half a day, maximum one day
> 3. At a more operational level planners or other professionals involved in the design of plans, programmes and policies, as well as EIA professionals are training targets as well
>
> C. Developing the training programme content and format:
>
> 1. Provide background of SEA – understanding the roots
> 2. Offer concepts, techniques and examples, in their right context, as relevant for the theme of the training course
> 3. Use local (or regional or national) examples and refer to local (or regional or national) institutions
> 4. Give good and bad examples, explain why
> 5. Be pragmatical, to the point and practical
> 6. Provide all perspectives and not only yours
> 7. Avoid giving recipes, and the notion of standard methodologies
> 8. Multiply or diversify communication tools and formats (theoretical exposure, individual interactions, groups interactions, exercises)
> 9. Apply professional training techniques (visual techniques, time control etc.)
>
> D. Logistics:
>
> 1. Check capacities of classroom (size, light, noise) for group meetings and display conditions
> 2. Provide background literature, guiding handouts
> 3. Provide the necessary breaks and relaxation moments
> 4. Refer to professional training manuals for additional general operational guidance on delivery and organization of the logistics of training.

Basically any SEA training must be tailor-made to the current needs of the recipients' target groups. SEA can be used in multiple forms, whether in sectoral contexts, or at different scales. Training for SEA must reflect such realities. Some authors criticize current approaches to SEA as being too process oriented, failing to deal with the substance of SEA. But what is the substance of SEA? Isn't it the decision-making processes to which SEA applies? Or is the substance of SEA the physical territory, or the social communities, upon which the programme, planning and policy options will be implemented? Perhaps, if not certainly, both!

44.6 Conclusions and Recommendations

Current practice of SEA is far from being close to a useful and efficient tool to enable integrated decision-making for sound and sustainable development, perhaps the key purpose of SEA. Even with respect to providing a better context for the development of Project EIA, SEA is falling short of this other major objective. Best guesses indicate that this happens because SEA is doing mainly what Project EIA was expected to deliver in its early days. Most examples of SEA show an SEA that covers for the deficiencies of Project EIA. As such SEA is trapped in a vicious cycle – the more SEA delivers what EIA was expected to deliver, the less EIA will deliver and the more SEA will replace EIA! And the real meaning of SEA remains to be fully fleshed.

Strategic decision-making is by nature difficult to pin down, and impossible to be rationalized into legally established straightjackets that set its structure and details. SEA was conceived to operate at strategic levels, therefore it should have adopted a strategic nature, yet structured and logic to ensure that all that matters is considered in development processes. Even though, as widely acknowledged, SEA shares the same roots of Project EIA, that should not necessarily mean that SEA and EIA share the same methodology, or the same rationale. One should be a strategic approach, while the other is clearly rational. However, many examples of SEA question this logic and confuse the expected roles of SEA and EIA. Because of this, when planning for SEA capacity-building, the model of SEA that is wanted must be clarified: whether it is in fact a strategic impact assessment approach or otherwise a grafted outcome, basically matching EIA, but certainly with identity problems.

Given this confusing context of SEA, hardly any sectoral decision-making will happily consider environmental or social concerns in strategic initiatives unless forced, or convinced, to do so. As with economic instruments, there are many non-legally based approaches that can be used as incentives to the use of strategic impact assessment. Public opinion is increasingly having a major impact in both public and private organizations. If well oriented, public opinion may act as a strong motivation to adhere to SEA. Licensing timings and approval fast tracks, tax waiving, financial benefits, seals and improved performance programmes are just a few examples that should be explored to enhance the capacity for improved delivery of SEA.

It appears therefore that capacity-building for SEA must explore, in each decisional context, what are these and other motivations that can enable the positive role of SEA as an integrative tool, or as a facilitator that acts in timely moments, using windows of opportunity. SEA is seldom explored in its positive sense, instead it is the barrier syndrome that takes over, similarly to what has happened with Project EIA, enforcing the negative role of SEA. SEA can not obviously deny its roots but it can chose to stay with them, or otherwise, to search for its own identity. Capacity-building for SEA has a major role to play in this respect.

References[2]

ADEPT – Community Development Agency (2004) Capacity-building. Internet address: http://www.adept.org.uk/chex/contents/background/definitions/definitions.htm, last accessed 09.08.2004

Bina O (2003) Re-conceptualising Strategic Environmental Assessment: theoretical overview and case study from Chile. PhD Thesis, University of Cambridge, Cambridge.

Clark R (2001) Making EIA Count in Decision-Making. In: Partidário MR, Clark R (eds) Perspectives on Strategic Environmental Assessment. CRC Press / Lewis Publishers, New York, pp 15-27

Devuyst D (1999) Sustainability assessment: the application of a methodological framework. J Environmental Assessment, Policy and Management 14:459-87

ELARD – Earth Link and Advanced Resources Development (2004) Review of SEA Practices Worldwide and SEA application in Lebanon. Final Report submitted to the United Nations Development Programme and the Lebanese Ministry of Environment, European Commission Life Third Countries Program

Fischer TB (2002) SEA performance criteria – the same requirements for every assessment? J Environmental Assessment Policy and Management 4(1): 83-99

Genter S (2004) Evaluating the Consideration of Biodiversity in NRM Policy Through PEA. MPhil dissertation, Murdoch University, Australia

George C (2001) Sustainability appraisal for sustainable development: integrating everything from jobs to climate change. Impact Assessment and Project Appraisal 19:95-106

George C, Nafti R, Curran J (2001) Capacity-building for trade impact assessment: lessons from the development of EIA. Impact Assessment and Project Appraisal 19(4):311-319

Gibson RB (2001) Specification of sustainability-based environmental assessment decision criteria and implications for determining "significance" in environmental assessment. Prepared under a contribution agreement with the Canadian Environmental Assessment Agency Research and Development Programme, Ottawa-Hull, Canada

HSHR – Human Strategies for Human Rights (2004) Capacity-building definition. Internet address: http:// www.hshr.org/cbdefined.htm, last accessed 09.08.2004

IAIA – International Association for Impact Assessment (2002) Strategic Environmental Assessment Performance Criteria. Special Publication Series nr. 1, IAIA. Internet ac-

---

[2] Note that legislation mentioned in the chapter is listed at the end of the handbook in the consolidated list of legislation (Appendix 2).

cess: http://www.iaia.org/Members/Publications/Special_Pubs/sp1.pdf, last accessed 09.08.2004

Jenkins B, Annandale D, Morrison-Saunders A (2003) Evolution of a sustainability assessment strategy for Western Australia, Environmental Policy and Law 201:56-65

Lee N, Kirkpatrick C (2001) Methodologies for sustainability impact assessments of proposals for new trade agreements. Environmental Assessment, Policy and Management 3:395-412

Noble B (2002) The Canadian experience with SEA and sustainability. Environmental Impact Assessment Review 22:3-16

Partidário MR (1996) SEA - key issues emerging from recent practice. Environmental Impact Assessment Review 16(1):31-55

Partidário MR (1999) Strategic Environmental Assessment – principles and potential. In: Petts J (ed) Handbook of Environmental Impact Assessment, vol. 1. Blackwell, Oxford 60-73

Partidário MR (2000) Elements of an SEA framework - improving the added-value of SEA. Environmental Impact Assessment Review 20(6):647-663

Partidário MR (2004) The contribution of Strategic Impact Assessment to Planning Evaluation. In: Miller D, Patassini D (ed), Accounting for non-market values in planning evaluation, Ashgate, Aldershot: ch. X (in press)

Pope J, Annandale D, Morris-Saunders A (2004) Conceptualising sustainability assessment, Environmental Impact Assessment Review 24:595-616

Sadler B, Verheem R (1996) Strategic Environmental Assessment – Status, Challenges and Future Directions. Ministry of Housing, Spatial Planning and the Environment of the Netherlands

Thissen W (2000) Criteria for evaluation of SEA. In: Partidario MP, Clark R (eds) Perspectives on Strategic Environmental Assessment. Lewis Publishers, London, pp 113-130

Vicente G, Partidário MR (2002) Environmental Communication at Strategic Levels of Decision Making- potencial role of SEA. Proceedings of the 22nd Annual Conference of the International Association for Impact Assessment, The Hague (in CD-ROM)

45 A Critique of SEA from the Point of View of the German Industry

Jürgen Ertel

Department of Industrial Sustainability, Brandenburg University of Technology (BTU), Cottbus, Germany

45.1 Introduction

The SEA Directive does not address industry directly. This statement seems to be strange because Art. 3 para 2 a) of the SEA Directive mentions some fields which are highly linked to industrial activities. This is obvious for the sector "industry" which is listed at the fifth position. The sectors "energy", "transport", "waste management" and "telecommunication", also listed in Art. 3 para 2 a) of the SEA Directive are further typical industrial fields. The reason for this apparent contradiction is, that Art. 2a) of the SEA Directive requires an SEA for plans and programmes which are:

- subject to preparation and/or adoption by an authority […] and
- which are required by legislative, regulatory or administrative provisions.

Therefore the typical addressees of this Directive are authorities. Only in certain cases will the work of authorities be done by private enterprises. This can be the case of waste management, in the telecommunication and energy sector or in respect to transport. Still, the second requirement, the legal obligation to plan, must be fulfilled. However, though industry prepares plans and programmes for internal and external purposes, there are only a few planning obligations required by law.

For this chapter business associations have been contacted and it was found, that SEA is considered of less significance compared to project EIA. This chapter explains and discusses this aspect with regard to the position of German industry, chosen as a representative example.

First, the chapter explains the role of industry in the legislative process and then evaluates the priority level of SEA for the industry. The chapter concludes with an analysis of the actual involvement of industry in the legislative process with particular regards to environmental matters.

Implementing Strategic Environmental Assessment. Edited by Michael Schmidt, Elsa João and Eike Albrecht. © 2005 Springer-Verlag

45.2 The Role of Industry in the Legislatorial Process

In the European Community, it is a legal requirement to involve industry in the course of legislation. This means, for example, the advanced notification of intended acts and invitations for giving statements and attendance to hearings. Hereby, the focus is not only on the rules themselves, but also on the problems presumably encountered during their implementation. In addition to this, the SEA Directive, like the majority of environmental laws, refers to Art. 6 of the Treaty of Amsterdam which demands to integrate environmental requirements into the definition of community policies and activities, in particular with a view to promoting sustainable development. (It should be kept in mind that the phrase 'sustainable development' includes the view of having an undisturbed free European market with no distortions in competition among the member states.) This aspect has also been subject to the famous Fifth Environment Action Programme: Towards sustainability – A European Community programme of policy and action in relation to the environment and sustainable development. This program represented, among other topics, the umbrella for several directives and ordinances that impacted directly on business companies. Therefore, it is a well established tradition of European enterprises to commit themselves quite vigorously to seizing the legal provisions of the process of legislation. This means that they took an active involvement on this process.

45.3 The Priority Level of SEA for the Industry

A good way to find out the general opinion of German enterprises about the SEA Directive is to contact the main business associations. This makes it possible to evaluate the general opinion of German enterprises on the importance, to the German industry, of the forthcoming SEA Directive. The main business associations[1] which were contacted in 2003/2004 are listed in Box 45.1.

Box 45.1. Main business associations, contacted about the SEA Directive

- BDI   Bundesverband der Deutschen Industrie e.V. (Federation of German Industries)
- ZVEI   Zentralverband Elektrotechnik- und Elektronikindustrie e.V. (German Electrical and Electronical Manufacturers' Association)
- DIHK   Deutscher Industrie- und Handelskammertag (Association of German Chambers of Industry and Commerce)
- VDMA   Verband der Deutschen Maschinen- und Anlagenhersteller (Federation of the German Engineering Industries)

---

[1] Not mentioned are others like the VDA (German Car Manufacturers' Association) and the VCI (Association of the German Chemical Industry) which are also members of the BDI.

It was from past experience that these associations were selected to be interviewed. At the first contact, surprisingly, there was no immediate response forthcoming; instead they had to be reminded of the SEA Directive. In fact, some remembered the Directive vaguely and figured out that the priority was considered rather low. The explanation for this is, that first and foremost, the current problems in industry are dealing with the implementation and measures for compliance with the new directives on product responsibility, such as:

- End-of-life Vehicles Directive
- WEEE Directive
- RoHS Directive
- Energy Using Products Directive (COM/2003/0453 final)
- EU Chemicals Policy (COM/2003/0644 final)

The possible problems encountered are the liabilities in the areas of establishing collection systems, recycling of such end-of-life equipment and providing the finance for these activities. In addition to these treatment requirements, the design of new products must take into account strict requirements, like meeting recycling quotas and the absence of diverse banned hazardous materials like hexavalent chromium, lead, polybrominated biphenyls and so forth (for details see Albrecht 2004). It is easily understood that these efforts currently have the highest priority and involve the search for proprietary solutions as well as the establishing of branch-wide operations. Compared to these tasks, the other environmental legislation draws less attention and is currently only in the status of being monitored. However, much attention is also spent at this time on the drafts like the Energy Using Products Directive. This new proposed Directive will interfere explicitly with product design and requires a company-based evaluation of the product impacts on the environment, thus, being also relevant in terms of public image and competitiveness and the new European Chemicals Policy with the REACH-programme (for details see SRU (2003) pp.1ff).

Second, the SEA Directive is only indirectly touching the interests of the economy. The main addressees are the administrational and legislatorial bodies in the member states. In essence, the Directive targets the adoption of environmental assessment procedures at the planning and programming level. This should benefit activities by providing a more consistent framework in which to include relevant environmental information into decision-making. The inclusion of wider set of factors in decision making should contribute to more sustainable and effective solutions (Recital 5 of the SEA Directive).

Such a goal of providing a more consistent framework is also in the best interest of business and has been pursued by the business associations. Since no new situation was detected with respect to the already strict environmental regulations in Germany, the main focus was put on the proper implementation of the Directive into German law. This more general concern also included the assumption that this new Directive could serve as an opportunity for general improvements like:

- Correct currently stricter rules in Germany
- Reduce time span for issuing permits

- Improve the efficiency for issuing permits
- Provide proper implementation into other legislation (alterations of existing laws)
- Ascertain a European wide framework, including the new members

45.4 Involvement of the Industry in the Legislative Process

These essential issues mentioned in Sect. 45.3 are already known from the process of designing the EIA Directive. This Directive was introduced into German law in 1990 and caused a series of new or amended laws (see Röhnert 1999). In a comprehensive approach the DIHK (2000), for instance, commented on the diverse measures to amend and harmonize those parts of the German environmental legislation that was affected by the EU legislation.

45.4.1 Activities in the EIA Process

Clearly, the German industry was taking primarily interest into the EIA process rather than into the later SEA, which was considered a mere extension towards policy, plans and programs (PPP). The current state is that the EIA procedure is considered as a necessary subpart of the general permitting process. Nothing else was to be expected from the introduction of the SEA, which is called in Germany also the *Plan-UVP*, which means Plan-EIA, used for the official term SEA. The SEA Directive's essentials are displayed in the following list:

- Applies only to plans and programs which are prepared or adopted by authorities and which are required by law
- Inclusion of environmental effect assessment at the earliest stage
- Not to be confused with "environmental impact assessment", also called Project EIA
- Designed for ensuring a procedure not a decision quality
- Decision criteria with respect to the need of employing SEA
- The Environmental Report, a consequence to the Aarhus Convention
- Promotion of international co-operation and information

45.4.2 Interferences with Other Legislation

It should finally be pointed out that SEA is also interrelated to other pieces of legislation. This is not only the already mentioned EIA Directive (see Chap. 3), but also the Habitats Directive (see Chap. 3) as well as the IPPC Directive (see Chap. 2). In particular, it must be mentioned that the scope of SEA might include subjects under the Annexes I and II of the EIA Directive. Art. 3 para 2a) of the SEA

Directive requires an SEA for plans and programmes in the certain sectors which set the framework for future development consent of projects listed in the annexes I and II of the EIA Directive (see Boxes 45.2 and 45.3). Therefore many more branches of industry than the interviewed ones are subject to the SEA legislation.

Box 45.2. Projects Subject to Annex I of the EIA Directive in conjunction with Art. 4 para 1 of the EIA Directive[2] (excerpts)

1. Crude oil refineries
2. Thermal power stations
   nuclear power stations incl. dismantling or decommissioning
3. Installations for the reprocessing of irradiated nuclear fuel
4. Integrated works for the initial smelting of cast-iron and steel
5. Installation for the extraction, processing and transformation of asbestos
6. Integrated chemical installations
7. Construction of lines for long-distance railway traffic, airports, motorways etc.
8. Inland waterways, trading ports, landing piers
9. Waste disposal and treatment facilities of hazardous waste
10. Waste disposal and treatment facilities of non-hazardous waste
11. Groundwater abstraction or artificial groundwater recharge schemes
12. Works for the transfer of water resources
13. Wastewater treatment plants
16. Pipelines for the transport of gas, oil or chemicals
18. Industrial plants for the production of pulp from timber, or paper and board

---

Box 45.3. Projects Subject to Annex II of the EIA Directive in conjunction with Art. 4 para 2 of the EIA Directive[3] (excerpts)

1. Agriculture, silviculture and aquaculture
2. Extractive industry
3. Energy industry
4. Production and processing of metals
5. Mineral industry
6. Chemical industry (Projects not included in Annex I)
7. Food industry
8. Textile, leather, wood and paper industries
9. Rubber industry
10. Infrastructure projects
11. Other projects
12. Tourism and leisure

45.4.3 Consideration of the Aarhus Convention

Finally a third document has to be considered, which is the Aarhus Convention of 1998 (for details see Chap. 3). The goal is the strengthening of the rights of the

---

[2] Mostly under certain conditions only, regarding for instance volume etc..
[3] Member States specify and determine projects as being subject to an assessment.

public in the areas of information and participation. The main items are to provide the public with access to information, to guarantee public participation in decision-making and to secure access to justice in environmental matters, as the name of the convention states. Environmental information should be made available to the public, within the framework of national legislation (Art. 4 of the Aarhus Convention) and exemption from disclosure (Art. 4 para 3 and 4 of the Aarhus Convention). Furthermore, the way of making environmental information available shall be transparent and information effectively accessible (Art. 5 of the Aarhus Convention) and the public concerned shall be informed early in an environmental decision and in an adequate, timely and effective manner inter alia of the fact that the activity is subject to a national or transboundary environmental impact assessment procedure (Art 6 of the Aarhus Convention). Finally appropriate practical and other provisions for the public to participate during the preparation of plans and programmes relating to the environment, within a transparent and fair framework are laid down in Art. 7 of the Aarhus Convention. The annex to this convention contains a list of activities.

With regards to EIA and SEA, industry was eager to ascertain the proper consideration and integration of the Aarhus Convention. The motivation was again, to provide compatible conditions for disclosure etc. in these Directives. Art 5 para 1 of the SEA Directive describes how an environmental report has to be prepared. Where an environmental assessment is required, an environmental report has to be prepared in which the likely significant environmental effects of implementing the plan or programme, and reasonable alternatives taking into account the objectives and the geographical scope of the plan or programme, are identified, described and evaluated. Further information about the content of an environmental report is given in Annex I of the SEA Directive (see Appendix 1).

45.4.4 Activities in the SEA Process

Obviously, industry could and did agree with the provisions of the SEA Directive. However, industry's main contribution to the legislatorial process took place during the discussion of the EIA Directive. Therefore it was considered that there was no necessity for major activities to influence the legislative process of the SEA Directive only a couple of years later.

This is even true if private undertakings like utility companies are required to perform an SEA in lieu of authorities (c.f. ruling of the European Court; for details see Chap. 3). Though, this issue is certainly a matter for law experts and maybe further court rulings, in view of industry, however, such obligations are already nowadays practiced in the course of EIA and would not make a difference to enterprises (see Chap. 46 for an opposite view point).

45.5 Conclusions and Recommendations

The SEA Directive and the implementation process into national law are not of high priority for the industry as described with regard to the example provided by the German industry. The main reason is that authorities are the typical addressees of the Directive's provisions. Only in certain fields, when there are planning obligation for private enterprises in specific sectors, mentioned in Art. 3 para 2 a) of the SEA Directive, could industry directly be affected by SEA provisions. It has been pointed out in this chapter that industry was mainly involved in the legislatorial process of project EIA, although industry also dealt with the SEA Directive. However, both project EIA and SEA do not lead to additional activities in industry since the EIA is considered to be a well established part of permitting. It is considered that a possible request for performing an SEA would not exceed the effort being done by industry.

Acknowledgements

The author is indebted to Dr. Hermann Hüvels, DIHK and Dr. Eike Albrecht for critical review and amendments to this article.

References[4]

Albrecht E (2004) Die europäische Elektro- und Elektronikschrott-Richtlinie – Grundzüge, Problembereiche und ihre Umsetzung in deutsches Recht (European Directive on waste electrical and electronic equipment – basics, difficulties and its implementation into German law). In: Knopp L, Busch G, Heinze A (eds) Brennpunkte der Abfallwirtschaft. Springer, Heidelberg

DIHK – Deutscher Industrie- und Handelskammertag (2000) Stellungnahme zum Entwurf für ein Gesetz zur Umsetzung der UVP-Änderungsrichtlinie, der IVU-Richtlinie und weiterer EG-Richtlinien zum Umweltschutz (Comments on the draft of a law for the implementation of the council Directive 97/11/EC of 3 March 1997 amending Directive 85/337/EEC on the assessment of the effects of certain public and private projects on the environment, of the Council Directive 96/61/EC of 24 September 1996 concerning integrated pollution prevention and control, and other EC-Directives on environmental protection). Stand: 30. Juni 2000, Internet address: www.dihk.de, last accessed 01.09.2004

Röhnert P (1999) Die EU-Richtlinien zur UVP und ihre Umsetzung in Deutschland, (The EU Directives for EIA and their transposition in Germany). Internet address: www.raumplanung.uni-dortmund.de/rgl/uvp.htm, last accessed 01.09.2004

---

[4] Note that legislation mentioned in the chapter is listed at the end of the handbook in the consolidated list of legislation (Appendix 2).

SRU – Sachverständigenrat für Umweltfragen (2003) Zur Wirtschaftsverträglichkeit der Reform der Europäischen Chemikalienpolitik – Stellungnahme. Internet address: www.umweltrat.de/03stellung/downlo03/stellung/Stellung_Reach_Juli2003.pdf, last accessed 01.09.2004

46 Best Practice Use of SEA – Industry, Energy and Sustainable Development

Ross Marshall[1] and Thomas B. Fischer[2]

1 National Environmental Assessment Service, Environment Agency, England
2 Department of Civic Design, University of Liverpool, England

46.1 Introduction

More than a decade after the Rio Summit in 1992, the challenge of moving economic development forwards, without compromising the availability of resources for future generations and at the same time protecting the environment and cultural heritage has become more urgent than ever. For industries, having to operate in progressively more globally oriented economies, with a greater awareness for environmental risks and stakeholder demands regarding corporate social responsibility, to date the concept of sustainable development has raised more questions than provided concrete answers. A fundamental problem is how to make a profit in today's marketplace whilst making decisions that customers in 5, 10 or even 50 years time would wish had been taken. In this context, strategic environmental assessment (SEA) has a tremendous potential. It is very likely to soon start taking on an increasingly important corporate role in the private sector.

This chapter is divided into several parts. Firstly, institutional links of SEA to industry are highlighted. Why SEA might be attractive to industry is examined. The question as to how SEA can be applied within a private sector organisation is addressed, referring to a regional transmission network SEA case study undertaken by a privatised electrical utility (ScottishPower). This case study shows how SEA frameworks can be established inside a private sector organisation. Furthermore, it portrays the practical issues that have to be addressed by corporate decision-makers. The following key questions are posed:

- Is SEA able to contribute to the final outcome of private sector led strategic actions for the development of essential utility infrastructure?
- Should organisations adopt 'tiered', 'integrated' or 'stand alone' approaches to SEA within existing decision-making frameworks?

Implementing Strategic Environmental Assessment. Edited by Michael Schmidt, Elsa João and Eike Albrecht. © 2005 Springer-Verlag

- Can a structured approach to decision making be supported by SEA and can SEA help to find a preferred course of action when planning regional transmission networks?

Finally, conclusions and recommendations are drawn.

46.2 Institutional Links between SEA and Industry

From the perspective of policy makers controlling the introduction of new industrial development, project EIA application in the public sector and in land use planning has resulted in a range of benefits. However, it has clearly failed to address strategic issues regarding capacity, cumulative effects and cross-sector synergies of developments. Taking the example of windfarms, EIA has been used to address the physical attributes of projects and has largely been applied in a site specific manner. This narrow focus has meant that so far the desire of planners and decision makers to comprehensively consider environmental impacts within public authority areas have not been satisfied. For example, the synergies and cumulative impacts of several windfarm projects have not been satisfactorily considered by only using EIA. Furthermore, proponents of windfarm developments have had to seek consent in an often antagonistic and reactive planning process. In the past, this has often resulted in direct argument between developer, pubic, local authority and statutory body and has failed to meet the needs of all parties.

SEA is a decision making support tool that is able to positively address these shortcomings. SEA allows for a greater scope of considering environmental issues and raises general awareness for strategic issues. Furthermore, it enables the anticipation of subsequent site-specific and reactive issues, the needs and concerns of potential objectors. Finally, SEA is able to show the cumulative benefits of individual strategies across industrial sectors.

On 21 July 2004, i.e. more than 10 years after SEA had been declared a cornerstone in the implementation of the Fifth Environmental Action programme, the sustainable development strategy for the European Union in the European Union, formal requirements were put in place for SEA of certain plans and programs. Whereas a large number of public sector plans and programmes are covered, the extent to which plans and programs of industry and private companies are affected has remained unclear for many prospective users (see Chap. 45). Furthermore, there is comparatively little reported SEA experience of industry on which to draw upon. Although policy-makers have now largely recognised SEA's theoretical potential to improve planning and decision making and few doubt the power of the SEA Directive to compel proponents to increasingly adopt an environmental governance and stewardship approach to future plan or programmes, to date the concept of SEA has not been clearly mapped out for industrial sectors, their players and marketplaces. As successful SEA applications of industry have been rare in the professional literature, voluntary SEA uptake amongst industrial proponents has remained low. Therefore, practice outside the governmental spectrum of activities has remained scant and inconsistent (Glasson et al 1999; Jay and Marshall

2004; Noble 2004). However, following implementation of SEA legislation in all EU member states, more environmentally aware companies have started to develop working frameworks in which to apply SEA. In this context, an important question is whether companies are able to meet statutory obligations for SEA and whether SEA can contribute towards their environmental performance. There are some indications that industry is slowly starting to examine whether SEA can be a practical tool for facilitating business objectives whilst meeting:

- business needs (profits, survival, growth, competitive advantage, etc),
- the requirements of customers for competitive goods or services,
- the selection of strategic actions, technology or markets that contribute to performance and market position, and
- the increasing desires of customers (and by default environmental pressure groups).

46.3 SEA Application in Industry Planning

Only specific sectors of industry have become familiar with SEA's more project-orientated predecessor - Environmental Impact Assessment (EIA). The correlation between organisations that have participated in EIA, and hence developed a capacity in EIA performance, and those that are likely to be obligated to undertake SEA has yet to be mapped out. Though broadly speaking, the methodology of SEA for plans and programmes is similar in structure and performance to that required for project EIA, SEA also includes some extra preliminary steps such as the determination of need, establishment of the geographic area and data streams to be analysed, and the identification of strategic alternatives (Thérivel et al. 1992). In doing so, SEA has the potential to encourage proponents to take a more proactive approach towards identifying potential environmental outcomes and implications for sustainable development.

If SEA is to fulfil its full potential beyond regulatory compliance, it must start to become relevant and responsive to the requirements of industrial organisations. To be successful in practice, SEA must bring forward or initiate change in the strategic management processes that are accepted as valid and worthwhile components of corporate planning or decision-making pathways (Marshall 2003). Ultimately this must enhance or contribute to the quality of projects and the deliverability of corporate plans or policy objectives. To be successful in this role, SEA frameworks require that an environmentally sound culture is embedded within the economic basis of an organisation. This culture will need to be prepared to embrace the concept that industry and has an important role to play in contributing to, rather than taking from, the societies that use or are reliant on its products and services. It is on these terms that SEA's focus on environmental parameters within strategic economic decision-making is able to contribute positively to the participation of industry within the wider social objectives for sustainable development.

The SEA Directive's requirements cover plan and programme making in forestry, energy, industry, transport, waste management, water management and telecommunications. In the UK, many of these sectors have now transferred from state responsibility to the control of private sector shareholders through privatisation programmes between 1980 and the mid 1990s. Formal governmental policy and guidance on how SEA should be applied to the strategic actions of privatised former public services, such as water, gas and electricity, is still awaited. On the one hand, their statutory duties may require them to plan for a set of obligated actions that may be subjected to SEA; on the other hand, their industrial/privatised status places them in a different organisational context to local authorities preparing land-use plans (Jay and Marshall 2004). It is therefore unclear to what extent private sector organisations are free to accrue commercial advantage or to pursue clear and transparent shareholder benefits through SEA whilst meeting their statutory obligations imposed through its introduction. It is within these boundaries that any benefit to sustainable development can be realised. There is therefore a vital need to consider how existing corporate decision-making frameworks, operating within commercially orientated and shareholder-driven organisations can and should apply SEA. Moreover, the global trend towards the privatisation of state-owned enterprises and an increasing reliance on these bodies for essential services suggests that society's self-interests may be best served by considering exactly how SEA should be applied in such industrial sectors. Set in this context, SEA methodologies for public authority land-use plans have already started to be used as the accepted model templates for SEA performance (ODPM 2003). However, it is far from clear if these are indeed the best possible or preferred approach for industrial, utility or agricultural sectors to adopt. The absence of guidance and a lack of clarity as to how exactly SEA should be applied presents a corporate dilemma for industry.

Disregarding obligated requirements for SEA, there is also the question of whether private companies might also wish to apply SEA voluntarily for their own benefit. In this context, the potential benefits that arise from SEA need to be made clear to industry. Whilst private companies appreciate that SEA is a systematic, procedural, objectives-led and participative decision support instrument for sustainable policy, plan and programme making. It is also often considered a dreary and resource intensive formality, a reluctant 'paper chase' that adds further to the impression that is a administrative burden that has to be paid for (Therivel 2004; IEMA 2003). However, experience has shown that SEA can actually lead to time and cost savings by establishing a strategic decision making framework, thus consolidating existing procedures and facilitating participation and consultation. Levels of public understanding and acceptance of plans and programmes can thus be raised (Dusik et al. 2002). Furthermore, SEA may assist in reducing negative environmental impacts or delays in corporate objectives that are ultimately costly for private companies.

46.4 SEA – A Strategic Decision-Making Framework for Businesses

In considering strategic decisions, corporate decision-makers have to ask two questions. Firstly, they need to know whether a decision might have a significant impact on other decisions of the organisation; and secondly, they need to consider whether other strategic decisions have a significant impact on what the strategic decision is trying to achieve. The value of SEA in this process is that it provides a strategic framework in which different ideas and concepts can become part of the evaluation process. To put it simple, SEA is a robust decision-making framework that is in line with good strategic business planning and good corporate decision-making. Fig. 46.1 summarises the fundamental stages of a typical strategic decision framework for businesses. Initially, a strategic vision should be developed, setting clear corporate aims and objectives. In this context, the questions of *why* development is needed and *what* it should involve are asked. Secondly, plans are designed that establish *how* and *where* development will or should occur. Thirdly, proto-implementation programmes are developed that specify (a) preferred option(s). In this context, *when* exactly development should occur is established. Finally, the criteria and outline of future projects are planned and implemented. The different stages of this hierarchy are iterative, i.e. they may not necessarily take place in a clear sequence, but feedback is possible at any stage.

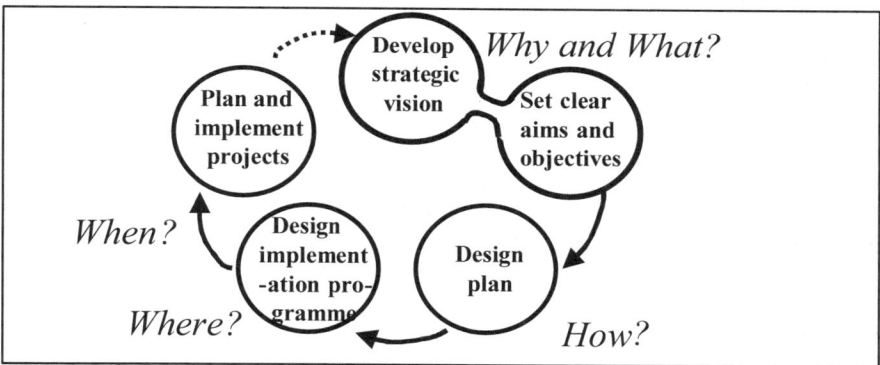

Fig. 46.1. Strategic decision making framework for businesses

Tiered environmental assessment systems are able to provide support for strategic company planning, providing it with a structured framework (see, for example Jansson 2000 and Fischer 2000). Within a tiered SEA framework, *why* and *what* questions are addressed at the highest tier, sometimes named policy-SEA (vision and objectives). *How* and *where* questions are addressed at the second SEA tier, which is sometimes named plan-SEA (design). *When* questions, finally, are addressed at the lowest SEA tier, which is sometimes called programme SEA, and ultimately by project EIA (implementation).

Large capital programmes of private companies require the co-operation, input and interaction of various business groups and professional disciplines. This demands the establishment of clear structures for communication, project management and participation. Whereas in established businesses, strategic decision-making structures for large-scale multi-million pound developments are often well established, they may not necessarily follow the structured strategic decision making framework presented in Fig. 46.1. SEA can create and support decision-making structures that can be understood by all participants – at least as far as this is possible – and enabling effective involvement and participation. SEA can thus aid the establishment of systems for corporate governance, distributing and specifying responsibilities in decision-making. Currently, the role of public involvement in corporate decision-making is still far from clear. However, participation is likely to be 'encouraged' by organisations where the results are likely to be beneficial to both parties and SEA can help to do this in an efficient and effective manner.

46.5 Electricity Provision and the UK Transmission Industry

Electricity is vital to the UK's development and well being. It must be readily available and affordable. Continuous massive capital spending is required in order to maintain and upgrade existing transmission networks. However, new infrastructure programmes often bring electricity utilities into dispute with those whose services they seek to ensure or whose energy uses they seek to provide for. The development or refurbishment of transmission infrastructures becomes even more difficult within democracies when local decision-makers or residents perceive no immediate benefit to themselves through a project's delivery, or where a plan or programme benefits members of society distant from the site of development (i.e. Not In My Backyard – NIMBY[1] attitudes arise).

Within the next decade, the issue of impact location versus perceived benefit area will become increasingly important, as many existing networks and their transmission structures (towers, transformers and conductors) reach the end of their engineered life and need replacement. Replacement will be problematic as, due to their strategic contribution to regional energy networks, few transmission structures can be decommissioned until their replacement systems are *in situ* and energised. Another challenge for the UK electricity industry is that the drive for increased renewable energy generation will require transmission companies to reinforce or re-develop existing networks to accommodate the predicted rise in renewables from within their licensed geographic areas. Within Scotland and Wales, over the next decade, an expansion of approximately seven hundred kilometres of new transmission overhead lines will be required to accommodate renewable en-

---

[1] For an explanation of NIMBY'ism see for example Petts 1999.

ergy development. Generally speaking, all UK transmission companies will be faced by the tri-part challenge of

a) continued growth in the UK's domestic energy market,
b) network ageing, and
c) the drive towards greater sustainability through renewable generation

It will be a test to their ingenuity to deliver new transmission systems, and of their ability to respond proactively to often conflicting environmental, energy and social demands. It is in this context that SEA has the potential to develop as an important and necessary strategic planning support instrument, either as a statutory obligation or through voluntary action.

Electricity provision – UK privatisation, the rise of private utilities and strategic network planning.

Prior to 1990, the UK electricity generation and transmission were the responsibility of a central nationalised Central Electricity Generating Board, with distribution and supply activities the responsibility of regional electricity boards. The Electricity Act 1989 (UK Government 1989) provided for the near-complete privatisation of the sector, and in England and Wales, vertical de-integration of the industry took place, giving rise to new privatised companies with responsibility for specific aspects of generation, transmission, or distribution and supply (Jay and Marshall 2004). ScottishPower is one of three UK companies who holds a statutory licence under the Electricity Act 1989 to 'develop and maintain an efficient, co-ordinated and economical transmission system of electricity supply'. The company's licensed transmission and distribution service area includes Southern Scotland, the North-West of England and North Wales.

Over the last seventy years, electricity networks have developed in response to urban and industrial demands, the required siting of generators and the needs of high-energy use industrial plants. Strategic network planning has traditionally concentrated on providing solutions to identified needs from within the confines and constraints of the existing network configuration. New networks have been considered where existing system capacity has been deemed inadequate. The objective when refurbishing, reinforcing or designing new network components is to identify the lowest cost feasible design, that meets accepted standards of system security, technical feasibility and economic viability with the least possible overall impact to people and the environment. In this context, the identification of a preferred transmission network option is a lengthy process. Firstly, the general need for future strategic action needs to be confirmed. Secondly, a preferred strategic solution needs to be chosen, and thirdly the transmission company needs to apply for consent to construct and operate the solution.

46.6 The Perceived Benefits of SEA to ScottishPower

ScottishPower has come to consider SEA as an appropriate tool to promote good corporate decision-making during the initial strategic planning of electricity transmission networks. Furthermore, it regards SEA as a means to promote better internal decision-making management. Initial interest in SEA was driven by an early recognition that strategic actions resulting in statutory transmission projects under the EIA Directive were likely to also fall under the scope of the SEA Directive. This initiated internal debate concerning whether strategic advantage resided in the voluntary up-take and examination of SEA prior to its statutory introduction in July 2004. A particular focus was put on the question as to how SEA and the existing processes for strategic action could be integrated. Currently, the company is interested in establishing whether SEA can create frameworks and agreed structures for communication, project management and decision-making that is understood by all participants. ScottishPower hopes that SEA is able to force the evaluation of the adequacy and reliability of current and future electrical supplies against a framework of strategic alternatives, the baseline environment and multi-criteria consequences.

Three factors played a major role in the debate within ScottishPower on how SEA should be applied. Firstly, the company was already familiar with project EIA integration in transmission planning and did not feel uneasy about SEA. Secondly, the company had sought to develop internal management systems to promote a corporate culture that included the consideration of the environment into its planning and decision-making frameworks, notably through ISO 14001 management systems and corporate environmental governance programmes (ISO 1999; PowerSystems 2003). Thirdly, the company already had developed existing processes prior to EIA that sought to include the consideration of strategic environmental factors into its existing overhead line routeing programmes (Marshall and Baxter 2002). Based on these experiences, the decision was taken to conduct SEA as an integrated rather than a stand-alone process. How ScottishPower subsequently took the first steps towards SEA implementation is shown by the following case study on the Mid-Wales regional transmission network planning SEA.

46.7 The Mid-Wales Case Study – ScottishPower's First Steps to SEA implementation

Background

Large parts of the electricity distribution system in Mid-Wales were first established prior to 1947. The system subsequently was adapted, replaced and refurbished to meet public electricity system obligations and serviced consistent with the needs of a rural area. The region is supplied from 400 to 132 kilovolt (kV) grid supply point substations at Trawsfynydd, Legacy and Swansea North (in South Wales) refer to Fig. 46.2.

There are two main area systems in the region, Aberystwyth and Maentwrog. The Aberystwyth area system is presently supplied from the Swansea North grid supply point and the Maentwrog area is supplied from the Trawsfynydd grid supply point. The Swansea North connections to Aberystwyth and at Rhydlydan are both supplied through single circuit 132 kV lines that are over 100 km long. This places the current network configuration at risk of major power failures following local faults, and if interrupted power supplies to the region could be lost for an extended period of time. The area under consideration has also attracted significant attention from windfarm developers with a number of sites constructed to date, increasing the demand on the system.

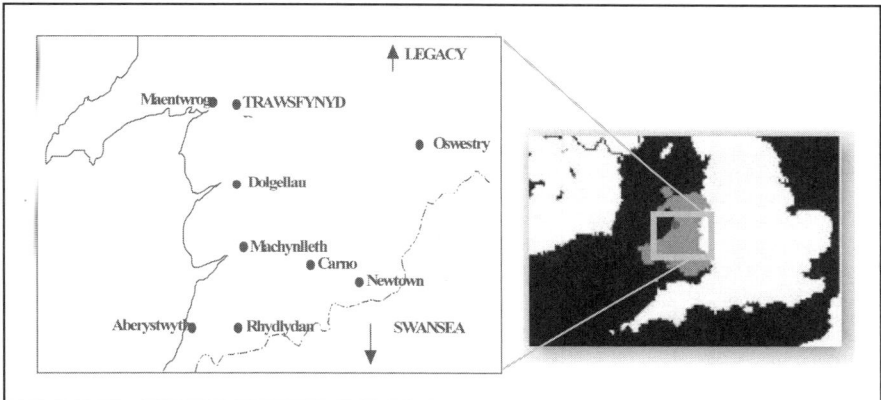

Fig. 46.2. Mid-Wales region

Although an original engineering solution at 132 kV had been first proposed in 1998, the decision was made to restart strategic planning for the programme anew, adopting a revised approach that sought to integrate SEA within the existing planning framework.

A tiered environmental assessment system

Initial thoughts on how to effectively consider the environment in strategic action planning for Mid-Wales led to the introduction of a tiered assessment system, consisting of three SEA tiers, an EIA tier and a follow-up tier. The three SEA tiers included a new 'preliminary establishment of need SEA' tier, a new 'regional transmission network SEA' tier and an established 'overhead line routeing methodology SEA' tier. The idea to apply SEA within a tiered system is not specific to ScottishPower, but has been advertised by various authors. It has probably developed furthest in transport planning, where three tiers with distinct tasks have been said to be the basis for effective SEA application (see Fischer 2002). Fig. 46.3 shows the integration of SEA into the existing regional network planning stages, with the strategic decision framework introduced in Fig. 46.1 also included. Subsequently, the three SEA tiers are explained in further detail.

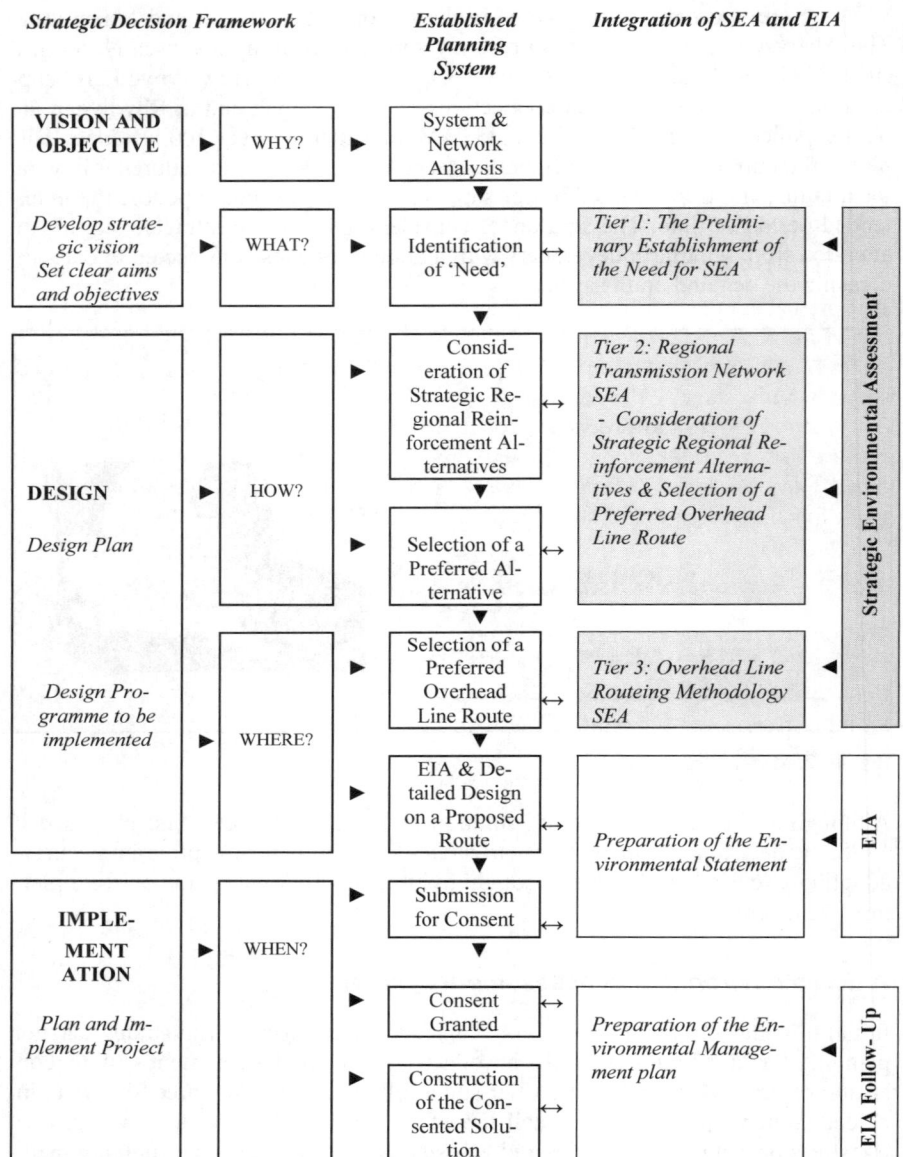

Fig. 46.3. Environmental Assessment integration in regional network planning (Marshall and Fischer 2004)

Tier 1 - The Preliminary Identification of 'Need' SEA

The establishment of 'need' is the initial assessment stage in transmission network planning. Ultimately, this stage is about justifying the granting of statutory consent to construct a preferred option. 'Need' arises from:

- existing conditions and forecasted energy demand on existing network systems;
- quality and security of supply to customers;
- age and condition of its infrastructure;
- demands of new generators seeking connection to electricity transmission grid.

An internal guidance procedure was developed to help staff conducting the first tier of the SEA/regional network programme. The guidance consisted of a simple series of questions and comparison tables that set the baseline context for the subsequent SEA stage - the identification of strategic alternatives (PowerSystems 2003). In practice, the procedure consisted of two distinct data streams, a 'data capture list', recording geographic and physical system evidence that encapsulated 'need', and the 'need definition review list' that built up into a reference framework for the assessment of strategic alternatives.

Tier 2 – Regional Transmission Network SEA

The objective of regional transmission network SEA is to allow the company to address the critical issue of 'what is the range of feasible strategic alternative solutions that should be considered in securing the electrical supply to or demand within a region or user group'. The SEA process used in this context is shown in Table 46.1. Its basis are the key principles and attributes laid out in Noble and Storey's (2001) generic seven-phase methodological framework for SEA in Canadian energy sector planning. The decision to use this methodological framework was taken based on the fact that the seven 'gateways' made rational sense to in-house personnel and senior management unfamiliar with the concept and practice of SEA.

The process starts with scoping the assessment issues, where problems are identified and the context is set within which the assessment will take place. Alternatives are described and scoping of the different assessment components takes place. Potential impacts are evaluated and impact significance is determined. After a final comparison of alternatives, a best practicable environmental option is identified. In addition to these seven procedural phases, SEA also makes use of a combination of methods and techniques for identifying strategic alternatives, evaluating those alternatives against specific assessment criteria, and determining a preferred course of strategic action. Whilst Noble and Storey's (2001) original framework included substantial statistical evaluation, ScottishPower were more interested in the procedural stages, as the input of an expert panel for determining preference in a mathematical model was regarded too costly and demanding for the purposes of regional network planning.

Table 46.1. Methodological framework for transmission network SEA (based on Noble and Storey 2001)

Phase	Attribute	Summary Description
1	Scoping the assessment issues	Problem identification, setting the context within which the assessment will take place.
2	Describing the alternatives	Identification of PPP alternatives, notably alternatives to identify as strategy for action
3	Scoping the assessment components	Specifying the criteria that will be used to evaluate the environmental implications of the various alternatives.
4	Evaluating the potential impacts	The evaluation and assessment of whether the effect of an alternative will be adverse or beneficial
5	Determining the impact significance	Determining the extent of the change within the context of identified impact, the cumulative intensity or severity of impacts across the scope of the alternative and their perceived importance to stakeholders
6	Comparing the alternatives	Determination of the preferred strategic option or PPP direction.
7	Identify the Best Practicable Environmental Option	The development of an overall strategy for action based on the possible alternatives and evaluative criteria assessed.

Tier 3 – Overhead Line Routeing Methodology SEA

The third SEA tier seeks to utilise an already existing in-house approach to the routeing of overhead transmission lines (ScottishPower 2001a; Marshall and Baxter 2002). Developed during the early 1990's, its objective is to strategically evaluate geographic route options in order to select a final preferred route for the transmission corridor, and following consultation with stakeholders, to identify a proposed route for EIA. The procedure, summarised in Fig. 46.4, is based on the simple premise that in the defence of its transmission network proposals, the company is best advised to have a robust and clearly defendable approach to routeing in place, based on which line placement may be justified.

The objective of the exercise is to bridge the gap between the selection of a preferred strategic alternative and the final design project submitted for EIA / developmental consent. The process is iterative and the steps may be re-visited several times before a balance is achieved between technical, economic and environmental considerations, using multi-criteria analysis (MCA). Critically, consultation is carried out throughout the process, with professional judgement being used to establish explicitly the balance between the various factors.

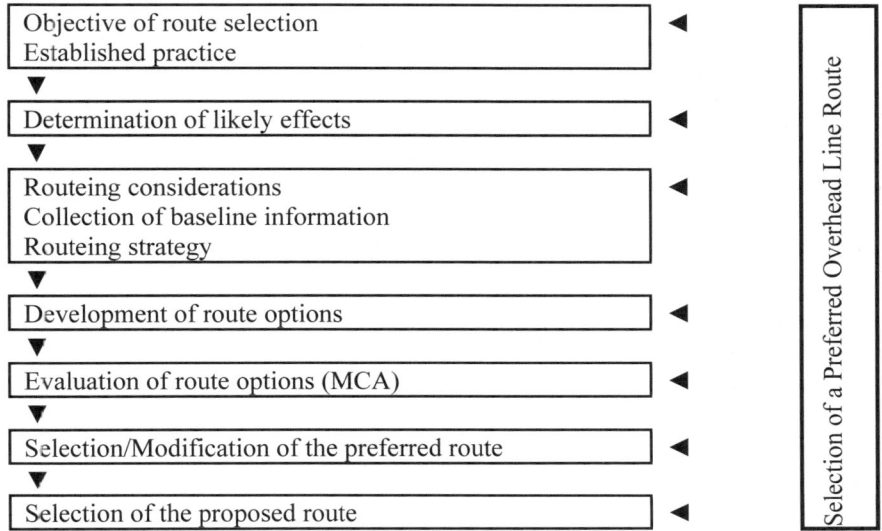

Fig. 46.4. Scope of ScottishPower's Strategic Routeing Methodology

46.8 Applying the Seven Phases of Regional Transmission Network SEA

This section describes how SEA was conducted for the regional transmission network plan in Mid-Wales. In this context, reference is made to the seven procedural stages introduced above.

Phase 1 – Scoping the Assessment Issues

In phase 1, an environmental reference framework was established, which consisted of data tables and constraint maps containing biophysical and socio-economic information. A key constraint identified early was the potential for adverse impacts on the setting and amenity of the Snowdonia National Park. The debate on the preliminary establishment of need was directed towards reviewing potential positive and adverse socio-environmental factors arising from reinforcement of the existing electrical systems and the scheme's relationship with the region's other wider existing environmental and social problems, and other plans and programmes. Subsequently, objectives for the next tier of the SEA were developed.

Phase 2 – Description of the Alternatives

All feasible alternatives open to the company within the constraints of its transmission licence were considered. The range of alternatives was enlarged to include all potential solutions that secured power supplies; for example, both internal and external sources to Mid-Wales were combined. Whilst the consideration of alternatives from outside the company's direct control and operations went contrary to established practice, this encouraged debates on the programme's needs and demands. Six overall reinforcement alternatives options were finally considered, some of them containing multi-criteria options (ScottishPower 2001b):

1. do-nothing
2. maintaining versus replacing the existing system
3. new embedded generation (such as thermal or renewable power plant)
4. overhead transmission solution (only company internal)
5. overhead transmission solution (also company external)
6. underground transmission solution (only localised)

Alternatives 1, 3 and 6 came into the process through SEA.

Phases 3 and 4 – Scoping Assessment Components and Evaluating Potential Impacts

On the basis of the programme's objectives set out in the initial stages, formal criteria were developed to evaluate the environmental implications of the various alternatives. In this context, parameters for nature conservation, cultural heritage, archaeological sites, amenity and landscape assessment criteria were set.

Table 46.2. Examples for programme objectives, criteria and indicators (adapted from ScottishPower 2002)

Programme Objectives	Criteria	Indicators
To take into account the impact of the programme on the natural and human environments	Potential for siting conflicts with recognised aspects of nature conservation, cultural heritage, archaeological sites, amenity and landscape importance; designations of international, national, regional, local importance considered	Existing registers and GIS-records from National and Welsh Statutory Bodies, Local Authority, NGOs, Wildlife Trusts, etc.
To minimise impact on the Snowdonia National Park	Potential for siting conflict with recognised features and aspects within the National Park and its boundaries	Snowdonia National Park Policy documents, Registers of Features

Furthermore, socio-economic criteria, reflecting wider goals for society were included that required objectives to be set against perceived positive or negative implications in terms of supply quality. Table 46.2 provides an example of the criteria and associated indicators used, referring to two objectives set in phase 1.

Phases 5 and 6 – Determining Impact Significance and Comparing Alternatives

Each strategic reinforcement alternative was evaluated individually regarding its potential impacts, cumulative intensity and severity against the indicators within its geographic reference area. For those impacts that were identified as being significantly adverse or beneficial, the perceived importance to stakeholders was also estimated, following the concept of 'issues attention' as brought forward by Downs (1972). Discussions comparing the different alternatives were limited to an internal project group, comprising asset managers, engineering specialists, environmental planners, the estates and legal departments, and a small group of independent environmental and landscape consultants. Each alternative was assessed using constraints mapping of biophysical factors, matrix analysis of compliance with other recorded plans and development policies, economic and engineering models, expert judgement and forecasting across future scenarios. Further assessments were performed and additional desk-based and site study programmes were commissioned to strengthen the data collection and analysis for each alternative. The results of these studies were fed back to the project management group and ultimately into the evaluation cycle.

Phase 7 – Identification of an Overall Strategy for Action

The final analysis, resulting from the assessment process, identified the preferred strategic environmental option as a new steel tower 132 kV connection from Legacy to Aberystwyth with localised distribution reinforcement to the North as required to support the network. Although this alternative did not result in the least number of negative environmental effects, it was the preferred alternative based on the avoidance of impact to Snowdonia National Park and a number of internationally recognised sites for nature conservation and national landscape designations. It also provided compliance with existing plans and programmes and the use of existing infrastructure as well as system reinforcement to those regions of Mid-Wales where windfarm development will be most extensive.

46.9 Conclusions and Recommendations

To date, few examples have demonstrated that the obligated (or voluntary) uptake of SEA can contribute value and input to industry. The example of the electricity company ScottishPower demonstrates that SEA, which is tailored to specific decision making situations, can not only address the environmental impact of future investment programmes, but can also enhance the environmental governance and stewardship of established corporate decision-making frameworks. SEA application to plan and programme making above the project level has shown to be able to widen the scope of examination from a purely technical viewpoint to discussions on the implications of socio-environmental aspects and the political acceptability of different alternatives. Furthermore, the ScottishPower example shows that SEA can stimulate debate on not only the positive and negative aspects of specific strategic alternatives, but also challenge established assumptions. In this context, the introduction of a tiered SEA approach with a clear allocation of tasks, gateways and objectives for each tier can create a discipline among the project's participants not previously known. Ultimately, the Mid-Wales SEA case study conducted by ScottishPower identified preferred alternatives that had not initially been perceived as the company's preferred choices.

If industry wants to achieve a more sustainable development, it must seek to evaluate environmental parameters on an equal footing with economic and technical parameters. In this context, SEA is a suitable instrument that is able to strengthen and improve corporate decision-making procedures. SEA which is integrated in decision-making can create an auditable trail and record of what decisions have been taken, at what point in time, on what basis, and by whom. This complements corporate governance structures developing in many major industries, enhancing the defensibility of management decisions. In order to succeed, the SEA process needs to be aligned well with an organisation's established decision-making culture. Isolating SEA from an organisations integrated decision-making structures runs the risk of only reactively looking at the environmental impacts of pre-determined choices and subsequently retrofitting it to the decision-making process. Through internal organisational management, the pace of SEA can complement the decision-making phases of the strategy under consideration.

Although the statutory application of SEA by obligated industries is likely to form the driving force for the majority of future private sector SEA initiatives, the necessity for compliance should not discourage the uptake of SEA and its application across non-regulated contexts where business advantage is perceived. Regardless of how EU Member States will use SEA in industrial sectors, the application of SEA principles can form a distinct business tool for the purposes of corporate decision-making within the decision-making structures of a company. Its voluntary uptake in a non-regulatory context can provide environmental governance and stewardship to established decision-making frameworks that can contribute positively to the strategic evolution of corporate organisations.

The way in which statutory obligations will be combined with the requirements of the marketplace in which organisations operate will ultimately determine SEA's success in contributing positively to sustainable development. In this context, more empirical evidence is needed on how to operate SEA efficiently in industrial sectors subjected to imposed constraints through statutory obligations, regulation, monopoly and competitive barriers.

Based on the evidence currently available, it is clear that only if the full range of alternatives is considered within a tiered system can SEA be effective. Therefore, the dispersed and fragmented delivery system of electricity in the UK through separate generators, transmission, distribution and supply companies will need to change. Otherwise, certain alternatives will remain unaddressed. Companies on their own will not be able to meet the challenge of increasing demand on its networks and the risks to consumers' loss of supply, by promoting sustainable strategies that ultimately reduce overall electricity demand in the network. Customers themselves can also have a significant impact by reducing their own energy demands and by using energy more efficiently, thus reducing the demand they themselves place on the systems and the need for new infrastructure – which they might otherwise oppose! However, currently government policy to encourage competition for customers works against the cumulative environmental impacts of cheaper electricity provision. There is no incentive for the transmission company to consider stated Government energy targets and show how such targets should be met without a 'trade off' between the objectives of national energy policy and other policies to protect the countryside and wider environment from necessary but potentially adverse infrastructure. In this context, SEA is limited in what can be considered and ultimately its contribution to industry's participation in sustainable development. It is to be hoped that the pragmatism for which industry is renowned will encourage effective SEA which might ultimately identify idiosyncrasies and barriers to progressing towards sustainable development.

References[2]

Downs A (1972) Up and down with ecology: the "issue attention cycle", The Public Interest, 28: 38-50.
Dusik J, Fischer TB, Sadler B (2003) Benefits of Strategic Environmental Assessment, REC, UNDP, 5 pages, www.ecissurf.org, last accessed 23.09.2004
European Community (1993) The Fifth Environment Action Programme: Towards sustainability – A European Community programme of policy and action in relation to the environment and sustainable development. Official Journal of the European Union, C 138, 17.5.1993, p 5
Fischer TB (2000) Lifting the fog on SEA – towards a categorisation and identification of some major SEA tasks: understanding policy-SEA, plan-SEA and programme-SEA (page 5). In: Bjarnadóttir H (ed) Environmental Assessment in the Nordic Countries: Nordregio, Stockholm, pp 39-46
Fischer TB (2002) Strategic Environmental Assessment in Transport and Land-Use Planning, Earthscan, London
Glasson J, Therivel R, Chadwick A (1999) Introduction to Environmental Impact Assessment, 2nd edn. UCL Press, London
IEMA – Institute for Environmental Management and Assessment (2003) Strategic Environmental Assessment: Report of the SEA Workshop, London, 5th November 2003

---

[2] Note that legislation mentioned in the chapter is listed at the end of the handbook in the consolidated list of legislation (Appendix 2).

ISO – International Organization for Standardization (1999) Environmental Management Systems – Specification with guidance for use. International Organization for Standardization

Jansson A (2000) Strategic Environmental Assessment for transport in four Nordic countries. In: Bjarnadottir H (ed) Environmental Assessment in the Nordic Countries: 39-46, Nordregio, Stockholm

Jay S, Marshall R (2004) The Place of Strategic Environmental Assessment in the Privatised Electricity Industry (in preparation)

Marshall R (2003) Embedding SEA in Industry – Identifying a Preferred Course of Action in Transmission Network Planning. Paper presented at the 23rd Annual Meeting of the International Association for Impact Aassessment, Building Capacity for Impact Assessment, 17-20th June 2003, Marrakesh, Morocco

Marshall R, Baxter R (2002) Strategic Routeing and Environmental impact Assessment for Overhead Electrical Transmission lines, Journal of Environmental Planning and Management 45(5):747-764

Marshall R, Fischer T (2004) Regional electricity transmission planning and tiered SEA in the UK - the case of ScottishPower

Noble B, Storey K (2001) Towards a structures approach to Strategic Environmental Assessment. Journal of Environmental Assessment Policy & Management, Vol. 3, No. 4, pp 483-508

Noble B (2004) Strategic Environmental Assessment benefits to industry;: a case study of integrated SEA in Saskatchewan's fortestry sector, Canada. Paper presented at the 24rd Annual Meeting of the International Association for Impact Aassessment, April, 2004, Vancouver, Canada

ODPM – Office of the Deputy Prime Minister (2003) The Strategic Environmental Assessment Directive: Guidance for Planning Authorities - Practical guidance on applying European Directive 2001/42/EC 'on the assessment of the effects of certain plans and programmes on the environment' to land use and spatial plans in England, Office of the Deputy Prime Minister: London, October 2003

Petts J (1999) Public participation and environmental impact assessment. In: Petts J, Handbook of Environmental Impact Assessment, Blackwell Science, Oxford

PowerSystems (2003) The Preliminary Establishment of Need. SP PowerSystems Ltd, ScottishPower

ScottishPower (2001a) 'Overhead Transmission Lines - Routeing and Environmental Impact Assessment – The ScottishPower Approach, ScottishPower, Glasgow

ScottishPower (2001b) Mid-Wales System Development Proposals, SP PowerSystems, Prenton, 29 January 2001

ScottishPower (2002) Proposed Mid-Wales Reinforcement – Environmental Review, Draft 1, SP PowerSystems, Prenton, February 2002

Thérivel R, Wilson E, Thompson S, Heaney D, Pritchard D (1992) Strategic Environmental Assessment, Earthscan, London

Thérivel R (2004) Strategic Environmental Assessment in Action, Earthscan, London.

UK Government (1989) The Electricity Act, HMSO, London

47 SEA Outlook – Future Challenges and Possibilities

Elsa João

Graduate School of Environmental Studies, University of Strathclyde, Scotland

47.1 Introduction

This chapter discusses barriers to a successful SEA. A successful SEA is considered here as one that contributes to a better strategic action and therefore complies with a key principle: that SEA must improve (and not just analyze) the policy, plan or programme (see Chap. 1). The litmus test for a successful SEA would be to compare the strategic action before and after the SEA is done, evaluating any environmental improvements that occurred. This will be clearer in the future as more and more SEA are carried out, and the consequences of the strategic actions (that can be long term) become evident. As the practice of SEA increases, it is also very likely that the quality of the SEA process will improve (similarly to what happened to Project EIA – see for example Lee et al. 1994). Meanwhile it is important that barriers for a successful SEA are evaluated in order to recommend the best course of action for future practice. This chapter concentrates on four main potential problems: bland alternatives, weak public participation, lack of the 'right data', and generally poor procedures and methodology. The chapter concludes with an analysis of the need for quality assurance mechanisms for SEA, and evaluates whether SEA should be integrated with sustainability appraisal or not.

47.2 SEA Barrier 1: Bland Alternatives

Alternatives (or options) in SEA are the different possible ways of achieving a future vision (Thérivel 2004). The quality and innovation of the SEA process overall depends to a large extent on the type of alternatives that are considered. If alternatives are bland and restricted, this will affect how SEA will be able to improve the strategic action.

Implementing Strategic Environmental Assessment. Edited by Michael Schmidt, Elsa João and Eike Albrecht. © 2005 Springer-Verlag

The ability to consider innovative alternatives, that might be very different from what has been done in the past, will depend to a large extent of the SEA team in charge of carrying out the SEA. As a way of bringing some lateral thinking into an SEA team, it is considered a good idea to involve 'outsiders' (either from a different department within the same organisation or from outside altogether) working in tandem with the people that are producing the strategic action (see Chap. 1). However, even with the best forward-thinking SEA team in the world, alternatives might be limited by choices made at an earlier, even more strategic level. The fact that the SEA Directive applies to only plans and programmes, and excludes policies, can have a negative impact in this respect. It can be argued that "un-SEAed national policies do not provide [local and regional level decision-makers] with an acceptable, sustainable framework for their decisions" (Thérivel 2004 p.209).

As SEA practice progresses, it will be very informative to compare the type of alternatives considered in the countries that have also legislated for SEA of policies versus the countries that have only implemented the restricted focus of the SEA Directive. A particularly interesting country to consider for this comparison is Scotland. Thanks to political devolution, the different countries in the UK (England, Scotland, Wales and Northern Ireland) are implementing the SEA Directive in their separate ways (see Chap. 6 for the case of England). Unlike the other UK countries, the Scottish Executive (the devolved government for Scotland) is planning to extend SEA to policies (which the Scottish Executive calls strategies). In other words, Scotland is planning to 'gold plate' the SEA Directive. The Scottish Executive's strong commitment to SEA is obvious in the Partnership Agreement that sets out the Executive's plans and priorities for Scotland (Scottish Executive 2003 p.47):

"We will legislate to introduce strategic environmental assessment to ensure that the full environmental impacts of all new strategies, programmes and plans developed by the public sector are properly considered."

The transposition of the SEA Directive in Scotland will have two stages. In stage 1, in July 2004, new SEA regulations implemented the requirements of the SEA Directive. Then in stage 2, probably by the end of 2005, an SEA Bill (primary legislation) will extend the 2004 requirements to cover a much broader range of strategic actions (Scottish Executive 2004). The Scottish approach to the implementation of the SEA Directive led Deasley (2003 p.17) to argue that "Scotland will have a more comprehensive and inclusive approach to SEA than perhaps any other part of Europe and may emerge as a European leader in this field." What remains to be seen is to what extent the more "comprehensive and inclusive approach to SEA" will lead to more innovative alternatives considered.

Time will also tell how, in the different Member States, will authorities with environmental responsibilities and the public react to 'bland alternatives' in the consultation process. Will responsible authorities be asked to go back to the drawing board and think about new alternatives? What about if the most innovative alternatives, that would be possible in theory, are not really possible in practice because the implementation of those would be the responsibility of another authority (e.g. from a different strategic level or from a different sector)? This all links with the important issue of quality assurance of the SEA process that is discussed in Sect. 47.5.

47.3 SEA Barrier 2: Weak Public Participation

The *minimum* amount of public participation that is required according to legislation is easy to determine. For example, according to the SEA Directive, only consultation of the authorities with environmental responsibilities is necessary when deciding on the scope and level of detail of the SEA, while the announcement of the draft plan and accompanying Environmental Report requires consultation of both authorities with environmental responsibilities and the public. What is more difficult to determine is what is the *best* public participation that should be carried out for SEA. The danger of having a weak public participation is that it can cause the SEA process to weaken as well. Public participation can be considered the ultimate test of the "utility and effectiveness of SEA, which can lead to increased public acceptance" of the strategic action (Fischer 2002 p.13).

There are four key issues related to trying to establish the best public participation possible: when public participation should take place, who exactly is the public, what is the level of participation that is required, and how to ensure effective participation (see also Chaps. 24, 29 and 30). According to the SEA Directive, only when the draft plan is announced, does a full public consultation need to take place. It can be argued that this is not enough. When asked if the public should be consulted at the scoping stage of SEA, a top UK SEA official replied "you would be a fool not to" (SEA Team Leader, SEA Consultation Authority, Scotland, pers. comm. 2004). The widespread perception is that generally more public consultation is required than the one prescribed by the SEA Directive. What is uncertain is how much more exactly.

Another difficult issue is to determine who exactly is the public (see Petts 1999). A strategic action might be of simultaneous interest to the national population as well as the local population, and these may have opposite interests. The conflict of interests between different types of population might be counterintuitive: the development of a new wind farm might be approved by the local people (e.g. if they get cheaper electricity) but might be opposed by the national population because the wind farm would ruin their perception of an unspoiled countryside.

Stakeholder analysis can be useful in determining who the stakeholders are. A stakeholder is any person, group or organisation with an interest or 'stake' in an issue either because they will be affected by it or may have some influence on its outcome. However, in the case of SEA the scope of the strategic action may be very difficult to define, both spatially and temporally (Ortolano and Shepherd 1995), and this makes determining who the public is difficult. Even when spatial boundaries can be delineated, the land areas involved in strategic decision making may be huge and involve many decision-making authorities. Again this would make it very cumbersome to carry out such a broad public participation exercise, with the added possible problem of the existence of many potentially opposite points of view.

The final two issues related to determining the best public participation possible is level of participation required and how to ensure effective participation. Wilcox (1994) suggests five levels of increased participation, and these can be useful for SEA:

- *Information* – The least you can do is tell people what is planned.
- *Consultation* – You offer a number of options and listen to the feedback.
- *Deciding together* – You encourage others to provide some additional ideas and options, and join in deciding the best way forward.
- *Acting together* – Not only do different interests decide together what is best, but they form a partnership to carry it out.
- *Supporting independent community initiatives* – You help others do what they want – perhaps within a framework of grants, advice and support provided by the resource holder.

The key idea behind these different levels of participation is that one level is not necessarily always better than another. They all serve different purposes, and different stages of SEA and strategic decision-making might require different levels of public participation. The last extreme level of public participation – supporting independent community initiatives – is quite extraordinary. It requires a change of attitude where stakeholders are seen as *partners* in plan making, and agencies (like environmental protection-type agencies) might become 'enablers' rather than 'doers'. In SEA this level of participation might be useful in the monitoring stage. An excellent example of this is the Ythan Project in Scotland that aims to involve local people in protecting and enhancing the Ythan River. This has included training farmers to do nutrient budgeting on their use of fertilizers and training local people to carry out river surveys (Ythan Project 2004). Finally, effective participation is only achieved with good feedback mechanisms. It is crucial that the public is informed on how the public participation changed decisions. Only by demonstrating that the public voice has been heard, will consultation fatigue be avoided and *strong* public participation can occur.

47.4 SEA Barrier 3: Lack of the 'Right Data'

There are two sides to the potential SEA barrier caused by lack of the 'right data'. On one hand, lack of data can prevent particular analyses and assessments being done. On the other hand, lack of data can be a poor excuse for certain analyses and assessments *not* being done. The reason for the last point has to do as much with the nature of SEA methodology as with the process of strategic decision-making. It is crucial that SEA is done irrespective of the data available. So for example, even if there is no specific health data, but the strategic action will potentially impact in human health, then this will still need to be assessed even if only qualitatively. The Environmental Report should then point any data gaps that in the future (e.g. in the monitoring stage) will need to be sorted out.

In addition, because SEA should be an "on-going process" (Partidário 1999), it is also "vital that the SEA process keeps pace with decision-making process, which is often very rapid" (Thérivel 2004 p.160). This means that "more data-hungry and specialist-intensive techniques are fine where there is reasonable time in which to carry out the SEA. But quick-and-dirty techniques may be the only ones that can keep up with the rapid decision-making process" (Thérivel 2004 p.162). It is more important that the SEA is done, even if quickly and with little data, than not done at all for lack of data. However, everything should be done to ameliorate this situation which is far from ideal.

If data needs to be selected for a particular assessment, it may necessary to determine what is the 'right data' for that assessment. A common criteria for evaluating data quality is to determine if the data is "fit for purpose". Depending what the data is being used for will affect its suitability and, therefore, how fit it is for that purpose. The notion of data quality is, to a large extent, a relative one. This means that the user will need to determine the suitability of the data in each case. In order to be able to do this, a quality report is normally required that can provide the basis for a user to decide if the data is good enough for the intended application. This quality report is sometimes called 'metadata' (i.e. data about data).

Fig. 47.1 shows six different parameters that can help determining the 'right data' for SEA. Starting with the fact that the data actually exists and is available for use (i.e. is not confidential and is available at an affordable price), and that there is metadata in order to make an informed decision about the suitability of the data. Next, it is important that the data is complete, current, reliable, unbiased and free of gross errors. Plus that the data needs to be at a correct and useful (spatial and temporal) scale. The choice of scale is important as it can affect the results of environmental assessments (João 2002).

Finally, in the case of SEA it is also important that a trend is known and a target has been set. A trend lets practitioners know how things have been evolving in the past. In environmental assessment it is important to have more than a snap shot at a particular point in time. It is critical to know for example if water quality is improving or population health getting worse. In addition, it is important to know what society is aiming for and a target is very useful in providing the size (rather than simply the direction) of the change expected. Targets establish a framework against which the efficacy of a strategic action can be judged, and simplify the monitoring and review of the effectiveness of the strategic action (ODPM et al. 2004). An important aspect about targets is that they must be *realistic*, "otherwise they become embarrassing symbols of defeat rather than positive goals" (Thérivel and Partidário 1996 p.32).

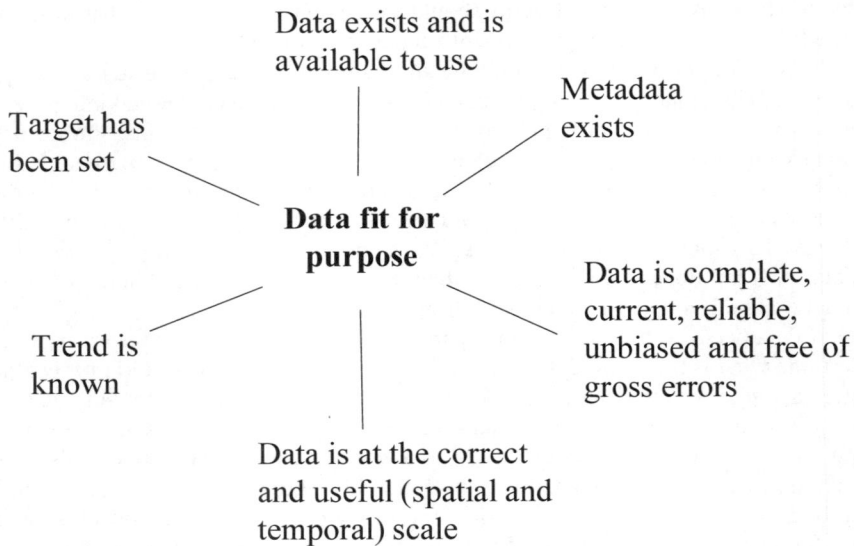

Fig. 47.1. Determining the 'right data' for SEA

An interesting case study carried in the East Midlands Region in England evaluated the availability of baseline environmental data for SEA. The study aimed to find out if, in particular, the East Midlands Observatory – a network of organisations established in 1999 with an interest and involvement in research and statistics about regional economic, social, environmental and spatial issues (see East Midlands Observatory 2004) – could provide the necessary data for SEA. The study found that the data directly available from the East Midlands Observatory was insufficient to establish the environmental baseline for SEA. Much of the required data was available but was often aggregated at the 'wrong level' (e.g. by Health Authority, Local Authority or Environment Agency Region, rather than by 'East Midlands Region'). Targets had not been developed for all regional indicators and, where they did exist, the data to assess performance was not always available. Data for certain indicators was hard to quantify and the methodology required to measure the variable had not been developed even at the National level (e.g. building functionality in terms of use, access, space).

While carrying out an SEA, all practitioners will come across data that is problematic to a lesser or greater extent. Choices will need to be made on whether to avoid using the data, collect further data, or make use of the data but include an explanation of its limitations (or indication of uncertainty) in the Environmental Report (ODPM et al. 2004).

47.5 SEA Barrier 4: Poor Procedures and Methodology

One of the difficulties of SEA is knowing how best to carry out SEA in practice, as methodologies are still evolving and best-practice examples are still rare. However, there is now useful guidance aimed specifically at practitioners (e.g. English Nature et al. 2004; ODPM et al. 2004), and as more SEA are done, best practice examples will start accumulating. A key aspect of this learning from existing practice is that an SEA review is carried out on how the SEA process went. This ensures that practitioners learn from their mistakes and their achievements and allows the SEA process to evolve with time.

However "more intractable than […] technical and information problems are those inherent in the policy-making process" (Glasson et al. 1999 p.404). At the end of the day, in order for SEA to improve the strategic action, the strategic-decision maker needs to allow that to happen. The crucial aspect in this is the necessity of the decision-maker to embrace the spirit of SEA. If SEA is seen as a *burden* rather than an *opportunity*, then it is unlikely that SEA can have much of an impact on the improvement of the strategic action. In the wrong hands SEA could be applied as a "dreary and resource-intensive formality, applied in a grudging minimalist fashion by people who just *hate* having to do it" (Thérivel 2004 p.3). Showing the value of SEA practice to decision-makers is crucial. In this respect there is a lot to be learned from the application of SEA based on *voluntary* approaches (rather than based only on formal legal frameworks) – see Chap. 46. It is key that decision-makers are involved but in an inspired way and from the start. Timing is crucial – if SEA interacts with strategic decision-making too late, then it might be too late to improve the strategic action and, therefore, too late to have a successful SEA.

The quality of the SEA process is paramount. ODPM et al. (2004 p.79) proposes a quality assurance list "intended to help test whether the requirements of the SEA Directive are met, identify any problems in the Environmental Report, and show how effectively the SEA has integrated environmental considerations into the plan-making process. The checklist is designed to be used by anyone involved in an SEA in any capacity: Responsible Authorities which carry out SEA, the organisations which they consult, inspectors, auditors, independent experts, and members of the public." The checklist contains 40 separate points organised according to the following topics:

- Objectives and context
- Scoping
- Alternatives
- Baseline information
- Prediction and evaluation of likely significant effects
- Mitigation measures
- The environmental report
- Consultation
- Decision-making and information on the decision
- Monitoring measures

Quality assurance procedures, such as the one described above and the IAIA (2002) SEA Performance Criteria (see Box 44.2), have an important role to play in ensuring SEA quality. For example, the Czech Republic, which has had SEA legislation since 1992, uses a system of licensed environmental assessment experts in order to help ensure that the SEA process is done with quality (see Chap. 13). The issue of bias, caused by the fact that a authority responsible for the strategic action is the same as the authority responsible for the SEA and could also be the same as one of the consultation authorities (e.g. the Environment Agency is going to be the organisation that will need to carry out the largest amount of SEA in the UK), has lead some practitioners in Scotland to suggest that the existence of a separate independent SEA organisation might be a good idea. This new SEA organisation could be a one-stop contact point for SEA in the country and it could deal with the administration of screening and scoping. It could also prepare guidance, record examples of good practice, be an arbiter in cases of dispute, and it could audit the quality of Environmental Reporting. Overall, it could monitor SEA effectiveness and procedures, and could support linkages between SEA and Project EIA. However, others disagree as they fear that a new SEA body will lead to more bureaucracy.

47.6 Conclusions and Recommendations

The quality of the SEA should be assured through the selection of innovative and appropriate alternatives, "choice of a good SEA team, the collection of appropriate information, the use of effective prediction techniques, consultation, and integration of the SEA findings into the strategic action" (ODPM et al. 2004 p.79). A good quality SEA that contributes to a better strategic action is a successful SEA.

One issue remains. Will SEA be considered more or less successful if it encompasses sustainability appraisal? How much should SEA include socio-economic issues, together with environmental issues? These questions divide most SEA researchers and practitioners at the moment. For example, at a conference on 13 September 2004, to launch the consultation for the Scottish SEA Bill (see Sect. 47.2), participants voted on this same issue. Of the 56 people attending, 37% agreed that SEA and sustainability appraisal should become a single tool, 44% disagreed, and 19% were not sure. For some, if SEA does not include socio-economic issues than that could be considered a 'step back' in relation to sustainability appraisal. Others, however, consider that this is a bad idea because there is the danger that the 'environment' will lose out in relation to the economic and social issues. The Scottish Executive takes very much this last point of view and prefers SEA to be a separate process from sustainability appraisal. In contrast, England has created new guidance that intends to ensure that sustainability appraisal meets the SEA Directive requirements plus widens the SEA Directive's approach to include social and economic as well as environmental issues (see ODPM 2004). The guidance acknowledges that SEA and sustainability appraisal are distinct, but argues that is possible to satisfy both through a single appraisal process. Future practice and comparison of the results of these two opposite approaches, should help elucidate what is the best way forward.

References[1]

Deasley N (2003) Strategic Environmental Assessment. SEPAView – The magazine of the Scottish Environment Protection Agency, pp. 16-17. SEPA, Stirling

East Midlands Observatory (2004) Web site of the East Midlands Observatory. Internet address: http://www.eastmidlandsobservatory.org.uk/index.asp, last accessed 09.08.2004

East Midlands Region (2003) East Midlands Region – SEA Baseline Environmental Data. Prepared by Colin Bush, East Midlands

English Nature, Countryside Council for Wales, Environment Agency, Royal Society for the Protection of Birds (2004) Strategic Environmental Assessment and Biodiversity: Guidance for Practitioners

Fischer TB (2002) Strategic Environmental Assessment in Transport and Land use Planning. Earthscan, London

Glasson J, Thérivel, R, Chadwick A (1999) Introduction to Environmental Impact Assessment, 2nd edn. UCL Press, London

IAIA – International Association for Impact Assessment (2002) Strategic Environmental Assessment Performance Criteria. Special Publication Series nr. 1, IAIA. Internet address: http://www.iaia.org/Members/Publications/Special_Pubs/sp1.pdf, last accessed 09.08.2004

João E (2002) How scale affects environmental impact assessment. Environmental Impact Assessment Review, 22 (4):287-306

Lee N, Walsh F, Reeder G (1994) Assessing the performance of the EA process. Project Appraisal, 9 (3):161-172

ODPM – Office of the Deputy Prime Minister (2004) Sustainability appraisal of regional strategies and local development frameworks – consultation paper, ODPM, Sept 2004, Internet address: www.planning.odpm.gov.uk, last accessed 30.09.2004

ODPM, Welsh Assembly Government, Scottish Executive, Northern Ireland Department of the Environment (2004) A Draft Practical Guide to the SEA Directive. Proposals by ODPM, the Scottish Executive, the Welsh Assembly Government and the Northern Ireland Department of the Environment for practical guidance on applying European Directive 2001/42/EC 'on the assessment of the effects of certain plans and programmes on the environment', Office of the Deputy Prime Minister, London

Ortolano L and Shepherd A (1995) Environmental Impact Assessment. In: Vanclay F and Bronstein D (eds.), Environmental and Social Impact Assessment, pp. 3-31. John Wiley & Sons, Chichester

Partidário MR (1999) Strategic Environmental Assessment – principles and potential. In: Petts J (ed) Handbook of Environmental Impact Assessment, vol. 1. Blackwell, Oxford, pp 60-73

Petts J (1999) Public participation and environmental impact assessment. In: Petts J (ed) Handbook of Environmental Impact Assessment, Blackwell Science, Oxford

---

[1] Note that legislation mentioned in the chapter is listed at the end of the handbook in the consolidated list of legislation (Appendix 2).

Scottish Executive (2003) A Partnership for a Better Scotland. A partnership coalition agreement between Labour and the Liberal Democrats, Internet address: //http://www.scotland.gov.uk/library5/government/pfbs-05.asp, last accessed 28.09.2004

Scottish Executive (2004) Strategic Environmental Assessment – A Consultation on the Proposed Environmental Assessment (Scotland) Bill. September 2004, paper 2004/12

Thérivel R (2004) Strategic Environmental Assessment in Action. Earthscan, London.

Thérivel R and Partidário MR (1996) The Practice of Strategic Environmental Assessment. Earthscan Publications Ltd, London

Wilcox D (1994) The Guide to Effective Participation, prepared for the Joseph Rowntree Foundation, Internet address: www.partnerhsips.org.uk/guide, last accessed 09.08.2004

Ythan Project (2004), Web site of the in Ythan Project (Scotland) that aims to involve local people in protecting and enhancing the Ythan River. Internet address: http://www.ythan.org.uk, last accessed 09.08.2004

Appendix 1 – The Full Text of the SEA Directive

Directive 2001/42/EC of the European Parliament and of the Council of 27 June 2001 on the assessment of the effects of certain plans and programmes on the environment

THE EUROPEAN PARLIAMENT AND THE COUNCIL OF THE EUROPEAN UNION,
Having regard to the Treaty establishing the European Community, and in particular Article 175(1) thereof,
Having regard to the proposal from the Commission(1),
Having regard to the opinion of the Economic and Social Committee(2),
Having regard to the opinion of the Committee of the Regions(3),
Acting in accordance with the procedure laid down in Article 251 of the Treaty(4), in the light of the joint text approved by the Conciliation Committee on 21 March 2001,

Whereas:
(1) Article 174 of the Treaty provides that Community policy on the environment is to contribute to, inter alia, the preservation, protection and improvement of the quality of the environment, the protection of human health and the prudent and rational utilisation of natural resources and that it is to be based on the precautionary principle. Article 6 of the Treaty provides that environmental protection requirements are to be integrated into the definition of Community policies and activities, in particular with a view to promoting sustainable development.

(2) The Fifth Environment Action Programme: Towards sustainability - A European Community programme of policy and action in relation to the environment and sustainable development(5), supplemented by Council Decision No 2179/98/EC(6) on its review, affirms the importance of assessing the likely environmental effects of plans and programmes.

(3) The Convention on Biological Diversity requires Parties to integrate as far as possible and as appropriate the conservation and sustainable use of biological diversity into relevant sectoral or cross-sectoral plans and programmes.

(4) Environmental assessment is an important tool for integrating environmental considerations into the preparation and adoption of certain plans and programmes which are likely to have significant effects on the environment in the Member States, because it ensures that such effects of implementing plans and programmes are taken into account during their preparation and before their adoption.

Implementing Strategic Environmental Assessment. Edited by Michael Schmidt, Elsa João and Eike Albrecht. © 2005 Springer-Verlag

(5) The adoption of environmental assessment procedures at the planning and programming level should benefit undertakings by providing a more consistent framework in which to operate by the inclusion of the relevant environmental information into decision making. The inclusion of a wider set of factors in decision making should contribute to more sustainable and effective solutions.

(6) The different environmental assessment systems operating within Member States should contain a set of common procedural requirements necessary to contribute to a high level of protection of the environment.

(7) The United Nations/Economic Commission for Europe Convention on Environmental Impact Assessment in a Transboundary Context of 25 February 1991, which applies to both Member States and other States, encourages the parties to the Convention to apply its principles to plans and programmes as well; at the second meeting of the Parties to the Convention in Sofia on 26 and 27 February 2001, it was decided to prepare a legally binding protocol on strategic environmental assessment which would supplement the existing provisions on environmental impact assessment in a transboundary context, with a view to its possible adoption on the occasion of the 5th Ministerial Conference "Environment for Europe" at an extraordinary meeting of the Parties to the Convention, scheduled for May 2003 in Kiev, Ukraine. The systems operating within the Community for environmental assessment of plans and programmes should ensure that there are adequate transboundary consultations where the implementation of a plan or programme being prepared in one Member State is likely to have significant effects on the environment of another Member State. The information on plans and programmes having significant effects on the environment of other States should be forwarded on a reciprocal and equivalent basis within an appropriate legal framework between Member States and these other States.

(8) Action is therefore required at Community level to lay down a minimum environmental assessment framework, which would set out the broad principles of the environmental assessment system and leave the details to the Member States, having regard to the principle of subsidiarity. Action by the Community should not go beyond what is necessary to achieve the objectives set out in the Treaty.

(9) This Directive is of a procedural nature, and its requirements should either be integrated into existing procedures in Member States or incorporated in specifically established procedures. With a view to avoiding duplication of the assessment, Member States should take account, where appropriate, of the fact that assessments will be carried out at different levels of a hierarchy of plans and programmes.

(10) All plans and programmes which are prepared for a number of sectors and which set a framework for future development consent of projects listed in Annexes I and II to Council Directive 85/337/EEC of 27 June 1985 on the assessment of the effects of certain public and private projects on the environment(7), and all plans and programmes which have been determined to require assessment pursuant to Council Directive 92/43/EEC of 21 May 1992 on

the conservation of natural habitats and of wild flora and fauna(8), are likely to have significant effects on the environment, and should as a rule be made subject to systematic environmental assessment. When they determine the use of small areas at local level or are minor modifications to the above plans or programmes, they should be assessed only where Member States determine that they are likely to have significant effects on the environment.

(11) Other plans and programmes which set the framework for future development consent of projects may not have significant effects on the environment in all cases and should be assessed only where Member States determine that they are likely to have such effects.

(12) When Member States make such determinations, they should take into account the relevant criteria set out in this Directive.

(13) Some plans or programmes are not subject to this Directive because of their particular characteristics.

(14) Where an assessment is required by this Directive, an environmental report should be prepared containing relevant information as set out in this Directive, identifying, describing and evaluating the likely significant environmental effects of implementing the plan or programme, and reasonable alternatives taking into account the objectives and the geographical scope of the plan or programme; Member States should communicate to the Commission any measures they take concerning the quality of environmental reports.

(15) In order to contribute to more transparent decision making and with the aim of ensuring that the information supplied for the assessment is comprehensive and reliable, it is necessary to provide that authorities with relevant environmental responsibilities and the public are to be consulted during the assessment of plans and programmes, and that appropriate time frames are set, allowing sufficient time for consultations, including the expression of opinion.

(16) Where the implementation of a plan or programme prepared in one Member State is likely to have a significant effect on the environment of other Member States, provision should be made for the Member States concerned to enter into consultations and for the relevant authorities and the public to be informed and enabled to express their opinion.

(17) The environmental report and the opinions expressed by the relevant authorities and the public, as well as the results of any transboundary consultation, should be taken into account during the preparation of the plan or programme and before its adoption or submission to the legislative procedure.

(18) Member States should ensure that, when a plan or programme is adopted, the relevant authorities and the public are informed and relevant information is made available to them.

(19) Where the obligation to carry out assessments of the effects on the environment arises simultaneously from this Directive and other Community

legislation, such as Council Directive 79/409/EEC of 2 April 1979 on the conservation of wild birds(9), Directive 92/43/EEC, or Directive 2000/60/EC of the European Parliament and the Council of 23 October 2000 establishing a framework for Community action in the field of water policy(10), in order to avoid duplication of the assessment, Member States may provide for coordinated or joint procedures fulfilling the requirements of the relevant Community legislation.

(20) A first report on the application and effectiveness of this Directive should be carried out by the Commission five years after its entry into force, and at seven-year intervals thereafter. With a view to further integrating environmental protection requirements, and taking into account the experience acquired, the first report should, if appropriate, be accompanied by proposals for amendment of this Directive, in particular as regards the possibility of extending its scope to other areas/sectors and other types of plans and programmes,

HAVE ADOPTED THIS DIRECTIVE:

Article 1
Objectives

The objective of this Directive is to provide for a high level of protection of the environment and to contribute to the integration of environmental considerations into the preparation and adoption of plans and programmes with a view to promoting sustainable development, by ensuring that, in accordance with this Directive, an environmental assessment is carried out of certain plans and programmes which are likely to have significant effects on the environment.

Article 2
Definitions

For the purposes of this Directive:

(a) "plans and programmes" shall mean plans and programmes, including those co-financed by the European Community, as well as any modifications to them:

- which are subject to preparation and/or adoption by an authority at national, regional or local level or which are prepared by an authority for adoption, through a legislative procedure by Parliament or Government, and

- which are required by legislative, regulatory or administrative provisions;

(b) "environmental assessment" shall mean the preparation of an environmental report, the carrying out of consultations, the taking into account of the environmental report and the results of the consultations in decision-making and the provision of information on the decision in accordance with Articles 4 to 9;

(c) "environmental report" shall mean the part of the plan or programme documentation containing the information required in Article 5 and Annex I;

(d) "The public" shall mean one or more natural or legal persons and, in accordance with national legislation or practice, their associations, organisations or groups.

Article 3
Scope

1. An environmental assessment, in accordance with Articles 4 to 9, shall be carried out for plans and programmes referred to in paragraphs 2 to 4 which are likely to have significant environmental effects.

2. Subject to paragraph 3, an environmental assessment shall be carried out for all plans and programmes,

> (a) which are prepared for agriculture, forestry, fisheries, energy, industry, transport, waste management, water management, telecommunications, tourism, town and country planning or land use and which set the framework for future development consent of projects listed in Annexes I and II to Directive 85/337/EEC, or

> (b) which, in view of the likely effect on sites, have been determined to require an assessment pursuant to Article 6 or 7 of Directive 92/43/EEC.

3. Plans and programmes referred to in paragraph 2 which determine the use of small areas at local level and minor modifications to plans and programmes referred to in paragraph 2 shall require an environmental assessment only where the Member States determine that they are likely to have significant environmental effects.

4. Member States shall determine whether plans and programmes, other than those referred to in paragraph 2, which set the framework for future development consent of projects, are likely to have significant environmental effects.

5. Member States shall determine whether plans or programmes referred to in paragraphs 3 and 4 are likely to have significant environmental effects either through case-by-case examination or by specifying types of plans and programmes or by combining both approaches. For this purpose Member States shall in all cases take into account relevant criteria set out in Annex II, in order to ensure that plans and programmes with likely significant effects on the environment are covered by this Directive.

6. In the case-by-case examination and in specifying types of plans and programmes in accordance with paragraph 5, the authorities referred to in Article 6(3) shall be consulted.

7. Member States shall ensure that their conclusions pursuant to paragraph 5, including the reasons for not requiring an environmental assessment pursuant to Articles 4 to 9, are made available to the public.

8. The following plans and programmes are not subject to this Directive:

- plans and programmes the sole purpose of which is to serve national defence or civil emergency,
- financial or budget plans and programmes.

9. This Directive does not apply to plans and programmes co-financed under the current respective programming periods(11) for Council Regulations (EC) No 1260/1999(12) and (EC) No 1257/1999(13).

Article 4
General obligations

1. The environmental assessment referred to in Article 3 shall be carried out during the preparation of a plan or programme and before its adoption or submission to the legislative procedure.

2. The requirements of this Directive shall either be integrated into existing procedures in Member States for the adoption of plans and programmes or incorporated in procedures established to comply with this Directive.

3. Where plans and programmes form part of a hierarchy, Member States shall, with a view to avoiding duplication of the assessment, take into account the fact that the assessment will be carried out, in accordance with this Directive, at different levels of the hierarchy. For the purpose of, inter alia, avoiding duplication of assessment, Member States shall apply Article 5(2) and (3).

Article 5
Environmental report

1. Where an environmental assessment is required under Article 3(1), an environmental report shall be prepared in which the likely significant effects on the environment of implementing the plan or programme, and reasonable alternatives taking into account the objectives and the geographical scope of the plan or programme, are identified, described and evaluated. The information to be given for this purpose is referred to in Annex I.

2. The environmental report prepared pursuant to paragraph 1 shall include the information that may reasonably be required taking into account current knowledge and methods of assessment, the contents and level of detail in the plan or programme, its stage in the decision-making process and the extent to which certain matters are more appropriately assessed at different levels in that process in order to avoid duplication of the assessment.

3. Relevant information available on environmental effects of the plans and programmes and obtained at other levels of decision-making or through other Community legislation may be used for providing the information referred to in Annex I.

4. The authorities referred to in Article 6(3) shall be consulted when deciding on the scope and level of detail of the information which must be included in the environmental report.

Article 6
Consultations

1. The draft plan or programme and the environmental report prepared in accordance with Article 5 shall be made available to the authorities referred to in paragraph 3 of this Article and the public.

2. The authorities referred to in paragraph 3 and the public referred to in paragraph 4 shall be given an early and effective opportunity within appropriate time frames to express their opinion on the draft plan or programme and the accompanying environmental report before the adoption of the plan or programme or its submission to the legislative procedure.

3. Member States shall designate the authorities to be consulted which, by reason of their specific environmental responsibilities, are likely to be concerned by the environmental effects of implementing plans and programmes.

4. Member States shall identify the public for the purposes of paragraph 2, including the public affected or likely to be affected by, or having an interest in, the decision-making subject to this Directive, including relevant non-governmental organisations, such as those promoting environmental protection and other organisations concerned.

5. The detailed arrangements for the information and consultation of the authorities and the public shall be determined by the Member States.

Article 7
Transboundary consultations

1. Where a Member State considers that the implementation of a plan or programme being prepared in relation to its territory is likely to have significant effects on the environment in another Member State, or where a Member State likely to be significantly affected so requests, the Member State in whose territory the plan or programme is being prepared shall, before its adoption or submission to the legislative procedure, forward a copy of the draft plan or programme and the relevant environmental report to the other Member State.

2. Where a Member State is sent a copy of a draft plan or programme and an environmental report under paragraph 1, it shall indicate to the other Member State whether it wishes to enter into consultations before the adoption of the plan or programme or its submission to the legislative procedure and, if it so indicates, the Member States concerned shall enter into consultations concerning the likely transboundary environmental effects of implementing the plan or programme and the measures envisaged to reduce or eliminate such effects.

Where such consultations take place, the Member States concerned shall agree on detailed arrangements to ensure that the authorities referred to in Article 6(3) and the public referred to in Article 6(4) in the Member State likely to be significantly affected are informed and given an opportunity to forward their opinion within a reasonable time-frame.

3. Where Member States are required under this Article to enter into consultations, they shall agree, at the beginning of such consultations, on a reasonable timeframe for the duration of the consultations.

Article 8
Decision making

The environmental report prepared pursuant to Article 5, the opinions expressed pursuant to Article 6 and the results of any transboundary consultations entered into pursuant to Article 7 shall be taken into account during the preparation of the plan or programme and before its adoption or submission to the legislative procedure.

Article 9
Information on the decision

1. Member States shall ensure that, when a plan or programme is adopted, the authorities referred to in Article 6(3), the public and any Member State consulted under Article 7 are informed and the following items are made available to those so informed:

(a) the plan or programme as adopted;

(b) a statement summarising how environmental considerations have been integrated into the plan or programme and how the environmental report prepared pursuant to Article 5, the opinions expressed pursuant to Article 6 and the results of consultations entered into pursuant to Article 7 have been taken into account in accordance with Article 8 and the reasons for choosing the plan or programme as adopted, in the light of the other reasonable alternatives dealt with, and

(c) measures decided concerning monitoring in accordance with Article 10.

2. The detailed arrangements concerning the information referred to in paragraph 1 shall be determined by the Member States.

Article 10
Monitoring

1. Member States shall monitor the significant environmental effects of the implementation of plans and programmes in order, inter alia, to identify at an early stage unforeseen adverse effects, and to be able to undertake appropriate remedial action.

2. In order to comply with paragraph 1, existing monitoring arrangements may be used if appropriate, with a view to avoiding duplication of monitoring.

Article 11
Relationship with other Community legislation

1. An environmental assessment carried out under this Directive shall be without prejudice to any requirements under Directive 85/337/EEC and to any other Community law requirements.

2. For plans and programmes for which the obligation to carry out assessments of the effects on the environment arises simultaneously from this Directive and other Community legislation, Member States may provide for coordinated or joint procedures fulfilling the requirements of the relevant Community legislation in order, inter alia, to avoid duplication of assessment.

3. For plans and programmes co-financed by the European Community, the environmental assessment in accordance with this Directive shall be carried out in conformity with the specific provisions in relevant Community legislation.

Article 12
Information, reporting and review

1. Member States and the Commission shall exchange information on the experience gained in applying this Directive.

2. Member States shall ensure that environmental reports are of a sufficient quality to meet the requirements of this Directive and shall communicate to the Commission any measures they take concerning the quality of these reports.

3. Before 21 July 2006 the Commission shall send a first report on the application and effectiveness of this Directive to the European Parliament and to the Council. With a view further to integrating environmental protection requirements, in accordance with Article 6 of the Treaty, and taking into account the experience acquired in the application of this Directive in the Member States, such a report will be accompanied by proposals for amendment of this Directive, if appropriate. In particular, the Commission will consider the possibility of extending the scope of this Directive to other areas/sectors and other types of plans and programmes. A new evaluation report shall follow at seven-year intervals.

4. The Commission shall report on the relationship between this Directive and Regulations (EC) No 1260/1999 and (EC) No 1257/1999 well ahead of the expiry of the programming periods provided for in those Regulations, with a view to ensuring a coherent approach with regard to this Directive and subsequent Community Regulations.

Article 13
Implementation of the Directive

1. Member States shall bring into force the laws, regulations and administrative provisions necessary to comply with this Directive before 21 July 2004. They shall forthwith inform the Commission thereof.

2. When Member States adopt the measures, they shall contain a reference to this Directive or shall be accompanied by such reference on the occasion of their official publication. The methods of making such reference shall be laid down by Member States.

3. The obligation referred to in Article 4(1) shall apply to the plans and programmes of which the first formal preparatory act is subsequent to the date referred to in paragraph 1. Plans and programmes of which the first formal preparatory act is before that date and which are adopted or submitted to the legislative procedure more than 24 months thereafter, shall be made subject to the obligation referred to in Article 4(1) unless Member States decide on a case by case basis that this is not feasible and inform the public of their decision.

4. Before 21 July 2004, Member States shall communicate to the Commission, in addition to the measures referred to in paragraph 1, separate information on the types of plans and programmes which, in accordance with Article 3, would be subject to an environmental assessment pursuant to this Directive. The Commission shall make this information available to the Member States. The information will be updated on a regular basis.

Article 14
Entry into force

This Directive shall enter into force on the day of its publication in the Official Journal of the European Communities.

Article 15
Addressees

This Directive is addressed to the Member States.
Done at Luxembourg, 27 June 2001.

For the European Parliament
The President
N. Fontaine

For the Council
The President
B. Rosengren

Footnotes

(1) OJ C 129, 25.4.1997, p. 14 and OJ C 83, 25.3.1999, p. 13.

(2) OJ C 287, 22.9.1997, p. 101.

(3) OJ C 64, 27.2.1998, p. 63 and OJ C 374, 23.12.1999, p. 9.

(4) Opinion of the European Parliament of 20 October 1998 (OJ C 341, 9.11.1998, p. 18), confirmed on 16 September 1999 (OJ C 54, 25.2.2000, p. 76), Council Common Position of 30 March 2000 (OJ C 137, 16.5.2000, p. 11) and Decision of the European Parliament of 6 September 2000 (OJ C 135, 7.5.2001, p. 155). Decision of the European Parliament of 31 May 2001 and Decision of the Council of 5 June 2001.

(5) OJ C 138, 17.5.1993, p. 5.

(6) OJ L 275, 10.10.1998, p. 1.

(7) OJ L 175, 5.7.1985, p. 40. Directive as amended by Directive 97/11/EC (OJ L 73, 14.3.1997, p. 5).

(8) OJ L 206, 22.7.1992, p. 7. Directive as last amended by Directive 97/62/EC (OJ L 305, 8.11.1997, p. 42).

(9) OJ L 103, 25.4.1979, p. 1. Directive as last amended by Directive 97/49/EC (OJ L 223, 13.8.1997, p. 9).

(10) OJ L 327, 22.12.2000, p. 1.

(11) The 2000-2006 programming period for Council Regulation (EC) No 1260/1999 and the 2000-2006 and 2000-2007 programming periods for Council Regulation (EC) No 1257/1999.

(12) Council Regulation (EC) No 1260/1999 of 21 June 1999 laying down general provisions on the Structural Funds (OJ L 161, 26.6.1999, p. 1).

(13) Council Regulation (EC) No 1257/1999 of 17 May 1999 on support for rural development from the European Agricultural Guidance and Guarantee Fund (EAGGF) and amending and repealing certain regulations (OJ L 160, 26.6.1999, p. 80).

ANNEX I
Information referred to in Article 5(1)

The information to be provided under Article 5(1), subject to Article 5(2) and (3), is the following:

(a) an outline of the contents, main objectives of the plan or programme and relationship with other relevant plans and programmes;

(b) the relevant aspects of the current state of the environment and the likely evolution thereof without implementation of the plan or programme;

(c) the environmental characteristics of areas likely to be significantly affected;

(d) any existing environmental problems which are relevant to the plan or programme including, in particular, those relating to any areas of a particular environmental importance, such as areas designated pursuant to Directives 79/409/EEC and 92/43/EEC;

(e) the environmental protection objectives, established at international, Community or Member State level, which are relevant to the plan or programme and the way those objectives and any environmental considerations have been taken into account during its preparation;

(f) the likely significant effects[1] on the environment, including on issues such as biodiversity, population, human health, fauna, flora, soil, water, air, climatic factors, material assets, cultural heritage including architectural and archaeological heritage, landscape and the interrelationship between the above factors;

(g) the measures envisaged to prevent, reduce and as fully as possible offset any significant adverse effects on the environment of implementing the plan or programme;

(h) an outline of the reasons for selecting the alternatives dealt with, and a description of how the assessment was undertaken including any difficulties (such as technical deficiencies or lack of know-how) encountered in compiling the required information;

(i) a description of the measures envisaged concerning monitoring in accordance with Article 10;

(j) a non-technical summary of the information provided under the above headings.

---

[1] These effects should include secondary, cumulative, synergistic, short, medium and long-term permanent and temporary, positive and negative effects.

ANNEX II

Criteria for determining the likely significance of effects referred to in Article 3(5)

1. The characteristics of plans and programmes, having regard, in particular, to

- the degree to which the plan or programme sets a framework for projects and other activities, either with regard to the location, nature, size and operating conditions or by allocating resources,
- the degree to which the plan or programme influences other plans and programmes including those in a hierarchy,
- the relevance of the plan or programme for the integration of environmental considerations in particular with a view to promoting sustainable development,
- environmental problems relevant to the plan or programme,
- the relevance of the plan or programme for the implementation of Community legislation on the environment (e.g. plans and programmes linked to waste-management or water protection).

2. Characteristics of the effects and of the area likely to be affected, having regard, in particular, to

- the probability, duration, frequency and reversibility of the effects,
- the cumulative nature of the effects,
- the transboundary nature of the effects,
- the risks to human health or the environment (e.g. due to accidents),
- the magnitude and spatial extent of the effects (geographical area and size of the population likely to be affected),
- the value and vulnerability of the area likely to be affected due to:
- special natural characteristics or cultural heritage,
- exceeded environmental quality standards or limit values,
- intensive land-use,
- the effects on areas or landscapes which have a recognised national, Community or international protection status.

Appendix 2 – Consolidated List of Legislation

The consolidated List of Legislation includes only the legislation that is mentioned in the handbook. The left column shows the terms used in the chapters and the right column gives the full information about the respective law. The list is organized as follows:

1 International Conventions 716

2 European Union Legislation 717
    2.1 Primary Treaties 717
    2.2 Secondary Legislation 718

3 National Legislation 721
    3.1 Austria 721
        Federal Acts 721
        The Acts of the Provinces 721
    3.2 Belgium 722
    3.3 Canada 722
    3.4 China 723
    3.5 Czech Republic 723
    3.6 England 724
    3.7 Estonia 724
    3.8 Finland 724
    3.9 Germany 725
        Federal Acts 725
        The Acts of the Laender 726
    3.10 Ghana 726
    3.11 Italy 727
    3.12 Kenya 728
    3.13 Latvia 728
    3.14 New Zealand 728
    3.15 Poland 729
    3.16 Sweden 731
    3.17 Ukraine 731
    3.18 United States 732

Implementing Strategic Environmental Assessment. Edited by Michael Schmidt, Elsa João and Eike Albrecht. © 2005 Springer-Verlag

1 International Conventions

Aarhus Convention	UN ECE Convention on Access to Information, Public Participation in Decision-making and Access to Justice in Environmental Matters; source: www.unece.org/env/pp/treatytext.htm (last access 03.09.04)
Agenda 21	www.un.org/esa/sustdev/documents/agenda21/english/agenda21toc.htm (last access 03.09.04)
Bern Convention	Convention on the Conservation of Wildlife and Natural Habitats, 1979; source: http://www.unep.ch/seas/main/legal/lbern.html (last access 03.09.04)
Convention on Biodiversity	Convention on Biological Diversity, 1992; source: www.biodiv.org/convention/articles.asp (last access 03.09.04)
Bonn Convention	Convention on the Conservation of Migratory Species of Wild Animals, 1979; source: http://www.unep.ch/seas/main/legal/lbonn.html (last access 03.09.04)
Cartagena Protocol	Cartagena Protocol on Biosafety, 2000; source: http://www.biodiv.org/biosafety/protocol.asp (last access 03.09.04)
CITES	Convention on International Trade in Endangered Species of Wild Fauna and Flora, 1973; source: http://www.cites.org/eng/disc/text.shtml (last access 03.09.04)
Espoo Convention	the UN ECE Convention on Environmental Impact Assessment in a Transboundary Context, 1991; source: www.unece.org/env/eia/eia.htm (last access 03.09.04)
Kiev SEA Protocol	Protocol on Strategic Environmental Assessment to the Convention on Environmental Impact Assessment in a Transboundary Context; source: www.unece.org/env/eia/sea_protocol.htm (last access 03.09.04)

Ramsar Convention	Convention on Wetlands of International Importance especially Waterfowl Habitat, 1971; source: www.ramsar.org/index_very_key_docs.htm (last access 03.09.04)
Rio Declaration	Rio Declaration on Environment and Development, 1992; source: http://www.un.org/documents/ga/conf151/aconf1512 6-1annex1.htm (last access 03.09.04)
Vienna Treaty Convention	Vienna Convention on the Law of Treaties, 1969; source: www.un.org/law/ilc/texts/treatfra.htm (last access 03.09.04)
World Heritage Convention	Convention concerning the Protection of the World Cultural and Natural Heritage, 1972; source: http://whc.unesco.org/nwhc/pages/doc/main.htm (last access 03.09.04)

2 European Union Legislation

2.1 Primary Treaties

EU Treaty	Treaty on European Union, OJ C 325 of 24.12.2002, p. 5
EC Treaty	Treaty establishing European Community, OJ C 325 of 24.12.2002, p. 33
Treaty of Maastricht	Treaty on European Union, OJ C 191 of 29.07.1992, p. 1 and Treaty establishing the European Community OJ C 224 of 31.08.1992, p. 1
Treaty of Amsterdam	Treaty of Amsterdam amending the Treaty on European Union, the Treaties establishing the European Communities and Related Acts, OJ C 340 of 10.11.1997, p. 1
Treaty of Nice	Treaty of Nice amending the Treaty on European Union, the Treaties establishing the European Communities and Certain related Acts, OJ C 80 of 10.03.2001, p. 1

2.2 Secondary Legislation

Access to Environmental Information Directive	Directive 2003/4/EC of the European Parliament and of the Council of 28 January 2003 on public access to environmental information and repealing Council Directive 90/313/EEC, OJ L 41 of 14.02.2003, p. 26
Air Quality Framework Directive	Council Directive 96/62/EC of 27 September 1996 on ambient air quality assessment and management, OJ L 296 of 21.11.1996, p. 55
Council Regulation 1782/2003	Council Regulations (EC) No. 1782/2003 of 29 September 2003 establishing common rules for direct support schemes under common agricultural policy and establishing certain support schemes for farmers and amending Regulations (EEC) No. 2019/93, (EC) No. 1452/2001, (EC) No. 1453/2001, (EC) No. 1454/2001, (EC) No.1868/94, (EC) No. 1251/1991, (EC) No. 1254/1999, (EC) No. 1673/2000, (EEC) No. 2358/71 and (EC) No. 2529/2001, OJ L 270 of 21.10.2003, p. 1, corr. OJ L 94 of 31.03.2004, p. 70
Directive on Farming in Mountain, Hill and Less-Favoured Areas	Council Directive 75/268/EEC of 28 April 1975 on mountain and hill farming and farming in certain less-favoured areas, OJ L 128 of 19.05.1975, p. 1, corr. OJ L 216 of 14.08.1975, p. 17, OJ L 206 of 05.08.1975, p. 1, OJ L 189 of 11.07.1975, p. 39 and OJ L 172 of 03.07.1975, p. 19
Groundwater Directive	Directive on Protection of Groundwater Against Pollution by Dangerous substances, Council Directive 80/68/EEC of 17 December 1979 on the protection of groundwater against pollution caused by certain dangerous substances, OJ L 20 of 26.01.1980, p. 43
Sewage Sludge Directive	Directive on Usage of Sewage Sludge in Agriculture, Council Directive 86/278/EEC of 12 June 1986 on the protection of the environment and in particular of the soil, when sewage sludge is used in agriculture, OJ L 181 of 04.07.1986, p. 6, corr. OJ L 191, 15.07.1986, p. 23
EIA Directive	Council Directive 85/337/EEC of 27 June 1985 on the assessment of the effects of certain public and private projects on the environment, OJ L 175 of 05.07.1985 p. 40, corr. OJ L 216 of 03.08.1991, p. 40, amended

	by Directive 97/11/EC of 03.03 1997, OJ L 73, p. 5, and Directive 2003/35/EC of 26.05.2003, OJ L 156, p. 17
EMAS Regulation	Regulation (EC) No 761/2001 of the European parliament and of the council of 19 March 2001 allowing voluntary participation by organisations in a Community eco-management and audit scheme (EMAS), OJ L 114 of 24.04.2001, p. 1, corr. OJ L 327 of 04.12.2002, p. 10
End of Life Vehicles Directive	Directive 2000/53/EC of the European Parliament and of the Council of 18 September 2000 on end-of life vehicles, OJ L 269 of 21.10.2000, p. 34
Habitats Directive	Council Directive 92/43/EEC of 21 May 1992 on the conservation of natural habitats and of wild fauna and flora, OJ L 206 of 22.07.1992, p. 7, corr. OJ L 031 of 06.02.1998, p. 30, OJ L 059 of 08.03.1996, p. 63 and OJ L 176, 20.07.1993, p. 29, as amended by Directive 97/62/EC of 27 October 1997, OJ L 305, p. 42
Hazardous Waste Directive	Council Directive 91/689/EEC of 12 December 1991 on hazardous waste, OJ L 377 of 31.12.1991, p. 20, corr. OJ L 023 of 30.01.1998, p. 39
IPPC Directive	Council Directive 96/61/EC of 24 September 1996 concerning integrated pollution prevention and control, OJ L 257 of 10.10.1996, p. 26, corr. OJ L 333 of 10.12.2002, p. 27, OJ L 140 of 30.05.2002, p. 39, OJ L 019 of 24.01.1998, p. 83, OJ L 082 of 22.03.1997, p. 63, OJ L 036 of 06.02.1997, p. 32 and OJ L 302 of 26.11.1996, p. 28
Nitrates Directive	Council Directive 91/676/EEC of 12 December 1991 concerning the protection of waters against pollution caused by nitrates from agricultural sources, OJ L 375 of 31.12.1991, p. 1, corr. OJ L 092 of 16.04.1993, p. 51
Packaging and Packaging Waste Directive	European Parliament and Council Directive 94/62/EC of 20 December 1994 on packaging and packaging waste, OJ L 365 of 31.12.1994, p. 10
Public Participation Directive	Directive 2003/35/EC of the European Parliament and of the Council of 26 May 2003 providing for public

720 Appendix 2 – Consolidated List of Legislation

	participation in respect of the drawing up of certain plans and programmes relating to the environment and amending with regard to public participation and access to justice Council Directives 85/337/EEC and 96/61/EC, OJ L 156 of 25.06.2003, p. 17
RoHS Directive	Directive 2002/95/EC of the European Parliament and of the Council of 27 January 2003 on the restriction of the use of certain hazardous substances in electrical and electronic equipment, OJ L 37 of 13.02.2003, p. 19
SEA Directive	Directive 2001/42/EC of the European Parliament and of the Council of 27 June 2001 on the assessment of the effects of certain plans and programs on the environment, OJ L 197 of 21.07.2001, p. 30
Structural Funds Regulation (expired)	Council Regulation (EEC) No 2081/93 of 20 July 1993 amending Regulation (EEC) No 2052/88 on the tasks of the Structural Funds and their effectiveness and on coordination of their activities between themselves and with the operations of the European Investment Bank and the other existing financial instruments, OJ L 193, p. 5
Structural Funds Regulation	Council Regulation (EC) No 1260/1999 of 21 June 1999 laying down general provisions on the Structural Funds, OJ L 161, p. 1, corr. OJ L 009 of 13.01.2000, p. 30, OJ L 271 of 21.10.1999, p. 47 and OJ L 194 of 27.07.1999, p. 68
Waste Framework Directive	Council Directive 75/442/EEC of 15 July 1975 on waste, OJ L 194 of 25.07.1975, p. 39
Water Framework Directive	Directive 2000/60/EC of the European Parliament and of the Council of 23 October 2000 establishing a framework for Community action in the field of water policy, OJ L 327 of 22.12.2000, p. 1, corr. OJ L 017 of 19.01.2001, p. 39
WEEE Directive	Directive 2002/96/EC of the European Parliament and of the Council of 27 January 2003 on waste electrical and electronic equipment (WEEE), OJ L 37 of 13.02.2003, p. 24, amended by Directive 2003/108/EC of 8 December 2003, OJ L 345 of 31.12.2003, p. 106

Wild Birds Directive	Council Directive 79/409/EEC of 2 April 1979 on the conservation of wild birds, OJ L 103 of 25.04.1979, p. 1, corr. OJ L 059 of 08.03.1996, p. 61, amended by Directive 97/49/EC of 29 July 1997, OJ L 223, p. 9

3 National Legislation

3.1 Austria

Federal Acts

EIA Act (2000)	Environmental Impact Assessment Act (2000) from 10 August 2000, amending the EIA Act, Federal Law Gazette 2000 No. I 89
Water Management Act (2003)	Water Management Act 2003 from 29 August 2003 amending the Water Management Act, Federal Law Gazette 2003 No. I 82
Federal Waste Management Act	(Federal) Waste Management Act 2002, Federal Law Gazette for the Republic of Austria (2002) No. I 102, 2002, last amended with Federal Law Gazette No. I 43 (2004), Vienna (SEA amendment currently proposed)

The Acts of the Provinces[1]

Spatial Planning Act (Province of Salzburg)	Spatial Planning Act 2003 from 17 December 2003, 27 February 2004, amending the Spatial Planning Act, Provincial Law Gazette of Salzburg Province No. 13/2004
Spatial Planning Act of Lower Austria	(Lower Austrian) Spatial Planning Act, Provincial Law Gazette of Lower Austria (1976) No. 8000 in its actual version, St. Poelten (SEA amendment currently proposed)
Styrian Spatial Planning Act	(Styrian) Spatial Planning Act, Provincial law Gazette of Styria (1974) No. 127 in its actual version, Graz

---

[1] There are 9 Provinces in Austria which have a similar role like the States in the USA

Carinthian Act on Environmental Planning	Currently, only a proposal exists (from 4 February 2004), Klagenfurt

3.2 Belgium

EIA decree	Decree of the Flemish Government of March 23rd 1989 concerning categories of works and activities for which an environmental impact statement is needed for the completion of the request for a building permit or an environmental permit, changed by the decrees of the Flemish Government of January 25th 1995, February 4th 1997, and March 10th 1998, Belgisch Staatsblad/Le Moniteur Belge – B.S. 17/05/1989
Environmental policy Decree	Decree of the Flemish Government of April 5th 1995, concerning general environmental policy, except for title IV, added to the Decree of December 18th 2002, Belgisch Staatsblad/Le Moniteur Belge – B.S. 03/06/1995
EIA and SEA Decree	Decree of the Flemish Government of December 18th 2002, to complete the Decree of April 5th 1995, with a title concerning environmental and safety reporting, Belgisch Staatsblad/Le Moniteur Belge – B.S. 13/02/2003
Port Decree	Decree concerning the policy and management of the sea ports of March 2nd 1999, Belgisch Staatsblad/Le Moniteur Belge – B.S. 08/04/1999
(Draft) Resolution to the EIA and SEA Decree	(Draft) Resolution of the Flemish Government of March 12th 2004, to determine the Flemish regulation for environmental and safety impact assessment, no official publication (31.8.2004).

3.3 Canada

Cabinet Directive on the Environmental Assessment of Policy, Plan and Program Proposals	1999, Ottawa: Government of Canada, Auditor General Act, 1995, c. 43, s. 4, Canadian Environmental Assessment Act, 1992, c.37, Farm Income Protection Act, 1991, c.22

3.4 China

EIA Law 2003	Law of the People's Republic of China on the Environmental Impact Assessment by the State Environmental Protection Administration of China (SEPA), 2003, from 28.10.2002, entering into force 01.09.2003, source: http://www.zhb.gov.cn/english/chanel-3/index.php3?chanel=3 (last access: 08.09.04)
Law on Water Pollution Control	The Law of the People's Republic of China on Water Pollution Control (October 1, 2002) The State Environmental Protection Administration of China (SEPA). Internet address: http://www.zhb.gov.cn/. Accessed 18/5/2004.

3.5 Czech Republic

Act No. 244/1992 Coll.	Act No. 244/1992 Coll., of 15. April 1992, on Environmental Impact Assessment, amended by Act No. 132/2000 Coll. and Act No. 100/2001 Coll., by which it was renamed "Act on Environmental Impact Assessment of Development Conceptions and Programs", and abolished by Act No. 93/2004 Coll.
Act No. 132/2000 Coll.	Act No. 132/2000 Coll., of 13. April 2000, Amending and Abolishing Some Acts Related to the Act on the Regions, the Act on the Municipalities, the Act on District Authorities and the Act on the Capital City of Prague
Act No. 100/2001 Coll.	Act No. 100/2001 Coll., of 20. February 2001, on Environmental Impact Assessment and Amending Some Related Acts (Act on Environmental Impact Assessment), amended by Act. No. 93/2004 Coll.
Act No. 93/2004 Coll.	Act No. 93/2004 Coll., of 29. January 2004, Amending Act No. 100/2001 Coll., on Environmental Impact Assessment and the Amendment of Some Related Acts (Act on Environmental Impact Assessment)
Act No. 50/1976 Coll.	Act No. 50/1976 Coll., of 27. April 1976, on Land-use Planning and Building (Building Act), as amended

3.6 England

Planning and Compulsory Purchase Act, 2004	Statutory Instrument 2004 No. 2097 (C. 89, Commencement No.1), HMSO, London

3.7 Estonia

EIA and EA Act	Act on Environmental Impact Assessment and Environmental Auditing, State Gazette, RT I 2000, 54, 348 http://www.envir.ee/oigusaktid/keskkonnaoigus/keskkonnamoju.html
EIA and EMS Act	Act on Environmental Impact Assessment and Environmental Management Systems, 2004 http://www.envir.ee/oigusaktid/eelnoud/Keskkonnamoju_hindamise_eelnou.pdf

3.8 Finland

Act on the Openness of Government Activities	Act on the Openness of Government Activities (621/1999), latest amendment 01.05.2004 (281/2004), http://www.finlex.fi
Environmental Impact Assessment Act	Environmental Impact Assessment Act (468/1994), latest amendment from 01.12.1999 (623/1999), http://www.finlex.fi
Forest Act	Forest Act (1093/1996), latest amendment 01.07.2004 (552/2004), http://www.finlex.fi
Land Use and Building Act	Land Use and Building Act (132/1999), latest amendment 01.09.2004 (476/2004), http://www.finlex.fi
Regional Development Act	Regional Development Act (602/ 2002), http://www.finlex.fi
Public Roads Act	Public Roads Act (243/1954), latest amendment 10.12.2001 (1059/2001), http://www.finlex.fi

3.9 Germany

Federal Acts

Basic Law	Basic Law for the Federal Republic of Germany (*Grundgesetz*) of 23 May 1949, Federal Law Gazette III, p. 1, last amendment of 26.07.2002, Fed. Law Gazette I, p. 2862
Building Code	(Federal) Building Code in the version of 27.08.1997, Fed. Law Gazette I, p. 2141, last amendment of 23.07.2002, Fed. Law Gazette I, p. 2850
Federal Nature Conservation Act	Federal Nature Conservation of 25.03.2002, Fed. Law Gazette I, p. 1193, amendment of 25.11.2003, Fed. Law Gazette I, p. 2304
EIA Act	Federal Act on Environmental Impact Assessment in the version of 05.09.2001, Fed. Law Gazette I, p. 2350, last amendment of 18.06.2002, Fed. Law Gazette I, p. 1914
Spatial Planning Act	Federal Spatial Planning Act of 18.08.1997, Fed. Law Gazette I, p. 2081, amendment of 15.12.1997, Fed. Law Gazette I, p. 2902
Federal Immissions Control Act	Federal Immissions Control Act in the version of 26.09.2002, Fed. Law Gazette I, p. 3830, last amendment of 06.01.2004, Fed. Law Gazette I, p. 2
Water Management Act	Federal Water Management Act in the version of 19.08.2002, Fed. Law Gazette I, p. 3245
Federal Highways Act	Federal Highways Act in the version of 20.02.2003, Fed. Law Gazette I, p. 286
Waste Management Act	Waste Management Act of 27.09.1994, Fed. Law Gazette I, p. 2703, last amendment of 25.01.2004, Fed. Law Gazette I, p. 82
Act to Adapt the Building Code to EU Directives	Act to Adapt the Building Code to EU Directives of 24.06.2004, Fed. Law Gazette I, p. 1359
Highway Development Act	Federal Highway Development Act in the version of 15.11.1993, Fed. Law Gazette I, p. 1878, amendment

	of 29.10.2001, Fed. Law Gazette I, p. 2785
Administrative Procedures Act	Federal Administrative Procedures Act in the version from 23.01.2003, Fed. Law Gazette I, p. 102, amendment of 05.05.2004, Fed. Law Gazette I, p. 718
Federal Soil Protection Act	Federal Soil Protection Act of 17.03.1998, Fed. Law Gazette I, p. 502, amendment of 09.09.2001, Fed. Law Gazette I, p. 2331
Federal Act on the improvement of the agrarian sector	Act on common task "improvement of the agrarian sector and the coastal protection" in the version of 21.07.1988, Fed. Law Gazette I, p. 1055, last amendment on 02.05.2002, Fed. Law Gazette I, p. 1527

The Acts of the Laender[2]

Waste Act of Baden Wuerttemberg	Waste Act of Baden-Wuerttemberg in the version of 15.10.1996, Law Gazette, p. 617, last amendment of 19.11. 2002, Law Gazette p. 428, corr. p. 531
Saxon Nature Protection and Conservation Act	Nature Protection and Conservation Act of Saxony in the version of 11.10.1994, Law and Ordinance Gazette, p. 1601, last amendment of 01.09.2003, Law and Ordinance Gazette, p. 418
Brandenburg Nature Conservation Act	Brandenburg Nature Conservation Act from 26 May 2004, Law Gazette I, p. 350

3.10 Ghana

Environmental Protection Agency Act	Environmental Protection Agency Act 1994, 490th Act of Parliament, date of assent: 30.12.1994, Ghana Government Gazette from 30.12.1994, p.1-17
Environmental Assessment Regulations	Environmental Assessment Regulations 1999, LI 1652, date of assent: 18.02.1999, Ghana Government Gazette from 26.02.1999, last amendment from 26.04.2002, LI 1703, Ghana Government Gazette from 17.05.2002, p 4.

---

[2] There are 16 *Laender* in Germany which have a similar role like the States in the USA

Energy Commission Act	Energy Commission Act 1997, 541st Act of Parliament, establishing the Energy Commission, date of assent: 31.12.1997, Ghana Government Gazette from 31.12.1997, p 1
Water Resources Act	Water Resources Act 1996, 522nd Act of Parliament establishing the Water Resources Commission, date of assent: 11.10.1996, Ghana Government Gazette from 20.10.1996
Local Government Act	Local Government Act 1993, 462nd Act of Parliament, legal basis to decentralization scheme, Consisting of various laws. Date of assent: 07.07.1993, Ghana Government Gazette from 18.07.1993, p.1

3.11 Italy

Law No. 20/2000	Urban/planning law of the Region of Emilia Romagna, concerning territorial protection and land use, from March 24th, 2000. Official Bulletin of the Emilia Romagna Region n.52, March 27th, 2000.
National Urban Act No. 1150, 1942	Urban National Act, from August 17th, 1942. Official Gazette n.244, October 16th, 1942.
Tuscany's Regional Urban/Planning Law No. 5/1995	Urban/planning law of the Region of Tuscany, concerning rules for territorial government, from January 16th, 1995. Official Bulletin of the Tuscany Region n.6, January 20th, 1995.
Law No. 5/2004	Conversion in law of the amended decree n.315 of November 14th, 2003, concerning urgent dispositions for the composition of the environmental impact assessment commission and for the authorizing procedures for infrastructures of electronic communication, from January 16th, 2004. Official Gazette n.13, January 17th, 2004.
Law No. 39/2002	Community Law 2001. Italy's dispositions for complying with the European Community, from March 1st 2002. Official Gazette n. 72, March 26th, 2002.
Law No. 284/2002	Conversion in law of the amended decree n.236 of October 25th 2002, concerning urgent dispositions for leg-

	islative deadlines, from December 27th. Official Gazette n. 303, December 28th, 2002.
Act of Instruction and Coordination (Dpr 1996)	Act of instruction and coordination for the implementation of art.40, comma 1, of law n.146, February 22nd 1994, concerning dispositions in environmental impact assessment, from April 12th, 1996. Official Gazette n.210, September 7th, 1996.
Constitution of the Italian Republic	Published in the Official Gazette n.7, December 1947. Amended by the following Constitutional Laws: n.2/1963; n.2/1967; n.3/2001; n.1/2002.

3.12 Kenya

Environmental Management and Coordination Act of 1999	Environmental Management and Coordination Act No. 8 of 1999, Kenya Gazette Supplement No. 3 (Acts No. 1), Kenya Gazette Supplement Act 2000, Nairobi, 14. January 2000

3.13 Latvia

EIA Act	Environmental Impact Assessment Act (Latvijas Vēstnesis 322/325 30.10.1998), latest amendments 26.02.2004 (L.V., 12.marts, nr.40)
Regulations on procedure for conducting SEA	Cabinet of Ministers' Regulations No 157 on procedure for conducting SEA, Riga 23.03.2004, (L.V. 53 05.04.2004)
Regulations on Rules of Procedure for Cabinet of Ministers'	Cabinet of Ministers' Regulations No 111 on Rules of Procedure for Cabinet of Ministers', Riga 12.03.2002, latest amendments 03.08.2004, Regulations No 684 (L.V. 6.aug., nr. 124)

3.14 New Zealand

Resource Management Act (RMA), 1991	Resource Management Act of 21.07.1991 (Act no. 69), 1991, last amendment 2004 (Act no. 46)
Town and Country Planning Act (TCPA), 1977	Government Printer, Wellington, New Zealand. Repealed by the Resource Management Act 1991.

Water and Soil Conservation Act (WSCA) 1967	Government Printer, Wellington, New Zealand. Repealed by the Resource Management Act 1991.
National Development Act (NDA), 1979	Government Printer, Wellington, New Zealand. Repealed by the Resource Management Act 1991.

3.15 Poland

Basic Law (Constitution)	The Constitution of the Republic of Poland of 2 April 1997, Dz.U. 78, p.483
EIA Act 2000	Act of 09.11.2000 on Public Access to Information and the Environmental Impact Assessment (EIA Act, 200) (*ustawa z 09.11.2000 r. o dostępie do informacji o środowisku i jego ochronie oraz ocenach oddziaływania na środowisko*), Dz. U. No 109, position 1157
Environmental Protection Act (EPA) 2001	The Environmental Protection Act of 27.04.2001 (EPA, 2001) (*ustawa z 27.04.2001 - Prawo ochrony środowiska*), Dz. U. No 62, position 621 with subsequent amendments
Land Use Planning and Management Act	Land Use Planning and Management Act of 27. 03. 2003 (*ustawa z 27.03.2003 r o planowaniu i zagospodarowaniu przestrzennym*), Dz. U. No 80, position 717
Act on the Sea Areas and Maritime Administration	Act on the Sea Areas and Maritime Administration of the Polish Republic of 21.03.1991 (*ustawa z 21.03.1991 o obszarach morskich Rzeczypospolitej Polskiej i administracji morskiej*), Dz. U. from 2003 final text after amendments, No 153, position 1502 with subsequent amendments
Ministers` Council Act	Ministers` Council Act of 08.08.1996 *(ustawa z 08.08.1996 o Radzie Ministrów)*, Dz. U. form 2003 final text after amendments No 24, position 199 with subsequent amendments
Resolution comprising Rules of Ministers` Council	Resolution No 49 of 19.03.2002 comprising Rules of Ministers` Council (*uchwała nr 49 Rady Ministrów z 19.03.2002 r Regulamin pracy Rady Ministrów*), MP No 13, position 221 with subsequent amendments

Act on Sections of Governmental Administration	Act on Sections of Governmental Administration of 04.09.1997 *(ustawa z 04.09.1997 o działach administracji państowej)*, Dz. U. from 2003 final text after amendments, No 159, position 1548 with subsequent amendments
Regional Local Government Act	Regional Local Government Act of 05.06.1998 *(ustawa z 05.06.1998 r. o samorządzie województwa)*, Dz. U. from 2001 final text after amendments, No 142, position 1590 with subsequent amendments
Water Law Act, 2001	Water Law Act from 27.04.2001 *(ustawa z 27.04.2001 Prawo wodne)*, Dz. U. No 115, position 1224 with subsequent amendments
Nature Conservation Act, 1991	Nature Conservation Act, 1991 *(ustawa z 16.10.1991 r. o ochronie przyrody)*, Dz. U. from 2001 final text after amendments No 99, position 1079 with subsequent amendments
Land Use Act, 1994	Land Use Act of 07.07.1994 *(ustawa z 07.07.1994 r. o zagospodarowaniu przestrzennym)*, original text Dz. U. 1994 r. No 89 position 415, final text Dz. U. 1999 r. No 15 position 139 with subsequent amendments
Environmental Protection and Management Act, 1980	Environmental Protection and Management Act of 31.01.1980 (ustawa z dnia 31.01.1980 r. o ochronie i ksztaltowaniu srodowiska), original text Dz. U. 1980 r. No 3 position 6, final text: Dz. U. 1994 r. No 49 position 196 with subsequent amendments
Toll Motorways Act, 1994	Toll Motorways Act of 27.10.1994 *(ustawa z dnia 27.10.1994 r. o autostradach platnych)*, original text Dz. U. 1994 r. No 127 position 627, final text Dz. U. 2001 r. No 110 position 1192 with subsequent amendments
Order on the requirements to be met by the environmental prognosis of land use plans	Order of the Minister of Environmental Protection of 09.03.1995 on the requirements to be met by the environmental prognosis of land use plans
Order on developments particularly harmful to the environment and human	Order of the Minister of Environmental Protection of 13.05.1995 on developments particularly harmful to the environment and human health and terms to be met by EIS

health and terms to be met by EIS	
Order on the terms to be met by the assessment of the impact of highways on the environment, agricultural and forest land and cultural heritage	Order of the Minister of Environmental Protection of 05.06.1995 on the terms to be met by the assessment of the impact of highways on the environment, agricultural and forest land and cultural heritage
Order on the types of the projects which may have significant impacts on the environments and detailed criteria for project screening for EIA	Order of the Minister of Environmental Protection of 24.09.2002 on the types of the projects which may have significant impacts on the environments and detailed criteria for project screening for EIA
Order on detailed criteria of SEA prognosis for land use plans	Order of the Minister of Environmental Protection of 14.11.2002 on detailed criteria of SEA prognosis for land use plans; Rozporzadzenie Ministra Srodowiska z dnia 14 listopada 2002 r. w sprawie szczegółowych warunków, jakim powinna odpowiadać prognoza oddziaływania na środowisko dotycząca projektów miejscowych planów zagospodarowania przestrzennego. (Dz. U. 02.197.1667 z dnia 27 listopada 2002 r.)

3.16 Sweden

The Environmental Code	Law (1998:808), last amended by (2004:667), published in English by Ministry publications series (Ds 2000:61)

3.17 Ukraine

Law About Ecological Expertise"	Law of Ukraine: "About Ecological Expertise" set into force 09.02.95 by decree BPN 46/95-BP, BVR (Bulletin of Verkhovna Rada) 1995, No. 8, p.54

Decree N 554 "About listing types of activities and objects, considered to have high ecological risk"	Decree N 554 "About listing types of activities and objects, considered to have high ecological risk" set into force 27.07.1995

3.18 United States

National Environmental Policy Act of 1969	National Environmental Policy Act of 1969, Pub. L. 91-190, 42 USC 4321-4327, January 1, 1970, as amended by Pub. L. 94-52, July 3, 1975, Pub. L. 94-83, August 9, 1975, and pub. L. 97-258, § 4(b), Sept. 13, 1982
Washington State Environmental Policy Act	Washington State Environmental Policy Act, RCW 43.21C.030 (2002). Washington Rules, WAS 197-11-330(1)(b)
Administrative Procedure Act	Administrative Procedure Act, 5 USC § 706.
40 CFR § 1508.7	Council on Environmental Quality, „Terminology and Index." US Code of Federal Regulations

Subject Index

Aarhus Convention 20, 33-34, 49-50, 58, 202, 435, 504, 669-670
accession countries 73, 193, 201, 217, 227, 437
accredited/authorised EIA/SEA experts 12, 138-140, 199, 229-234
acquis communautaire 63
adaptive management 249
additive effects 387, 399, 400
    see also cumulative impacts
Agenda 21 19, 131, 163, 191, 314, 422, 431, 515
agricultural/agriculture 38-39, 100, 599-619
    agricultural land use 602-605
    agricultural plans and programmes 605-608
    agricultural policy 600-601
    good agricultural and environmental condition framework 614
    see also Common Agricultural Policy
Air Quality Framework Directive 37-38
alternatives/options 7-8, 13, 163, 181, 312, 372-380, 575, 578, 592, 686-687, 691-693
    danger of bland alternatives 691
    for road corridors 375-380
    for waste dumping ground sites 372-375
    overhead line routeing methodology 684-685
archaeological heritage 342
Austria 17, 149-158, 621-623
    case studies 150-155, 621-623
    EIA Act 149
    federal vs. provincial level 149
    Weiz land use plan 152-155
    waste management 622
Azerbaijan 498, 501

b-test *see* business, business test
baseline data *see* data
baseline-led assessment 88
Belgium 16, 57, 59, 137-148
bias 71-72, 114-115, 698
biodiversity 111, 523-538, 539-554
    action plan 394
    assessment 167
    between diversity 526
    definition 524-525
    evaluation/measurement 525-527, 532-534, 547-548, 550
    gain 604
    guidance 87
    inadequately dealt with 111
    indicators/indices 401, 404, 525-527, 531-533, 542, 548-549, 550-551, 618
    information and assessment center 402
    loss of 602, 604
    value of 528 528-532, 547-549
    within diversity 526
building schemes 105-107
Bulgaria 63
business 27, 42, 272, 281, 677-678
    business impact assessment 27
    business test (b-test) 28
    small business 642
    see also industry

Canada 251-268
    analysing environmental effects 258
    appropriate level of effort 259-260
    Canadian Environmental Assessment Act 251-253, 257
    case studies 264-266
    documentation and reporting 261-262
    failures of SEA system 263
    quality assurance 263-264
    preliminary scan 256
    public participation 260
CAP *see* Common Agricultural Policy
capacity-building 63-64, 313, 649-663
    definition 650
    for improved compliance 63-64
    needs and key drivers 655-658
    see also training
case by case approach 221
case studies *see* named countries; cumulative impacts; industry/energy; methods/methodologies; public consultation/participation
categorical exclusion (USA) 243-244

causal effect diagrams 352, 361-362, 459-469
CEE (Central and Eastern European Countries) 63
China 331-345
   environmental situation 331-335
   Three Gorges Dam Project 336-341
CIT see Countries in Transition
civil emergency 43, 204
clean/cleaner technology 444-445
climate 165
co-decision procedure 21, 24, 27
Committee of Regions 23, 24
Common Agricultural Policy (CAP) 602
   CAP reform 611
communal statute 61-62
communication skills 654-655
compatibility appraisal 354
competence (of the European Community) 31-33, 57
conceptions 3, 194, 220
   definition 220
concurrant programme appraisal (CPA) 496, 506, 508
conference "Environment for Europe" 28, 33, 34
consumer impact assessment 85
Convention on Biodiversity 542-543
cost-benefit analysis 93, 280, 376
Council on Environmental Quality 240, 398
   see also NEPA; United States
Countries in Transition (CIT) 321
   see also accession countries
critique of SEA/SEA Directive 69-79, 665-672
cross compliance 613
cross-sectoral integration 295
cultural/culture 340-343, 409-420
   concept/components 412-416
   conservation 340-343
   cultural heritage 342
   cultural impact assessment 416-418
   cultural integrity 410-411
cumulative impacts 6, 249, 385-395, 397-408
   case studies 404-406
   definition 386-387, 398-399
   Environmental Monitoring and Assessment Program (EMAP) 401
   methods 388-393, 402

   mitigation/enhancement 265-266, 393, 406-407
   monitoring 404
   scoping 388-389, 400
   temporal accumulation 399
   transboundary issues 405
   see also additive effects; neutralising effects; synergistic effects
Czech Republic 3-4, 193-200, 698
   information system 197
   pilot project 436-441
   quality assurance 199
   Waste Management Plan 196

dams see China, Three Gorges Dam Project; Ghana, Volta and Kpong dams
data 110-111, 156, 201, 210-211, 311, 317-318, 353, 380, 390, 400-401, 487-489, 500-504, 546, 623-625, 694-696
   common environmental database 489
   data appropriateness 500-504, 694-696
   data quality issues 503-504
   data needs of SEA tools 353
   for biodiversity 546
   information system 197
   metadata 503, 695
   representativeness 500
decision-making 51, 168-169, 239, 365, 592-594
   improvement of 9-12, 239, 467
   multistage decision-making process 377
   place-based approach 249
   relationship with SEA 205, 214, 653
   see also business
defence 43, 83, 86, 204
Denmark 16-17, 70, 157
developing countries 291, 306, 640
documentation see environmental report
DPSIR-Framework see Driving Forces Pressures State Impacts Responses Framework
Driving Forces Pressures State Impacts Responses (DPSIR) Framework 603

e-commerce impact appraisal 85
e-test see environment test
Earth Summit 19, 674

Ecological Expertise (EE) *see* State Ecological Expertise (SEE)
ecological footprint 446-447
Economic and Social Committee 23
economic development planning 83, 95
economy/economic evaluation 371, 529-530, 547-548
EIA *see* Environmental Impact Assessment (EIA)
EIA Directive 6, 16, 20-21, 25-26, 36, 45-46, 58, 71, 75, 100, 102, 117, 202, 608, 668, 670
  Annexes I and II 45, 73, 373, 521, 607-608, 669
  competent authority 75
  implementation problems 70, 179
  legislative competence 16, 20
  relation to the SEA Directive 35-36, 187, 475
  see also Environmental Impact Statement (EIS)
electricity *see* industry/energy
England 83-98
  case studies 89-91
  land use planning 86-89
  local and structure plans 86, 87
  North West England 87, 89
  transport planning 91-96
  see also environmental appraisal
EMAS (Eco Management and Audit Scheme) 42, 156
energy *see* industry/energy
environmental action programmes 20
environmental appraisal 84-85, 87-88, 96
environmental impact assessment (EIA) 4, 34, 138-140, 178, 275, 308-309, 312, 366-369, 497, 607, 682
  difference/link to SEA 4-5, 138-140, 312, 366-369, 590, 604, 682
  learning from past experience 495, 505
  main stages 497
  methodology 109-110
  project-level actions 366
  tiering from SEA *see* tiering
  see also named countries
Environmental Impact Statement (EIS) 138, 242, 244
environmental law (EU) 16-21

environmental management/planning 201, 202, 209
Environmental Monitoring and Assessment Program *see* cumulative impacts, EMAP
Environmental Protection Agency (EPA) *see* United States
Environmental Protection and Enhancement Procedures (EPEP) *see* New Zealand
environmental report 10-11, 46-48, 72-73, 105-107, 113, 166, 205, 206, 208, 214, 221-222, 231, 261, 560-561, 588-589
  legal value 72
  see also named countries
environmental test (E-test) 27
EPEP *see* New Zealand, Environmental Protection and Enhancement Procedures
Espoo Convention 28, 34, 49-50, 160, 202, 299, 504
Estonia 227-235
  project EIA and SEA in 228
  quality criteria 233-234
European Atomic Energy Community (ECSC) 16
European Coal and Steel Community (ECSC) 16
European Commission 16-17, 21-24, 26-27
  communication on impact assessment 39
  Environmental Technologies Action Plan 544
  MOLAND project 361
  SEA guidance 112
  strategy for soil protection 474
European Community (EC) 17
  agricultural policy 600-601
European Court of Justice 19, 33, 57, 60-64, 65, 70
European Directive 2001/42/EC *see* SEA Directive
European Economic Community (EEC) 16
European Parliament 21, 23, 24
European Single Act 17
European Union 17
EUROSTAT 482
exclusion zoning 371, 375

Subject Index

exclusive economic zones 108
exergy
  concept 447
  cumulative exergy consumption 450
  exergetic life cycle analysis (ELCA) 450
  loss 452
expert judgement 350

federal system
  in Austria 149, 721-722
  in Germany 66, 99-100, 566, 725-726
Finland 159-175
  case study 165
  challenges 170-172
  EIA Act 160
  forest programme 2010 164, 167
financial/budget plans/programmes 44
Flanders, Belgium 137-148
  case studies 142-148
forest/forestry 45, 74, 83, 156, 210
Former Soviet Union (FSU) 63, 64, 496, 499, 504
FSU *see* Former Soviet Union
Functional Strategy Analysis 462

gAEP *see* preliminary agrarian structure planning
gender impact assessment 85
General Agreement of Tariffs and Trade (GATT) 405
geographical information systems (GIS) 88, 166, 351, 356-360, 402, 404, 488, 686
  advantages of 357
  case study 358-359
  disadvantages of 357
Georgia 498, 501
Germany 16, 59, 99-116, 155, 558
  Amendment to Spatial Planning Act 108
  arcane culture 49
  Building Code 36, 105, 572-573, 575
  case studies 113
  critique of SEA/SEA Directive 23 665-672
  current SEA Practice 99
  EIA Act 64, 100-101, 475
  environmental assessment 581
  environmental legislation 475

  Federal Transport Infrastructure Plan 378
  implementation of SEA 99, 109, 111, 114
  industry 665-672
  Laender 59, 66, 557-559, 566, 573-575
  landscape planning 558-559
  local development planning 99, 106, 107, 112
  Nature Conservation Act 476, 558, 566
  participation of NGO 49
  plan approval procedure 101, 102
  planning legislation/system 475, 609
  Preliminary agrarian structure planning (gAEP) 608-616
  spatial planning 99
  Spatial Planning Act 108
  Soil Protection Act 476
  subsidiarity clause 100
  transport planning 585-595
Ghana 305-319
  constraints to EA in 309
  environmental assessment in 308
  Volta and Kpong Dams 306-307
  water resource management 305, 310, 312, 313, 317
GIS *see* Geographical Information Systems
globalisation 299
GOMMMS (guidance on multi-modal methodologies) 92
Great Britain *see* United Kingdom
groundwater issues 44, 153, 398, 406-407
guidance for SEA/SEA Directive 8, 25, 85-86, 222-223, 535

Habitats Directive 38, 45, 77, 543
Health Impact Assessment (HIA) 27, 85
health impacts 27, 34, 85, 123, 153
  inadequately dealt with 111
HIV/Aids 299
Hong Kong 335
housing sector 464-466

impartiality *see* bias
implementation of SEA
  minimalistic approach 177, 187
  problems of 69, 73, 132

support of the 111
see also named countries
indicators 11, 153-154, 417, 433, 444, 446, 447, 449, 451, 452, 454-456, 686
  based on a thermodynamic approach 443-458
  cultural indicators 417
  for biodiversity *see* biodiversity, indicators
indirect effects 143, 155, 168, 398, 402, 404
  of legislation 41
  see also cumulative impacts
industrial metabolism 445, 449, 452-454
industry/energy 632-640, 665-672, 673-690
  case study 680-681
  electricity 678-679
  institutional links 674
  mining 632-640
  offshore oil and gas facilities 86
  transmission industry/network 678, 683-684, 685-688
information system for SEA 197
infrastructure 41, 131, 142-143, 156, 179-180, 211, 251, 300, 352, 579, 604, 636, 673
  farm infrastructure 602
  improvement areas 86
  master plan mobility for Antwerp 144-145
  planning 163, 181
  telecommunications 84
  transport infrastructure plan (Finland) 164
  transport infrastructure plan (Germany) 558
  transportation projects 93
  village infrastructure project (Ghana) 310
  see also transport
infringement procedure 59, 66, 70
integrative environmental protection 19, 20, 32
inter-agency coordination 247, 248
interdisciplinary 244, 247, 249
international plans and programmes 42
Italy 16, 117-135
  Emilia Romagna Region 117-134

Environmental and territorial sustainability assessment (VALSAT) 126-131
  planning in 119, 120, 122-129
  planning instruments, levels of 124
  SEA implementation 117, 119, 131

Kenya 291-303
  history of EA of strategic decisions 293
  institutional framework 297
  need for SEA 298
  physical planning 300
  SEA elements and tools 294, 295
Kiev SEA Protocol 3, 28, 34, 41, 49-50, 326
  inclusion of health impacts 34

land use partitioning analysis 352
land use planning/plans 76, 78, 84, 86, 95, 139, 207, 365-367, 371, 378, 380, 382
landscape planning/plans 366, 371, 372, 378, 380-382, 557-570
  environmental report 560-561
  integrated procedure 566-567
  procedure 559
  relation to urban planning 573-575
  stages 560
Latvia 217-225
  case study 223-224
  competent authority 220
  development plan 223
  EIA Act 219, 224
  planning document 217-222, 224
  preliminary assessment 221
  SEA application form 221
legislation, consolidated list 715-732
level of effort *see* scale/detail
life cycle analysis/assessment 371, 446
Laender *see* Germany, Laender
Local Agenda 21 *see* Agenda 21

mad cow disease 19
malaria 299
marine environment 76-77, 83
Material Flow Analysis (MFA) 447
matrix/matrices 89, 153-154, 350, 354, 402, 405
methodology/methods 8, 12, 109, 131, 133, 141-142, 161-164, 213, 256, 279,

349-363, 365-383, 459-469, 487, 623, 697
case studies 358-359, 378,
EIA methodology 109-110
for elaborating urban sustainability 134
for public participation 427-428, 435
for risk assessment 484-486
for soil evaluation 478
for transposition European directives 33, 59-60, 63, 109
in landscape planning 566
Multi Modal Methodology 84-85, 93
of measurements 70
report 142
see also named countries
Mexico 406
mining industry *see* industry/energy
mitigation 10, 13, 52, 90, 93, 134, 153-154, 393
modeling 144, 352-353, 402
physical and biological modeling 407
monitoring 35, 47, 52, 65, 84, 90-96, 100, 107, 110, 112, 122, 156, 169, 204, 205, 249, 295, 312, 382, 401-403, 508, 595, 634
harmonisation 35
in the IPPC Directive 35
in the Water Framework Directive 35
included into SEA Directive 24
indirect monitoring 52
information of monitoring measures 51, 123
integrated monitoring 52
of the Building and Land Use Act (Finland) 162
of transport plans 595
Vital Signs monitoring network (USA) 403-404
multi-criteria/attribute analysis 353, 371

NATA (new approach to trunk roads appraisal) *see* transport planning
National Environmental Policy Act (NEPA) 15, 179, 239-246, 397, 398, 407
cumulative impacts 398
decision-making process 243
implementing government bodies 241
procedural steps 242
purpose 240

task force 240
triggering 241
Natura 2000 109, 211, 378, 394
Natural Resources Inventory and Monitoring Program (NRIMP) 404
nature conservation/preservation 139, 149, 209-211, 373, 380, 382
neutralising effects 387
see also cumulative impacts
Netherlands 3, 27, 621, 623, 624
New Zealand 269-287
EIA development 272-279
Environmental Protection and Enhancement Procedures (EPEP) 272, 274
Resource Management Act 1991 (RMA) 269-272, 276-280, 282-284, 286
sustainable management 276, 279
Newly Independent States (NIS) 321
NGO *see* non-governmental organisation
NIMBY (not in my backyard) 678
nibbling 399
see also cumulative impacts
Nitrates Directive 37
non-governmental organisation (NGO) 47, 50-51, 58, 221, 223, 230, 232, 298, 308, 311, 313, 327-329, 422-423, 427, 434-435, 437, 539, 577, 627, 637, 640-641, 687
Nordic Triangle 164
North American Free Trade Agreement (NAFTA) 405
Northern Ireland 86, 692

objectives-led process 88
obviating development 7
OVOS (Assessment of Environmental Impacts) 64, 322, 496

penalty *see* infringement procedure
performance criteria of SEA (from IAIA) 651-652, 698
plans 39, 72, 220
definition 4, 220, 605
Poland 73-78, 201-215
administrative structure and SEA procedure 206
area of application 206
case studies 211-213

Subject Index 739

competent authority 207
compliance with SEA Directive 204
developments of SEA in 201-202
EIA Act 202
environmental data 210
history of EIA and SEA 202, 203
implementaion 73
Land Use Act 2003 202, 204-205, 209
legal situation 74-78
National Development Plan (NDP) 212
national energy plan 72, 76
procedure for local land use plans 209
procedure for plans, programmes, policies, strategies (PPPS) 208
sectoral plans 207
Police and Judicial Cooperation in Criminal Matters 17-18
policy appraisal 85
policies 39, 220
definition 4, 220
PPP *see* policies, plans and programmes
precautionary principle 18-22, 32, 38, 104
preliminary agrarian structure planning (gAEP) 608-616
principle of precaution and prevention *see* precautionary principle
principle of proportionality 59
principle of subsidiarity 60
private plans/programmes 42, 670, 676
privatisation 679
procedural law 53
process design 625
production and consumption chain 450, 452-454
programmes 38, 72, 220
definition 4, 220, 605
programme of measures 36, 43, 517
project EIA *see* Environmental Impact Assessment (EIA)
public consultation/participation 20, 49-51, 72, 77, 90, 93-94, 103, 106, 139, 141, 143, 146-147, 167-168, 184, 197, 202-206, 208-209, 213-215, 229-230, 248, 260, 292, 295, 312-313, 350, 407, 421-432, 433-441, 567, 577, 625-628, 638-640, 652, 657, 661, 693-694
case studies 429-430, 436-440
definition of the public 49-50, 423
individual rights 61

internet 198
levels of participation 424
link to Water Framework Directive 520
methods 427-428, 435
opportunities and obstacles 422
"participation paradox" 427
problems of weak participation 693
procedural requirements 425-426
SEA Round Table 627-628
transboundary consultation 51, 77, 436
see also named countries

quality assurance/control (of SEA) 72, 111-112, 181-182, 190, 199-200, 213-214, 222-223, 233-234, 263-264, 459, 697-698
of the strategic action 459
see also performance criteria
quality of life assessment 351

Regional Environmental Center for Central and Eastern Europe (REC) 211, 212
regulatory impact assessment 27, 85, 262
renewability coefficient 451
reporting *see* environmental report
responsible authority 47
reverse impacts 499
risk assessment/analysis 85, 354, 371, 376, 378-379, 484-486, 566
River Basin Management Plan 36, 43, 513-521, 548
see also Water Framework Directive
RMA *see* New Zealand, Resource Management Act 1991 (RMA)
Romania 63
rural planning 84, 95

scale/detail 92, 145, 205, 254, 407, 417, 502, 516, 604, 610-611
level of effort 259-260
scenarios 368-369, 372, 381
scenario/sensitivity analysis 353
scoping 46, 129, 140-141, 153, 208, 587-588, 605, 685-686
of transport plans 586-587
Scotland 3, 86, 242, 692, 698
ScottishPower 680

screening 25, 46, 65, 102, 106, 155, 183, 207-208, 586, 589-590
  in EIA 65
  test 106
  see also categorical exclusion
SEA Directive 15-29, 31-56, 57-67, 69-79, 701-713
  addressees of the 59
  area of application 39-45
  basic provisions 73-74
  competence norm 18, 32-33
  connection to member states' laws 57-58, 69-70
  content 25-26
  "detailed directive" 33
  development 23-25
  "effet utile" 33
  European context 15-21
  history of 22-25
  links with other legislation 33-39, 108-109
  mistakes, consequences of 52
  objectives/purpose 26-27, 31
  outlook 27-28
  possibilities 57
  problems 69-79, 692
  procedural provisions/requirements 25, 27, 33, 45-46, 71
  scope 71, 160-161, 183-184
  see also named countries; implementation of SEA Directive; public consultation/participation; transposition
SEA round table 627
SEA-type assessments 84, 87
SEE *see* State Ecological Expertise
SIA *see* sustainability impact assessment
significance of impacts 73, 243-244, 687
Silesia see public consultation/participation, case studies
skeleton law 558
socio and economic impacts 154, 166, 637-638
soil/soil resources 473-493
  evaluation 478
  functions 479, 484
  impact on by chemicals 481
  impact on by mechanical stress and precipitation 480
  maps 488
  methods for soil evaluation 478

Protection Act (Germany) 476
  properties 480-482
  protective goals 490
  resources 473
Soviet Union *see* Former Soviet Union
Spain 70
spatial planning 23, 107, 112, 149, 166, 178-181, 183, 184, 187-189
  for offshore wind power utilization 112
  Land-wide spatial plan 107
  Spatial Planning Act (Germany) 34, 105, 108, 113
  Spatial Planning Act (Styria) 154
spatial resistance 378
spatial sensitivity 378
stakeholders *see* public consultation/participation
State Ecological Expertise (SEE) 63, 322-326
  comparison with SEA and EIA 323
State Ecological Review (SER) 63
strategic actions/decisions 4, 9, 292, 366
strategic assessment instruments 84
strategic environmental assessment (SEA)
  barriers of 691-700
  best practice of 673-690
  definition of 4, 22, 219, 261
  difference/link to project EIA 4, 138-140, 312, 366-369, 590, 604, 682
  future of SEA 48, 691
  key principles 3-14
  information system 197
  integration with planning process 9-12, 625-626
  need/advantages of 13, 79, 680, 683
  problems 69-79, 179, 184, 186-189
  review 508
  SEA team 12, 627, 698
  stages/steps 11, 366, 370, 389, 507, 685
  transparency 182
  voluntary application 676
strategic environmental and social impact assessment (SESA) 508
strategic planning 83, 121, 247
strategies 4, 40, 220
strategy analysis 459-469
  contribution of 466
  method of 462

structural and functional analysis 459, 464
structural planning 83, 95
subsidiarity clause 52-53, 60, 105
sustainability/sustainable development 18, 37, 38, 83, 88, 104, 114, 140, 443, 444, 452, 455, 673, 688
 and mobility 142
 and public participation 20
 and SEA Directive 21, 26, 32
 definition 19
 in European activities 19-20, 27, 32
 framework 279
 global discourse 271
 management of natural and physical resources 276
 sustainable agricultural land use 611
 weak sustainability 272
sustainability appraisal 28, 83-84, 88-91, 156
sustainability impact assessment (SIA) 15, 27, 156
Sweden 177-191
 comprehensive planning 180, 181, 183, 188, 189
 environmental assessment 178
 legislation implementing the SEA Directive 181
 minimalistic approach 177, 187
 municipal planning 189
 resistance to EA 179
 SEA implementation 177, 185
 SEA objectives 182
synergistic effects 48, 387, 399-400
 see also cumulative impacts

Tariff-Union 16
Technology Assessment 446
team *see* strategic environmental assessment (SEA), SEA team
technology, environmental impact 444
terminology issues 4, 39, 70-1, 186, 232
thermodynamics/thermodynamic approach to SEA 443, 447, 448, 450
tiering 44-5, 90, 94, 95, 104, 107, 110, 145-146, 161-162, 204, 205, 295, 681-685
 requirements 107
 tiered EA system 681-685
 tiered reporting 166
timing 8, 161-162, 187

 of assessment 576
 of consultation 426
Total Material Requirement (TMR) 447
tools for SEA *see* metods/methodologies
tourism 45, 74, 83, 179, 181, 194, 196, 207, 242, 337, 514, 640, 670
town and country planning 83, 93
training (of SEA) 659-661
 see also capacity-building
transboundary cumulative effects 218, 397-398, 405, 406
transboundary effects 218, 241, 253, 633
transboundary issues 49, 51, 72, 77-78, 433-441
transparency 122, 182, 504
 lack of 179, 447, 641
 of decision making 281, 421, 431, 459, 558
transport planning 83, 84, 91-95, 112, 585-597
 assessment of alternatives 592-593
 decision making 593-595
 duplication of assessment 590
 environmental effects 591
 Federal Transport Infrastructure Plan (Germany) 378, 558
 in Germany 585-595
 monitoring 595
 multi-modal studies 84, 85, 92, 94, 95
 new approach to trunk roads appraisal (NATA) 84, 85, 92
 Trans-European Transport Networks 23, 91
 transport scheme appraisal 85
transposition 57-67
 contracts and agreements 61, 62
 in accession states 73
 instruments 60
 into the National Legal System 63
 measures 60-62
 parliamentary acts 61
 ordinances 61, 64, 65
 special SEA Act 64-65
 supplement to EIA Act 65
 transposing provisions, content of the 59
 see also named countries; SEA Directive
Treaties of Rome 16
Treaty of Amsterdam 18, 21, 24, 514, 666

Treaty of Maastricht 17
Treaty of Nice 18, 32
trends analysis 402
trilateral border region 438
Turkey 498, 631
  case study 639-640
  mining industry 631-643

uncertainty 368-369
UK *see* United Kingdom
Ukraine 321-330
  comparison of SEA and EIA 323
  obstacles for SEA 326-327
unitary development plan 84, 86
United Nations Environmental Programme (UNEP) 402, 444, 542, 546
United Kingdom 17, 27-28, 157, 275, 692
United States 239-250, 397-408
  Bureau of Land Management 241
  case studies 245, 404, 406
  Clean Air Act 240, 241
  Clean Water Act 240, 241
  cumulative effects analysis 399
  Endangered Species Act 240
  Environmental Protection Agency (EPA) 401
  Federal Water Pollution Control Act 240
  fish and wildlife 241
  Forest Service 240, 241, 245
  Marine Protection, Research, and Sanctuaries Act 240
  National Park System 241, 406
  NEPA *see* National Environmental Policy Act (NEPA)
  Resource Conservation and Recovery Act 240
  Safe Drinking Water Act 240
  State Environmental Policy Act 241
  US-Mexico Borderlands 406
  Vital Signs monitoring network 403-404
  Washington State Department of Transportation 243
  see also categorical exclusion
usability 467
urban planning 118, 123, 571-584
  assessment of alternatives 575
  German system 571
  quality of 574
  relation to landscape planning 573-575
  statutes 32
  urbanisation 299, 579

VALSAT (Environmental and territorial sustainability assessment) *see* Italy
Vienna 622-623
Vienna Treaty Convention 58

Wales 86-87, 94, 692
Waste Framework Directive 37
waste management 7, 10, 45, 74, 101, 112, 194, 196, 205, 207, 242, 298, 376, 621-630, 665, 676
  procedure for the assessment of sites 373
  Viennese waste management plan 622
  waste dumping ground sites 372
  waste management plans 37, 214, 373
  waste treatment 373
Water Framework Directive 36, 46, 50, 150, 222, 382, 495, 500, 503-505, 508, 513-522, 539, 548-550
water impact assessment (WIA) 504-505
water management 45-46, 74-76, 83, 86-87, 95, 98, 101, 139, 196, 211, 242, 317-318, 520, 676
water management plan 72, 78, 99, 101, 112, 194, 267
water resources 43, 315-318, 332-333, 495-496, 498-499, 501, 505, 509, 513, 518, 669
  see also dams
Wild Birds Directive 38, 106, 204, 222, 543, 614
win-win opportunities 655
wind energy 86, 87, 108, 112, 359, 361, 448
  wind farms 108
  offshore wind power utilisation 86, 87, 112, 359, 361
World Conservation Monitoring Centre (WCMC) 402
World Trade Organization (WTO) 405

zoning plans 105-107, 559-561, 566-567

Printing and Binding: Strauss GmbH, Mörlenbach